Waves

In SI Units

Waves

Berkeley Physics Course – Volume 3

In SI Units

The preparation of this course was supported by a grant
from the National Science Foundation to
Education Development Center

Frank S. Crawford, Jr.

Professor of Physics
University of California, Berkeley

McGraw Hill Education (India) Private Limited

NEW DELHI

McGraw Hill Education Offices
New Delhi New York St Louis San Francisco Auckland Bogotá Caracas
Kuala Lumpur Lisbon London Madrid Mexico City Milan Montreal
San Juan Santiago Singapore Sydney Tokyo Toronto

McGraw Hill Education (India) Private Limited

Waves, Berkeley Physics Course – Volume 3

Adapted in India by arrangement with McGraw-Hill Global Education Holdings, LLC, New York.

Sales territories: India, Pakistan, Nepal, Bangladesh, Sri Lanka and Bhutan

ISBN (13 digit): 978-0-07-070217-2
ISBN (10 digit): 0-07-070217-9

Managing Director: *Kaushik Bellani*
Head—Higher Education Publishing and Marketing: *Vibha Mahajan*
Publishing Manager—SEM & Tech Ed. *Shalini Jha*
Editorial Executive: *Smruti Snigdha*
Development Editor: *Renu Upadhyay*
Executive—Editorial Services: *Sohini Mukherjee*
Assistant Manager Production: *Anjali Razdan*
General Manager—Production: *Rajender P Ghansela*

Published by McGraw Hill Education (India) Private Limited
P-24, Green Park Extension, New Delhi 110 016, typeset at ACE PRO India Private Limited, Ghaziabad 201 014 and printed at A P Offset Pvt. Ltd., Delhi 110 095

Visit us at: www.mheducation.co.in

Contents

Note from the Publishers

SI METRICATION—WHY?

SI (Système Internationale) units are more popular and convenient amongst all users across the Indian subcontinent. Hence to make this classic Berkeley volume more user-friendly, there was a need to SI metricate it.

CONTRIBUTIONS TO THE SI METRICATION

We would like to express our sincere gratitude to the SI metricator of this Berkeley Volume:

Geetanjali Sethi

Faculty
St. Stephen's College
University of Delhi
New Delhi

Also noteworthy is the contribution of the following accuracy checker:

Dr R S Tiwari

Assistant Professor
Apollo Institute of Technology
Kanpur

Request for feedback and suggestions

Tata McGraw-Hill invites comments, suggestions and feedback from readers, all of which can be sent to *tmh.sciencemathsfec dback@gmail.com*. Please feel free to report piracy-related issues.

Preface to the Berkeley Physics Course

This is a two-year elementary college physics course for students majoring in science and engineering. The intention of the writers has been to present elementary physics as far as possible in the way in which it is used by physicists working on the forefront of their field. We have sought to make a course which would vigorously emphasize the foundations of physics. Our specific objectives were to introduce coherently into an elementary curriculum the ideas of special relativity, of quantum physics, and of statistical physics.

This course is intended for any student who has had a physics course in high school. A mathematics course including the calculus should be taken at the same time as this course.

There are several new college physics courses under development in the United States at this time. The idea of making a new course has come to many physicists, affected by the needs both of the advancement of science and engineering and of the increasing emphasis on science in elementary schools and in high schools. Our own course was conceived in a conversation between Philip Morrison of Cornell University and C. Kittel late in 1961. We were encouraged by John Mays and his colleagues of the National Science Foundation and by Walter C. Michels, then the Chairman of the Commission on College Physics. An informal committee was formed to guide the course through the initial stages. The committee consisted originally of Luis Alvarez, William B. Fretter, Charles Kittel, Walter D. Knight, Philip Morrison, Edward M. Purcell, Malvin A. Ruderman, and Jerrold R. Zacharias. The committee met first in May 1962, in Berkeley; at that time it drew up a provisional outline of an entirely new physics course. Because of heavy obligations of several of the original members, the committee was partially reconstituted in January 1964, and now consists of the undersigned. Contributions of others are acknowledged in the prefaces to the individual volumes.

The provisional outline and its associated spirit were a powerful influence on the course material finally produced. The outline covered in detail the topics and attitudes which we believed should and could be taught to beginning college students of science and engineering. It was never our intention to develop a course limited to honors students or to students with advanced standing. We have sought to present the principles of physics from fresh and unified viewpoints, and parts of the course may therefore seem almost as new to the instructor as to the students.

The five volumes of the course as planned will include:

I. Mechanics (Kittel, Knight, Ruderman)
II. Electricity and Magnetism (Purcell)
III. Waves (Crawford)
IV. Quantum Physics (Wichmann)
V. Statistical Physics (Reif)

The authors of each volume have been free to choose that style and method of presentation which seemed to them appropriate to their subject.

The initial course activity led Alan M. Portis to devise a new elementary physics laboratory, now known as the Berkeley Physics Laboratory. Because the course emphasizes the principles of physics, some teachers may feel that it does not deal sufficiently with experimental physics. The laboratory is rich in important experiments, and is designed to balance the course.

The financial support of the course development was provided by the National Science Foundation, with considerable indirect support by the University of California. The funds were administered by Educational Services Incorporated, a nonprofit organization established to administer curriculum improvement programs. We are particularly indebted to Gilbert Oakley, James Aldrich, and William Jones, all of ESI, for their sympathetic and vigorous support. ESI established in Berkeley an office under the very competent direction of Mrs. Mary R. Maloney to assist the development of the course and the laboratory. The University of California has no official connection with our program, but it has aided us in important ways. For this help we thank in particular two successive chairmen of the Department of Physics, August C. Helmholz and Burton J. Moyer; the faculty and nonacademic staff of the Department; Donald Coney, and many others in the University. Abraham Olshen gave much help with the early organizational problems.

Eugene D. Commins   Edward M. Purcell
Frank S. Crawford, Jr.   Frederick Reif
Walter D. Knight   Malvin A. Ruderman
Philip Morrison   Eyvind H. Wichmann
January, 1965   Alan M. Portis   Charles Kittel, *Chairman*

A Further Note

Volumes I, II, and V were published in final form in the period from January 1965 to June 1967. During the preparation of Volumes III and IV for final publication some organizational changes occurred. Education Development Center succeeded Educational Services

Incorporated as the administering organization. There were also some changes in the committee itself and borne redistribution of responsibilities. The committee is particularly grateful to those of our colleagues who have tried this course in the classroom and who, on the basis of their experience, have offered criticism and suggestions for improvements.

As with the previously published volumes, your corrections and suggestions will always be welcome.

June, 1968 Frank S. Crawford, Jr. Frederick Reif
Berkeley,   Charles Kittel   Malvin A. Ruderman
California   Walter D. Knight   Eyvind H. Wichmann
            Alan M. Portis   A. Carl Helmholz ⎫ *Chairmen*
                                   Edward M. Purcell ⎭

Preface to Volume III

This volume is devoted to the study of waves. That is a broad subject. Everyone knows many natural phenomena that involve waves—there are water waves, sound waves, light waves, radio waves, seismic waves, de Broglie waves, as well as other waves. Furthermore, perusal of the shelves of any physics library reveals that the study of a single facet of wave phenomena—for example, *supersonic sound waves in water*—may occupy whole books or periodicals and may even absorb the complete attention of individual scientists. Amazingly, a professional "specialist" in one of these narrow fields of study can usually communicate fairly easily with other supposedly narrow specialists in other supposedly unrelated fields. He has first to learn their slang, their units (like what a parsec is), and what numbers are important. Indeed, when he experiences a change of interest, he may become a "narrow specialist" in a new field surprisingly quickly. This is possible because scientists share a common language due to the remarkable fact that many entirely different and apparently unrelated physical phenomena can be described in terms of a common set of concepts. Many of these shared concepts are implicit in the word *wave*.

The principal objective of this book is to develop an understanding of basic wave concepts and of their relations with one another. To that end the book is organized in terms of these concepts rather than in terms of such observable natural phenomena as sound, light, and so on.

A complementary goal is to acquire familiarity with many interesting and important examples of waves, and thus to arrive at a concrete realization of the wide applicability and generality of the concepts. After each new concept is introduced, therefore, it is illustrated by immediate application to many different physical systems: strings, slinkies, transmission lines, mailing tubes, light beams, and so forth. This may be contrasted with the different approach of first developing the useful concepts using one simple example (the stretched string) and then considering other interesting physical systems.

By choosing illustrative examples having geometric "similitude" with one another I hope to encourage the student to search for similarities and analogies between different wave phenomena. I also hope to stimulate him to develop the courage to *use* such analogies in "hazarding a guess" when confronted with new phenomena. The use of analogy has well-known dangers and pitfalls, but so does everything. (The guess that light waves might be "just like" mechanical waves, in a sort of jelly-like "ether" was very fruitful; it helped guide Maxwell in his attempts to guess his famous equations. It yielded interesting predictions. When experiments—especially those of Michelson and Morley—indicated that this mechanical model could not be entirely correct, Einstein showed how to discard the model yet keep Maxwell's equations. Einstein preferred to guess the equations directly—what might be called "pure" guesswork. Nowadays, although most physicists still use analogies and models to help them guess new equations, they usually publish only the equations.)

The home experiments form an important part of this volume. They can provide pleasure—and insight—of a kind not to be acquired through the ordinary lecture demonstrations and laboratory experiments, important as these are. The home experiments are all of the "kitchen physics" type, requiring little or no special equipment. (An optics kit is provided. Tuning forks, slinkies, and mailing tubes are not provided, but they are cheap and thus not "special.") These experiments really *are* meant to be done at home, not at the lab. Many would be better termed *demonstrations* rather than experiments.

Every major concept discussed in the text is demonstrated in at least one home experiment. Besides illustrating concepts, the home experiments give the student a chance to experience close personal "contact" with phenomena. Because of the "home" aspect of the experiments, the contact is intimate and leisurely. This is important. There is no lab partner who may pick up the ball and run with it while you are still reading the rules of the game (or sit on it when you want to pick it up); no instructor, explaining the meaning of *his* demonstration, when what you really need is to perform *your* demonstration, with your own hands, at your own speed, and as often as you wish.

A very valuable feature of the home experiment is that, upon discovering at 10 P.M. that one has misunderstood an experiment done last week, by 10:15 P.M. one can have set it up once again and repeated it. This is important. For one thing, in real experimental work no one ever "gets it right" the first time. Afterthoughts are a secret of success. (There are others.) Nothing is more frustrating or more inhibiting to learning than inability to pursue an experimental afterthought because "the equipment is torn down," or "it is after 5 P.M.," or some other stupid reason.

Finally, through the home experiments I hope to nurture what I may call "an appreciation of phenomena." I would like to beguile

the student into creating with his own hands a scene that simultaneously surprises and delights his eyes, his ears, and his brain ...

      Clear-colored stones
         are vibrating in the brook-bed ...
         or the water is.
                   —SOSEKI†

† Reprinted from *The Four Seasons* (tr. Peter Beilenson), copyright © 1958, by The Peter Pauper Press, Mount Vernon, N.Y., and used by permission of the publisher.

Acknowledgments

In its preliminary versions, Vol. III was used in several classes at Berkeley. Valuable criticisms and comments on the preliminary editions came from Berkeley students; from Berkeley professors L. Alvarez, S. Parker, A. Portis, and especially from C. Kittel; from J. C. Thompson and his students at the University of Texas; and from W. Walker and his students at the University of California at Santa Barbara. Extremely useful specific criticism was provided by S. Pasternack's attentive reading of the preliminary edition. Of particular help and influence were the detailed criticisms of W. Walker, who read the almost-final version.

Luis Alvarez also contributed his first published experiment, "A Simplified Method for Determination of the Wavelength of Light," *School Science and Mathematics* **32,** 89 (1932), which is the basis for Home Exp. 9.10.

I am especially grateful to Joseph Doyle, who read the entire final manuscript. His considered criticisms and suggestions led to many important changes. He also introduced me to the Japanese haiku that ends the preface. He and another graduate student, Robert Fisher, contributed many fine ideas for home experiments. My daughter Sarah (age 4½) and son Matthew (2½) not only contributed their slinkies but also demonstrated that systems may have degrees of freedom nobody ever thought of. My wife Bevalyn contributed her kitchen and very much more.

Publication of early preliminary versions was supervised by Mrs. Mary R. Maloney. Mrs. Lila Lowell supervised the last preliminary edition and typed most of the final manuscript. The illustrations owe their final form to Felix Cooper.

I acknowledge gratefully the contributions others have made, but final responsibility for the manuscript rests with me. I shall welcome any further corrections, complaints, compliments, suggestions for revision, and ideas for new home experiments which may be sent to me at the Physics Department, University of California, Berkeley, California, 94720. Any home experiment used in the next edition will show the contributor's name, even though it may first have been done by Lord Rayleigh or somebody.

F. S. Crawford, Jr.

Teaching Notes

Traveling waves have great aesthetic appeal, and it would be tempting to begin with them. In spite of their aesthetic and mathematical beauty, however, waves are physically rather complicated because they involve interactions between large numbers of particles. Since I want to emphasize physical systems rather than mathematics, I begin with the simplest physical *system,* rather than with the simplest *wave.*

Chapter 1 Free Oscillations of Simple Systems: We first review the free oscillations of a one-dimensional harmonic oscillator, emphasizing the physical aspects of inertia and return force, the physical meaning of ω^2, and the fact that for a real system the oscillation amplitude must not be too large if we are to get simple harmonic motion. Next, we consider free oscillations of two coupled oscillators and introduce the concept of normal mode. We emphasize that the mode is like a single "extended" harmonic oscillator, with all parts throbbing at the same frequency and all in phase, and that, for a given mode, ω^2 has the same physical meaning as it does for a one-dimensional oscillator.

What to omit: Throughout the book, several physical systems recur repeatedly. The teacher should not discuss all of them, nor the student study all of them. Examples 2 and 8 are longitudinal oscillations of mass and springs for one (Ex. 2) and two (Ex. 8) degrees of freedom. In later chapters this system is extended to many degrees of freedom, to continuous systems (rubber rope and slinky undergoing longitudinal oscillations) and is used as a model to assist comprehension of sound waves. A teacher who wishes to omit sound may also wish to omit all longitudinal oscillations from the beginning. Similarly, Examples 4 and 10 are *LC* circuits for one and two degrees of freedom. In later chapters they are extended to *LC* networks and then to continuous transmission lines. A teacher who wishes to omit the study of electromagnetic waves in transmission lines, therefore, can omit all *LC* circuit examples from the very beginning. (He can do this and still give a thorough discussion of electromagnetic waves, starting in Chap. 7 with Maxwell's equations.) Do *not* omit transverse oscillations (Examples **3** and 9).

Home experiments: We strongly advocate Home Exp. 1.24 (Sloshing mode in a pan of water) and the related Prob. 1.25 (Seiches), to get the student started "doing it himself." Home Exp. 1.8 (Coupled cans of soup) makes a good class demonstration. Of course, you may already have available such a demonstration (coupled pendulums). Nevertheless, I advocate slinky and soup cans for its crudity, even as a class demonstration, since it may encourage the student to get his own slinky and soup.

Chapter 2   *Free Oscillations of Systems with Many Degrees of Freedom:* We extend the number of degrees of freedom from two to a very large number and find the transverse modes—the standing waves—of a continuous string. We define k and introduce the concept of a dispersion relation, giving ω as a function of k. We use the modes of the string to introduce Fourier analysis of periodic functions in Sec. 2.3. The exact dispersion relation for beaded springs is given in Sec. 2.4.

What to omit: Sec. 2.3 is optional—especially if the students already know some Fourier analysis. Example 5 (Sec. 2.4) is a linear array of coupled pendulums, the simplest system having a low-frequency cutoff. They are used later to help explain the behavior of other systems that have a low-frequency cutoff. A teacher who does not intend to discuss at a later time systems driven below cutoff (waveguide, ionosphere, total reflection of light in glass, barrier penetration of de Broglie waves, high-pass filters, etc.) need not consider Example 5.

Chapter 3   *Forced Oscillations:* Chapters 1 and 2 started with free oscillations of a harmonic oscillator and ended with free standing waves of closed systems. In Chaps. 3 and 4 we consider forced oscillations, first of *closed* systems (Chap. 3) where we find "resonances," and then in *open* systems (Chap. 4) where we find traveling waves. In Sec. 3.2 we review the damped driven one-dimensional oscillator, considering its transient behavior as well as its steady-state behavior. Then we go to two or more degrees of freedom, and discover that there is a resonance corresponding to every mode of free oscillation. We also consider closed systems driven below their lowest (or above their highest) mode frequency and discover exponential waves and "filtering" action.

What to omit: Transients (in Sec. 3.2) can be omitted. Some teachers may also wish to omit everything about systems driven beyond cutoff.

Home experiments: Home Exps. 3.8 (Forced oscillations in a system of two coupled cans of soup) and 3.16 (Mechanical bandpass filter) require phonograph turntables. They make excellent class demonstrations, especially of exponential waves for systems driven beyond cutoff.

Chapter 4 Traveling Waves: Here we introduce traveling waves resulting from forced oscillations of an *open* system (contrasted with the *standing* waves resulting from forced oscillations of a *closed* system that we found in Chap. 3). The remainder of Chap. 4 is devoted to studying phase velocity (including dispersion) and impedance in traveling waves. We contrast the two "traveling wave concepts," *phase velocity* and *impedance*, with the "standing wave concepts," *inertia* and *return force*, and also contrast the fundamental difference in phase relationships for standing versus traveling *waves*.

Home experiments: We recommend Home Exp. 4.12 (Water prism). This is the first optics kit experiment; it uses the purple filter (which passes red and blue but cuts out green). We strongly recommend Home Exp. 4.18 (Measuring the solar constant at the earth's surface) with your face as detector.

Chapter 5 Reflection: By the end of Chap. 4 we have at our disposal both standing and traveling waves (in one dimension). In Chap. 5 we consider general super positions of standing and traveling waves. In deriving reflection coefficients we make a very "physical" use of the superposition principle, rather than emphasizing boundary conditions. (Use of boundary conditions is emphasized in the problems.)

What to omit: There are many examples, involving sound, transmission lines, and light. Don't do them all! Chapter 5 is essentially the "application", of what we have acquired in Chaps. 1–4. Any or all of it can be omitted.

Home experiments: Everyone should do Home Exp. 5.3 (Transitory waves on a slinky). Home Exps. 5.17 and 5.18 are especially interesting.

Chapter 6 Modulations, Pulses, and Wave Packets: In Chaps. 1–5 we work mainly with a single frequency ω (except for Sec. 2.3 on Fourier analysis). In Chap. 6 we consider superpositions, involving different frequencies, to form pulses and wave packets and to extend the concepts of Fourier analysis (developed in Chap. 2 for periodic functions) so as to include nonperiodic functions.

What to omit: Most of the physics is in the first three sections. A teacher who has omitted Fourier analysis in Sec. 2.3 will undoubtedly want to omit Secs. 6.4 and 6.5, where Fourier integrals are introduced and applied.

Home experiments: No one believes in group velocity until they have watched water wave packets (see Home Exp. 6.11). Everyone should also do Home Exps. 6.12 and 6.13.

Problems: Frequency and phase modulation are discussed in problems rather than in the text. So are such interesting recent developments as Mode-locking of a laser (Prob. 6.23), Frequency multiplexing (Prob. 6.32). and Multiplex Interferometric Fourier Spectroscopy (Prob. 6.33).

Chapter 7   Waves in Two and Three Dimensions: In Chaps. 1–6 the waves are all one-dimensional. In Chap. 7 we go to three dimensions. The propagation vector k is introduced. Electromagnetic waves are studied using Maxwell's equation as the starting point. (In earlier chapters there are many examples of electromagnetic waves in transmission lines, evolving from the *LC*-circuit example.) Water waves are also studied.

What to omit: Sec. 7.3 (Water Waves) can be omitted, but we recommend the home experiments on water waves whether or not Sec. 7.3 is studied. A teacher mainly interested in optics could actually start his course at Sec. 7.4 (Electromagnetic Waves) and continue on through Chaps. 7,8, and 9.

Chapter 8   Polarization: This chapter is devoted to study of polarization of electromagnetic waves and of waves on slinkies, with emphasis on the physical relation between partial polarization and coherence.

Home experiments: Everyone should do at least Home Exps. 8.12, 8.14, 8.16, and 8.18 (Exp. 8.14 requiring slinky; the others, the optics kit).

Chapter 9   Interference and Diffraction: Here we consider superpositions of waves that have traveled different paths from source to detector. We emphasize the physical meaning of coherence. Geometrical optics is treated as a wave phenomenon—the behavior of diffraction-limited beam impinging on various reflecting and refracting surfaces.

Home experiments: Everyone should do at least one each of the many home experiments on interference, diffraction, coherence, and geometrical optics. We also strongly recommend 9.50 (Quadrupole radiation from a tuning fork.)

Problems: Some topics are developed in the problems: Stellar interferometers, including the recently developed "long-base-line interferometry" (Prob. 9.57); the analogy between the phase-contrast microscope and the conversion of AM radio wave to FM is discussed in Prob. 9.59.

Home Experiments

General remarks: At least one home experiment should be assigned per week. For your convenience we list here all experiments involving water waves, waves in slinkies, and sound waves. We also later describe the optics kit.

Water waves: Discussed in Chap. 7; in addition they form a recurring theme developed in the following series of easy home experiments:

Slinkies: Every student should have a Slinky (about $1 in any toy store). Four of the following experiments require a record-player turntable and are therefore outside the "kitchen physics" cost range. However, many students already have record players. (The experiments involving record players make good lecture demonstrations.)

Sound: Many home experiments on sound involve use of two identical tuning forks, preferably C523.3 or A440. The cheapest kind (about $1.25 each), which are perfectly adequate, are available in music stores. Mailing tubes can be purchased for about 25¢ each in stationery or art-supply stores. The following home experiments involve sound:

2.5   Piano as Fourier-analyzing machine—insensitivity of ear to phase

2.6   Piano harmonics—equal-temperament scale

3.27   Resonant frequency width of a mailing tube

4.6   Measuring the velocity of sound with wave packets

4.15   Whiskey-bottle resonator (Helmholtz resonator)

4.16   Sound velocity in air, helium, and natural gas

4.26   Sound impedance

5.15   Effective length of open-ended tube for standing waves

5.16   Resonance in cardboard tubes

5.17   Is your sound-detecting system (eardrum, nerves, brain) a phase-sensitive detector?

5.18   Measuring the relative phase at the two ends of an open tube

5.19   Overtones in tuning fork

5.31   Resonances in toy balloons

6.13   Musical trills and bandwidth

9.50   Radiation pattern of tuning fork—quadrupole radiation

Optics Kit

Components: Four linear polarizers, a circular polarizer, a quarter-wave plate, a half-wave plate, a diffraction grating, and four color filters (red, green, blue, and purple). The components are described in the text (linear polarizer on p.411; circular polarizer, p. 433; quarter-and half-wave retardation plates, p. 435; diffraction grating, p. 496). Some experiments also require microscope slides, a showcase-lamp line source, or a flashlight-bulb point source as described in Home Exp. 4.12, p. 217. Aside from Exp. 4.12, all experiments requiring the optics kit are in Chaps. 8 and 9. They are too numerous to list here.

Home Experiment

The first experiment involving the complete optics kit should be identification of all the components by the student. (Components are listed on the envelope container glued to the inside back cover.) Label the components in some way for future reference. For example, use scissors to round off slightly the four corners of the circular polarizer, and then scratch "IN" near one edge of the input face or stick a tiny piece of tape on that face. Clip *one* corner of the one-quarter-wave retarder; clip *two* corners of the two-quarter- (half-) wave retarder. Scratch a line along the axis of easy transmission on the linear polarizers. (This axis is parallel to one of the edges of the polarizer.)

We should remark that the "quarter-wave plate" gives a spatial retardation of 1400 ± 200 Å, nearly independent of wavelength (for visible Light). Thus the wavelength for which it is a quarter-wave retarder is 5600 ± 800 Å. The ± 200 Å is the manufacturer's tolerance. A manufactured batch that gives retardation 1400 Å is a quarter-wave retarder for green (5600 Å), but it retards by less than one quarter wave for longer wavelengths (red) and more for shorter (blue). Another batch that happens to give retardation 1400 + 200 = 1600 Å is a quarter-wave retarder only for red (6400 Å). One that retards by 1400 − 200 is a quarter-wave plate only for blue (4800 Å). Similar remarks apply to the circular polarizer, since it consists of a sandwich of quarter-wave plate and linear polarizer at 45 deg, and the quarter-wave plate is a retarder of 1400 ± 200 Å. Thus there may be slightly distracting color effects when using white light. The student must be warned that in any experiment where he is supposed to get "black," i.e., extinction, he will always have some "non-extinguished" light of the "wrong" color leaking through. For example, I was naive when I wrote Home Exp. 8.12. You should perhaps strike out everything after the word "band" in the sentences "Do you see the dirk band at green? That is the color of 5600 Å!"

Use of Complex Numbers

Complex numbers simplify algebra when sinusoidal oscillation or waves are to be superposed. They may also obscure the physics. For that reason I have avoided their use, especially in the first part of the hook. All the trigonometric identities that are needed will be found inside the front cover. In Chap. 6 I do make use of the complex representation exp $i\omega t$, so as to use the well-known graphical or "phasor diagram" method of superposing vibrations. In Chap. 8 (Polarization) I use complex quantities extensively. In Chap. 9 (Interference and Diffraction) I do not make much use of complex quantities, even though it would sometimes simplify the algebra. Many teachers may wish to make much more extensive use of complex numbers than I do, especially in Chap. 9. In the sections on Fourier series (Sec. 2.3) and Fourier integrals (Secs. 6.4 and 6.5). I make no use of complex quantities. (I especially wanted to avoid Fourier integrals involving "negative frequencies"!)

A Note on the MKS System of Electrical Units†

Most textbooks in electrical engineering, and many elementary physics texts, use a system of electrical units called the *rationalized MKS system.* This system employs the MKS mechanical units based on the *meter,* the *kilogram,* and the *second.* The MKS unit of force is the *newton,* defined as the force which causes a 1-kilogram mass to accelerate at a rate of 1 meter/sec². Thus a newton is equivalent to exactly 10^5 dynes. The corresponding unit of energy, the newton-meter, or *joule,* is equivalent to 10^7 ergs.

The electrical units in the MKS system include our familiar "practical" unit—coulomb, volt, ampere, and ohm—along with some new ones. Someone noticed that it was possible to assimilate the long-used practical units into a complete system devised as follows. Write Coulomb's law as we did in Eq. 1.1:

$$\mathbf{F}_2 = k\,\frac{q_1 q_2 \hat{\mathbf{r}}_{21}}{r_{21}^{\,2}} \tag{1}$$

† Reprinted from Berkeley Physics Course, Vol. II, Electricity and Magnetism, by Edward M. Purcell, © 1963, 1964, 1965 by Education Development Center, Inc., successor by merger to Educational Services Incorporated.

Instead of setting k equal to 1, give it a value such that F_2 will be given in newtons if q_1 and q_2 are expressed in coulombs and r_{21} in meters. Knowing the relation between the newton and the dyne, between the coulomb and the esu, and between the meter and the centimeter, you can easily calculate that k must have the value 0.8988×10^{10}. (Two 1-coulomb charges a meter apart produce quite a force—around a million tons!) It makes no difference if we write $1/(4\pi\varepsilon_0)$ instead of k, where the constant ε_0 is a number such that $1/(4\pi\varepsilon_0) = k = 0.8988 \times 10^{10}$. Coulomb's law now reads:

$$F = \frac{1}{4\pi\varepsilon_0}\frac{q_1 q_2}{r^2} \tag{2}$$

with the constant ε_0 specified as

$$\varepsilon_0 = 8.854 \times 10^{-12} \text{ coulomb}^2/\text{newton-m}^2 \tag{3}$$

Separating out a factor $1/4\pi$ was an arbitrary move, which will have the effect of removing the 4π that would appear in many of the electrical formulas, at the price of introducing it into some others, as here in Coulomb's law. That is all that "rationalized" means. The constant ε_0 is called the dielectric constant (or "permittivity") of free space.

Electric potential is to be measured in volts, and electric field strength E in volts/meter. The force on a charge q, in field **E**, is:

$$\mathbf{F} \text{ (newtons)} = q\mathbf{E} \text{ (coulombs} \times \text{volts/meter)} \tag{4}$$

An ampere is 1 coulomb/sec, of course. The force per meter of length on each of two parallel wires, r meters apart, carrying current I measured in amperes, is:

$$f \text{ (newtons/meter)} = \left(\frac{\mu_0}{4\pi}\right)\frac{2I^2}{r}\frac{(\text{amp}^2)}{(\text{meters})} \tag{5}$$

Recalling our CGS formula for the same situation,

$$f(\text{dynes/cm}) = \frac{2I^2}{rc^2}\frac{(\text{esu/sec})^2}{(\text{cm}^3/\text{sec}^2)} \tag{6}$$

we compute that $(\mu_0/4\pi)$ must have the value 10^{-7}. Thus the constant μ_0, called the permeability of free space, must be:

$$\mu_0 = 4\pi \times 10^{-7} \text{ newtons/amp}^2 \text{ (exactly)} \tag{7}$$

The magnetic field **B** is defined by writing the Lorentz force law as follows:

$$\mathbf{F} \text{ (newtons)} = q\mathbf{E} + q\mathbf{v} \times \mathbf{B} \tag{8}$$

where **v** is the velocity of a particle in meters/sec, q its charge in coulombs. This requires a new unit for **B**. The unit is called a *tesla*, or a *weber/m²*. One tesla is equivalent to precisely 10^4 gauss. In this system the auxiliary field **H** is expressed in different units, and is related to **B**, in free space, in this way:

$$\mathbf{B} = \mu_0 \mathbf{H} \text{ (in free space)} \tag{9}$$

The relation of **H** to the free current is

$$\int \mathbf{H}.ds = I_{\text{free}} \tag{10}$$

I_{free} being the free current, in amperes, enclosed by the loop around which the line integral on the left is taken. Since ds is to be measured in meters, the unit for *H* is called simply, *ampere/meter.*
Maxwell's equations for the fiels in free space look like this, in the rationalized **MKS** system:

$$\begin{aligned}
\text{div } \mathbf{E} &= \rho & \text{curl } \mathbf{E} &= -\frac{\partial \mathbf{B}}{\partial t} \\
\text{div } \mathbf{B} &= 0 & \text{curl } \mathbf{B} &= \mu_0 \varepsilon_0 \frac{\partial \mathbf{E}}{\partial t} + \mu_0 \mathbf{J}
\end{aligned} \tag{11}$$

If you will compare this with our Gaussian CGS version, in which c appears out in the open, you will see that Eqs. 11 imply a wave velocity $1/\sqrt{\varepsilon_0 \mu_0}$ (in meters/sec, of course). That is:

$$\varepsilon_0 \mu_0 = \frac{1}{c^2} \tag{12}$$

In our Gaussian CGS system the unit of charge, esu, was established by Coulomb's law, with $k \equiv 1$. In this **MKS** system the coulomb is defined, basically, not by Eq. 1 but by Eq. 5, that is, by the force between currents rather than the force between charges. For in Eq. 5 we have $\mu_0 \equiv 4\pi \times 10^{-7}$. In other words, if a new experimental measurement of the speed of light were to change the accepted value of c, we should have to revise the value of the constant ε_0, not that of μ_0.

A partial list of the **MKS** units is given below, with their equivalents in Gaussian CGS units.

Quantity	Symbol	Unit, in rationalized MKS System	Equivalent, in Gaussian CGS units
Distance	s	meter	10^2 cm
Force	**F**	newton	10^5 dynes
Work energy	W	joule	10^7 ergs
Charge	q	coulomb	2.998×10^9 esu
Current	I	ampere	2.998×10^9 esu/sec

Quantity	Symbol	Unit, in rationalized MKS System	Equivalent, in Gaussian CGS units
Electric potential	φ	volt	(1/299.8) statvolts
Electric field	**E**	volt/meter	(1/29980) statvolts/cm
Resistance	R	ohm	1.139×10^{-12} sec/cm
Magnetic field	**B**	tesla	10^4 gauss
Magnetic flux	Φ	weber	10^8 gauss-cm^2
Auxiliary field H	**H**	amperes/meter	$4\pi \times 10^{-3}$ oersted

This **MKS** system is convenient in engineering. For a treatment of the fundamental physics of fields and matter, it has one basic defect. Maxwell's equations for the vacuum fields, in this system, are symmetrical in the electric and magnetic field only if **H**, not **B**, appears in the role of the magnetic field. (Notice that Eqs. 11 above are not symmetrical, even in the absence of **J**.) On the other hand, as we showed in Chapter 10, **B**, not **H**, is the fundamental magnetic field inside matter. That is not a matter of definition or of units, but a fact of nature, reflecting the absence of magnetic charge. Thus the **MKS** system, as it has been constructed, tends to obscure either the fundamental electromagnetic symmetry of the vacuum, or the essential asymmetry of the sources. That was one of our reasons for preferring the Gaussian CGS system in this book. The other reason is that Gaussian CGS units, augmented by practical units on occasion, are still the units used by most working physicists.

Chapter 1

Free Oscillations of Simple Systems

Chapter 1   Free Oscillations of Simple Systems

1.1 Introduction

The world is full of things that move. Their motions can be broadly categorized into two classes, according to whether the thing that is moving stays near one place or travels from one place to another. Examples of the first class are an oscillating pendulum, a vibrating violin string, water sloshing back and forth in a cup, electrons vibrating (or whatever they do) in atoms, light bouncing back and forth between the mirrors of a laser. Parallel examples of traveling motion are a sliding hockey puck, a pulse traveling down a long stretched rope plucked at one end, ocean waves rolling toward the beach, the electron beam of a TV tube; a ray of light emitted at a star and detected at your eye. Sometimes the same phenomenon exhibits one or the other class of motion (i.e., standing still on the average, or traveling) depending on your point of view: the ocean waves travel toward the beach, but the water (and the duck sitting on the surface) goes up and down (and also forward and backward) without traveling. The displacement pulse travels down the rope, but the material of the rope vibrates without traveling.

We begin by studying things that stay in one vicinity and oscillate or vibrate about an average position. In Chaps. 1 and 2 we shall study many examples of the motion of a closed system that has been given an initial excitation (by some external disturbance) and is thereafter allowed to oscillate freely without further influence. Such oscillations are called *free* or *natural oscillations*. In Chap. 1 study of these simple systems having one or two moving parts will form the basis for our understanding of the free oscillations of systems with many moving parts in Chap. 2. There we shall find that the motion of a complicated system having many moving parts may always be regarded as compounded from simpler motions, called *modes*, all going on at once. No matter how complicated the system, we shall find that each one of its modes has properties very similar to those of a simple harmonic oscillator. Thus for motion of any system in a single one of its modes, we shall find that each moving part experiences the same return force per unit mass per unit displacement and that all moving parts oscillate with the same time dependence $\cos(\omega t + \varphi)$, i.e., with the same frequency ω and the same phase constant φ.

Each of the systems that we shall study is described by some physical quantity whose displacement from its equilibrium value varies with position in the system and time. In the mechanical examples (involving moving parts which are point masses subject to return forces), the physical quantity is the displacement of the mass at the point x,y,z from its equilibrium position. The displacement is described by a vector $\psi(x,y,x,t)$. Sometimes we call this vector function of x, y, z, t a *wave function*. (It is only a continuous function of x, y, and z when we can use the continuous approximation, i.e., when near neighbors have essentially the same motion.) In some of the electrical examples, the physical quantity may be the current in a coil or the charge on a capacitor. In others, it may be the electric field $\mathbf{E}(x,y,z,t)$ or the magnetic field $\mathbf{B}(x,y,z,t)$. In the latter cases, the waves are called electromagnetic waves.

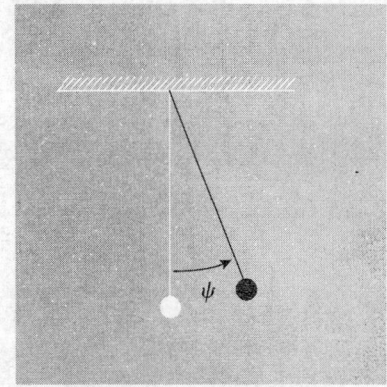

1.2 Free Oscillations of Systems with One Degree of Freedom

We shall begin with things that stay in one vicinity, oscillating or vibrating about an average position. Such simple systems as a pendulum oscillating in a plane, a mass on a spring, and an LC circuit, whose configuration at any time can be completely specified by giving a single quantity, are said to have one degree of freedom—loosely speaking, one moving part (see Fig. 1.1). For example, the swinging pendulum can be described by the angle that the string makes with the vertical, and the LC circuit by the charge on the capacitor. (A pendulum free to swing in any direction, like a bob on a string, has not one but two degrees of freedom; it takes two co-ordinates to specify the position of the bob. The pendulum on a grandfather clock is constrained to swing in a plane, and thus has only one degree of freedom.)

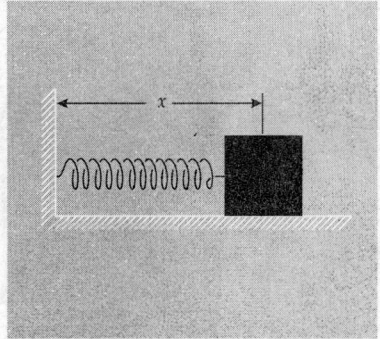

For all these systems with one degree of freedom, we shall find that the displacement of the "moving part" from its equilibrium value has the same simple time dependence (called *harmonic oscillation*),

$$\psi(t) = A \cos (\omega t + \varphi). \tag{1}$$

For the oscillating mass, ψ may represent the displacement of the mass from its equilibrium position; for the oscillating LC circuit, it may represent the current in the inductor or the charge on the capacitor. More precisely, we shall find Eq. (1) gives the time dependence provided the moving part does not move too far from its equilibrium position. [For large-angle swings of a pendulum, Eq. (1) is a poor approximation to the motion; for large extensions of a real spring, the return force is not proportional to the extension, and the motion is not given by Eq. (1); a large enough charge on a capacitor will cause it to "break down" by sparking between the plates, and the charge will not satisfy Eq. (1).]

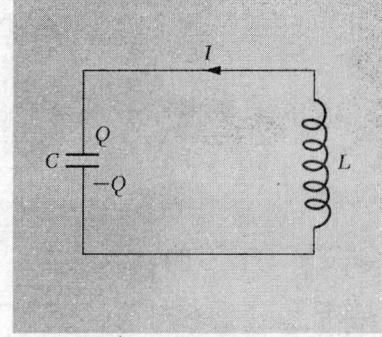

Fig. 1.1 *Systems with one degree of freedom. (The pendulum is constrained to swing in a plane.)*

Nomenclature. We use the following nomenclature with Eq. (1): A is a positive constant called the *amplitude*; ω is the *angular frequency*, measured in radians per second; $v = \omega / 2\pi$ is the *frequency*, measured in cycles per second, or hertz (abbreviated cps, or Hz). The inverse of v is called the *period T*, which is given in seconds per cycle:

$$T = \frac{1}{v}. \tag{2}$$

The *phase constant φ* corresponds to the choice of the zero of time. Often we are not particularly interested in the value of the phase constant. In these cases we can always "reset the clock," so that φ becomes zero, and then we write $\psi = A \cos \omega t$ or $\psi = A \sin \omega t$, instead of the more general Eq. (1).

Return force and inertia.   The oscillatory behavior represented by Eq. (1) always results from the interplay of two intrinsic properties of the physical system which have opposite tendencies: *return force* and *inertia*. The "return force" tries to return ψ to zero by imparting a suitable "velocity" $d\psi/dt$ to the moving part. The greater ψ is, the stronger the return force. For the oscillating *LC* circuit, the return force is due to the repulsive force between the electrons, which makes the electrons prefer not to crowd onto one of the capacitor plates, but rather to distribute themselves equally on each plate, giving zero charge. The second property, "inertia," "opposes" any change in $d\psi/dt$. For the oscillating *LC* circuit, the inertia is due to the inductance *L*, which opposes any change in the current $d\psi/dt$ (letting ψ stand for the charge on the capacitor).

Oscillatory behavior.   If we start with ψ positive and $d\psi/dt$ zero, the return force gives an acceleration which induces a negative velocity. By the time ψ returns to zero, the negative velocity is maximum. The return force is zero at $\psi = 0$, but the negative velocity now induces a negative displacement. Then the return force becomes positive, but it must now overcome the inertia of the negative velocity. Finally, the velocity $d\psi/dt$ is zero, but by that time the displacement is large and negative, and the process reverses. This cycle goes on and on: the return force tries to restore ψ to zero; in so doing, it induces a velocity; the inertia preserves the velocity and causes ψ to "overshoot." The system oscillates.

Physical meaning of ω^2.   The angular frequency of oscillation ω is related to the physical properties of the system in every case (as we shall show) by the relation

$$\boxed{\omega^2 = \text{return force per unit displacement per unit mass.}} \tag{3}$$

Sometimes, as in the case of the electrical examples (*LC* circuit), the "inertial mass" may not actually be mass.

Damped oscillations. If left undisturbed, an oscillating system will continue to oscillate forever in accordance with Eq. (1). However, in any real physical situation, there are "frictional," or "resistive," processes which "damp" the motion. Thus a more realistic description of an oscillating system is given by a "damped oscillation." If the system is "excited" into oscillation at $t = 0$ (by giving it a bump or closing a switch or something), we find (see Vol. I, Chap. 7, page 209)

$$\psi(t) = Ae^{-t/2\tau} \cos(\omega t + \varphi), \tag{4}$$

for $t \geq 0$, with the understanding that ψ is zero for $t < 0$. For simplicity we shall use Eq. (1) instead of the more realistic Eq. (4) in the examples that follow. We are thus neglecting friction (or resistance in the case of the *LC* circuit) by taking the decay time τ to be infinite.

Example 1: Pendulum

A simple pendulum consists of a massless string or rod of length l attached at the top to a rigid support and at the bottom to a "point" bob of mass M (see Fig. 1.2). Let ψ denote the angle (in radians) that the string makes with the vertical. (The pendulum swings in a plane; its configuration is given by ψ alone.) The displacement of the bob, as measured along the perimeter of the circular arc of its path, is $l\psi$. The corresponding instantaneous tangential velocity is $l\,d\psi/dt$. The corresponding tangential acceleration is $l\,d^2\psi/dt^2$. The return force is the tangential component of force. The string does not contribute to this force component. The weight Mg contributes the tangential component $-Mg \sin \psi$. Thus Newton's second law (mass times acceleration equals force) gives

$$Ml\frac{d^2\psi}{dt^2} = -Mg \sin \psi(t). \tag{5}$$

We now use the Taylor's series expansion [Appendix, Eq. (4)]

$$\sin \psi = \psi - \frac{\psi^3}{3!} + \frac{\psi^5}{5!} - \cdots, \tag{6}$$

where the ellipsis (. . .) denotes the rest of the infinite series. We see that for sufficiently small ψ (in radians, remember), we can neglect all terms in Eq. (6) except the first one, ψ. You may ask, "How small

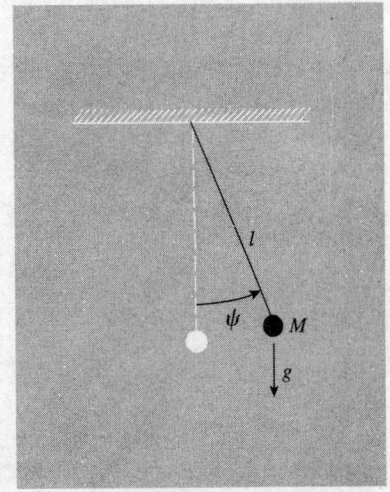

Fig. 1.2 Simple pendulum.

is 'sufficiently small'?" That question has no universal answer—it depends on how accurately you can determine the function $\psi(t)$ in the experiment you have in mind (this is physics, remember—nothing is perfectly measurable) and on how much you care. For example, for $\psi = 0.10$ rad (5.7 deg), sin ψ is 0.0998; in some problems "0.0998 = 0.1000" is a poor approximation. For $\psi = 1.0$ rad (57.3 deg), sin ψ is 0.841; in some problems "0.8 = 1.0" is an adequate approximation.

If we retain only the first term in Eq. (6), then Eq. (5) takes on the form

$$\frac{d^2\psi}{dt^2} = -\omega^2\psi, \qquad (7)$$

where

$$\omega^2 = \frac{g}{l}. \qquad (8)$$

The general solution of Eq. (7) is the harmonic oscillation given by

$$\psi(t) = A \cos{(\omega t + \varphi)}.$$

Note that the angular frequency of oscillation, given by Eq. (8), can be written

ω^2 = return force per unit displacement per unit mass:

$$\omega^2 = \frac{Mg\psi}{(l\psi)M} = \frac{g}{l},$$

using the approximation that sin ψ equals ψ.

The two constants A and φ are determined by the initial conditions, i.e., by the displacement and velocity at time $t = 0$. (Since ψ is an angular displacement, the corresponding "velocity" is the angular velocity $d\psi/dt$.) Thus we have

$$\psi(t) = A \cos{(\omega t + \varphi)},$$
$$\dot{\psi}(t) \equiv \frac{d\psi(t)}{dt} = -\omega A \sin{(\omega t + \varphi)}$$

so that

$$\psi(0) = A \cos\varphi,$$
$$\dot{\psi}(0) = -\omega A \sin\varphi.$$

These two equations may be solved for the positive constant A and for sin φ and cos φ (which determine φ).

Example 2: Mass and springs—longitudinal oscillations

Mass M slides on a frictionless surface. It is connected to rigid walls by means of two identical springs, each of which has zero mass, spring constant K, and relaxed length a_0. At the equilibrium position, each spring is stretched to length a, and thus each spring has tension $K(a - a_0)$ at equilibrium (see Fig. 1.3*a* and *b*). Let z be the distance of M from the left-hand wall. Then its distance from the right-hand wall is $2a - z$ (see Fig. 1.3*c*). The left-hand spring exerts a force $K(z - a_0)$ in the $-z$ direction. The right-hand spring exerts a force $K(2a - z - a_0)$ in the $+z$ direction. The total force F_z in the $+z$ direction is the superposition (sum) of these two forces:

$$F_z = -K(z - a_0) + K(2a - z - a_0)$$
$$= -2K(z - a).$$

Newton's second law then gives

$$\frac{M\,d^2z}{dt^2} = F_z = -2K(z - a). \qquad (9)$$

The displacement from equilibrium is $z - a$. We designate this by $\psi(t)$:

$$\psi(t) \equiv z(t) - a.$$

then

$$\frac{d^2\psi}{dt^2} = \frac{d^2z}{dt^2}.$$

Now we can write Eq. (9) in the form

$$\frac{d^2\psi}{dt^2} = -\omega^2\psi, \qquad (10)$$

with

$$\omega^2 = \frac{2K}{M}. \qquad (11)$$

The general solution of Eq. (10) is again the harmonic oscillation $\psi = A \cos(\omega t + \varphi)$. Note that Eq. (11) has the form ω^2 = force per unit displacement per unit mass, since the return force is $2K\psi$ for a displacement ψ.

Example 3: Mass and springs–transverse oscillations

The system is shown in Fig. 1.4. Mass M is suspended between rigid supports by means of two identical springs. The springs each have zero mass, spring constant K, and unstretched length a_0. They each have length a at the equilibrium position of M. We

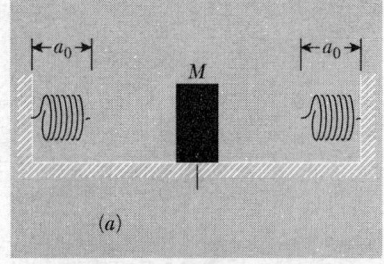

(a)

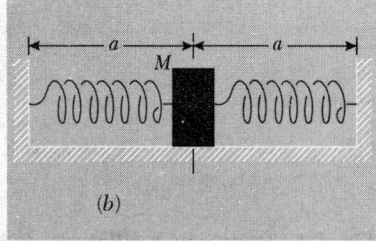

(b)

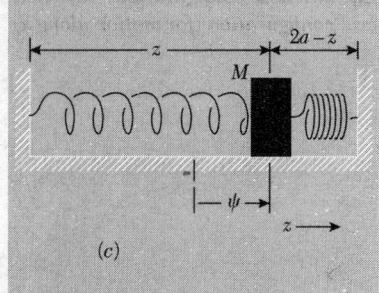

(c)

Fig. 1.3 Longitudinal oscillations. (a) Springs relaxed and unattached, (b) Springs attached, M at equilibrium position. (c) General configuration.

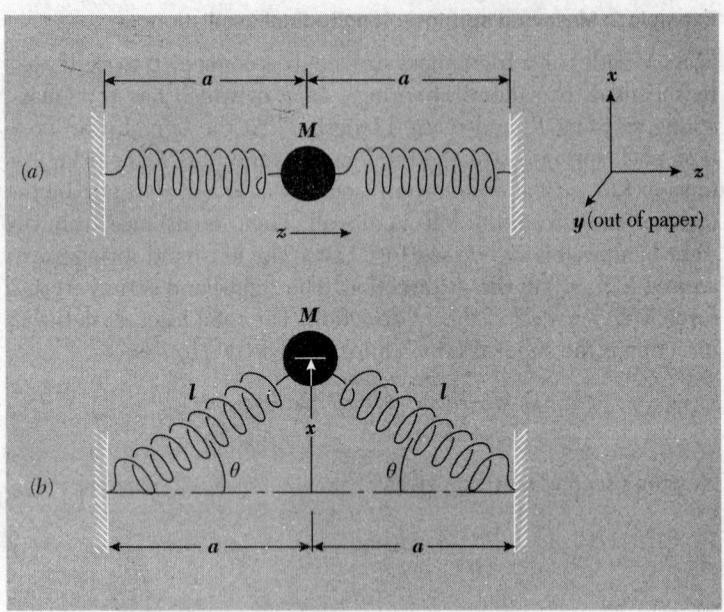

Fig. 1.4 Transverse oscillations. (a) Equilibrium configuration, (b) General configuration (for motion along x).

neglect the effect of gravity. (Gravity does not produce any return force in this problem. It does cause the system to "sag," but that does not affect the results in the order of approximation that we are interested in.) Mass M now has three degrees of freedom: It can move in the z direction (along the axis of the springs) to give "longitudinal" oscillation. That is the motion we considered above, and we need not repeat those considerations. It can also move in the x direction or in the y direction to give "transverse" oscillations. For simplicity, let us consider only motion along x. We may imagine that there is some frictionless constraint that allows complete freedom of motion in the transverse x direction but prevents motion along either y or z. (For example, we could drill a hole through M and arrange a frictionless rod passing through the hole, rigidly attached to the walls, and oriented along x. However, you can easily convince yourself that such a constraint is unnecessary. From the symmetry of Fig. 1.4, you can see that if at a given time the system is oscillating along x, there is no tendency for it to acquire any motion along y or z. The same circumstance holds true for each of the other two degrees of freedom: no unbalanced force along x or y is developed due to oscillation along z, nor along x or z due to oscillation along y.)

At equilibrium (Fig. 1.4a), each of the springs has length a and exerts a tension T_0, given by

$$T_0 = K(a - a_0).\qquad(12)$$

In the general configuration (Fig. 1.4*b*), each spring has length *l* and tension

$$T = K(l - a_0). \tag{13}$$

This tension is exerted along the axis of the spring. Taking the *x* component of this force, we see that each spring contributes a return force $T \sin \theta$ in the $-x$ direction. Using Newton's second law and the fact that $\sin \theta$ is x/l, we find

$$M \frac{d^2x}{dt^2} = F_x = -2T \sin \theta$$

$$= -2K(l - a_0) \frac{x}{l} = -2Kx\left(1 - \frac{a_0}{l}\right). \tag{14}$$

Equation (14) is exact, under our assumptions (including the assumption, expressed by Eq. (13), that the spring is a "linear" or "Hooke's law" spring). Notice that the spring length *l* which appears on the right side of Eq. (14) is a function of *x*. Therefore Eq. (14) is not exactly of the form that gives rise to harmonic oscillations, because the return force on *M* is not exactly linearly proportional to the displacement from equilibrium, *x*.

Slinky approximation. There are two interesting ways in which we can obtain an approximate equation with a linear restoring force. The first way we shall call the *slinky approximation*, in which we neglect a_0/a compared to unity. Hence, since *l* is always greater than *a*, we neglect a_0/l in Eq. (14). [A slinky is a helical spring with relaxed length a_0 about 7.5 cm. It can be stretched to a length *a* of about 4.5 m without exceeding its elastic limit. That would give $a_0/a < 1/60$ in Eq. (14).] Using this approximation, we can write Eq. (14) in the form

$$\frac{d^2x}{dt^2} = -\omega^2 x, \tag{15}$$

with

$$\omega^2 = \frac{2K}{M} = \frac{2T_0}{Ma} \quad \text{(for } a_0 = 0). \tag{16}$$

This has the solution $x = A \cos (\omega t + \varphi)$, i.e., harmonic oscillation. Notice that there is no restriction on the amplitude *A*. We can have "large" oscillations and still have perfect linearity of the return force. Notice also that the frequency for transverse oscillations, as given by Eq. (16), is the same as that for longitudinal oscillations, as given by Eq. (11). That is not true in general. It holds only in the slinky approximation, where we effectively take $a_0 = 0$.

Small-oscillations approximation. If a_0 cannot be neglected with respect to *a* (as is the case, for example, with a rubber rope under the conditions ordinarily met in lecture demonstrations), the slinky approximation does not apply. Then F_x in Eq. (14) is not linear in x. However, we shall show that if the displacements x are small compared with the length a, then l differs from a only by a quantity of order $a(x/a)^2$. In the *small-oscillations approximation*, we neglect the terms in F_x which are nonlinear in x/a. Let us now do the algebra: We want to express l in Eq. (14) as $l = a +$ something, where "something" vanishes when $x = 0$. Since l is larger than a, whether x is positive or negative, "something" must be an even function of x. In fact we have from Fig. 1.4

$$l^2 = a^2 + x^2$$
$$= a^2(1 + \epsilon), \qquad \epsilon \equiv \frac{x^2}{a^2}.$$

Thus

$$\frac{1}{l} = \frac{1}{a}(1 + \epsilon)^{-(1/2)}$$
$$= \frac{1}{a}\left[1 - \left(\frac{1}{2}\epsilon\right) + \left(\frac{3}{8}\epsilon^2\right) - \cdots\right], \qquad (17)$$

where we have used the Taylor's series expansion [Appendix Eq. (20)] for $(1 + x)^n$ with $n = -\frac{1}{2}$ and $x = \epsilon$. Next we make the small-oscillations approximation. We assume we have $\epsilon \ll 1$ and discard the higher-order terms in the infinite series of Eq. (17). (Eventually we shall drop everything except the first term, $1/a$.) Thus we have

$$\frac{1}{l} \approx \frac{1}{a}\left[1 - \left(\frac{1}{2}\epsilon\right)\right]$$
$$= \frac{1}{a}\left[1 - \left(\frac{1}{2}\frac{x^2}{a^2}\right)\right]. \qquad (18)$$

Inserting Eq. (18) into Eq. (14), we find

$$\frac{d^2x}{dt^2} = -\frac{2Kx}{M}\left(1 - \frac{a_0}{l}\right)$$
$$= -\frac{2Kx}{M}\left\{1 - \frac{a_0}{a}\left[1 - \left(\frac{1}{2}\frac{x^2}{a^2}\right) + \cdots\right]\right\}$$
$$= -\frac{2K}{Ma}(a - a_0)x + \frac{K}{M}a_0\left(\frac{x}{a}\right)^3 + \cdots. \qquad (19)$$

Discarding the cubic and higher-order terms, we obtain

$$\frac{d^2x}{dt^2} \approx -\frac{2K}{Ma}(a - a_0)x = -\frac{2T_0x}{Ma}. \qquad (20)$$

[In the second equality of Eq. (20), we used T_0 as given by Eq. (12).]
Equation (20) is of the form

$$\frac{d^2x}{dt^2} = -\omega^2 x,$$

with

$$\omega^2 = \frac{2T_0}{Ma}. \tag{21}$$

Therefore $x(t)$ is given by the harmonic oscillation

$$x(t) = A \cos(\omega t + \varphi).$$

Notice that ω^2 given by Eq. (21) is the return force per unit displacement per unit mass: for small oscillations, the return force is the tension T_0 times $\sin\theta$, which is x/a, times two (two springs). The displacement is x; the mass is M. Thus the return force per unit displacement per unit mass is $2T_0(x/a) / xM$.

Notice that the frequency for transverse oscillations is given by $\omega^2 = 2T_0/Ma$ for both the case of the slinky approximation ($a_0 = 0$) and the small-oscillations approximation ($x/a \ll 1$), as we see by comparing Eqs. (16) and (21). In the slinky approximation, the longitudinal oscillation also has this same frequency, as we see from Eqs. (11) and (16). If the slinky approximation does not hold (i.e., if a_0/a cannot be neglected), then the longitudinal oscillations and (small) transverse oscillations do not have the same frequency, as we see from Eqs. (11), (12), and (21). In this case,

$$(\omega^2)_{\text{long}} = \frac{2Ka}{Ma}, \tag{22}$$

$$(\omega^2)_{\text{tr}} = \frac{2T_0}{Ma}, \qquad T_0 = K(a - a_0). \tag{23}$$

Thus for small oscillations of a rubber rope (where a_0/a cannot be neglected), the longitudinal oscillations are more rapid than the transverse oscillations:

$$\frac{\omega_{\text{long}}}{\omega_{\text{tr}}} = \frac{1}{\left[1 - \dfrac{a_0}{a}\right]^{1/2}}.$$

Example 4: *LC* circuit

(For a more complete discussion of *LC* circuits, see Vol. 2, Chap. 8.) Consider the series *LC* circuit of Fig. 1.5. The charge displaced from the bottom to the top plate of the left-hand capacitor is Q_1. That displaced from bottom to top of the right-hand capacitor is Q_2. The electromotive force (emf) across the inductance is equal to

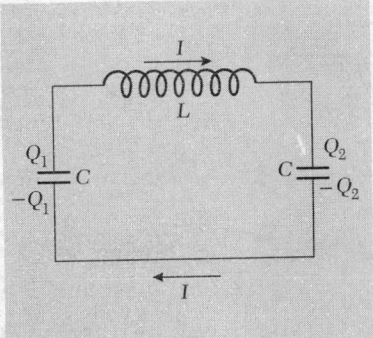

Fig. 1.5 *Series LC circuit. The sign conventions for Q and I are indicated. Q_1 (or Q_2) is positive if the upper plate is positive with respect to the lower plate; I is positive if positive charge flows in the direction of the arrows.*

the "back emf," $L \, dI/dt$. Charge Q_1 provides an electromotive force equal to $C^{-1} Q_1$, such that positive Q_1 drives current in the direction of the arrow in Fig. 1.5. Thus positive Q_1 gives positive $L \, dI/dt$. Similarly, from Fig. 1.5, positive Q_2 gives negative $L \, dI/dt$. Thus we have

$$L \frac{dI}{dt} = C^{-1}Q_1 - C^{-1}Q_2. \qquad (24)$$

At equilibrium there is no charge on either capacitor. The charge Q_2 is built up by the current I at the expense of the charge Q_1. Thus, using charge conservation and the sign conventions of Fig. 1.5, we have

$$Q_1 = -Q_2, \qquad (25)$$

$$\frac{dQ_2}{dt} = 1. \qquad (26)$$

Because of Eqs. (25) and (26), there is only one degree of freedom. We can describe the instantaneous configuration of the system by giving Q_1, or Q_2, or I. The current I will be most convenient in our later work (when we go to systems having more than one degree of freedom), and we shall use it here. We first use Eq. (25) to eliminate Q_1 from Eq. (24); then we differentiate with respect to t and use Eq. (26) to eliminate Q_2:

$$L \frac{dI}{dt} = C^{-1}Q_1 - C^{-1}Q_2 = -2C^{-1}Q_2;$$

$$L \frac{d^2I}{dt^2} = -2C^{-1}\frac{dQ_2}{dt} = -2C^{-1}I.$$

Thus the current $I(t)$ obeys the equation

$$\frac{d^2I}{dt^2} = -\omega^2 I,$$

with

$$\omega^2 = \frac{2C^{-1}}{L}, \qquad (27)$$

and $I(t)$ undergoes harmonic oscillation:

$$I(t) = A \cos(\omega t + \varphi).$$

We can think of Eq. (27) as an illustration of the fact that ω^2 is always the "return force" per unit "displacement" per unit "inertia." We can take the "return force" to be the electromotive force $2C^{-1}Q$, where Q is the "charge displacement" Q_2. We then take the self-inductance L to be the "charge inertia." Then the return force per unit displacement per unit inertia is $(2C^{-1}Q)/QL$.

You may have noticed a mathematical parallelism between Examples 2, 3, and 4. We purposely gave these examples the same spatial symmetry ("inertia" in the center, "driving forces" located symmetrically on either side) so as to produce the parallelism. Such parallelisms are often useful as mnemonic devices.

1.3 Linearity and the Superposition Principle

In Sec. 1.2 we solved for the oscillations of the pendulum and of the mass and springs only for the cases where we could assume the return force to be proportional to $-\psi$, with (for example) no dependence on ψ^2, ψ^3, etc. A differential equation that contains no higher than the first power of ψ, of $d\psi/dt$, of $d^2\psi/dt^2$, etc., is said to be *linear* in ψ and its time derivatives. If, in addition, no terms independent of ψ occur, the equation is said to be *homogeneous* as well. If any higher powers of ψ or its derivatives occur in the equation, the equation is said to be *nonlinear.* For example, Eq. (5) is nonlinear, as we can see from the expansion of $\sin \psi$ given by Eq. (6). Only when we neglect the higher powers of ψ do we obtain a linear equation.

Nonlinear equations are generally difficult to solve. (The nonlinear pendulum equation is solved exactly in Volume I, pp. 225 ff.) Fortunately, there are many interesting physical situations for which linear equations give a very good approximation. We shall deal almost entirely with linear equations.

Linear homogeneous equations. Linear homogeneous differential equations have the following very interesting and important property: *The sum of any two solutions is itself a solution.* Nonlinear equations do not have that property. The sum of two solutions of a nonlinear equation is not itself a solution of the equation.

We shall prove these statements for both cases (linear and nonlinear) at once. Suppose that we have found the differential equation of motion of a system with one degree of freedom to be of the form

$$\frac{d^2\psi(t)}{dt^2} = -C\psi + a\psi^2 + \beta\psi^3 + \gamma\psi^4 + \cdots, \qquad (28)$$

as we found, for example, for the pendulum [Eqs. (5) and (6)] or for the transverse oscillations of a mass suspended by springs [Eq. (19)]. If the constants a, β, γ, etc. are all zero or can be taken to be zero as a sufficiently good approximation, then Eq. (28) is linear and homogeneous. Otherwise, it is nonlinear. Now suppose that $\psi_1(t)$ is a solution of Eq. (28) and that $\psi_2(t)$ is a different solution. For example, ψ_1 may be the solution corresponding to a particular initial displacement and initial velocity of a pendulum bob, and ψ_2

may correspond to different initial displacement and velocity. By hypothesis ψ_1 and ψ_2 each satisfy Eq. (28). Thus we have

$$\frac{d^2\psi_1}{dt^2} = -C\psi_1 + a\psi_1{}^2 + \beta\psi_1{}^3 + \gamma\psi_1{}^4 + \cdots, \tag{29}$$

and

$$\frac{d^2\psi_2}{dt^2} = -C\psi_2 + a\psi_2{}^2 + \beta\psi_2{}^3 + \gamma\psi_2{}^4 + \cdots. \tag{30}$$

The question of interest to us is whether or not the *superposition* of ψ_1 and ψ_2, defined as the sum $\psi(t) = \psi_1(t) + \psi_2(t)$, satisfies the same equation of motion, Eq. (28). Do we have

$$\frac{d^2(\psi_1 + \psi_2)}{dt^2} = -C(\psi_1 + \psi_2) + a(\psi_1 + \psi_2)^2 + \beta(\psi_1 + \psi_2)^3 + \cdots ? \tag{31}$$

The Eq. (31) has the answer "yes" if and only if the constants a, β, etc. are zero. That is easily shown as follows. Add Eqs. (29) and (30).

The sum gives Eq. (31) if and only if all the following conditions are satisfied:

$$\frac{d^2\psi_1}{dt^2} + \frac{d^2\psi_2}{dt^2} = \frac{d^2(\psi_1 + \psi_2)}{dt^2}, \tag{32}$$

$$-C\psi_1 - C\psi_2 = -C(\psi_1 + \psi_2), \tag{33}$$

$$a\psi_1{}^2 + a\psi_2{}^2 = a(\psi_1 + \psi_2)^2, \tag{34}$$

$$\beta\psi_1{}^3 + \beta\psi_2{}^3 = \beta(\psi_1 + \psi_2)^3, \text{etc.} \tag{35}$$

Equations (32) and (33) are both true. Equations (34) and (35) are not true unless a and β are zero. Thus we see that the superposition of two solutions is itself a solution if and only if the equation is linear.

The property that a superposition of solutions is itself a solution is unique to homogeneous linear equations. Oscillations that obey such equations are said to obey the *superposition principle*. We shall not study any other kind.

Superposition of initial conditions. As an example of the applications of the concept of superposition, consider the motion of a simple pendulum under small oscillations. Suppose that one has found a solution ψ_1 corresponding to a certain set of initial conditions (displacement and velocity) and another solution ψ_2 corresponding to a different set of initial conditions. Now suppose we prescribe a third set of initial conditions as follows: We *superpose* *the initial conditions* corresponding to ψ_1 and ψ_2. That means that we give the bob an initial displacement that is the algebraic sum of the initial displacement corresponding to the motion $\psi_1(t)$ and that corresponding to $\psi_2(t)$, and we give the bob an initial velocity that is the algebraic sum of the two initial velocities corresponding to ψ_1 and ψ_2. Then there is no need to do any more work to find the new motion, described by $\psi_3(t)$. The solution ψ_3 is just the superposition $\psi_1 + \psi_2$. We let you finish the proof. This result holds *only* if the pendulum oscillations are sufficiently small so that we can neglect the nonlinear terms in the return force.

Linear inhomogeneous equations.   Linear inhomogeneous equations (i.e., equations containing terms independent of ψ) also give rise to a superposition principle, though of a slightly different sort. There are many physical situations analogous to a driven harmonic oscillator, which satisfies the equation

$$\frac{M \, d^2\psi(t)}{dt^2} = -C\psi(t) + F(t), \tag{36}$$

where $F(t)$ is an "external" driving force that is independent of $\psi(t)$. The corresponding superposition principle is as follows: Suppose a driving force $F_1(t)$ produces an oscillation $\psi_1(t)$ (when F_1 is the only driving force), and suppose another driving force $F_2(t)$ produces an oscillation $\psi_2(t)$ [when $F_2(t)$ is present by itself]. Then, if both driving forces are present simultaneously [so that the total driving force is the superposition $F_1(t) + F_2(t)$], the corresponding oscillation [i.e., corresponding solution of Eq. (36)] is given by the superposition $\psi(t) = \psi_1(t) + \psi_2(t)$. We leave it to you to show that this is true for the linear inhomogeneous Eq. (36) and not true for an equation nonlinear in $\psi(t)$. (See Prob. 1.16.)

The systems we dealt with in Sec. 1.2 and our illustrations of the superposition principle in this section have all been systems of one degree of freedom. However, the superposition principle is applicable to systems of any number of degrees of freedom (when the equations are linear), and we shall be using it very often, usually without mentioning its name.

Example 5: Spherical pendulum

To illustrate the application of the superposition principle when we have two degrees of freedom, we consider the motion of a pendulum consisting of a bob of mass M on a string of length l. The pendulum is free to swing in any direction and is called a *spherical pendulum*. At equilibrium the string is vertical, along z, and the bob is at $x = y = 0$. For displacements x and y that are sufficiently small, you can easily show that $x(t)$ and $y(t)$ satisfy the differential equations

$$M \frac{d^2x}{dt^2} = -\frac{Mg}{l}\,x \tag{37}$$

$$M \frac{d^2y}{dt^2} = -\frac{Mg}{l}\,y. \tag{38}$$

These two equations are "uncoupled," by which we mean that the x component of force depends only on x, not on y, and vice versa. Thus Eq. (37) does not contain y, and similarly Eq. (38) does not contain x. Equations (37) and (38) can be solved independently to give

$$x(t) = A_1 \cos(\omega t + \varphi_1) \tag{39}$$

$$y(t) = A_2 \cos(\omega t + \varphi_2), \tag{40}$$

with

$$\omega^2 = \frac{g}{l},$$

where the constants A_1, A_2, φ_1, and φ_2 are determined by the initial conditions of displacement and velocity in the x and y directions. The complete motion can now be thought of as a *superposition* of the motion $\hat{\mathbf{x}}x(t)$ and the motion $\hat{\mathbf{y}}y(t)$, where $\hat{\mathbf{x}}$ and $\hat{\mathbf{y}}$ are unit vectors. The power of the superposition principle lies in the fact that we can solve for the x and y motions separately and then merely superpose the two motions to get the complete motion involving both degrees of freedom.

1.4   Free Oscillations of Systems with Two Degrees of Freedom

In nature there are many fascinating examples of systems having two degrees of freedom. The most beautiful examples involve molecules and elementary particles (the neutral K mesons especially); to study them requires quantum mechanics. Some simpler examples are a double pendulum (one pendulum attached to the ceiling, the second attached to the bob of the first); two pendulums coupled by a spring; a string with two beads; and two coupled LC circuits. (See Fig. 1.6.) It takes two variables to describe the configuration of such a system, say ψ_a and ψ_b. For example, in the case of a simple pendulum free to swing in any direction, the "moving parts" ψ_a and ψ_b would be the positions of the pendulum in the two perpendicular horizontal directions; in the case of coupled pendulums, the moving parts ψ_a and ψ_b would be the positions of the pendulums; in the case of two coupled LC circuits, the "moving parts" ψ_a and ψ_b would be the charges on the two capacitors or the currents in the circuits.

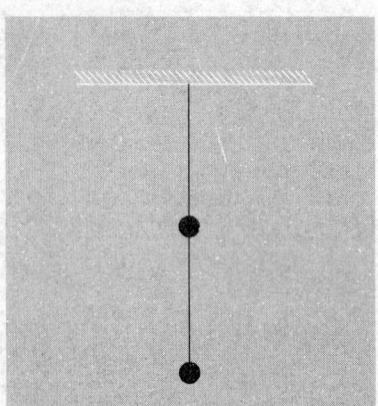
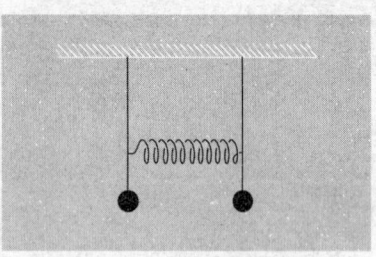
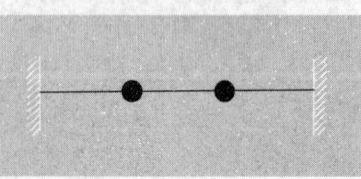
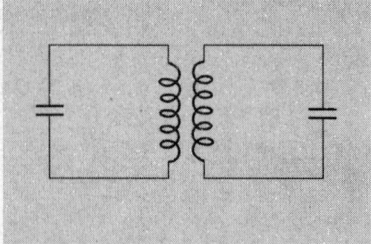

Fig. 1.6   Systems with two degrees of freedom. (The masses are constrained to remain in the plane of the figure.)

The general motion of a system with two degrees of freedom can have a very complicated appearance; no part moves with simple harmonic motion. However, we will show that for two degrees of freedom and for linear equations of motion the most general motion is a *superposition* of two independent simple harmonic motions, both going on simultaneously. These two simple harmonic motions (described below) are called *normal modes* or simply *modes*. By suitable starting conditions (suitable initial values of ψ_a, ψ_b, $d\psi_a/dt$, and $d\psi_b/dt$), we can get the system to oscillate in only one mode or the other. Thus the modes are "uncoupled," even though the moving parts are not.

Properties of a mode. When only one mode is present, each moving part undergoes simple harmonic motion. All parts oscillate with the same frequency. All parts pass through their equilibrium positions (where ψ is zero) simultaneously. Thus, for example, one never has in a single mode, $\psi_a(t) = A \cos \omega t$ and $\psi_b(t) = B \sin \omega t$ (different phase constants) or $\psi_a(t) = A \cos \omega_1 t$ and $\psi_b(t) = B \cos \omega_2 t$ (different frequencies). Instead one has, for one mode (which we call mode 1),

$$\psi_a(t) = A_1 \cos (\omega_1 t + \varphi_1),$$
$$\psi_b(t) = B_1 \cos (\omega_1 t + \varphi_1) = \frac{B_1}{A_1} \psi_a(t), \tag{41}$$

with the *same frequency and phase constant* for both degrees of freedom (moving parts). Similarly, for mode 2, the two degrees of freedom a and b move according to

$$\psi_a(t) = A_2 \cos (\omega_2 t + \varphi_2),$$
$$\psi_b(t) = B_2 \cos (\omega_2 t + \varphi_2) = \frac{B_2}{A_2} \psi_a(t). \tag{42}$$

Each mode has its won characteristic frequency: ω_1 for mode 1, ω_2 for mode 2. In each mode the system also has a characteristic "configuration" or "shape," given by the ratio of the amplitude of motion of the moving parts: A_1/B_1 for mode 1 and A_2/B_2 for mode 2. Note that in a mode the ratio $\psi_a(t)/\psi_b(t)$ is constant, independent of time. It is given by the appropriate ratio A_1/B_1 or A_2/B_2, which can be either positive or negative.

The most general motion of the system is (as we will show) simply a superposition with both modes oscillating at once:

$$\psi_a(t) = A_1 \cos (\omega_1 t + \varphi_1) + A_2 \cos (\omega_2 t + \varphi_2),$$
$$\psi_b(t) = B_1 \cos (\omega_1 t + \varphi_1) + B_2 \cos (\omega_2 t + \varphi_2). \tag{43}$$

Let us consider some specific examples.

Example 6: Simple spherical pendulum

This example is almost too simple, for it does not reveal the full richness of complexity of the general motion that corresponds to Eqs. (43) because the two modes, corresponding respectively to oscillation in the x and in the y direction, have the same frequency, given by $\omega^2 = g/l$. Rather than the superpositions of Eq. (43), corresponding to two different frequencies, we have the simpler results obtained in Eqs. (30) and (40)

$$x(t) \equiv \psi_a(t) = A_1 \cos(\omega_1 t + \varphi_1), \quad \omega_1 = \omega,$$
$$y(t) \equiv \psi_b(t) = B_2 \cos(\omega_2 t + \varphi_2), \quad \omega_2 = \omega_1 = \omega, \quad (44)$$

where we have forced Eqs. (44) to appear to resemble Eqs. (43). For the two modes to have the same frequency is unusual; the two modes are then said to be "degenerate."

Example 7: Two-dimensional harmonic oscillator

In Fig. 1.7 we show a mass M that is free to move in the xy plane. It is coupled to the walls by two unstretched massless springs of spring constant K_1 oriented along x and by two unstretched massless springs of spring constant K_2 oriented along y. In the small-oscillations approximation, where we neglect x^2/a^2, y^2/a^2, and xy/a^2, we shall show that the x component of return force is due entirely to the two springs K_1. Similarly, the y component of return force is entirely due to the springs K_2. You can prove this by writing out the exact F_x and F_y and then discarding nonlinear terms. Here is an easier way to see it: Start at the equilibrium position of Fig. 1.7a. Mentally make a small displacement x of M in the $+x$ direction. The return force at this stage in the argument is given by inspection of Fig. 1.7:

$$F_x = -2K_1 x, \qquad F_y = 0.$$

Next make a second small displacement y (starting at the terminus of the first displacement), this time in the $+y$ direction. The question of interest is whether F_x changes. The K_1 springs get longer by a small amount proportional to y^2. We neglect that. The K_2 springs change their length by an amount proportional to y (one gets shorter, the other longer), but the projection of their force on the x direction is proportional also to x. We neglect the product yx. Thus F_x is unchanged. A similar argument applies to F_y. Thus we obtain the two linear equations

$$M \frac{d^2 x}{dt^2} = -2K_1 x, \quad \text{and} \quad M \frac{d^2 y}{dt^2} = -2K_2 y, \quad (45)$$

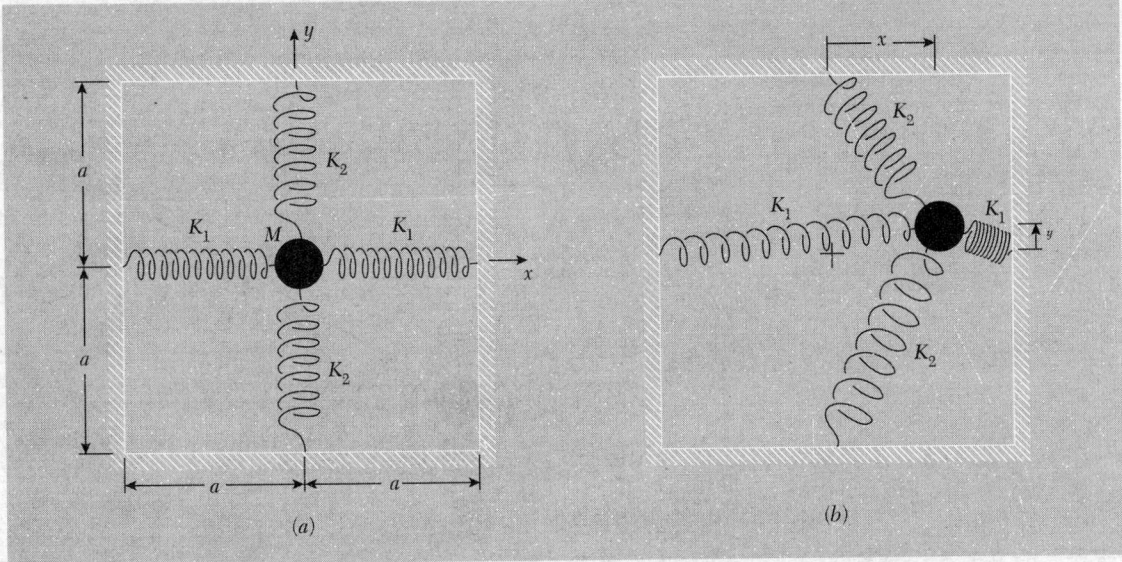

Fig. 1.7 *Two-dimensional harmonic oscillator. (a) Equilibrium. (b) General configuration.*

which have the solutions

$$x = A_1 \cos(\omega_1 t + \varphi_1), \qquad \omega_1{}^2 = \frac{2K_1}{M},$$

$$y = B_2 \cos(\omega_2 t + \varphi_2), \qquad \omega_2{}^2 = \frac{2K_2}{M}. \qquad (46)$$

We see that the x motion and y motion are uncoupled, and that each is a harmonic oscillation with its own frequency. Thus the x motion corresponds to one normal mode of oscillation, the y motion to the other. The x mode has amplitude A_1 and phase constant φ_1 that depend only on the initial values $x(0)$ and $\dot{x}(0)$, i.e., the x displacement and velocity at time $t = 0$. Similarly the y mode has amplitude B_2 and phase constant φ_2 that depend only on the initial values $y(0)$ and $\dot{y}(0)$.

Normal coordinates.   Notice that our solution (46), which is completely general, is still not as general in appearance as Eqs. (43). That is because we were lucky! Our natural choice for x and y along the springs gave us the uncoupled Eq. (45), each of which corresponds to one of the modes. In terms of Eq. (43), we came out with ψ_a luckily chosen so that A_2 came out identically zero and with ψ_b chosen so that B_1 came out identically zero. Our fortunate choice of coordinates gave us what are called *normal coordinates;* in this example the normal coordinates are x and y.

Suppose we had not been so lucky or so wise. Suppose we had used a coordinate system x' and y' related to x and y by a rotation through angle a, as shown in Fig. 1.8. By inspection of the figure

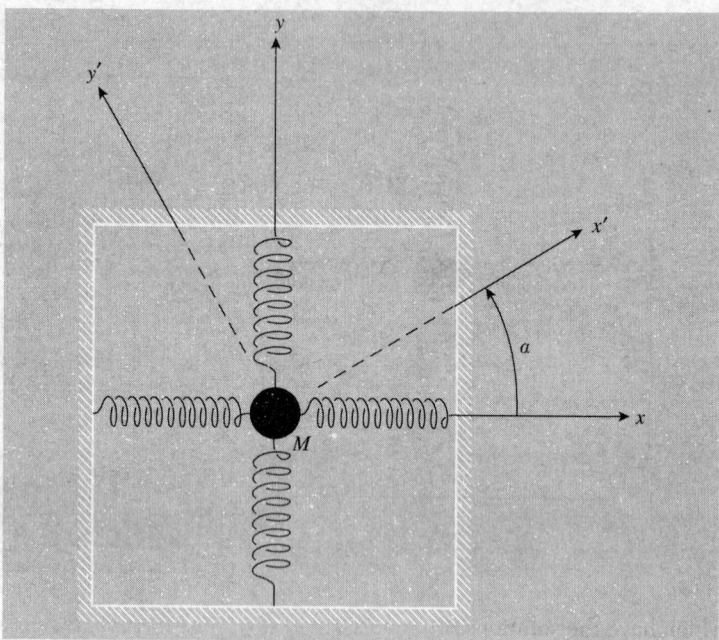

Fig. 1.8   *Rotation of coordinates.*

we see that the normal coordinate x is a linear combination of the coordinates x' and y', as is the other normal coordinate, y. If we had used the "dumb" coordinates x' and y' instead of the "smart" coordinates x and y, we would have obtained two "coupled" differential equations, with both x' and y' appearing in each equation, rather than the uncoupled Eq. (5).

In most problems involving two degrees of freedom it is not easy to find the normal coordinates "by inspection," as we did in the present example. Thus the equations of motion of the different degrees of freedom are usually coupled equations. One method of solving these two coupled differential equations is to search for new variables that are linear combinations of the original "dumb" coordinates such that the new variables satisfy uncoupled equations of motion. The new variables are then called "normal coordinates." In the present example we know how to find the normal coordinates, given the "dumb" coordinates x' and y'. Simply rotate the coordinate system so as to obtain x and y, each of which is a linear combination of x' and y'. In a more general problem, we would have to use a more general linear transformation of coordinates than can be obtained by a simple rotation. That would be the case if, for example, the pairs of springs in Fig. 1.7 were not orthogonal.

Systematic solution for modes.   Without considering any specific physical system, we assume that we have found two coupled first-order linear homogeneous equations in the "dumb" coordinates x and y:

$$\frac{d^2x}{dt^2} = -a_{11}x - a_{12}y \tag{47}$$

$$\frac{d^2y}{dt^2} = -a_{21}x - a_{22}y. \tag{48}$$

Now we simply *assume* that we have osculation in a single normal mode. That means we assume that both degrees of freedom, namely x and y, oscillate with harmonic motion with the *same frequency and same phase constant*. Thus we *assume* we have

$$x = A\cos(\omega t + \varphi), \qquad y = B\cos(\omega t + \varphi), \tag{49}$$

with ω unknown and B/A unknown at this stage. Then we have

$$\frac{d^2x}{dt^2} = -\omega^2 x, \qquad \frac{d^2y}{dt^2} = -\omega^2 y. \tag{50}$$

Substituting Eq. (50) into Eqs. (47) and (48) and rearranging, we obtain two homogeneous linear equations in x and y:

$$(a_{11} - \omega^2)x + a_{12}y = 0, \tag{51}$$

$$a_{21}x + (a_{22} - \omega^2)y = 0. \tag{52}$$

Equations (51) and (52) each give the ratio y/x:

$$\frac{y}{x} = \frac{\omega^2 - a_{11}}{a_{12}}, \tag{53}$$

$$\frac{y}{x} = \frac{a_{21}}{\omega^2 - a_{22}}. \tag{54}$$

For consistency, we need to have Eqs. (53) and (54) give the same result. Thus we need the condition

$$\frac{\omega^2 - a_{11}}{a_{12}} = \frac{a_{21}}{\omega^2 - a_{22}},$$

i.e.,

$$(a_{11} - \omega^2)(a_{22} - \omega^2) - a_{21}a_{12} = 0. \tag{55}$$

Another way to write Eq. (55) is to say that the determinant of coefficients of the linear homogeneous Eqs. (51) and (52) must vanish:

$$\begin{vmatrix} a_{11} - \omega^2 & a_{12} \\ a_{21} & a_{22} - \omega^2 \end{vmatrix} \equiv (a_{11} - \omega^2)(a_{22} - \omega^2) - a_{21}a_{12} = 0. \tag{56}$$

Equation (55) or (56) is a quadratic equation in the variable ω^2. It has two solutions, which we call ω_1^2 and ω_2^2. Thus we have found that if we assume we have oscillation in a single mode, there are exactly two ways that that assumption can be realized. Frequency ω_1

is the frequency of mode 1; ω_2 is that of mode 2. The shape or configuration of x and y in mode 1 is obtained by substituting $\omega^2 = \omega_1{}^2$ back into either one of Eqs. (53) and (54). [They are equivalent, because of Eq. (56).] Thus

$$\left(\frac{y}{x}\right)_{\text{mode 1}} = \left(\frac{B}{A}\right)_{\text{mode 1}} = \frac{B_1}{A_1} = \frac{\omega_1{}^2 - a_{11}}{a_{12}}. \quad (57a)$$

Similarly,

$$\left(\frac{y}{x}\right)_{\text{mode 2}} = \left(\frac{B}{A}\right)_{\text{mode 2}} = \frac{B_2}{A_2} = \frac{\omega_2{}^2 - a_{11}}{a_{12}}. \quad (57b)$$

Once we have found the mode frequencies ω_1 and ω_2 and the amplitude ratios B_1/A_1 and B_2/A_2, we can write down the most general superposition of the two modes as follows:

$$x(t) = x_1(t) + x_2(t) = A_1 \cos{(\omega_1 t + \varphi_1)} + A_2 \cos{(\omega_2 t + \varphi_2)}, \quad (58)$$

$$y(t) = \frac{B_1}{A_1} A_1 \cos{(\omega_1 t + \varphi_1)} + \frac{B_2}{A_2} A_2 \cos{(\omega_2 t + \varphi_2)}$$

$$= B_1 \cos{(\omega_1 t + \varphi_1)} + B_2 \cos{(\omega_2 t + \varphi_2)}. \quad (59)$$

Notice that, whereas we chose A_1, φ_1, A_2, and φ_2 with complete freedom in Eq. (58), we had no freedom at all left when we came to write the constants in Eq. (59), because φ_1 and φ_2 were already fixed and because we had to satisfy Eqs. (57).

The most general solution of Eqs. (47) and (48) consists of a superposition of any two independent solutions which satisfies the four initial conditions given by $x(0)$, $\dot{x}(0)$, $y(0)$, and $\dot{y}(0)$. A superposition of the two normal modes, with the four constants A_1, φ_1, A_2, and φ_2 determined by the four initial conditions, is such a solution. Thus the general solution can be (although it need not be) written as a superposition of the modes.

Example 8: Longitudinal oscillations of two coupled masses

The system is shown in Fig. 1.9. The two masses M slide on a frictionless table. The three springs are massless and identical, each with spring constant K. We will let the reader do the systematic solution (Prob. 1.23), but here let us try to *guess* the normal modes. We know there must be two modes, since there are two degrees of freedom. In a mode, each moving part (each mass) oscillates with harmonic motion. This means that each moving part oscillates with the same frequency, and thus *the return force per unit displacement per unit mass is the same for both masses*. (We learned in Sec. 1.2 that ω^2 is the return force per unit displacement per unit mass. That holds for each moving part, whether it is a single isolated system

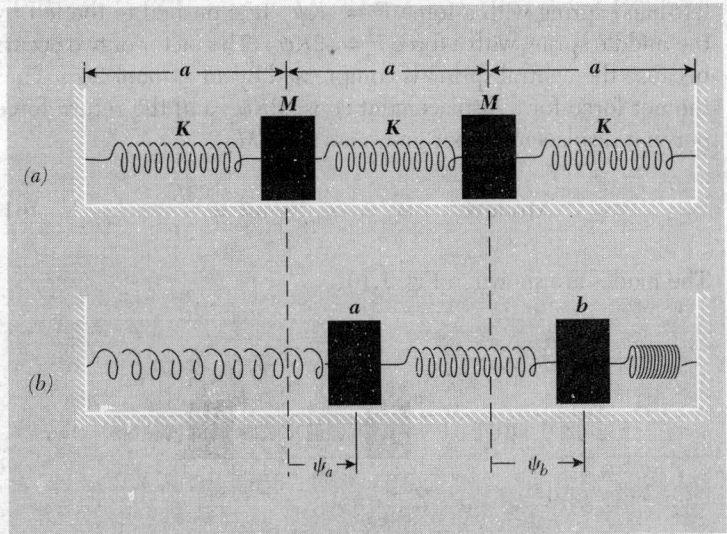

Fig. 1.9 Longitudinal oscillations. (a) Equilibrium. (b) General configuration.

with one degree of freedom or is part of a larger system. The only requirement is that the motion be harmonic motion with a single frequency.)

In the present example the masses are equal. We need therefore only search for configurations that have the same return force per unit displacement for both masses. Let us guess that the displacements may be the same, and see if that works: Suppose we start at the equilibrium position and then displace both masses by the same amount to the right. Is the return force the same on each mass? Notice that the central spring has the same length as it had at equilibrium, so that it exerts no force on either mass. The left-hand mass is pulled to the left because the left-hand spring is extended. The right-hand mass is pushed to the left with the *same* force, because the right-hand spring is compressed by the same amount. We have therefore discovered one mode!

$$\text{Mode 1:} \quad \psi_a(t) = \psi_b(t), \quad\quad \omega_1^2 = \frac{K}{M}. \quad\quad (60)$$

The frequency $\omega_1{}^2 = K/M$ in Eq. (60) follows from the fact that each mass oscillates just as it would if the central spring were removed.

Now let us try to guess the second mode. From the symmetry we guess that if a and b move oppositely we may have a mode. If a moves a distance ψ_a to the right and b moves an equal distance to the left, each has the same return force. Thus the second mode has $\psi_b = -\psi_a$. The frequency ω_2 can be found by considering a single mass and finding its return force per unit displacement per unit mass. Consider the left-hand mass a. It is pulled to the left by the left-hand spring with a force $F_z = -K\psi_a$. It is pushed to the left by the middle spring with a force $F_z = -2K\psi_a$. (The factor of two occurs because the central spring is compressed by an amount $2\psi_a$.) Thus the net force for a displacement ψ_a is $-3K\psi_a$, and the return force per unit displacement per unit mass is $3K/M$:

$$\text{Mode 2:} \qquad \psi_a = -\psi_b, \qquad \omega_2^2 = \frac{3K}{M}. \qquad (61)$$

The modes are shown in Fig. 1.10.

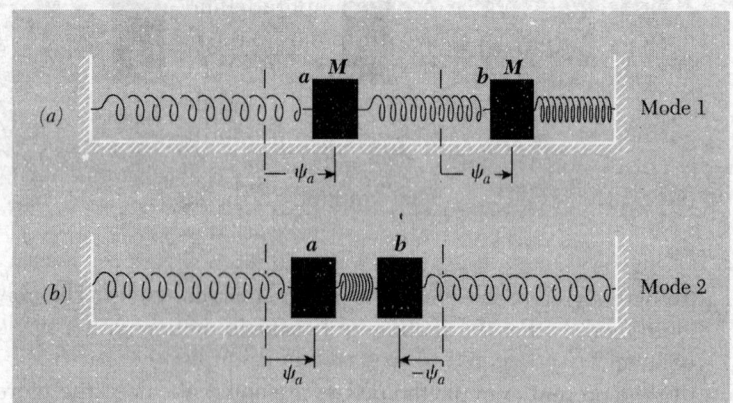

Fig. 1.10 Normal modes of longitudinal oscillation. (a) Mode with lower frequency. (b) Mode with higher frequency.

We shall solve this problem once more, using the method of searching for normal coordinates, i.e., "smart" coordinates. The "smart" coordinates are always a linear combination of ordinary "dumb" coordinates, such that instead of two coupled linear equations, one obtains two uncoupled equations. From Fig. 1.9b, we easily see that the equations of motion for a general configuration are

$$M \frac{d^2\psi_a}{dt^2} = -K\psi_a + K(\psi_b - \psi_a), \qquad (62)$$

$$M \frac{d^2\psi_b}{dt^2} = -K(\psi_b - \psi_a) - K\psi_b. \qquad (63)$$

By inspection of these equations of motion, we see that alternately adding and subtracting these equations will produce the desired uncoupled equations. Adding Eqs. (62) and (63), we obtain

$$M \frac{d^2}{dt^2}(\psi_a + \psi_b) = -K(\psi_a + \psi_b). \qquad (64)$$

Subtracting Eq. (63) from Eq. (62), we obtain

$$M \frac{d^2(\psi_a - \psi_b)}{dt^2} = -3K(\psi_a - \psi_b). \qquad (65)$$

Equations (64) and (65) are uncoupled equations in the variables $\psi_a + \psi_b$ and $\psi_a - \psi_b$. They have the solutions

$$\psi_a + \psi_b \equiv \psi_1(t) = A_1 \cos(\omega_1 t + \varphi_1), \qquad \omega_1^2 = \frac{K}{M}, \qquad (66)$$

$$\psi_a - \psi_b \equiv \psi_2(t) = A_2 \cos(\omega_2 t + \varphi_2), \qquad \omega_2^2 = \frac{3K}{M}, \qquad (67)$$

where A_1 and φ_1 are the amplitude and phase constant of mode 1 and where A_2 and φ_2 are the amplitude and phase constant of mode 2. We see that $\psi_1(t)$ corresponds to the motion of the center of mass, since $\frac{1}{2}(\psi_a + \psi_b)$ is the position of the center of mass. (We could have divided Eq. (64) by 2 and defined ψ_1 to be the position of the center of mass. The proportionality factor of $\frac{1}{2}$ is not of much interest.) We see that ψ_2 is the compression of the central spring, or (what amounts to the same thing) it is the relative displacement of the two masses. If we had been smart enough, we might have chosen ψ_1 and ψ_2 to start with, since the motion of the center of mass and the "internal motion" (relative motion of the two particles) are physically interesting variables. In many cases it is not so easy to find a simple physical meaning for the normal coordinates. Thus we shall usually stick with our original "dumb" coordinates even after finding the modes, simply because we understand them best.

In the present problem we have found the normal coordinates ψ_1 and ψ_2. Let us go back to our more familiar coordinates ψ_a and ψ_b. Solving Eqs. (66) and (67), we find

$$2\psi_a = A_1 \cos(\omega_1 t + \varphi_1) + A_2 \cos(\omega_2 t + \varphi_2) \qquad (68)$$

$$2\psi_b = A_1 \cos(\omega_1 t + \varphi_1) - A_2 \cos(\omega_2 t + \varphi_2). \qquad (69)$$

Notice that if we have a motion that is purely mode 1, then A_2 is zero, and, according to Eqs. (68) and (69), we have $\psi_b = \psi_a$. Similarly, in mode 2 we have $A_1 = 0$ and $\psi_b = -\psi_a$. That is what we found before [in Eqs. (60) and (61)].

Example 9: Transverse oscillations of two coupled masses

The system is shown in Fig. 1.11. The oscillations are assumed to be confined to the plane of the paper. Therefore there are just two degrees of freedom. The three identical massless springs have a relaxed length a_0 that is less than the equilibrium spacing a of the masses. Thus they are all stretched. When the system is at its equilibrium configuration (Fig. 1.11a), the springs have tension T_0.

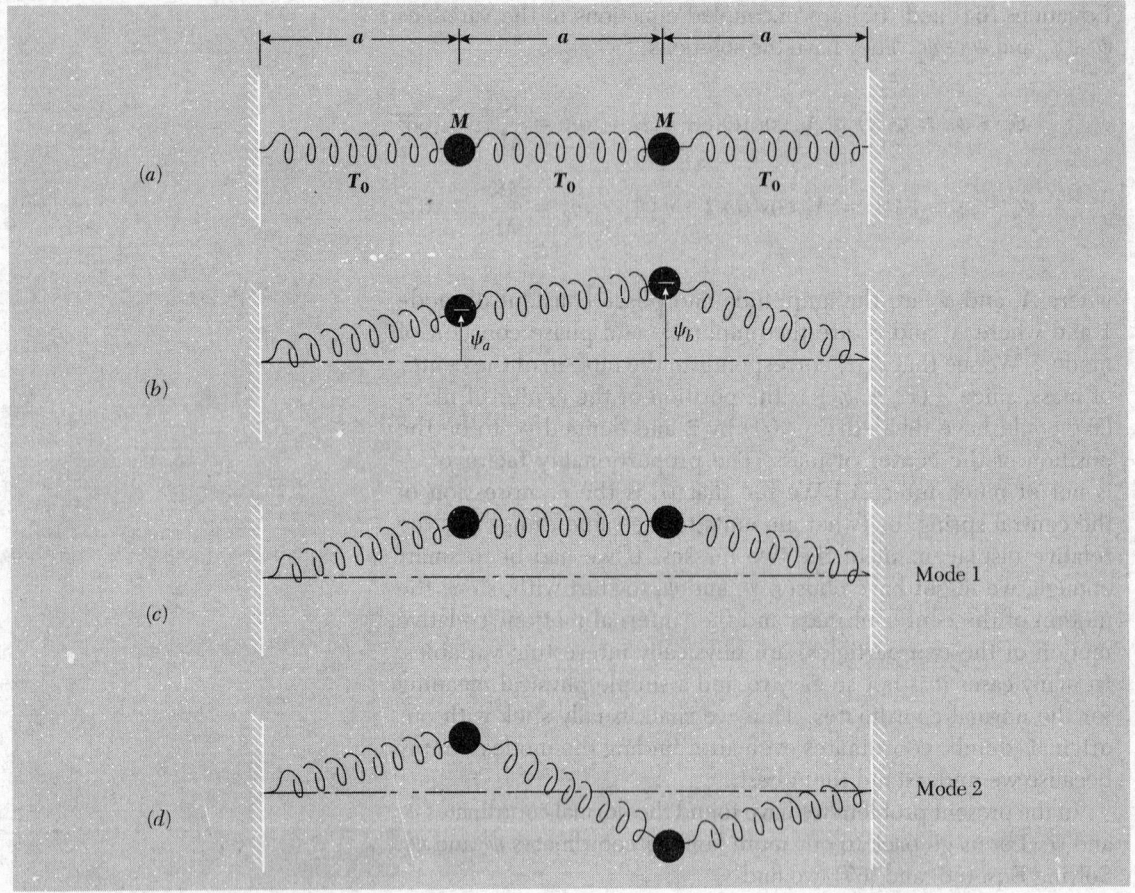

Fig. 1.11 Transverse oscillations.
(a) Equilibrium. (b) General configuration. (c) Mode with lower frequency.
(d) Mode with higher frequency.

Because of the symmetry of the system, the modes are easy to
guess. They are shown in Fig. 1.11. The lower mode (the one with
the lower frequency, i.e., the one with the smaller return force
per unit displacement per unit mass for each of the masses) has a
shape (Fig. 1.11c) such that the center spring is never compressed
or extended. The frequency is thus obtained by considering either
one of the masses separately, with the return force provided only by
the spring that connects it to the wall. For either the slinky approximation (unstretched spring length of zero) or the small-oscillations
approximation (displacements very small compared with the
spacing a), we shall show presently that a displacement ψ_a of the
left-hand mass causes the left-hand spring to exert a return force
of $T_0(\psi_a/a)$. Hence, in this mode the return force per unit displacement per unit mass, ω_1^2, is given by

$$\text{Mode 1:} \quad \omega_1^2 = \frac{T_0}{Ma}, \quad \frac{\psi_b}{\psi_a} = +1. \tag{70}$$

We see this as follows. First consider the slinky approximation (Sec. 1.2). In this approximation, the tension T is larger than T_0 by the factor l/a, where l is the spring length and a is the length at equilibrium (Fig. 1.11a). The spring exerts a transverse return force equal to the tension T times the sine of the angle between the spring and the equilibrium axis of the springs, i.e., the return force is $T(\psi_a/l)$. But $T = T_0(l/a)$. Thus the return force is $T_0(\psi_a/a)$, and this gives Eq. (70). Next consider the small-oscillations approximation (Sec. 1.2). In that approximation, the increase in length of the spring is neglected, because it differs from the equilibrium length a only by a quantity of order $a(\psi_a/a)^2$, and therefore the increase in tension also is neglected. The tension is thus T_0 when the displacement is ψ_a. The return force is equal to the tension T_0 times the sine of the angle between the spring and the equilibrium axis. This angle may be taken to be a "small" angle, since the oscillations are small. Then the angle (in radians) and its sine are equal, and both are equal to ψ_a/a. Thus the return force is $T_0(\psi_a/a)$. This gives Eq. (70).

Similarly, we can obtain the frequency for mode 2 (Fig. 1.11d) as follows: Consider the left-hand mass. The left-hand spring contributes a return force per unit displacement per unit mass of T_0/Ma, as we have just seen in considering mode 1. In mode 2 the center spring is "helping" the left-hand spring, and in fact it is providing twice as great a return force as is the left-hand spring. This is easily seen in the small-oscillations approximation: The spring tension is T_0 for both springs, but the center spring makes twice as large an angle with the axis as does the end spring, so that it gives twice as large a transverse force component. The total return force per unit displacement per unit mass, ω_2^2, is thus given by

$$\text{Mode 2:} \qquad \omega_2^2 = \frac{T_0}{Ma} + \frac{2T_0}{Ma} = \frac{3T_0}{Ma}, \qquad \frac{\psi_b}{\psi_a} = -1. \quad (71)$$

Notice that in the slinky approximation, where the relation $T_0 = K(a - a_0)$ becomes $T_0 = Ka$, the frequencies of the modes of transverse oscillation [Eqs. (70) and (71)] are the same as those for longitudinal oscillation [Eqs. (60) and (61)]. Thus we have a form of degeneracy. This degeneracy does not occur for the small-oscillation approximation, where a_0 is not negligible compared with a.

If the modes had not been so easy to guess, we would have written down the equations of motion of the two masses a and b and then proceeded with the equations, rather than with a mental picture of the physical system itself. We shall let you do that (Prob. 1.20).

Example 10: Two coupled *LC* circuits

Consider the system shown in Fig. 1.12. Let us find the equations of "motion"—motion of the charges in this case. The electromotive

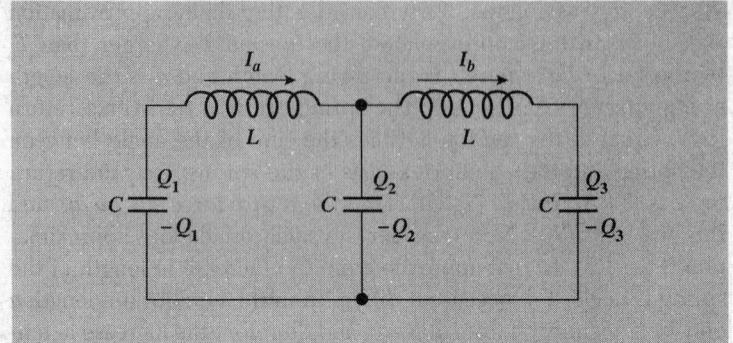

Fig. 1.12 Two coupled LC circuits. General configuration of charges and currents. The arrows give sign conventions for positive currents.

force (emf) across the left-hand inductance is $L\, dI_a/dt$. A positive charge Q_1 on the left-hand capacitor gives an emf $C^{-1}Q_1$ that tends to increase I_a (with our sign conventions). A positive charge Q_2 on the middle capacitor gives an emf $C^{-1}Q_2$ that tends to decrease I_a. Thus we have for the complete contribution to $L\, dI_a/dt$

$$L\frac{dI_a}{dt} = C^{-1}Q_1 - C^{-1}Q_2. \tag{72}$$

Similarly,

$$L\frac{dI_b}{dt} = C^{-1}Q_2 - C^{-1}Q_3. \tag{73}$$

As in Sec. 1.2, we will express the configuration of the system in terms of currents rather than charges. To do this, we differentiate Eqs. (72) and (73) with respect to time and use conservation of charge. Differentiating gives

$$L\frac{d^2I_a}{dt^2} = C^{-1}\frac{dQ_1}{dt} - C^{-1}\frac{dQ_2}{dt}, \tag{74}$$

$$L\frac{d^2I_b}{dt^2} = C^{-1}\frac{dQ_2}{dt} - C^{-1}\frac{dQ_3}{dt}. \tag{75}$$

Charge conservation gives

$$\frac{dQ_1}{dt} = -I_a, \qquad \frac{dQ_2}{dt} = I_a - I_b, \qquad \frac{dQ_3}{dt} = I_b. \tag{76}$$

Substituting Eqs. (76) into Eqs. (74) and (75), we obtain the coupled equations of motion

$$L\frac{d^2I_a}{dt^2} = -C^{-1}I_a + C^{-1}(I_b - I_a) \tag{77}$$

$$L\frac{d^2I_b}{dt^2} = -C^{-1}(I_b - I_a) - C^{-1}I_b. \tag{78}$$

Now that we have the two equations of motion we want to find the two normal modes. These can be found by searching for normal coordinates, by guessing, or by the systematic method (see Prob. 1.21). One finds

$$\text{Mode 1:} \quad I_a = I_b, \qquad \omega_1{}^2 = \frac{C^{-1}}{L}.$$

$$\text{Mode 2:} \quad I_a = -I_b, \qquad \omega_2{}^2 = \frac{3C^{-1}}{L}. \tag{79}$$

Notice that in mode 1 the center capacitor never acquires any charge, and it could be removed without affecting the motion of the charges. Also, in mode 1 the charges Q_1 and Q_3 are always equal in magnitude and opposite in sign. In mode 2 the charges Q_1 and Q_3 are always equal in both magnitude and sign, and Q_2 has twice that magnitude, but opposite sign.

We purposely chose the three Examples (8–10) of longitudinal oscillations (Fig. 1.9), transverse oscillations (Fig. 1.11), and coupled *LC* circuits (Fig. 1.12) to have the same spatial symmetry and to give equations of motion and normal modes with the same mathematical form. We also chose these examples to be the natural extensions (to two degrees of freedom) of the similar systems with one degree of freedom that we considered in Examples (2–4) in Sec. 1.2, as shown in Figs. 1.3, 1.4, and 1.5. In Chap. 2 we shall extend these same three examples to an arbitrarily large number of degrees of freedom.

1.5 Beats

There are many physical phenomena where the motion of a given moving part is a superposition of two harmonic oscillations having different angular frequencies ω_1 and ω_2. For example, the two harmonic oscillations may correspond to the two normal modes of a system having two degrees of freedom. As a contrasting example, the two harmonic oscillations may be due to driving forces produced by two independently oscillating uncoupled systems. This sort of situation is illustrated by two tuning forks of different frequencies. Each produces its own "note" by causing harmonically oscillating pressure variations at the fork, which radiate through the air as sound waves. The motion induced in your eardrum is a superposition of two harmonic oscillations.

In all these examples, the mathematics is the same. For simplicity we assume that the two harmonic oscillations have the same amplitude. We also assume that the two oscillations have the same phase constant, which we take to be zero. Then we write the superposition ψ of the two harmonic oscillations ψ_1 and ψ_2:

$$\psi_1 = A \cos \omega_1 t, \quad \psi_2 = A \cos \omega_2 t, \tag{80}$$

$$\psi = \psi_1 + \psi_2 = A \cos \omega_1 t + A \cos \omega_2 t. \tag{81}$$

Modulation. We shall now recast Eq. (81) into an interesting form. We define an "average" angular frequency ω_{av} and a "modulation" angular frequency ω_{mod}:

$$\omega_{av} \equiv \tfrac{1}{2}(\omega_1 + \omega_2), \qquad \omega_{mod} \equiv \tfrac{1}{2}(\omega_1 - \omega_2). \qquad (82)$$

The sum and difference of these give

$$\omega_1 = \omega_{av} + \omega_{mod}, \qquad \omega_2 = \omega_{av} - \omega_{mod}. \qquad (83)$$

Then we may write Eq. (81) in terms of ω_{av} and ω_{mod}:

$$\psi = A \cos \omega_1 t + A \cos \omega_2 t$$
$$= A \cos(\omega_{av}t + \omega_{mod}t) + A \cos(\omega_{av}t - \omega_{mod}t)$$
$$= [2A \cos \omega_{mod}t] \cos \omega_{av}t,$$

i.e.,

$$\psi = A_{mod}(t) \cos \omega_{av}t, \qquad (84)$$

where

$$A_{mod}(t) = 2A \cos \omega_{mod}t. \qquad (85)$$

We can think of Eqs. (84) and (85) as representing an oscillation at angular frequency ω_{av}, with an amplitude A_{mod} that is not constant but rather varies with time according to Eq. (85). Equations (84) and (85) are exact. However, it is most useful to write the superposition Eq. (81), in the form of Eqs. (84) and (85) when ω_1 and ω_2 are of comparable magnitude. Then the modulation frequency is small in magnitude compared with the average frequency:

$$\omega_1 \approx \omega_2; \qquad \omega_{mod} \ll \omega_{av}.$$

In that case, the modulation amplitude, $A_{mod}(t)$, varies only slightly during several of the so-called "fast" oscillations of $\cos \omega_{av}t$, and therefore Eq. (84) corresponds to "almost harmonic" oscillation at frequency ω_{av}. Of course, if A_{mod} is exactly constant, Eq. (84) represents exact harmonic oscillation at angular frequency ω_{av}. Then $\omega_{av} = \omega_1 = \omega_2$, since A_{mod} is only constant if ω_{mod} is zero. If ω_1 and ω_2 differ only slightly, the superposition of the two (exactly harmonic) oscillations ω_1 and ω_2 is called an "almost harmonic" or "almost monochromatic" oscillation of frequency ω_{av} with a slowly varying amplitude.

Almost harmonic oscillation. This is our first example of a very important and very general result that we will encounter many times: A linear superposition of two or more exactly harmonic oscillations having different frequencies (and different amplitudes and phase constants), with all the frequencies lying in a relatively narrow range or "band" of frequencies, gives a resultant oscillation that is "almost" a harmonic oscillation, with a frequency ω_{av} that lies somewhere in the band of the "component" oscillations that make up the superposition. The resultant motion is not exactly a harmonic oscillation because the amplitude and phase constant are not exactly constant, but only "almost constant." Their variation is negligible during one cycle of oscillation at the average "fast" frequency ω_{av}, provided that the frequency range or "bandwidth" of the component harmonic oscillations is small compared with ω_{av}. (We shall prove these remarks in Chap. 6.)

Some physical examples of beats follow:

Example 11: Beats produced by two tuning forks

When a sound wave reaches your ear, it produces a variation in air pressure at the eardrum. Let ψ_1 and ψ_2 represent the respective contributions to the gauge pressure produced outside your ear-drum by two tuning forks, numbered 1 and 2. (The gauge pressure is just the pressure on the outer surface of your eardrum minus the pressure on the inner surface; the pressure on the inner surface is normal atmospheric pressure. This pressure difference provides the driving force to drive the eardrum.)

If both forks are struck equally hard at the same time and are held at the same distance from the eardrum, the amplitudes and phase constants for the gauge pressures ψ_1 and ψ_2 are the same, and thus Eq. (80) correctly represents the two pressure contribu-tions. The total pressure (which gives the total force on the drum) is the superposition $\psi = \psi_1 + \psi_2$ of the contributions from the two forks. It is given either by Eq. (81) or by Eqs. (84) and (85). If the frequencies of the two forks, v_1 and v_2, differ by more than about 6% of their average value, then your ear and brain ordinarily pre-fer Eq. (81). That is, you "hear" the total sound as two separate notes with slightly different pitches. For example, if v_2 is $\frac{5}{4}$ times v_1, you hear two notes with an interval of a "major third." If v_2 is $1.06v_1$, you hear v_2 as a note "one half-tone higher" in pitch than v_1. However, if v_1 and v_2 differ by less than about 10 cps, your ear (plus brain) no longer easily recognizes them as different notes. (A musician's trained ear may do much better.) Then a superposi-tion of the two is not heard as a "chord" made up of the two notes v_1 and v_2, but rather as a single pitch of frequency v_{av} with a slowly varying amplitude A_{mod}, just as given by Eqs. (84) and (85).

Square-law detector. The modulation amplitude A_{mod} oscillates at the modulation angular frequency ω_{mod}. Whenever $\omega_{mod}t$ has increased by an amount 2π (radians of phase), the amplitude A_{mod} has gone through one complete cycle of oscillation (i.e., the "slow" oscillation at the modulation frequency) and has returned to its original value. At two times during one cycle, A_{mod} is zero. At those times, the ear doesn't hear anything—there is no sound. In between the silences, you hear a sound at the average pitch. Since $\cos \omega_{\text{mod}} t$ goes from zero to +1, to zero, to –1, to zero, to +1, etc., we see that A_{mod} has opposite signs at successive loud times. Nevertheless, your ear does not recognize "two kinds" of loud times, as you will discover if you perform the experiment with tuning forks. Thus your ear (plus brain) does not distinguish positive from negative values of A_{mod}. It only distinguishes whether the magnitude of A_{mod} is large ("loud") or small ("soft") that is, whether the *square* of A_{mod} is large or small. For that reason, your ear plus brain) is sometimes said to be a *square-law detector*. Since A_{mod}^2 has *two* maxima for every modulation cycle (during which $\omega_{\text{mod}} t$ increases by 2π), the repetition rate for the Sequence "loud, soft, loud, soft, loud, soft, . . ." is twice the modulation frequency. This repetition rate of large values of A_{mod}^2 is called the *beat frequency*:

$$\omega_{\text{beat}} = 2\omega_{\text{mod}} = \omega_1 - \omega_2. \qquad (86)$$

We can see this algebraically as follows:

$$A_{\text{mod}}(t) = 2A \cos \omega_{\text{mod}} t.$$
$$\left[A_{\text{mod}}(t) \right]^2 = 4A^2 \cos^2 \omega_{\text{mod}} t;$$

but

$$\cos^2 \theta = \tfrac{1}{2}\left[\cos^2 \theta + \sin^2 \theta + \cos^2 \theta - \sin^2 \theta \right] = \tfrac{1}{2}\left[1 + \cos 2\theta \right].$$

Thus

$$\left[A_{\text{mod}}(t) \right]^2 = 2A^2 \left[1 + \cos 2\omega_{\text{mod}} t \right],$$

i.e.,

$$(A_{\text{mod}})^2 = 2A^2 \left[1 + \cos \omega_{\text{beat}} t \right]. \qquad (87)$$

Thus A_{mod}^2 oscillates about its average value at twice the modulation frequency, i.e., at the beat frequency, $\omega_1 - \omega_2$.

The superposition of two harmonic oscillations with nearly equal frequencies to produce beats is illustrated in Fig. 1.13.

Example 12: Beats between two sources of visible light

In 1955, Forrester, Gudmundsen, and Johnson performed a beautiful experiment showing beats between two independent sources of visible light with nearly the same frequency.† The light sources were gas discharge tubes containing freely decaying mercury atoms with an average frequency of $\nu_{\text{av}} = 5.49 \times 10^{14}$ cps, corresponding to the bright "green line" of mercury. The atoms were placed in a magnetic

†A. T. Forrester, R. A. Gudmundsen, and P. O. Jhonson, "Photoelectric mixing of incoherent light," *Phys. Rev.* **99**, 1691 (1955).

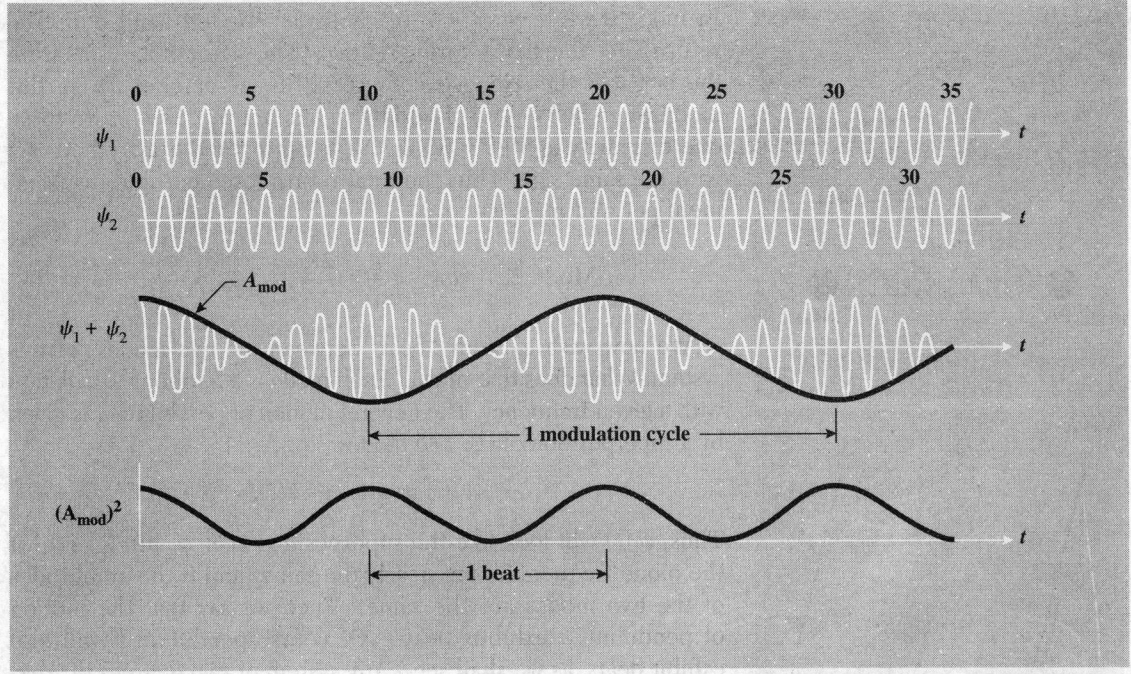

field. This caused the green radiation to "split" into two neighboring frequencies, with the frequency difference proportional to the magnetic field. The beat frequency was $\nu_1 - \nu_2 \approx 10^{10}$ cps. This is a typical "radar" or "microwave" frequency. Their detector used the photoelectric effect to give an output electric current proportional to the square of the modulation amplitude of the resultant electric field in the light wave. Thus the detector was a square-law detector. The output of their detector showed a time variation similar to the "loudness," A_{mod}^2, in Fig. 1.13.

Fig. 1.13 Beats. ψ_1 and ψ_2 are the pressure variations at your ear produced by two tuning forks with frequency ratio $\nu_1/\nu_2 = 10/9$. The total pressure is the superposition $\psi_1 + \psi_2$, which is an "almost harmonic" oscillation at frequency ν_{av} with slowly varying amplitude A_{mod} (t). The loudness is proportional to $(A_{mod})^2$ and consists of a constant (average value) plus a sinusoidal variation at the beat frequency. The beat frequency is twice the modulation frequency.

Example 13: Beats between the two normal modes of two weakly coupled identical oscillators

Consider the system of two identical pendulums coupled by a spring shown in Fig. 1.14. The normal modes are easily guessed by analogy with the longitudinal oscillations of the identical masses studied in Sec. 1.4. In mode 1 we have $\psi_a = \psi_b$. The coupling spring could just as well be removed; the return force is entirely due to gravity. The return force per unit displacement per unit mass (assuming small-oscillation amplitudes, for which we have a linear restoring force) is $Mg\theta/(l\theta)M = g/l$:

$$\text{Mode 1:} \quad \omega_1^2 = \frac{g}{l}, \quad \psi_a = \psi_b. \quad (88)$$

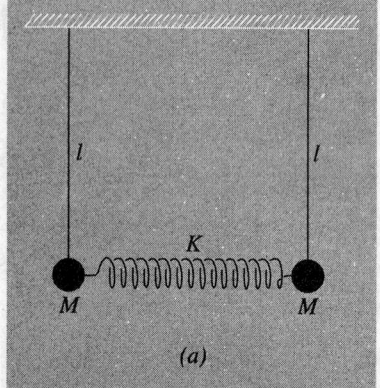

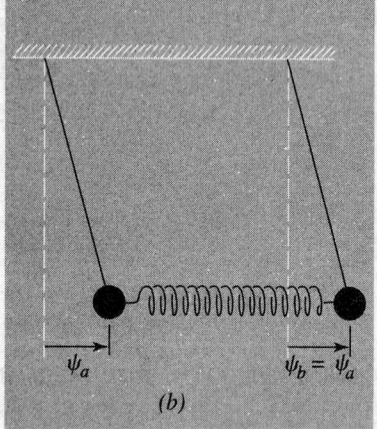

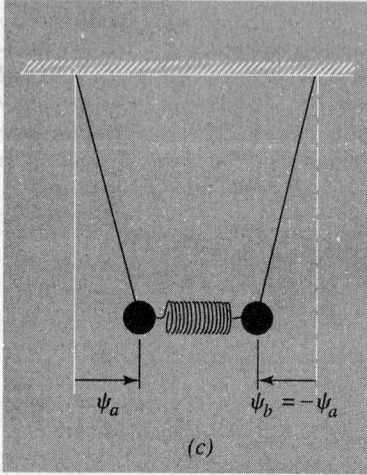

Fig. 1.14 *Coupled identical pendulums. (a) Equilibrium configuration. (b) Mode with lower frequency, (c) Mode with higher frequency.*

In mode 2 we have $\psi_a = -\psi_b$. Consider the left-hand bob. The return force due to the spring is $2K\psi_a$. (The factor of 2 results from the fact that the spring is compressed by an amount $2\psi_a$ in this mode when bob a is displaced by an amount ψ_a.) The return force due to gravity is $Mg\theta = Mg\psi_a/l$. The spring and gravity both act with the same sign. Thus the total return force per unit displacement per unit mass is

$$\text{Mode 2:} \qquad \omega_2^{\,2} = \frac{g}{l} + \frac{2K}{M}, \qquad \psi_a = -\psi_b. \qquad (89)$$

We now wish to study "beats between the two modes" of this system. What does that mean? Each mode is a harmonic oscillation with a given frequency. The general motion of pendulum a is given by a superposition of the two modes:

$$\psi_a(t) = \psi_1(t) + \psi_2(t).$$

Thus, $\psi_a(t)$ will look like the superposition $\psi_1 + \psi_2$ in Fig. 1.13 if the mode frequencies are nearly the same (and if the amplitudes of the two modes are the same). Then we say that the motion of pendulum a exhibits beats. (Of course pendulum b will also exhibit beats, as we shall see.) Any system of two degrees of freedom can exhibit beats, but the system we have chosen is convenient because we can easily make the beat frequency $\nu_1 - \nu_2$ small compared with the average frequency by using a sufficiently weak spring or by making the mass M large. [To see this, compare Eqs. (88) and (89).]

What do the beats look like? According to our discussion in Sec. 1.4, the displacements of the bobs, ψ_a and ψ_b, can be expressed in terms of the normal coordinates ψ_1 and ψ_2 by the general superposition

$$\psi_a = \psi_1 + \psi_2 = A_1 \cos(\omega_1 t + \varphi_1) + A_2 \cos(\omega_2 t + \varphi_2),$$

$$\psi_b = \psi_1 - \psi_2 = A_1 \cos(\omega_1 t + \varphi_1) - A_2 \cos(\omega_2 t + \varphi_2). \qquad (90)$$

By analogy with the tuning forks, we will get the largest beat effect if the two modes are present with equal amplitudes. (If either A_1 or A_2 is nearly zero compared to the other, there is virtually no beat effect, since (approximately) only one harmonic oscillation is present. Both oscillations should have approximately equal amplitudes to produce strong beats.) Therefore we take $A_1 = A_2 = A$. The choice of phase constants φ_1 and φ_2 corresponds to the initial conditions, as we shall see. By analogy with our example of the tuning forks, we take $\varphi_1 = \varphi_2 = 0$. With these choices for A_1, A_2, φ_1 and φ_2, Eqs. (90) give

$$\psi_a(t) = A \cos \omega_1 t + A \cos \omega_2 t, \qquad \psi_b(t) = A \cos \omega_1 t - A \cos \omega_2 t. \qquad (91)$$

The velocities of the bobs are given by

$$\dot{\psi}_a(t) \equiv \frac{d\psi_a}{dt} = -\omega_1 A \sin \omega_1 t - \omega_2 A \sin \omega_2 t,$$

$$\dot{\psi}_b(t) \equiv \frac{d\psi_b}{dt} = -\omega_1 A \sin \omega_1 t + \omega_2 A \sin \omega_2 t. \tag{92}$$

In order to see how to excite the two modes in just such a way as to get oscillations corresponding to Eq. (91), let us consider the *initial conditions* at time $t = 0$. According to Eqs. (91) and (92), the initial displacements and velocities of the bobs are given by

$$\psi_a(0) = 2A, \qquad \psi_b(0) = 0; \qquad \dot{\psi}_a(0) = 0, \qquad \dot{\psi}_b(0) = 0.$$

Therefore we hold bob a at displacement $2A$, bob b at zero, and release both bobs from rest at the same time, which we call $t = 0$.

After that we just watch. (You should do this experiment yourself. You need two cans of soup, a slinky, and some string. See Home Experiment 1.8.) A fascinating process unfolds. Gradually the oscillation amplitude of pendulum a decreases and that of pendulum b increases, until eventually pendulum a is resting and pendulum b is oscillating with the amplitude and energy that pendulum a started out with. (We are neglecting frictional forces.) The vibration energy is transferred completely from one pendulum to the other. By the symmetry of the system we see that the process continues. The vibration energy slowly flows back and forth between a and b. One complete round trip for the energy from a to b and back to a is a beat. The beat period is the time for the round trip and is the inverse of the beat frequency.

All of this is predicted by Eqs. (91) and (92). Using $\omega_1 = \omega_{av} + \omega_{mod}$ and $\omega_2 = \omega_{av} - \omega_{mod}$ in Eqs. (91), we get the "almost harmonic" oscillations

$$\begin{aligned} \psi_a(t) &= A \cos (\omega_{av} + \omega_{mod})t + A \cos (\omega_{av} - \omega_{mod})t \\ &= (2A \cos \omega_{mod}t) \cos \omega_{av}t \\ &\equiv A_{mod}(t) \cos \omega_{av}t \end{aligned} \tag{93}$$

and

$$\begin{aligned} \psi_b(t) &= A \cos(\omega_{av} + \omega_{mod})t - A \cos(\omega_{av} - \omega_{mod})t \\ &= (2A \sin \omega_{mod}t) \sin \omega_{av}t \\ &\equiv B_{mod}(t) \sin \omega_{av}t. \end{aligned} \tag{94}$$

Let us find an expression for the energy (kinetic plus potential) of each pendulum. We think of the oscillation amplitude $A_{mod}(t)$ as essentially constant over one cycle of the "fast" oscillation, and we also neglect the energy that is transferred between the weak coupling spring and the pendulum. (If the spring is very weak, it never has a significant amount of stored energy.) Thus during one fast oscillation cycle we think of pendulum *a* as a harmonic oscillator of frequency ω_{av} with constant amplitude, A_{mod}. The energy is then easily seen to be given by twice the average value of the kinetic energy (averaged over one "fast" cycle). This gives

$$E_a = \tfrac{1}{2} M\omega_{av}{}^2 A_{mod}{}^2 = 2MA^2\omega_{av}{}^2 \cos^2 \omega_{mod}t. \tag{95}$$

Similarly,

$$E_b = \tfrac{1}{2} M\omega_{av}{}^2 B_{mod}{}^2 = 2MA^2\omega_{av}{}^2 \sin^2 \omega_{mod}t. \tag{96}$$

The total energy of both pendulums is constant, as we see by adding Eqs. (95) and (96):

$$E_a + E_b = (2MA^2\omega_{av}{}^2) = E. \tag{97}$$

The energy difference between the two pendulums is

$$E_a - E_b = E\ (\cos^2 \omega_{mod}t - \sin^2 \omega_{mod}t)$$
$$= E \cos 2\omega_{mod}t = E \cos (\omega_1 - \omega_2)t. \tag{98}$$

Combining Eqs. (97) and (98) gives

$$E_a = \tfrac{1}{2} E\left[1 + \cos(\omega_1 - \omega_2)t\right], \tag{99a}$$
$$E_b = \tfrac{1}{2} E\left[1 - \cos(\omega_1 - \omega_2)t\right]. \tag{99b}$$

Equations (99) show that the total energy E is constant and that it flows back and forth between the two pendulums at the beat frequency. In Fig. 1.15 we plot $\psi_a(t)$, $\psi_b(t)$, E_a, and E_b.

Esoteric examples

In the study of microscopic systems—molecules, elementary particles—one encounters several beautiful examples of systems that are mathematically analogous to our mechanical example of two identical weakly coupled pendulums. One needs quantum mechanics to understand these systems. The "stuff" that "flows" back and forth between the two degrees of freedom, in analogy to the energy transfer between two weakly coupled pendulums, is *not* energy but probability. Then energy is "quantized"—it cannot "subdivide" to flow. Either one "moving part" or the other has *all* the energy. What "flows" is the probability to *have* the excitation energy. Two examples, the ammonia molecule (which is the "clockworks" of the ammonia clock) and the neutral *K* mesons, are discussed in Supplementary Topic 1.

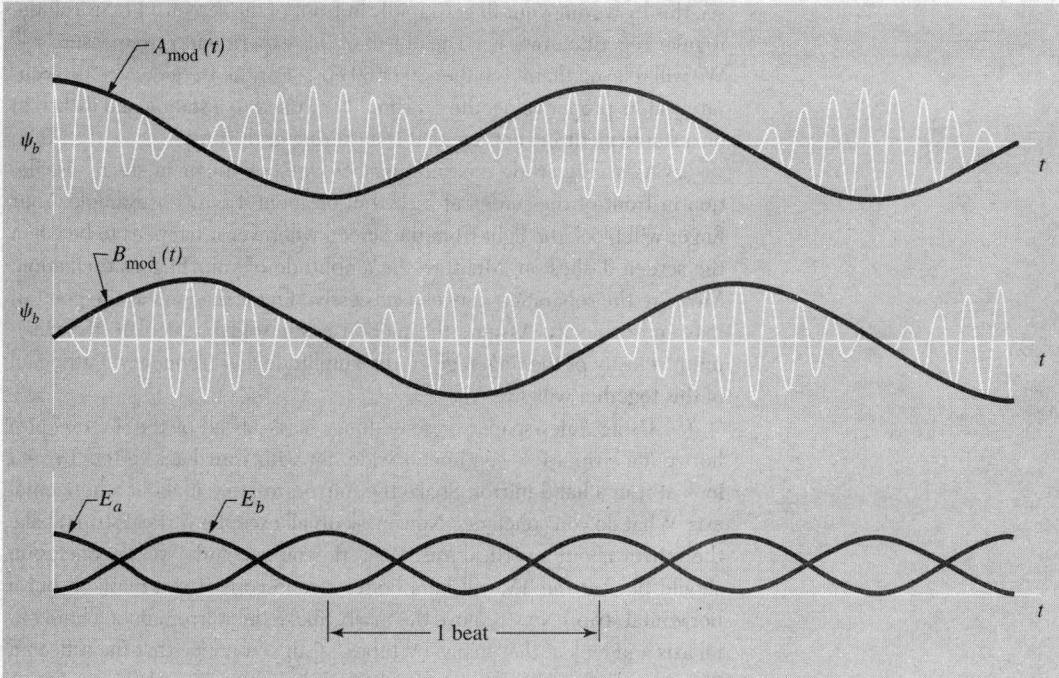

Fig. 1.15 *Energy transfer between two weakly coupled identical pendulums. Energy flows back and forth from a to b at the frequency* $|\nu_1 - \nu_2|$, *the beat frequency of the two modes.*

Problems and Home Experiments

1.1 Find the two mode frequencies in cps (cycles per second) for the *LC* network shown in Fig. 1.12, with $L = 10$ H (henrys) and $C = 6\ \mu$F (microfarads). Also, sketch the current configuration for each mode.

$$Ans.\ \ n1 \approx 20\ cps,\ n2 \approx 35\ cps.$$

1.2 If you set a small block of wood (or something) on a record player turntable and look at it from the side as the turntable goes around, using only one eye so as to get rid of your depth perception, the apparent motion (i.e., motion projected perpendicular to your line of sight) is harmonic, i.e., of the form $x = x_0 \cos \omega t$. (*a*) Prove the foregoing statement. (*b*) Make a simple pendulum by suspending a small weight (like a nut or bolt) from a string hung over the back of a chair. Adjust the length of string until you can get your pendulum to swing in synchronization with the projected motion of the block on the turntable when the record player is set at 45 rpm. This gives you a nice demonstration of the fact that the projection of a uniform circular motion is a harmonic oscillation. It is also a nice way to measure *g*. If *g* has the standard "textbook value" of 980 cm/sec², show that $l \approx 45$ cm for $\nu = 45$ rpm. That should be easy to remember!

1.3 TV set as a stroboscope. The light emitted by a TV set makes a good stroboscope. A given point on the screen is actually dark most of the time; it is lit a small fraction of the time at a regular repetition rate. (You can

Home experiment

see this by waving your finger rapidly in front of die screen.) Let us call the regular repetition rate v_{TV}. The object of this experiment is to measure v_{TV}. We will tell you that it is either 30 or 60 cps. (For the frequency to be accurately at its proper value, the set should be tuned to a station and locked in on a stable picture—not one that is flickering or drifting.)

(*a*) As a very crude measurement, wave your finger in steady oscillation in front of the screen at a rhythm of about 4 cps, for example. Your finger will block the light from the screen wherever it happens to be when the screen flashes on. Measure the amplitude of your finger's oscillation. Measure the separation between successive finger shadows at the point of maximum velocity. Assume the motion is sinusoidal. Calculate the maximum velocity of the finger, given the amplitude and frequency. Putting all of this together will give you v_{TV}.

(*b*) Using a newspaper or something, mask off all of the TV except a horizontal strip a few centimeter wide. Sit with your back to the TV and look at it in a hand mirror. Shake the mirror, rotating it about a horizontal axis. What do you conclude? Now mask off all except a vertical strip. Shake the mirror about a vertical axis. What do you conclude? (One conclusion should be that the TV will be a better stroboscope if you mask all but a horizontal strip.) Now remove the mask. Shake the mirror about a horizontal axis and look at the "many TV tubes." Can you notice that the reflected TV's seen in the oscillating mirror have only half as many horizontal lines per unit vertical distance as does the stationary tube seen when you don't shake the mirror?

(*c*) Here is an accurate way to measure v_{TV} using a record player turntable. Make a stroboscopic disk by drawing a circle on a piece of white paper with the edge of a protractor. Make pencil marks at angular intervals which will produce stroboscopic superposition of successive marks; mark $\frac{1}{3}$ of the circle for strobing at 120 cps, $\frac{1}{3}$ for 60 cps, and $\frac{1}{3}$ for 30 cps. Punch a hole in the center and put it on the turntable like a record. Then illuminate it with the TV and see which sector of the circle has the appearance of the original pencil marks. (If you want a very accurate turntable strobe disk, you can get one at a high fidelity phonograph shop or from Audiotex Co., Los Angeles 18, Calif., or Rockford, Ill., Catalogue No. 30–228, list price 55 cents.)

Home experiment   **1.4 Measuring the frequency of vibrations.** (*a*) *Piano strings.*   Now that you know v_{TV} (Home Exp. 1.3), use the TV set to measure the frequency of vibration of piano strings. Illuminate the lowest two octaves of strings with the TV (at night with other lights out). Hold down the damper pedal and strum all of these strings with your hand near the middle of the strings. (If you use the piano hammers, as in playing, the vibration amplitudes are too small.) You can quickly see which string "stands still." Note the exact string; then strum the string one octave down. If you were correct, the lower string should appear to stand still but to be "double." (Why?) You have now found the piano string (and corresponding note on the keyboard) that has the frequency v_{TV}. You can obtain the frequency of each successive octave of that note by multiplying by two. Look up the answer in the *Handbook of Chemistry and Physics* (indexed under "musical scales") to see if your piano is in tune. (The equal-temperament scale with A440 is the standard tuning.)

(*b*) *Guitar strings.* An analogous experiment can be done with a guitar. Suppose the lowest string, the E string, is in tune. Strobe it with the TV. It does not stand still. Loosen it. After you have gone down an interval of about a fourth, i.e., to the B below that E, it will stand still. Go down another octave to see whether the string "doubles." (At this lowest note the string is very loose, but it still works fairly well for strobing.) Finally, use your results to tell us the pitch of the low E string on a guitar. Is it E82 or E164?

(*c*) *Hacksaw blade.* Another nice experiment is to strobe a vibrating hacksaw blade with the TV. Clamp the blade to a table with a C clamp. Vary the length of the blade to vary the pitch.

1.5 Consider the energy transfer between two weakly coupled identical oscillators (Sec. 1.5). At $t = 0$, when oscillator a has all the oscillation energy and b has none, it is easy to see which is the "driven" oscillator (it is b) and which the "driving force" (it is a). Now consider the time $t = \frac{1}{4} T_{beat}$, a quarter of a beat cycle after $t = 0$. At that time pendulum a has lost half its energy, and b has gained that energy; the pendulums have the same oscillation amplitude. How do they "know" which one is the driven and which the driver? How do they know which way the energy should flow? To put it differently, suppose you are allowed to look at the system and follow it through one oscillation (a fast oscillation of frequency about ω_1 or ω_2) at a time when both have the same energy. How can you predict whether the energy partition will (*a*) stay the same; (*b*) change so as to increase the energy of b; (*c*) change the other way? Try not to use the formulas; that is too easy. Look at the system itself—what pulls on what, when, etc. (*Hint:* The phase relationships are crucial.)

1.6 Devise a damping mechanism ("friction") that will damp only mode 1 of the coupled pendulums of Fig. 1.14. Devise another that will damp only mode 2. Notice that friction at the supports (hinges) damps *both* modes. So does air resistance. These will not work. See Supplementary Topic 1.

1.7 Coupled hacksaw blades. Clamp two hacksaw blades (about 25 cents apiece at a hardware store) to a table with C clamps, leaving about four inches free to vibrate. One way to adjust them to the same frequency is to shorten the protruding part of one blade until it vibrates at a recognizable pitch and then tune the other to sound the same pitch. Another way is to "strobe" each of them, using the light from a TV tube as a convenient stroboscope. (See Home Exp. 1.3.) When the two blades are in reasonably close tune, couple them with a rubber band. Strum one of them and watch the beats between the modes. Vary the coupling by moving the rubber band in or out along the blades. If the two blades are not in tune, do you get beats?

Home experiment

Here are some other examples of coupled identical oscillators that give nice beats:

(i) two identical magnets hung so they can swing on a piece of iron—the magnets are coupled by their fields; (ii) two clotheslines or strings tied to the same flexible post at one end and tied independently at the other ends; (iii) two strings on a guitar tuned to the same pitch.

Home experiment **1.8 Coupled cans of soup.** One of the standard sizes for a can of soup has an outer diameter of about 6.67 cm and fits perfectly into the end of a slinky. (A slinky is available in any toy store for about $1.) Get a slinky and two cans of soup. Use the cans for pendulum bobs, suspended by strings about 50 cm long tied and taped to the cans. Couple the bobs with the slinky (tape helps). Measure the frequencies of the two longitudinal modes and the frequency for energy transfer. (Start with one pendulum at its equilibrium position and the other displaced.) Does your experiment show this frequency to be the beat frequency $\nu_1 - \nu_2$? From the frequency of the lowest mode, the beat frequency, and the number of turns of the slinky that you are using, calculate the inverse spring constant per turn of the slinky K^{-1}/a.

This system actually has four degrees of freedom. In addition to the two longitudinal degrees and corresponding modes studied above, there are two transverse modes with the bobs oscillating perpendicular to the spring. Find these two modes and measure the two mode frequencies. Compare these frequencies with those for the longitudinal modes. Explain.

1.9 Suppose one pendulum consists of a 1-meter string with a bob that is an aluminum sphere 5.0 cm in diameter. A second pendulum consists of a 1-meter string with a bob that is a brass sphere 5 cm in diameter. The two pendulums are set into oscillation at the same time and with the same amplitude A. After 5 minutes of undisturbed osculation, the aluminum pendulum is oscillating with one-half of its initial amplitude. What is the oscillation amplitude of the brass pendulum? Assume that the friction is due to the relative velocity of bob and air and that the instantaneous rate of energy loss is proportional to the square of the velocity of the bob. Show that the energy decays exponentially. (Show that for any other velocity dependence, say v^4, the energy does not decay exponentially.) Show that for the assumed exponential decay the mean decay time is proportional to the mass of the bob. The final answer is 0.81 A for the amplitude of the brass pendulum.

1.10 A massless spring with no mass attached to it hangs from the ceiling. Its length is 20 cm. A mass M is now hung on the lower end of the spring. Support the mass with your hand so that the spring remains relaxed, then suddenly remove your supporting hand. The mass and spring oscillate. The lowest position of the mass during the oscillations is 10 cm below the place it was resting when you supported it. (a) What is the frequency of oscillation? (b) What is the velocity when the mass is 5 cm below its original resting place? *Ans.* (a) 2.2 cps; (b) 70 cm/sec.

A second mass of 300 gm is added to the first mass, making a total of $M + 300$ gm. When this system oscillates, it has half the frequency of the system with mass M alone. (c) What is M? (d) Where is the new equilibrium position?

Ans. (c) 100 gm; (d) 15 cm below old position.

1.11 Find the modes and their frequencies for the coupled springs and masses sliding on a frictionless surface shown below. At equilibrium the springs are relaxed. Take $M_1 = M_2 = M$.

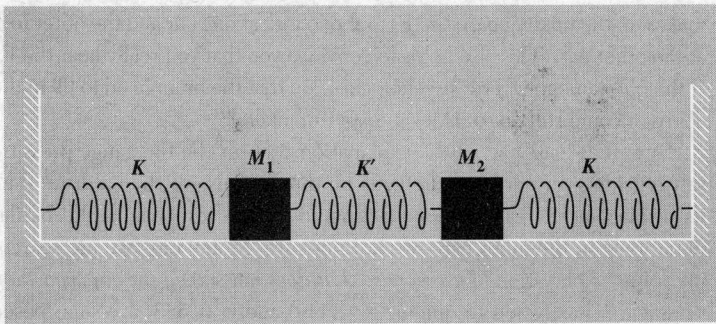

Problem 1.11

1.12 Beats from two tuning forks. Use two forks, each having the same nominal frequency. C523.3 and A440 forks are available in many music stores for about $1. [Forks for C517 and A435 (International pitch) are available for about 85 cents each from Central Scientific Co., Irving Park Road, Chicago, Ill., 60613, Cenco Nos. 8455-1 and 2.]

Home experiment

Strike one fork against the other at equal distances from the ends of the prongs. Hold both forks near one ear and make fine adjustments in the positions of the forks until you hear beats. "Load" one prong of one fork by wrapping it with a rubber band. Vary the beat frequency by pushing the rubber band nearer to or farther from the end of the prong.

Some ordinary dinner forks make good tuning forks, as do some carving forks (if the handle does not damp the vibrations). You should be able to find two forks that give nearly the same pitch and give beats. Some wine glasses also give clear tones (they usually vibrate in several modes at once). In listening for beats between bells (or brandy snifters or pot lids), you will hear beats coming from a single bell! When this happens, it is because the bell has two modes that are closely spaced in frequency. When you hit it on one edge, you excite both modes.

1.13 Nonlinearities in your ear—combination tones. For this experiment you need one A440 and one C523 tuning fork. (Other combinations also work well.) You also need a quiet environment. Strike the forks together. Bring C523 up to your ear, then A440 (at the same time removing C523); then, keeping A440 at your ear, bring back C523. But this time don't

Home experiment

focus your attention on either the A440 or C523. Listen for a note about a major third *below* the A440. (The technique of listening to first the C, then the A, then both, is to help get your attention to progress downward in pitch with successive configurations.) After some trying, you may hear the F below A440 when the A and C are both present. (Many people don't hear it. Most violin players hear it immediately. If you don't know what to listen for, try the notes on a piano to find out.) Altogether one has a pleasant F-major triad, i.e., F, A, C. To prove that the phenomenon happens in your eardrum (or perhaps your basilar membrane) and not in your brain (i.e., to prove that it is not heard merely because your brain just likes to hear major triads and makes up the missing part, the F), put one fork at one ear and the other fork at the other ear. (This also helps to convince you that you really hear the F.) If the phenomenon were "psychological," in that the brain liked to fill in the chord, it could still do so. Does it (experimentally)?

Here is at least part of the explanation: Let $p(t)$ be the gauge pressure just outside your eardrum. Let $q(t)$ be the response of the eardrum (i.e., its displacement), or perhaps $q(t)$ should be the response of the basilar membrane in the inner ear—we are not sure. At any rate, we are searching for an explanation of *a response that does not satisfy the superposition principle*. Thus, when frequency v_1 (A440) and v_2 (C523) are superposed at the ear, the response includes not only v_1 and v_2, but also a third frequency, v_3 ($\approx$F349). That suggests a nonlinearity. (We already know that linear responses obey the superposition principle, and we shall see it again below.) Let us assume that $q(t)$ is a nonlinear function of $p(t)$:

$$q(t) = \alpha p(t) + \beta p^2(t) + \gamma p^3(t).$$

Now let $p(t)$ be a superposition of two different harmonic oscillations (produced by the two tuning forks). For simplicity, let us take the amplitudes equal and the phase constants zero. Also, let us use units in which each amplitude is unity, so that we don't have so much to write. Thus we take

$$p(t) = \cos \omega_1 t + \cos \omega_2 t.$$

The response $q(t)$ of the eardrum (or basilar membrane?) is then given by

$$q(t) = \alpha \left[\cos \omega_1 t + \cos \omega_2 t \right] + \beta \left[\cos \omega_1 t + \cos \omega_2 t \right]^2 + \gamma \left[\cos \omega_1 t + \cos \omega_2 t \right]^3.$$

If β and γ are zero, then $q(t)$ is said to be linear in its response. (It responds like a perfectly linear Hooke's law spring to an impressed force.) In that case, $q(t)$ is just a superposition of harmonic oscillations at frequencies ω_1 and ω_2. (Then you don't hear the F!) The term with β is a quadratic nonlinearity; the term with γ is a cubic nonlinearity.

We want to express $q(t)$ as a superposition of harmonic oscillations. We need some trigonometric identities; we shall derive them now. Let $f(x) \equiv \cos x$. We already know the identity $\cos (x+y) + \cos (x-y) = 2 \cos x \cos y$, i.e.,

$$f(x)f(y) = \tfrac{1}{2} f(x + y) + \tfrac{1}{2} f(x - y).$$

We use this result to derive the identity (needed for the cubic nonlinearity)

$$
\begin{aligned}
\left[f(x)f(y) \right] f(z) &= \left[\tfrac{1}{2} f(x + y) + \tfrac{1}{2} f(x - y) \right] f(z) \\
&= \tfrac{1}{2} f(x + y)f(z) + \tfrac{1}{2} f(x - y)f(z) \\
&= \tfrac{1}{4} f(x + y + z) + \tfrac{1}{4} f(x + y - z) + \tfrac{1}{4} f(x - y + z) \\
&\quad + \tfrac{1}{4} f(x - y - z).
\end{aligned}
$$

Now let us find the quadratic response term. Letting $\theta_1 \equiv \omega_1 t$, $\theta_2 \equiv \omega_2 t$, we have (for the quadratic nonlinearity)

$$
\begin{aligned}
(\cos \omega_1 t + \cos \omega_2 t)^2 &\equiv \left[f(\theta_1) + f(\theta_2) \right]^2 \\
&= \left[f(\theta_1)f(\theta_1) \right] + \left[2 f(\theta_1)f(\theta_2) \right] + \left[f(\theta_2)f(\theta_2) \right] \\
&= \left[\tfrac{1}{2} f(\theta_1 + \theta_1) + \tfrac{1}{2} f(\theta_1 - \theta_1) \right] \\
&\quad + \left[f(\theta_1 + \theta_2) + f(\theta_1 - \theta_2) \right] \\
&\quad + \left[\tfrac{1}{2} f(\theta_2 + \theta_2) + \tfrac{1}{2} f(\theta_2 - \theta_2) \right].
\end{aligned}
$$

Thus the quadratic response term includes frequencies $2\omega_1$, 0, $\omega_1 + \omega_2$, $\omega_1 - \omega_2$, and $2\omega_2$. These are called *combination tones* or *combination frequencies*.

The cubic nonlinear response term has

$$
\begin{aligned}
(\cos \omega_1 t + \cos \omega_2 t)^3 &= \left[f(\theta_1) + f(\theta_2) \right]^3 \\
&= f^3(\theta_1) + 3 f^2(\theta_1)f(\theta_2) + 3 f(\theta_1) f^2(\theta_2) + f^3(\theta_2).
\end{aligned}
$$

Using the identity for $f(x)f(y)f(z)$, we see that the term $f^3(\theta_1)$ is a superposition of harmonic oscillations of frequencies $3\omega_1$ and ω_1; the term $f^2(\theta_1)f(\theta_2)$ is a superposition of frequencies $2\omega_1 + \omega_2$, $2\omega_1 - \omega_2$, and ω_2; $f(\theta_1) f^2(\theta_2)$ is a superposition of $2\omega_2 + \omega_1$, $2\omega_2 - \omega_1$, and ω_1; $f^3(\theta_2)$ is a superposition of $3\omega_2$ and ω_2. Thus the cubic response term is a superposition of harmonic oscillations with combination tones $3\omega_1$, ω_1, $2\omega_1 \pm \omega_2$, $2\omega_2 \pm \omega_1$, ω_2, and $3\omega_2$.

Back to the tuning fork experiment! A little arithmetic shows that our F does not result from a quadratic nonlinearity. In fact, it is given by the cubic contribution $2\omega_1 - \omega_2$:

$$
\begin{aligned}
\nu_1 &= A440 \\
\nu_2 &= C523 \\
2\nu_1 - \nu_2 &= 880 - 523 = 357.
\end{aligned}
$$

According to the handbook, F is 349 and F# is 370. Thus $2\nu_1 - \nu_2$ is a rather "sharp" F; it is $\frac{8}{21}$ of the way from F to F#. (It also *sounds* a bit sharp.) (If you use tuning forks tuned to the scientific or "just" scale, then you get an exact F. It also sounds exact.)

Now comes the interesting part. Is the cubic nonlinearity in the eardrum? Or is it perhaps in the resonating basilar membrane? I believe it is not in the drum, for the following reasons: when I move the two forks away from my ear, so that the perceived intensity from each one decreases, I still hear the nonlinear term. If it were due to nonlinear response in my eardrum, it should fall off in loudness much faster with distance than the v_1 and v_2 terms, but it doesn't. Also, the nonlinear contribution $2v_2 - v_1 = 1046 - 440 = 606 \approx$ halfway between D and D# should be present, but I don't hear it. None of this proves that the basilar membrane *is* responsible, but just that the drum does *not* seem to be responsible. Does this leave only the basilar membrane or its nerve endings? I don't know the answer. (I discovered the effect accidentally while inventing home experiments. Perhaps it is well known and under-stood already.)

Optical harmonics   It is possible to produce optical harmonics (and sum and difference frequencies, i.e., combination frequencies) by making use of the small nonlinear contribution to the dielectric constant of a transparent substance. The cover of the magazine *Scientific American* for July 1963 has a beautiful photograph showing a beam of red light of wavelength 6940 angstrom units (1 angstrom $= 10^{-8}$ cm) incident on a crystal. Emerging from the opposite side of the crystal is a beam of blue light of wavelength 3470 angstroms. Halving the wavelength is equivalent to doubling the frequency. Therefore the nonlinearity must be a quadratic one. See also "The Interaction of Light with Light," by J. A. Giordmaine, *Scientific American* (April 1964).

1.14   Superposition of initial conditions gives superposition of corresponding motions.   Suppose *a* and *b* are two coupled oscillators. Consider three different initial conditions:

(i) a and b are released from rest with amplitudes 1 and –1, respectively;

(ii) they are released from rest with amplitudes 1 and 1;

(iii) they are released from rest with amplitudes 2 and 0, respectively. Thus the initial conditions for case (iii) are a superposition of those for cases (i) and (ii).

Show that the motion in case (iii) is a superposition of the motions for cases (i) and (ii).

1.15   Prove the general case corresponding to the example of Prob. 1.14. (Include the velocities as well as the displacements in the initial conditions.)

1.16   Prove the superposition principle for inhomogeneous linear equations of motion given after Eq. (36). Prove that it does not apply to nonlinear inhomogeneous equations.

1.17   Write down the three equations for a system of three degrees of freedom analogous to the general Eqs. (47) and (48). Show that if one assumes a mode, one gets a determinantal equation analogous to Eq. (56), except that it is a three-by-three determinant. Show that this gives a cubic equation in the variable ω^2. Since a cubic has three solutions, there are three modes. Generalize to N degrees of freedom. This constitutes a proof that N modes exist for a system of N degrees of freedom. They must exist, because here you have a prescription for finding them.

1.18 Beats between weakly coupled nonidentical guitar strings. Borrow a Home experiment guitar. Tune the two lowest strings to the same frequency. Pluck one string and watch the other closely. (They should be tuned as exactly as possible to the same frequency. The most exact tuning is in fact obtained by maximizing the beats that you see.) Now pluck the other and watch. Is the energy transferred completely from one string to the other during the beating process? Can you get the energy to transfer completely by improving the tuning? Describe what you observe. What is the explanation? See Prob. 1.19.

Home experiment

1.19 Nonidentical coupled pendulums. Consider two pendulums, a and b, with the same string length l, but with different bob masses, M_a and M_b. They are coupled by a spring of spring constant K which is attached to the bobs. Show that the equations of motion (for small oscillations) are

$$M_a \frac{d^2\psi_a}{dt^2} = -M_a \frac{g}{l}\psi_a + K(\psi_b - \psi_a),$$

$$M_b \frac{d^2\psi_b}{dt^2} = -M_b \frac{g}{l}\psi_b - K(\psi_b - \psi_a).$$

Solve these two equations for the two modes by the method of searching for normal coordinates. Show that $\psi_1 \equiv (M_a\psi_a + M_b\psi_b)/(M_a + M$ and $\psi_2 = \psi_a - \psi_b$ are normal coordinates. Find the frequencies and configurations of the modes. What is the physical significance of ψ_1? Of ψ_2? Find a superposition of the two modes which corresponds to the initial conditions at time $t = 0$ that both pendulums have zero velocity, that bob a have amplitude A, and that bob b have amplitude zero. Let E be the total energy of bob a at $t = 0$. Find an expression for $E_a(t)$ and for $E_b(t)$. Assume weak coupling. Does the energy of bob a transfer completely to bob b during a beat? Is it perhaps the case that if the pendulum which initially has all the energy is the heavy one, the energy is not completely transferred, but if it is the light one, the energy is completely transferred?

Ans. $\qquad \omega_1^2 = \frac{g}{l}, \qquad \omega_2^2 = \frac{g}{l} + K\left(\frac{1}{M_a} + \frac{1}{M_b}\right).$

$\psi_a = A\left(\frac{M_a}{M}\cos\omega_1 t + \frac{M_b}{M}\cos\omega_2 t\right), \quad \psi_b = A\frac{M_a}{M}(\cos\omega_1 t - \cos\omega_2 t),$

where $M = M_a + M_b$.

After defining $\omega_{\text{mod}} = \frac{1}{2}(\omega_2 - \omega_1)$ and $\omega_{\text{av}} = \frac{1}{2}(\omega_2 + \omega_1)$, one finds

$$\psi_a = \left(A \cos \omega_{\text{mod}} t \right) \cos \omega_{\text{av}} t + \left(A \frac{M_a - M_b}{M} \sin \omega_{\text{mod}} t \right) \sin \omega_{\text{av}} t,$$

$$\psi_b = \left(2A \frac{M_a}{M} \sin \omega_{\text{mod}} t \right) \sin \omega_{\text{av}} t.$$

The energy of each pendulum is easily found in the weak-coupling approximation, where we neglect the time variation of the sine or cosine of $\omega_{\text{mod}} t$ during one cycle of the fast oscillation at frequency ω_{av}, because we assume $\omega_{\text{mod}} \ll \omega_{\text{av}}$. We also neglect the energy stored in the spring at any instant. Then you should find

$$E_b = E \left(\frac{2M_a M_b}{M^2} \right) \left[1 - \cos (\omega_2 - \omega_1)t \right]$$

$$E_a = E \left[\frac{M_a{}^2 + M_b{}^2 + 2M_a M_b \cos (\omega_2 - \omega_1)t}{M^2} \right].$$

Thus the energy of pendulum a (the one with all the energy at time zero) varies sinusoidally at the beat frequency, oscillating between a maximum value of E and a minimum value of $[(M_a - M_b)/M]^2 E$.

The energy of pendulum b oscillates at the beat frequency between a minimum value of zero and a maximum value of $(4M_a M_b/M^2)E$. The total energy $E_a + E_b$ is constant (since we neglect damping). Now look at Home Exp. 1.18. Also, give a *qualitative* explanation of why the energy transfer does not continue to completion, so to speak, when the masses are unequal. [*Hint:* Consider the two extreme cases (i) M_a is huge compared with M_b and (ii) M_a is tiny compared with M_b.]

1.20 Transverse oscillations of two coupled masses. Using either the slinky approximation or the small-oscillations approximation, find the two coupled equations of motion for the transverse displacements ψ_a and ψ_b, of Fig. 1.11. (*a*) Use the systematic method to find the frequencies and amplitude ratios for the two normal modes, (*b*) Find linear combinations of ψ_a and ψ_b that give uncoupled equations; i.e., find the normal coordinates, and find the frequencies and amplitude ratios for the two modes.

Ans. See Eqs. (70) and (71).

1.21 Oscillations of two coupled LC circuits. Find the two normal modes of oscillation of the coupled *LC* circuits shown in Fig. 1.12, with equations of motion given by Eqs. (77) and (78). (*a*) Use the systematic method, (*b*) Use the method of finding normal coordinates.

Ans. See Eq. (79).

1.22 A heavy object placed on a rubber pad that is to be used as a shock absorber compresses the pad by 1 cm. If the object is given a vertical tap, it will oscillate. (The oscillations will be damped; we neglect the damping.)

Estimate the oscillation frequency. (*Hint:* Assume the pad acts like a Hooke's law spring.)

Ans. About 5 cps.

1.23 **Longitudinal oscillations of two coupled masses.** The system is shown in Fig. 1.9. The equations of motion are given by Eqs. (62) and (63). Use the systematic method given in Eqs. (47) through (59) to find the modes. You should *not* simply plug into these equations, however, you should go through the analogous steps "without looking."

Ans. See Eqs. (60) and (61).

1.24 **Sloshing mode in a pan of water.** The lowest mode of oscillation in a closed body of liquid can be called the "sloshing" mode. It is easily excited, as anyone knows who has ever tried to carry a pan of water without sloshing.

Home experiment

Partly fill a rectangular pan with water. Push the pan a little. It sloshes. A better method is to lay the pan on a flat horizontal surface, fill it to the brim, and then overfill it so that water bulges above the level of the brim. Gently nudge the pan. After higher modes have damped out, you will be left with the sloshing mode, which will then oscillate with very little damping. (It *is* a gravitational mode, even though you are using surface tension to hold the water "above the walls"; this is done so as to minimize damping.) The water surface remains practically flat. (It is flat after the higher modes have damped out.) Assume it is flat throughout the motion—horizontal when it passes through the equilibrium position, and tilted at the extremes of its oscillation. Let x be along the horizontal oscillation direction and let y be vertical upward. Let $\bar{x}$ and $\bar{y}$ be the horizontal and vertical coordinates of the center of gravity of the water. Let $\bar{x}_0$ and $\bar{y}_0$ be the equilibrium values of $\bar{x}$ and $\bar{y}$. Find a formula for $\bar{y} - \bar{y}_0$ as a function of $\bar{x} - \bar{x}_0$. (A convenient parameter to work with is the water level at one end of the pan, relative to its equilibrium level.) The potential energy increase of the body of water is $mg(\bar{y} - \bar{y}_0)$. You will find that $\bar{y} - \bar{y}_0$ is proportional to $(\bar{x} - \bar{x}_0)^2$. Thus the center of gravity has a potential energy like that of a harmonic oscillator. Use Newton's second law as if the entire mass m were at the center of gravity. Find a formula for the frequency.

Ans. $\omega^2 = 3gh_0/L^2$, where h_0 is the equilibrium depth of water, $g = 980$ cm/sec^2, and L is *half* the length of the pan along the direction of wave motion, i.e., along x. Try out this formula on your experiment with the pan of water, i.e., measure ω, h_0, and L, and see how they agree with the formula. Now see Prob. 1.25.

1.25 **Seiches.** According to an encyclopedia the average depth of Lake Geneva is about 150 meters. The length is about 60 kilometers (including the narrower western end). If we approximate the lake by a rectangular pan, we can use the formula for ω^2 obtained in Home Exp. 1.24. What does it predict, under those assumptions, for the period of the seiches (sloshing modes) that go in the long direction of the lake? (The observed period is of the order of an hour.) The seiches are probably excited by sudden differences in atmospheric pressure at one part of the lake relative to that at another. The observed amplitudes are up to 150 cm.

In June 1954, a seiche of amplitude about ten feet that occurred in Lake Michigan swept away a number of people fishing from piers.

According to *Time* (Nov. 17, 1967), shock waves from the great Alaskan earthquake of Good Friday 1964 produced seiches in rivers, lakes, and harbors along the Gulf Coast of the United States and caused water to slosh over the top of a swimming pool in an Atlantic City hotel in New Jersey.

Chapter 2

Free Oscillations of Systems with Many Degrees of Freedom

2.1 Introduction

In Chap. 1 we studied oscillations of systems having one or two degrees of freedom. In this chapter we shall study systems having N degrees of freedom, where N can range up to some very large number, which we shall loosely call "infinity."

For a system with N degrees of freedom, there are always exactly N modes (see Prob. 1.17). Each mode has its own frequency ω and its own "shape" given by the amplitude ratios $A : B : C : D : \ldots$ etc., corresponding to the degrees of freedom $a, b, c, d, \ldots$, etc. In each mode, all moving parts go through their equilibrium positions simultaneously; that is, every degree of freedom oscillates in that mode with the same phase constant. Thus there is a single phase constant for the entire mode, which is determined by the initial conditions. Since each degree of freedom oscillates in a given mode with the same frequency ω, each moving part experiences the same return force per unit displacement per unit mass, given by ω^2.

As an example, suppose we have a system with four degrees of freedom a, b, c, d. Then there are four modes. Suppose that in mode 1 the amplitude ratios are

$$A : B : C : D = 1 : 0 : -2 : 7.$$

Then the motions of a, b, c, and d (if mode 1 is the only excited mode) are given by

$$\psi_a = A_1 \cos (\omega_1 t + \varphi_1), \qquad \psi_b = 0, \qquad \psi_c = -2\psi_a, \qquad \psi_d = 7\psi_a,$$

where A_1 and φ_1 depend on the initial conditions.

If a system contains a very large number of moving parts, and if these parts are distributed within a limited region of space, the average distance between neighboring moving parts becomes very small. As an approximation, one may wish to think of the number of parts as becoming infinite and the distance between neighboring parts as going to zero. One then says that the system behaves as if it were "continuous." Implicit in this point of view is the assumption that the motion of near neighbors is nearly the same. This assumption allows us to describe the vector displacement of all the moving parts in a small neighborhood of a point x, y, z, with a single vector quantity $\psi(x, y, z, t)$. Then the "displacement" $\psi(x, y, z, t)$ is a continuous function of position,

x, y, z, and of time t. It replaces the description using the displacements $\psi_a(t)$, $\psi_b(t)$, etc., of the individual parts. We then say we are dealing with *waves*.

Standing waves are normal modes. The modes of a continuous system are called *standing waves*, or *normal modes*, or simply *modes*. According to the discussion above, a truly continuous system has an infinite number of independent moving parts, although they occupy a finite space. There are therefore an infinite number of degrees of freedom, and hence an infinite number of modes. This is not literally true for a real material system. One liter of air does not contain an infinite number of moving parts, but only 2.7×10^{22} molecules, each of which has three degrees of freedom (for motion along x, y, and z directions). Thus a bottle containing 1 liter of air does not have an infinite number of possible vibrational modes of the air, but only 8×10^{22} at most. Anyone who has practiced blowing a bottle or a flute knows that it is not easy to excite more than the first few modes. (We usually distinguish the modes by calling the one with the lowest frequency number 1, the next higher number 2, etc.) In practice we are often concerned only with the first few (or few dozen or few thousand) modes. As we shall see, it turns out that the lowest modes behave as if the system were continuous.

The most general motion of a system can be written as a superposition of all its modes, with the amplitude and phase constant of each mode set by the initial conditions. The appearance of the vibrating system in such a general situation is very complicated, simply because the eye and brain cannot contemplate several things at once without confusion. It is not easy to look at the complete motion and "see" each mode separately when many are present.

Modes of beaded string. We study first the transverse oscillations of beaded strings. By "strings" we shall really mean springs. We will assume that we have linear (i.e., Hooke's law) massless springs connecting point masses M. (In our figures, we will draw the springs as straight lines rather than as helices.)

In Fig. 2.1 we exhibit a sequence of systems of beaded strings. The first system has $N = 1$ (one degree of freedom), the next $N = 2$, etc. In each case, we exhibit without proof the configurations of the normal modes. Later we shall derive the exact configuration and frequency for each mode.

It should already be possible for you to see (assuming the configurations shown are those of the modes) that we have correctly ordered the configurations in order of ascending mode frequency. That is because the strings make increasingly large angles with the

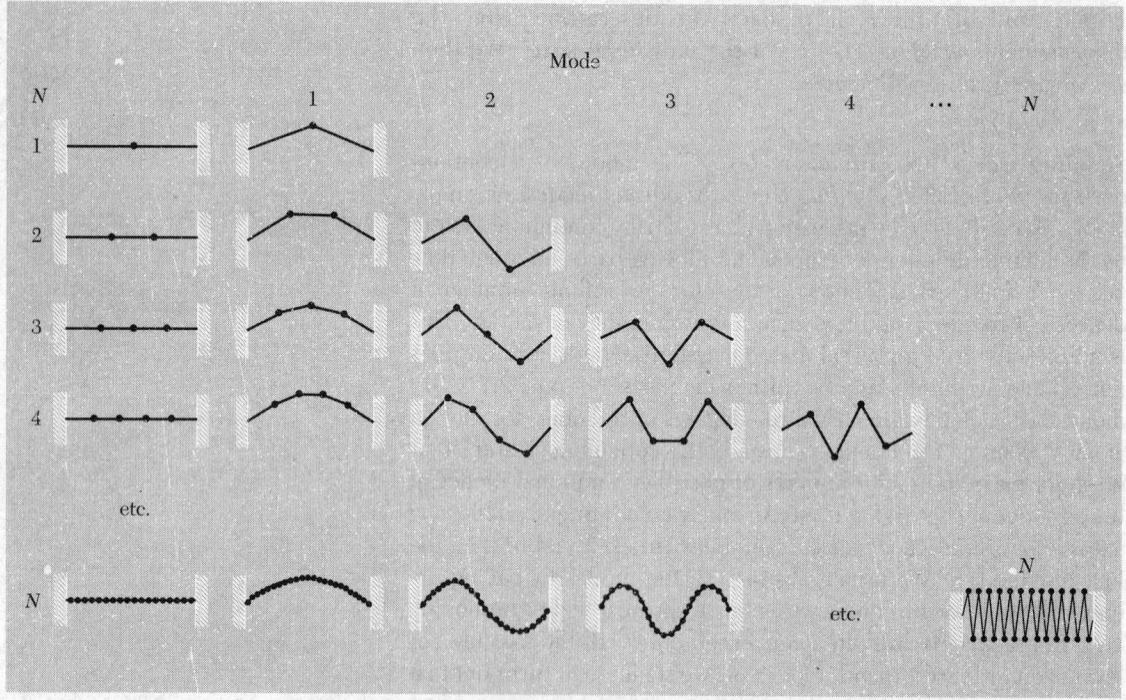

Fig. 2.1 Transverse vibrational modes of a beaded string. A string with N beads has N modes. In mode m the string crosses the equilibrium axis m – 1 times and has m half-wavelengths. The highest frequency mode is the "zigzag" configuration shown.

equilibrium axis as we increase the mode number (taking the displacement of a given bead to be the same). Consequently the return force per unit displacement per unit mass for a given bead in a given system increases when we go from one configuration to the next, and therefore so does the mode frequency.

Another thing that is apparent is that our sequence of assumed mode shapes always gives exactly N configurations: the first mode always has zero "nodes" (places where the string crosses the axis, excluding the end points), the second has one node, etc. The highest mode always has the largest possible number of nodes, namely $N - 1$, which is achieved by "zigzagging" up and down, i.e., crossing the axis once between each two successive masses.

2.2 Transverse Modes of Continuous String

We now consider the case where N is huge, say $N = 1,000,000$ or so. Then for the lowest modes (say the first few thousand), there are a very large number of beads between each node. Thus the displacement varies slowly from one bead to the next. [We shall *not* consider here the highest modes, since they approach the "zigzag limit," where a description using a continuous function $\psi(x, y, z, t)$ is not possible. Therefore, in accordance with the remarks above, we shall not describe the instantaneous configuration by the list of displacements $\psi_a(t)$, $\psi_b(t)$, $\psi_c(t)$, $\psi_d(t)$, etc., of each bead. Instead we consider all the particles with *equilibrium* positions in the neighborhood of the point x, y, z (a neighborhood being an infinitesimal cube, if you wish, with edges of length Δx, Δy, and Δz) as having the same instantaneous vector displacement $\psi(x, y, z, t)$:

$$\psi(x, y, z, t) = \hat{\mathbf{x}}\psi_x(x, y, z, t) + \hat{\mathbf{y}}\psi_y(x, y, z, t) + \hat{\mathbf{z}}\psi_z(x, y, z, t), \quad (1)$$

where $\hat{\mathbf{x}}, \hat{\mathbf{y}}$, and $\hat{\mathbf{z}}$, are unit vectors and ψ_x, ψ_y, ψ_z are the components of the vector displacement ψ. It is important to realize that x, y, z label the *equilibrium* position of the particles in that neighborhood. Thus x, y, z are not functions of time.

Longitudinal and transverse vibration.   Equation (1) is of a much more general form than we need in order to study the vibrations of a string. Suppose that at equilibrium the string is stretched along the z axis. Then the coordinate z is sufficient to label the equilibrium position of each bead (to an accuracy Δz) and Eq. (1) can be written in the simpler form

$$\psi(z, t) = \hat{\mathbf{x}}\psi_x(z, t) + \hat{\mathbf{y}}\psi_y(z, t) + \hat{\mathbf{z}}\psi_z(z, t). \quad (2)$$

Vibrations along the z direction are called *longitudinal* vibrations. Vibrations along the x and y directions are called *transverse* vibrations. At present we wish to consider only the transverse vibrations of the string. Therefore we assume ψ_z is zero:

$$\psi(z, t) = \hat{\mathbf{x}}\psi_x(z, t) + \hat{\mathbf{y}}\psi_y(z, t). \quad (3)$$

Linear polarization.   As a further simplification, we assume that the vibrations are entirely along $\hat{\mathbf{x}}$ (i.e., $\psi_y \equiv 0$). The vibrations are then said to be *linearly polarized* along $\hat{\mathbf{x}}$. (In Chap. 8 we shall study general states of polarization.) Now we can drop the unit vector $\hat{\mathbf{x}}$ and the subscript on ψ_x from the notation:

$$\psi(z, t) = \text{instantaneous transverse displacement of}$$
$$\text{particles having equilibrium position } z. \quad (4)$$

Now consider a very small segment of the continuous string. At equilibrium, the segment occupies a small interval of length Δz centered at z. The mass ΔM of the segment divided by the length Δz is defined as the *mass density* ρ_0, measured in units of mass per unit length:

$$\Delta M = \rho_0 \, \Delta z. \quad (5)$$

The mass density is assumed to be uniform along the string. The string tension at equilibrium, denoted by T_0, is also assumed to be uniform.

For a general (non equilibrium) situation, the segment has a transverse displacement $\psi(z, t)$, averaged over the segment. (See Fig. 2.2.). The segment is no longer exactly straight; it has (generally) a slight curvature. This is indicated in Fig. 2.2 by the fact that θ_1 and θ_2 are not equal. The tension in the segment is no longer T_0, since the segment is longer than its equilibrium length Δz. Let us find the net force F_x on the segment at the instant shown. At its left end, the segment is pulled downward with a force $T_1 \sin \theta_1$. At its right end, it is pulled upward with a force $T_2 \sin \theta_2$. Thus the net force upward is

$$F_x(t) = T_2 \sin \theta_2 - T_1 \sin \theta_1. \tag{6}$$

We want to express $F_x(t)$ in terms of $\psi(z, t)$ and its space derivative

$$\frac{\partial \psi(z, t)}{\partial z} = \text{slope of string at position } z \text{ at time } t. \tag{7}$$

According to Fig. 2.2, the string slope at z_1 is $\tan \theta_1$, and the slope at z_2 is $\tan \theta_2$. Also, $T_1 \cos \theta_1$ is the horizontal component of the string tension at z_1, and $T_2 \cos \theta_2$ is the horizontal component at z_2. Now, we want eventually to obtain a *linear* differential equation of motion. To this end, we shall assume that we can use either the slinky approximation or the small oscillations approximation. In the slinky approximation, T is larger than T_0 by a factor $1/\cos \theta$, because the segment is longer than Δz by a factor $1/\cos \theta$. Therefore $T \cos \theta = T_0$. In the small-oscillations approximation, we neglect the increase in length of the segment, and we also approximate $\cos \theta$ by 1. Thus we have $T \cos \theta = T_0$ in that case also. Then Eq. (6) gives

$$\begin{aligned} F_x(t) &= T_2 \, \sin \theta_2 - T_1 \, \sin \theta_1 \\ &= T_2 \, \cos \theta_2 \tan \theta_2 - T_1 \, \cos \theta_1 \tan \theta_1 \\ &= T_0 \, \tan \theta_2 - T_0 \, \tan \theta_1 \\ &= T_0 \left(\frac{\partial \psi}{\partial z} \right)_2 - T_0 \left(\frac{\partial \psi}{\partial z} \right)_1. \end{aligned} \tag{8}$$

Now consider the function $f(z)$ defined by

$$f(z) = \frac{\partial \psi(z, t)}{\partial z}, \tag{9}$$

where we have suppressed the variable t in writing $f(z)$ because we intend to hold t constant. We expand $f(z)$ in a Taylor's series around z_1 and then set $z = z_2$. [See Appendix Eq. (3)]:

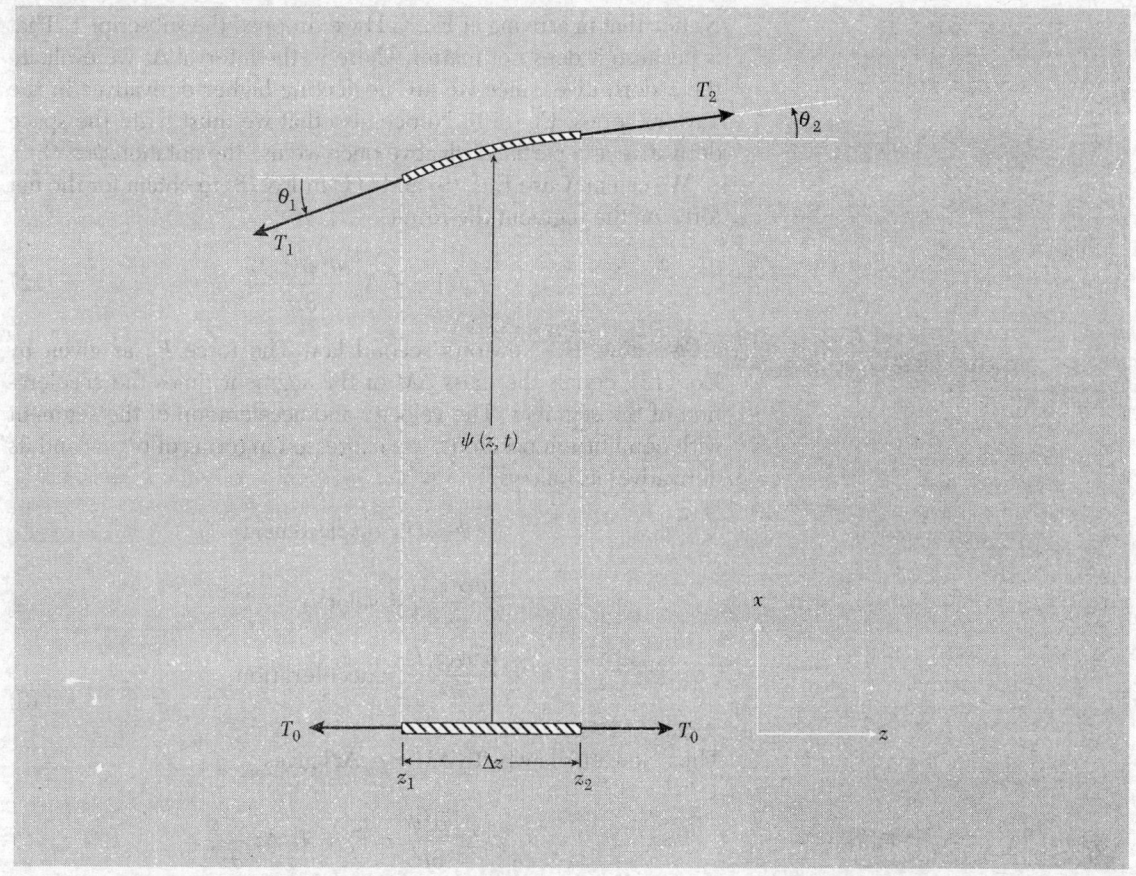

Fig. 2.2 *Transverse oscillations of a continuous string. At bottom is the equilibrium position of an infinitesimal segment along the z axis. Above is a general position and configuration of the same segment.*

$$f(z_2) = f(z_1) + (z_2 - z_1)\left(\frac{df}{dz}\right)_1$$

$$+ \frac{1}{2}(z_2 - z_1)^2 \left(\frac{d^2 f}{dz^2}\right)_1 + \cdots, \qquad (10)$$

where $z_2 - z_1 = \Delta z$, according to Fig. 2.2. We now go to the limit in which Δz is small enough so that we can neglect quadratic and higher terms in Eq. (10). Then we write

$$f(z_2) - f(z_1) = \Delta z \left(\frac{df}{dz}\right)_1 = \Delta z \frac{d}{dz}\left(\frac{\partial \psi(z, t)}{\partial z}\right)$$

$$= \Delta z \frac{\partial}{\partial z}\left(\frac{\partial \psi(z, t)}{\partial z}\right)$$

$$= \Delta z \frac{\partial^2 \psi(z, t)}{\partial z^2}. \qquad (11)$$

Notice that in arriving at Eq. (11) we dropped the subscript 1. That is because it does not matter where in the interval Δz we evaluate the z derivative, since we are neglecting higher derivatives in the Taylor's series, Eq. (10). Notice also that we must write the space derivative as a partial derivative once we use the notation $\psi(z, t)$.

We can now use Eqs. (9) and (11) in Eq. (8) to obtain for the net force on the segment the result

$$F_x(t) = T_0 \Delta_z \frac{\partial^2 \psi(z, t)}{\partial z^2}. \tag{12}$$

We now use Newton's second law. The force F_x, as given by Eq. (12), equals the mass ΔM of the segment times the acceleration of the segment. The velocity and acceleration of the segment with equilibrium position z are expressed in terms of $\psi(z, t)$ and its derivatives as follows:

$$\psi(z, t) = \text{displacement}$$

$$\frac{\partial \psi(z, t)}{\partial t} = \text{velocity}$$

$$\frac{\partial^2 \psi(z, t)}{\partial t^2} = \text{acceleration.} \tag{13}$$

Thus Newton's law [with $\Delta M = \rho_0 \, \Delta z$] gives

$$\rho_0 \Delta z \frac{\partial^2 \psi}{\partial t^2} = F_x = T_0 \, \Delta z \frac{\partial^2 \psi}{\partial z^2},$$

i.e.,

$$\boxed{\frac{\partial^2 \psi(z, t)}{\partial t^2} = \frac{T_0}{\rho_0} \frac{\partial^2 \psi(z, t)}{\partial z^2}} \tag{14}$$

Classical wave equation.   Equation (14) is a very famous second-order linear partial differential equation. It is called the *classical wave equation*. We will encounter it often and will eventually know many of the properties of its solutions and the physical situations where it occurs. (Of course the positive constant T_0/ρ_0 is special to the string. In other physical applications, some other positive constant appears in its place in the wave equation.)

Standing waves.   We are trying to find the normal modes—the standing waves—of a continuous string. Therefore we *assume* that we have a mode. We assume that all parts of the string oscillate in harmonic motion at the same angular frequency ω and with the same phase constant φ. Thus $\psi(z, t)$, which is the displacement of string particles with equilibrium position z, should have the same time dependence, $\cos(\omega t + \varphi)$ for all particles, i.e., for all z. As usual, the phase constant φ corresponds to the "turn-on time" of the mode. The "shape" of a mode made up of discrete degrees of freedom labeled a, b, c, etc., is given by the relative vibration amplitudes, A, B, C, etc. In the present case of a continuous string, where the (infinitely many) degrees of freedom are labeled by the parameter z, the amplitude of vibration of the degrees of freedom at z (i.e., in a small neighborhood of z) can be written as a continuous function of z denoted by $A(z)$. The shape of $A(z)$ as a function of z depends on the mode; that is, each mode has a different $A(z)$. Thus we can write down the *general expression for a standing wave:*

$$\psi(z, t) = A(z) \cos (\omega t + \varphi). \qquad (15)$$

The acceleration corresponding to Eq. (15) is

$$\frac{\partial^2 \psi}{\partial t^2} = -\omega^2 \psi = -\omega^2 A(z) \cos (\omega t + \varphi) \qquad (16)$$

The second partial derivative of Eq. (15) with respect to z is

$$\frac{\partial^2 \psi}{\partial z^2} = \frac{\partial^2 [A(z) \cos (\omega t + \varphi)]}{\partial z^2}$$

$$= \cos (\omega t + \varphi) \frac{d^2 A(z)}{dz^2}, \qquad (17)$$

where we have an ordinary derivative with respect to z rather than a partial derivative because $A(z)$ has no time dependence. Inserting Eqs. (16) and (17) into Eq. (14) and canceling the common factor $\cos (\omega t + \varphi)$, we obtain

$$\frac{d^2 A(z)}{dz^2} = -\omega^2 \frac{\rho_0}{T_0} A(z). \qquad (18)$$

Equation (18) governs the shape of the mode. Since each mode has a different angular frequency ω, and since ω^2 appears in Eq. (18), we see that different modes have different shapes, as expected.

Equation (18) is of the form of the differential equation for harmonic oscillation, but for oscillation in space rather than in time. The general form of a harmonic oscillation in space can be written

$$A(z) = A \sin \left(2\pi \frac{z}{\lambda} \right) + B \cos \left(2\pi \frac{z}{\lambda} \right), \qquad (19)$$

where the constant λ represents the distance over which one complete oscillation occurs. Thus it is called the *wavelength*. It is the parameter for oscillations in space analogous to the period T for oscillations in time. The wavelength λ is measured in units of centimeters per cycle (i.e., per cycle of spatial oscillation along z), or simply in centimeters.

To see how to adapt this solution to Eq. (18), differentiate Eq. (19) twice:

$$\frac{d^2 A(z)}{dz^2} = -\left(\frac{2\pi}{\lambda}\right)^2 A(z). \tag{20}$$

Then comparing Eqs. (18) and (20), we see that we need to have

$$\left(\frac{2\pi}{\lambda}\right)^2 = \omega^2 \left(\frac{\rho_0}{T_0}\right) = (2\pi v)^2 \frac{\rho_0}{T_0}, \tag{21}$$

i.e.,

$$\boxed{\lambda v = \sqrt{\frac{T_0}{\rho_0}} \equiv v_0 = \text{constant.}} \tag{22}$$

Wave velocity.   Equation (22) gives the relation between wavelength and frequency for transverse standing waves on a continuous homogeneous string. The constant $(T_0/\rho_0)^{1/2}$ has the dimensions of velocity, since λv has dimensions length/time. The velocity $v_0 \equiv (T_0/\rho_0)^{1/2}$ is called the "phase velocity for traveling waves," for this system. (We will study traveling waves in Chap. 4.) In our present study of standing waves, the concept of phase velocity is not needed, because standing waves do not "go anywhere." They "stand and wave" like a big "distributed" harmonic oscillator. Hereafter in this chapter we shall avoid calling $(T_0/\rho_0)^{1/2}$ a velocity, because we want your mental picture to be that of standing waves.

The general solution for the displacement $\psi(z, t)$ of the string in a single mode (standing wave) is obtained by combining Eqs. (15) and (19):

$$\psi(z, t) = \cos(\omega t + \varphi)[A \sin (2\pi z/\lambda) + B \cos (2\pi z/\lambda)]. \tag{23}$$

Boundary conditions.   Equation (23) is slightly *too* general. It does not manifest the important boundary conditions. Our vibrating string is *fixed at both ends*, but we have not yet incorporated that bit of information into the solution. We do so as follows. Suppose the string has total length L. Let us choose the origin of coordinates so that the left-hand end of the string is at $z = 0$. The right-hand end is then at $z = L$. Consider $z = 0$. The string is fixed there, so $\psi(0, t)$ must be zero for all t. This condition requires that $B = 0$, since, for all times t,

$$\psi(0, t) = \cos (\omega t + \varphi)[0 + B] = 0. \tag{24}$$

Thus we have

$$\psi(z, t) = A\cos(\omega t + \varphi)\sin\frac{2\pi z}{\lambda}. \qquad (25)$$

The other boundary condition is that the string be fixed at $z = L$, so $\psi(L, t)$ must be zero for all t. We certainly do not want to choose $A = 0$ in Eq. (25), since that corresponds to the uninteresting situation of a string permanently at rest. The only way we can satisfy the boundary condition at L is to have

$$\sin\frac{2\pi L}{\lambda} = 0. \qquad (26)$$

The only wavelengths λ that can satisfy this boundary condition are those for which the number of half-wavelengths, L, is an integer. Thus the acceptable wavelengths must satisfy one of the following possibilities:

$$\frac{2\pi L}{\lambda} = \pi, 2\pi, 3\pi, 4\pi, 5\pi, \ldots. \qquad (27)$$

(Why did we exclude the case $2\pi L/\lambda = 0$?). This sequence of possible ways to satisfy the boundary conditions corresponds to all the possible modes of the string. We number the modes according to the sequence, beginning with the first term in the sequence as number 1. Then according to Eq. (27), we have the wavelengths of the modes given by

$$\lambda_1 = 2L, \quad \lambda_2 = \tfrac{1}{2}\lambda_1, \quad \lambda_3 = \tfrac{1}{3}\lambda_1, \quad \lambda_4 = \tfrac{1}{4}\lambda_1, \ldots. \quad (28)$$

Harmonic frequency ratios. The corresponding frequencies of the modes are found by using Eq. (22):

$$\nu_1 = \frac{\upsilon_0}{\lambda_1}, \quad \nu_2 = 2\nu_1, \quad \nu_3 = 3\nu_1, \quad \nu_4 = 4\nu_1, \ldots. \quad (29)$$

The frequencies $2\nu_1$, $3\nu_1$, etc., are called the second, third, etc., *harmonics* of the *fundamental* frequency ν_1. The fact that the mode frequencies ν_2, ν_3, etc., consist of a sequence of *harmonics* of the lowest mode frequency ν_1 is a result of our assumption that the string is perfectly uniform and flexible. Most real physical systems have mode frequencies that do not follow this harmonic sequence of frequency ratios. For example, the mode frequencies for a string of nonuniform mass density do not form a sequence of harmonics of the fundamental. Instead one might have, for example, $\nu_2 = 2.78\nu_1$, $\nu_3 = 4.62\nu_1$, etc. For a real piano

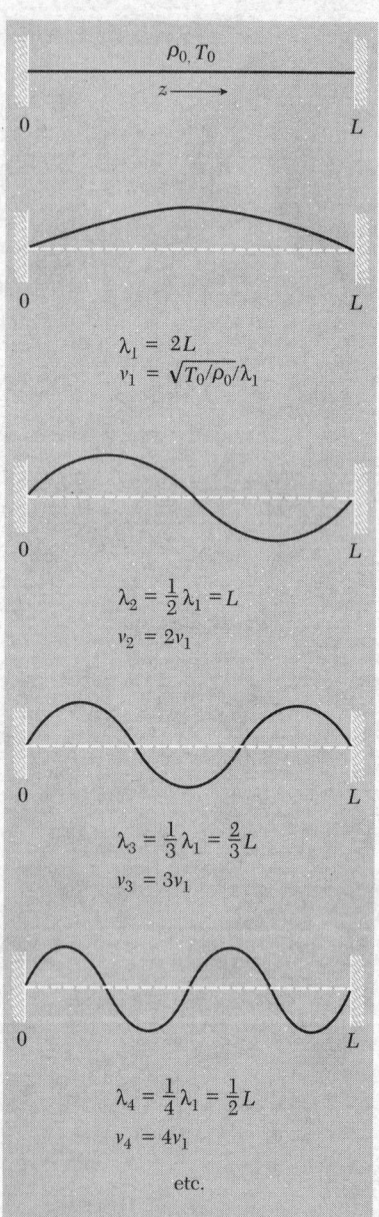

$$\rho_0, T_0$$

$$z \longrightarrow$$

0   L

0   L

$$\lambda_1 = 2L$$
$$\nu_1 = \sqrt{T_0/\rho_0}/\lambda_1$$

0   L

$$\lambda_2 = \tfrac{1}{2}\lambda_1 = L$$
$$\nu_2 = 2\nu_1$$

0   L

$$\lambda_3 = \tfrac{1}{3}\lambda_1 = \tfrac{2}{3}L$$
$$\nu_3 = 3\nu_1$$

0   L

$$\lambda_4 = \tfrac{1}{4}\lambda_1 = \tfrac{1}{2}L$$
$$\nu_4 = 4\nu_1$$

etc.

**Fig. 2.3   Modes of continuous homo-
geneous string with fixed ends.**

or violin string, the mode frequencies follow approximately, but not exactly, the harmonic sequence. That is because they are not perfectly flexible. (For a qualitative argument that shows how these "harmonic" frequency ratios are due to the uniformity of the string, see Prob. 2.7.)

The modes of the string are shown in Fig. 2.3. The equilibrium configuration would correspond to the missing first term, $2\pi L/\lambda = 0$, in the sequence given by Eq. (27). The corresponding frequency is zero. There is no motion, and the equilibrium state is not called a mode.

Wavenumber.   The inverse of the wavelength λ. is called the *wavenumber* σ. Its units are cycles per centimeter or, more often, "inverse centimeters." It is the parameter for oscillations in space analogous to the frequency ν for oscillations in time.

$$\sigma = \frac{1}{\lambda} = \text{wavenumber (cycles per cm)}. \tag{30}$$

The wavenumber times 2π is called the *angular wavenumber k*. Its units are radians of phase per centimeter. It is the quantity for oscillations in space analogous to the angular frequency ω for oscillations in time.

$$k = \frac{2\pi}{\lambda} = \text{angular wavenumber (radians per cm)}. \tag{31}$$

We can illustrate the use of these quantities by writing the same standing wave in several equivalent forms:

$$\begin{aligned}
\psi(z, t) &= A \sin 2\pi\, \frac{t}{T} \sin 2\pi\, \frac{z}{\lambda} \\
&= A \sin 2\pi\nu t \sin 2\pi\sigma z \\
&= A \sin \omega t \sin kz.
\end{aligned} \tag{32}$$

As another illustration, we can describe the sequence of normal modes given by Eqs. (27), (28), and (29) as follows:

$$k_1 L = \pi \text{ rad}, \qquad k_2 L = 2\pi \text{ rad}, \qquad k_3 L = 3\pi \text{ rad, etc.} \tag{33}$$

$$\sigma_1 L = \tfrac{1}{2} \text{ cycle}, \qquad \sigma_2 L = 1 \text{ cycle}, \qquad \sigma_3 L = \tfrac{3}{2} \text{ cycle, etc.} \tag{34}$$

Dispersion relation.   Equation (22) gives the relation between frequency and wavelength for the normal modes of the uniform flexible string:

$$\nu = \sqrt{\frac{T_0}{\rho_0}} \cdot \frac{1}{\lambda} = \sqrt{\frac{T_0}{\rho_0}} \cdot \sigma,$$

or (multiplying by 2π)

$$\omega = \sqrt{\frac{T_0}{\rho_0}}\, k. \tag{35}$$

Equation (35) gives the relation between frequency and wave-number for the normal modes of the string. (Note that we dropped the adjective "angular" from the designations "angular frequency" and "angular wavenumber." This is common practice, but the symbols and the units always remove any ambiguity.) Such a relation, giving ω as a function of k, is called a *dispersion relation*. It is a convenient way of characterizing the wave behavior of a system.

Dispersion law for real piano string. The dispersion relation given by Eq. (35) is extremely simple, but we shall find more complicated ones later. For a more complicated dispersion relation, the quantity $\lambda v = \omega/k$ is *not* constant, i.e., it is not independent of wavelength. For example, it turns out that the dispersion law for a real piano string is given approximately by

$$\frac{\omega^2}{k^2} = \frac{T_0}{\rho_0} + ak^2 \tag{36}$$

where a is a small positive constant that would be zero if the string were perfectly flexible [In that case Eq. (36) reduces to Eq. (35).] The modes of a real piano string have the same spatial dependence as those of a perfectly flexible string, i.e., $\lambda_1 = 2L$, $\lambda_2 = \frac{1}{2}\lambda_1$, $\lambda_3 = \frac{1}{3}\lambda_1$, etc., because the boundary conditions are the same. But the mode frequencies do *not* satisfy the "harmonic" sequence $v_2 = 2v_1$, $v_3 = 3v_1$, etc., because the dispersion relation Eq. (36) does not give that sequence. The harmonic sequence is obtained only in the idealized limit where a is zero, i.e., where we have $\lambda v = $ constant. For a real piano string the frequencies of the higher modes are slightly "sharper" (i.e., have slightly higher frequencies) than the frequencies given by the harmonic sequence.

Nondispersive and dispersive waves. Waves satisfying the simple dispersion relation $\omega/k = $ constant are called "nondispersive waves." When ω/k depends on the wavelength (and hence on the frequency), the waves are called "dispersive." For dispersive waves, it is customary to make a plot of ω versus k. In the present example of the flexible string this plot is just a straight line passing through the point $\omega = k = 0$ and having slope $(T_0/\rho_0)^{1/2}$, as shown in Fig. 2.4.

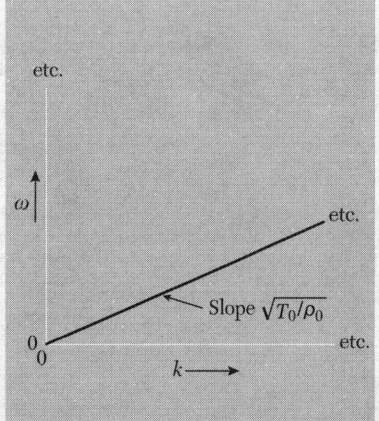

Fig. 2.4   *Dispersion relation for continuous, homogeneous flexible string.*

2.3   *General Motion of Continuous String and Fourier Analysis*

The most general state of motion of the continuous string (with both ends fixed and for transverse vibrations along x) is given by a super-position of all the modes, numbered 1, 2, 3,. . ., with amplitudes $A_1, A_2, A_3, \ldots$, and phase constants $\varphi_1, \varphi_2, \varphi_3, \ldots$:

$$
\begin{aligned}
\psi(z, t) &= A_1 \sin k_1 z \cos (\omega_1 t + \varphi_1) \\
&+ A_2 \sin k_2 z \cos (\omega_2 t + \varphi_2) + \ldots,
\end{aligned} \tag{37}
$$

where k_n are chosen as described in the preceding section to satisfy the boundary conditions at $z = 0$ and $z = L$, and where ω_n are related to k_n by the dispersion relation $\omega(k)$. The amplitudes A_n and phase constants φ_n, which complete the description of the motion for all positions z and times t, are determined by specifying the *initial conditions*, namely, the instantaneous displacement $\psi(z, t)$ and the corresponding instantaneous velocity $v(z, t) = \partial\psi(z, t)/\partial t$ for each point z at $t = 0$.

Motion of string fixed at both ends.   Suppose that for $t < 0$ we constrain the string to follow a prescribed shape $f(z)$ by means of some sort of template. Then, at $t = 0$, we let the string go by suddenly removing the template. Thus at $t = 0$ each part of the string has its displacement $\psi(z, 0)$ equal to $f(z)$ and has velocity $v(z, 0)$ equal to zero. Now, the nth term in the velocity [which is the time derivative of Eq. (37)] is proportional to $\sin (\omega_n t + \varphi_n)$, which reduces to $\sin \varphi_n$ at $t = 0$. Thus we can make $v(z, 0) = 0$ for all z simply by setting each phase constant φ_n equal either to zero or to π. However, the phase constant $\varphi_1 = \pi$ (for example) is just equivalent to a minus sign affixed to A_1. Therefore we can satisfy these initial conditions if we set all the phase constants to zero but allow the amplitudes A_1, A_2, etc., to be either positive or negative. Thus we have, for $v(z, 0) = 0$,

$$
\psi(z, t) = A_1 \sin k_1 z \cos \omega_1 t + A_2 \sin k_2 z \cos \omega_2 t + \ldots, \tag{38}
$$

and, at $t = 0$,

$$
\psi(z, 0) = f(z) = A_1 \sin k_1 z + A_2 \sin k_2 z + \ldots, \tag{39}
$$

As we shall see below, Eq. (39) determines the amplitudes $A_1, A_2, \ldots$.

Fourier series for function with zeros at both ends.   Now, the function $f(z)$ can be a very general function of z. The only condition we specified was that it was to constrain the string. Therefore, virtually all we require of $f(z)$ is that we have $f(z) = 0$ at $z = 0$ and $z = L$. We also require that $f(z)$ not be "jagged" on a "small" scale, since our wave function $\psi(z, t)$ is supposed to be a slowly varying function of z.

Therefore, $f(z)$ must be reasonably smooth in order for us to be able to use it to constrain the string and still have the string obey the differential equation that we obtained in the "continuous" approximation. Thus we have found that *any* reasonable function $f(z)$ that vanishes at $z = 0$ and L can be expanded in a series of the form of Eq. (39), i.e., as a sum of sinusoidal oscillations. Equation (39) is called a *Fourier series* or *Fourier expansion*. It is a special example of a Fourier series in that it applies only to functions $f(z)$ that vanish at $z = 0$ and L. However, a much broader class of functions can be expressed in appropriate Fourier expansions. We shall now find this broader class of functions.

Our function $f(z)$ was used to constrain the string, and therefore it was defined only between $z = 0$ and L. However, the functions $\sin k_1 z$, $\sin 2k_1 z$, $\sin 3k_1 z$, etc., that make up the infinite series of Eq. (39) are defined for all z from $-\infty$ to $+\infty$. Also, we notice that $\sin k_1 z$ is *periodic* in z with period λ_1. This means it satisfies the *periodicity condition*, namely, that for any given z, it must have the same value at $z + \lambda_1$ as it does at z. (The period λ_1 is $2L$, in our example.) We notice that the function $\sin 2k_1 z$ is also periodic in z with period λ_1. (Of course it goes through two cycles in distance λ_1; it is thus periodic with period $\frac{1}{2}\lambda_1$, as well as periodic with period λ_1). In fact, all the sinusoidal functions in the expansion, Eq. (39), are periodic in z with period λ_1. Therefore, the expansion itself is periodic with period λ_1. Thus we can broaden the class of functions which have a Fourier expansion of the form of Eq. (39): all periodic functions $F(z)$ with period λ_1 that vanish at $z = 0$ and at $z = \frac{1}{2}\lambda_1$ can be expanded in a Fourier series of the form of Eq. (39). Given a function $f(z)$ defined only between $z = 0$ and L and vanishing at those points, we can construct a periodic function $F(z)$ which will have the same Fourier expansion as $f(z)$ by the following procedure: Between $z = 0$ and L, we let $F(z)$ coincide with $f(z)$. Between L and $2L$, we construct $F(z)$ by making an "inverted mirror image" of $f(z)$ in a "mirror" located at $z = L$. Now that we have defined $F(z)$ between $z = 0$ and $2L$, we simply repeat it in successive intervals of length $2L$ to define $F(z)$ for all z. The construction is shown in Fig. 2.5.

Fourier analysis of a periodic function of z. We now broaden the class of functions for which we can write Fourier expansions once more, as follows: Equation (39) corresponds only to functions that are periodic with period λ_1 and that vanish at $z = 0$ and $\frac{1}{2}\lambda_1$. However, the condition that the function vanish at $z = 0$ and $\frac{1}{2}\lambda_1$ was the result of our particular choice of boundary conditions, namely that the string have both ends fixed. Without those particular boundary conditions, we would have obtained solutions for the string vibrations which included not only the terms in $\sin mk_1 z$ but

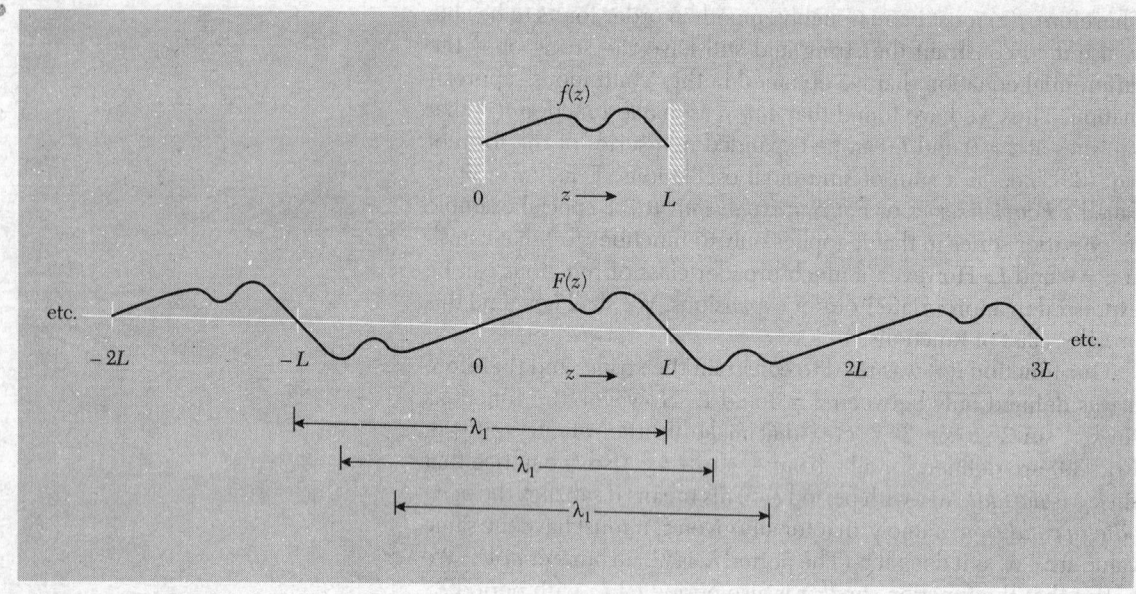

Fig. 2.5 *Construction of a periodic function F(z) with period* $\lambda_1 = 2L$ *from a function f(z) that vanishes at z = 0 and L. Note that F(z) satisfies the periodicity condition.*

also terms in $\cos m k_1 z$. These functions are also periodic in z with period, λ_1, but they do not vanish at $z = 0$ and $\frac{1}{2}\lambda_1$. (They correspond to string vibrations with a free end or ends.) By including them in the series, we finally arrive at a very general class of functions for which we can write Fourier series: all (reasonable) periodic functions $F(z)$ with period λ_1, i.e., functions such that $F(z + \lambda_1) = F(z)$ for all z, can be expanded in a Fourier series of the form

$$F(z) = \sum_{n=0}^{\infty} \left[A_n \sin n \frac{2\pi}{\lambda_1} z + B_n \cos n \frac{2\pi}{\lambda_1} z \right]$$

$$= B_0 + \sum_{n=1}^{\infty} A_n \sin n \frac{2\pi}{\lambda_1} z + \sum_{n=1}^{\infty} B_n \cos n \frac{2\pi}{\lambda_1} z$$

$$= B_0 + \sum_{n=1}^{\infty} A_n \sin n k_1 z + \sum_{n=1}^{\infty} B_n \cos n k_1 z. \tag{40}$$

Finding Fourier coefficients.   The process of finding the amplitudes or *Fourier coefficients* B_0, A_n, and B_n (for all n) for a given periodic function $F(z)$ is called *Fourier analysis*. We shall now show you how to find these coefficients.

First we find B_0 as follows: We integrate both sides of Eq. (40) over any complete period of $F(z)$; i.e., we integrate from $z = z_1$ to $z = z_2$, where z_1 is any value of z and where $z_2 = z_1 + \lambda_1$. The function $F(z)$ is assumed to be known; therefore its integral from z_1 to z_2, which is the integral of the left side of Eq. (40), can be found. Now consider the integral of the right side of Eq. (40). There are an infinite number of terms and therefore an infinite number of integrals to consider. The first term is B_0; it produces the integral

$$\int_{z_1}^{z_2} B_0 \, dz = B_0(z_2 - z_1) = B_0\lambda_1. \tag{41}$$

All the other terms give zero when integrated over one period. That is because $\sin nk_1 z$ and $\cos nk_1 z$ are as often negative as positive in any complete period, and therefore they integrate to zero:

$$\int_{z_1}^{z_2} \sin nk_1 z \, dz = 0; \qquad \int_{z_1}^{z_2} \cos nk_1 z \, dz = 0$$

Thus we have found B_0. It is given by

$$B_0\lambda_1 = \int_{z_1}^{z_2} F(z) \, dz. \tag{42}$$

Next we show you how to find A_m, where m is any particular value of n in Eq. (40) from 1 to infinity. The trick is to multiply both sides of Eq. (40) by $\sin mk_1 z$ and integrate both sides over one complete period of $F(z)$. The integral of the left-hand side can be evaluated since $F(z)$ is known. Now consider the integral of the right-hand side. The first term is the integral of B_0 times $\sin mk_1 z$; that integrates to zero because it includes m complete periods of $\sin mk_1 z$. That leaves us with the integrals of $\sin nk_1 z \sin mk_1 z$ and of $\cos nk_1 z \sin mk_1 z$ for $n = 1, 2, \ldots$. Consider the particular term that has $n = m$. The square of $\sin mk_1 z$ averages to $\frac{1}{2}$ over one period of $F(z)$ of length λ_1 (which is m complete periods of the function $\sin mk_1 z$). This gives a contribution $\frac{1}{2}A_m\lambda_1$ to the integral of the right side of Eq. (40). All other terms contribute zero. We see that as follows: Consider for example the integrand $\sin nk_1 z \sin mk_1 z$, for m not equal to n. This can be written in the form

$$\sin nk_1 z \sin mk_1 z = \tfrac{1}{2} \cos (n - m)k_1 z - \tfrac{1}{2} \cos (n + m)k_1 z. \tag{43}$$

Since $n - m$ and $n + m$ are integers, each of the two terms on the right side of Eq. (43) is as often positive as negative in any complete period of $F(z)$ of length λ_1. Therefore both terms integrate to zero (except for the case $n = m$ which we have already considered). Similarly, the terms of the form $\cos nk_1 z \sin mk_1 z$ integrate to zero because of the identity

$$\cos nk_1 z \sin mk_1 z = \tfrac{1}{2} \sin (m + n)k_1 z + \tfrac{1}{2} \sin (m - n)k_1 z.$$

Thus we find that

$$\frac{1}{2} A_m \lambda_1 = \int_{z_1}^{z_2} \sin mk_1 z \, F(z) \, dz. \tag{44}$$

Similarly, we can find the coefficients B_m by multiplying both sides of Eq. (40) by $\cos mk_1 z$ and integrating over one period of length λ_1. The only nonzero contribution to the integral of the right side comes from the term with coefficient B_m. Thus we find that

$$\frac{1}{2} B_m \lambda_1 = \int_{z_1}^{z_2} \cos mk_1 z \, F(z) \, dz. \tag{45}$$

Fourier coefficients. Our results are given by Eqs. (40), (42), (44), and (45), which we collect in one place for convenience of future reference:

$$
\begin{aligned}
F(z) &= B_0 + \sum_{m=1}^{\infty} A_m \sin mk_1 z + \sum_{m=1}^{\infty} B_m \cos mk_1 z, \\[2mm]
B_0 &= \frac{1}{\lambda_1} \int_{z_1}^{z_1 + \lambda_1} F(z) \, dz, \\[2mm]
A_m &= \frac{2}{\lambda_1} \int_{z_1}^{z_1 + \lambda_1} F(z) \sin mk_1 z \, dz, \\[2mm]
B_m &= \frac{2}{\lambda_1} \int_{z_1}^{z_1 + \lambda_1} F(z) \cos mk_1 z \, dz,
\end{aligned}
\tag{46}
$$

where z_1 is any value of z. Equations (46) tell us how to Fourier-analyze $F(z)$, any periodic function of z having period λ_1.

Square wave. Here is an illustrative example, the Fourier analysis of a "square wave." Let $f(z)$ be zero at the points $z = 0$ and $z = L$, but let it equal $+1$ for $0 < z < L$. (This function has a discontinuity at $z = 0$ and another at $z = L$, so that it does not satisfy the assumption in our discussion above that it be "smooth" everywhere. Therefore we cannot reasonably expect the Fourier series to give a perfect representation of a square wave: It turns out that there is a sharp "overshoot spike" at $z = 0$ and at $z = L$ for every partial sum of the series. As more and more terms are added, the spike gets sharper, but its height does not go to zero.)

The periodic function $F(z)$ that we construct according to the prescription of Fig. 2.5 is given as follows: $F(z) = 0$ for $z = 0$; $+1$ for $0 < z < L$; 0 for $z = L$; -1 for $L < z < 2L$; etc.; as shown in Fig. 2.6.

Using Eqs. (46), one can easily obtain the results (Prob. 2.11) $B_0 = 0$; $B_m = 0$ for all m; $A_m = 0$ for $m = 2, 4, 6, 8. . .$(even integers); $A_m = 4/m\pi$ for $m = 1, 3, 5, 7,. . .$(odd integers). Thus $F(z)$ is given by

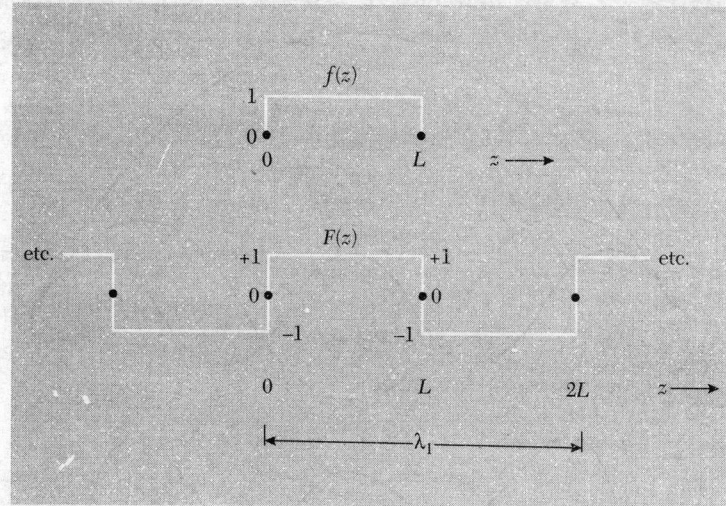

Fig. 2.6 Square wave f(z). Periodic square wave F(z).

$$F(z) = B_0 + \sum_{m=1}^{\infty} B_m \cos mk_1z + \sum_{m=1}^{\infty} A_m \sin mk_1z$$

$$= \frac{4}{\pi}\left\{ \sin k_1z + \frac{1}{3} \sin 3k_1z + \frac{1}{5} \sin 5k_1z + \cdots \right\}$$

$$= 1.273 \sin \frac{\pi z}{L} + 0.424 \sin \frac{3\pi z}{L} + 0.255 \sin \frac{5\pi z}{L} + \cdots \tag{47}$$

In Fig. 2.7 are shown the square wave $f(z)$, the first three contributing terms given by Eq. (47), and the superposition of these first three terms.

Suppose that instead of trying to force a slinky into the configuration of the sharp-cornered function $f(z)$ which we have been considering, we constrain it at time zero to follow exactly the function

$$g(z) = 1.273 \sin \frac{\pi z}{L} + 0.424 \sin \frac{3\pi z}{L} + 0.255 \sin \frac{5\pi z}{L}. \tag{48}$$

This corresponds to the first three terms of Eq. (47) and is plotted in Fig. 2.7b. Now we let the slinky go at $t = 0$. What is $\psi(z, t)$? Does the shape remain constant as t increases? (See Prob. 2.16.)

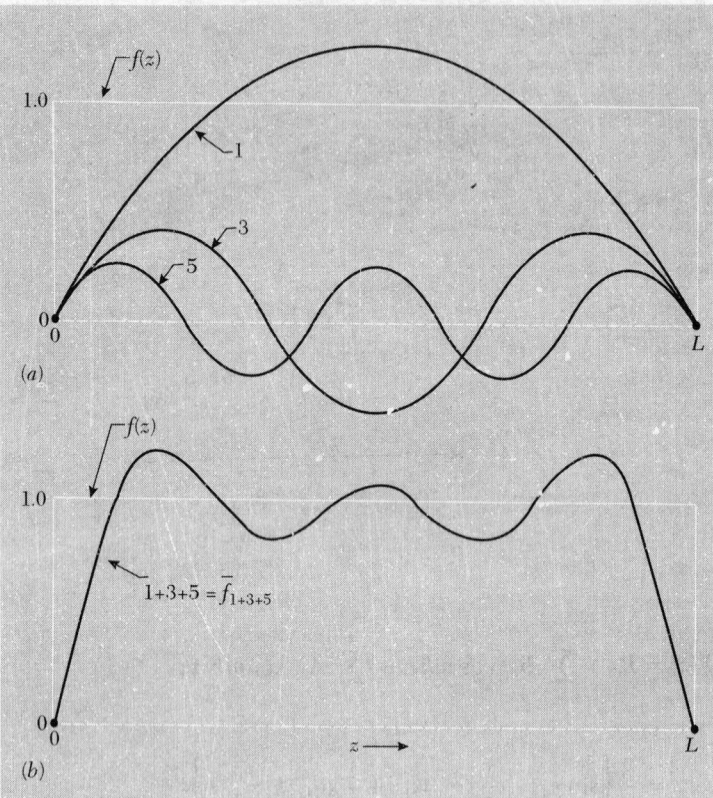

(a)

(b)

Fig. 2.7 *Fourier analysis of square wave f(z). (a) Square wave f(z) and the first three contributions to its Fourier decomposition. The labels 1, 3, and 5 refer to the normal modes 1, 3, 5. (b) Square wave f(z) and the superposition f_{1+3+5} of its first three Fourier components.*

Fourier analysis of a periodic function of time.   Suppose we are given a function $F(t)$ that is defined for all t and that is periodic in t with period T_1:

$$F(t + T_1) = F(t) \qquad \text{for any } t. \tag{49}$$

We assume that $F(t)$ can be expanded in the Fourier series

$$F(t) = B_0 + \sum_{m=1}^{\infty} A_n \sin n\omega_1 t + \sum_{m=1}^{\infty} B_n \cos n\omega_1 t, \tag{50}$$

with

$$\omega_1 = 2\pi \nu_1 = \frac{2\pi}{T_1} \tag{51}$$

The Fourier coefficients can be obtained directly from our results for the Fourier analysis of a spatially periodic function $F(z)$, which we studied above. The mathematical analysis cannot distinguish the variable $\theta = \omega_1 t$ from the variable $\theta = k_1 z$. Thus we obtain the results for the coefficients in Eqs. (50) directly from Eqs. (46):

$$B_0 = \frac{1}{T_1} \int_{t_1}^{t_1+T_1} F(t)\, dt,$$

$$B_n = \frac{2}{T_1} \int_{t_1}^{t_1+T_1} F(t) \cos n\omega_1 t\, dt, \qquad (52)$$

$$A_n = \frac{2}{T_1} \int_{t_1}^{t_1+T_1} F(t) \sin n\omega_1 t\, dt,$$

where the time t_1 is any convenient time.

Sound of a piano chord. We shall illustrate this with a superposition of known ingredients, rather than by a Fourier analysis of a known function $F(t)$. Suppose you have a piano that is tuned to the "scientific scale." (See Home Exp. 2.6 if you want to know more about musical scales.) Let $v_1 = 128$ cps. That is the note C one octave (i.e., a factor of 2 in frequency) below middle C. Now let $v_3 = 3v_1 = 384$ cps. That is the G above middle C. Let $v_5 = 5v_1 = 640$ cps. That is the E above the G above middle C. Now strike all three notes at the same time. One hears a nice "open" chord. If you strike them at exactly the same time, and if you adjust your striking force so that the gauge pressure of air produced at your ear by the C128 string is (in appropriate units) $1.273 \sin 2\pi v_1 t$, pressure by the G384 string is $0.424 \sin 2\pi v_3 t$, and pressure by the E640 string is $0.255 \sin 2\pi v_5 t$, then the total air pressure $p(t)$ at your ear is the superposition

$$p(t) = 1.273 \sin 2\pi v_1 t + 0.424 \sin 2\pi v_3 t + 0.255 \sin 2\pi v_5 t. \quad (53)$$

But Eq. (53) is very similar to Eq. (48), which is plotted in Fig. 2.7b. All we have to do to obtain a plot of $p(t)$ is to change variables from $k_1 z$ to $\omega_1 t$ and extend the plot shown in Fig. 2.7b. Thus we get the result shown in Fig. 2.8.

If we do not strike all the keys at exactly the same time (i.e., to within an accuracy of much less than $\frac{1}{128}$ sec), the relative phases of the three notes will not be those of Eq. (53), and the superposition will not look like Fig. 2.8. But your ear does not notice this! Your ear (plus brain) performs a Fourier analysis on the total pressure. That must be so, because you "hear" the individual notes of the chord and recognize them. But the information as to relative phase of the notes is apparently discarded or perhaps not obtained. Otherwise you would notice a difference in the sound depending on the relative phases.

The pitch-detecting device in the ear is called the *basilar membrane*. It is enclosed in a fluid-filled, spiral-shaped organ in the inner ear called the *cochlea*. The cochlea is mechanically coupled to the eardrum. The end of the basilar membrane nearest the eardrum resonates at about 20,000 cps; the end farthest from the drum resonates at about 20 cps. Thus the extreme range of audible

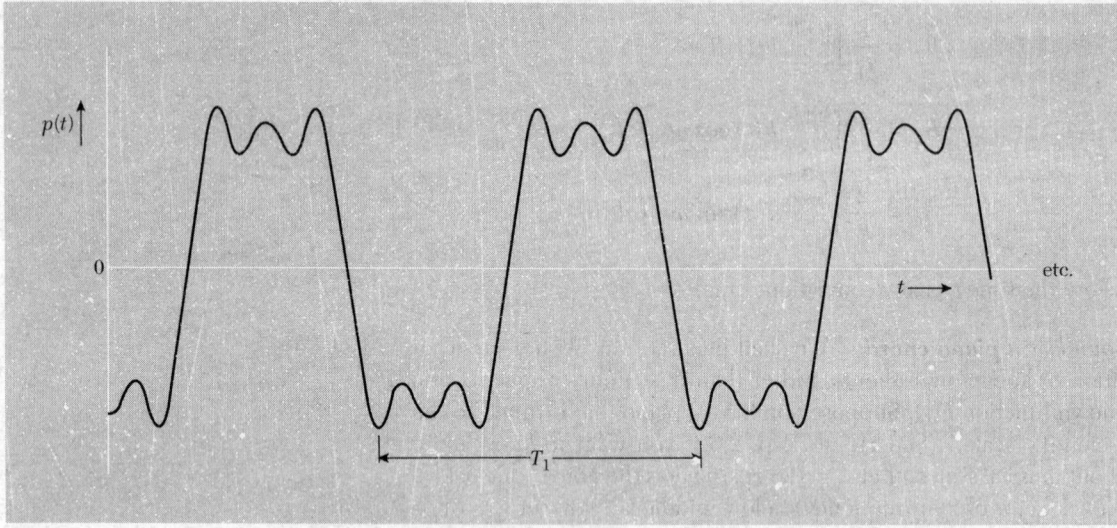

Fig. 2.8   Gauge pressure at ear due to superposition of the notes C128, G384, and E640 with the relative amplitudes and phases of Eq. (53). The period T_1 is (1/128) sec.

frequencies is about 20 cps to 20 kc. The cochlear nerve has sensors in the basilar membrane and "transduces" the mechanical vibrations into electrical signals that are carried to the brain, where they are somehow processed to become our hearing sensations. By doing the experiment of hitting the chord over and over and seeing that our sensation is the same [even though $p(t)$ must have a very different shape depending on the relative phases], we have learned that somewhere the information as to the relative phase of the vibrations of different parts of the basilar membrane is lost. Perhaps this information is never picked up. Perhaps the transducer is a *square-law detector*, i.e., one that puts out an electrical signal proportional to the square of the amplitude of vibration of the membrane. Or perhaps the nerve signal does carry phase information [i.e., perhaps the signal does give $\psi(z, t)$ rather than $\psi^2(z, t)$], but the brain does not use the phase information, i.e., it does not form a superposition of $\psi(z, t)$ from different nerve signals. Apparently there is not much survival value in the phase information; otherwise in our evolutionary development we surely would have acquired some phase-detecting mechanism.

Other boundary conditions.   In the general problem of transverse vibrations of a continuous string, it is not necessary that the string be fixed at both ends. One or both ends can be "free," at least as far as transverse oscillations are concerned. The tension and equilibrium configuration of the string can be maintained by a constraint in the form of a massless, frictionless ring sliding on a fixed rod oriented along x, i.e., transverse to the equilibrium axis of the string (which we always take along z). The normal modes will then have different configurations from those we obtained for the

string with both ends fixed. The shapes of the modes are still sinusoidal functions of z, as given by Eq. (19). The dispersion relation
between frequency and wavelength is still that given by Eq. (22) In
fact our entire discussion preceding Eq. (23), the general solution
for the displacement of the string in a single mode, is independent
of the boundary conditions. It was only in the discussion following
Eq. (23), that we specialized the solution to the case of the string
fixed at $z = 0$ and L.

At a free end of a vibrating string, there is (by definition) no
transverse force exerted on the end of the string, i.e., the frictionless rod exerts no transverse force on the frictionless ring.
Then (by Newton's third law) the string and frictionless ring
exert no transverse force on the frictionless rod. That means the
string must be horizontal. *The slope of the string at a free end is
zero at all times.* If one tries to exert a transverse force on the
free end of a string, the string moves in such a way as to reduce
the force to zero even as it is being applied. It never becomes
different from zero, and the string remains horizontal, but of
course not motionless. (The moral is that you cannot push on
something that refuses to push back, but you can move it where
you please.)

In Fig. 2.9 we show the modes of a string with one end fixed and
the other free. We have labeled the successive modes according to
the number of quarter-wavelengths contained in the string length L.
Notice that the even harmonics with frequencies $2v_1$, $4v_1$, etc. are
missing. The Fourier analysis of functions $f(z)$ with value zero at
$z = 0$ and slope zero at $x = L$ is discussed in Prob. 2.29.

Dependence of tone quality on method of excitation. When
a piano string is struck by its hammer, the fundamental (v_1), the
second harmonic or octave ($2v_1$), the octave plus a fifth ($3v_1$), the second octave ($4v_1$), the second octave plus a major third ($5v_1$), and the
second octave plus a fifth ($6v_1$) are all excited to some extent, as are
higher harmonics of the fundamental tone v_1. The amount and phase
of each Fourier component (each harmonic) depend on the initial
configuration and velocity of all parts of the string at the instant just
after it has been struck by the hammer. These depend to a great
extent on the location of the hammer, i.e., on its distance from the
end of the string. No mode that has a node (a permanently motionless point) at the striking point will be excited by the hammer blow,
since the hammer imparts an initial velocity to the part of the string
it hits. For example, if the string is plucked at its center, the modes
with a node at the center are not excited. Inspection of Fig. 2.3
shows that in that case all the even harmonics are missing. Thus if
we pluck the string for C128 in the middle, we expect it to vibrate in
a superposition of C128, G384, E640, etc. The "tone quality" is then

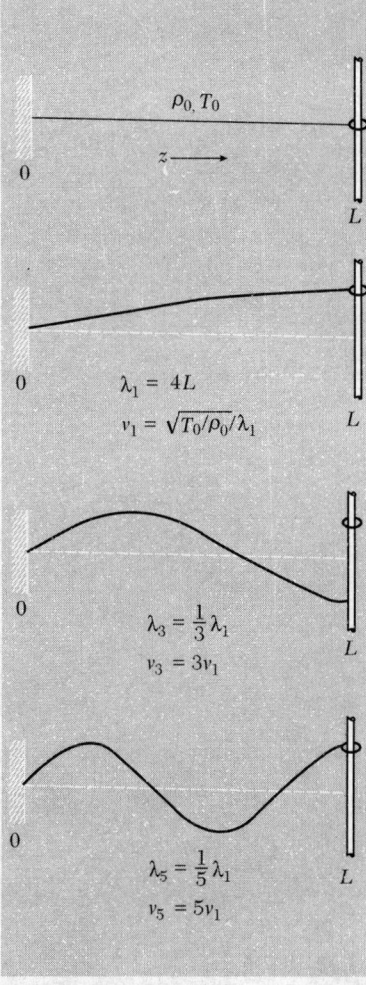

**Fig. 2.9 Modes of continuous string
with one fixed end and one free end.**

appreciably different from that produced when the string is struck near one end and vibrates in a superposition of C128, C256, G384, CSI2, E640, G768, etc.

Modes of homogeneous string form complete set of functions. Starting with string fixed at both ends, e discovered that *any* reasonable unction $f(z)$ that is defined between $z = 0$ and $z = L$ and that is zero at $z = 0$ and L can be expanded in the Fourier series

$$f(z) = \sum_{n=1}^{\infty} A_n \sin nk_1 z; \qquad k_1 L = \pi. \qquad (54)$$

For that reason, the functions $\sin nk_1 z$, with $n = 1, 2, 3, \ldots$, are said to be a *complete set* of functions [with respect to functions $f(z)$ that vanish at $z = 0$ and L]. A complete set of functions is defined as a set such that *any* (reasonable) function $f(z)$ can be written as a superposition of functions from the set by choosing suitable constant coefficients.

In homogeneous string.   Besides the sinusoidal functions that constitute a Fourier series, are there other complete sets? Yes, infinitely many sets! We can see this as follows. Suppose that the string is not homogeneous, i.e., that either its mass density or its tension (or both) is a continuous function of position z. (An example of a "string" with varying density and tension is provided by a vertically hanging slinky with fixed top and bottom ends. The tension at the bottom is less than that at the top by the weight Mg, where M is the total mass of the slinky.) Then the equation of motion of a small segment of string does not again lead to the classical wave equation, which is

$$\frac{\partial^2 \psi(z, t)}{\partial t^2} = \frac{T_0}{\rho_0} \frac{\partial^2 \psi(z, t)}{\partial z^2}.$$

Instead, if we have equilibrium tension $T_0(z)$ and density $\rho_0(z)$, we easily find (Prob. 2.10) that we have

$$\frac{\partial^2 \psi(z, t)}{\partial t^2} = \frac{1}{\rho_0(z)} \frac{\partial}{\partial z} \left[T_0(z) \frac{\partial^2 \psi(z, t)}{\partial z} \right], \qquad (55)$$

which reduces to the classical wave equation only if $T_0(z)$ and $\rho_0(z)$ are constants, independent of z. In a normal mode of this *inhomogeneous string*, just as in a mode of the homogeneous string, every part of the string vibrates in harmonic motion with the same frequency and phase constant:

$$\psi(z, t) = A(z) \cos (\omega t + \varphi). \qquad (56)$$

Thus

$$\frac{\partial^2 \psi}{\partial t^2} = -\omega^2 A(z) \cos(\omega t + \varphi), \tag{57}$$

$$\frac{\partial \psi}{\partial z} = \cos(\omega t + \varphi)\frac{dA(z)}{dz}. \tag{58}$$

Substituting these in Eq. (55) and canceling the common factor cos $(\omega t + \varphi)$ yields the equation for the shape of the mode:

$$\frac{1}{\rho_0(z)}\frac{d}{dz}\left[T_0(z)\frac{dA(z)}{dz}\right] = -\omega^2 A(z). \tag{59}$$

Sinusoidal shape of standing waves is characteristic of homogeneous systems.   The shape of the mode is given by $A(z)$, which is obtained by solving the differential equation Eq. (59) with the appropriate boundary conditions that $A(z) = 0$ at $z = 0$ and L. The function $A(z)$ is not sinusoidal in shape unless T_0 and ρ_0 are constants. Thus *sinusoidal* oscillations in space are only characteristic of the shapes of the normal modes of a *homogeneous* system.

Modes of inhomogeneous string form complete set of functions.   We shall tell you without proof the characteristics of the normal modes for an inhomogeneous string with ends fixed at $z = 0$ and L. The lowest mode corresponds to a solution of Eq. (59), $A_1(z)$, which is zero only at $z = 0$ and L. (That is like one half-wavelength of a "distorted sine wave," which has no nodes between 0 and L.) This mode has frequency ω_1. The next mode has *one node* between $z = 0$ and L and thus resembles one full wavelength of a distorted sine wave. It has characteristic frequency ω_2. The mth mode has $m - 1$ nodes between $z = 0$ and L and resembles m half-wavelengths of a distorted sine wave. There are an infinite number of modes (for a continuous string). The functions $A_1(z)$, $A_2(z)$, $A_3(z)$, ..., which give the space dependence of the modes, form a complete set with respect to any reasonable function $f(z)$ that vanishes at $z = 0$ and L. A reasonable function $f(z)$ is defined to be one which the string or slinky can follow without violating any of our assumptions. In that case we can make a template that has the shape $f(z)$, fit the inhomogeneous string to the template, and let it go from rest at $t = 0$. The string will vibrate in an infinite superposition of its modes:

$$\psi(z, t) = \sum_{m=1}^{\infty} c_m A_m(z) \cos \omega_m t \tag{60}$$

Then at $t = 0$ we have

$$\psi(z, 0) = f(z) = \sum_{m=1}^{\infty} c_m A_m(z). \tag{61}$$

Equation (61) shows that $f(z)$ (subject to our assumptions) can be expanded in the set of functions $A_m(z)$. Thus $A_m(z)$ form a complete set of functions. This argument is exactly analogous to the one that convinced us that the sinusoidal functions of a Fourier series form a complete set with respect to functions $f(z)$ that vanish at $z = 0$ and L.

Eigenfunctions.   There are an infinite number of different ways that we can construct a string with nonuniform mass density and tension. Therefore, there are an infinite number of different complete sets $A_m(z)$. Sinusoidal functions of z are thus not the only complete set of functions for expanding functions $f(z)$. But they are a very important set, because they are very simple and easy to understand. Furthermore, they give the shapes of the modes whenever we have a system that is spatially *homogeneous.* When that is not the case, the sinusoidal functions are not very useful. Instead one tries to find and use the appropriate functions $A_m(z)$ that correspond to the normal modes of the system. These functions $A_m(z)$, or, more generally, $A_m(x, y, z)$ for a three-dimensional system, are called *eigenfunctions.* They give the *space* dependence of the normal modes.

For every position x, y, z, the *time* dependence of a mode is *always* given by $\cos(\omega t + \varphi)$. Thus a mode is essentially nothing but the simultaneous small oscillation (small enough to give linear equations) of all the moving parts, all parts oscillating with the same frequency and same phase constant. When the entire system is in a single mode, it pulsates and throbs like one big oscillator. Each mode has its own "shape," i.e., its own eigenfunction. The relation between mode frequency and shape is called the dispersion relation, $\omega(k)$, when the shapes of the eigenfunctions are sinusoidal. When they are not sinusoidal, there is, of course, no such thing as wavelength or wave number k. Then the relation between mode frequency and shape is not usually called by the name "dispersion relation."

We will not study inhomogeneous systems further. When you study quantum physics, you will study the eigenfunctions (shapes) of de Broglie standing waves in systems with nonconstant potential energy. These are analogous to the standing waves of an inhomogeneous string. See Supplementary Topic 2.

2.4 Modes of a Noncontinuous System with N Degrees of Freedom

In Sec. 2.2 we considered a continuous string, which is a system with infinitely many degrees of freedom. No real mechanical system has an infinite number of degrees of freedom, and we are interested in real systems. In this section we will find the exact solution for the modes of a uniform beaded string having N beads and with fixed ends. In the limit that we take the number of beads N to be infinite (and maintain the finite length L), we shall find the standing waves that we studied in Sec. 2.2. Our purpose is not merely that, however. Rather, we shall find that, in going to the limit of a continuous string, we discarded some extremely interesting behavior of the system. Remember that in order to use the smooth function $\psi(z, t)$ to describe the displacement when N is huge but not infinite, we have to prohibit ourselves from considering the highest modes, i.e., the modes $m = N, N - 1, N - 2$, etc. We must confine ourselves to, values of m much less than N. That is because mode N has the zigzag configuration shown in Fig. 2.1, and thus neighboring beads do *not* have nearly the same displacement.

The most interesting new result we shall obtain in this section is that the dispersion law which we obtained for the *continuous* string, namely "ω equals a constant times k," does not generally hold. This relationship between frequency and wavelength, which implies that the frequency doubles when the wavelength is halved (i.e., which gives the harmonic frequency ratios), is an approximation that holds for the flexible string only in the continuous limit. The fact that it does not hold for a "lumpy" (but otherwise uniform) string provides an example of an interesting physical phenomenon called *dispersion*. A medium that satisfies the simple dispersion relation above, "ω equals constant times k," is called *nondispersive* (for the appropriate waves). If any other dispersion relation holds, the medium is called *dispersive*. Now consider the example:

Example 1: Transverse oscillations of beaded string

The system is shown in Fig. 2.10. There are N beads. The string is fixed at $z = 0$ and L. The beads are located at $z = a, 2a, \ldots,$ Na. The total length L is $(N + 1)a$. The beads each have mass M. The string (or spring) segments are identical. They are massless and are perfect Hooke's law springs. The equilibrium tension is T_0. If the springs (strings) satisfy the slinky approximation (tension proportional to length), the oscillations may have arbitrarily large amplitude and still give us linear equations of motion. If the springs are not slinkies, we shall confine ourselves to small amplitude oscillations, so as to get linear equations.

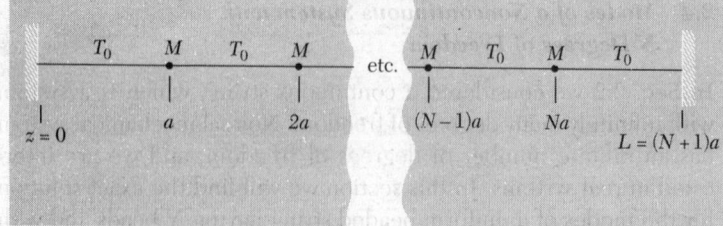

Fig. 2.10 Equilibrium configuration of beaded string.

Now we consider the general configuration shown in Fig. 2.11. [It is not completely general, in that we are now considering only transverse oscillations along x. Later we shall consider the longitudinal oscillations along z. The general motion is of course a superposition of longitudinal oscillation along z and transverse oscillations along x and y]. The displacement of bead n upward (in the figure) from its equilibrium position is $\psi_n(t)$, with $n = 1, 2, 3, \ldots, N-1, N$. We focus our attention on an arbitrary bead n and its neighbors $n-1$ (to the left) and $n+1$ (to the right).

Equation of motion.   We want the equation of motion for bead n. We have already solved a problem very similar to this one (in Sec. 1.2, for one degree of freedom, and in Sec. 1.4 for two degrees of freedom). Therefore, we leave it to you to show that, for either the slinky approximation or the small-oscillations approximation, Newton's law applied to the motion of bead n gives

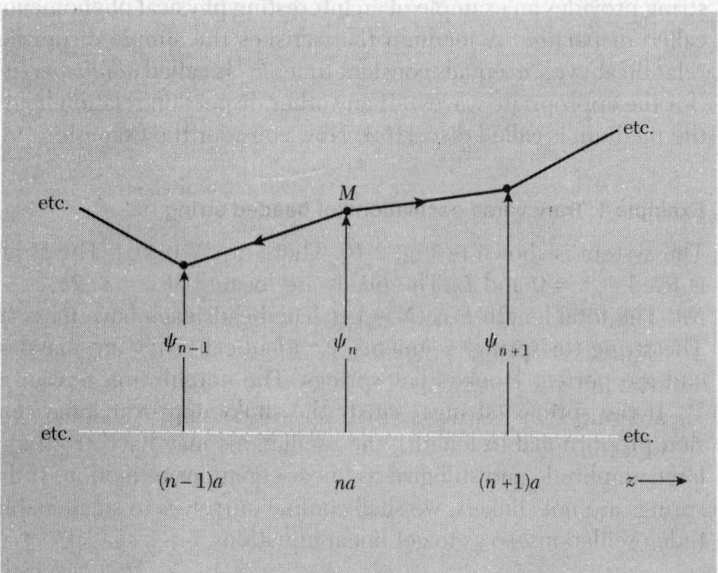

Fig. 2.11 General configuration of beaded string for transverse oscillations along x.

$$M \frac{d^2\psi_n(t)}{dt^2} = T_0 \left[\frac{\psi_{n+1}(t) - \psi_n(t)}{a} \right] - T_0 \left[\frac{\psi_n(t) - \psi_{n-1}(t)}{a} \right]. \quad (62)$$

Equation (62) is perfectly general; it holds for an arbitrary motion of the freely oscillating system, i.e., for an arbitrary superposition of the N different modes.

Normal modes. We want to find the frequencies and configurations of the individual modes. Therefore, we assume that we have a single mode of frequency ω. Each bead oscillates harmonically with the same frequency ω and with the same phase constant φ. The shape of the mode is given by the ratios of amplitudes of the beads. We let A_n designate the vibration amplitude for bead n in the mode we are considering. Thus we have, in a single mode,

$$\psi_1(t) = A_1 \cos(\omega t + \varphi); \quad \psi_2(t) = A_2 \cos(\omega t + \varphi);$$

$$\psi_{n-1}(t) = A_{n-1} \cos(\omega t + \varphi); \quad \psi_n(t) = A_n \cos(\omega t + \varphi);$$

$$\psi_{n+1}(t) = A_{n+1} \cos(\omega t + \varphi); \quad (63)$$

Because of Eq. (63), we have

$$\frac{d^2\psi_n(t)}{dt^2} = -\omega^2 \psi_n(t) = -\omega^2 A_n \cos(\omega t + \varphi). \quad (64)$$

Now use Eq. (64) on the left side of Eq. (62), and Eqs. (63) on the right side. Then cancel the common time-dependent factor $\cos(\omega t + \varphi)$. This gives

$$-M\omega^2 A_n = \frac{T_0}{a}(A_{n+1} - 2A_n + A_{n-1});$$

i.e.,

$$A_{n+1} + A_{n-1} = A_n \left(2 - \frac{Ma}{T_0}\omega^2 \right). \quad (65)$$

Equation (65) looks formidable. It gives the shape of the mode with angular frequency ω. Let us solve it by bold guesswork. We are guided in our guessing by our previous solution for the modes of a continuous string with fixed ends at $z = 0$ and L. In that problem we found the shape of the modes to be given by

$$A(z) = A \sin \frac{2\pi z}{\lambda} = A \sin kz. \quad (66)$$

Our solution for A_n must of course reduce to Eq. (66) in the limit of infinitely many beads (the continuous limit). Let us try a solution obtained by Simply setting $z = na$ in Eq. (66):

$$A_n = A \sin \frac{2\pi na}{\lambda} = A \sin kna. \tag{67}$$

Then

$$A_{n+1} = A \sin k(n+1)a = A \sin(kna + ka)$$
$$= A (\sin kna \cos ka + \cos kna \sin ka).$$
$$A_{n-1} = A \sin k(n-1)a = A \sin(kna - ka)$$
$$= A (\sin kna \cos ka - \cos kna \sin ka).$$
$$A_{n+1} + A_{n-1} = 2A \sin kna \cos ka = 2A_n \cos ka \tag{68}$$

Inserting Eq. (68) into Eq. (65), we obtain

$$2A_n \cos ka = A_n \left(2 - \frac{Ma}{T_0} \omega^2 \right). \tag{69}$$

Exact dispersion relation for beaded string.   Equation (69) is supposed to hold for every bead n, whether or not A_n happens to be zero for that particular bead n in a particular mode. Therefore, we can take n to correspond to a bead that is not at a node, i.e., one for which A_n is not zero. Then we cancel A_n and obtain the condition that our guess for a solution must meet to be an actual solution:

$$2 \cos ka = 2 - \frac{Ma}{T_0} \omega^2,$$

i.e.,

$$\omega^2 = \frac{2T_0}{Ma} (1 - \cos ka)$$
$$= \frac{2T_0}{Ma} \left[1 - \left(\cos^2 \frac{ka}{2} - \sin^2 \frac{ka}{2} \right) \right],$$
$$\omega^2 = \frac{4T_0}{Ma} \sin^2 \frac{ka}{2} = \frac{4T_0}{Ma} \sin^2 \frac{\pi a}{\lambda}. \tag{70}$$

Equation (70), which relates frequency to wavelength (or wavenumber) for a mode with angular frequency ω, is the dispersion relation for the beaded string.

Boundary conditions.   We have not yet specified the boundary conditions completely. When we wrote Eq. (67) instead of the more general expression

$$A_n = A \sin kna + B \cos kna, \tag{71}$$

we had already satisfied the boundary condition at $z = 0$, namely that the string displacement be zero there for any mode. Setting $z = na = 0$ in Eq. (71) and demanding $A_0 = 0$ gives $B = 0$. We must still satisfy the boundary condition at $z = L$, namely that the string displacement be zero there also. The wall at $z = L$ corresponds to the "fixed bead $N + 1$." Thus we need $A_{N+1} = 0$:

$$A_{N+1} = A \sin k(N + 1)a = A \sin kL = 0. \qquad (72)$$

There are N possible solutions for Eq. (72). Each solution corresponds to a single mode m, with $m = 1, 2, \ldots, N$. We number the modes so that $m = 1$ has the longest wavelength. Thus we have

$$k_1 L = \pi, \qquad k_2 L = 2\pi, \ \ldots, \ k_m L = m\pi, \ \ldots, \ k_N L = N\pi. \quad (73)$$

The reason that there are only N solutions [the modes specified by Eq. (73)] is that the last term in the sequence of Eq. (73) corresponds to a completely zigzag configuration: starting at $z = 0$, the first string segment "zigs" up to the first bead, the second string "zags" down to bead 2, ..., string $N + 1$ zags (or zigs) from bead N to the wall. Equation (72) does have the further solutions $k_{N+1}L = (N + 1)\pi$, $k_{N+2}L = (N + 2)\pi$, etc., but to do all the zigzagging such solutions imply, takes more string segments than we have.

The equation for the shape of the modes, Eq. (65), was obtained without consideration of boundary conditions. (Figure 2.11 contains no boundary conditions.) The most general solution of this equation is given by Eq. (71), with B/A and k determined by the boundary conditions. If you substitute Eq. (71) into Eq. (65), you find the dispersion relation Eq. (70), independent of the boundary conditions, i.e., independent of the values of A, B, and k, as you can easily show (Prob. 2.19). For our particular boundary conditions (string fixed at $z = 0$ and L), we obtain the mode configurations of Eq. (72) with k_m given by Eq. (73). Then the frequencies ω_m are given by Eq. (70).

Notice that the mode configurations of Eq. (73) are exactly the same as those that we obtained for the continuous string, the only difference being that for the continuous string we had $N = \infty$ and thus no highest mode. Also, for the beaded string, the string segments are straight and do not follow the smooth sinusoidal function A_m that passes through the beads.

We illustrate the modes by showing the case $N = 5$ in Fig. 2.12. In Fig. 2.13 we plot the dispersion relation given by Eq. (70):

$$\omega(k) = 2\sqrt{\frac{T_0}{Ma}} \sin \frac{ka}{2}. \qquad (74)$$

The five labeled points give k and ω for the five modes of the string of five beads with both ends fixed. If there were a different number

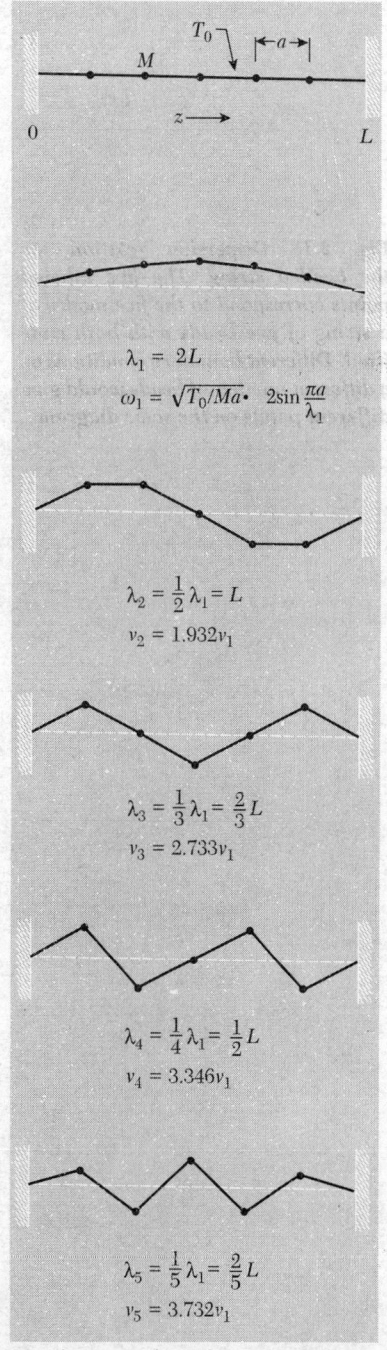

Fig. 2.12 *Modes of a string with five beads.*

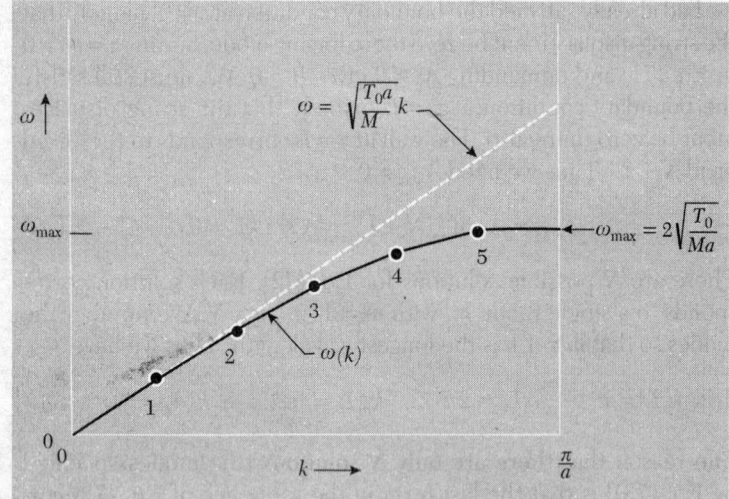

Fig. 2.13 Dispersion relation for the beaded string. The five labeled points correspond to the five modes of a string of five beads with both ends fixed. Different boundary conditions or a different number of beads would give different points on the same diagram.

of beads or different boundary conditions (for example we could have a free end at $z = L$), the points representing the modes would lie at different places on the *same curve* $\omega(k)$. Thus Fig. 2.13 holds for *every* beaded string.

Continuous or long wavelength limit. In the continuous approximation, we assume an infinite number of beads in the region between $z = 0$ and $z = L$. Thus the bead separation a goes to zero. It is interesting to look at the properties of our exact dispersion relation, Eq. (74), for bead separations a that are "very small" but not *exactly* zero, in order to see that the dispersion relation approaches that of the continuous string. We have to know what we mean by "small", i.e., small with respect to what? The continuous approximation is a good one when bead spacing a is small compared with the wavelength λ:

$$a \ll \lambda; \qquad ka = 2\pi \frac{a}{\lambda} \ll 1.$$

We now use the Taylor's series expansion [Appendix Eq. (4)],

$$\sin x = x - \tfrac{1}{6} x^3 + \cdots .$$

We insert this series with x equal to $\tfrac{1}{2} ka$ into Eq. (74):

$$\omega(k) = 2\sqrt{\frac{T_0}{Ma}} \left[\frac{1}{2} ka - \frac{1}{48} (ka)^3 + \cdots \right]$$

$$= \sqrt{\frac{T_0 a}{M}} k \left[1 - \frac{1}{24} (ka)^2 + \cdots \right],$$

i.e.

$$\omega(k) \approx \sqrt{\frac{T_0 a}{M}} k. \tag{75}$$

Equation (75) is the "nondispersive" dispersion law that we obtained for the continuous string in Sec. 2.3, with $M/a = \rho_0$.

Dispersion relation for real piano string. We have discovered that the modes of a noncontinuous string do not satisfy the nondispersive wave dispersion relation of Eq. (75). Therefore, we expect that the overtones of a piano string, for example, the string with fundamental tone C128, are not given exactly by the octave C256, twelfth G384, double octave C512, etc. That is correct; they are not. According to Eq. (74) or, more easily, its plot in Fig. 2.13, an increase in k does not produce a proportional increase in frequency, but rather slightly less. Therefore, you might expect the overtones of the piano string to be slightly "flat" with respect to the predictions of the continuous-string theory, i.e., you might expect the second harmonic to have $v_2 < 256$, the third to have $v_3 < 384$, etc. That is incorrect! The overtones of a piano string are not flat; they are *sharp* with respect to the simple "harmonic-overtone" prediction of Eq. (75)! That is because, although the model of a perfectly continuous and perfectly flexible string is not a perfect description of a piano string, neither is the beaded-string model. In fact the beaded-string model is worse than the continuous-string model, since it gives to the result of the latter a "correction" with the wrong sign! The trouble with the continuous-string model for a piano string is not that it needs some beads, but rather that a real piano string is not perfectly flexible. When you bend it, it wants to straighten out again, even if there is no tension helping to pull it straight. Consequently, the return force on a small curved segment of string (i.e., the force that tends to straighten the string out—it is straight at equilibrium) is slightly greater than that predicted by the "completely flexible" model. The mode frequency is of course given by ω^2 = return force per unit displacement per unit mass. The higher modes have shorter wavelengths, so they are bent more. The stiffness is thus more important for the higher modes than for the lower, and hence the frequency increases *faster* than expected from the flexible-string model.

There is an interesting "fine point" to this explanation. The return force due to the tension and that due to the stiffness *both* increase with k. Therefore, if the stiffness is to play a relatively larger role at higher values of k than at its lower values, the return force due to stiffness must increase with k at a rate greater than the return force due to tension. The return force due to tension is proportional to k^2. That due to stiffness turns out to be proportional to k^4.

Thus the dispersion relation for a real piano string is given by

$$\omega^2 \approx \frac{T_0}{\rho_0} k^2 + a k^4, \qquad (76)$$

where a is a positive constant that is due to the stiffness. If the stiffness term were also proportional to k^2, we would still get the "nondispersive" dispersion relation, Eq. (75), with T_0/ρ_0 merely replaced by $(T_0/\rho_0) + a$. Then the frequency ratios would still be the "harmonic" ones $\nu_2 = 2\nu_1$, $\nu_3 = 3\nu_1$, etc. Now we shall consider more examples:

Example 2: Longitudinal oscillations of a system of springs and masses

This is an important example because it will later provide us with a very simple model to help us understand sound waves. (Sound waves consist of longitudinal vibrations; i.e., the vibrations are perpendicular to the "wavefronts".)

We have already studied the cases $N = 1$ and 2 in Secs. 1.2 and 1.4, respectively. We now consider the general case of N masses coupled by springs as shown in Fig. 2.14.

The equation of motion of bead n is very easily derived. (If you have difficulty, review the derivation for $N = 2$ in Sec. 1.4.) One finds

Fig. 2.14 Longitudinal oscillations of N masses and $N + 1$ springs. (a) Equilibrium configuration. (b) General configuration.

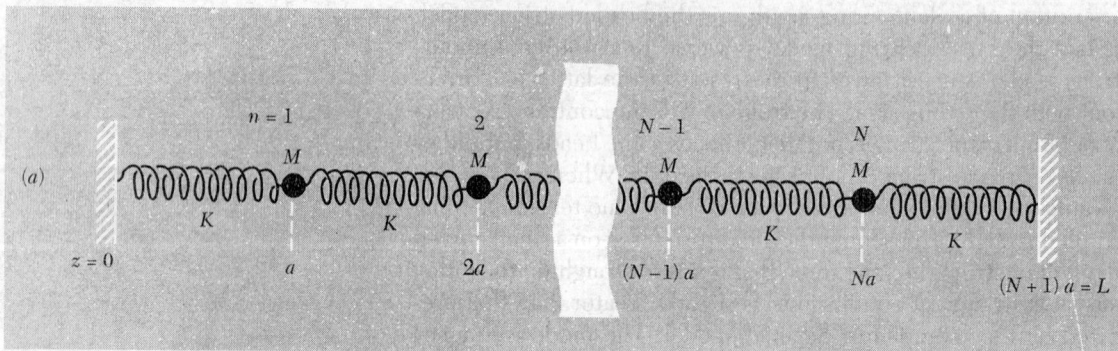

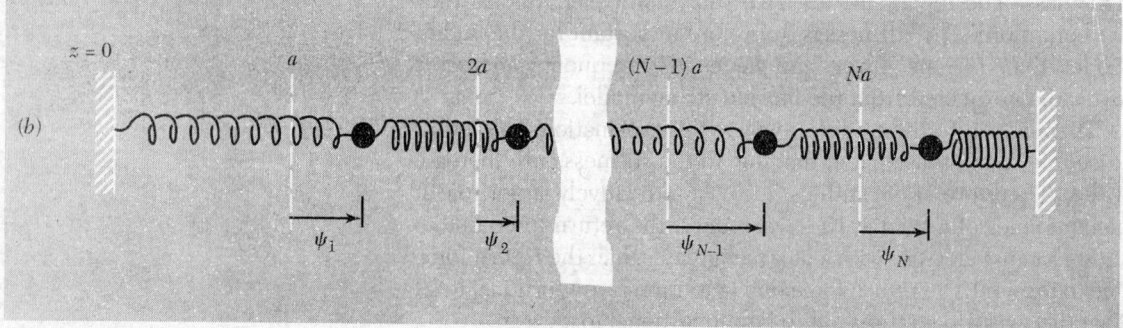

$$M \frac{d^2\psi_n}{dt^2} = K(\psi_{n+1} - \psi_n) - K(\psi_n - \psi_{n-1}). \tag{77}$$

The mathematical form of Eq. (77) is the same as that of the equation of motion for transverse displacements, Eq. (62), except for the replacement of constant T_0/a by the spring constant K. Therefore, all of our former mathematical steps can be repeated. Thus we get the dispersion relationship [obtained by replacing T_0/a by K in Eq. (74)]

$$\omega(k) = 2\sqrt{K/M}\ \sin \frac{ka}{2} = 2\sqrt{K/M}\ \sin \frac{\pi a}{\lambda}. \tag{78}$$

In the mode with wavenumber k, the motion of mass n is given by

$$\psi_n(t) = A \sin nka \cos[\omega(k)t + \varphi], \tag{79}$$

with the N different possibilities for k given by

$$k_1 L = \pi, \qquad k_2 L = 2\pi, \cdots, k_N L = N\pi. \tag{80}$$

The dispersion relation plotted in Fig. 2.13 need only be suitably relabeled to represent Eq. (78).

Lumped parameters and distributed parameters. When we considered the transverse vibrations of a beaded string, we went to the continuous limit by letting the bead spacing a go to zero (keeping L constant). Once one has gone to sufficiently small a/λ, so that the continuous approximation is a good one, then one can use a different physical model of the system. Instead of letting the spacing a continue to go to zero while maintaining the mental picture of massless springs alternating with point masses, we can distribute the mass uniformly along the spring. Then there are no longer lumped masses and massless springs (lumped spring constants). Instead there is just one long spring with the mass distributed along it. A good example is provided by a slinky. For the "repeat length" a we can now take the length along z of a single turn of the helical spring. The meanings of the parameters M and K become respectively the mass of one turn (of the helix) and the spring constant of one turn. If one has a total of N turns (N is not the number of degrees of freedom, now), then the total mass is NM. The total spring constant (i.e., the spring constant of the complete spring of length $L = Na$) is K/N. (That is because two identical springs hooked in series make one longer spring with half the spring constant of the two component springs.)

Instead of retaining the repeat length a (of one turn of the helix), we can eliminate it completely from the notation (in the continuous approximation) by replacing M/a by the mass per unit length (linear mass density) $\rho_0 = M/a$. Similarly, one can eliminate K, the spring constant of one turn, replacing it by a quantity that is characteristic of the material of the spring and its construction. That quantity is the inverse spring constant per unit length, K^{-1}/a. That is easily seen as follows. For a spring of total length $L = Na$, the spring constant K_L is N times smaller than K:

$$K_L = \frac{1}{N} K = \frac{a}{L} K. \tag{81}$$

Thus we have $K_L \cdot L = Ka$, which is independent of L; i.e., Ka is a property of the "springiness" of the material and is independent of the length of the spring. Since we like to deal with quantities that have dimensions of "something per unit length," we write the relation

$$K_L \cdot L = Ka$$

in the form

$$\frac{K^{-1}}{a} = \frac{K_L^{-1}}{L}. \tag{82}$$

Now we can express the result by saying that the inverse spring constant per unit length is a property of the spring which is independent of its length.

Example 3: Slinky

A slinky is a helical spring having $N \approx 100$ turns, with each turn about 7 cm in diameter. The unstretched length is about 6 cm. When stretched to a length L of about a meter, it satisfies the slinky approximation very well. A convenient "repeat length" a is given by the length per turn, $a = L/N$. Then K is the spring constant for one turn, and K^{-1}/a is independent of the length L. (The mass is of course distributed, not lumped at intervals of length a.) The dispersion relation for longitudinal oscillations is obtained by going to the continuous limit, starting with Eq. (78):

$$\omega(k) = 2\sqrt{\frac{K}{M}} \sin \frac{ka}{2}$$

$$= 2\sqrt{\frac{K}{M}} \left[\frac{ka}{2} - \frac{1}{6}\left(\frac{ka}{2}\right)^2 + \cdots \right]$$

$$\approx \sqrt{\frac{Ka^2}{M}} k$$

$$= \sqrt{\frac{Ka}{(M/a)}} k. \tag{83}$$

The dispersion relation for transverse oscillations is [see Eq. (75)]

$$\omega(k) \approx \sqrt{\frac{T_0}{M/a}}k$$

$$\approx \sqrt{\frac{Ka}{M/a}}k, \tag{84}$$

since $T_0 = Ka$ in the slinky approximation. Thus the slinky has the same dispersion relation for longitudinal and for transverse oscillations. Therefore, *if* the boundary conditions are the same (for example, both ends rigidly fixed for vibrations along x, y, or z), the modes for x, y, and z vibrations have the same sequence of wavelengths and frequencies. You can easily verify for yourself that the longitudinal and transverse modes have the same frequencies. We strongly recommend that you now perform some of the Home Experiments involving slinkies. There is no better way to understand waves than to make your own. Get a slinky. (Made by James Industries, Inc., Hollidaysburg, Pa.; available in any toy store for about $1.)

Example 4: *LC* network

Consider the sequence of coupled inductances and capacitances shown in Fig. 2.15. From Fig. 2.15*b* (and our discussion involving the same network for the case $N = 2$, in Sec. 1.4), we easily find that the equation for the electromotive force across the nth inductance is given by

$$L\frac{dI_n}{dt} = -C^{-1}Q' + C^{-1}Q.$$

Then

$$L\frac{d^2I_n}{dt^2} = -C^{-1}\frac{dQ'}{dt} + C^{-1}\frac{dQ}{dt}.$$

Using conservation of charge to eliminate dQ'/dt and dQ/dt, we get

$$L\frac{d^2I_n}{dt^2} = -C^{-1}[I_n - I_{n+1}] + C^{-1}[I_{n-1} - I_n] \tag{85}$$

$$= C^{-1}[I_{n+1} - I_n] - C^{-1}[I_n - I_{n-1}].$$

Equation (85) has the same mathematical form as Eq. (77), the equation of motion governing the longitudinal oscillations of a sequence of masses and springs. Therefore, without yet worrying about the boundary conditions, we can write down the dispersion relation and the general solution for the currents in the inductances. The dispersion relation is obtained by replacing K/M by C^{-1}/L in Eq. (78):

$$\omega(k) = 2\sqrt{\frac{C^{-1}}{L}} \sin\frac{ka}{2}. \tag{86}$$

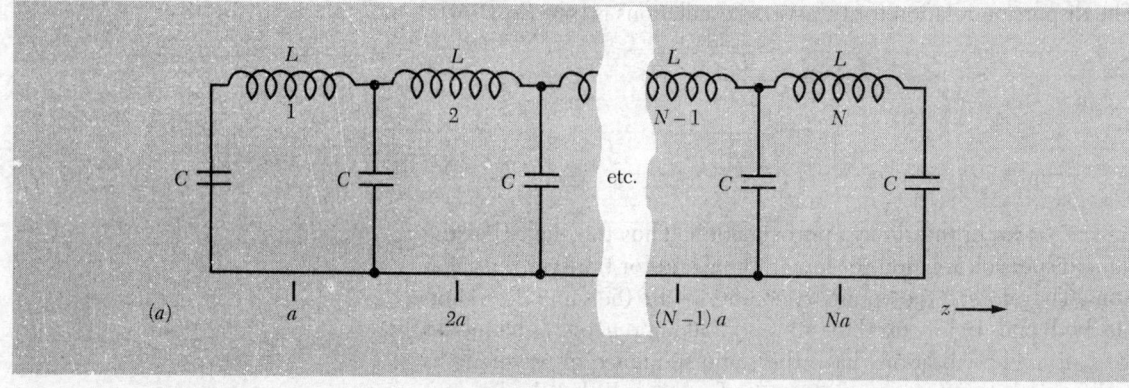

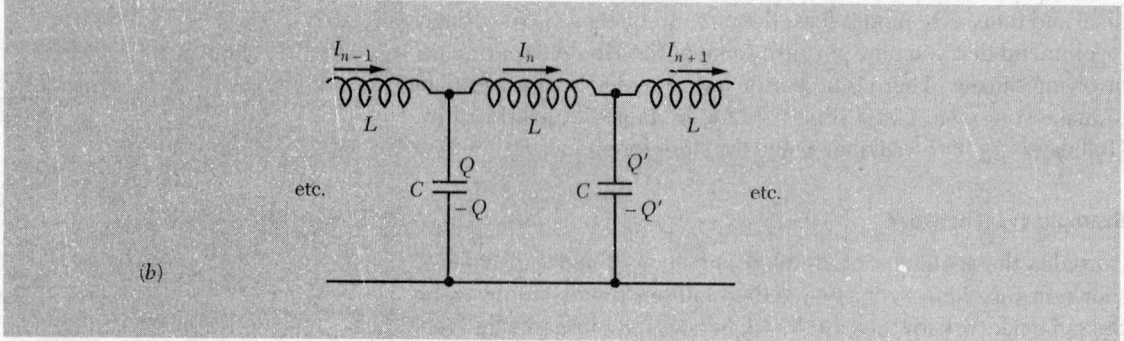

Fig. 2.15 *Network of coupled inductances and capacitances. (a) The lumped parameters. (b) General current and charge configuration at nth inductance.*

The general solution of Eq. (85) for a single mode, without regard to boundary conditions, is

$$I_n(t) = [A \sin nka + B \cos nka] \cos [\omega(k)t + \varphi], \qquad (87)$$

where the constants A and B and the sequence of values of k corresponding to modes depend on the boundary conditions at the ends of the system.

The meaning of ka. You may have noticed that the equation governing the behavior of the LC circuit, Eq. (85), does not contain the distance a. We labeled such a distance in Fig. 2.15, but there was no need to do so, since a circuit diagram is not a spatial diagram, and the behavior of the circuit does not depend on its spatial configuration. What then do we mean by ka in the dispersion relation, Eq. (86), and in the general solution for the currents, Eq. (87)? When the length along z really has an important physical significance, as in the oscillating string, then we know that k means the increase per unit length along z of the phase of the function $A \sin kz + B \cos kz$ which gives the shape of the mode. When we have lumped parameters, as in the beaded string, we write $z = na$, where

$n = 1, 2, \ldots$ is the bead index. Then, the quantity ka appearing in the shape function $A \sin nka + B \cos nka$ is the product of radians of phase (of the shape function) per unit distance times the distance between lumped masses. Thus ka is the number of radians of phase increase from the lumped mass n to the succeeding lumped mass $n + 1$. In the case of the system of lumped inductances and capacitances, the quantity ka is similarly the increase in phase of the "shape" function $A \sin nka + B \sin nka$ when we go from one lumped inductance to the next. We do not really need to specify the distance a by which the inductances are separated. We could merely replace ka by some symbol, say θ, where θ would then denote the increase of phase when n is increased by 1 in the shape function $A \sin n\theta + B \cos n\theta$. That notation is too abstract for us and abandons the mathematical similarity to the mechanical examples, so we shall retain the idea that the lumped inductances are separated by distance a.

Other forms of dispersion relation.   You may have noticed that every lumped-parameter system we have considered so far in this section has an exact dispersion relation of the form

$$\omega(k) = \omega_{\text{max}} \sin \frac{ka}{2}, \qquad (88)$$

as plotted in Fig. 2.13, with ω_{max} a constant depending on the physical system. That is only because of our choice of systems. In every case we considered, we chose a system in which the return force on a given mass (or inductance) is entirely the result of the coupling of that mass to neighboring masses and is proportional to the relative displacement of that mass and its neighbors. Such systems are numerous, but there are many other interesting and important forms of dispersion relations. For example, there are systems having the property that the return force on a given moving part has the following two independent contributions. One is due to the force arising from coupling with similar neighboring moving parts. If this were the only contribution, the dispersion relation would take the form of Eq. (88). The other is due to its coupling to some "external" force. This external contribution depends only on the displacement of the moving part from its equilibrium position and not on the displacements of the neighboring parts. If it were the only contribution, then the moving parts would be uncoupled, and their displacements would be the normal coordinates of the entire system. This kind of system is illustrated by the following example.

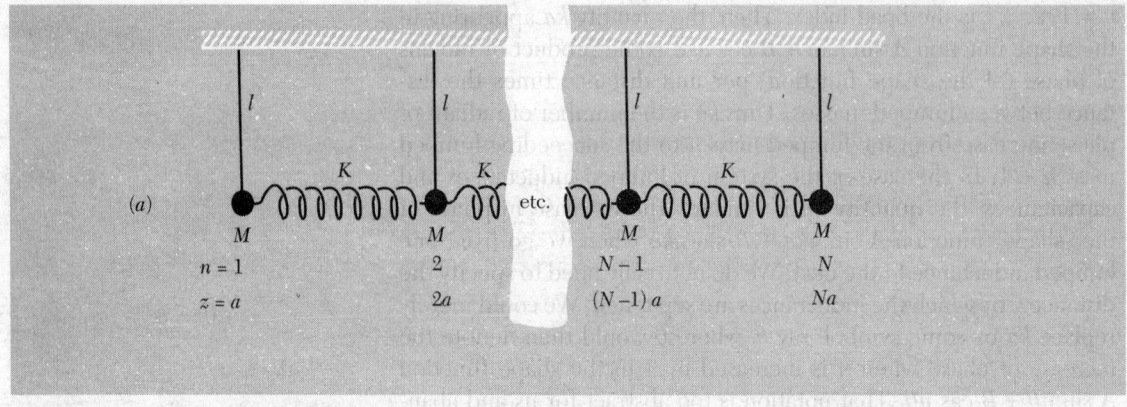

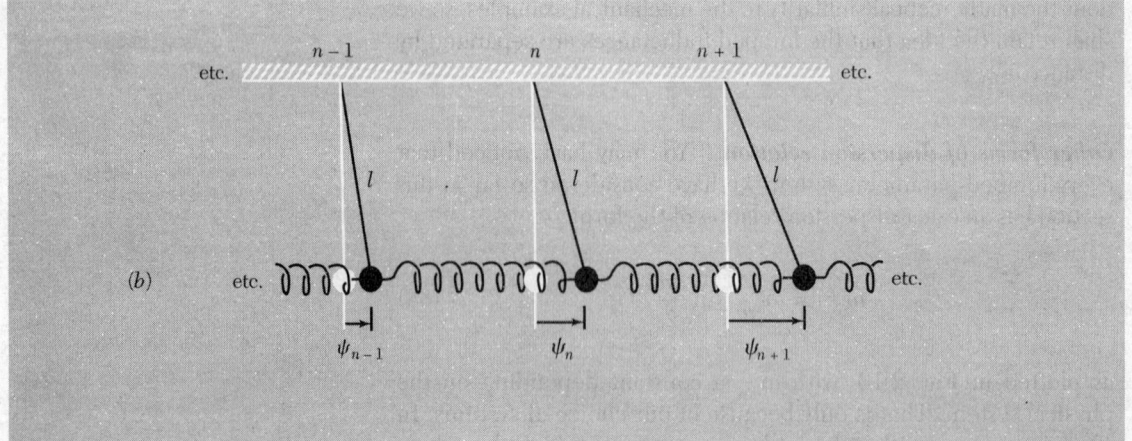

Fig. 2.16 Coupled pendulums. (a) Equilibrium. (b) General configuration.

Example 5: Coupled pendulums

The system is shown in Fig. 2.16. Each mass has a return force made up of two contributions. The "external" contribution is that due to gravity. It is proportional to the displacement of the mass from its equilibrium position and is independent of the displacement of its neighbors. The second independent force contribution is due to the coupling of a given mass to its neighbors by means of the springs. This contribution depends on the displacement of the neighboring masses.

Let us try to guess the dispersion relation. If we had only the coupling between masses, i.e., if g were zero, then we would have the dispersion relation for longitudinal oscillations of coupled masses. Thus the return force per unit displacement per unit mass, ω^2, would be given by Eq. (78):

$$\omega^2 = 4 \frac{K}{M} \sin^2 \frac{ka}{2}, \qquad \text{if } g = 0. \qquad (89)$$

Now suppose that (with $g = 0$) we have oscillation in a single mode, with a definite shape determined by the definite value of k, which is set by the boundary conditions. Imagine that we can gradually increase g from zero up to its final value of 980 (cgs units), using a "gravity knob." (A more practical method can be invented. What else could you vary?) When we increase g from zero to the very small value g', the return force per unit displacement per unit mass for *each* particle increases by the *same* amount, the contribution from g':

Contribution of g' to ω^2 is $\dfrac{g'}{l}$ for every mass.

That means that the masses will continue to oscillate with the same configuration, same k, and same linear combination of $\sin kz$ and $\cos kz$, but they will merely oscillate faster. That is because they *had* the same return force per unit displacement per unit mass, ω^2, when g' was zero, and now we have added the *same amount* to the frequency squared of every mass. Therefore, all masses still have the same ω^2 and hence are still in a mode. Thus by gradually turning on g, we preserve the modes without mixing them up. The shapes and wavelengths are the same as for $g = 0$, and the total return force per unit mass per unit displacement is now

$$\omega^2(k) = \frac{g}{l} + \frac{4K}{M}\sin^2\frac{ka}{2}. \tag{90}$$

If you prefer a less qualitative derivation, of this relation, see Prob. 2.26; there *you* will find the equation of motion of mass n, verify the dispersion relation Eq. (90), and find the shapes of the modes. (Can you already see that for the boundary conditions of Fig. 2.16, the lowest mode has $k = 0$?)

Later we shall encounter more examples of dispersion laws of the form of Eq. (90), which we can write in the general form

$$\omega^2(k) = \omega_0^2 + \omega_1^2 \sin^2\frac{ka}{2}. \tag{91}$$

In the continuous limit, where we have $ka \ll 1$, this relation becomes

$$\omega^2(k) = \omega_0^2 + v_0^2 k^2, \tag{92}$$

where v_0^2 is the constant $\omega_1^2 a^2/4$.

We shall encounter dispersion laws of the form of Eq. (92) when we study electromagnetic radiation in a waveguide and electromagnetic waves in the earth's ionosphere. (That is also the form

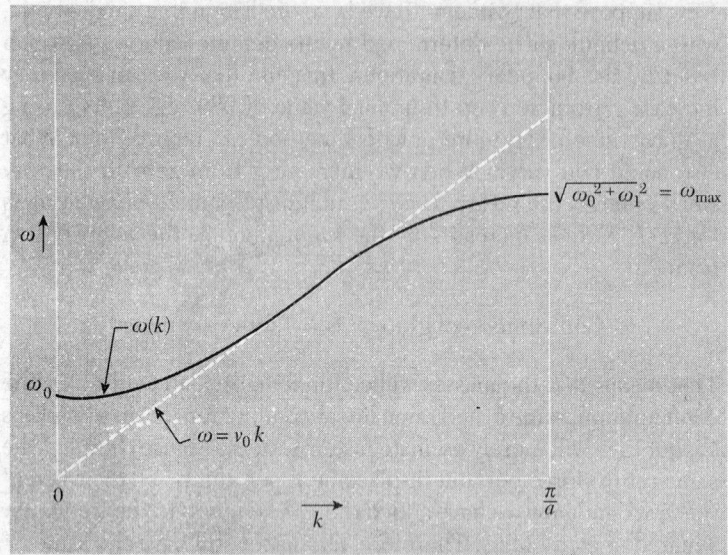

Fig. 2.17 Dispersion relation for coupled pendulums.

of dispersion law for relativistic de Broglie waves in the quantum description of particles.) We plot Eq. (91) in Fig. 2.17.

Example 6: Plasma oscillations

This is an interesting example of a system with a dispersion relation like that of coupled pendulums. In Chap. 4 we shall derive the dispersion relation for electromagnetic waves in the earth's ionosphere. It turns out to have the form of Eq. (92):

$$\omega^2(k) = \omega_p{}^2 + c^2 k^2, \tag{93}$$

where c is the velocity of light and where ω_p, called the "plasma oscillation frequency," is given by

$$\omega_p{}^2 = \frac{4\pi N e^2}{m}. \tag{94}$$

Here N is the number density of electrons (in electrons per cm³), e is the charge on the electron, and m is the electron mass. From Fig. 2.17 we see that the mode of lowest possible frequency for a system with a dispersion relation like Eq. (91) or (92) is a mode with $k = 0$, i.e., with infinite wavelength. That just means all the pendulums oscillate with the same phase constant and same amplitude. The pendulum frequency is then given by $\omega^2 = g/l$. The frequency of that lowest mode in the present example is the plasma oscillation frequency ω_p, as we see by setting $k = 0$ in Eq. (93). We shall now consider that mode and derive Eq. (94) as its frequency.

A neutral plasma consists of a gas of neutral molecules with some of the molecules ionized. Every singly ionized molecule consists of one positive ion which has set free one (negative) electron. The earth's ionosphere is a layer of air (actually several layers with somewhat differing properties) that contains many ionized air molecules (N_2 and O_2 molecules). The ionization of an air molecule usually occurs by means of absorption of an ultraviolet- light quantum emitted by the sun. The ion and free-electron density is greatest at about 200 to 400 km above the earth's surface. Higher than that, the electron (and ion) density decreases because the density of neutral air molecules available to be ionized decreases. Lower than that, the electron density decreases because the ultraviolet radiation has already been mostly absorbed. (We would soon die of sunburn without the protective air layer above us.)

Since the plasma is neutral (on the average), it does not act as a source of electrostatic field. However, one region of the plasma may have at any one instant a slight excess of charge, with a corresponding deficit in some other neighboring region. This creates an electric field in the plasma. Under the influence of the electric field, the ions are accelerated in one direction (i.e., along the field), and the electrons in the other. The charges move in directions that tend to cancel those charge excesses and deficits which created the electric field. Thus we have a "return force." By the time the charge excess has been canceled to zero and the corresponding electric field reduced to zero, the ions and electrons have acquired velocities. Their inertia causes them to "overshoot," and we get a new charge excess and deficit with opposite sign from the original one. We have. here the typical situation that sustains oscillations, once they are excited.

If we are interested only in the net motion of charge back and forth from one region to another, we can forget about the positive ions and consider the entire motion of charge to be due to the motion of the electrons. This is because the acceleration of an electron is larger than that of a singly charged ion by the ratio of their respective masses (about 3×10^4), since the electric force is the same on each.

Let us study a simplified situation where the plasma is contained between confining walls. We neglect motion of the ions compared with that of the electrons. At any instant there may.be an excess of charge Q on one wall and a corresponding deficit on the other wall. This produces a spatially uniform electric field in the plasma (Vol. II, Sec. 3.5) given by

$$E_x = -4\pi \frac{Q}{A}, \tag{95}$$

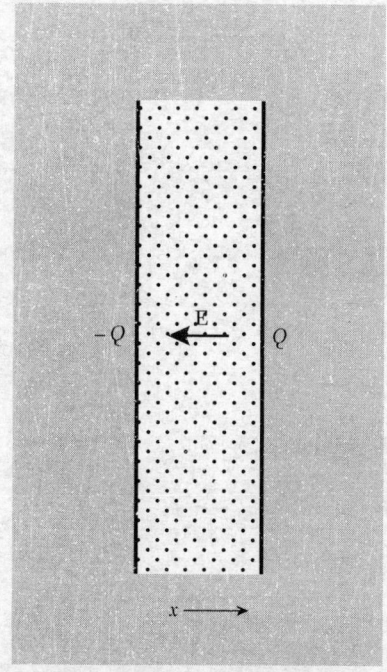

Fig. 2.18 Oscillations in a confined plasma.

Where A is the area of the wall, and where the minus sign signifies that E_x tends to return Q to zero. There is no other source of electric field. (The plasma between the walls is neutral, because every electron that moves to the right out of a given region is replaced by another moving in from the left.) A single electron has mass m and charge q. Newton's second law gives

$$\frac{m \, d^2x}{dt^2} = qE_x \tag{96}$$

for every electron in the plasma. (We are neglecting other forces on the electrons which arise from collisions among the electrons and ions; these forces average to zero and give no net motion of charge.) Now suppose that there are N free electrons per cubic centimeter and that each electron is displaced from its average (equilibrium) position by a distance x. Then the net charge deposited on one wall (and removed from the other) is given by

$$Q = NqAx. \tag{97}$$

Differentiating Eq. (97) twice with respect to time and inserting Eqs. (96) and (95) yields

$$\frac{d^2Q}{dt^2} = -\frac{4\pi Nq^2}{m} Q. \tag{98}$$

This has the solution

$$Q = Q_0 \cos(\omega t + \varphi),$$

with

$$\omega^2 = \frac{4\pi Ne^2}{m} \equiv \omega_p^{\ 2}. \tag{99}$$

The quantity ω_p is called the plasma oscillation frequency.

The free-electron density N in the earth's ionosphere varies with height and with time. The recombination of ions and electrons to form neutral molecules continues after sunset, but the forming of new ions ceases. The electron density therefore decreases at night. Typical daytime plasma oscillation frequencies $\nu_p \, (= \omega_p/2\pi)$ are

$$\nu_p = 10 \text{ to } 30 \text{ Mc (megacycle per second)}, \tag{100}$$

which corresponds to $N \approx 10^6$ to 10^7 free electrons per cm³.

Esoteric examples

If one combines De Broglie's hypothesis, which says that a particle of momentum p has a wave number k given by $p = \hbar k$, with the "Bohr frequency condition," which says that a particle of energy E has a wave frequency ω given by $E = \hbar \omega$, one can then find a dispersion relation between ω and k for particles, given the relation between E and p. Examples are given in Supplementary Topic 2.

Problems and Home Experiments

2.1 Slinky–dependence of frequency on length. Hold the first turn of your slinky in your left hand and the last turn in your right, with your hands about 1 m apart. Measure the frequency for vertical transverse oscillations. (Don't worry about the sag.) Next stretch the slinky out as far as you can reach. Measure the frequency. Now fasten each end to something so that the total length is 2.4 m to 3 m. Measure the frequency. Explain your result. Use your frequency measurement to determine the inverse spring constant per turn. Suppose that N_0 is the total number of turns of the slinky. Hold or fasten the slinky so that only N turns are free. Before doing the experiment, predict the frequency dependence on N/N_0. Then do the experiment.

Home experiment

2.2 Slinky as a continuous system. Fasten each end to something fixed. (Tape, string, "eyes" (as in hook and eye), and C clamps may possibly be useful.) A convenient length is about 2.4 m to 3 m. Don't worry about the sag. Excite the lowest transverse mode in each transverse direction. Measure the frequencies for both modes. Also excite the lowest longitudinal mode and measure its frequency. (There are two good ways to excite a desired mode. One is to constrain the slinky in an appropriate way and then let it go; the other is to hold it near one end and gently shake it at the right frequency until it builds up a decent amplitude, then let go. Use both methods.) Next learn how to excite the second mode, in which the length L is two half-wavelengths. Do this for all three directions, x, y, and z. Measure the frequencies. With some practice you should also be able to excite the third modes.

Home experiment

Now excite the lowest vertical mode and the second longitudinal mode simultaneously. (This can easily be done by a suitable initial constraint.) Look at the system and measure the beat frequency between the longitudinal (second) mode and twice the lowest vertical mode. This is easy once you get the idea and practice for a few minutes. This is a good way to see by inspection that one has an accurate factor of two in frequency in going from the "fundamental" to the "first octave." Similarly you can easily excite the lowest vertical mode and the second horizontal mode simultaneously.

2.3   Null measurements.   Read Home Exp. 2.2 (though you need not perform it to work this problem). Suppose you measure slinky frequencies by counting oscillations for about ten seconds and then dividing the number of complete cycles by the time. Suppose that you read the watch to an accuracy of ± 1 sec and that you can estimate a "complete" oscillation to an accuracy of about $\pm \frac{1}{4}$ cycle. The frequency v_1 of the lowest mode is about 1 cps. The frequency v_2 of the second mode is about 2 cps.

(*a*) Roughly, what is the fractional (or percent) accuracy that your measurement would give for v_1? For v_2? (We want an answer like "$v_1 = 1.0 \pm 0.1$, $v_2 = 2.0 \pm 0.2$," or whatever it is.)

(*b*) Next, suppose you were to excite both modes simultaneously and measure the beat frequency between $2v_1$ and v_2 as described in Home Exp. 2.2. This could be done by watching for beats for about 10 sec, i.e., about ten cycles of v_1. Suppose that you could detect, to an accuracy of $\frac{1}{4}$ beat, no beat between $2\,v_1$ and v_2 in that time. Thus your experimental result would be $v_2 - 2v_1 = 0$. What is the experimental accuracy? (We want an answer like "$v_2 - 2\,v_1 = 0 \pm 0.10\,v_1$", or whatever it is.) What is the accuracy of an estimate of the quantity $v_2 - 2v_1$ (expressed in the same way) obtained by combining the results of your independent measurements of v_1 and v_2 from part (*a*)? Can you see an experimental advantage in the method of counting beats? Explain why you would do much better with that method. Try to generalize this into a statement on "how to make a measurement, if possible."

Home experiment   **2.4   "Tone quality" of a slinky.**   The tone quality of a musical instrument depends on what harmonics are excited. [For example, the even harmonics are missing (or nearly so) from a clarinet; only v_1, $3v_1$, $5v_1$, etc., are present.] Locate the center of your slinky (suspended as in Home Exp. 2.2). Excite the slinky by giving it a sudden push at the center with your hand. Try different degrees of abruptness. You should soon be able to see that the even harmonics are always missing and the more impulsive your excitation, the larger the number of (odd) modes you excite. Can you devise a way to excite only the *even* modes?

Try plucking a guitar or piano string at different places–in the middle or near one end—and see if you can hear a difference in "tone quality."

Home experiment   **2.5   Piano as Fourier-analyzing machine—insensitivity of ear to phase.**   Find a piano. Hold down the damper pedal. Shout "hey" into the region of the strings and sounding board. Listen. Shout "ooh." Try all vowels. The piano strings are picking up (in somewhat distorted form) and then preserving the Fourier analysis of your voice! Notice that the recognizable vowel sound persists for several seconds. What does that tell you about the importance to your ear and brain of the relative phases of the Fourier components that make up the sound?

2.6 Piano harmonics–equal-temperament scale. Look in the
Handbook of Chemistry and Physics under "musical scales" for convenient
tables of pitches for the three common scales:

Home experiment

> American Standard pitch (A440) equal-temperament chromatic scale
> International pitch (A435) equal-temperament chromatic scale
> Scientific or Just scale (based on C256, which gives A426.67)

First we shall explain the scientific scale. Let 256 cps equal one unit of frequency, $v = 1$. The *harmonics* of this fundamental note are then $v = 2, 3, 4$, etc.;
the *subharmonics* are defined to be $\frac{1}{2}$, $\frac{1}{3}$, $\frac{1}{4}$, etc. Middle C on the piano
is C256 (if the piano is tuned that way). It is called C_4. (The subscript refers
to the octave. The subscript increases by one at each higher octave of C.)
Suppose that piano strings perfectly obeyed the "continuous, perfectly flexible string" dispersion law. Then the mode frequencies of a given string
would consist of the harmonic sequence v_1, $2v_1$, $3v_1$, etc. The names and
frequencies of the first 16 harmonics of string C_4, and also its first two subharmonics would be as follows (we underline C_4 and its octaves):

Name: F_3 C_3 $\underline{C_4}$ $\underline{C_5}$ G_5 $\underline{C_6}$ E_6 G_6 Bb_6 $\underline{C_7}$ D_7 E_7 $F\#_7$ G_7 $G\#_7$ Bb_7 B_7 $\underline{C_8}$

v: $\frac{1}{3}$ $\frac{1}{2}$ $\underline{1}$ $\underline{2}$ 3 $\underline{4}$ 5 6 7 $\underline{8}$ 9 10 11 12 13 14 15 $\underline{16}$

An octave is always higher by a factor of two in frequency (compare G_6 and G_7).
Now let us construct a scale within the single octave between C_4 and, C_5 by
dividing or multiplying the harmonics and subharmonics of C_4 by appropriate powers of 2. We then get the Scientific or Just diatonic scale in C major
(diatonic means we have only the "white notes," none of the "black notes"
on the piano keyboard):

Name: C   D   E   F   G   A   B   C

v: 1 $\frac{9}{8}$ $\frac{5}{4}$ $\frac{4}{3}$ $\frac{3}{2}$ $\frac{5}{3}$ $\frac{15}{8}$ 2

(We sneaked in the A. It is $\frac{5}{4}$ F.) The note C is called the tonic in this scale.
  The smallest musical interval in this diatonic scale is called a *minor second*. The frequency ratio for a minor second is F/E = C/B = $\frac{16}{15}$ = 1.067.
The next larger ratio is called a major second. There are two kinds of major
seconds: D/C = G/F = B/A = $\frac{9}{8}$ = 1.125; E/D = A/G = $\frac{10}{9}$ = 1.111. There
are also two kinds of the next larger ratio, the minor third: F/D = $\frac{3}{2}\frac{2}{7}$ =
1.185; G/E = C/ A = $\frac{6}{5}$ = 1.200. There is only one kind of major third:
E/C = A/F = B/G = $\frac{5}{4}$ = 1.250. Now comes the difficulty, musically. Suppose
that, in the course of composing for a piano tuned to this scale, you suddenly decide that you want to change to a new "key signature," that is, to a
diatonic scale with a different note as tonic. For example, you might want to
change from C major to D major. You want the same kind of scale, i. e., the
same frequency ratios as before. Thus you want the first major second in the
new scale, E/D, to be a "D/C type of major second" with ratio 1.125. Unfortunately you can't use the E that you already have, because that gives E/D

= 1.111. So you need a new string, E′, with E′/C = (1.125)(D/C) = 1.265, whereas E/C = 1.250. The next note after E′ also demands a new string, called F#. We make it in the ratio F#/D = E/C, so that F# = $\left(\frac{5}{4}\right)\left(\frac{9}{8}\right)$ = 1.407. (This is a "black key" on the piano.) Notice that the piano has now acquired a *new kind of minor second:* F#/F = 1.0555. As you complete the scale, you must add more and more keys. Then, if you want to play in still other keys, the situation just gets worse and worse. (Try completing the D scale. You have to add the "black key" C# to get the note corresponding to B on the C scale. But what other "primed" strings do you need?

The equal-temperament scale gets around all this by making all notes equally spaced on a logarithmic scale. The octave is divided into 12 minor seconds ("half tones"), each of which has the frequency ratio $2^{1/12}$ = 1.059. Then all major seconds have the ratio $2^{2/12}$ =. 1.122; all minor thirds have $2^{3/12}$, etc. None of the intervals are "just" except the octaves, but all intervals are close to the just values in diatonic scales based on any note as tonic.

Try the following experiments:

(i) Steadily hold down the key for, say, Bb_6, so as to lift its damper without sounding the note. Now strike a lower note sharply, hold it for a few seconds, and release the lower key so as to damp the lower note. If you now hear the Bb_6 string sounding, it must have been excited by one of the harmonics present in the mode pattern of the lower vibrating string. Try various lower strings. The note an octave down should work. So should the 12th down, E*b*, since B*b* is its third harmonic. So should the C string for which Bb_6 is the 7th harmonic, provided that that harmonic is present in the vibrating string C. Another way to work this experiment is to hit the same lower note, say C_4, while silently holding down different higher notes to see if they are excited. When you have found a note that is excited, try the neighboring note one minor second away. Is it excited?

(ii) This time hold the key of a low note down silently and strike a higher note sharply. If the higher note is one of the overtones of the lower string, you will excite that overtone in the lower string without exciting the lowest mode (the fundamental) of the lower string. Thus you get to hear what the harmonics of the lower string sound like when they are not masked by the loud fundamental.

(iii) Use the method of (ii) to learn what the first 6 or 7 harmonics of C_4 (or a lower C) sound like. Then learn how to hear a particular harmonic in the tone pattern when the lower key is struck in the nomal manner. For example, to learn how to hear the 7th harmonic, Bb_6, when C_4 is struck, hold down C_4 silently and hit Bb_6 sharply. That tells you what Bb_6 sounds like when it is the 7th harmonic of C_4. Then, while your memory is fresh, hit string C_4 and concentrate on picking Bb_6 out of the sound (dominated by the fundamental of C_4). Note that the frequency of this note when it occurs as the 7th harmonic of C_4, i.e., on the C_4 *string*, will not be exactly the same as the frequency of the fundamental note of the Bb_6 string. It will be close enough so that it can be excited, probably, but as soon as the Bb_6 string is damped and the C_4 string has had a few seconds to forget how it was excited, it will oscillate at its own (7th harmonic) frequency, not the exciting frequency. Thus it sounds slightly different from the exciting note. (Of course if the piano is out of tune, it may sound *very* different.) Because of this small frequency difference, you can hear beats as follows:

(iv) Hold down C_4 silently. Hit C_5 sharply. This excites the second mode of the C_4 string. Now, before this dies away, damp the C_5 string and then hit C_5 again, gently, trying to match its loudness to what is left of the second harmonic of C_4. Listen for beats. (This works better on some pianos than on others. It should be done in an otherwise quiet room.)

(v) The lowest two notes on the piano are $A_0 27.5$ and $A\#_0 29.1$. Their beat frequency is thus 1.6 cps, which is easily detectable. Hit both notes together, gently. Once you think you hear beats, let one key up, but not the other. Do the beats go away? (Is the piano in tune?)

2.7 Why an ideal continuous string gives exactly "harmonic" frequency ratios but a beaded string does not.   Consider a beaded string of very many beads (say 100) with both ends fixed. We will think of this string as being essentially continuous. Suppose it is oscillating in its lowest mode; then the length L is one half-wavelength of a sine wave. Now consider the second mode: The length L is two half-wavelengths, so the first half of L is one half-wavelength. Now compare the 50 beads in the first half of the string when it is in its second mode, with the entire 100 beads when the string is in its lowest mode. In each case the beads follow a curve that is one half-wavelength of a sine wave. Compare bead 1 (in mode 2) with the average of beads 1 and 2 (in mode 1); compare bead 2 (in mode 2) with beads 3 and 4 (mode 1), etc. Thus, in mode 2, bead 17 has the same amplitude as the average of beads 33 and 34 in mode 1 (if the sine waves have the same amplitude). But in mode 2 the string at bead 17 makes twice as great an angle to the equilibrium axis as does the string at beads 33 and 34 in mode 1(using the small-angle approximation). Thus the return force per unit displacement on bead 17 is twice that on the two beads 33 and 34. Also, the mass of bead 17 is only half that of the two beads 33 and 34. Thus *the return force per unit displacement per unit mass* is *four times greater for bead 17 in mode two than it is for the combination of beads 33 and 34 in mode one*. Thus we find, in the "nearly continuous" approximation (implied by our "large number" of beads), $\omega_2 = 2\omega_1$.

This argument does not work if the number of beads becomes small. Explain why not. Then you can see why you get the "harmonic" ratios $v_2 = 2v_1$, $v_3 = 3v_1$, etc. in the continuous limit, but not when there are only a few beads, as shown for example in Fig. 2.12.

2.8   How many years does it take to double your money if you invest it at 5.9% interest (per year, compounded annually)? [*Hint*: Consider the equal-tempered scale (Home Exp. 2.6).]

2.9   Complete the "just" diatonic scale for D major which we started in Home Exp. 2.6. We found there that we had to add a new E string, which we called E'. We needed our first "black note," F#. We will also need another black note, C#. What about G, F, A, and B? Can we use the ones we have, or do we need G', F', A', and B'?

2.10   Derive Eq. (55), the wave equation for a nonuniform string.

2.11   Obtain the result Eq. (47) for the Fourier coefficients of $F(z)$ plotted in Fig. 2.6.

2.12   Find the mode configurations and frequencies for the first three modes of transverse vibration of a continuous string with tension T_0, mass density ρ_0, and length L, given the boundary conditions that *both ends are free*. (They slide on frictionless rods that pass through massless rings at each end of the string.) Show that the lowest mode has the peculiar property of having infinite wavelength and zero frequency. In this mode, the string translates with uniform velocity. (That includes the possibility of its remaining at rest with arbitrary displacement.)

2.13   Find the three mode configurations and frequencies for transverse vibrations of a uniformly beaded string having three beads and four string segments, given the boundary conditions that both ends are free. (The end string segments have massless rings at their ends and slide on frictionless rods.) Compare the lowest mode with that in Prob. 2.12.

2.14   Consider an *LC* network of three inductances and four capacitances, arranged as in Fig. 2.15 for $N = 3$, except that *the outer two capacitors are short-circuited*. Find the three modes–current configurations and frequencies. Compare the physical significance of the "peculiar" lowest mode in this problem with that in Prob. 2.13. Compare the boundary conditions with those in Prob. 2.13.

2.15   Consider the piano string which sounds middle C, C256 (Scientific scale). The density of the steel of the string is about 9 gm/cm³. (This is *not* the linear mass density ρ_0. Why not?) Suppose the string diameter is $\frac{1}{2}$ mm and its length is 100 cm. What is its tension in Newtons and in kg-wt 9.8 N = 1 kg-wt;

Ans. $T_0 \approx 460$ N.

2.16   Find $\psi(z, t)$ for a slinky constrained at $t = 0$ to follow the function $g(z)$ given by Eq. (48). Plot $\psi(z, t_0)$, where $\omega_1 t_0 = \pi/3$. Compare the shape of $\psi(z, t)$ with that of $\psi(z, 0)$, which is shown in Fig. 2.7.

2.17   Compare the tension in a steel guitar string with that in a catgut string of the same length, diameter, and pitch (of lowest mode). The density of steel is about 9 gm/cm^3; the density of catgut is not much more than 1 gm/cm^3. Are the string diameters actually the same for steel-string guitars as for catgut-string guitars? Look at guitars and find out. Once you have looked and estimated the ratio of diameters, recalculate the ratio of tensions in the two cases.

2.18   Derive the classical wave equation [Eq. (14)] in the following way: Start with the exact Eq. (62). Now go to the continuous approximation. Replace subscript n by location z, taking into account the bead separation is length a. Use the Taylor's series expansions of the right side of Eq. (62). *Include one more term than* is *necessary to obtain the classical wave equation.* Give a criterion for neglecting this and higher-order terms.

2.19   Show that by taking Eq. (71) as the solution to the equation of motion for transverse vibrations of the beaded string, Eq. (65), one gets the dispersion relation Eq. (70). Show that this fact is independent of the choices for the constants A, B, and k, which depend only on initial conditions and boundary conditions.

2.20   Use Eqs. (73) and (70) to obtain the frequency ratios shown in Fig. 2.12 for $N = 5$.

2.21   Find the mode configurations and frequencies for transverse oscillations of a beaded string of 5 beads with one end fixed and one free. Plot the five corresponding points on the dispersion relation $\omega(k)$ as in Fig. 2.13.

2.22   By inspection of Fig. 2.13 and a sketch of the system, show an easy way to add 6 more points to Fig. 2.13, so that it gives the modes for a beaded string of 11 beads with both ends fixed.

2.23   Show that Eqs. (73) and (74) give the same answers for the frequencies for $N = 1$ and $N = 2$ as those we obtained in Sec. 1.2 and Sec. 1.4.

2.24   Sketch the configurations of the five modes of a 5-beaded string corresponding to Eqs. (78) to (80).

2.25   Plot the dispersion relation for the system shown in Fig. 2.15.

2.26   Show that, for the system of coupled pendulums shown in Fig. 2.16, the equation of motion for the nth pendulum bob is given (for small oscillations) by

$$\frac{d^2\psi_n}{dt^2} = -\frac{g}{l}\psi_n + \frac{aK}{M}\left(\frac{\psi_{n+1} - \psi_n}{a}\right) - \frac{aK}{M}\left(\frac{\psi_n - \psi_{n-1}}{a}\right)$$

Show that the general solution for a mode, without regard to boundary conditions, is

$$\psi_n(t) = \cos(\omega t + \varphi)[A \sin nka + B \cos nka]$$

Show that the dispersion relation is

$$\omega^2 = \frac{g}{l} + \frac{4K}{M} \sin^2 \frac{ka}{2}$$

Show that, for the boundary conditions shown in Fig. 2.16 (i.e., no springs coupling the end bobs to the wall), the above solution reduces to

$$\psi_n(t) = \cos(\omega t + \varphi) B \cos nka,$$

with the nth bob located at $z = (n - \frac{1}{2})a$. Show that the lowest mode has $k = 0$. Sketch its configuration. What would be the behavior of the system in this configuration if the gravitational constant were gradually reduced to zero? Sketch the configurations for $N = 3$ and give the frequencies for the three modes.

2.27 Find the system of coupled capacitances and inductances which is "analogous" to the system of coupled pendulums of Fig. 2.16, in the sense that the equation of motion for the nth inductance has the same form as the equation of motion for the nth pendulum bob, which you found in Prob. 2.26. Find the dispersion relation.

2.28 Go to the continuous limit in the coupled-pendulum problem (Prob. 2.26). Show that the equation of motion becomes a wave equation of the form

$$\frac{\partial^2 \psi}{\partial t^2} = -\omega_0^2 \psi + v_0^2 \frac{\partial^2 \psi}{\partial z^2}.$$

2.29 Prove each of the following numbered statements using two methods: (*a*) the "physical" method that makes use of the normal modes of a continuous string with appropriate boundary conditions, and (*b*) the method of Fourier analysis of a periodic function of z.

(i) Any (reasonable) function $f(z)$ defined between $z = 0$ and $z = L$ and having value zero at $z = 0$ and slope zero at $z = L$ can be expanded in a Fourier series of the form

$$f(z) = \sum_n A_n \sin nk_1 z; \quad n = 1, 3, 5, 7, \ldots; \quad k_1 L = \frac{\pi}{2}.$$

(*Note*: In using the method of Fourier analysis you must first construct a periodic function from $f(z)$ so that you can use the formulas of Fourier analysis.)

(ii) Any (reasonable) function $f(z)$ defined between $z = 0$ and $z = L$ and having zero slope at $z = 0$ and zero slope at $z = L$ can be expanded in a Fourier series of the form

$$f(z) = B_0 + \sum_n B_n \cos nk_1 z; \quad n = 1, 2, 3, 4, \ldots; \quad k_1 L = \pi.$$

(iii) Any (reasonable) function $f(z)$ defined between $z = 0$ and $z = L$ and having slope zero at $z = 0$ and value zero at $z = L$ can be expanded in a Fourier series of the form

$$f(z) = \sum_n B_n \cos nk_1 z; \quad n = 1, 3, 5, 7, \ldots; \quad k_1 L = \frac{\pi}{2}.$$

2.30 Fourier analysis of a periodically repeated square pulse.

When you clap your hands periodically, the resulting air pressure at your ear can be approximated by a periodically repeated square pulse. Let $F(t)$ represent the gauge pressure at your ear. Take $F(t)$ to be $+ 1$ unit for the short time interval Δt, and zero before and after that interval. This "square pulse" of unit height and width Δt [on a plot of $F(t)$ versus t] is repeated periodically at time intervals of length T_1. The short interval Δt gives the duration of each clapping sound. The period T_1 is the time between successive claps. The frequency $v_1 = T_1^{-1}$ is the clapping frequency. You are to Fourier analyze $F(t)$.

(*a*) Show that you can choose the time origin so that only cosines of $n\omega_1 t$ appear, i.e., so that

$$F(t) = B_0 + \sum_{n=1}^{\infty} B_n \cos n\omega_1 t$$

(*b*) Show that $B_0 = \Delta t/T_1$, which is just the fractional "on" time. Show that

$$B_n = \frac{2}{n\pi} \sin (n\pi v_1 \Delta t), \quad \text{for } n = 1, 2, \ldots$$

(*c*) Show that for $\Delta t \ll T_1$ the "fundamental" tone v_1 and the low harmonics $2v_1, 3v_1, 4v_1$, etc. all have essentially the same value for their Fourier amplitudes B_n.

(*d*) Sketch B_n versus nv_1, going up to sufficiently high n so that B_n has gone through zero two or three times.

(*e*) Show from part (*d*) that the "most important" frequencies (i.e., those with relatively large values for B_n) go from the fundamental, v_1, to a frequency of the order of $1/\Delta t$. Thus we may call $1/\Delta t$ by the name v_{max}. Actually, of course, there is no maximum frequency since the Fourier series extends to $n = \infty$. However, the most important frequencies lie between zero and v_{max}. The "frequency band" of dominant frequencies has "bandwidth" equal to about $v_{max} = 1/\Delta t$. The important frequencies are thus

$$v = 0, v_1, 2v_1, 3v_1, 4v_1, \ldots v_{max} = \frac{1}{\Delta t}$$

The bandwidth of dominant frequencies can be given the name Δv. Then your result can be written

$$\Delta v \Delta t \approx 1.$$

This is a very important relation. It holds not only for our assumed $F(t)$, a repeated "square" pulse of width Δt, but for any pulse shape that can be characterized as being zero most of the time and nonzero for a time of duration about Δt. If (as in our example) the pulse is repeated at intervals T_1, then the dominant frequencies are zero, $\nu_1, 2\nu_1, 3\nu_1$, etc. up to about $1/\Delta t$. If the pulse is not repeated but only occurs once, then it turns out (as we shall show in Chap. 6) that the "Fourier spectrum" of important frequencies still occupies the frequency band extending from zero to about $1/\Delta t$, but it is a continuous spectrum, including all frequencies in the band rather than just the fundamental, ν_1, and its harmonics.

The present problem can help you to understand the frequency spectrum of the electromagnetic radiation called *synchrotron radiation* which is emitted by a relativistic electron undergoing uniform circular motion. It can be shown (Chap. 7) that a nonrelativistic electron undergoing uniform circular motion at frequency ν_1 emits electromagnetic radiation having the single frequency ν_1. That is because the electric field in the radiation for nonrelativistic electron velocities is just proportional to the component of acceleration of the charge perpendicular to the line of sight from the charge to the observer. For circular motion this projected acceleration is simple harmonic motion. Therefore for a nonrelativistic electron the radiated field is proportional to the cosine (or sine) of $\omega_1 t$. For a relativistic electron, the radiated field does not have the time dependence $\cos \omega_1 t$. Instead the radiation intensity is very strongly concentrated in direction, along the instantaneous direction of the velocity of the charge. When the electron is headed directly toward the observer, it is emitting radiation that will later be detected by the observer. At other times, the radiation it is emitting will not reach the observer. Thus the electric field measured by the observer is strong for a short time interval Δt once every period T_1, and nearly zero at other times. Thus the frequency spectrum observed consists of $\nu_1 = 1/T_1$ and its harmonics $2\nu_1, 3\nu_1$), etc., up to a maximum (important) frequency of about $1/\Delta t$ Show that the time interval Δt is given roughly by $\Delta t/T_1 \approx \Delta\theta/2\pi$, where $\Delta\theta$ is the "angular full-width" of the radiation pattern.

2.31 Sawtooth shallow-water standing waves. Shallow-water waves are waves in which the amplitude of motion of the water at the bottom of the pan, lake, or ocean is comparable in magnitude with that at the surface. The sloshing mode (Home Exp. 1.24) is a shallow-water wave. Show this experimentally by stirring some coffee grounds into the water so that some grounds are near the bottom. Excite the sloshing mode (the one where the surface remain sessentially flat) and look, at the motion of coffee grounds at the bottom and at the surface near the center of the pan. Look also near the end walls.

Now consider the following idealized *sawtooth shallow-water standing wave.* Suppose you have two independent pans of the same shape with water of the same equilibrium depth, h, undergoing oscillation in the sloshing mode. The pans are adjacent so that, if there were no partition, the two would form one long pan along the horizontal oscillation direction. Suppose the phases of the oscillations are such that water in one pan always moves horizontally opposite to that in the other, causing the water to pile up to its maximum height in both pans at the same time at the walls separating the two. Now imagine that you remove the walls separating the two pans. The water at the boundary surface had no horizontal motion when the walls were in place. It still has none because of the symmetry of the motion of the two bodies of water (now joined to form one long body). The motion should continue unchanged! We can continue joining pans if we wish. We have here a standing wave with a sawtooth shape. Let us approximate that shape by a sine wave. Then we see that *one pan length equals one half-wave-length. (Note:* If you Fourier-analyze this periodic function of z, the first and most dominant term in the Fourier expansion will be the one that we are using to approximate the sawtooth.) Use that approximation in the formula that gives the frequency of the sloshing mode (see Home Exp. 1.24). Show that one thus obtains

$$\lambda v = \frac{2\sqrt{3}}{\pi} \sqrt{gh} = 1.10\sqrt{gh}.$$

We see that *these waves are nondispersive. (Note:* It turns out that the exact dispersion relation for sinusoidal shallow-water waves is $\lambda v = \sqrt{gh}$. Our sawtooth approximation gives a propagation velocity 10% too high.)

For *deep-water waves* (waves where the equilibrium water depth is large compared with the wavelength), the wave amplitude falls off exponentially with depth below the surface, with a decrease in amplitude by a factor of $e = 2.718\ldots$ for every increase in depth by an amount $\lambdabar \equiv \lambda/2\pi$, where $\lambdabar$ (pronounced "lambda bar") is called the reduced wavelength. To a crude approximation we can say that a deepwater wave is something like a shallow-water wave from the surface down to an effective depth $h = \lambdabar$, since in that region the amplitude is relatively large and roughly constant, whereas for depths much greater than $\lambdabar$ the amplitude is very small. Thus we guess that the dispersion relation for deep-water waves may be obtained from that for shallow-water waves by replacing the equilibrium depth h for shallow-water waves by the mean amplitude attenuation length $\lambdabar$ for deep-water waves. That conjecture turns out to be correct, as we shall show in Chap. 7. Thus the dispersion law for deep-water waves is given by $\lambda v = \sqrt{g\lambdabar}$.

2.32 Fourier analysis of symmetrical sawtooth. By a symmetrical sawtooth we mean one where the front and back edges of each tooth have the same slope. Let $z = 0$ occur at one of the crests (tooth points). Show that the periodic sawtooth $f(z)$ has a Fourier series given by

$$f(z) = 0.82A \left[\cos k_1 z + \tfrac{1}{4} \cos 2k_1 z + \tfrac{1}{9} \cos 3k_1 z + \ldots\right],$$

where $k_1 = 2\pi/\lambda_1$, where λ_1 is the tooth length (from one crest to the next), and where A is the tooth amplitude, i.e., $2A$ is the vertical distance from the minimum (at a trough) to the maximum (at the crest). Thus we see that the contribution from the nth term has amplitude proportional to $1/n^2$. This tells you something about how good our approximation was in Prob. 2.31, where we approximated the sawtooth by its first Fourier component to obtain the dispersion relation $\lambda v = 1.10 \sqrt{gh}$.

Home experiment

2.33 Surface tension modes. Circular surface-tension standing waves can be .clearly demonstrated as follows: Fill a paper or styrofoam cup to the brim and then add a little more, so that the water bulges above the top of the cup. Tap the cup gently. Watch! To see the waves easily look at the reflection of the sky. Alternatively, use a small bright light held several feet above the surface, and look at the pattern on the bottom caused by the lenslike effect of the waves. To see that surface tension is involved, try adding a little detergent to the water.

2.34 Boundary conditions at free end of string. Consider the four different systems shown in the figure.

(i) Show that all four systems have the same frequency in the modes shown.

(ii) Suppose you want to use the same formula for the constraint on wavenumber in cases (c) and (d) as that which is used in case (a). Then show that L in your formula must equal $\left(\frac{3}{2}\right) a$ in those cases. Give the formula.

2.35 A flexible string of length L is stretched with equilibrium tension T between fixed supports. Its mass per unit length is ρ, so that its total mass is $M = \rho L$. The string is set into vibration with a hammer blow which impulsively imparts a transverse velocity v_0 to a small segment of length a at the center. Evaluate the amplitudes of the lowest three harmonics excited.

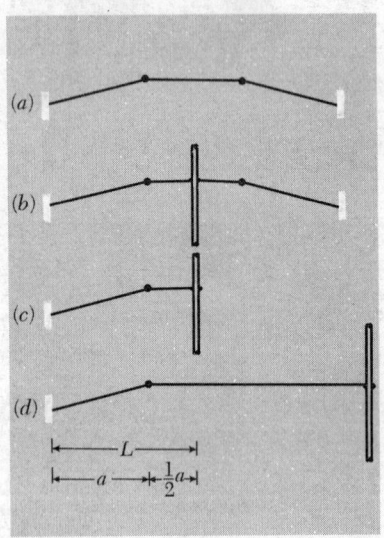

Problem 2.34

Chapter 3

Forced Oscillations

Chapter 3   Forced Oscillations

3.1   Introduction

In Chaps. 1 and 2 we studied the free oscillations of various systems. In the present chapter we shall study *forced* oscillations of these systems. This means that we shall investigate the behavior of the systems when a given external time-dependent force is applied to the system in some way. With no loss of generality, we shall specialize to harmonically oscillating driving forces and investigate the response of the system as a function of frequency.

In Sec. 3.2 we shall review the free oscillations of a damped one-dimensional oscillator. Then we shall look at the transient response when the damped oscillator starts at rest and is driven by a harmonically oscillating force. We shall find the interesting phenomenon of "transient beats" between the driving force and the free oscillation "transient." Then we shall study the steady-state oscillations which remain after the transients have decayed to zero. We shall examine the resonant response of the driven oscillator as we slowly vary the driving frequency. In Sec. 3.3 we shall study a system with two degrees of freedom and shall find that each of the modes of free oscillation contributes to the driven motion of a given moving part. In fact, we shall derive the very simple result that the motion of a given moving part is a superposition of independent contributions, one from each mode. In Sec. 3.4 we shall discover the remarkable behavior of a system of several degrees of freedom which is driven at a frequency above or below its lowest mode frequency. In Sec. 3.5 we shall study the behavior of a driven system of many coupled pendulums. We will thus be led to the discovery of exponential waves.

All of the phenomena discussed in this chapter can be studied experimentally in simple home experiments with coupled pendulums, using a slinky for coupling spring, standard sized (≈ 300 gm) cans of soup (which fit nicely inside the slinky) for bobs, and a phonograph turntable for a driving force.

3.2   Damped Driven One-dimensional Harmonic Oscillator

This section is partly a review of Vol. 1, Chap. 7, where you studied the free oscillations and the steady-state forced oscillations of a damped oscillator. We will also consider the "transient response" of the oscillator subject to a harmonic driving force when the oscillator is initially at rest at its equilibrium position.

Consider a point mass M oscillating in the x direction. Its displacement from equilibrium is $x(t)$. The mass M experiences a *return force* $-M\omega_0^2 x(t)$ due to a spring of spring constant $K = M\omega_0^2$.

If there were no other force, the mass would therefore undergo harmonic oscillation with angular frequency ω_0. The mass also experiences a *frictional drag* force $-M\Gamma\dot{x}(t)$, where Γ is a constant that we can call the *damping constant per unit mass,* or simply the damping constant. Finally, the mass is subject to an *external force* $F(t)$. Newton's second law then gives for the equation of motion of M the inhomogeneous second-order linear differential equation

$$M\ddot{x}(t) = -M\omega_0^2 x(t) - M\Gamma\dot{x}(t) + F(t). \tag{1}$$

First we consider the special case where is no external force.

Transient decay of free oscillations. The equation of motion, Eq. (1), becomes

$$\ddot{x}(t) + \Gamma\dot{x}(t) + \omega_0^2 x(t) = 0. \tag{2}$$

We try a solution $x_1(t)$ of the form

$$x_1(t) = e^{-(1/2)t/\tau} \cos(\omega_1 t + \theta) \tag{3}$$

where τ, ω_1, and θ are unknown. By direct substitution, we find that Eq. (3) gives a solution to Eq. (2) for *any* value of the phase constant θ, provided that we choose

$$\tau = \frac{1}{\Gamma} \tag{4}$$

and

$$\omega_1^2 = \omega_0^2 - \tfrac{1}{4}\Gamma^2. \tag{5}$$

The most general solution of Eq. (2) is a superposition of two linearly independent solutions with two "arbitrary" constants that may be chosen to fit the initial displacement and velocity $x_1(0)$ and $\dot{x}_1(0)$. Two independent solutions can be obtained by setting θ equal (for example) to zero or $-\tfrac{1}{2}\pi$. Thus the general solution can be written in the form

$$x_1(t) = e^{-(1/2)\Gamma t}(A_1 \sin \omega_1 t + B_1 \cos \omega_1 t). \tag{6}$$

The constants A_1 and B_1 are easily seen to be given by $B_1 = x_1(0)$ and by $\omega_1 A_1 = \dot{x}_1(0) + \tfrac{1}{2}\Gamma x_1(0)$. Then Eq. (6) gives

$$x_1(t) = e^{-(1/2)\Gamma t} \left\{ x_1(0) \cos \omega_1 t + \left[\dot{x}_1(0) + \frac{1}{2}\Gamma x_1(0) \right] \frac{\sin \omega_1 t}{\omega_1} \right\} \tag{7}$$

When $\tfrac{1}{2}\Gamma$ is small compared with ω_0, the oscillation is said to be *weakly damped.* When $\tfrac{1}{2}\Gamma$ equals ω_0, the motion is called *critically damped.* In that case Eq. (5) gives that ω_1 is zero. In the solution Eq. (7) we then replace $\cos \omega_1 t$ by 1 and $(1/\omega_1) \sin \omega_1 t$ by t, since the limit as ω_1 goes to zero of $(1/\omega_1) \sin \omega_1 t$ is just t.

When $\frac{1}{2}\Gamma$ is greater than ω_0, the oscillator is said to be *overdamped*. In that case, Eq. (5) gives ω_1^2 as negative. This means that ω_1 is given by

$$\omega_1 = \pm i\,|\omega_1|, \qquad |\omega_1| = \sqrt{\tfrac{1}{4}\Gamma^2 - \omega_0^2}, \qquad (8)$$

where i is the square root of -1. The solution Eq. (7) is still valid and can be written in the form (Prob. 3.25)

$$x_1(t) = e^{-(1/2)\Gamma t}\left\{ x_1(0)\cosh|\omega_1|t \right.$$

$$\left. + \left[\dot{x}_1(0) + \frac{1}{2}\Gamma x_1(0)\right]\frac{\sinh|\omega_1|t}{|\omega_1|}\right\} \qquad (9)$$

We shall only be concerned with the case where $\frac{1}{2}\Gamma$ is less than ω_0, in which case the oscillator is said to be *underdamped*. That includes the case of weak damping, where we have $\frac{1}{2}\Gamma \ll \omega_0$. In the case of weak damping, we may consider the exponential factor $e^{-(1/2)\Gamma t}$ to be essentially constant during any one cycle of oscillation. Then the velocity is given to a sufficiently good approximation by taking the time derivative of Eq. (6) with $e^{-(1/2)\Gamma t}$ regarded as a constant. It is then easily shown that the energy (kinetic plus potential) is essentially constant during any one cycle but decays exponentially over a time interval that includes many cycles:

$$E(t) = \tfrac{1}{2}M\dot{x}_1^2(t) + \tfrac{1}{2}M\omega_0^2 x_1^2(t)$$

$$= E_0 e^{-\Gamma t} = E_0 e^{-t/\tau}, \qquad (10)$$

where

$$E_0 = \tfrac{1}{2}M(\omega_1^2 + \omega_0^2)(\tfrac{1}{2}A_1^2 + \tfrac{1}{2}B_1^2). \qquad (11)$$

We now turn to the case of an underdamped oscillator subject to an external force $F(t)$ that is not zero.

Steady-state oscillation under harmonic driving force. A very large class of functions $F(t)$ can be written as a Fourier expansion over various frequencies ω:

$$F(t) = \sum_\omega f(\omega)\cos[\omega t + \varphi(\omega)]. \qquad (12)$$

For example, as we saw in Sec. 2.3, any (reasonable) periodic function $F(t)$ can be expanded in this way. In addition, many nonperiodic functions can also be expanded in Fourier series or Fourier integrals, as we shall see in Chap. 6. Consider a single Fourier component of such a force:

$$F(t) = F_0 \cos \omega t, \tag{13}$$

where we have chosen the zero of time so as to set the phase constant to zero. Once we know how to find $x(t)$ for a harmonic driving force like that of Eq. (13), we can find $x(t)$ for a superposition like that of Eq. (12). That is because, according to our discussion in Sec. 1.3, the inhomogeneous linear equation satisfies a superposition principle such that the solution corresponding to a superposition of different external forces is just the superposition of the individual solutions. Therefore we need consider only the inhomogeneous equation with an external force of a single harmonic component:

$$M\ddot{x}(t) + M\Gamma\dot{x}(t) + M\omega_0{}^2 x(t) = F_0 \cos \omega(t). \tag{14}$$

We want to find the *steady-state solution* of Eq. (14). The steady-state solution gives the motion of the oscillator after the harmonic driving force has been applied for a time that is very long compared with the decay time τ. Then the "transient oscillations," which describe the behavior of the oscillator during the first few mean decay times after the initial driving force is applied, have decayed away to zero. The oscillator then undergoes harmonic oscillations at the driving frequency ω. There are no adjustable or "arbitrary" constants. The oscillation amplitude is proportional to the amplitude F_0 of the driving force. The phase constant has a definite relationship to the phase constant of the driving force.

Absorptive and elastic amplitudes. Rather than describe the oscillation in terms of its amplitude and phase constant, we can describe it in terms of two amplitudes A and B that give the oscillation component $A \sin \omega t$ that is 90 degrees out of phase with the driving force $F_0 \cos \omega t$, and the oscillation component $B \cos \omega t$ that is in phase with the driving force. Thus the steady-state solution $x_s(t)$ can be written as

$$x_s(t) = A \sin \omega t + B \cos \omega t, \tag{15}$$

with suitable choices for A and B. By direct substitution you may verify that $x_s(t)$ satisfies Eq. (14) if and only if A and B are given by

$$A = \frac{F_0}{M} \frac{\Gamma\omega}{[(\omega_0{}^2 - \omega^2)^2 + \Gamma^2\omega^2]} \equiv A_{\mathrm{ab}}, \tag{16}$$

$$B = \frac{F_0}{M} \frac{(\omega_0{}^2 - \omega^2)}{[(\omega_0{}^2 - \omega^2)^2 + \Gamma^2\omega^2]} \equiv A_{\mathrm{el}}. \tag{17}$$

The constant A_{ab} is called the *absorptive amplitude*. The constant A_{el} is called the *elastic amplitude*. (Sometimes the elastic amplitude is called instead, the "dispersive" amplitude.) These names are chosen because the time-averaged input power absorption is entirely due to the term $A_{ab} \sin \omega t$. The term $A_{el} \cos \omega t$ contributes to the instantaneous power absorption $P(t)$, but averages to zero over one cycle of steady-state oscillation. These results follow from the fact that the instantaneous power $P(t)$ is the force $F_0 \cos \omega t$ times the velocity $\dot{x}(t)$. The instantaneous velocity has a contribution that is in phase with the force and a contribution that is 90 degrees out of phase with the force. Only that velocity contribution which is in phase with the force contributes to the *time-averaged power, P.* This "in phase" velocity is contributed by the "out of phase" displacement, $A_{ab} \sin \omega t$. We see these relations algebraically as follows:

$$F(t) \quad = F_0 \cos \omega t,$$
$$x_s(t) \quad = A_{ab} \sin \omega t + A_{el} \cos \omega t,$$
$$\dot{x}_s(t) = \omega A_{ab} \cos \omega t - \omega A_{el} \sin \omega t.$$

Then the steady-state instantaneous input power is given by

$$P(t) = F(t)\,\dot{x}_s(t) = F_0 \cos \omega t [\omega A_{ab} \cos \omega t - \omega A_{el} \sin \omega t]. \quad (18)$$

Denoting the time average over one cycle by brackets $\langle \ \rangle$, we find

$$P = F_0 \omega A_{ab} \langle \cos^2 \omega t \rangle - F_0 \omega A_{el} \langle \cos \omega t \sin \omega t \rangle.$$

But

$$\langle \cos^2 \omega t \rangle \equiv \frac{1}{T} \int_{t_0}^{t_0 + T} \cos^2 \omega t \, dt = \frac{1}{2}, \quad (19)$$

where T is the oscillation period. Similarly

$$\langle \cos \omega t \sin \omega t \rangle = \tfrac{1}{2} \langle \sin 2\omega t \rangle = 0. \quad (20)$$

Thus we obtain the time-averaged steady-state input power,

$$P = \tfrac{1}{2} F_0 \omega A_{ab}. \quad (21)$$

In Eq. (21) we have verified that the time-averaged input power is proportional to the amplitude A_{ab} of that part of the steady-state displacement $x_s(t)$ which is 90 deg out of phase with the driving force. This result is independent of the phase convention whereby we chose the force to be proportional to $\cos \omega t$, rather than to the more general expression $\cos(\omega t + \varphi)$.

At steady state the time-averaged power must equal the time average of the power dissipated by friction. The instantaneous frictional force is $-M\Gamma \dot{x}(t)$. The instantaneous frictional power is the frictional force times the velocity. You can easily show that the time-averaged power loss to friction is given by

$$P_{\text{fr}} = M\Gamma \langle \dot{x}_s{}^2 \rangle$$
$$= \tfrac{1}{2} M\Gamma \omega^2 [A_{ab}{}^2 + A_{el}{}^2], \quad (22)$$

and can show that this is in fact equal to the time-averaged input power P given by Eq. (21). (See Prob. 3.6.)

At steady state the energy stored in the oscillator is not perfectly constant, because the instantaneous power input $F(t)\,\dot{x}_s(t)$ given by Eq. (18) does not equal the instantaneous frictional power loss, $M\Gamma\,\dot{x}_s{}^2(t)$ It is only when we average over one cycle that the power input and power loss due to friction are equal. We are interested in the time average of the stored energy. You can easily show that for steady-state oscillation the *time-averaged stored energy E* is given by

$$E = \tfrac{1}{2}M\langle \dot{x}_s{}^2 \rangle + \tfrac{1}{2}M\omega_0{}^2\langle x_s{}^2\rangle$$
$$= \tfrac{1}{2}M(\omega^2 + \omega_0{}^2)\left(\tfrac{1}{2}A_{ab}{}^2 + \tfrac{1}{2}A_{el}{}^2\right). \tag{23}$$

(See Prob. 3.10.) Notice that the term with ω^2 is the time-averaged kinetic energy and that that with $\omega_0{}^2$ is the time-averaged potential energy. These are equal only if we have $\omega = \omega_0$. (Recall that for a weakly damped *free* oscillator the time-averaged kinetic and potential energies are equal.) That fact can be understood qualitatively as follows: If ω is large compared with ω_0, the velocity of M gets reversed before it has a chance to acquire a large displacement and hence before it can acquire a large potential energy stored in the spring. On the other hand, if ω is small compared with ω_0, the velocity never gets very large, and then the time-averaged potential energy dominates.

Notice that for $\omega = \omega_0$ the stored energy E given by Eq. (23) is equal to the product of the steady-state power dissipation [given by Eq. (22)] times the decay time τ for free oscillation. This is qualitatively understandable: If we turned off the driving force, friction would cause the oscillator energy to "decay" exponentially with a mean decay time τ, as shown by Eq. (10). When we drive the oscillator at its natural frequency, which is essentially ω_0 for weak damping, the oscillation amplitude continues to build up until, at steady state, the power input is matched by the power loss due to friction. Since friction dissipates most of the energy in a time τ, the steady-state stored energy is equal to that which has been supplied "recently" by the driving force, i.e., within a time τ. Thus we expect that at equilibrium the stored energy will be approximately equal to the input power times τ, which is equal to the frictional power times τ. We have seen that that is indeed the case, for $\omega = \omega_0$ (For ω not equal to ω_0, the relationship between input power and stored energy is less easily guessed.)

Resonance.   Next we shall consider the variation in response of the oscillator when we slowly vary the driving frequency, always maintaining an essentially constant frequency during any time interval of duration equal to many mean decay times τ, so that we are always essentially at steady state. The time-averaged input power P is given by [Eqs. (21) and (16)]

$$P = P_0 \frac{\Gamma^2 \omega^2}{(\omega_0{}^2 - \omega^2)^2 + \Gamma^2 \omega^2}, \qquad (24)$$

where P_0 is the value of P "at resonance," i.e., at $\omega = \omega_0$. The maximum value of P occurs at resonance. The "half-power points" are defined as those values of ω for which P is half of its maximum value. You may show that the half-power points are given by (Prob. 3.11)

$$\omega^2 = \omega_0{}^2 \pm \Gamma\omega, \qquad (25)$$

which is equivalent to

$$\omega = \sqrt{\omega_0{}^2 + \tfrac{1}{4}\Gamma^2} \pm \tfrac{1}{2}\Gamma. \qquad (26)$$

[Notice that Eq. (25) is two separate quadratic equations in ω. Each of these quadratics has one positive and one negative solution. The two positive solutions give Eq. (26).] The frequency interval between the two half-power points is called the *full-frequency width at half-maximum power*, or simply the *resonance full width*, written $(\Delta\omega)_{\text{fwhm}}$ or simply $\Delta\omega$. According to Eq. (26),

$$(\Delta\omega)_{\text{fwhm}} = \Gamma. \qquad (27)$$

Now, we have already found [Eq. (4)] that the free oscillations have a mean decay lifetime τ given by $\tau = 1/\Gamma$. Thus we have found a very important relation between the resonance full-width for *forced* oscillations and the mean decay time for *free* oscillations:

$$\boxed{(\Delta\omega)_{\text{res}}\,\tau_{\text{free}} = 1.} \qquad (28)$$

In words, *the frequency width of the resonance curve for driven oscillations equals the inverse decay lifetime for free oscillations*. This is a result of great generality. It holds not only for a system of one degree of freedom, but also for systems of many degrees of freedom, as we shall later show. In those cases it turns out that the resonances occur at the frequencies of the normal modes for undamped free oscillation, just as they do for the one-dimensional oscillator. (The resonance frequency ω_0 is only equal to the oscillation frequency ω_1 for free oscillation provided that the damping constant Γ is zero. In the damped free oscillations the frequency is "pulled" from ω_0 to ω_1, because of the damping factor $e^{-(1/2)\Gamma t}$. In the forced oscillations the amplitude is constant and the resonant frequency is what the free oscillation frequency would be if there were no damping.) When there are several degrees of freedom the resonant width and free-decay time for each mode satisfy Eq. (28), provided the resonances are sufficiently well separated in frequency so that they do not "overlap."

Equation (28) has very important consequences for experiment. Often it is easier experimentally to observe the resonant response of a system than it is to observe its free decay. In that case, one can obtain the mean decay time for free oscillation by studying instead the resonant response so as to obtain $\Delta\omega$. The decay time is then obtained from Eq. (28).

Example 1: Decay time of a mailing tube

Here is an illustrative application of Eq. (28) to a system having many degrees of freedom. Take a cardboard mailing tube. Excite it suddenly and then let it decay freely. To do this merely tap the tube on your head. The tap excites mainly the lowest mode, in which the tube length is one half-wavelength. The system oscillates. It radiates sound energy out the ends of the tube and also loses some sound energy by "friction" of the air rubbing against the rough walls of the tube (i.e., sound energy gets converted into "heat"). Thus we have damped oscillations. The question is, what is the mean decay time? Your ear easily recognizes that there is a dominant frequency, the same dominant frequency you hear when you blow steadily across the end of the tube. However the decay time is too rapid to measure with the unaided ear. You have two choices. You can get a microphone, audio amplifier, and oscilloscope; trigger the oscilloscope sweep at the same time that you excite the oscillations; and put the amplifier output on the vertical plates. (This is done most easily with a good oscilloscope that has an "internal trigger," so that the beginning of the output of the amplifier can be used to trigger the sweep.) Photograph the oscilloscope trace and measure τ directly—or alternatively, indirectly as follows: Get an audio signal generator. Let it drive a small loudspeaker at the entrance to the tube. This drives the tube in steady-state oscillations at the signal frequency. Pick up the tube's output radiation at the other end with your microphone, measure the wave amplitude on the oscilloscope. Now vary the driving frequency. [Experimentally, it may be easier to hold the driving frequency constant and vary the length of the tube with a trombone. Plot the square of the amplitude versus inverse tube length (why inverse?).] Find the half-output power points. This gives $\Delta\omega$. Then use Eq. (28) to find τ.

Without this equipment, you can still do fairly well. Use one tuning fork and five or six tubes that are identical except for their lengths. Move the tuning fork quickly along the row of tubes and try to decide on the "full width at half-maximum power output." You might be able to devise a way to recognize a factor of two in intensity at a given pitch. In any case, you can estimate $\Delta\omega$ (and hence the decay time) to within a factor of two. I found typical decay times for mailing tubes are 20 to 50 millisec. (See Home Exp. 3.27.)

Frequency dependence of elastic amplitude.   The term $A_{el} \cos \omega t$ in the steady-state oscillation $x_s(t)$ is the part of $x_s(t)$ which is *in phase* with the driving force $F_0 \cos \omega t$. As discussed earlier, this "elastic" term makes no contribution to the time-averaged energy absorption. Furthermore, at resonance (i.e., at $\omega = \omega_0$), A_{el} is zero. Does that mean that A_{el} is unimportant? No. Indeed, at driving frequencies that are *far from resonance, the elastic term is the dominant one*. We see that as follows: The elastic amplitude is given by [Eq. (17)]

$$A_{el} = \frac{F_0}{M} \frac{(\omega_0^2 - \omega^2)}{(\omega_0^2 - \omega^2)^2 + \Gamma^2 \omega^2}. \tag{29}$$

The ratio of elastic to absorptive amplitudes is given by [See Eqs. (16) and (17)]

$$\frac{A_{el}}{A_{ab}} = \frac{\omega_0^2 - \omega^2}{\Gamma \omega}. \tag{30}$$

For ω less than ω_0, A_{el}/A_{ab} is positive and can be made as large as you please by choosing ω sufficiently small. For ω greater than ω_0, A_{el}/A_{ab} is negative and can be made as large in magnitude as you please, by choosing ω sufficiently large. For either of these cases we have $\Gamma \omega \ll |\omega_0^2 - \omega^2|$, and may neglect the contribution $A_{ab} \sin \omega t$ to $x_s(t)$ provided we are willing to neglect the very small time-averaged power. (When we are far from resonance, the power absorption is very small compared with that at resonance.) Thus, far from resonance, the steady-state solution is given by $A_{el} \cos \omega t$, with A_{el} given by neglecting the term $\Gamma^2 \omega^2$ in the denominator of Eq. (29):

$$x_s(t) \approx A_{el} \cos \omega t \approx \frac{F_0 \cos \omega t}{M(\omega_0^2 - \omega^2)}. \tag{31}$$

Notice that the damping constant Γ has completely disappeared from the result, Eq. (31). In fact it is easy to see that Eq. (31) gives the *exact* steady-state solution to the equation of motion, Eq. (14), if we set $\Gamma = 0$ in that equation (Prob. 3.13).

In Fig. 3.1 we plot the absorptive and elastic amplitudes in the vicinity of the resonance.

Other "resonance curves."   The behavior of the forced harmonic oscilla–tor is described by several different quantities, all of which exhibit similar (but not identical) "resonance shapes" when plotted against frequency. These quantities are the absorptive amplitude A_{ab}, the squared magnitude of the amplitude, $|A|^2 \equiv A_{el}^2 + A_{ab}^2$, the input power P (which also equals the dissipated power), and the stored energy E. In this paragraph we will write them all down for the sake of comparison. From Eqs. (16), (17), (22), and (23), we have

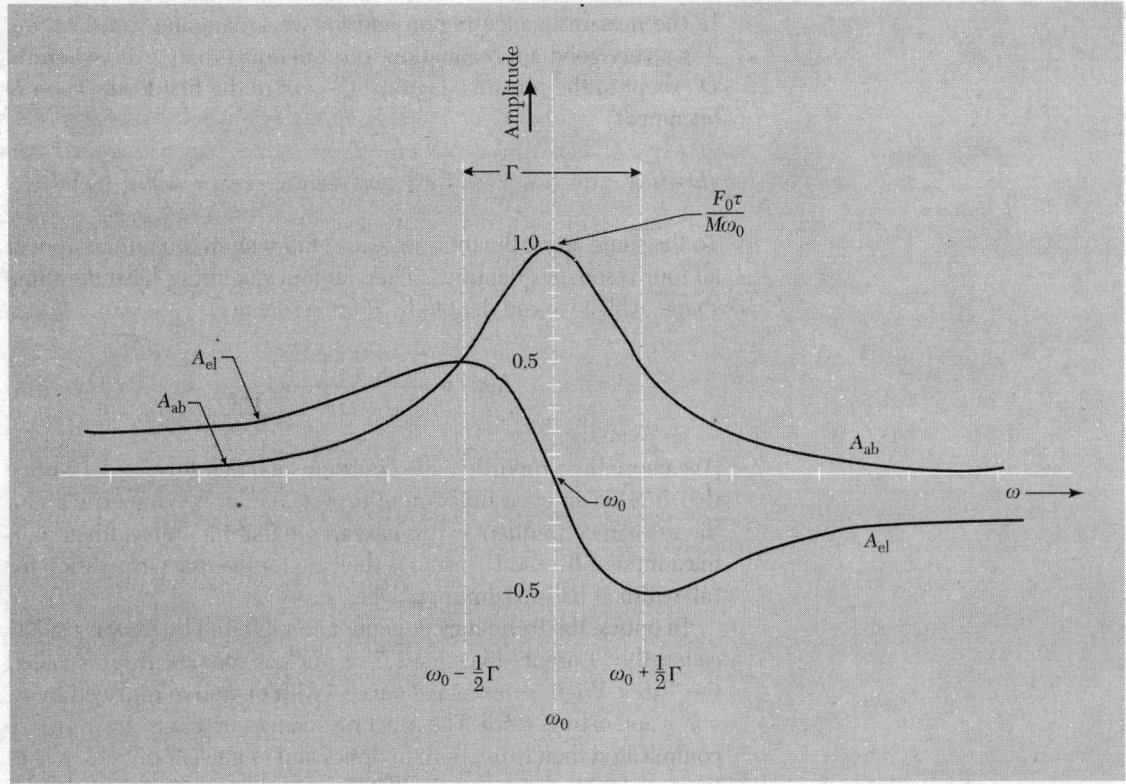

Fig. 3.1 *Resonance in driven oscilla-tor. When the oscillator is subject to the external force $F_0 \cos \omega t$, the steady-state oscillation is $x_s(t) = A_{ab} \sin \omega t + A_{el} \cos \omega t$.*

$$A_{ab}(\omega) = A_{ab}(\omega_0) \frac{\Gamma^2 \omega_0 \omega}{(\omega_0{}^2 - \omega^2)^2 + \Gamma^2 \omega^2}, \tag{32}$$

$$|A(\omega)|^2 = |A(\omega_0)|^2 \frac{\Gamma^2 \omega_0{}^2}{(\omega_0{}^2 - \omega^2)^2 + \Gamma^2 \omega^2}, \tag{33}$$

$$P(\omega) = P(\omega_0) \frac{\Gamma^2 \omega^2}{(\omega_0{}^2 - \omega^2)^2 + \Gamma^2 \omega^2}, \tag{34}$$

$$E(\omega) = E(\omega_0) \frac{\frac{1}{2}\Gamma^2(\omega^2 + \omega_0{}^2)}{(\omega_0{}^2 - \omega^2)^2 + \Gamma^2 \omega^2}. \tag{35}$$

All of these quantities have in common the "resonance denomina-tor" D, given by

$$D \equiv (\omega_0{}^2 - \omega^2)^2 + \Gamma^2 \omega^2 = (\omega_0 - \omega)^2 (\omega_0 + \omega)^2 + \Gamma^2 \omega^2.$$

Near resonance (i.e., near $\omega = \omega_0$), the rapid variation of D with ω is almost entirely due to the factor $(\omega_0 - \omega)^2$ in the first term. The occurrence of ω elsewhere in D and in the numerators of the four quantities above contributes only much slower variation. Now, the absorptive amplitude and the other quantities given above are rela-tively important only "near" resonance. (We can very loosely define "near" as meaning, say, within the range $\omega_0 - 10\Gamma < \omega < \omega_0 + 10\Gamma$.)

In the near-resonance region and for weak damping, i.e., $\Gamma \ll \omega_0$, it is a very good approximation to set ω equal to ω_0 everywhere in D except in the sensitive factor $(\omega_0 - \omega)^2$ of the first term. Then D becomes

$$D \approx (\omega_0 - \omega)^2 (\omega_0 + \omega_0)^2 + \Gamma^2 \omega_0^2 = 4\omega_0^2 \left[(\omega_0 - \omega)^2 + \left(\tfrac{1}{2} \Gamma \right)^2 \right].$$

To the same approximation we can set $\omega = \omega_0$ in the numerators of all four resonant quantities. Then all four quantities have the same shape, which we can denote by R for resonance:

$$R(\omega) \equiv \frac{\left(\tfrac{1}{2} \Gamma \right)^2}{(\omega_0 - \omega)^2 + (\tfrac{1}{2} \Gamma)^2}. \tag{36}$$

[We chose the proportionality constant such that $R(\omega_0) = 1$.] Notice that $R(\omega)$ is an even function of $\omega_0 - \omega$, i.e., it is symmetric about the resonance frequency. It is easy to see that the full width at half-maximum of $R(\omega)$ is Γ, just as is the case for the exact expression for full width at half-maximum power.

In optics, the frequency dependence exhibited by $R(\omega)$ is usually called the "Lorentz line shape." In nuclear physics, $R(\omega)$ is called the "Breit-Wigner resonance curve," with ω_0 and ω replaced by $E_0 = \hbar\omega_0$ and ω by $E = \hbar\omega$. The exact resonance curves are always more complicated than $R(\omega)$, both in optics and in nuclear physics, just as they are for the harmonic oscillator.

Transient forced oscillations. We wish to find the solution of Eq. (14), the differential equation of a harmonically driven damped harmonic oscillator, for given arbitrary initial conditions $x(0)$ and $\dot{x}(0)$. For that we need the general solution. The general solution is given by a superposition of the steady-state solution $x_s(t)$ and the general solution $x_1(t)$ of the homogeneous equation of motion (the equation for free oscillations):

$$x(t) = x_s(t) + x_1(t)$$

$$= A_{ab} \sin \omega t + A_{el} \cos \omega t + e^{-(1/2)\Gamma t}[A_1 \sin \omega_1 t + B_1 \cos \omega_1 t], \tag{37}$$

where A_1 and B_1 are arbitrary constants that can be chosen so as to satisfy the initial conditions of displacement and velocity. We can see that Eq. (37) is the general solution as follows: First, it satisfies the given second-order differential equation. Second, it can be made to match arbitrary initial conditions $x(0)$ and $\dot{x}(0)$ by suitable choices for A_1 and B_1. That is all that is required for it to be a unique solution, according to the theory of differential equations.

Initially undisturbed oscillators. Let us specialize our general solution to the interesting situation where at $t = 0$ the *oscillator is practically at rest at its equilibrium position*. We choose $B_1 = -A_{el}$, since that will give the initial condition that $x(0) = 0$. Now let us choose A_1 as simply as we can, yet provide that the initial velocity $\dot{x}(0)$ is essentially zero. We are only interested in weak damping; therefore we take $e^{-(1/2)\Gamma t}$ to be essentially constant during any given cycle. With that approximation you can show $\dot{x}(0) \approx \omega A_{ab} + \omega_1 A_1$. Since we are interested in driving frequencies not too far from resonance we simply take $A_1 = -A_{ab}$. Then

$$\dot{x}(0) \approx (\omega - \omega_1)A_{ab}, \tag{38}$$

which is zero for $\omega = \omega_1$ or for $A_{ab} = 0$ (which implies $\Gamma = 0$). With these choices we have $x(0) = 0$ and $\dot{x}(0) \approx 0$. Then Eq. (37) becomes

$$x(t) = A_{ab}[\sin \omega t - e^{-(1/2)\Gamma t} \sin \omega_1 t] + A_{el}[\cos \omega t - e^{-(1/2)\Gamma t} \cos \omega_1 t]. \tag{39}$$

Some interesting special cases follow.

Case 1: Driving frequency equals natural oscillation frequency

Setting $\omega = \omega_1$ in Eq. (39) we obtain

$$x(t) = [1 - e^{-(1/2)\Gamma t}][A_{ab} \sin \omega t + A_{el} \cos \omega t] = [1 - e^{-(1/2)\Gamma t}]x_s(t), \tag{40}$$

where $x_s(t)$ is the steady-state solution. Thus when the driving frequency is exactly equal to the natural oscillation frequency ω_1 the steady-state solution is "present from the beginning." Its oscillation amplitude builds up smoothly from zero to its final steady-state value.

Case 2: Zero damping and interminable beats

Setting $\Gamma = 0$ gives $A_{ab} = 0$ and

$$A_{el} = \frac{F_0/M}{\omega_0^2 - \omega^2}.$$

Then Eq. (39) gives

$$x(t) = \frac{F_0}{M} \frac{[\cos \omega t - \cos \omega_0 t]}{\omega_0^2 - \omega^2}. \tag{41}$$

Equation (41) is similar to the superposition of two harmonic oscillations, encountered when we studied the phenomenon of beats between two tuning forks in Sec. 1.5. We recall that we can write $x(t)$ either as a linear superposition of two exactly harmonic oscillations, as in Eq. (41), or, alternatively, as an "almost harmonic" oscillation at the "fast" average frequency $\omega_{av} = \frac{1}{2}(\omega_0 + \omega)$ with a "slowly varying" amplitude that oscillates harmonically at the "slow" modulation frequency $\omega_{\text{mod}} = \frac{1}{2}(\omega_0 - \omega)$. The latter procedure gives (Prob. 3.22)

$$x(t) = A_{\text{mod}}(t) \sin \omega_{\text{av}}t, \tag{42}$$

where

$$A_{\text{mod}}(t) = \frac{F_0}{M} \frac{2 \sin \frac{1}{2}(\omega_0 - \omega)t}{(\omega_0{}^2 - \omega^2)}. \tag{43}$$

Thus the oscillation amplitude oscillates forever at the modulation frequency $\frac{1}{2}(\omega_0 - \omega)$. The stored energy oscillates about its average value, going from zero to its maximum value E_0 according to the relation

$$\begin{aligned} E(t) &= E_0 \sin^2 \tfrac{1}{2}(\omega_0 - \omega)t \\ &= \tfrac{1}{2}E_0[1 - \cos(\omega_0 - \omega)t]. \end{aligned} \tag{44}$$

Thus the energy oscillates forever at the beat frequency between the driving frequency and the natural oscillation frequency.

In order to observe almost-interminable beats you may hang a can of soup or something on a string of length about 45 cm. Couple this pendulum by means of several feet of rubber bands to a record player turning at 45 rpm.

For the special case $\omega = \omega_0$ Eq. (43) implies that the amplitude of the fast oscillations increases linearly with time, forever, corresponding to an infinite beat period:

$$x(t) = \left[\frac{1}{2} \frac{F_0 t}{M\omega_0} \right] \sin \omega_0 t. \tag{45}$$

After an infinite time the amplitude is infinite.

Case 3: Transient beats

For weak damping and for ω near ω_1, it is not difficult (but it is tedious) to show that the stored energy is given approximately by (Prob. 3.24)

$$E(t) = E[1 + e^{-\Gamma t} - 2e^{-(1/2)\Gamma t} \cos (\omega - \omega_1)t], \tag{46}$$

where E is the steady-state energy. (If we take $\omega = \omega_1$, we obtain Case 1 above. If we take $\Gamma = 0$, we obtain Case 2 above.) Thus we see that if we start at $t = 0$ with no energy in the oscillator, the energy $E(t)$ does not build up smoothly to its steady-state value unless the driving frequency ω is exactly equal to the free-oscillation frequency ω_1. Instead, the energy undergoes oscillations at the beat frequency $\omega - \omega_1$. These beats are caused by the fact that the oscillator "likes" to oscillate at its natural frequency ω_1, whereas it is being pushed at frequency ω. Therefore the driving force is sometimes pushing with a relative phase which helps build up the oscillation amplitude, but it is sometimes (one-half of a beat period later) pushing with the opposite phase, thus diminishing the oscillations. If there were no damping, these beats would go on and on forever as in Case 2. However, because of the damping, the oscillator gradually adjusts its phase with respect to that of the driving force. After a sufficiently long time, the oscillator settles into a steady state of vibration with no beats, oscillating exactly at the driving frequency ω, with the relative phase between the oscillator and the driving force stabilized so that the amount of energy delivered to the oscillator in each push (each cycle) of the driving force is exactly equal to the energy lost by the oscillator in one cycle due to the frictional drag. Then the oscillator energy remains constant, and the relative phase of the oscillator and driving force remains constant. The transient buildup of energy is shown in Fig. 3.2.

Fig. 3.2   Transient beats. (We chose the beat period equal to the decay time τ.) The stored energy $E(t)$ rises from zero, undergoes damped oscillations at the beat frequency between the driving force and the natural oscillations, and eventually settles at the steady-state value E.

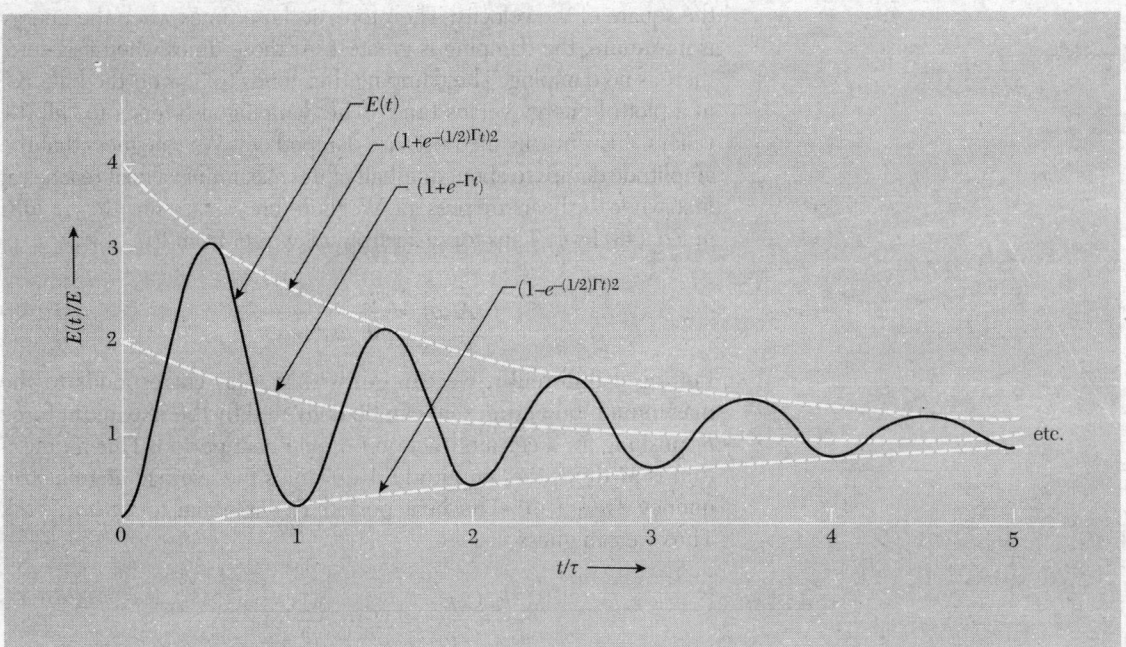

Qualitative derivation of resonance shape.   Let us use the insight
gained from studying the transient response of the oscillator in order
to guess the ratio of the steady-state oscillation amplitude exactly at
resonance to that at other frequencies. Let us start with the oscillator
at rest and drive it exactly at its resonance frequency. If there were no
damping, the oscillation amplitude would build up linearly with time
forever [see Eq. (45)]. Actually it *starts* to build up linearly with time
because at first, when the average speed is small, the damping is negligible. However, because of the damping, it eventually levels off to a
value equal to that amplitude which it can "acquire" in a time of order τ.
Due to the damping it can only retain the motion it has acquired
"recently," i.e., within a time of order τ. We can guess this amplitude
by thinking of the maximum force F_0 as pushing for a time τ so as to
impart a maximum momentum $F_0\tau$ to the mass. But the maximum
momentum of the oscillating mass is M times the maximum velocity,
which is $\omega_0 A(\omega_0)$. Thus $F_0\tau \approx M\omega_0 A(\omega_0)$ and we have

$$A(\omega_0) \approx \frac{F_0\tau}{M\omega_0}, \tag{47}$$

as our guess for the steady-state amplitude at $\omega = \omega_0$.

Now let us drive the oscillator at a frequency ω that differs greatly
from the resonance frequency ω_0. If there were no damping, the amplitude would oscillate forever at the modulation frequency $\frac{1}{2}(\omega_0 - \omega)$,
and the oscillator energy would oscillate at the beat frequency, $\omega_0 - \omega$.
We now "turn on" the damping. The energy loss to damping goes as
the square of the velocity. Therefore, at those times when the energy
is maximum, the damping is greatest. At those times when it is zero,
there is no damping. The damping thus tends to "cut off the hilltops"
in a plot of energy versus time. (The damping also tends to "fill the
valleys.") Eventually the beats are damped out. We can guess that the
amplitude damps to about one-half of the maximum value it reaches at
first, when the beats are present. We therefore replace $\sin\frac{1}{2}(\omega_0 - \omega)t$
in Eq. (43) by $\frac{1}{2}$. Thus for ω far from ω_0 we get from Eq. (43)

$$A(\omega) \approx \frac{F_0}{M}\frac{1}{\omega_0{}^2 - \omega^2}. \tag{48}$$

Putting it differently, we can guess that $A(\omega)$ corresponds to the
maximum momentum that can be delivered by the maximum force
F_0 pushing for a certain fraction f of one beat period. This momentum is M times the amplitude $A(\omega)$ times the average angular fre-
quency $\frac{1}{2}(\omega_0 + \omega)$. The beat period T_{beat} is equal to $2\pi/(\omega_0 - \omega)$.
Thus we can guess

$$\frac{F_0 f 2\pi}{(\omega_0 - \omega)} \approx MA(\omega)\frac{1}{2}(\omega_0 - \omega).$$

This gives Eq. (48) if we are smart enough to guess $f = 1/4\pi$. (I wasn't.)

We know from our exact solution that exactly at resonance the oscillation amplitude is $A_{ab}(\omega_0)$, since A_{el} is zero at resonance. Indeed, our guessed amplitude $A(\omega_0)$ is equal to $A_{ab}(\omega_0)$ as you can verify by comparing Eqs. (47) and (16). We know that far from resonance the exact solution gives that the oscillation amplitude is essentially $A_{el}(\omega)$. Our guessed amplitude $A(\omega)$ for ω far from ω_0 is indeed equal to $A_{el}(\omega)$ far from resonance, as you can verify by comparing Eqs. (48) and (17).

3.3 *Resonances in System with Two Degrees of Freedom*

In Chap. 1 we found that each mode of a freely oscillating system with more than one degree of freedom behaves very much like a simple harmonic oscillator. The main difference is that the system occupies a finite region of space, and thus the "harmonic oscillator" is spread throughout the region occupied by the system, rather than being confined to a point mass. Thus each mode has a characteristic "shape," a concept not needed for a one-dimensional oscillator.

In Chap. 1 we neglected damping in studying the modes of freely oscillating systems. When damping is taken into account, one finds (as we shall see) that each mode is similar to a damped one-dimensional oscillator. Thus each mode has its own characteristic damping mechanism and damping constant Γ, and each therefore has its own characteristic decay time τ. For some systems, the damping mechanisms may be associated with the individual "moving parts," and then all modes may have roughly the same damping constants and decay times. An example is that of a system of two identical pendulums coupled by a spring in which the damping is provided by something rubbing either on each of the two supporting strings or on each bob. Since both bobs move equally in each mode, the two modes have the same decay times. For other systems, the damping mechanisms are associated with the modes. For example, the spring that couples two pendulums may have some stretchy tape stuck on it, so that it experiences a frictional damping when it is extended or compressed. If that is essentially the only damping mechanism, then mode 2 (the mode in which the spring is stretched and compressed) has a much larger damping constant than mode 1, i.e., $\Gamma_2 \gg \Gamma_1$, and thus mode 2 has a much shorter decay time than mode 1: $\tau_2 \ll \tau_1$.

When one drives a system having several modes, one finds a resonance whenever the driving frequency is near a mode frequency. It turns out that *the absorptive and elastic amplitudes for a given moving part are simply superpositions of the ampli-tude contributions arising from each resonance* (each mode of the undriven system). Each of these contributions has a form similar to that which we found in Sec. 3.2 for a system with a single degree of freedom.

If we (slowly) vary the driving frequency and plot the rate of energy absorption by a given moving part as a function of the driving frequency ω, we find a resonance each time ω passes through the neighborhood of a mode frequency. (We shall use "resonance frequency" and "mode frequency" as interchangeable expressions, even though the one refers to forced oscillations and the other to free oscillations.) Each resonance exhibits a frequency full width given by [see Eq. (28)]

$$\Delta\omega = \Gamma = \frac{1}{\tau},$$

where $\Delta\omega$ is the full width at half-maximum power absorption, and where Γ rand τ are the damping constant and decay time for free oscillations of that particular mode. This relation holds if the damping is sufficiently weak and if the individual resonances are separated by frequency intervals large compared to their half-widths. In that case we have at most one mode contributing to the absorptive amplitude at any given point on the frequency plot. On the other hand, it turns out that we cannot ordinarily neglect *any* of the elastic contributions. (See Prob. 3.20.)

Example 2: Forced oscillations of two coupled pendulums

The system is shown in Fig. 3.3, and is also described in Home Exp. 3.8 (where the pendulum bobs are cans of soup, the spring is a slinky, the driving force is provided by a phonograph turntable coupled to the system by ten feet or so of rubber bands, and damping is provided by having the strings rub on something). For Simplicity, we assume. that each pendulum has the same damping constant, Γ. The equations of motion are easily seen to be

$$M\ddot{\psi}_a = -\frac{Mg}{l}\,\psi_a - K(\psi_a - \psi_b) - M\Gamma\dot{\psi}_a + F_0 \cos\omega t, \qquad (49)$$

$$M\ddot{\psi}_b = -\frac{Mg}{l}\,\psi_b + K(\psi_a - \psi_b) - M\Gamma\dot{\psi}_b. \qquad (50)$$

We have already studied the free oscillations of this system in the absence of damping. Thus we know that if F_0 and Γ are both zero, the modes are given by

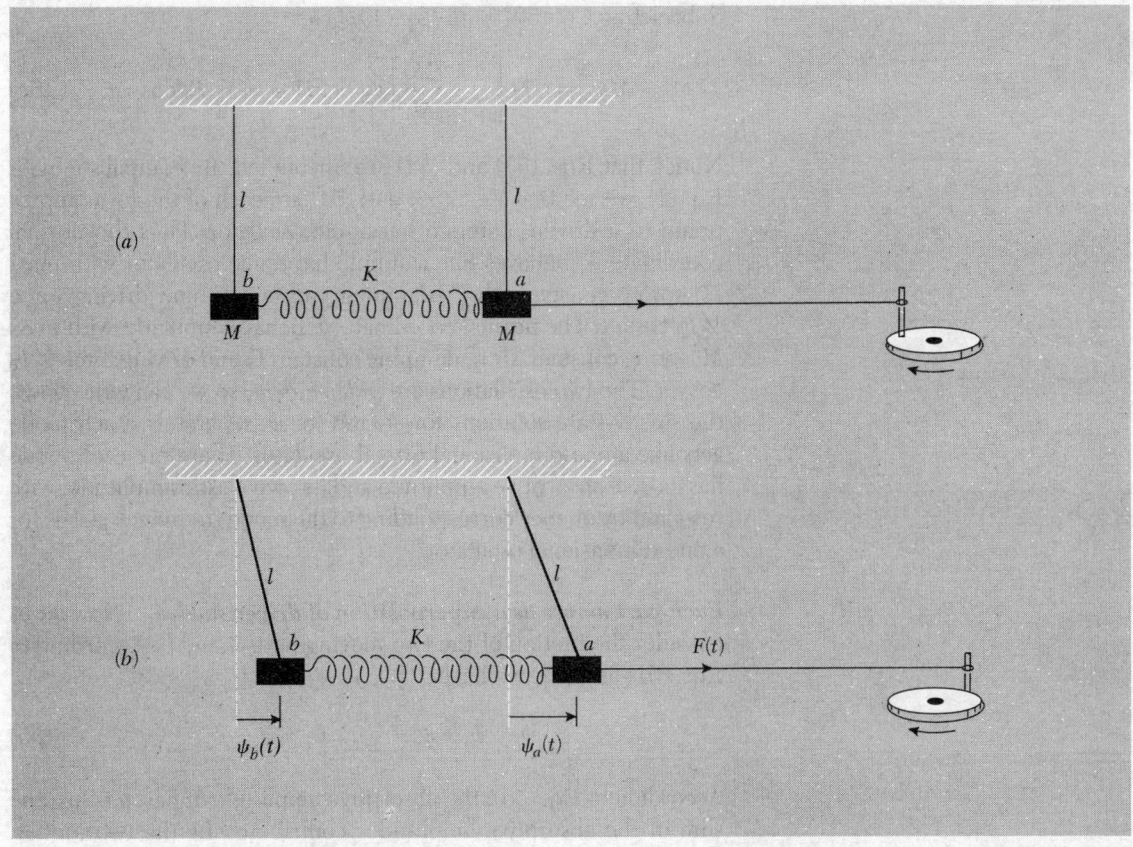

Fig. 3.3 Forced oscillations of coupled pendulums. (a) Equilibrium. (b) General configuration.

Mode 1: $\psi_a = \psi_b,$ $\omega_1{}^2 = \dfrac{g}{l},$ $\psi_1 = \dfrac{1}{2}\big(\psi_a + \psi_b\big),$ (51)

Mode 2: $\psi_a = -\psi_b,$ $\omega_2{}^2 = \dfrac{g}{l} + \dfrac{2K}{M},$ $\psi_2 = \dfrac{1}{2}\big(\psi_a - \psi_b\big),$ (52)

where the superpositions ψ_1 and ψ_2 are the normal coordinates.

Each mode acts like driven oscillator. Let us transform to the normal coordinates ψ_1 and ψ_2. If we add Eqs. (49) and (50), we get

$$M\ddot{\psi}_1 = -\frac{Mg}{l}\psi_1 - M\Gamma\dot{\psi}_1 + \frac{1}{2}F_0 \cos \omega t. \qquad (53)$$

Subtracting Eq. (50) from Eq. (49) gives

$$M\ddot{\psi}_2 = -M\left[\frac{g}{l} + \frac{2K}{M}\right]\psi_2 - M\Gamma\dot{\psi}_2 + \frac{1}{2}F_0\cos\omega t. \qquad (54)$$

Notice that Eqs. (53) and (54) are uncoupled. By compalison with Eq. (1), we see that Eqs. (53) and (54) are each of the form appropriate to a driven, damped harmonic oscillator. Thus the normal coordinate ψ_1 behaves like a simple harmonic oscillator with mass M, spring constant $M\omega_1^2$, damping constant Γ, and driving force $\frac{1}{2}F_0\cos\omega t$. The normal coordinate ψ_2 behaves similarly, with mass M, spring constant $M\omega_2^2$, damping constant Γ, and driving force $\frac{1}{2}F_0$ $\cos\omega t$. The two oscillations are *independent,* so we can write down the steady-state solutions for ψ_1 and for ψ_2 separately. Each mode acts like a one-dimensional forced oscillator. Therefore each mode has its own absorptive amplitude and its own elastic amplitude, with resonant frequency corresponding to the mode frequency just as for a one-dimensional oscillator.

Each part moves as a superposition of driven modes. Now let us consider the motion of the two moving parts a and b. According to Eqs. (51) and (52) we have

$$\psi_a = \psi_1 + \psi_2, \qquad \psi_b = \psi_1 - \psi_2. \qquad (55)$$

According to Eq. (55), the absorptive amplitude of part a is just the sum of the absorptive amplitudes contributed by the two modes. The absorptive amplitude of part b is the difference of the absorptive amplitudes of the two modes. Similarly the elastic amplitude of part a is the sum of the elastic amplitudes of the two modes, and that of part b is their difference.

When the driving frequency is equal to one of the mode frequencies then the motion of a and b is essentially what it would be in that mode (for free oscillations).

In Fig. 3.4 we plot the absorptive and elastic amplitudes for ψ_a and ψ_b.

In this example we found the general result that the steady-state amplitude for each moving part can be written as a superposition of contributions from each resonance, i.e., from each mode of the freely oscillating system. Each contribution to the superposition has the form appropriate to the driven oscillator corresponding to that mode. The contribution from each mode depends on how the driving force is coupled to the system. For the configuration of Fig. 3.3, we found that each moving part receives equal contributions (except for sign) from each of the two modes. However, if we had tied the rubber band to the center of the spring instead of to one of the bobs, we would have found

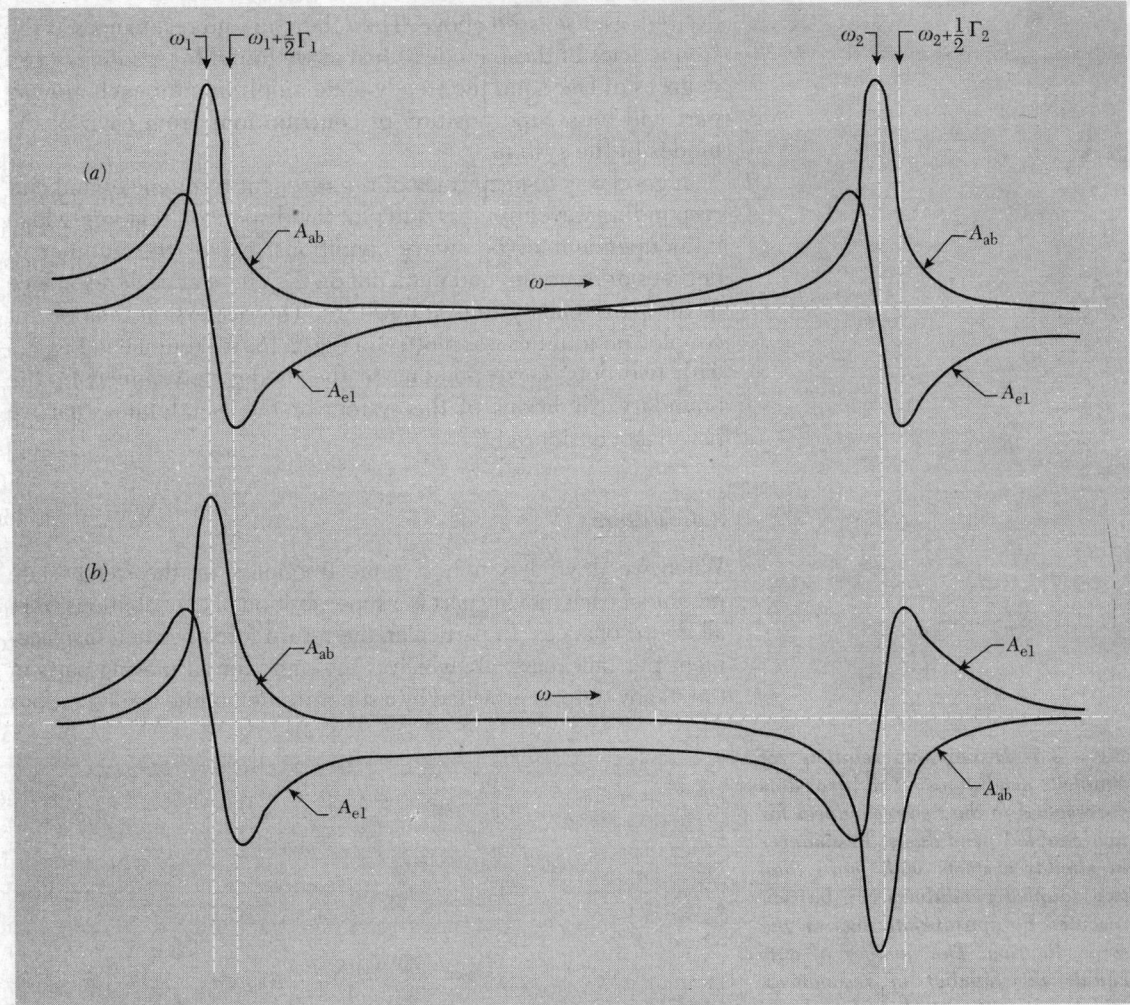

Fig. 3.4 *Resonances in a system with two degrees of freedom. Plots of absorptive and elastic amplitudes versus frequency for (a) the pendulum coupled directly to the driving force and (b) the pendulum farther from the driving Force. The angular frequency separation $\omega_2 - \omega_1$ is taken to be thirty times $\frac{1}{2}\Gamma$, the resonance half-width of either mode.*

very different proportions of the two modes. Thus the relative contribution from each mode depends on the details of how the driving force is applied.

Forced oscillations of system of many coupled pendulums. Suppose that instead of just two pendulums we have many, coupled in a linear array. If we apply a harmonic driving force to the system and vary the driving frequency (but vary it slowly enough so that we are always at "steady state"), we will obtain a resonance whenever the driving frequency passes through one of the mode frequencies. (Of course, the driving force may be coupled in such a way that some modes are not excited, as discussed above. Then there are no resonances at the frequencies of those modes.) Just as we found for systems of two degrees of freedom, the steady-state amplitude for each moving part will be a superposition of contributions from each of the modes of the system.

A good way to keep track of the resonant frequencies and corresponding wavenumbers is to plot the dispersion relation (which is independent of boundary conditions and of the number of degrees of freedom) and put a dot on the curve at each resonance of the particular system in question. The dispersion relation for coupled pendulums was plotted in Fig. 2.18. We replot it in Fig. 3.5 with two dots, corresponding to the modes determined by the boundary conditions of the system of two pendulums that we have just considered.

3.4 Filters

When we drive a system at some frequency ω, the steady-state motion of each moving part is a superposition of contributions from all the resonances. In particular, the return force per unit displacement per unit mass, ω^2, which is the same for all moving parts in the steady state, is provided by a superposition of the configurations

Fig. 3.5 Dispersion relation of coupled pendulums. The two dots correspond to the two resonances for two coupled pendulums. Resonances in similar systems with more than two coupled pendulums can be represented by appropriate dots on the same diagram. The number of dots equals the number of resonances, which equals the number of modes of free oscillation.

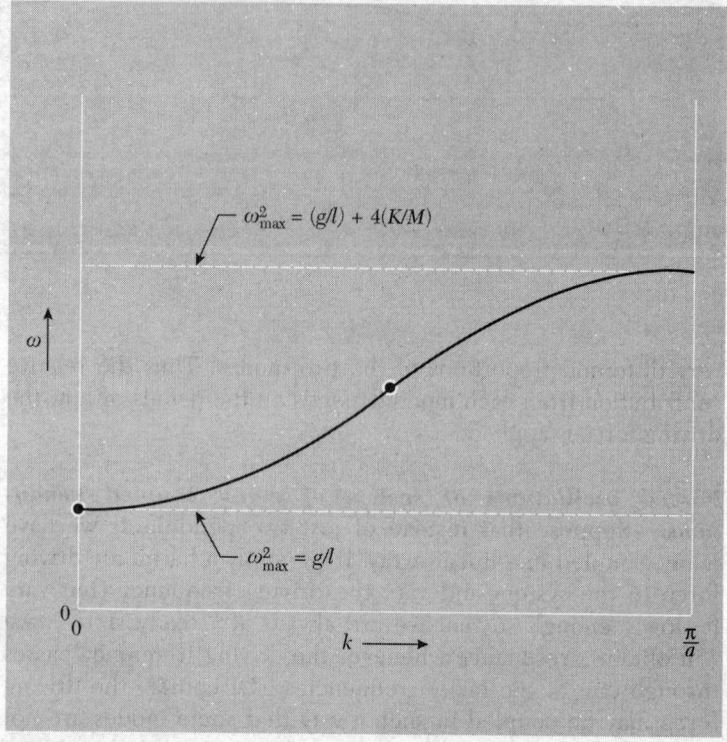

corresponding to the various modes. Consider qualitatively what happens as we vary ω^2. Suppose first that ω lies somewhere between the minimum and maximum mode frequencies, but that it does not lie near any particular resonance. Then the amplitude of a given moving part is essentially just the elastic contributions from all the modes. Different modes contribute with different signs, depending on what moving part we are considering. [See Eqs. (55), and compare the contributions of mode 2 to ψ_a and to ψ_b.] As we increase ω^2, we may approach a resonance frequency. When we pass from below the resonance to above it, the sign of the elastic contribution from that mode reverses. As we continue to increase the frequency, various moving parts increase or decrease their amplitudes in some more or less complicated way, as we pass one mode frequency after another. Finally, we pass the highest mode frequency. After that, there are no more sign changes in the contributions; i.e., the sign of each contribution to the elastic amplitude of a given moving part keeps the same sign as we increase the frequency beyond the highest resonance. Therefore the moving parts maintain more or less the shape of the highest mode (but not exactly, of course). Something very interesting happens. If the system is a Slinear array and we drive it from one end at a frequency above the highest mode frequency, the moving part nearest the driven end has the largest amplitude, its neighbor has a smaller amplitude, the third moving part has amplitude smaller still, etc. The amplitude is *attenuated* with increasing distance from the *input* end of the system. The system is then called a *filter*.

Example 3: Two coupled pendulums as a mechanical filter

Consider, for example, the two coupled pendulums of Fig. 3.3. Suppose we drive the input end (pendulum a) at a frequency higher than the mode frequency ω_2. Now, pendulum a is coupled directly to the driving force. Therefore, in the steady state the return force on pendulum a is provided in part by the driving force. That is not the case for pendulum b. Its return force is provided only by the spring and gravity, just as is the case for free oscillations. Now, in a mode of free oscillation the largest return force per unit displacement that the spring (and gravity) can provide is that corresponding to the highest mode configuration, where the bobs move oppositely to one another. But that is now insufficient to equal ω^2 if the configuration is precisely that of the highest mode, where the magnitude $|B|$ of the oscillation of pendulum b is the same as the magnitude $|A|$ of the oscillation of pendulum a. The only way that pendulum b can have the same return force per unit mass per unit displacement as pendulum a is for b to have smaller displacement: $|B| < |A|$. The higher ω goes above ω_2, the smaller must be the corresponding displacement of pendulum b relative to that of a. To put it differently, b cannot keep up with a unless it travels less.

If we have three or more coupled pendulums in a linear array instead of two, and if we drive them at one end at a frequency exceeding the highest mode frequency, an analogous situation occurs. The configuration in the steady state resembles that of the highest mode; i.e., each bob moves with a phase opposite to that of its neighbor on either side. This provides the greatest possible return force per unit mass per unit displacement for each bob. But that is still insufficient to equal ω^2 unless each successive bob has a smaller displacement than that of the next bob nearer to the input end (the driven end). Thus the amplitude of motion of successive bobs decreases as one progresses farther from the driven end of the system.

High-frequency cutoff.   Thus we have an example of a *mechanical filter* If you push on the input end with a force $F_0 \cos \omega t$, the amplitude of motion at the output end (the end farthest from the driving force) is much smaller than that at the input end, provided that ω is much greater than the highest mode frequency of the system. The configuration of the system is like that of the highest mode, except for the progressive decrease of amplitude as one travels farther from the driven end. *The highest mode frequency (for free oscillation) is called a cutoff frequency (for forced oscillation).* A driving force at one end with frequency above the cutoff frequency gives motion that is not "passed" through the filter—it is "cut off." We say that the system is being "driven above cutoff." In Fig. 3.6 we show a system

Fig. 3.6 Mechanical filter. The driving frequency exceeds the highest mode frequency. The configuration is such that the relative phases of the bobs are the same as those in the highest mode. The "output" amplitude (that of bob c) is less than the "input" amplitude (that of bob a).

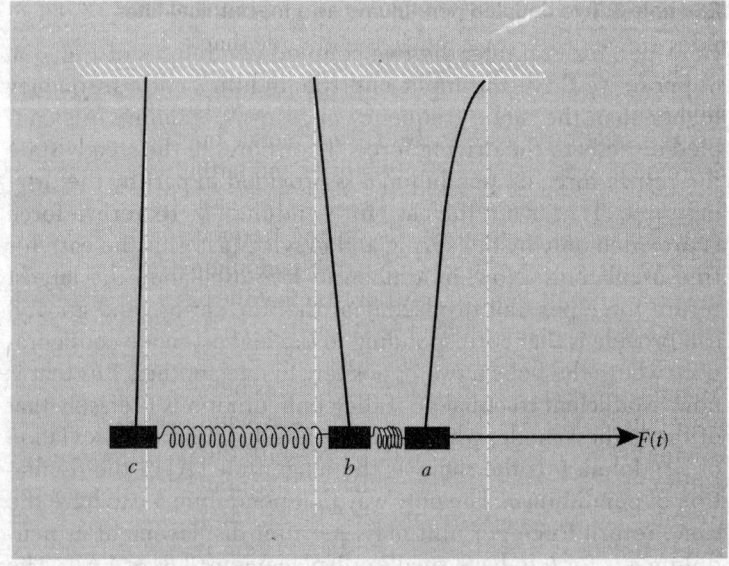

of three coupled pendulums driven above cutoff. (This situation can easily be set up with one slinky and three cans of soup. See Home Exp. 3.16.)

Low-frequency cutoff.   Now let us consider what happens if we drive the system at a frequency less than the lowest natural frequency (i.e., the lowest mode frequency for free oscillations). We shall show that if the driving frequency is much less than the lowest natural frequency, then the output amplitude (i.e., the amplitude of the bob farthest from the driving force) is much less than the input amplitude (the amplitude of the driven bob). Thus *the lowest mode frequency is also a cutoff frequency.*

Consider our system of two coupled pendulums (Fig. 3.3). The lowest mode corresponds to a configuration in which all the pendulums oscillate in phase with one another and with the same amplitude. The springs are not stretched or compressed; the return force is due to gravity alone. Thus the frequency is $\omega_1 = \sqrt{g/l}$. Now suppose we drive the system at a frequency ω that is less than ω_1. Then in the steady state the return force per unit displacement per unit mass, ω^2, must be less than g/l for every bob. The bob at the input end has part of its return force provided by the direct coupling to the driving force. The second bob has to get along with only the force provided by gravity and by the spring. The only way its return force per unit displacement per unit mass can be reduced below g/l is for the spring to contribute a force with opposite sign to that provided by gravity. It is easy to see that the displacement of bob b must therefore be less than that of bob a, but in the same direction. (Then the spring is stretched.) Thus the relative phases are the same as those of the lowest mode, but the relative amplitudes are not. (Bob b has smaller amplitude than bob a.)

The same result holds if we have three or more coupled pendulums driven below the lowest mode frequency. The relative phases are the same as those of the lowest mode, but the amplitudes decrease progressively as we get farther and farther from the input end. This situation is shown in Fig. 3.7. The easiest way to understand Fig. 3.7 is to think of the driving frequency as being zero. Then you have just a steady force, the pendulums are motionless, and your intuition immediately tells you that the pendulum configuration does indeed resemble Fig. 3.7.

Nomenclature.   The band of frequencies between the low- and high-cutoff frequencies is called the *pass band* of the filter. For driving frequencies lying in the pass band, the amplitude at the output end is comparable with that at the input end. For driving frequencies lying outside the pass band, the output amplitude is less than the input amplitude. The system is therefore called

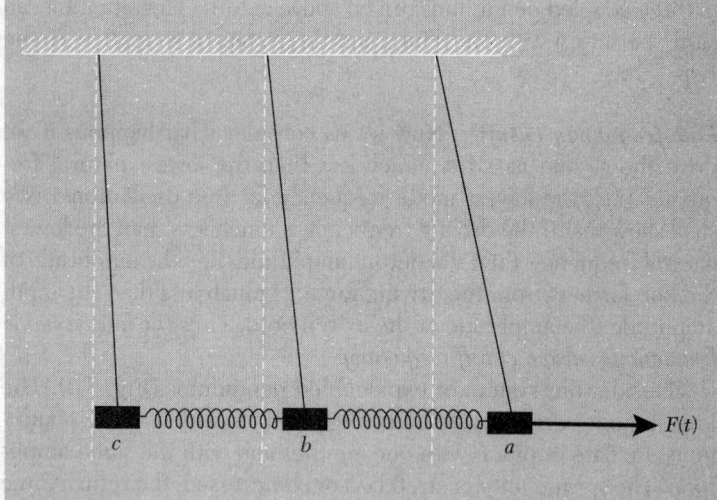

Fig. 3.7 Mechanical filter. The driving frequency is less than the lowest mode frequency. The configuration is such that the relative phases are the same as those of the lowest mode. The output amplitude (that of bob c) is less than the input amplitude (that of bob a).

a *bandpass filter.* If the low-frequency cutoff frequency is zero (i.e., if the lowest mode has frequency zero), the system is called a *low-pass filter.* For example, if in the system of coupled pendulums we allow the length of the pendulum strings to become infinite, then the strings are always vertical and never give any return force. (The strings are then equivalent to supporting the bobs on a "frictionless table.") Then the lowest mode frequency is zero (corresponding to translational motion). If we drive the system at one end, we have a low-pass filter which passes frequencies from zero to the high-frequency cutoff.

If the lowest mode frequency is greater than zero but the highest mode frequency is "infinite," the system is called a *high-pass filter.* For example, if we let K/M go to infinity in the system of coupled pendulums, we get a high-pass filter. The springs are then so stiff (or the masses so small) that the springs can always provide sufficient return force per unit mass per unit displacement without requiring successively decreasing amplitudes, no matter how high the driving frequency.

A system of two or three (or more) slinky-coupled soup-can pendulums driven at one end by a record player can nicely demonstrate the properties of a bandpass filter. (See Home Exp. 3.16.)

Example 4: Mechanical bandpass filter

Two coupled pendulums driven at one end as in Fig. 3.3 form a simple mechanical bandpass filter. We will let you show (Problem 3.28) that the ratio of output amplitude to input amplitude (neglecting damping) is given by

$$\frac{\psi_b}{\psi_a} \approx \frac{\omega_2{}^2 - \omega_1{}^2}{\omega_2{}^2 + \omega_1{}^2 - 2\omega^2}, \tag{56}$$

with

$$\omega_1{}^2 = \frac{g}{l}, \qquad \omega_2{}^2 = \frac{g}{l} + 2\frac{K}{M}. \qquad (57)$$

Notice that when ω is equal to either of the resonance values ω_1 or ω_2 the amplitude ratio is the same as it would be in the corresponding mode: ψ_b/ψ_a is +1 for $\omega = \omega_1$ and is –1 for $\omega = \omega_2$. As ω is decreased below the lowest mode frequency, ω_1, the amplitude ratio remains positive as it decreases from +1, its value at $\omega = \omega_1$, to $(\omega_2{}^2 - \omega_1{}^2)/(\omega_2{}^2 + \omega_1{}^2)$, its value at $\omega = 0$. Thus oscillations with driving frequencies well below the low-frequency cutoff are strongly attenuated in passing through the filter, provided the frequency range of the pass band is small compared with the average frequency in the pass band. As ω is increased above the highest mode frequency, ω_2, the amplitude ratio remains negative. It decreases in magnitude with increasing ω and becomes $-(\omega_2{}^2 - \omega_1{}^2)/2\omega^2$ for sufficiently high frequencies. Thus driving frequencies well above the high frequency cutoff are strongly attenuated.

Example 5: Mechanical low-pass filter

Suppose we start with the two coupled pendulums of Fig. 3.3. Now increase the height of the string support and lengthen the strings (so the bobs remain where they were). When the strings become "infinitely long," they remain vertical under any finite displacements of the bobs. Therefore gravity exerts no return force, and the strings act merely as a support, equivalent to a frictionless table. Then the lowest mode frequency, $\omega_1{}^2 = g/l$, goes to zero. Thus we have a low-pass filter, which passes frequencies between zero and the high-frequency cutoff, $\omega_2{}^2 = 2K/M$. (That is also the result for two spring-coupled masses resting on a frictionless table and driven at one end by a harmonic driving force.) The amplitude attenuation ratio ψ_b/ψ_a is given by Eq. (56) with ω_1 set equal to zero:

$$\frac{\psi_b}{\psi_a} = \frac{\omega_2{}^2}{\omega_2{}^2 - 2\omega^2} = \frac{K/M}{(K/M) - \omega^2}. \qquad (58)$$

We see that the attenuation ratio is +1 at zero frequency. It is infinite (meaning that ψ_a is zero) at $\omega_2 = \frac{1}{2}\omega_2{}^2$. It is –1 at the high-frequency cutoff. It becomes very small (and negative) for very high frequencies.

Here is an application of Eq. (58). Suppose we have a delicate piece of apparatus that will not work if it is subject to horizontal jiggling; however, vertical jiggling does no harm. Therefore we mount it on a flat plate which rests on a flat, frictionless, horizontal table (the frictionless support being provided by a thin film of air, perhaps). In order to keep the plate from drifting off the table, we must provide some horizontal support. Suppose the walls, floor, and ceiling are all vibrating with frequency components of 20 cps or more, the worst offender being at 20 cps. Suppose that if the plate is fastened to the walls through rigid supports, the vibration is 100 times greater (in amplitude) than we can tolerate. Suppose the apparatus plus plate weighs 10 kg. What should we do? Let us couple the apparatus and its plate to the walls through a low-pass filter, consisting of a spring in the x direction and another spring in the y direction. Each spring has spring constant K (to be determined). The x and y degrees of freedom are independent, so we need only consider the x motion. We designate the wall at the connection of the spring as "moving part a" and the apparatus as "moving part b." Now, in obtaining Eq. (58) we considered two masses coupled by a spring, with mass a driven by a force $F_0 \cos \omega t$. But of course mass b doesn't know what is pushing on mass a; it only knows that it is coupled to a moving part through a spring K and that at steady state there is a certain relation between its motion and that of a, the relation being given by Eq. (58). Therefore we can use Eq. (58) even when mass a is replaced by a jiggling wall, with ψ_a giving the motion of the end of the spring attached to the wall. We want ψ_b/ψ_a to be less than 10^{-2} for frequencies 20 cps or higher:

$$\frac{\psi_a}{\psi_b} = 1 - \frac{\omega^2}{K/M} = -100,$$

i.e.,

$$\frac{K}{M} = \frac{\omega^2}{101}, \quad \sqrt{\frac{K}{M}} \approx \frac{\omega}{10}.$$

For a fixed wall, the natural oscillation angular frequency of the apparatus and plate is $\sqrt{K/M}$. We see that if we want to attenuate frequencies v and above by at least a factor $f = 10^{-2}$, then the spring constant K must be weak enough so that the natural oscillation frequency of the apparatus is less than $f^{1/2}v$. In our example, the natural frequency must be less than $\frac{20}{10} = 2$ cps.

Here is another example. Suppose you are uncomfortable sitting on the floor because the floor is vibrating vertically at 20 cps. (Perhaps it is the floor of an airplane or something.) You therefore sit on a cushion. The cushion attenuates the vertical jiggling amplitude by a factor of 100. (You are comfortable.) How far does the top of the cushion sink down when you sit on it? (Prob. 3.12.)

Example 6: Electrical bandpass filter

Let us find an electrical analog for the mechanical example of the two coupled pendulums of Fig. 3.3. For each mass M we substitute an inductance L. For the coupling spring of spring constant K we substitute a capacitor with inverse capacitance C^{-1}. Now, the gravitational return force on each pendulum depends on the displacement of that pendulum but not on its coupling to the other pendulum. Similarly we want to provide an emf to drive each inductance independently of its coupling to the other. We do this by breaking the inductance into two halves and inserting a capacitor C_0 in series in the middle of the inductance. Finally, we neglect the fact that each inductance has resistance R (from the coil of wire that makes up the inductance). All other resistance is neglected. The system is shown in Fig. 3.8.

We shall let you work out the equations of motion and find the normal coordinates and modes (Prob. 3.29). Here we simply guess the results by analogy with the coupled pendulums:

$$\textbf{Mode 1: } I_a = I_b, \qquad \omega_1^2 = \frac{C_0^{-1}}{L} \qquad (59)$$

$$\textbf{Mode 2: } I_a = -I_b, \qquad \omega_2^2 = \frac{C_0^{-1}}{L} + \frac{2C^{-1}}{L} .$$

The attenuation provided by the bandpass filter is given by Eq. (56), if we neglect damping (i.e., neglect the resistance of the coils):

$$\frac{I_b}{I_a} = \frac{\omega_2^2 - \omega_1^2}{\omega_2^2 + \omega_1^2 - 2\omega^2} = \frac{1/LC}{\left(1/LC_0\right) + \left(1/LC\right) - \omega^2} . \qquad (60)$$

Example 7: Electrical low-pass filter

If we short-circuit the capacitor C_0 in Fig. 3.8, its capacitance becomes effectively infinite. The lowest mode frequency becomes zero, corresponding to steady DC (direct current). Then we have a

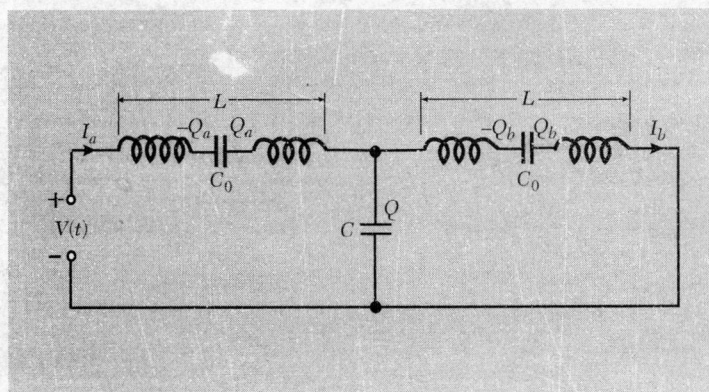

Fig. 3.8   Coupled LC circuits driven at one end by voltage V(t). This circuit is the electrical analog of the two coupled pendulums of Fig. 3.3.

low-pass filter, as shown in Fig. 3.9. The current attenuation ratio is given by Eq. (60) with I/C_0 set equal to zero:

$$\frac{I_b}{I_a} = \frac{\omega_2{}^2}{\omega_2{}^2 - 2\omega^2} = \frac{1/LC}{\left(1/LC\right) - \omega^2}. \tag{61}$$

Example 8: Low-pass filter for a DC power supply

This is a very practical application of Eq. (61). In a typical direct-current (DC) power supply, one starts with alternating current (AC) from a wall receptacle that furnishes power at a frequency of 60 cps and at an rms (root-mean-square) voltage of about 110 volts. This voltage is applied to the input winding of a transformer. The output winding of the transformer may have more turns than the input winding (step-up transformer) or less turns (step-down transformer), depending on what DC voltage we eventually want to obtain. The output winding is connected across a diode, which passes current in only one direction. That would give "half-wave rectified" DC current. In practice two diodes are used with a center-tapped output winding to obtain "full-wave rectification." This current is used to charge up a capacitor, which then acts as a source of steady voltage. The charge (and hence the voltage) on the capacitor is not exactly constant, however. To a good approximation it is given by a constant plus a small "ripple" oscillating at a frequency of 120 cps (for full-wave rectification). (*Question*: Why does the ripple frequency equal twice the AC frequency that we started with?) If this charged capacitor is used as a source of DC voltage to power radio or phonograph tubes, the output of the radio will include an annoying 120 cps "hum." (To hear this hum, listen to a radio just after it is turned on, before the tubes warm up. Of course, a battery-operated radio doesn't have this hum. Alternatively, find an electric clock or a "high-intensity lamp" that uses 12-volt auto headlight bulbs. Both of these have inductive

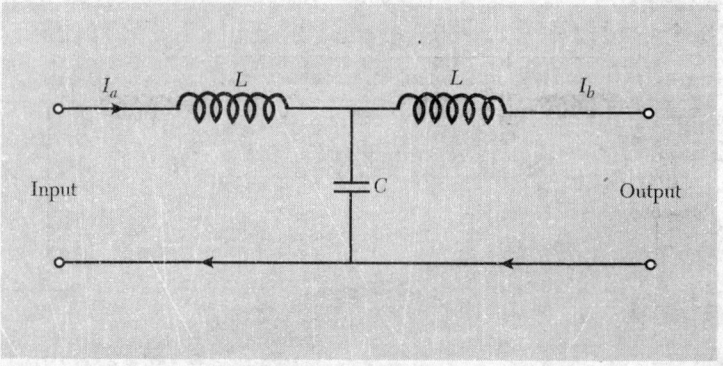

Fig. 3.9 Low-pass electrical filter.

windings, and you can hear the hum due to the mechanical stress in the windings. Why is *this* hum at 120 cps instead of at 60 cps?)

To get rid of the 120 cps hum, we connect the capacitor across the input terminals of the low-pass filter of Fig. 3.9, and use the output of the filter as a source of DC voltage. Values used for L and C in a typical filter are (see any handbook for radio amateurs) $L = 10h$, $C = 6\mu f$. Then the high-frequency cutoff is given by

$$\nu_2 = \frac{1}{2\pi}\sqrt{\frac{2}{LC}} = \frac{1}{6.28}\sqrt{\frac{2}{10 \times 6 \times 10^{-6}}} = 29.1 \text{ cps.}$$

The attenuation factor for the 120-cps component of the current is given by Eq. (61):

$$\frac{I_b}{I_a} = \frac{\nu_2^2}{\nu_2^2 - 2\nu^2} = \frac{(29.1)^2}{(29.1)^2 - 2(120)^2} = -0.030.$$

Thus the ripple component is reduced by a factor of about 30. The DC component is not affected by the filter.

3.5 *Forced Oscillations of Closed System with Many Degrees of Freedom*

In this section we will study the behavior of a system of coupled identical pendulums experiencing steady-state oscillations under the influence of a driving force of arbitrary frequency ω. At first we will not specify the boundary conditions; neither will we specify which one or ones of the moving parts are directly coupled to the driving force. (The latter specifications can be thought of as being part of the boundary conditions.) We shall look only at the equation of motion of a pendulum bob that does not happen to be coupled directly to any driving force. By doing this, we shall be able to find the general solution for the motion of the bob with unspecified boundary conditions. Of course, in any specific situation it is necessary to specify conditions of the system completely—whether the ends are fixed or free (or neither), where the forces are applied, etc.

Neglect of damping. We shall omit damping terms from the equations of motion. Will that limit the generality of our results? Yes, but not very seriously. Recall that in Sec. 3.3 we found that, as long as ω is not near any resonance (any mode frequency for free oscillations of the system), the displacement of each moving part is a superposition of only the elastic contributions, one from each mode. The absorptive amplitudes do not contribute. That is because the absorptive amplitude falls off much more rapidly as the frequency changes than does the elastic amplitude. As long as ω is at least 5 or 10 half-widths away from any resonance frequency, we can neglect the absorptive terms. That is equivalent to setting $\Gamma = 0$ in the results. We shall instead set $\Gamma = 0$ in the equations of motion, but we shall nevertheless assume that there is some damping, so that the system reaches a steady state of oscillation at the driving frequency ω. We know that if there actually were no damping, then the system would not reach any steady state but would continue forever to experience "interminable beats." We are assuming that there *is* damping, but we shall avoid describing the behavior when ω is near a resonance. (We already know what that behavior is from the results of Sec. 3.3.)

Relative phases of moving parts. An important consequence of our neglecting the absorptive amplitudes contributed by the different modes is that every mode gives a contribution (to the displacement of a given moving part) that is either in phase or 180 deg out of phase with the driving force $F_0 \cos (\omega t + \varphi_0)$. That is because the elastic amplitude is a positive or negative constant times $\cos (\omega t + \varphi_0)$, as we showed in Sec. 3.3. Another way to obtain that result without looking back at Sec. 3.3 is as follows: Assume that there is no damping, but that nevertheless you have somehow gotten the system into a steady state of oscillation at the driving frequency ω. Since there is no damping, there is no energy dissipation. Therefore the driving force must not be doing any work, positive or negative, on *any* moving part. (Otherwise the driving force would perform some net work in each oscillation cycle.) That implies that the displacement of every moving part is either in phase or 180 deg out of phase with the driving force; i.e., we have purely elastic amplitudes.

We thus have the important result that *at steady state (and for ω not near a resonance) every moving part has the same phase constant*, and this phase constant is the same as that of the driving force. (We let the amplitude of each moving part be either positive or negative, so that we do not have to talk about 180-deg phase constants.) We also have the result that *the return force per unit displacement per unit mass, ω^2, is the same for all moving parts*, since each moving part is oscillating at the same frequency. (Notice that these are just the conditions that also hold for a single normal mode of a freely oscillating undamped system!)

We are now ready to consider a specific system.

Example 9: Coupled pendulums

By now, it will not surprise you to learn that, by merely changing names (for example, changing "string length" to "capacitance" and "mass" to "inductance") and drawing new sketches, one can obtain the results for very many different physical systems from the final results for the coupled pendulums, without repeating the mathematical details. (We did this often in Chap. 2.) For now, we shall pretend we are interested only in coupled pendulums.

Three of the identical coupled pendulums in a linear array (with the total number of pendulums left unspecified and with unspecified boundary conditions) are shown in Fig. 3.10. The equation of motion for the displacement $\psi_n(t)$ of bob n is (for small oscillations)

$$M\ddot{\psi}_n = -M\omega_0^2\psi_n + K(\psi_{n+1} - \psi_n) - K(\psi_n - \psi_{n-1}), \qquad (62)$$

where

$$\omega_0{}^2 = \frac{g}{l}.$$

Before we attempt the exact solution of Eq. (62), we shall study its solutions in the continuous approximation. That means that we will not obtain any information about the motion for configurations like that of the highest mode of free oscillation, in which the successive bobs have a "to and fro" configuration. ("To and fro" is the longitudinal analog of the transverse "zigzag.") We will thus have to limit ourselves (for the present) to frequencies well below the upper cutoff frequency. Only when we come back and solve the equation exactly will we be able to discuss the motion for driving frequencies in the upper part of the pass band and above the upper cutoff frequency.

Fig. 3.10 Coupled pendulums with unspecified boundary conditions.

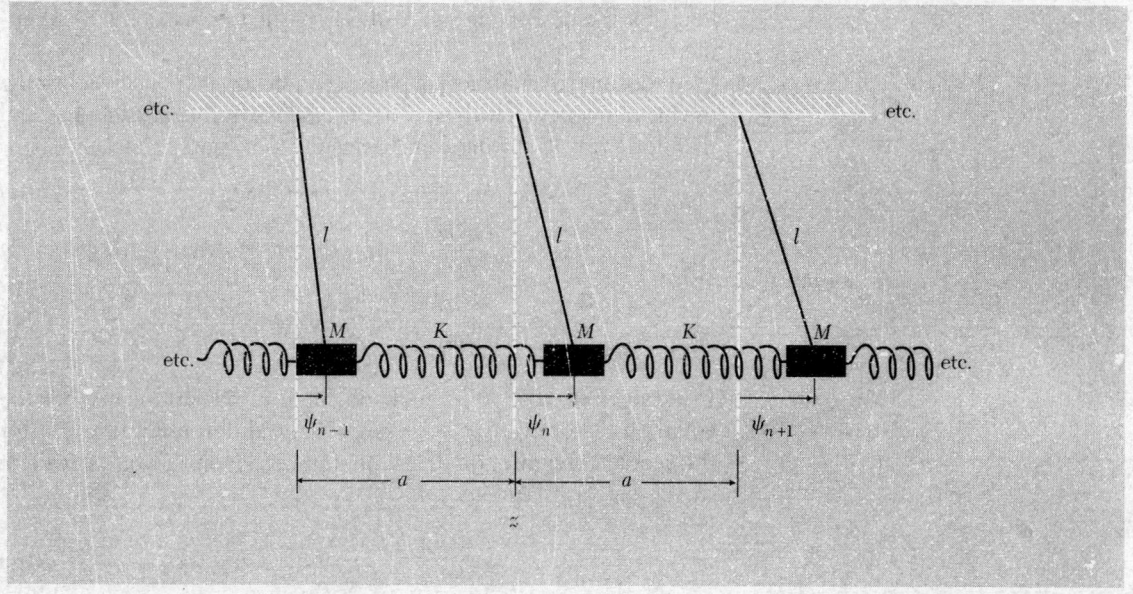

Continuous approximation. We assume $\psi_n(t)$ varies slowly with increasing n. This means that we assume that all the bobs in a small vicinity of bob n, which has equilibrium position at z, have nearly the same motion as bob n, so that the motion of a bob at z can be described by a continuous function $\psi(z, t)$. We expand the appropriate terms in Eq. (62) in a Taylor's series:

$$\psi_n(t) = \psi(z, t).$$

$$\psi_{n+1}(t) = \psi(z + a, t) = \psi(z, t) + a \frac{\partial \psi(z, t)}{\partial z} + \frac{1}{2} a^2 \frac{\partial^2 \psi(z, t)}{\partial z^2} + \cdots,$$

$$\psi_{n-1}(t) = \psi(z - a, t) = \psi(z, t) - a \frac{\partial \psi(z, t)}{\partial z} + \frac{1}{2} a^2 \frac{\partial^2 \psi(z, t)}{\partial z^2} + \cdots.$$

Thus

$$\psi_{n+1} - \psi_n = a \frac{\partial \psi}{\partial z} + \frac{1}{2} a^2 \frac{\partial^2 \psi}{\partial z^2} + \cdots,$$

$$\psi_n - \psi_{n-1} = a \frac{\partial \psi}{\partial z} - \frac{1}{2} a^2 \frac{\partial^2 \psi}{\partial z^2} + \cdots.$$

Then inserting these expressions (and also $\ddot{\psi}_n(t) = \partial^2 \psi(z, t) / \partial t^2$) into Eq. (62), we get

$$\boxed{\frac{\partial^2 \psi(z, t)}{\partial t^2} = -\omega_0^2 \psi(z, t) + \frac{Ka^2}{M} \frac{\partial^2 \psi(z, t)}{\partial z^2}.} \tag{63}$$

Klein-Gordon wave equation. Equation (63) is a famous wave equation. It is not the classical wave equation, except when ω_0 is zero. It is sometimes called the "Klein-Gordon wave equation." (It holds for the de Broglie waves of relativistic free particles. See Supplementary Topic 2.)

We assume that all moving parts are in steady-state oscillation at the driving frequency ω, with no work being done by the driving force. and therefore with all moving parts having the same phase constant. Then

$$\psi(z, t) = \cos (\omega t + \varphi) A(z), \tag{64}$$

$$\frac{\partial^2 \psi}{\partial t^2} = -\omega^2 \cos (\omega t + \varphi) A(z), \tag{65}$$

$$\frac{\partial^2 \psi}{\partial z^2} = \cos (\omega t + \varphi) \frac{d^2 A(z)}{dz^2}. \tag{66}$$

Inserting Eqs. (64), (65), and (66) in Eq. (63) and canceling the common factor $\cos (\omega t + \varphi)$, we get the differential equation for the spatial configuration of the pendulums when they are driven at steady state at frequency ω:

$$\frac{d^2 A(z)}{dz^2} = \frac{M}{Ka^2} (\omega_0^2 - \omega^2) A(z). \tag{67}$$

The solutions of Eq. (67) have very different dependence on z for the two cases $\omega^2 > \omega_0^2$ and $\omega^2 < \omega_0^2$. When ω^2 is greater than ω_0^2, we obtain sinusoidal waves of the type studied earlier (Sec. 2.2) for the continuous string:

$\omega^2 > \omega_0^2$: *Sinusoidal waves.* For $\omega^2 > \omega_0^2$, Eq. (67) has the form

$$\frac{d^2A(z)}{dz^2} = -k^2 A(z), \tag{68}$$

where k^2 is the positive constant

$$k^2 = (\omega^2 - \omega_0^2)\frac{M}{Ka^2}. \tag{69}$$

Equation (69) is the dispersion relation for waves in the system with $\omega^2 > \omega_0^2$. The general solution of Eq. (68) is

$$A(z) = A \sin kz + B \cos kz, \tag{70}$$

where A and B are constants determined by the boundary conditions. Depending on the boundary conditions, there will be certain wavelengths (and corresponding driving frequencies) that correspond to a "resonance." The resonance frequencies are the same as the frequencies of the normal modes (standing waves) of the freely oscillating system.

Now we come to something new and important:

$\omega^2 < \omega_0^2$: *Exponential waves.* If ω^2 is less than ω_0^2, we define the positive constant κ (kappa) as the positive square root of the positive quantity

$$\kappa^2 = (\omega_0^2 - \omega^2)\frac{M}{Ka^2}. \tag{71}$$

(Do not confuse kappa with the similar-appearing capital K.) Equation (71) is the dispersion relation for the case $\omega^2 < \omega_0^2$. Then Eq. (67) has the form

$$\frac{d^2A(z)}{dz^2} = \kappa^2 A(z). \tag{72}$$

The presence of a plus sign on the right-hand side of Eq. (72) gives its solutions a completely different shape from that of the sinusoidal solutions of the similar-appearing Eq. (68). Because of the minus sign in Eq. (68), its solution, the sinusoidal function $A(z)$ given by Eq. (70), is always bending toward the z axis. Therefore it always eventually crosses the z axis. After it crosses, it bends back and eventually crosses it again, etc., thus giving oscillations in space. By contrast, the plus sign on the right-hand side of Eq. (72) means that its solution $A(z)$ always bends *away* from the z axis. Therefore if $A(z)$ happens to have positive value and positive slope (or negative value and slope), it will never return to the z axis. If it has positive value and negative slope, it will approach the z axis more and more slowly with increasing z. If it does eventually cross the z axis with negative slope (which it may, but need not), it will continue on to more and more negative values of $A(z)$ with increasing z and will never again cross the z axis.

The general solution of Eq. (72) is a superposition of two exponential functions:

$$A(z) = Ae^{-\kappa z} + Be^{+\kappa z}. \tag{73}$$

To see that this $A(z)$ is a solution, differentiate it:

$$\frac{dA(z)}{dz} = -\kappa Ae^{-\kappa z} + \kappa Be^{+\kappa z},$$

$$\frac{d^2A(z)}{dz^2} = -(-\kappa)^2 Ae^{-\kappa z} + (\kappa)^2 Be^{+\kappa z} = \kappa^2 A(z),$$

so that Eq. (73) satisfies Eq. (72). The constants A and B are determined by the boundary conditions. Thus for $\omega^2 < \omega_0^2$, the general solution $\psi(z, t)$, is

$$\psi(z, t) = (Ae^{-\kappa z} + Be^{+\kappa z}) \cos(\omega t + \varphi). \tag{74}$$

Coupled pendulums as high-pass filter. Equation (74) gives the general form of an *exponential wave*. The frequency $\omega_0^2 = g/l$ acts as a *lowfrequency cutoff frequency*. This was to be expected. since we found that same value for the low-frequency cutoff for the simple system of two coupled pendulums. At that lowest mode frequency, all the pendulums are swinging in phase with one another, with return force provided only by gravity. The springs are not stretched or compressed. The wavelength is "infinite", i.e., k is zero. If the system is driven below cutoff, it cannot sustain sinusoidal space dependence for the relative amplitudes of the oscillating bobs. Instead, the relative amplitudes of the bobs depend exponentially on the distance, as given by Eq. (74). Thus the system acts like a high-pass filter. (Actually it is a bandpass filter, but we are only treating the system in the continuous approximation, and we must avoid considering the response of the system near the highest modes with their "zigzag" configuration.)

Suppose that the system is driven at $z = 0$ and extends from $z = 0$ to $z = L$, at which point the last spring is tied to a rigid wall. It should be intuitively evident that if we drive the system below cutoff, the amplitude $A(z)$ must decrease with increasing distance z from the driven end. If the system is very long, i.e., if L is large, the amplitude must become very small by the time we reach the wall at $z = L$. In the limit that L is "infinite," the amplitude must be zero at $z = L$. That implies that the contribution $Be^{+\kappa z}$ in Eq. (74) must vanish, i.e., that B must be zero. That guess is correct. (See Prob. 3.30.)

In Fig. 3.11 we show an example of this situation. Notice that for the example shown in Fig. 3.11, it makes very little difference whether or not we actually tie down the end at $z = L$. If we have $\kappa L \gg 1$, the oscillation amplitude is essentially zero before we reach $z = L$. Thus we can experimentally simulate an "infinite" length with a finite length L that is large compared with $1/\kappa$. (See Home Exp. 3.16.)

Nomenclature for exponential waves. The constant κ (kappa) is called the *amplitude attenuation constant* or simply the *attenuation constant*. Its units are *fractional amplitude attenuation per unit length* or simply *attenuation per unit length*. These

Fig. 3.11 *Coupled pendulums driven at left end at a frequency below the cutoff frequency ω_0. (a) Instantaneous configuration of system. (b) Plot of A(z).*

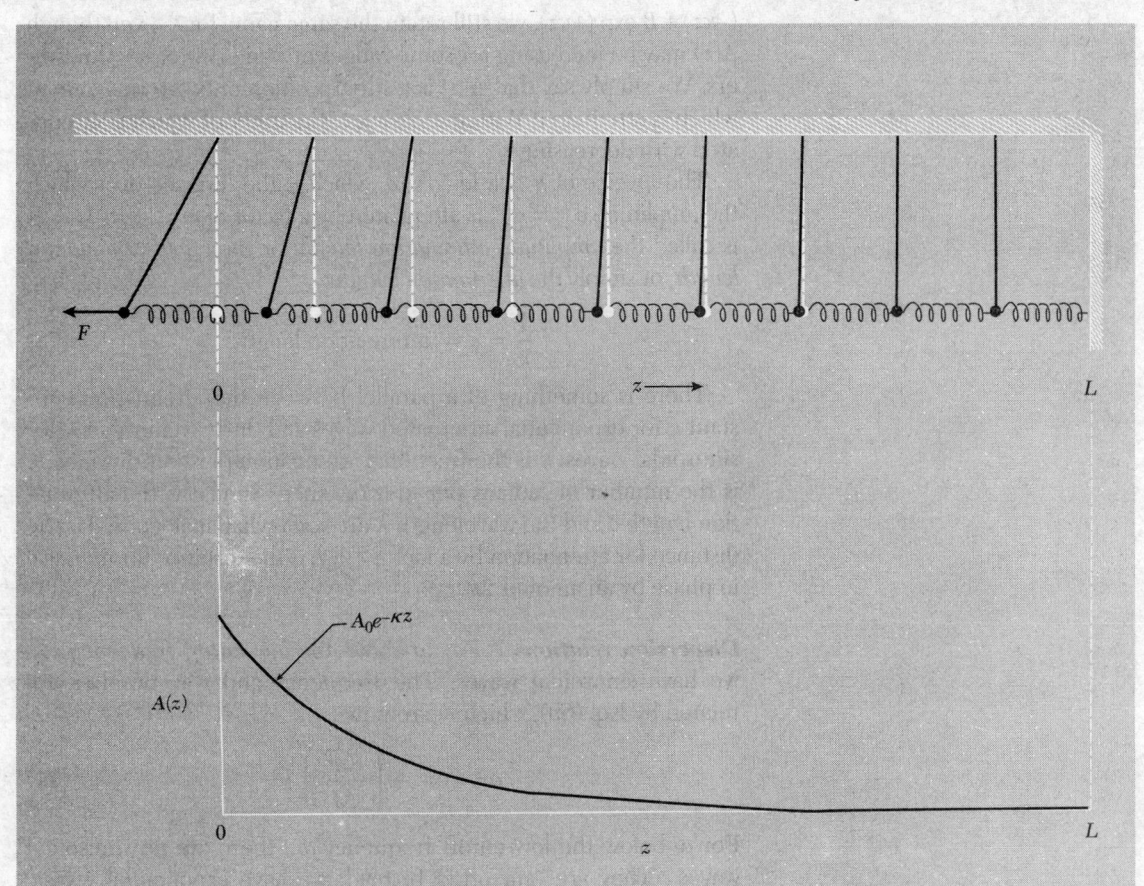

units are arrived at by considering an amplitude $A(z)$ produced by a driving force at the left end of a long system, i.e., long enough so that we have only the decreasing exponential:

$$\psi(z, t) = A(z) \cos \omega t, \tag{75}$$

with

$$A(z) = Ae^{-\kappa z}. \tag{76}$$

The fractional amplitude attenuation per unit length of the amplitude $A(z)$ is defined as

$$-\frac{1}{A(z)} \frac{dA(z)}{dz} = \text{fractional amplitude attenuation per unit length} \tag{77}$$

This is equal to κ in the case that $A(z)$ is given by Eq. (76). On the other hand, when $A(z)$ is given by $A(z) = B \exp (+\kappa z)$, the amplitude is attenuated when z decreases rather than when it increases. However, that causes no confusion, and we can continue to call κ the attenuation constant. When we have the general solution $A \exp (-\kappa z) + B \exp (+\kappa z)$, we still retain the same name for κ, even though $A(z)$ may be increasing for some ranges of z and decreasing for others. We simply say that $A(z)$ is a superposition of two terms, one of which is attenuated with increasing z, the other of which is attenuated with decreasing z.

The inverse of κ is a length, δ, which is the distance over which the amplitude $e^{-\kappa z} = e^{-z/\delta}$ is attenuated by a factor of $e = 2.718 \ldots$. It is called the *amplitude attenuation length*, or the *e-fold attenuation length*, or simply the *attenuation length*:

$$\frac{1}{\kappa} = \delta = \text{attenuation length.} \tag{78}$$

There is something of a parallel between the attenuation constant κ for exponential attenuated waves and the wavenumber κ for sinusoidal waves: κ is the fractional attenuation per unit distance; κ is the number of radians per unit distance. Similarly, the attenuation length δ and the wavelength λ are somewhat analogous: δ is the distance for attenuation by a factor e^{-1}; λ is the distance for increase in phase by an amount 2π.

Dispersion relations. For ω above the *low-cutoff frequency* ω_0, we have sinusoidal waves. The frequency and wavenumber are related by Eq. (69), which we rewrite:

$$\omega^2 = \omega_0^2 + \left(\frac{Ka^2}{M} \right) k^2. \tag{79}$$

For ω below the low-cutoff frequency ω_0, there are no sinusoidal waves. (They are "cut off.") Instead, we have exponential waves.

The frequency ω and attenuation constant κ are related by Eq. (71), which we rewrite:

$$\omega^2 = \omega_0^2 - \left(\frac{Ka^2}{M}\right)\kappa^2. \tag{80}$$

Equations (79) and (80) constitute the complete dispersion relation for the system (in the continuous approximation).

For the frequency range in which the forced oscillations are sinusoidal waves, the dispersion relation Eq. (79) for *forced* oscillations is identical with that which we found for the modes of *free* oscillation. [See Sec. 2.4, Eqs. (2.90), (2.91), and (2.92).] That is no accident. In both derivations, we found the equation of motion of bead n and then inserted the assumption that all the moving parts moved in harmonic oscillation with the same frequency ω (a mode frequency in one case, the steady-state oscillation frequency in the other) and the same phase constant. The dispersion relation followed from these assumptions. That is generally the case: *the dispersion relation for forced sinusoidal oscillations is the same as that for free oscillations.*

Dispersive or reactive medium.   In the example we are considering, the "medium" in which the waves occur consists of the system of coupled pendulums. A medium which can support sinusoidal waves is said to be a *dispersive* medium. That just means that ω is not below the cutoff frequency ω_0. A medium which cannot support sinusoidal waves but which instead gives exponential waves (without energy dissipation) is called a *reactive* medium. The same medium can of course be reactive at some frequencies and dispersive at others, as is the case for our coupled pendulums.

Example 10: The ionosphere

The earth's ionosphere is an example of a medium (for electromagnetic waves) that is dispersive for frequencies above a cutoff frequency (called the plasma oscillation frequency ν_p) and reactive for frequencies below that cutoff. The dispersion relation for driven oscillations in the earth's ionosphere is very similar to that for the coupled pendulums. It turns out that it is given by

$$\omega^2 = \omega_p^2 + c^2 k^2, \quad \omega > \omega_p, \tag{81}$$

and

$$\omega^2 = \omega_p^2 - c^2 \kappa^2, \quad \omega < \omega_p.$$

The plasma oscillation frequency is the frequency of the lowest mode of vibration of the "free" electrons, which we derived in Eq. (2.99), Sec. 2.4. Typical daytime plasma oscillation frequencies

ν_p $(=\omega_p/2\pi)$ are 10 to 30 Mc. If the ionosphere is driven "at one end" by a radio station emitting typical AM broadcast frequencies, say $\nu = 1000$ Kc, it behaves as a reactive medium, since $\nu \ll \nu_p$. The waves are exponentially attenuated, just as is the case for our coupled pendulums in Fig. 3.11. No work is done on the ionosphere in this process, since (it turns out) the velocity of every electron is $\pm 90°$ out of phase with the electric field in its neighborhood. Now, in the case of the driven pendulums of Fig. 3.11, the average energy output of the driving force is zero (neglecting the damping). Energy that is instantaneously given to the pendulums is returned to the driving force later in the cycle. In the case of the three-dimensional problem of the radio station and the ionosphere, that is not the case: The radio station gets back very little of the energy it broadcasts. The ionosphere absorbs no energy, but the waves are reflected back to earth over a broad region, not just back to the transmitter. This *total reflection* from the "underside" of the ionosphere provides a technique used to broadcast to distant receivers that are "out of sight" around the curve of the earth's surface: One simply bounces the signals off the ionosphere. This works whenever ω is below the cutoff frequency ω_p.

Typical FM radio and TV broadcasting frequencies are around 100 Mc. This frequency is greater than the cutoff frequency of 10 to 30 Mc of the ionosphere. Therefore the ionosphere behaves like a dispersive medium at FM and TV frequencies. That means it is "transparent." There is no exponential attenuation of the waves; they are sinusoidal instead. Thus there is no total reflection of the electromagnetic waves back to the earth, and one cannot use the ionosphere as an aid to transmitting signals in the way that one can at AM frequencies. Thus transmission is restricted to "line of sight."

The ionosphere is also a dispersive medium for electromagnetic waves of frequency $\nu \approx 10^{15}$ cps, the order of 'magnitude of frequency of *visible light*. We know that the ionosphere is not reactive at 10^{15} cps, because otherwise we would not see the stars or the sun. [However, our vision evolved to see whatever frequencies got through and lit things up, so we might see ultraviolet stars instead.] In a later chapter we shall derive the dispersion relation of the ionosphere, Eq. (81).

Wave penetration into a reactive region. When the ionosphere is driven by a radio station at a frequency below cutoff, the radio waves are *totally reflected* back to earth. But this does not happen all all. one place, so to speak. Let us consider an analogous problem for coupled pendulums (which have the same form of dispersion relation as the ionosphere) in the continuous approximation. Suppose the first bob, at $z = 0$, is driven by whatever force is necessary to produce the motion $\psi_1(t) = A_0 \cos \omega t$. In the region between $z = 0$ and $z = L$, we have a number of coupled pendulums, each with a length l_1 such that

$$\omega_0^2 = \frac{g}{l_1} < \omega^2. \tag{82}$$

Thus this region (we will call it region 1) is dispersive. (The driving force is the "radio station." The region from $z = 0$ to $z = L$ is "ordinary air," not "plasma.") At $z = L$, the pendulum strings suddenly became shorter, each having a length l_2 such that

$$\omega_0^2 = \frac{g}{l_2} > \omega^2. \tag{83}$$

Thus this region (called region 2) is reactive. (Region 2 is the "plasma.") This region extends to $z = \infty$. The system is shown in Fig. 3.12.

Let us find $\psi(z, t)$, given that at $z = 0$ it is equal to $A_0 \cos \omega t$. For any z we will have

$$\psi(z, t) = A(z) \cos \omega t, \tag{84}$$

where $A(z)$ is to be determined. In region 2, the reactive region between $z = L$ and infinity, $A(z)$ must be given by

$$A_2(z) = C e^{-\kappa(z-L)}, \tag{85}$$

with C an unknown constant and with κ given by

$$\kappa^2 = \frac{M}{K a^2}\left(\frac{g}{l_2} - \omega^2\right), \tag{86}$$

assuming ω^2 is less than the cutoff g/l_2. In the dispersive region between $z = 0$ and $z = L$, $A(z)$ is given by

$$A_1(z) = A \sin k(z - L) + B \cos k(z - L), \tag{87}$$

with A and B unknown constants and with k given by

$$k^2 = \frac{M}{K a^2}\left(\omega^2 - \frac{g}{l_1}\right), \tag{88}$$

assuming ω^2 is greater than g/l_1. We now impose the boundary conditions: At $z = L$ the functions $A_1(z)$ and $A_2(z)$ must join smoothly, i.e., their values and slopes must be equal at $z = L$. Equating their values at $z = L$ gives $B = C$. Equating their slopes at $z = L$ gives $kA = -\kappa C$. Thus we have in region 1

$$A_1(z) = C\left[\frac{-\kappa}{k} \sin\ k(z - L) + \cos\ k(z - L)\right]. \tag{89}$$

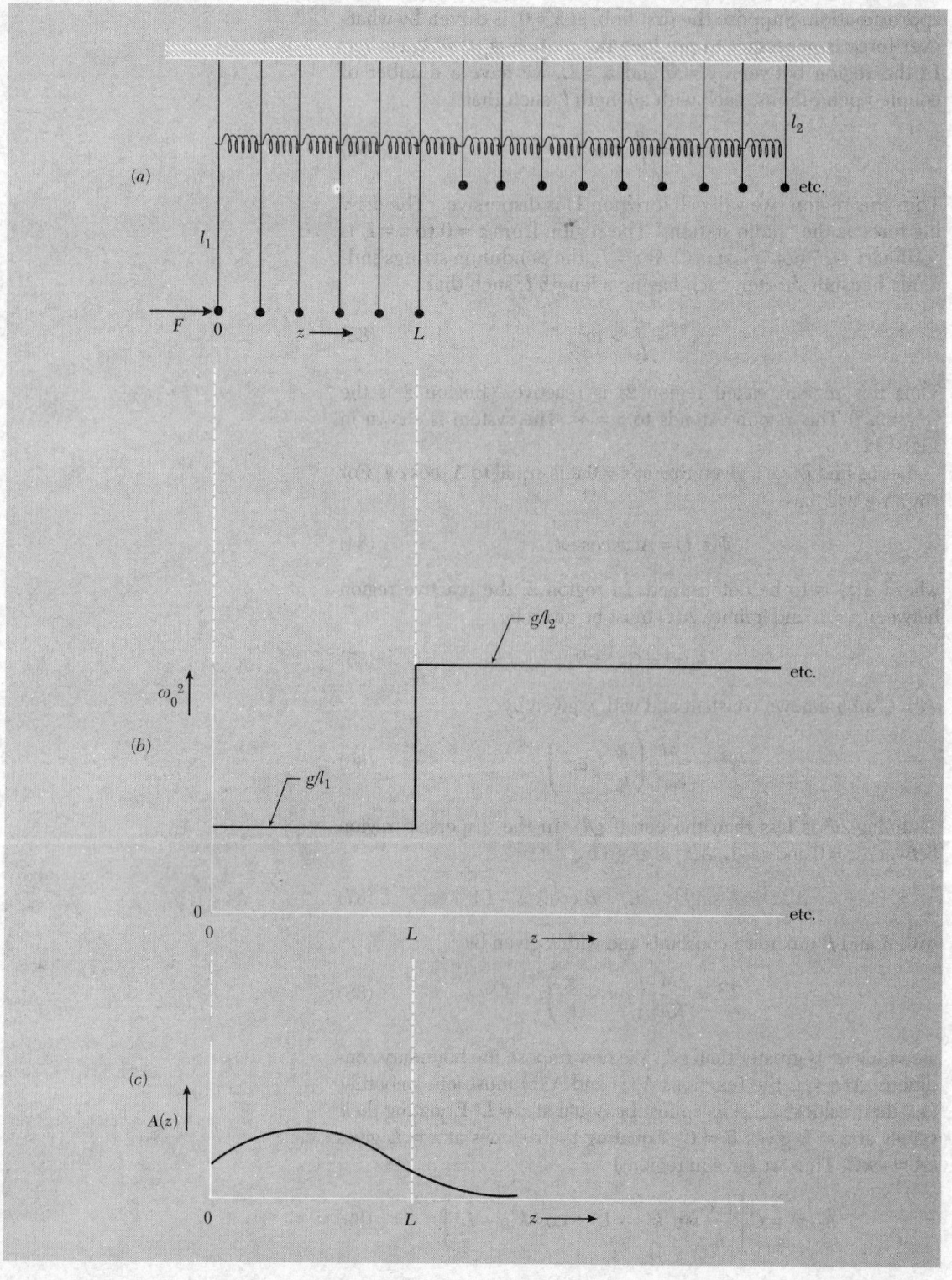

The boundary condition at $z = 0$ is that $A_1(z) = A_0$ at $z = 0$. Then Eq. (89) gives

$$C = \frac{A_0}{\dfrac{\kappa}{k} \sin kL + \cos kL}. \qquad (90)$$

The complete solution is given by Eqs. (84), (85), (89), and (90) plus the dispersion relations Eqs. (86) and (88).

Resonance.   The denominator of Eq. (90) goes to zero fm certain values of kL, giving "infinite" amplitude C. (When damping is not neglected, we do not get any infinite amplitudes.) These values of kL determine the resonant frequencies for the system. To find the resonant frequencies, one must use the dispersion relations, as well as Eq. (90). (See Prob. 3.31.) The amplitude $A(z)$ for ω near the lowest resonance is plotted in Fig. 3.12. with C taken to be large but not infinite.

Bound modes.   We see from Fig. 3.12c that a reactive region that extends a long distance (to $L' = \infty$ in our example) acts something like a "gradual wall." The bob at $z = L$ is not held fixed, as at a wall; nevertheless the bob motion is negligible beyond a few attenuation distances δ past $z = L$. This suggests that if we enclose a dispersive region by surrounding it with an infinitely thick reactive region on either side, we can have *modes* (of free oscillation) of the pendulums in the dispersive region, somewhat as if they were contained between two walls. That conjecture is correct. The modes are called *bound modes.* They occur approximately at the resonant frequencies of the system of Fig. 3.12.

One interesting feature of the bound modes is that there are only a limited number of bound modes for a given system, even if there are an "infinite" number of pendulums in the dispersive region. That is because the frequency increases as we progress from one bound mode to the next higher one, until finally we reach a bound mode for which the frequency is greater than $\sqrt{g/l_2}$. For $\omega^2 > g/l_2$ the outer regions are dispersive, and they no longer serve to "contain" oscillations. in the central region.

In quantum physics, one finds that the de Broglie waves of electrons in atoms behave like bound modes of coupled pendulums. The modes of oscillation of the electrons are called *bound states.*

Fig. 3.12   *(See Page 142) Coupled pendulums with sudden change-in ω_0^2 at $z = L$. (a) The system. The pendulum at $z = 0$ is coupled to an external driving force. (b) Plot of ω_0^2 versus z. For driving frequencies ω between $\sqrt{g/l_1}$ and $\sqrt{g/l_2}$, region 1 (from $z = 0$ to L) is dispersive, and region 2 (from $z = L$ to ∞) is reactive. (c) Plot of amplitude A(z) versus z for driving frequency ω near the lowest resonance frequency of the system.*

An example of a quantum system with bound states is given in Supplementary Topic 3.

Exact solution for the forced oscillations of a system of coupled pendulums. We have been examining the properties of forced oscillations of coupled pendulums in the continuous approximation. Now let us find the *exact* solution of the equation of motion of a pendulum in the linear array, Eq. (62), which we recopy here:

$$\ddot{\psi}_n = -\omega_0^2\psi_n + \frac{K}{M}(\psi_{n+1} - 2\psi_n + \psi_{n-1}). \tag{91}$$

We assume that all moving parts oscillate in harmonic motion with the same frequency and phase constant:

$$\psi_n = A_n \cos \omega t. \tag{92}$$

Inserting Eq. (92) into Eq. (91) and canceling the factor cos ωt, we get

$$-\omega^2 A_n = -\omega_0^2 A_n - \frac{2K}{M}A_n + \frac{2K}{M}\left(\frac{A_{n+1} + A_{n-1}}{2}\right),$$

i.e.,

$$\omega^2 = \omega_0^2 + \frac{2K}{M}\left(1 - \frac{\frac{1}{2}(A_{n+1} + A_{n-1})}{A_n}\right). \tag{93}$$

Dispersive frequency range. (This is the "pass band," in filter terminology.) In the dispersive region, the oscillations are sinusoidal in space. Thus, let us assume a solution of the form

$$A_n = A \sin kna + B \cos kna. \tag{94}$$

Then

$$A_{n+1} = A \sin (kna + ka) + B \cos (kna + ka),$$
$$A_{n-1} = A \sin (kna - ka) + B \cos (kna - ka). \tag{95}$$

Thus

$$A_{n+1} + A_{n-1} = 2A \sin kna \cos ka + 2B \cos kna \cos ka$$
$$= 2 \cos ka(A \sin kna + B \cos kna) = 2 \cos kaA_n \tag{96}$$

Insertion of this result into Eq. (93) gives

$$\omega^2 = \omega_0^2 + \frac{2K}{M}(1 - \cos ka), \tag{97}$$

i.e.,

$$\omega^2 = \omega_0^2 + \frac{4K}{M}\sin^2\frac{ka}{2}. \tag{98}$$

Equation (98) is the dispersion law for the dispersive frequency range. It gives frequencies from $\omega^2 = \omega_0^2$ to $\omega^2 = \omega_0^2 + 4K/M$, corresponding to values of ka from $ka = 0$ to $ka = \pi$. Equation (98) is exactly the same dispersion law as that which we obtained for freely oscillating coupled pendulums in Eq. (2.90), Sec. 2.4.

Lower reactive range.   Using our experience with the continuous approximation, let us guess that the general solution for frequencies below the lowfrequency cutoff ω_0 has the form of an *exponential wave:*

$$A_n = Ae^{-\kappa na} + Be^{+\kappa na}. \tag{99}$$

Then

$$A_{n+1} + A_{n-1} = (e^{\kappa a} + e^{-\kappa a})A_n. \tag{100}$$

Then Eq. (93) gives the dispersion law

$$\omega^2 = \omega_0{}^2 + \frac{2K}{M}\left[1 - \frac{1}{2}(e^{\kappa a} + e^{-\kappa a})\right]. \tag{101}$$

Equation (101) can be put into forms that resemble Eqs. (97) and (98). Using the definitions of the hyperbolic sine and hyperbolic cosine [Appendix Eqs. (11) and (12)], we find

$$\omega^2 = \omega_0{}^2 + \frac{2K}{M}(1 - \cosh \kappa a) \tag{102}$$

or

$$\omega^2 = \omega_0{}^2 - \frac{4K}{M}\sinh^2 \frac{1}{2}\kappa a. \tag{103}$$

At $\omega = \omega_0$, the dispersive solution Eq. (98) gives $k = 0$, and the reactive solution Eq. (103) gives $\kappa = 0$. These are both "flat waves," and thus they agree.

Upper reactive range.   This range consists of all frequencies above the high-frequency cutoff ω_{max}, where $\omega_{max}{}^2 = \omega_0{}^2 + 4K/M$. For this range, we are guided by our study of filters of two degrees of freedom. There we found that oscillations driven at a frequency above the high-frequency cutoff have a zigzag shape, like that of the highest mode, but that they also have an attenuation of amplitude with increasing distance from the input end. (See Fig. 3.6.) Let us make the guess that the shape of A_n is given by the *exponential zigzag wave*

$$A_n = (-1)^n(Ae^{-\kappa na} + Be^{+\kappa na}). \tag{104}$$

Then we get (by the same steps that gave Eq. (100), except for the minus signs)

$$A_{n+1} + A_{n-1} = -A_n(e^{\kappa a} + e^{-\kappa e}).$$

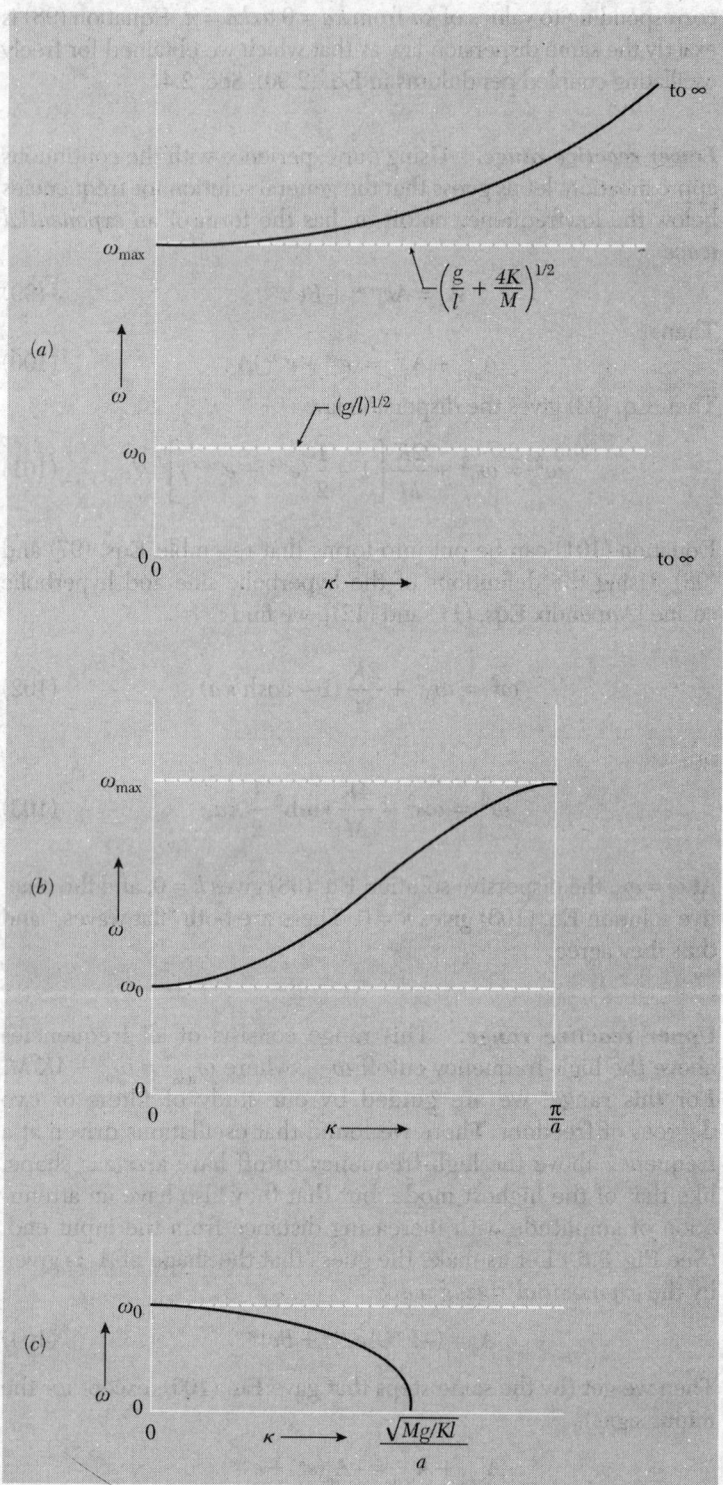

Fig. 3.13 *Complete dispersion relation for coupled pendulums. (a) Above highfrequency cutoff: the waves are zigzag exponential waves. (b) Dispersive frequency range: sinusoidal waves. (c) Below low-frequency cutoff: exponential waves.*

Than Eq. (93) gives the dispersion law

$$\omega^2 = \omega_0{}^2 + \frac{2K}{M}\left\{1 + \frac{1}{2}(e^{\kappa a} + e^{-\kappa a})\right\}$$

$$= \omega_0{}^2 + \frac{2K}{M}\left\{1 + \cosh \kappa a\right\} \tag{105}$$

$$= \omega_0{}^2 + \frac{4K}{M}\cosh^2\frac{1}{2}\kappa a. \tag{106}$$

At $\kappa = 0$, ω^2 is $\omega_0{}^2 + 4K/M = \omega_{max}{}^2$. Thus exactly at the high-frequency cutoff ω_{max} we have no attenuation.

In Fig. 3.13 we plot the exact dispersion law for all frequencies as given by Eqs. (98), (103), and (106).

Problems and Home Experiments

3.1 See Eq. (10) and fill in the algebraic steps omitted in obtaining the result $E = E_0 e^{-t/\tau}$.

3.2 Show by direct substitution that $x_1(t)$ as given by Eq. (3) is a solution of the damped harmonic oscillator equation of motion, Eq. (2).

3.3 Show that if $x_1(t)$ is a solution of Eq. (1) for a driving force $F_1(t)$, and if $x_2(t)$ is a solution for a different driving force $F_2(t)$, then the force $F(t) = F_1(t) + F_2(t)$ gives the solution $x(t) = x_1(t) + x_2(t)$, provided that the initial conditions $x(0)$ and $\dot{x}(0)$ for the superposition are also the corresponding sums of the initial conditions, i.e., provided $x(0) = x_1(0) + x_2(0)$ and $\dot{x}(0) = \dot{x}_1(0) + \dot{x}_2(0)$.

3.4 show by substitution that Eqs. (15), (16), and (17), give a solution to Eq. (14).

3.5 Transient beats. For this and some later experiments you will need a phonograph record turntable. In the present experiment, you are to drive a pendulum by means of the turntable. For a pendulum bob you can use a can of soup or something. A convenient turntable speed is 45 rpm. (What is the corresponding length of the pendulum string?) Tape a light cardboard box to the turntable and tape a pencil upright to the box. Slip a loop of string over the pencil. To the loop fasten one end of a 2 m to 2.5 m length of tied-together rubber bands, and fasten the other end to the string of the pendulum. Measure the frequency of the pendulum for free oscillations (an ordinary watch with a second hand is all you need).

Home experiment

Measure the beat frequency when the pendulum is driven. Try different string lengths. Add some damping by placing a book or board or something for the pendulum string to rub against. (A slot is best, i.e., a board or book on either side.) The reason for having such a long coupling rubber band is to make it a sufficiently weak spring. Also, it is best to couple the rubber band near enough to the top of the pendulum string so that the amplitude of motion of the string at that point is considerably less than the amplitude of motion of the pencil (on the turntable), even for large pendulum amplitude. This ensures a driving force independent of the pendulum amplitude.

3.6   Verify Eq. (22) for the power loss due to friction. Verify that it is equal to the input power as given by Eq. (21).

Home experiment   **3.7   Resonance in a damped slinky.**   Stretch the slinky out to 2.0 m or so and support it at both ends. One end should be clamped in such a way that the slinky can be easily released and reclamped with a different number of turns between the clamps. Drive the slinky with the phonograph turntable of Home Exp. 3.5, using a long coupling rubber band. Use the 45-rpm speed. Measure the free-oscillation frequency of the slinky. (Units of rpm are most convenient.) This frequency can be varied by changing the number of helical turns between the fixed supports. (See Home Exp. 2.1.) *Measure the mean decay time* τ. Increase the damping by adding a long strip of tape (the stretchy kind) along the slinky, so as to get a convenient damping time (say 10 to 20 sec). *Plot a resonance curve*, i.e., plot $|A|^2$ versus ω_0, with ω fixed at 45 rpm. Notice the phase relations and make sure you understand them. One way to measure $|A|$ is to use a light that casts sharp shadows (clear bulb instead of frosted bulb, i.e., a point source). Measure the positions on the wall or floor of the shadow of a piece of tape on the slinky. Calculate the expected full width at half-maximum, measured in slinky turns, *while* you are doing the experiment. (You may decide to shorten the damping time if you find it will give an inconveniently narrow resonance or too long a time for transient beats to die.)

Possible sources of trouble: If the rubber band ever relaxes completely and then "snaps" taut, the Fourier analysis of the force exerted by the rubber band will contain harmonics of 45 rpm as well as 45 rpm itself. These will excite harmonics of the slinky. It is an interesting trouble at least. *Another*: Twang the rubber band and watch it oscillate. Make sure that its oscillations are quite fast compared with one or two times 45 rpm; otherwise crazy things happen. You may find other problems. See if you can see the disappearance of the elastic amplitude and appearance of the absorptive amplitude when you are exactly at resonance, i.e., look at the relative phase of phonograph (pencil) and slinky. What do you get for the product of the resonant full width times the mean decay time? Does your result agree (within experimental errors) with Eq. (28)?

3.8 Forced oscillations in a system of two coupled cans of soup. The setup is shown in Fig. 3.3 and the theory is given in Secs. 3.3 and 3.4. The strings should be attached to sticks on which the strings can be rolled up or let down to vary the frequencies. Clamp the sticks to a bookcase or table or something. The string lengths should be variable over a range of 30 to 70 cm. As you vary the string lengths, you vary both ω_1^2 and ω_2^2 in such a way that their difference remains constant. Thus varying the string while holding the driving frequency constant is almost equivalent to varying the driving frequency while holding ω_1^2 and ω_2^2 constant. For given string lengths, measure the frequencies of the two modes (with the phonograph uncoupled). Then drive the system at 45 rpm. Drive the longitudinal oscillations (mainly) by aligning the coupling direction to be along the slinky. The longitudinal and transverse modes have the same sets of frequencies, as you can easily determine. This can lead to distracting, though interesting, effects—especially near a resonance. There are five interesting frequencies to hit, namely the two resonance frequencies and the regions far below, halfway between, and far above the resonances. Look at the filter characteristics above and below cutoff. Study the phase relationships—just look, and see if you understand. Transient beats can last a very long time if no damping is provided. It is best to damp the strings by having them rub on something. It is probably too time consuming to trace out the resonant curves. Don't bother. (You did that once, we hope. in Home Exp. 3.7.) Instead, measure the damping times of the two modes and calculate the expected resonant full widths Γ, using $\Delta\omega\tau = 1$. How close is your situation to that described in Fig. 3.4? Do the mechanical filter equations of Sec. 3.4 work?

An alternative way to vary the relative frequencies is, of course, to use the other turntable settings of 78, 33, and 16 rpm. Unfortunately these do not provide a continuous variable.

3.9 A jackhammer pounds the pavement at about 20 cps. The handle pounds the operator's hands at the same frequency. Design a low-pass filter for incorporation into the handle to reduce the vibration amplitude in the handle by a factor of 10. One way (called "brute force") is simply to increase the mass of the body of the tool (the part the hammer blade recoils against) by a factor of 10. Since the tool already weighs about 50 lbs, however, try an arrangement of springs and masses.

3.10 Verify that the time-averaged stored energy E for steady-state oscillation is given by Eq. (23).

3.11 Verify that the half-power points for the steady-state resonance curve are given by Eqs. (25) and (26).

3.12 Mechanical filter. (See Sec. 3.4.) A piece of delicate apparatus sits on a floor which has a vertical vibration of about 20 cps. You wish to attenuate this jiggling by a factor of 100, so you set the apparatus on a cushion. About how far down should the top of the cushion sink when the apparatus is placed

Home experiment

on it? (*Hint:* See the example following Eq. (58), Sec. 3.4. Also, approximate the cushion by a perfect Hooke's law spring.)   *Ans.* About 6 cm.

3.13 Show that Eq. (31) gives the exact steady-state solution to the driven oscillator equation (14), for the case where the damping constant Γ is zero.

3.14 Show that if the pendulums of Fig. 3.10 are coupled by slinkies, they have the same equations of motion for transverse horizontal oscillation as they do for the longitudinal motion shown.

3.15 Sketch a system of inductances and capacitances that has equations of motion similar in form to Eq. (62), and derive their equations of motion.

Home experiment **3.16 Mechanical bandpass filter.** With only two coupled pendulums, one cannot see the exponential character of the filtering be avior— any curve can be passed through two points. Put a third can of soup inside your slinky, halfway between the other two cans, and suspend it so as to have a system like that in Fig. 3.6 and 3.7. Drive the system above cutoff and below cutoff with the phonograph. Measure the ratios ψ_a/ψ_b and ψ_b/ψ_c. Are they equal? Should they be?

3.17 Assume the ionosphere starts suddenly at a boundary, at which the cutoff frequency v_p suddenly increases from zero to 20 Mc. Find the amplitude attenuation distance δ for AM radio waves of frequency 1000 Kc.

Ans. About 2.5 meters, independent of frequency, as long as the frequency is far below cutoff.

3.18 Using the coupled pendulums as a guide, write down the complete dispersion relation for the analogous system of coupled inductances and capacitances. We want the dispersion law in the pass band and in the two cutoff regions of frequency.

3.19 Show that, if we use the weak-damping approximation and if we stay reasonably near a resonance, the absorptive and elastic amplitudes can be written (with a suitable choice of units) in the form

$$A_{ab} = \frac{1}{x^2+1}, \quad A_{el} = \frac{-x}{x^2+1},$$

where $x = (\omega - \omega_0)/\frac{1}{2}\Gamma$.

3.20 Suppose we have a system with two resonances at frequencies ω_1 and ω_2 which make equal contributions to the elastic amplitude of some moving part. For ω far from both ω_1 and ω_2, we can write (in some units or other)

$$A_{el} = \frac{1}{\omega_1^2 - \omega^2} + \frac{1}{\omega_2^2 - \omega^2}.$$

Show that, if ω differs from ω_1 and ω_2 by much more than their difference $\omega_2 - \omega_1$, then A_{el} is (to a good approximation) just twice as large as either of the two contributions. That is, show that

$$A_{\text{el}} = \left(\frac{2}{\omega_{\text{av}}^2 - \omega^2} \right) \left\{ 1 + \epsilon^2 + \cdots \right\},$$

where

$$\omega_{\text{av}}^2 = \frac{1}{2} \left(\omega_1^2 + \omega_2^2 \right), \quad \epsilon = \frac{1}{2} \frac{\left(\omega_1^2 - \omega_2^2 \right)}{\omega_{\text{av}}^2 - \omega^2}.$$

3.21 Start with the exact dispersion law for the coupled pendulums given by Eqs. (98), (103), and (106). Assume you have $a/\lambda \ll 1$ and $a/\delta \ll 1$. Then the continuous approximation should be a good one. (Why?) Expand the dispersion formulas in a Taylor's series and keep only the first significant terms. Compare your result with that obtained in Sec. 3.5 in the continuous approximation.

3.22 Interminable transient beats. (See Sec. 3.2.) Verify the "amplitude-modulated almost harmonic-oscillation" form of the oscillator displacement for transient oscillation with zero damping, i.e., verify Eq. (43). Show that for zero damping and for driving frequency exactly at resonance the modulated amplitude builds up linearly with time [Eq. (45)].

3.23 Exponential penetration into reactive region. Setup a system of soup cans and slinky as in Fig. 3.12. Couple your phonograph drive system to one end of the dispersive region. Design the lengths so that 78 rpm is above the upper cutoff frequency, 45 rpm is in the pass band, and 33 rpm (and 16 rpm) are below cutoff. If you can design a quick and easy way to change all the string lengths at once by the same amount, you can vary ω_0^2 (and hence all the resonance frequencies)continuously, keeping the driving frequency constant, and look for resonances.

Home experiment

3.24 Transient beats. Verify Eq. (46), which gives the time dependence of the stored energy in a driven oscillator that starts with zero energy at time zero. Assume weak damping. Assume the driving frequency is near (but not exactly equal) to ω_1. Thus take $\omega/\omega_1 = 1$, where appropriate. (It is *not* appropriate to take $\omega = \omega_1$ in an expression like $\cos \omega t - \cos \omega_1 t$, because whatever small difference there is between ω and ω_1 eventually leads to a large effect, i.e., to a large relative shift in phase.)

3.25 Show that the result for the "overdamped" oscillator given by Eq. (9) follows from Eqs. (7) and (8). (*Hint:* First verify the identities $\cos ix = \cosh x$, $\sin ix = i \sinh x$; then use them.)

3.26 Critical damping. Starting with the equation for the underdamped free oscillations, Eq. (7), show that for critical damping the solution becomes

$$x_1(t) = e^{-(1/2)\Gamma t} \{ x_1(0) + [\dot{x}_1(0) + \tfrac{1}{2}\Gamma x_1(0)]t \}.$$

Show that this same result is obtained if you start instead with the equation for overdamped oscillations, Eq. (9).

Home experiment   **3.27   Resonant frequency width of a mailing tube.**   Read the paragraphs following Eq. (28). In the lowest normal mode of sound waves in a mailing tube open at both ends, the tube length is essentially one half-wavelength. (There is a small "end correction," so that the tube length is actually about one tube diameter less than a half-wavelength.) The velocity of sound is about 330 m/sec. If your tuning fork is C523 cps, the mailing tube will resonate loudest if its length is about 32 cm.

(*a*) Verify that statement. The resonant frequency v_0 of a tube of length L is thus given by

$$v_0 = \frac{523}{(L/L_0)} = \frac{\omega_0}{2\pi},$$

with L_0 about 32 cm. (L_0 will not be exactly 32 cm, because of the end effect mentioned above.)

(*b*) Verify that formula. Now cut 5 or 6 mailing tubes with values of L judiciously chosen so as to "cover" the peak of the resonance curve and the two half-power points on either side of the peak. The sound intensity I is expected to have a "resonant shape" given by

$$I = \frac{\left(\frac{1}{2}\Gamma\right)^2}{(\omega_0 - \omega)^2 + \left(\frac{1}{2}\Gamma\right)^2},$$

where we have normalized I to equal 1.0 at $\omega = \omega_0$. In your experiment the driving frequency v is that of the tuning fork and is therefore constant. The resonant frequency v_0 is varied by changing the tube length. You are to find the tube length L_0 for resonance, including the end correction (this is most easily done by ear, tapping the tube on your head and comparing the pitch with that of the tuning fork), and you are to find the two tube lengths that correspond to the half-power points. Thus you are to find .the full-width Γ. That will give you, indirectly, the decay time for free oscillation. Your main experimental problem is to devise a reasonably simple method of estimating a decrease in sound intensity by a factor of two.

3.28 Two coupled pendulums as a mechanical bandpass filter.   Consider the system shown in Fig. 3.3 and described in Sec. 3.3. Neglect damping. Show that

$$\psi_a \approx \frac{F_0}{2M}\cos\,\omega t\left\{\frac{1}{\omega_1{}^2 - \omega^2} + \frac{1}{\omega_2{}^2 - \omega^2}\right\},$$

$$\psi_b \approx \frac{F_0}{2M}\cos\,\omega t\left\{\frac{1}{\omega_1{}^2 - \omega^2} + \frac{1}{\omega_2{}^2 - \omega^2}\right\},$$

and

$$\frac{\psi_b}{\psi_a} \approx \frac{\omega_2{}^2 - \omega_1{}^2}{\omega_2{}^2 + \omega_1{}^2 - 2\omega^2},$$

where ω_1 is; the lower of the two mode frequencies, ω_2 is the higher, and ω is the driving frequency

3.29 Electrical bandpass filter. Consider the filter shown in Fig. 3.8. Find the differential equation for I_a and for I_b. Show that the normal coordinates are $I_a + I_b$ and $I_a - I_b$, and that the modes are given by Eqs. (59).

3.30 Coupled pendulums. Consider a linear array of coupled pendulums driven below cutoff at $z = 0$ and attached to a rigid wall at $z = L$, as shown in Fig. 3.11. Show that if $\psi(z, t)$ equals $A_0 \cos \omega t$ at $z = 0$, then $\psi(z, t) = A(z) \cos \omega t$, where

$$A(z) = A_0 \frac{[e^{-\kappa z} - e^{-\kappa L} e^{-\kappa(L-z)}]}{1 - e^{-2\kappa L}}.$$

Notice that for $L \to \infty$ this become simply $A_0 e^{-\kappa z}$.

3.31 Resonance in a system of coupled pendulums. Read the discussion following Eq. (90). Find the resonant values of ω^2 as follows. (*a*) Show that at resonance one has
$$k \cot kL = -\kappa.$$
That shows that resonant values of $\theta \equiv kL$ must lie in quadrant 2 (90 deg to 180 deg), quadrant 4 (270 deg to 360 deg), quadrant 6, quadrant 8, etc. (*b*) Let Ka^2/ML^2 equal "one unit" of return force per unit displacement per unit mass, i.e., of ω^2. Let $g/l_1 = \omega_1^2$, $g/l_2 = \omega_2^2$. Then show that the resonance values of ω^2 are obtained by plotting versus θ the two function

$$\omega^2 = \omega_1^2 + \theta^2$$
$$\omega^2 = \omega_2^2 - \theta^2 \cot^2 \theta.$$

The resonances are given by half of the points of intersection of the two curves. Why only half? (*Note:* ω^2, ω_1^2, and ω_2^2 are dimensionless in the above equations; i.e., they are given in units Ka^2/ML^2.) Make a sketch showing a typical plot that gives the resonance frequencies. What happens at very high frequency?

3.32 Total reflection of visible light from a silvered mirror. Assume that the "valence" electron of a silver atom becomes a "free" electron in solid silver. Look up (in *Handbook of Physics and Chemistry*) the valence of silver, its atomic weight, and its mass density. Thus find the number density N of free electrons per unit volume in solid silver. Assume that the dispersion relation for light in silver has the same form as for light (or other electromagnetic waves) in the io losphere, i.e., assume

$$\omega^2 = \omega_p^2 + c^2 k^2, \qquad \text{if } \omega^2 \gtreqqless \omega_p^2,$$
$$\omega^2 = \omega_p^2 - c^2 \kappa^2, \qquad \text{if } \omega^2 \lesseqqgtr \omega_p^2,$$

where $\omega_p^2 = 4\pi Ne^2/m$, and e and m are the charge and mass of the electron.

(*a*) Calculate the cutoff frequency v_p for solid silver. Show that for visible light the frequency v of the light is below cutoff. Therefore we expect that a sufficiently thick layer of silver will give total reflection for normally incident visible light. That is what gives the "silvery" appearance of a sil-vered mirror.

(*b*) Calculate the mean attenuation distance δ for red light of vacuum wave-length 0.65×10^{-4} cm and for blue light of vacuum wavelength 0.45×10^{-4} cm. A "halfsilvered" mirror is a slab of glass with a silver layer which is somewhat thinner than an attenuation distance, so that about half of the light gets through (i.e., the reflection is not total). Suppose you look at a "white" light bulb through a piece of half-silvered mirror. (The "white" light actually contains all visible colors.) Would you expect the transmitted light to look white? Have a bluish tinge? Have a reddish tinge? What about the reflected light?

(*c*) How thick must the silver layer be in order to reduce the intensity (which is proportional to the square of the amplitude) of blue light at the back side of the layer by a factor of l00? Such a mirror should reflect 99% of the incident light. (Actually the reflectivity is more like 9.5%, for visible light. We have neglected energy loss due to the resistivity of the silver. Also, the surface may tarnish, acquiring a layer of silver oxide with properties quite different from those of metallic silver.)

(*d*) For what frequencies should the silver layer become transparent? (Give the vacuum wavelength also. This is called "ultraviolet" light.)

Home experiment   **3.33 Sawtooth shallow-water standing waves.** These are described in Prob. 2.31. We wish now to learn how to excite the low-est sawtooth modes in a pan of water. The lowest sawtooth mode is the sloshing mode; there is only one-half a "tooth." The surface is flat. The pan length is one half-wavelengt. The next sawtooth mode would have one complete tooth, i.e., the pan length would be one wavelength (of the lowest Fourier component of the sawtooth). This mode is *not* excited if you excite the oscillations by pushing the pan back and forth. *Explain why it is* not *excited.* The next mode has $1\frac{1}{2}$ teeth, i.e., three flat regions. The pan length is thus three halfwavelengths. You *can* excite this mode. Use trial and error, gently shaking the pan. When you think you have it, release the pan and let it oscillate freely. After some practice, you can recognize the mode and can excite it. Here is a more systematic method: Borrow a metronome, or make one by hanging a weight on a string (a pendulum) with the bob hitting a piece of paper or something to make a noise. For a given metronome setting, gently shake the pan at the metronome tempo and wait for a steady state. Vary the metronome tempo to search for resonance. When you are near the resonance, watch the *transient beats!* They are not only beautiful; they also tell you about how far you are from resonance. Calculate the expected resonance frequency, using $v = \sqrt{gh}$. Do this *while* you are doing the experiment, so as to zero in rapidly on the resonance (i.e., so you don't search in the wrong ballpark) When you hit the resonance, let go of the pan and let it oscillate freely; time the free oscillations.

If you use a sufficiently lightweight pan so that the mass is mostly due to the water and not to the pan, and if you have a total mass that is big enough to give a significant reactive push on your hands, you can zero in on the resonance by a combination of feeling in your hands and watching. Then you don't need a metronome.

You may see cusplike "points" if you excite sawtooth waves in both horizontal directions. When these break and toss water into the air, you can be sure that no linear wave theory could ever explain them!

3.34 Rectangular two-dimensional standing surface waves on water. Obtain an icebox dish of rectangular shape (not one of stiff Lucite but the more pliable polyethylene kind). Fill it to the brim with water, and then overfill it so that the water surface bulges over the top. (This reduces the damping by the sides.) Tap gently and seethe grid of freely oscillating standing waves. Obtain a toy gyroscope (at any toy store, or, e.g., see the catalog of Edmund Scientific Co., Barrington, New Jersey 08007) Hold the spinning gyroscope against the side of the dish (or, for example, couple through an inverted pie pan on which both are set). You can watch the wavelength of the forced oscillations (the standing waves) gradually lengthen as the gyro slows down. You will probably also notice the effect of passing through resonance.

Home experiment

3.35 Standing waves in water. (*a*) Dip a vibrating tuning fork in water and look at the waves, especially between the prongs.

(*b*) Hold a vibrating tuning fork flat on a water surface (like two parallel floating logs) and look between the prongs. (Some of the modes of the fork are rapidly damped. There is one which persists for several seconds.) Try illuminating with a small light source at various angles (parallel and perpendicular to the prongs) to see the amazing amount of structure.

Home experiment

3.36 Harmonics and subharmonics. Given a harmonic oscillator with natural oscillation frequency $\nu_0 = 10$ cps and a very long decay time. If this oscillator is driven with a harmonically oscillating force at frequency 10 cps, it will acquire a large amplitude, i.e., it will "resonate" at the driving frequency. No other *harmonically* oscillating driving force will produce a large amplitude (a resonance).

(*a*) Justify the preceding statement. Next, suppose the oscillator is subject to a force that is a periodically repeated square pulse of duration 0.01 sec, repeated once per second.

(*b*) Describe qualitatively the Fourier analysis of the repeated square pulse.

(*c*) Will the harmonic oscillator "resonate" (acquire a large amplitude) under the influence of that driving force?

(*d*) Suppose the driving force is the same square pulse (of width 0.01 sec) but repeated twice per second. Will the oscillator resonate? Answer the same question for repetition rates of 3/sec, 4/sec, 5/sec, 6/sec, 7/sec, 8/sec and 9/sec.

(*e*) Now we come to something new. What if we drive the same oscillator with the same square pulse at a repetition rate 20/sec? Will the oscillator resonate? Note that the oscillator frequency is in this case a *subharmonic* of the fundamental repetition rate of the driving square pulse.

(*f*) Similarly consider driving forces consisting of repeated square pulses at 3, 4, etc., times the oscillator frequency. Does the oscillator resonate? Explain.

(g) Now we come again to something different. Suppose that the driving force is only coupled to the oscillator when the oscillator displacement from equilibrium is positive. For example, that is the case when you push a child on a swing. You only push when her displacement puts her within reach of your arms (the driving force). Reconsider the question of whether you can excite subharmonics in this case of "asymmetric coupling." Suppose the "swing" oscillates at 1 cps. If you push (blindly, whether or not the swing is there) at 2 cps, will the swing resonate? What if you push at 3 cps? at 3.5 cps? Explain. Now explain how it is that a high-frequency driving force from (for example) an airplane engine can excite a resonance at a much lower frequency that is a subharmonic of the driving frequency, i.e., is $\frac{1}{2}$, $\frac{1}{3}$, etc., of the driving frequency. Would you expect the excitation of subharmonics to be common in systems that can shake and rattle? Explain.

Traveling Waves

Chapter 4 Traveling Waves

4.1 Introduction

The systems we considered in Chaps. 1, 2, and 3 were *closed systems*, i.e., systems enclosed by definite boundaries, so that all the energy remains within the confines of the system. We found that the free oscillations of a closed system can be described in terms of a superposition of standing waves, i.e., modes, and that the steady-state forced oscillations can be described in terms of a superposition of standing-wave contributions arising from the modes. The character of the modes present is determined by the boundary conditions.

Open systems.   In Chap. 4, we shall consider forced oscillations of *open systems*, i.e., systems which have *no outer boundary*. For example, if a man playing a trumpet is suspended by a long rope from the gondola of a balloon high above the earth, the air acts like an open system or open medium for sound waves, at least to an extent that enables us to neglect echoes, reflections from the earth back to the trumpeter. If the same trumpeter plays instead in a closed room having hardwood floor, walls, and ceiling, the effect will be quite different. In that case, the air in the room acts like a closed system and will resonate at its mode frequencies, if suitably driven. However, if the walls of the room are covered by perfectly sound-absorbing material, so that no waves are ever reflected back to the transmitter (the trumpet), then the room behaves as if it were a completely open system with no outer boundaries. Thus it is not necessary that the medium actually extend to infinity for it to be an open system.

The waves produced by a driving force that is coupled to an open medium are called *traveling waves*—they travel away from the disturbance that created them. Traveling waves have the important property that they transport energy and momentum. Thus if you drop a rock into a quiet pond, the expanding circular waves that propagate outward from the splash may later impart kinetic energy to a distant floating bug or may increase the gravitational potential energy of a twig lying half in and half out of the water on a sandy beach by washing it up onto the beach.

If a driving force (coupled to the open medium) oscillates with harmonic motion, the traveling waves it produces are called *harmonic traveling waves*. At steady state, all moving parts of the system oscillate with harmonic motion at the driving frequency.

Amplitude relations.   If the waves are spreading out in two or three dimensions, the amplitude of the motion is smaller the farther the moving part is from the source of the waves (assuming the source is small). On the other hand, if the medium is one-dimensional (as, for example, a stretched string driven at one end and extending to infinity or ter minating in a wave-absorbing device), then the amplitude of harmonic motion of the moving parts does not decrease with distance from the source (assuming the medium is homogeneous). That is the case not only for one-dimensional waves (as on a string), but also two-dimensional "straight waves" (ocean swells from a distant storm) and three-dimensional "plane waves" (radio waves from a distant star).

Phase relations. The relative phase between two different moving parts in an open medium carrying a harmonic traveling wave is very different from that for a standing wave in a closed system. In the case of a standing wave, which may be either a normal mode of free oscillation of a closed system or a forced oscillation of a closed system, ail the moving parts oscillate in phase with one another (except for possible minus signs). In a traveling wave, that is not the case. Instead, if moving part b is farther from the driving force than moving part a, then b goes through the same motion as a but at a later time, because of the time that the wave takes to travel from a to b. Thus b has a phase constant which differs from that of a by an amount equal to the frequency times the time delay.

4.2 *Harmonic Traveling Waves in One Dimension and Phase Velocity*

Suppose we have a one-dimensional system consisting of a continuous, homogeneous string stretching from $z = 0$ to infinity. The string is attached at $z = 0$ to the output terminal of some device ("transmitter") that can shake the string and thus "radiate" traveling waves along the string. Suppose that the displacement $D(t)$ of the output terminal is given by the harmonic oscillation

$$D(t) = A \cos \omega t. \tag{1}$$

We wish to find the displacement $\psi(z, t)$ of a moving part located at position z, where z is anywhere between $z = 0$ and infinity. We can easily find $\psi(z,t)$ at $z = 0$. Since the string is tied directly to the output terminal of the transmitter, the displacement of the string at $z = 0$ is equal to $D(t)$:

$$\psi(0, t) = D(t) = A \cos \omega t. \tag{2}$$

Phase velocity. Now, from our common experience of watching traveling water waves, we know that they travel with constant velocity as long as the properties of the medium (water depth, for example) remain constant. When the waves are harmonic traveling waves, this velocity is called the *phase velocity v_φ*. We also recognize that the motion of a moving part at position z at time t is the same as that of the moving part at $z = 0$ at the earlier time t', where t' is earlier than t by the time that the wave takes to travel the distance z at velocity v_φ:

$$t' = t - \frac{z}{v_\varphi}.\tag{3}$$

Thus we have the form of a traveling sinusoidal wave,

$$\psi(z, t) = \psi(0, t')$$

$$= A \cos \omega t'$$

$$= A \cos \omega \left(t - \frac{z}{v_\varphi} \right)$$

$$= A \cos \left(\omega t - \frac{\omega z}{v_\varphi} \right).\tag{4}$$

Notice that at fixed z, $\psi(z, t)$ is a harmonic oscillation in time. Notice also that at fixed t, $\psi(z, t)$ is a sinusoidal oscillation in space. Of course, both of these statements also hold for a sinusoidal standing wave, which can have the form, for example,

$$\psi(z, t) = B \cos \omega t \cos (a - kz),\tag{5}$$

where a is a constant. For fixed time, the space dependence of the traveling wave given by Eq. (4) has the same form as that of the standing wave of Eq. (5). Thus, if we write the traveling wave in the form

$$\psi(z, t) = A \cos (\omega t - kz),\tag{6}$$

then we can use the same concept of wavenumber k (and wavelength λ) for a sinusoidal traveling wave at fixed time as we have been using for a standing wave. By comparison of Eqs. (4) and (6) we see that for a sinusoidal traveling wave at a fixed time the rate of increase of phase angle per unit length, k, is given by

$$k = \frac{\omega}{v_\varphi},\tag{7}$$

i.e., the phase velocity is given by

$$\boxed{v_\varphi = \frac{\omega}{k},}\tag{8a}$$

or, since $\omega = 2\pi\nu$ and $k = 2\pi/\lambda$,

$$\boxed{v_\varphi = \lambda\nu,}\tag{8b}$$

or, since $v = 1/T$,

$$\boxed{v_\varphi = \frac{\lambda}{T}} \quad . \tag{8c}$$

The phase velocity of a sinusoidal traveling wave is an extremely important quantity. We give the various forms of Eqs. (8) and urge you to learn each of them forward, upside down, and backward. In Fig. 4.1 we show a sinusoidal traveling wave.

   Equations (8) are so important that we now give an alternative derivation. We define the *phase function* $\varphi(z, t)$ of a sinusoidal traveling wave propagating in the $+z$ direction to be the argument of the wave function $\cos (\omega t - kz)$:

$$\varphi(z, t) = \omega t - kz. \tag{9}$$

[We have suppressed a possible phase constant in Eq. (9).] At a given z, the phase increases linearly with time in the term ωt. At a given time t, the phase decreases linearly with z in the term $-kz$. Going to greater z decreases the phase because it corresponds to waves emitted at earlier times. (Our sign convention for positive phase is not universal; some people prefer to call $kz - \omega t$ the phase.) If we wish to follow a given wave crest [maximum of $\cos \varphi(z, t)$] or trough [minimum of $\cos \varphi(z, t)$] while the wave propagates, we must keep looking at different z as t changes, in order to keep the phase $\varphi(z, t)$ constant. Thus by taking the total differential of $\varphi(z, t)$ and setting the result equal to zero, we can find the relation between z and t for a point of constant phase. The total differential of $\varphi(z, t)$ is given by

$$d\varphi = \left(\frac{\partial \varphi}{\partial t}\right) dt + \left(\frac{\partial \varphi}{\partial z}\right) dz = \omega \, dt - k \, dz, \tag{10}$$

which is zero provided dt and dz are related by

$$v_\varphi \equiv \left(\frac{dz}{dt}\right)_{[d\varphi = 0]} = \frac{\omega}{k}, \tag{11}$$

which is Eq. (8a).

Do traveling waves have the same dispersion relation as standing waves?   In Chap. 2 we found that the dispersion relation giving ω as a function of k (or k as a function of ω) for freely oscillating standing waves in a given medium does not depend on the boundary conditions, although the particular values of k do. In Chap. 3 we found that the standing waves that result from forced oscillations of a closed system satisfy the very same dispersion law as freely

Fig. 4.1 Driving force at z = 0 describes harmonic motion of period T. Sinusoidal traveling wave propagates in +z direction. The wavelength is λ. The phase velocity is λ/T = ω/k = λν. Every point on the string undergoes the same harmonic motion as that at z = 0, but at a later time.

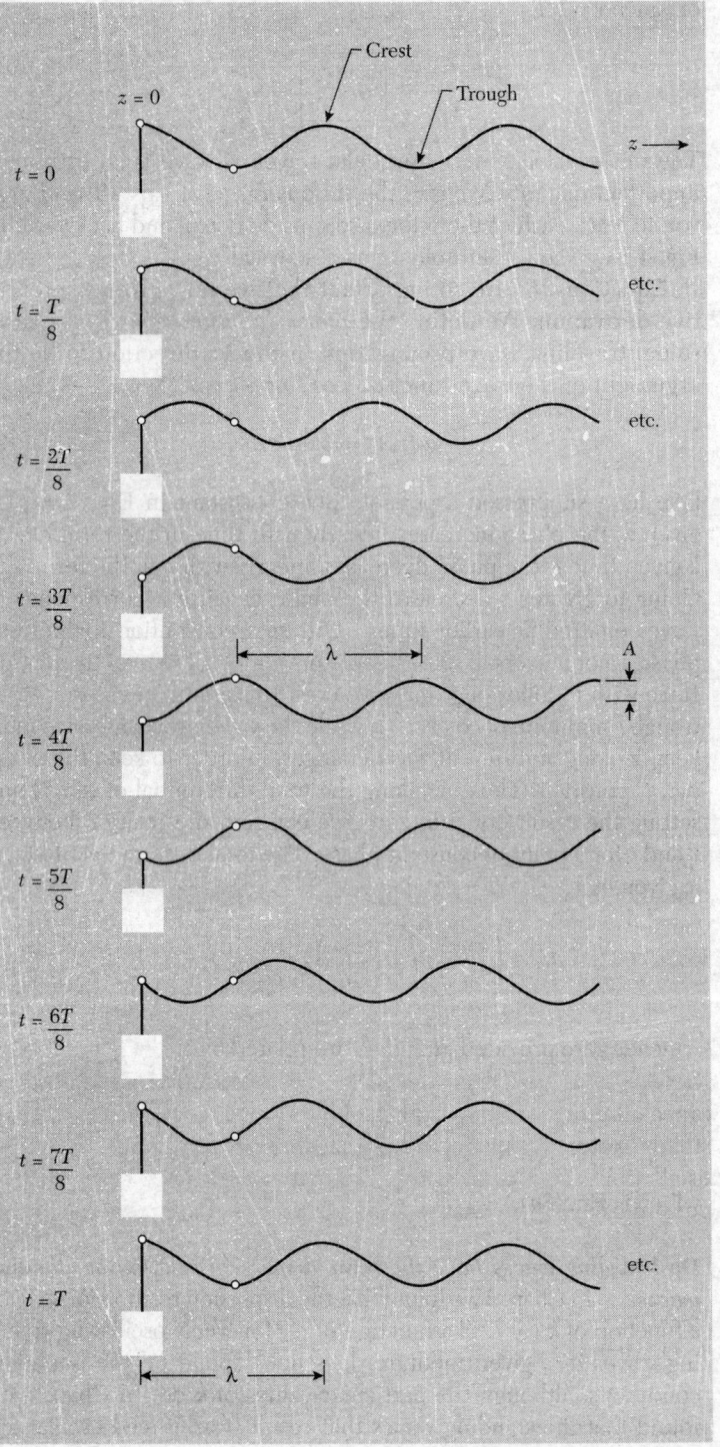

oscillating standing waves, with particular values of k that depend on boundary conditions. (We also discovered a new kind of wave, the exponential wave, for a system driven above or below its maximum and minimum mode frequencies.) In our present study of traveling waves in open systems, there are no boundary conditions other than that at the end coupled to the transmitter. We expect that the dispersion relation will (as before) be independent of boundary conditions. However, there is one thing about traveling waves that is quite different from standing waves due either to free or to forced oscillations of a closed system, and that is the relative phase of different moving parts. In the case of both the free and the forced oscillations (with damping neglected), all moving parts have the same phase. That is not so for traveling waves. Might not that affect the dispersion relation? No, as we shall now demonstrate.

Dispersion law for linear array of coupled pendulums. Let us consider a specific example, but one that will be sufficiently general to convince us that, indeed, all dispersion relations have the same form for traveling waves as for standing waves. Now, in introducing the concept of a traveling wave, we used the continuous string as a simple example. But of course we can have traveling waves in lumped parameter systems as well as in continuous systems, just as is the case for standing waves . Therefore, in order to obtain a very general result, we consider that wonderfully rich system, the coupled pendulums. We shall find the *exact dispersion law for an infinite linear array of coupled pendulums* which is driven at $z = 0$. We ask you to look at Fig. 3.10, Sec. 3.5, which shows a general configuration of three successive coupled pendulums, and to convince yourself that the exact equation of motion of bob n is as given in Eq. (3.62), Sec. 3.5, which we recopy here:

$$\ddot{\psi}_n = -\frac{g}{l}\,\psi_n + \frac{K}{M}\,(\psi_{n+1} - \psi_n) - \frac{K}{M}\,(\psi_n - \psi_{n-1}). \qquad (12)$$

Since all moving parts should oscillate with harmonic motion for steady state traveling waves, as well as for steady-state forced oscillations of a closed system, we know that, whatever the phase constant for ψ_n may be, we must have

$$\ddot{\psi}_n = -\omega^2 \psi_n. \qquad (13)$$

Inserting Eq. (13) into Eq. (12), collecting terms, and dividing by $\psi_n(t)$, we find

$$\omega^2 = \frac{g}{l} + \frac{2K}{M} - \frac{K}{M}\frac{(\psi_{n+1} + \psi_{n-1})}{\psi_n}. \qquad (14)$$

Sinusoidal traveling wave. Now we assume that we have a *sinusoidal traveling wave* of the form

$$\psi_n = A \cos(\omega t + \varphi - kz), \qquad z = na;$$

then, as you can easily show,

$$\psi_{n+1} + \psi_{n-1} = 2\psi_n \cos ka.$$

Thus Eq. (14) becomes

$$\omega^2 = \frac{g}{l} + \frac{2K}{M}(1 - \cos ka), \qquad (15)$$

i.e.,

$$\omega^2 = \frac{g}{l} + \frac{4K}{M}\sin^2 \tfrac{1}{2}ka. \qquad (16)$$

This is the very same dispersion law that we found in Sec. 3.5, Eqs. (3.91) through (3.98), for forced oscillations. We see that the frequency range for sinusoidal waves is the same for traveling as for standing waves; it extends from ω_{min} to ω_{max}, where

$$\omega_{min}^2 = \frac{g}{l} \equiv \omega_0^2, \qquad \omega_{max}^2 = \frac{g}{l} + \frac{4K}{M}. \qquad (17)$$

Exponential waves in an open system. For driving frequencies below the low-frequency cutoff ω_0, we may guess that the dispersion law for a driven open system will again be the same as that for a closed one. This is correct. Thus for an open system of coupled pendulums extending from $z = 0$ to $+\infty$ and driven at $z = 0$ at frequency $\omega < \omega_0$, we have

$$\psi(z, t) = Ae^{-\kappa z}\cos \omega t, \qquad z = na, \qquad (18)$$

$$\omega^2 = \omega_0^2 - \frac{4K}{M}\sinh^2 \frac{1}{2}\kappa a. \qquad (19)$$

Exponential zigzag waves. Similarly, for a driving frequency above the upper cutoff frequency, we obtain the *exponential zigzag waves*

$$\psi(z, t) = A(-1)^n e^{-\kappa z}\cos \omega t, \qquad z = na, \qquad (20)$$

$$\omega^2 = \dot{\omega}_0^2 + \frac{4K}{M}\cosh^2 \frac{1}{2}\kappa a. \qquad (21)$$

Thus an exponential wave in a driven open system differs from that of the general case of the driven closed system only in that we must discard the solution $e^{+\kappa z}$ that goes to infinity at $z = +\infty$. Notice that in an exponential wave all moving parts oscillate with the same phase constant [see Eqs. (18) and (20)]; thus there is no such thing as a phase velocity, because there is no waveform that propagates without change of shape, nor even a waveform that propagates with change of shape but with recognizable wave crests and troughs.

Thus we have demonstrated with the example of the coupled pendulums that for a given medium the dispersion law, which relates ω and k, is the same for traveling waves as it is for standing waves due to free oscillations or to steady-state forced oscillations of a closed system.

Dispersive and nondispersive sinusoidal waves. When the dispersion law has the simple form

$$v(k) = \frac{\omega(k)}{k} = \text{constant, independent of } k, \qquad (22)$$

the waves are said to be *nondispersive*; otherwise they are called *dispersive*. (The use of the symbol k implies in either case that they are sinusoidal.) A dispersive wave that is a superposition of traveling waves with different wavenumbers will change its shape as the superposition progresses in space, because different wavelength components travel at different speeds. The different frequency components of the superposition thus become "dispersed." *Dispersive waves are sinusoidal waves for which the phase velocity $v_{\varphi} = \omega/k$ varies with wavelength.*

Reactive exponential waves. When the driving frequency ω is not in the "pass band" between the low-frequency cutoff (which may be at zero frequency, in some examples) and the high-frequency cutoff (which may be at infinite frequency, in some examples), then, as we have seen, the waves are exponential (not sinusoidal) in their space dependence. This kind of exponential wave is sometimes said to be "reactive," whereas a sinusoidal wave is said to be "dispersive." Sometimes one speaks of a "dispersive medium" or a "reactive medium." Of course the same medium can be dispersive in one frequency range (the pass band) and reactive in another range (outside the pass band).

In the following examples we deal with phase velocities of dispersive waves.

Example 1: Transverse waves on a beaded string

The dispersion relation† for transverse waves on a beaded string with equilibrium tension T_0, bead mass M, and bead spacing a is [see Eq. (2.70), Sec. 2.4]

$$\omega^2 = \frac{4T_0}{Ma} \sin^2 \frac{1}{2} ka, \qquad 0 \leq k \leq \frac{\pi}{a}. \qquad (23)$$

---

†A very fine experimental demonstration of this dispersion relation, Eq. (23), is is given by J. M. Fowler, J. T. Brooks, and E. D. Lambe. "One-dimensional Wave Demonstration," *Am J. Phys.* **35**, 1065 (1967).

Therefore the phase velocity for transverse traveling waves is given by

$$v_\varphi^{\ 2} = \frac{\omega^2}{k^2} = \frac{4T_0}{Ma} \frac{\sin^2 \frac{1}{2}ka}{k^2}, \tag{24}$$

for $0 \le ka \le \pi$. For frequencies above the high-frequency cut-off, which is $\omega_0 = \sqrt{4T_0/Ma}$, the waves are zigzag exponential waves and there is no such thing as a phase velocity. For frequencies between zero and ω_0, the waves are dispersive waves, since the phase velocity is not a constant but depends on k. In the long-wavelength (or small bead spacing) limit, where we have $a/\lambda \ll 1$, the phase velocity becomes essentially independent of wavelength, so that the waves become nondispersive. We can see that by expanding $\sin \frac{1}{2}ka$ in a Taylor's series:

$$v_\varphi = \sqrt{\frac{T_0 a}{M}} \frac{\sin(\frac{1}{2}ka)}{(\frac{1}{2}ka)}$$

$$= \sqrt{\frac{T_0 a}{M}} \frac{(\frac{1}{2}ka) - \frac{1}{6}(\frac{1}{2}ka)^3 + \cdots}{(\frac{1}{2}ka)}$$

$$= \sqrt{\frac{T_0 a}{M}} \left[1 - \frac{1}{24}(ka)^2 + \cdots \right]. \tag{25}$$

Then, defining ρ_0 as the average mass per unit length at equilibrium, i.e., $\rho_0 \equiv M/a$, we have for the *continuous string* that

$$v_\varphi = \sqrt{\frac{T_0}{\rho_0}}. \tag{26}$$

Thus the phase velocity for transverse traveling waves on a continuous string is a constant, independent of frequency. Equation (26) is identical with the result for ω/k that we obtained in Chap. 2 for the dispersion law of standing waves on the continuous string [Eq. (2.22), Sec. 2.2].

Example 2: Longitudinal waves on a beaded spring

The dispersion law can be obtained from that for the transverse waves by simply replacing the tension T_0 by the spring constant K times the bead spacing a [see Eq. (2.78), Sec. 2.4]. In the continuous limit, we obtain [substituting Ka for T_0 in Eq. (26)]

$$v_\varphi = \sqrt{\frac{Ka}{\rho_0}} = \sqrt{\frac{K_L L}{\rho_0}}, \tag{27}$$

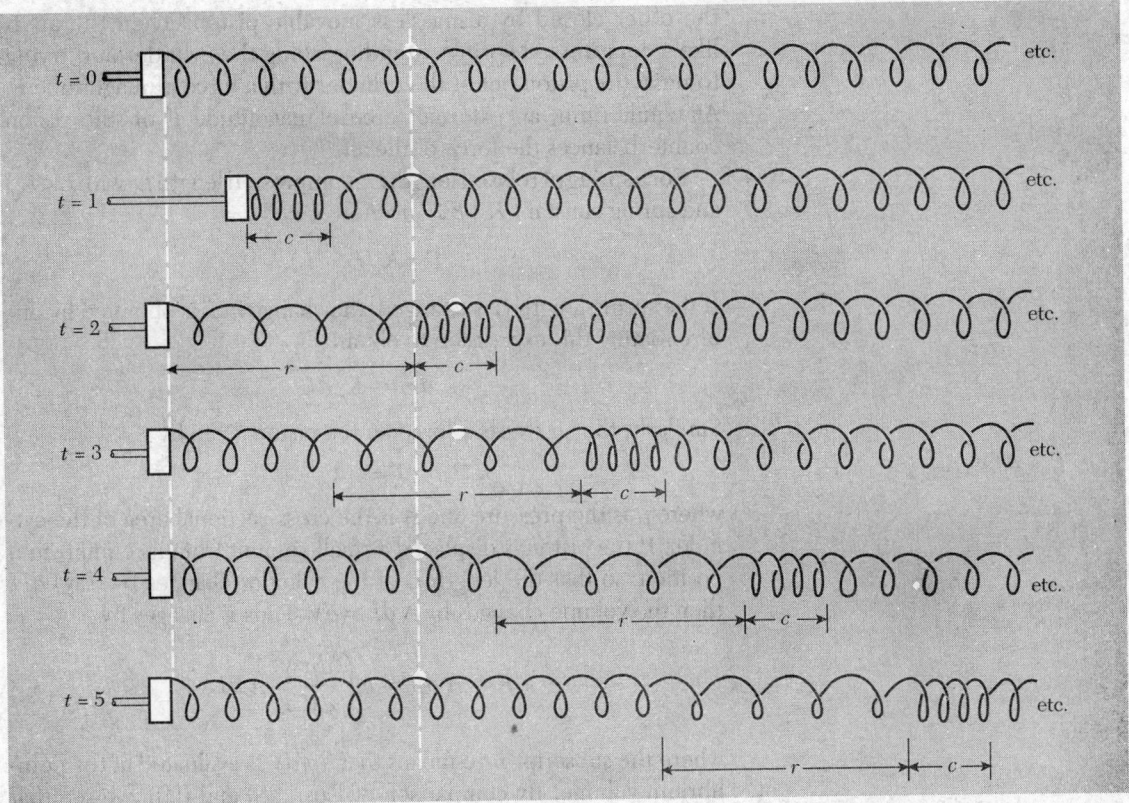

Fig. 4.2 Longitudinal traveling wave, consisting of a single compression c and a single rarefaction r, propagating on a spring. The sixth helical turn has a marker on it so you can follow its motion.

where we wrote $Ka = K_L L$ to remind you that if you add springs in series to make a long spring of total length L, the total spring constant K_L is just a/L times the spring constant K of one spring segment of length a. According to Eq. (27), longitudinal waves on a continuous spring are nondispersive. In Fig. 4.2 we show a traveling "wave packet," consisting of one "compression" and one "rarefaction," traveling on a spring.

Phase velocity of sound—Newton's model. Newton was the first to derive an expression predicting the velocity of sound waves in air. Newton's formula gives the wrong answer; it predicts a velocity of about 280 m/sec, whereas the observed velocity is 332 m/sec [at STP (standard temperature and pressure), i.e., at 1 atmosphere pressure and at a temperature of 0°C]. His derivation is very simple, and the reason it gives the wrong answer is quite interesting. Here is his derivation.

If air is confined in a closed container, it exerts an outward pressure on the walls of the container. Thus the air acts like a *compressed* spring which would like to extend itself. Suppose that the container is a long cylinder with one end closed by a wall and the other closed by a massless movable piston. Then the air is like a compressed spring extending along the cylinder and trying to push the piston out of the cylinder with a force of magnitude F. At equilibrium, an external force of magnitude F on the piston counterbalances the force of the air.

For a spring of relaxed length L_1, compressed length L (with $L < L_1$) and spring constant K_L, F is given by

$$F = K_L(L_1 - L).$$

If the spring length L is changed, the change in F is obtained by differentiating this expression to obtain

$$dF = -K_L \, dL. \tag{28}$$

Similarly, the air exerts a force on the piston given by

$$F = pA,$$

where p is the pressure and A is the cross-sectional area of the cylinder. If the piston is displaced a small amount from its equilibrium position, so that the length L of the cylinder changes by (say) dL, then the volume changes by $A \, dL = dV$. Thus F chaages by

$$dF = A \, dp = A\left(\frac{dp}{dV}\right)_0 A \, dL, \tag{29}$$

where the subscript zero means that dp/dV is evaluated at the equilibrium volume. By comparison of Eqs. (28) and (29), we see that the "equivalent spring constant" of the air in the tube is given by

$$K_L = -A^2\left(\frac{dp}{dV}\right)_0. \tag{30}$$

Suppose we have a compressed spring of spring constant K_L held at equilibrium with length L_0 and linear mass density ρ_0 (linear). Then the phase velocity for longitudinal waves is given by [see Eq. (27)]

$$v^2 = \frac{K_L L_0}{\rho_0(\text{linear})}. \tag{31}$$

In adapting Eq. (31) for sound, we use Eq. (30) for K_L. We also have that $AL_0 = V_0$, the equilibrium volume, and that the linear mass density is given by

$$\rho_0(\text{linear}) L_0 = \rho_0(\text{volume}) AL_0, \tag{32}$$

where ρ_0(volume) is the volume mass density at equilibrium. Inserting Eqs. (30) and (32) into Eq. (31) and dropping the qualifier "volume" from the volume mass density at equilibrium, ρ_0, we get for the *velocity of sound*

$$v^2 = -\frac{V_0 \left(dp/dV\right)_0}{\rho_0}.$$ (33)

We must still find dp/dV, the rate of change of pressure with volume. Here Newton used *Boyle's law*, which says that at constant temperature the product of pressure times volume is constant:

$$pV = p_0 V_0, \quad p = \frac{p_0 V_0}{V},$$ (34)

where p_0 is the equilibrium pressure. Differentiating gives

$$\frac{dp}{dV} = -\frac{p_0 V_0}{V^2},$$

i.e., at equilibrium, with $V = V_0$, we have

$$V_0 \left(\frac{dp}{dV}\right)_0 = -p_0.$$ (35)

Thus Eq. (33) becomes *Newton's result*

$$v_{\text{Newton}} = \sqrt{\frac{p_0}{\rho_0}}.$$ (36)

For air at STP, we have

$$p_0 = 1\,\text{atm} = 1.01 \times 10^5\,\text{N/m}^2$$ (37)

$$\rho_0 = \frac{29\,\text{gm/mole}}{22.4\,\text{liter/mole}} = 1.29 \times 10^{-3}\,\text{gm/cm}^3.$$

Thus Newton finds for the velocity of sound

$$v_{\text{Newton}} = \sqrt{\frac{1.01 \times 10^6}{1.29 \times 10^{-3}}} = 2.80 \times 10^4\,\text{cm/sec} = 280\,\text{meter/sec}.$$ (38)

The *experimental* velocity (which you should memorize) is, for air at STP,

$$v = 332\,\text{meter/sec}$$
$$= 745\,\text{miles/hr}$$
$$= 1\,\text{mile/4.8 sec}.$$ (39)

[You are perhaps familiar with the common method of estimating the distance to a lightning flash by counting the number of seconds between the flash and the thunder. Then "one mile equals five seconds," approximately. Similarly, you can measure the velocity of sound using a stopwatch and a firecracker (set off by a helper).]

Correcting Newton's mistake. Now comes the interesting question: How could Newton come so close to the right answer (which shows that something is right with his derivation) and yet miss it by 15% (which shows something is wrong with his derivation)? The trouble came from assuming Boyle's law, which holds only at constant temperature. The temperature in a sound wave does *not* remain constant. The air located (at a given instant) in a region of compression has had work done on it. It is slightly hotter than its equilibrium temperature. The neighboring regions one half-wavelength away are regions of rarefaction. They have cooled slightly in expanding. (Energy is conserved; the excess energy at a compression equals the energy deficit at a rarefaction.) Because of the increase in temperature in a compression, the pressure in the compression is *larger* than predicted by Boyle's law, and the pressure in a rarefaction is *less* than that predicted. This effect produces a larger return force than expected and hence a larger phase velocity.

It turns out that instead of Boyle's law (which holds at constant temperature) we should use the *adiabatic gas law*, which gives the relation between p and V when no heat is allowed to flow. (There is not sufficient time for heat to flow from the compressions to the rarefactions so as to equalize the temperature. Before that can happen, a half-cycle has elapsed, and a former region of compression has become a region of rarefaction. Thus the result is the same as if there were "walls" preventing the heat from flowing from one region to another.) This relation can be shown to be given by

$$pV^\gamma = p_0 V_0^\gamma, \qquad p = p_0 V_0^\gamma V^{-\gamma}, \qquad (40)$$

where γ is a constant called "the ratio of specific heat at constant pressure to specific heat at constant volume" and has the numerical value

$$\gamma = 1.40 \text{ for air at STP.}$$

Differentiating Eq. (40) and then setting $V = V_0$ gives

$$\frac{dp}{dV} = -\gamma p_0 V_0^\gamma V^{-\gamma-1},$$

$$V_0 \left(\frac{dp}{dV} \right)_0 = -\gamma p_0.$$

Inserting this into Eq. (33) gives the correct result for the velocity of sound:

$$v_{\text{sound}} = \sqrt{\frac{\gamma p_0}{\rho_0}}$$

$$= \sqrt{1.40}\, v_{\text{Newton}} = 332 \text{ meter/sec.} \qquad (41)$$

Let us examine why the heat does not have time to flow from a compression to a rarefaction and thus to equalize the temperature. In order for the heat flow to keep the temperature everywhere constant, the heat would have to flow a distance of one half-wavelength (from a compression to a rarefaction) in a time which is short compared with one-half of a period of oscillation (after half a period, the compressions and rarefactions will have exchanged places). Thus for the heat flow to be fast enough, one would need

$$v(\text{heat flow}) \gg \frac{\frac{1}{2}\lambda}{\frac{1}{2}T} = v_{\text{sound}}. \tag{42a}$$

It turns out that the heat flow is mostly due to conduction, i.e., due to the transfer of translational kinetic energy from one air molecule to another via collisions. For an air molecule of mass M in air at absolute temperature T, the rms thermal velocity (translational velocity due to heat energy) in a given direction z turns out to be

$$v_{\text{rms}} = \langle v_z{}^2 \rangle^{1/2} = \sqrt{\frac{kT}{M}}, \tag{42b}$$

where k is a constant called Boltzmann's constant. The velocity of sound can be also expressed in terms of T and M. It is given by

$$v_{\text{sound}} = \sqrt{\frac{\gamma p_0}{\rho_0}} = \sqrt{\frac{\gamma kT}{M}}. \tag{42c}$$

Thus, aside from the constant $\sqrt{\gamma}$, the velocity of sound equals the rms thermal velocity of a molecule along z. Thus *if* the molecules traveled in straight lines for distances of order $\frac{1}{2}\lambda$ before making collisions, they would "just make it" in time to transfer heat. They would not on the average satisfy Eq. (42a), but some of the exceptionally fast ones would. There could thus be a significant amount of heat transfer in one half-period. But instead of traveling in straight lines for distances of order $\frac{1}{2}\lambda$, the molecules zigzag their way in a random fashion, only going distances between collisions of the order of 10^{-5} cm (for air at STP). As long as the wavelength is long compared with 10^{-5} cm, the adiabatic law is therefore a very good approximation. (The shortest wavelength for audible sound Waves corresponds to $v \approx 20{,}000$ cps, so $\lambda = v/\nu \approx 3.32 \times 10^4/2 \times 10^4 = 1.6$ cm.)

Example 3: Electromagnetic waves in the earth's ionosphere and phase velocities that exceed *c*

The dispersion relation for electromagnetic waves in the ionosphere turns out to be (approximately)

$$\omega^2 = \omega_p{}^2 + c^2 k^2, \tag{43}$$

where c is the velocity of light and $\omega_p = 2\pi\nu_p$ is the angular frequency of natural oscillations of the electrons of plasma. For driving frequencies ω above the cutoff frequency ω_p, the ionosphere is a dispersive medium, and thus the electromagnetic waves are sinusoidal. That is the case for typical FM or TV frequencies of 100 Mc or so. The phase velocity for a traveling wave of frequency ω is given by

$$v_\varphi{}^2 = \frac{\omega^2}{k^2} = c^2 + \frac{\omega_p{}^2}{k^2}. \tag{44}$$

But this velocity exceeds c, the vacuum velocity of light (and of all other electromagnetic waves, including the 100-Mc TV waves we are now considering).

Indeed, the phase velocity *does* exceed c, but that does *not* mean that it is in conflict with the theory of relativity. Remember that a phase velocity v_φ merely gives the phase relation between the *steady-state* harmonic oscillation of a moving part (an electron in the ionosphere) at position z_1 and that of another moving part at position z_2. In a steady state of harmonic oscillation, there is no telling which oscillation at z_2 is the "result" of a particular oscillation at z_1. None is. The whole system is in a steady state, which has been attained after a long time in which the transients were dying out. We will find out (in Chap. 6) that if you *modulate* the wave by varying its amplitude, thereby sending information (e.g., a TV show) via the electromagnetic waves, then the *modulations do not propagate at the phase velocity*. They propagate at a different velocity, called the group velocity. The group velocity is always less than c, the velocity of light in vacuum.

Let us try to understand how we can obtain a phase velocity greater than c. Notice that the source of the "trouble" (if you are troubled) is in the constant $\omega_p{}^2$ that appears in the dispersion relation. If $\omega_p{}^2$ were zero, the phase velocity would equal c and thus would not exceed c. This constant is the return force on an electron per unit displacement per unit mass which leads to the free oscillations of the electrons in the plasma:

$$\omega_p{}^2 = \frac{Ne^2}{M\epsilon_0} \tag{45}$$

where ϵ_0 is permittivity of free space.

As such, it is analogous to the gravitational contribution to the return force of coupled pendulums. The coupled pendulums have the dispersion relation (in the long-wavelength approximation)

$$\omega^2 = \frac{g}{l} + \frac{Ka^2}{M}k^2, \tag{46}$$

which is of the same general form as that for the ionosphere, Eq. (43). Now suppose that we cut all the springs that couple the linear array of pendulums, i.e., we let $K = 0$. (We cannot so easily imagine how to set $c = 0$ in Eq. (43). In that respect the coupled pendulums are more convenient.) Then the dispersion relation for the array of pendulums gives the phase velocity

$$v_\varphi^2 = \frac{\omega^2}{k^2} = \frac{g}{lk^2}, \tag{47}$$

which can be made greater than the velocity of *light* in vacuum by taking lk^2 sufficiently small! Physically, we see how to go about it. There is absolutely no coupling between pendulums. We simply arrange along array of pendulums so that they all oscil late with the same amplitude and with the phase constant between one pendulum and the next steadily increasing in such a way that the wavelength (the distance over which the phase constant has increased by 2π) is larger than c times the common period of the pendulums. Then the phase velocity exceeds c! This is not a joke; it is a phase velocity and it *does* exceed c.

On the other hand, if we decide to *change* the amplitude of motion of one of the down stream pendulums by some means, we will find that it cannot be done in such a hurry. If we couple the pendulums together, so that there is a way to modify a down stream pendulum's behavior by changing the motion of an upstream pendulum (other than by walking down there), then we will find that we cannot send a modulation down the array at the phase velocity, since, to a large extent, the phase velocity has nothing to do with the coupling between pendulums. Instead, the modulation travels at the group velocity, which is less than c.

Example 4: Transmission line—low-pass filter

The system is shown in Fig. 4.3. The transmission line is driven at the input end ($z = 0$) by a harmonically oscillating voltage. We neglect resistance. In Sec. 2.4 we found that the equations of motion for this system are identical in form with those for longitudinal oscillation of a system of masses and springs, provided we replace K by C^{-1}/a, and M by L/a. The dispersion relation was found to be

$$\omega^2 = \frac{4C^{-1}}{L} \sin^2 \frac{1}{2} ka$$

in the dispersive frequency range (the pass band) from zero to $\omega_0 = 2\sqrt{C^{-1}/L}$. In the low-frequency limit ($k \approx 0$) or continuous limit ($a \approx 0$), we can replace $\sin \frac{1}{2}ka$ by $\frac{1}{2}ka$. Then the phase velocity is given by

$$v_\varphi^2 = \frac{\omega^2}{k^2} = \frac{1}{(C/a)(L/a)}, \tag{48}$$

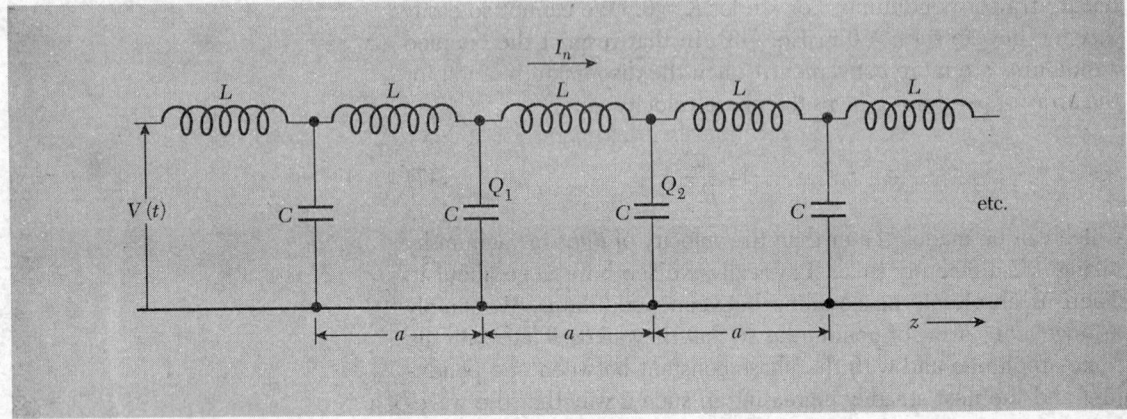

Fig. 4.3 *Transmission line driven at z = 0 and extending to infinity.*

where C/a is the shunt capacitance per unit length and L/a is the series inductance per unit length. Thus for a continuous transmission line (any pair of parallel conductors) in vacuum, the phase velocity is the inverse of the square root of the capacitance per unit length times the inductance per unit length and is a constant, independent of frequency. Thus the waves of voltage and current are nondispersive waves.

Can the phase velocity of this low-pass transmission line exceed c?
We know from Example 3 (of the ionosphere) that it is possible to have a phase velocity which exceeds c without violating relativity. But, at least in that example, we could have any phase velocity we pleased for a good reason: there was a low cutoff frequency ω_p. We saw that we could even make a system of coupled pendulums with phase velocity exceeding c. But in the present low-pass filter, there is no corresponding cutoff frequency. That is, there is no "return force" on the currents in the inductances except that which is provided by coupling to the adjacent capacitors. Therefore we should not expect to be able to find a phase velocity that exceeds c. Now consider Eq. (48). Let us try to make the phase velocity as large as we possibly can. That means we want the series inductance per unit length and the shunt capacitance per unit length to be as small as possible. By inspection of Fig. 4.3, we see that we can make the inductance per unit length a minimum by replacing each lumped inductance by a straight wire. We can minimize the shunt capacitance by simply removing all the lumped capacitances. You might at first suppose that now both C/a and L/a would be zero, so that Eq. (48) would give infinity for v_φ. That is wrong. We must not forget that two straight wires (one carrying the current out, the other carrying it back) have a nonzero self-inductance per unit length. They also have a nonzero shunt capacitance per unit length. In fact, you can show (perhaps after some reviewing of Vol. II) that the shunt capacitance

per unit length and the series inductance per unit length for *two infinitely long, straight, parallel wires* are given by (Prob. 4.8)

$$\frac{C}{a} = \frac{\epsilon_0}{4 \ln\left(\dfrac{D+r}{r}\right)} \quad \text{(esu)}, \qquad (49)$$

$$\frac{L}{a} = \frac{\mu_0}{\pi} \ln\left(\frac{D+r}{r}\right) \quad \text{(esu)}, \qquad (50)$$

where r is the radius of each wire and D is the distance between the wires (from the closest surface of one wire to the closest surface of the other). Taking the product of Eqs. (49) and (50), we obtain the remarkable result

$$\frac{C}{a}\frac{L}{a} = \mu_0 \epsilon_0 = \frac{1}{C^2}. \qquad (51)$$

where $C = \dfrac{1}{\sqrt{\mu_0 \epsilon_0}}$ is velocity of light in vacuum and μ_0 and ϵ_0 are magnetic permeability and electrical permittivity of free space respectively.

Thus, by Eq. (48), *the phase velocity for traveling waves of current (or voltage) on a transmission line consisting of two straight parallel wires is c, the velocity of light in vacuum.*

Phase velocity of straight and parallel transmission lines. Suppose now that we construct other transmission lines by building them up of "pairs of parallel wires," one carrying a current down the line, the other carrying it back. We shall call these *straight and parallel transmission lines.* It should not surprise you to learn that in these cases again, although C/a and L/a depend strongly on the geometry of the arrangement, their product is always $1/c^2$, as in Eq. (51). That can be understood by thinking of what happens if you suddenly change the voltage across the transmission line at one end. Each "pair of wires" carries a voltage pulse at velocity c. The pulse from one pair of wires cannot disturb that on any other pair of wires, because the waves are moving as fast as they can—nothing can pass them to disturb them.

Example 5: Parallel-plate transmission line

The system consists of two parallel conducting plates of width w in the y direction, separated at their inner surfaces by a gap g in the x direction, and carrying current in the z direction, as shown in Fig. 4.4. We wish to calculate the capacitance and the inductance per unit length along z. For that purpose, we can take the potential $V(t)$

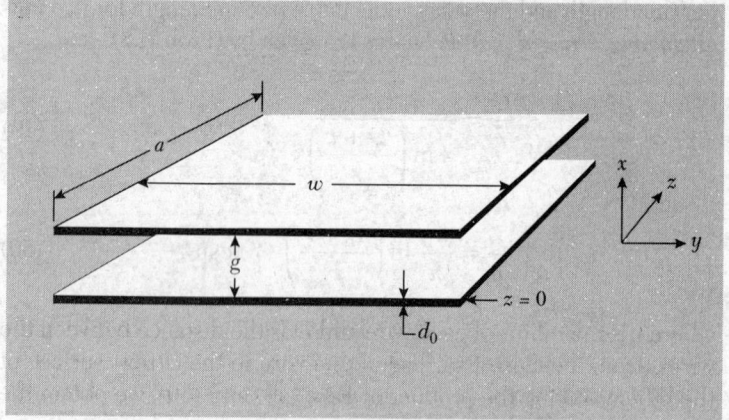

Fig. 4.4 Parallel-plate transmission line. The driving force (not shown) provides a potential difference V(t) between the plates at z = 0 and furnishes a current I(t) which (at any instant) is going out in the +z direction onto one plate and returning in the −z direction on the other. The dimension a is an arbitrary length along the z direction, taken to be small compared with the wavelength of the traveling waves.

between the plates at $z = 0$ to be constant. Then we have steady current. (We can assume that the two plates are joined at $z = \infty$ so as to ensure that current going out has a return path. Alternatively, we could simply assume the two plates extend to infinity and never join—the results are the same.)

Let us take the bottom plate positive and the top plate negative. Then *the electric field is in the +x direction* (see Fig. 4.4). Assume w is large compared with g, so that there is no "edge effect." Let Q be the charge displacement on an area of the plates (indicated in Fig. 4.4) of width w along y and of length a along z. (The length a is arbitrary, but it helps in the derivation if we include it explicitly.) Let C be the capacitance of this area of the plates. Then we have the relations (see Vol. II, Sec. 3.5, if you need review)

$$Q = CV, \tag{52}$$

$$V = gE_x, \tag{53}$$

$$E = \frac{Q}{wa\,\epsilon_0}, \tag{54}$$

where Eqs. (52) and (53) hold in either esu or in MKS units, and where Eq. (54) is 4π times the charge per unit area and gives the electric field in esu (statvolts per cm). Solving these three equations for C, we find the capacitance per unit length to be

$$\frac{C}{a} = \frac{w\epsilon_0}{g}. \tag{55}$$

Now let us find the inductance per unit length, L/a. The bottom plate is connected to the positive terminal of the power supply, the top plate to the negative terminal. Therefore a positive current

I flows in the $+z$ direction on the bottom plate and in the $-z$ direction on the top plate. Using your right hand and Fig. 4.4, you can convince yourself that *the magnetic field between the plates is in the $+y$ direction*. The magnetic field is zero in the region outside the plates, as you can easily convince yourself. Let L be the self-inductance of the part of the plates indicated in Fig. 4.4. The magnetic flux Φ through the area ga is given by

$$\Phi = B_y ga. \tag{56}$$

The magnetic flux B_y is given by

$$wB_y = \mu_0 I. \tag{57}$$

where μ_0 is defined as permeability of free space. (See Vol. II, Sec. 6.6; the "sheet current density" defined there is the same as our I/ω.) The self-inductance L is defined by [see Vol. II, Sec. 7.8, Eqs. (7.53) and (7.54)]

$$L\frac{dI}{dt} = \frac{d\Phi}{dt} ;$$

i.e., for steady current I,

$$LI = \Phi. \tag{58}$$

Solving Eqs. (56), (57), and (58) for L, we find the self-inductance per unit length to be

$$\frac{L}{a} = \frac{\mu_0 g}{w} . \tag{59}$$

Perhaps you are worried about the fact that we have calculated the self-inductance using a steady current, whereas you know that the Maxwell equation which gave us Eq. (57) for steady current is (Vol. II, Sec. 7.13)

$$\vec{\nabla} \times \mathbf{B} = \mu_0 \vec{\mathbf{J}} + \mu_0 \epsilon_0 \frac{\partial \vec{\mathbf{E}}}{\partial t} . \tag{60}$$

Thus we have neglected the contribution from the, "displacement current" term, $\mu_0 \epsilon_0 \dfrac{\partial \mathbf{E}}{\partial t}$. It turns out (Prob. 4.10) that this omission is justifiable provided that the thickness d_0 of each plate satisfies

$$d_0 \ll \lambda. \tag{61}$$

We shall assume that this condition holds.

The phase velocity v_φ for traveling waves is given by [using Eqs. (48), (55), and (59)]

$$v_\varphi = \frac{1}{\sqrt{\mu_0\,\epsilon_0}}. \tag{62}$$

Thus we have found that the phase velocity equals c for two quite different examples of straight and parallel transmission lines. It should be plausible that this is a general result: *the phase velocity for any transmission line consisting of two isolated, identical, straight, parallel conductors in vacuum is c.*

4.3 *Index of Refraction and Dispersion*

If all the space between the plates of a parallel-plate transmission line is filled with a dielectric material having relative electrical permittivity ϵ_r, the capacitance is thereby increased by a factor ϵ_r (See Vol. II, Sec. 9.9). (That is also the case for the parallel-wire transmission line, except that in that case we must fill all space with the dielectric. For a parallel-plate capacitance, the electric field outside of the region between the plates is zero, and it doesn't matter whether or not there is dielectric material there.) Similarly, if the material inserted has relative magnetic permeability μ_r, then the self-inductance is increased by a factor μ_r. [We shall only be considering materials like glass, water, air, or similar materials, for which the magnetic permeability is essentially unity. Therefore, there is no need for you to review at this time the physics of magnetic materials (Vol. II, Chap. 10). We shall carry the constant μ_r with us in what follows but shall always set it equal to unity when we consider an actual example.] Therefore, the phase velocity of traveling waves of current and voltage propagating along a parallel-plate transmission line (or any other straight and parallel transmission line), with all the space filled with a material of dielectric constant ϵ_r and of magnetic permeability μ_r, is given by

$$v_\varphi = \sqrt{\frac{a}{L}\frac{a}{C}} = \frac{v_\varphi}{\sqrt{\mu_r\epsilon_r}}(\text{vacuum}),$$

i.e.,

$$\boxed{v_\varphi = \frac{c}{\sqrt{\mu_r\epsilon_r}}.} \tag{63}$$

Equation (63), which we have obtained for the special case of traveling waves of current and voltage on a transmission line, is actually a very general result. It holds for any kind of electromagnetic wave propagating through matter. Thus, for example, Eq. (63) holds for visible light propagating through a piece of glass or other dielectric material.

Let us make the generality of Eq. (63) plausible. We have seen that it holds for current and voltage waves propagating on a transmission line. Now, the space between the plates of the line is filled with electric and magnetic fields. (The electric field corresponds to the voltage across the plates; the magnetic field corresponds to the current along the plates.) Thus the electric and magnetic field patterns must propagate with the same velocity as the current and voltage waves. (The field patterns are themselves waves, of course—they vary in space and time, and that characteristic is what constitutes a wave.) When the space is vacuum, the velocity is c. But we know that c is the velocity of *all* electromagnetic waves in vacuum, whether or not they happen to be between the plates of a transmission line. When the space is filled with a material of constants ϵ and μ, the velocity of the electric and magnetic field waves (accompanying the voltage and current waves) is $c/\sqrt{\epsilon\mu}$. It seems plausible that that is the velocity of all electromagnetic waves in such material, whatever their source, i.e., whether they are the electromagnetic waves accompanying voltage and current waves on the plates of a transmission line or are, for example, electromagnetic waves produced by a distant light bulb or a radio antenna or a star. One of the things that we have tried to drive home in Chaps. 1–3 is that *the dispersion relation is independent of the boundary conditions*. It depends only on the inherent properties of the waves and the medium. Electromagnetic waves can be produced by means of voltages applied at the end of a transmission line or by means of a transmitter or antenna, without benefit of a transmission line. These merely represent different boundary conditions, i.e., different ways to drive the system. (The system is the medium, consisting of the material with constants ϵ and μ.) The dispersion law, Eq. (63), is independent of these conditions. We have not proved it, but we hope we have made it plausible. (We shall prove it in Chap. 7.)

Equation (63) holds for all electromagnetic radiation and in particular for light. (We will study electromagnetic radiation in more detail in Chap. 7.) The factor $\sqrt{\dfrac{\mu\epsilon}{\mu_0\epsilon_0}}$, called the *index of refraction*, is designated by n. You should learn all the following ways to express the effect of the index of refraction. It helps to keep things straight if you remember the example of glass, which has an index of refraction for visible light of about 1.5. Then, in all the following expressions, you should form a mental image of what quantities get bigger and what quantities get smaller in glass as compared with vacuum:

$$n = \frac{c}{v_\varphi} = \sqrt{\frac{\mu\epsilon}{\mu_0\epsilon_0}} = \sqrt{\mu_r\epsilon_r}, \quad (64)$$

$$\lambda = \frac{1}{n}\frac{c}{v} = \frac{1}{n}\lambda(\text{vacuum}), \quad (65)$$

$$k = n\frac{\omega}{c} = nk(\text{vacuum}). \quad (66)$$

where μ_r and ϵ_r are called relative permeability and permittivity (dielectric constant) respectively.

Of course the frequency of the driving force is not affected by the medium, and c means the velocity of light in vacuum. Therefore if you want to designate the wavelength in vacuum, you can just call it c/v, instead of, for example, λ(vacuum). Similarly k(vacuum) = ω/c. In glass, the wavelength of visible light is only about $\frac{2}{3}$ as large as its value in vacuum. The number of waves per centimeter, $\sigma = 1/\lambda$, is larger by a factor of 1.5 in glass than in vacuum.

Table 4.1 gives values of the index of refraction of common materials for yellow sodium light of wavelength $\lambda = 5893$ Å (1 Å = 1 Angstrom unit = 10^{-8} cm). You should memorize the approximate values $n = \frac{3}{2}$ for glass and plastics, $n = \frac{4}{3}$ for water, and $n = 1 + 0.3 \times 10^{-3} = 1 + 0.3$ mil for air.

Variation of index of refraction with color—dispersion. A prism (which is a wedge-shaped piece of glass or other transparent material) bends a beam of incident light by an amount that depends on the color, i.e., on the wavelength of the light. The different colors in a parallel beam of "white" light are bent through different angles an are thus "dispersed"; i.e., they emerge from the prism at different angles and give a rainbow like colored pattern on a screen situated behind the prism. This is shown in Fig. 4.5.

Refraction and Snell's law. A beam of light of a given color is bent or *refracted* whenever it encounters a surface where the phase velocity takes on a new value, i.e., where the index of refraction n changes. The amount of refraction depends on the ratio n_1/n_2 of the index of refraction in medium 1 (from which the beam is incident) to that in medium 2 (into which it passes). It also depends on the *angle of incidence*, which is defined as the angle that the incident light beam makes with the normal to the surface. The *angle of refraction* is defined as the angle that the refracted beam makes with the normal to the surface. (We shall always take

Table 4.1 *Indices of refraction of common materials*

Material	Index, 5893 Å
Air (STP)	1.0002926
Water (20°C)	1.33
Zinc crown glass	1.52
Heavy lead glass	1.90
Lucite	1.50
Transparent Scotch tape	1.50

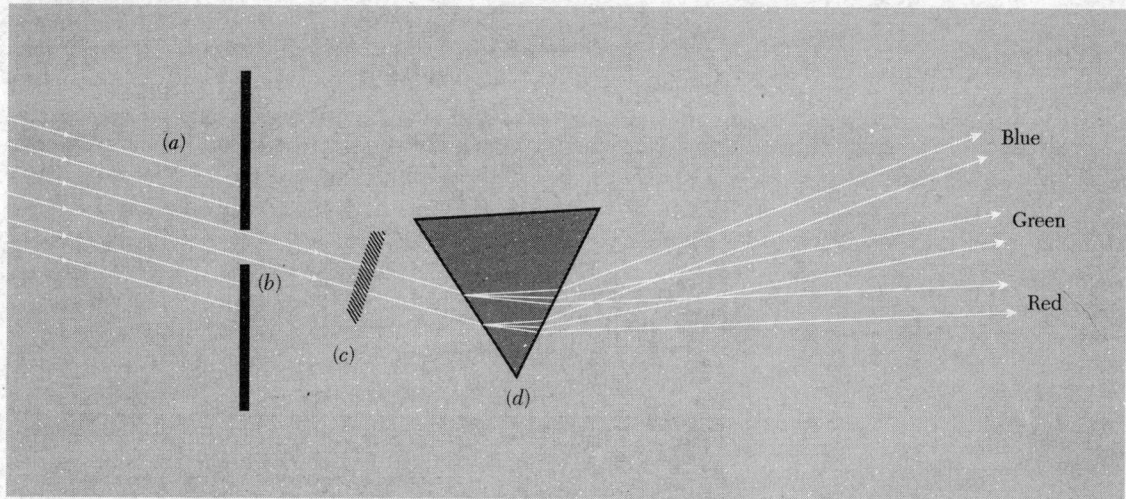

the angle of incidence and angle of refraction to be positive angles between 0° and 90°). These definitions are illustrated in Fig. 4.6.

We can easily derive the relation between n_1/n_2, θ_1, and θ_2 as follows. The "wave crests" of the light beam, or "wavefronts," as they are called in the three-dimensional wave we have here, are perpendicular to the direction of travel of the light beam. When a given wavefront reaches the boundary where the index increases (as in going from air to glass), one end of the wavefront reaches the boundary before the other end. Thus the phase velocity decreases at one end before it does at the other. Therefore the angle of the wavefront changes, somewhat as the angle of a row of marchers changes if one end of the row slows down while the other does not. The geometrical relations are shown in Fig. 4.7.

Consider the two right triangles having the common hypotenuse x in Fig. 4.7. From the figure, we see that

$$l_1 = x \sin \theta_1, \qquad l_2 = x \sin \theta_2. \tag{67}$$

Let t be the time it takes for the traveling wave to progress a distance l_1 in medium 1 or a distance l_2 in medium 2. Then

$$l_1 = \frac{ct}{n_1}, \qquad l_2 = \frac{ct}{n_2}. \tag{68}$$

Thus

$$ct = n_1 l_1 = n_2 l_2.$$

Then, using Eq. (67), we get

Fig. 4.5 Dispersion. Sunlight (a) is incident on an opaque screen having a slit (b) perpendicular to the paper. The beam of white light formed by the slit passes first through a filter (c), which transmits light of only one color, and then through a glass prism (d), which bends the light by an amount that depends on the color. Blue is bent more than red. With no filter, all the colors are present, spread out in the same order as that seen in the rainbow.

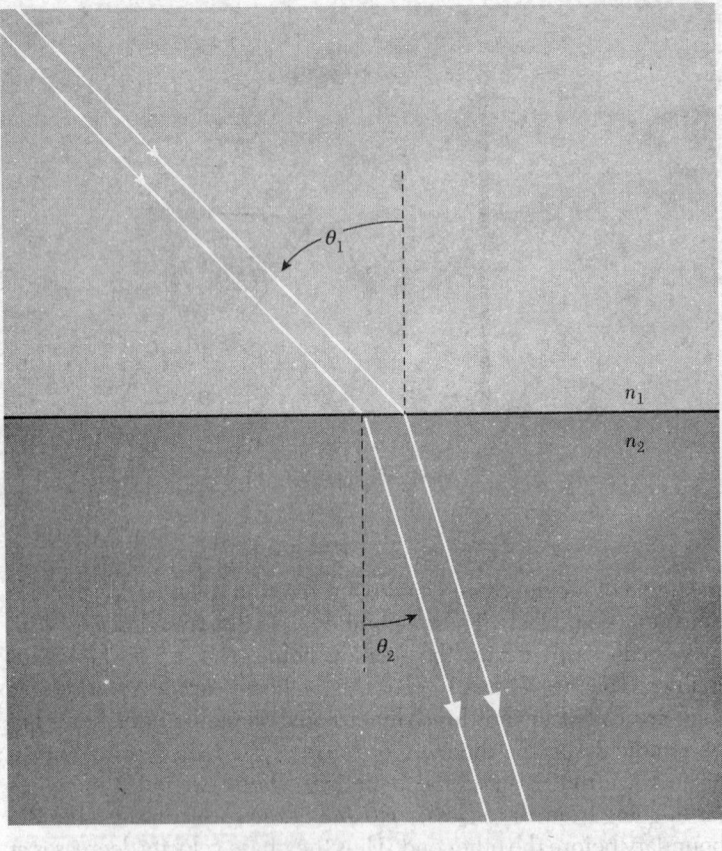

Fig. 4.6 Nomenclature. For a light beam traveling in the direction indicated by the arrows, θ_1 is called the angle of incidence, and θ_2 is called the angle of refraction.

$$\boxed{n_1 \sin \theta_1 = n_2 \sin \theta_2.}$$ (69)

Equation (69) is called *Snell's law* of refraction.

Dispersion of glass.   We now see that the dispersion of the prism results from the fact that the *index of refraction* is *greater for blue than for red light.* Here are some values for zinc crown glass taken from the *Handbook of Physics and Chemistry.* The wavelengths are given in units of angstroms (10^{-8} cm) and microns (10^{-4} cm). The frequencies ($v = c/\lambda$) are given in units of 10^{14} Hz (1 hertz = 1 cycle per second).

Table 4.2 can be summarized crudely by saying that the index of refraction of glass is about 1.5 over the entire visible range of frequencies and that the *dispersion*, i.e., the rate of change of n with λ, gives an *increase* in the index n of about six mils (i.e., about 0.006) for every 1000-Å *decrease* in wavelength.

You can study the dispersion of water with a simple prism made from two microscope slides (plus putty and tape) and your purple filter, which absorbs green but passes red and blue.

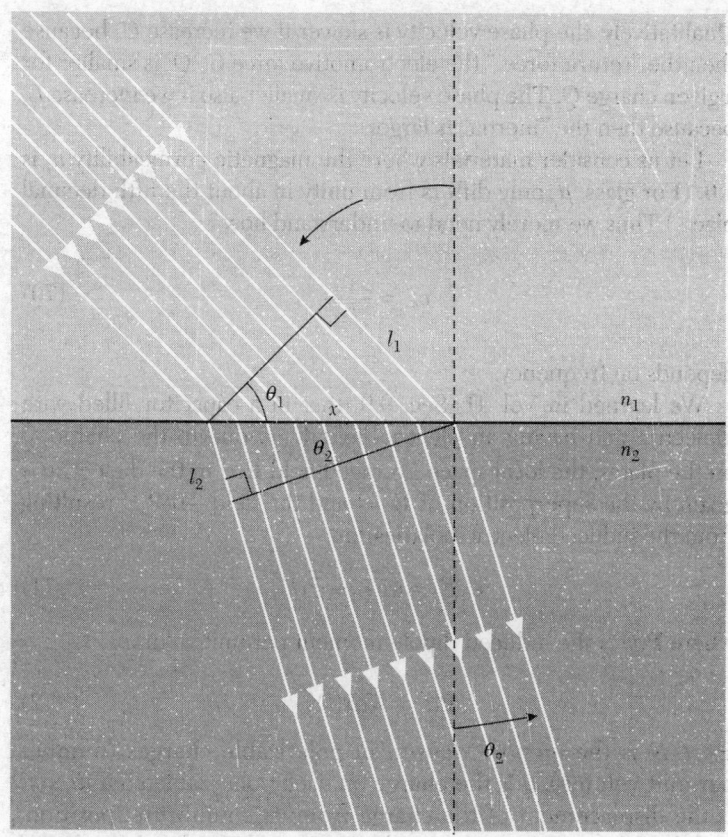

Fig. 4.7 *Refraction. If n_2 is greater than n_1, the end of the wavefront to the right (looking along the beam's direction of travel) travels a distance l_2, which is less than the distance l_1 traveled by the left end. Therefore the beam bends toward the normal, as shown.*

Table 4.2 *Dispersion of index of refraction of glass*

Color name	$\lambda(\text{Å})$	$\lambda(\mu)$	$\nu(10^{14}\,Hz)$	n
Near ultraviolet	3610	0.361	8.31	1.539
Dark blue	4340	0.434	6.92	1.528
Blue-green	4860	0.486	6.18	1.523
Yellow	5890	0.589	5.10	1.517
Red	6560	0.656	4.57	1.514
Very dark red	7680	0.768	3.91	1.511
Infrared	12,000	1.20	2.50	1.505
Far infrared	20,000	2.00	1.50	1.497

(See Home Exp. 4.12.)

Why does the index of refraction vary with frequency?   Let us go back to our transmission line. The phase velocity is

$$v_\varphi = \frac{1}{\sqrt{(C/a)(L/a)}}.$$

Qualitatively, the phase velocity is slower if we increase C, because then the "return force," the electromotive force $C^{-1}Q$, is smaller for a given charge Q. The phase velocity is smaller also if we increase L, because then the "inertia" is larger.

Let us consider materials where the magnetic permeability μ_r is 1.0. (For glass, μ_r only differs from unity in about the fifth decimal place.) Thus we merely need to understand how

$$v_\varphi = \frac{c}{\sqrt{\epsilon_r}} \tag{70}$$

depends on frequency.

We learned in Vol. II, Sec. 9.9, that in a capacitor filled with dielectric and having an electric field $\mathbf{E}_Q(t)$ due to the charge Q on the plates, the local space-averaged field $\mathbf{E}(t)$ in the dielectric is given by the superposition of $\mathbf{E}_Q(t)$ and the field $-4\pi\mathbf{P}(t)$ resulting from the induced electric polarization:

$$\epsilon_0 \vec{E} = \epsilon \vec{E}_Q - \vec{P}(t), \tag{71}$$

where $\mathbf{P}(t)$ is the induced dipole moment per unit volume:

$$\mathbf{P}(t) = Nqx(t)\hat{\mathbf{x}}. \tag{72}$$

Here N is the *number density* of polarizable charges (number per unit volume), q is the charge on each polarizable charge, $x(t)$ is the displacement of the charge from its equilibrium position, and $\hat{\mathbf{x}}$ is a unit vector. Let us take $\mathbf{E}_Q$, $\mathbf{P}$, and $\mathbf{E}$ to be along $\hat{\mathbf{x}}$ and drop the vector symbols. Since the capacitance C is defined by $C = Q/V$, where V is the potential difference between the plates, we see that, upon insertion of the dielectric, the reduction of electric field due to induced polarization (with a proportional decrease in V) gives an increase in C. The factor by which C increases is called the dielectric constant, ϵ. Thus according to Eqs. (71) and (72)

$$\epsilon_r = \frac{E_Q}{E} = 1 + \frac{P}{\epsilon_0 E} = 1 + \frac{Nqx(t)}{\epsilon_0 E}. \tag{73}$$

Example 6: Simple model of "glass molecule"

Despite the simplicity of the model we shall now make, it will manifest essentially all the successful features of any classical (i.e., pre-quantum mechanical) models describing the microscopic interaction of light with matter. Nor are these negligible successes; classical mechanics predicts many of the observed features of these phenomena, as we shall see. The reason is that a quantum-mechanical description supersedes but does not necessarily contradict the classical description; it includes the classical description as a limiting case, applicable under conditions which are present in a wide range of ordinary phenomena.

We assume that a "glass molecule" consists of a massive motionless nucleus to which is attached a charge q having a relatively small mass M. The charge is attached by means of a spring of spring constant $M\omega_0^2$. The motion of the charge is damped with damping constant Γ. The equation of motion of q is thus given by

$$M\ddot{x} = -M\omega_0^2 x - M\Gamma\dot{x} + qE(t). \qquad (74)$$

Now we suppose that the external field $E_0(t)$ is harmonically varying at angular frequency ω. Then $P(t)$ and $E(t)$ will also vary at frequency ω. Thus we can take the field at a given "average" molecule to be

$$E(t) = E_0 \cos \omega t. \qquad (75)$$

But then Eq. (74) is just the harmonically driven oscillator that we considered in Sec. 3.2, with $F_0 = qE_0$. The solution $x(t)$ for steady-state oscillations is given by

$$x(t) = A_{el} \cos \omega t + A_{ab} \sin \omega t, \qquad (76)$$

where A_{el} and A_{ab} are the elastic and absorptive amplitudes. Now, in the case of a "colorless" transparent substance like clear glass or water, there are no important resonances of the glass molecules in the visible frequency range. (That is why it is transparent and "colorless.") In the case of substances like colored glass or the gelatin filters in your optics kit, there are resonances in the visible range. In fact, it is the absorption of radiation energy due to the term $A_{ab} \sin \omega t$ at these resonances that subtracts part of the color spectrum from incident white light and leaves the transmitted color that you see. (You should now look at a "white" source, such as an incandescent lamp, using your diffraction grating and filters.) We do not wish to consider the behavior of colored filters at frequencies near absorbing resonances. Therefore we shall neglect the term $A_{ab} \sin \omega t$ in Eq. (76). We know from Chap. 3 that that is a good approximation as long as we are not near a resonance. The general case (including absorption) is discussed in Supplementary Topic 9.

The index of refraction is thus given by

$$n^2 = \epsilon_r = 1 + \frac{Nqx}{\epsilon_0 E} = 1 + \frac{Nq}{\epsilon_0} \frac{A_{el}}{E_0}. \qquad (77)$$

Assuming that we are far from resonance, i.e., taking $\Gamma = 0$ in Eq. (74), we have [see Eq. (3.17), Sec. 3.2],

$$A_{el} = \frac{F_0}{M} \frac{1}{\omega_0^2 - \omega^2} = \frac{qE_0}{M} \frac{1}{\omega_0^2 - \omega^2}.$$

Thus we obtain

$$\boxed{\frac{c^2 k^2}{\omega^2} = n^2 = \epsilon_r^2 = 1 + \frac{Nq^2}{\epsilon_0 M} \cdot \frac{1}{{\omega_0}^2 - \omega^2}.} \tag{78}$$

To adapt this result, based on a simple model with a single resonance, to a real piece of glass, we should sum the contributions to $n^2 - 1$ from all the important resonances. In that case, ${\omega_0}^2$ in Eq. (78) can be taken to mean a rough "average" resonance frequency. (See Prob. 3.20.) For N we should take the number of glass molecules per cubic centimeter times the average number of contributing resonances per molecule. The number of electrons that make substantial contributions is about equal to the number of "outer shell" or "valence" electrons.

When ω is in the range of frequencies of visible light, the most important resonances in glass turn out to be at "ultraviolet" frequencies, corresponding to wavelengths $\lambda = c/v$ of order 1000 Å (i.e., 10^{-5} cm) or less. Wavelengths of visible light are about 5 times greater than that; frequencies ω of visible light are correspondingly about 5 times smaller than the average resonance frequency ω_0. Then $n^2 - 1$ is positive, according to Eq. (78). That agrees with experiment, for visible light in glass. Notice also that as ω increases (always remaining less than ω_0), the denominator ${\omega_0}^2 - \omega^2$ in Eq. (78) becomes smaller, whence $n^2 - 1$ becomes larger. Thus blue light (higher frequency) has a larger index than red light. That agrees with the experimental result that a prism bends blue more than it does red.

Phase velocities which are greater than c. When the driving frequency ω of the electromagnetic radiation (the light) is less than the resonance frequency ω_0, we obtain the results given above, namely that the phase velocity is less than c, that the wavelength is less than the wavelength in vacuum, and that increasing frequency leads to increasing index of refraction. That is called "normal" dispersion. When the driving frequency is greater than the resonance frequency, as is the case for "extreme ultraviolet" light in glass, then we have from Eq. (78) that $n^2 - 1$ is negative; i.e., n^2 is less than 1. If n^2 lies between zero and 1, we again have what is called normal dispersion. But in that case the phase velocity is greater than c, the wavelength is greater than the wavelength in vacuum, and increasing the frequency again leads to increasing index of refraction. (When the frequency finally becomes very large, n finally increases to 1, and the light behaves as in vacuum.) In the frequency range $\omega_0 - \frac{1}{2}\Gamma < \omega < \omega_0 + \frac{1}{2}\Gamma$ it turns out that the index of refraction decreases with increasing ω. That is called "anomalous" dispersion.

The physical origin of phase velocities that are greater than c lies in the crucial phase relation existing between the driving force $qE(t)$ and the oscillation $x(t)$ of the driven charge q. We know that if the driving frequency is below the resonance frequency, then $x(t)$ can "follow" the force $qE(t)$. Then the charges are displaced in the same direction as the force and build up a field that tends to cancel the original field. (This holds for q either positive or negative.) This reduced field gives a reduced return force and hence a reduced phase velocity. On the other hand, when the charge is driven above its resonance frequency, it "cannot keep up," and the displacement $x(t)$ is always in the opposite direction from the instantaneous force $qE(t)$. Thus, for example, when you push an otherwise free ball back and forth from one hand to the other, you exert the maximum force to the left when the ball is in contact with your right hand at its farthest distance to the right. The displacement at a given time is mostly due to the force exerted one half-cycle earlier. The field due to the relative displacement of the charges thus tends to increase the original field. That gives an increased return force and hence a phase velocity greater than that in vacuum.

We can conclude that there is nothing more mysterious about a phase velocity's being greater than c than there is about a ball's being at the right in spite of the fact that it is being pushed to the left.

Exponential waves—reactive frequency range. When the driving frequency ω exceeds the resonance frequency ω_0, then n^2 is less than 1, according to Eq. (78). As long as n^2 lies between zero and 1, we have sinusoidal waves, i.e., k^2 comes out to be a positive number. That will certainly be the case for sufficiently large ω (always assuming $\omega > \omega_0$), since, for huge ω, n^2 is only slightly less than 1. But between $\omega = \omega_0$ plus several Γ [so that we can use the approximate form for A_{el} that gave Eq. (78)] and $\omega = \infty$, there is a region where Eq. (78) gives n^2 to be negative. That will be the case in the frequency range such that

$$\frac{Nq^2}{\epsilon_0 M} > \omega^2 - \omega_0^2, \tag{79}$$

where we assume $\omega^2 - \omega_0^2 \gg \Gamma\omega_0$, to ensure that we are safely above the resonance and can use the approximate expression for A_{el}. When Eq. (79) holds, then Eq. (78) gives that n^2 is negative, i.e., that k^2 is negative. That merely means that the differential equation for the spatial dependence of the waves is *not*

$$\frac{\partial^2 \psi(z,t)}{\partial z^2} = -k^2 \psi(z,t), \qquad k^2 > 0, \tag{80a}$$

corresponding to sinusoidal waves, but is instead

$$\frac{\partial^2 \psi(z,t)}{\partial z^2} = +\kappa^2 \psi(z,t), \qquad \kappa^2 > 0, \qquad (80b)$$

corresponding to exponential waves. That is a situation we have encountered before, for example with a system of coupled pendulums. When the dispersion relation for k^2 in terms of ω^2 gives a negative value for k^2, we merely change the name from k^2 to $-\kappa^2$ and recognize that the waves are exponential rather than sinusoidal.

We shall give a qualitative derivation of the condition for exponential waves, Eq. (79), after we have considered the special case where ω_0 is zero. That special case gives the dispersion law for the ionosphere, as we shall now show:

Example 7: Dispersion of the ionosphere

In Sec. 2.4 (Example 6), we gave a simple model of the plasma of the earth's ionosphere and derived the frequency ω_p of free vibrations in what could be called the "sloshing mode" of the ionosphere (the mode with infinite wavelength, similar to the sloshing mode in a pan of water, where the water surface remains flat as the water sloshes). In that model, we neglect the motion of the positive ions and also neglect any damping of the motion of the "free" electrons. (There actually is damping due to the collisions of electrons with ions, with a consequent transfer of energy from the oscillation to random "thermal" energy.) The equation of motion of a single electron of charge q and mass M is then

$$M\ddot{x} = qE(t), \qquad (81)$$

where $E(t)$ is the electric field at the position of the electron. *For free oscillations*, $E(t)$ is entirely due to the polarization per unit volume:

$$\epsilon_0 \vec{E} = -\vec{P} = -Nqx(t). \qquad (82)$$

Thus for free oscillations Eqs. (81) and (82) give

$$\ddot{x} = \frac{Nq^2}{\epsilon_0 M} x \equiv -\omega_p^2 x. \qquad (83)$$

Thus we have repeated (in briefer form) the earlier derivation of the equation of motion for oscillations at the plasma frequency ω_p. Now suppose that the plasma is driven at one end by a radio or TV transmitter. (Assume that the geometry is of the straight and parallel kind, as in a parallel-plate transmission line, so that our problem is as simple as possible.) Then $E(t)$ is the superposition [analogous to Eq. (71)]

$$\epsilon_0 \vec{E} = \epsilon \vec{E}_{tr} - \vec{P}(t) \tag{84}$$

where E_{tr} (subscript means transmitter) is the field that *would* be present if there were no free electrons contributing. The equation of motion of the electron is similar to that of the electron in the "glass molecule," provided we set both the "spring constant," $K = M\omega_0^2$, and the damping constant Γ to zero [see Eq. (74)]. Thus the free electron has "zero resonance frequency," $\omega_0 = 0$. The result for the index of refraction, i.e., for the dispersion relation, is therefore obtained by merely setting $\omega_0 = 0$ in Eq. (78):

$$\frac{c^2 k^2}{\omega^2} = n^2 = \epsilon_r = 1 - \frac{\omega_p^2}{\omega^2}, \tag{85}$$

with

$$\omega_p^2 = \frac{Ne^2}{\epsilon_0 M}.$$

Multiplying Eq. (85) through by ω^2 puts it in the form we gave earlier:

$$\omega^2 = \omega_p^2 + c^2 k^2, \qquad \omega^2 \geqq \omega_p^2. \tag{86}$$

For the reactive frequency region, we have exponential waves:

$$\omega^2 = \omega_p^2 - c^2 \kappa^2, \qquad \omega^2 \leqq \omega_p^2. \tag{87}$$

It is only fair to mention that our model of the ionosphere is not exact. Some of our physical assumptions break down at various frequencies for various interesting reasons, and the exact dispersion relation is actually considerably more complicated than that indicated by Eqs. (86) and (87). For example, at sufficiently low frequencies, an electron makes several collisions with ions per oscillation cycle, on the average. The damping force is then dominant; we neglected damping in our model. Also, at certain frequencies there are resonances other than that which occurs at the plasma oscillation frequency ω_p. For example, the plasma oscillations of the slower and heavier positive ions become important for low frequencies. (Their plasma oscillation frequency is about 100 kc.) Similarly, the "cyclotron frequency" ω_c which corresponds to circular motion of the electrons in the earth's magnetic field (about $\frac{1}{2}$ gauss) is important. For an interesting discussion of experimental results, see "Ionosphere Explorer I Satellite: First Observations from the Fixed-Frequency Topside Sounder," by W. Calvert, R. Knecht, and T. Van Zandt, *Science* **146**, 391 (Oct. 16, 1964).

Qualitative explanation for low-frequency cutoff. We know for any system (for example, for a system of coupled pendulums) that the frequency of the lowest possible mode for free oscillations is also the lowest possible frequency for sinusoidal waves under a harmonic driving force. Thus the lowest mode frequency is a low-frequency cutoff, for forced oscillations. For driving frequencies below the cutoff frequency, the waves are exponential rather than sinusoidal.

Exactly at the cutoff frequency the wavelength for sinusoidal waves is infinitely long, as is the attenuation distance for exponential waves. (For coupled pendulums, the pendulums all swing in phase.) Thus if we wish to look for a low-frequency cutoff in any dispersion law, we merely set $k = 0$ in the dispersion relation. The frequency obtained from the dispersion law with $k = 0$ is then the cutoff frequency, which we can call $\omega_{c.o.}$. In our example of the index of refraction, we have [see Eq. (78)]

$$n^2 = \frac{c^2 k^2}{\omega^2} = 1 + \frac{Nq^2}{\epsilon_0 M} \frac{1}{\omega_0^2 - \omega^2}.$$

Setting $k = 0$ gives the low-frequency cutoff frequency:

$$\omega_{c.o.}^2 = \omega_0^2 + \frac{Nq^2}{\epsilon_0 M}. \tag{88}$$

Now, as always, ω^2 is the return force per unit mass per unit displacement. According to our discussion (above) of the ionosphere, this return force (per unit mass per unit displacement) for free oscillations of the electrons in the ionosphere is $\omega_p^2 = 4\pi Ne^2/M$. This is the lowest normal mode of oscillation for the electrons and has infinite wavelength, i.e., all electrons oscillate in phase. It is clear that if we now add to each oscillating charge a binding force by means of a spring of spring constant $M\omega_0^2$, then we are merely adding a return force (per unit mass per unit displacement) ω_0^2 to each charge. The charges can still oscillate all in phase, so that k is still zero, and the system is still in its lowest mode of free oscillation. Thus we see that the right-hand side of Eq. (88) gives the return force per unit mass per unit displacement for the lowest mode of free oscillation. It is therefore a low-frequency cutoff. Thus we find Eq. (88), as well as the inequality Eq. (79), both of which hold for the "reactive" frequency range where the waves are exponential.

Here is another more physical explanation for the existence of the lowfrequency cutoff in the dispersion law for the index of refraction. For simplicity, we set $\omega_0 = 0$. Then our "model" is the ionosphere. The question is, why is there a low-frequency cutoff, Eq. (89)?

$$\omega_{c.o.}^2 = \frac{Nq^2}{\epsilon_0 M}. \tag{89}$$

First we point out that in many respects the ionosphere (or our model of it) is something like an ordinary metallic conductor.

In each case there are "free" electrons which carry current if an electric field is maintained in the medium. Now, a metallic conductor in a "static" electric field (one where the charges are at rest and the fields are all constant in time) has zero electric field inside the conductor. The reason the field is zero is not that the metal has somehow "blocked" the external driving field or gobbled it up. The external field is in fact still there, inside the metal. But it is "canceled" by superposition with another field, the field produced by the charges that have been driven to the surface of the metal. If the driving field is suddenly turned on from zero, the electrons in the metal require some time to move because they have inertia, and the field inside is at first not zero, but rather just that due to the external driving field. After the charges move and come into equilibrium, they produce a field which, superposed with the driving field, gives a resultant of zero. (If that is not the case, then they are not yet in equilibrium. They keep moving until it is the case.) Let us call the time it takes for equilibrium the "mean relaxation time," designated by τ. If the driving field is reversed in a time short compared with τ, then the flow of charge will not have time to set up a canceling field before it has to start flowing back in the other direction. Thus *the cutoff frequency will be of order* τ^{-1}. For incident electromagnetic radiation with frequency high compared with the cutoff frequency τ^{-1}, the electrons will not have time to move so as to cancel the field to zero. The medium will therefore be "transparent" for frequencies above cutoff. At "infinite" frequency, the electrons will not have time to move at all, and the radiation will go through as if in vacuum. If the system is driven at one end at frequencies below the cutoff frequency τ^{-1}, it will act like a high-pass filter driven below cutoff. The field at points very close to the driven end will essentially equal the driving field. At points farther downstream, the motion of the electrons has time to cancel the incident field, and we get progressively greater cancellation—exponential attenuation— with increasing distance.

Let us estimate the relaxation time τ. Suppose the field E_0 is turned on at time zero. It produces an acceleration given by force/$M = qE_0/M$. During a time t, if this acceleration remained constant, the electrons would travel a distance $\frac{1}{2}at^2$, where a is the acceleration. Dropping the factor of $\frac{1}{2}$ for our crude estimate, we get a displacement x in time t given by

$$x \approx \frac{qE_0}{M}t^2. \tag{90}$$

Suppose the motion of the charges is limited by the "surfaces" of the plasma (the ionosphere) or of the metal. The total charge added to one surface and subtracted from the other is

$$Q = NqxA, \tag{91}$$

where N is the number density, A is a cross-sectional area, and x is the displacement. The charges Q on one surface and $-Q$ on the other produce a uniform field E given by

$$E = \frac{Q}{A\epsilon_0} = \frac{Nqx}{\epsilon_0} \approx \frac{Nq}{\epsilon_0} \frac{q\mathrm{E}_0 t^2}{M}. \qquad (92)$$

If the time t is long enough so that E (the canceling field) can build up so as to equal E_0 (the driving field), then we shall have reached equilibrium. Therefore the relaxation time τ is obtained by setting $E \approx E_0$ and $t \approx \tau$ in Eq. (92):

$$\omega^2_{\text{c.o.}} \approx \tau^{-2} \approx \frac{Nq^2}{\epsilon_0 M},$$

which agrees with the exact result, Eq. (89).

Qualitative discussion of index of refraction in dispersive frequency range.   An isolated charged particle oscillating in vacuum emits electromagnetic waves that travel in vacuum at the velocity of light. Therefore when an incident light wave drives a single charged particle in steady-state oscillation, the oscillating charge emits radiation that travels in vacuum at velocity c. The fields radiated by the oscillating charge superpose with the incident field to give a resultant field. When there are many charges, as in a piece of glass (or in the ionosphere), each charge is driven by the local electric field in the vicinity of the charge. This field is in turn the result of a superposition of the field that would be present if there were no charges (the "incident field") plus the fields radiated by all the oscillating charges.

Each oscillating charge (in a piece of glass, for example) radiates waves that travel with the velocity of light *in vacuum*, c, even though the waves are "going through glass." How is it possible for a superposition of waves all having the same velocity c, same frequency ν, and hence same wavelength c/ν, to give a resultant that has a wavelength λ that is not c/ν and a phase velocity that is different from c? The clue is in that word "phase." Everything depends on the relative phase between the field radiated by a single oscillating charge and the field that is driving it. If the field radiated by the driven charge were exactly in phase with the driving radiation, then at some downstream observation point it would increase the total field (by so called "constructive interference"), but it would not produce any shift in phase of the total field and therefore would not affect the phase velocity. Similarly, if the radiated field were 180 deg out of phase with the driving field, the superposition of radiated and driving fields would give a resultant less than the incident field (by "destructive interference"), but it would not shift the phase. In order for the radiation from the charges to shift the phase of the resultant, it must include a contribution that is either +90 deg or –90 deg out of phase with the driving field. The phase constant of the resultant is mostly determined by the driving field (because the driving field is larger than the infinitesimal contribution from the single charge that we are considering), but the phase constant is slightly "pulled" by the contribution from the oscillating charge.

Thus, for example, suppose that at a fixed downstream point from the driven charge the field due to the incident radiation is $E_0 \cos \omega t$. That is the electric field at the observation point when there is no glass present. It is due to the oscillating electrons in some distant light bulb, for example. When the glass is inserted between the distant light bulb and the observer, the field contributed by the oscillating light-bulb electrons is *still* given by $E_0 \cos \omega t$ and *still* travels (through the glass and all) with velocity c. Now suppose that a small contribution from some of the oscillating glass molecules is given by the field $\varepsilon \sin \omega t$, where ε is very small and is (for example) positive. This radiation also travels through the rest of the glass at velocity c, but it has, by hypothesis, a 90-deg phase shift relative to the driving radiation. The superposition gives the resultant oscillating field at the observation point:

$$E(t) = E_0 \cos \omega t + \varepsilon \sin \omega t.$$

For $\varepsilon \ll E_0$, this is equivalent to

$$E(t) = E_0 \cos (\omega t - \delta), \qquad \delta \equiv \frac{\varepsilon}{E_0} \ll 1,$$

as you can easily see (use $\cos \delta \approx 1$ and $\sin \delta \approx \delta$). Thus we see that at a downstream point the resultant $E(t)$ has a phase shift δ when the glass is present. The observer at the downstream point has to "wait longer" for the phase of $E(t)$ to acquire a given value, i.e., he has to wait for $\omega t - \delta$ to reach the value that ωt would reach if there were no glass. He therefore says that the phase velocity is less than c. Note that if the contribution of the glass were proportional to $\cos \omega t$, then there would be no phase shift, since then the resultant would be

$$E(t) = (E_0 + \varepsilon) \cos \omega t,$$

and the phase velocity would still be the vacuum value, c. Instead, experiment shows that the phase velocity of the resultant is different from c, in spite of the fact that every contribution to the superposition travels at velocity c. That means the radiation from glass molecules that arrives at a given time t must be ±90 deg out of phase with the radiation from the light bulb that arrives at the same time.

The only thing that remains is to show that an infinitesimal contribution from radiating glass molecules is indeed ± 90 deg out of phase with the driving field. We do that as follows. Suppose the incident field is $E_0 \cos \omega t$. Then the oscillating charge has displacement $x(t) = A_{el} \cos \omega t$, for ω not near a resonance. In Chap. 7 we shall learn that the radiation from an oscillating charge is proportional to the "retarded acceleration." That means that the radiated field a distance z downstream is proportional to the acceleration of the charge at the earlier time $t - (z/c)$ when the radiation was being emitted. For harmonic motion the acceleration is $-\omega^2$ times the displacement. Thus we arrive at the horrible result that the radiation contributed by each of the oscillating charges is proportional to cos ωt, whereas we have decided that it *must* be proportional to sin ωt if we are to get a phase velocity different from c! The explanation is this: Suppose we have a "plane wave" of radiation propagating in the z direction. Then at a given instant we should consider not only the contribution from the one molecule directly upstream from the observation point but all the contributions from a thin slab of the glass perpendicular to the direction of propagation of the wave. As we have just seen, the molecule closest to the observation point contributes an infinitesimal contribution that is in phase with the driving field (neglecting plus or minus signs), but other molecules in the slab are farther away. Their contributions take longer to arrive (always traveling at velocity c). When we integrate over an infinitely wide slab, it turns out (as we shall show in Chap. 7) that the net contribution from the slab has a phase that is 90 deg behind that from the closest molecule. In other words, the average molecule in the slab is effectively one quarter-wavelength farther from the observation point than is the closest molecule. Thus we have found the source of the 90-deg phase shift, and we can see how it is that many waves, all traveling at velocity c, can superpose to give a resultant with phase velocity not c. The question of whether the phase velocity is greater than or less than c hinges only on whether the driven oscillations are in phase or are 180 deg out of phase with the driving radiation. That in turn depends, as we have seen, on whether the driving frequency is below or above the resonance frequency. Since all the molecules are at steady state, there is no need to "worry" about the fact that the phase velocity can exceed c.

Nomenclature: Why do we always consider E and neglect B?   We don't always, but we often do. Part of the reason we usually express the effect of electromagnetic waves in terms of **E** and suppress **B** from the formulas is the following: When electromagnetic waves interact with a charged particle of charge q and velocity **v**, the force on the particle is given by the Lorentz force (Vol. II, Sec. 5.2)

$$\mathbf{F} = q\vec{\mathbf{E}} + q\vec{\boldsymbol{v}} \times \vec{\mathbf{B}}.$$

In an electromagnetic traveling wave in vacuum, it turns out that **E** and **B** have the same instantaneous magnitude. Therefore, the magnitude of the force contributed by **B** is smaller than that contributed by **E** by a factor of order $|\mathbf{v}/c|$. Now, it turns out that when **E** and **B** are due to ordinary light sources or even due to a powerful laser, the fields **E** and **B** are sufficiently weak so that the maximum velocity $|\mathbf{v}|$ attained in the steady-state motion of driven electrons in a piece of ordinary materials is tiny compared with c. Thus, there are a large number of physical situations where we can neglect the force due to **B**. That is why we emphasize **E**.

Sometimes, however, the effects of **B** can dominate, small as they are according to the above discussion. And, of course, if **E** and **B** are not due to radiation (traveling waves) but are (for example) static fields due to independent charges and currents, then **B** and **E** are not constrained to have the same magnitude. For example, we could then have $|\mathbf{E}| = 0$ and $|\mathbf{B}| = 100$ kilogauss.

4.4 *Impedance and Energy Flux*

In studying modes and standing waves, we found that a continuous medium can be characterized by two parameters, one denoting a "return force," the other "inertia." For a continuous string, the equilibrium tension T_0 gives the return force, and the mass density ρ_0 gives the inertia. For the low-pass transmission line, the corresponding parameters are $(C/a)^{-1}$, the inverse of the shunt capacitance per unit length, and L/a, the inductance per unit length. For longitudinal waves on a spring, the return force parameter is Ka, and the inertial parameter is $M/a = \rho_0$. For sound waves, the "springiness" is given by $\gamma \rho_0$, the inertia by the volume mass density ρ_0. In all cases, the standing wave modes behave in a way that is analogous to a simple harmonic oscillator. (For coupled pendulums or a band-pass transmission line, we need another parameter, the low-frequency cutoff.)

Traveling waves behave very differently from standing waves; they transport energy and momentum. The phase relations are different from those for standing waves. An extended system carrying traveling waves does not behave like "one big harmonic oscillator," as it does when it carries standing waves. Thus the harmonic oscillator parameters, return force and inertia, are not the best physical parameters to describe a medium carrying traveling waves. One quantity that does characterize a medium carrying traveling waves is the phase velocity v_φ. For transverse waves on a string, this is given by

$$v_\varphi = \sqrt{\frac{T_0}{\rho_0}}, \tag{93}$$

which is just a certain combination of the return force and inertia parameters T_0 and ρ_0. An independent combination of T_0 and ρ_0 is given by

$$Z = \sqrt{\rho_0 T_0}. \tag{94}$$

This quantity is called the *characteristic impedance*, or simply *impedance*, for transverse waves on a continuous string. As we shall show, the impedance determines the rate at which energy is radiated onto the string by a given driving force. Thus it turns out that the phase velocity and the impedance are the two natural parameters to describe traveling waves in a given medium.

Example 8: Transverse traveling waves on continuous string

Suppose we have a continuous string stretching from left to right, with the left end at $z = 0$ driven transversely by a harmonically oscillating force. The system is shown in Fig. 4.8. Let us designate the connection by which the driving force is imparted to the string—the "transmitter output terminal"—by the letter L (for left), and let us designate the string immediately in contact with the terminal. by the letter R (for right). At equilibrium (Fig. 4.8a), we have no transverse component of force on L. The force along z is the equilibrium tension T_0. For the general configuration of Fig. 4.8b, the string tension is T. The transverse component of force exerted by the string on the transmitter output terminal, $F_x(R \text{ on } L)$, is given by

$$F_x(R \text{ on } L) = T \sin \theta$$

$$= (T \cos \theta) \frac{\sin \theta}{\cos \theta}$$

$$= T_0 \tan \theta$$

$$= T_0 \frac{\partial \psi}{\partial z}. \tag{95}$$

The result (95) holds exactly for an ideal slinky, which has $T = T_0/\cos \theta$.
It also holds for any spring for small angles θ.

Characteristic impedance. Now let us assume that the transmitter is driving a completely open medium (the string) at steady state, so that *it is emitting traveling waves in the +z direction*. Then $\psi(z, t)$ has the form

$$\psi(z, t) = A \cos (\omega t - kz). \tag{96}$$

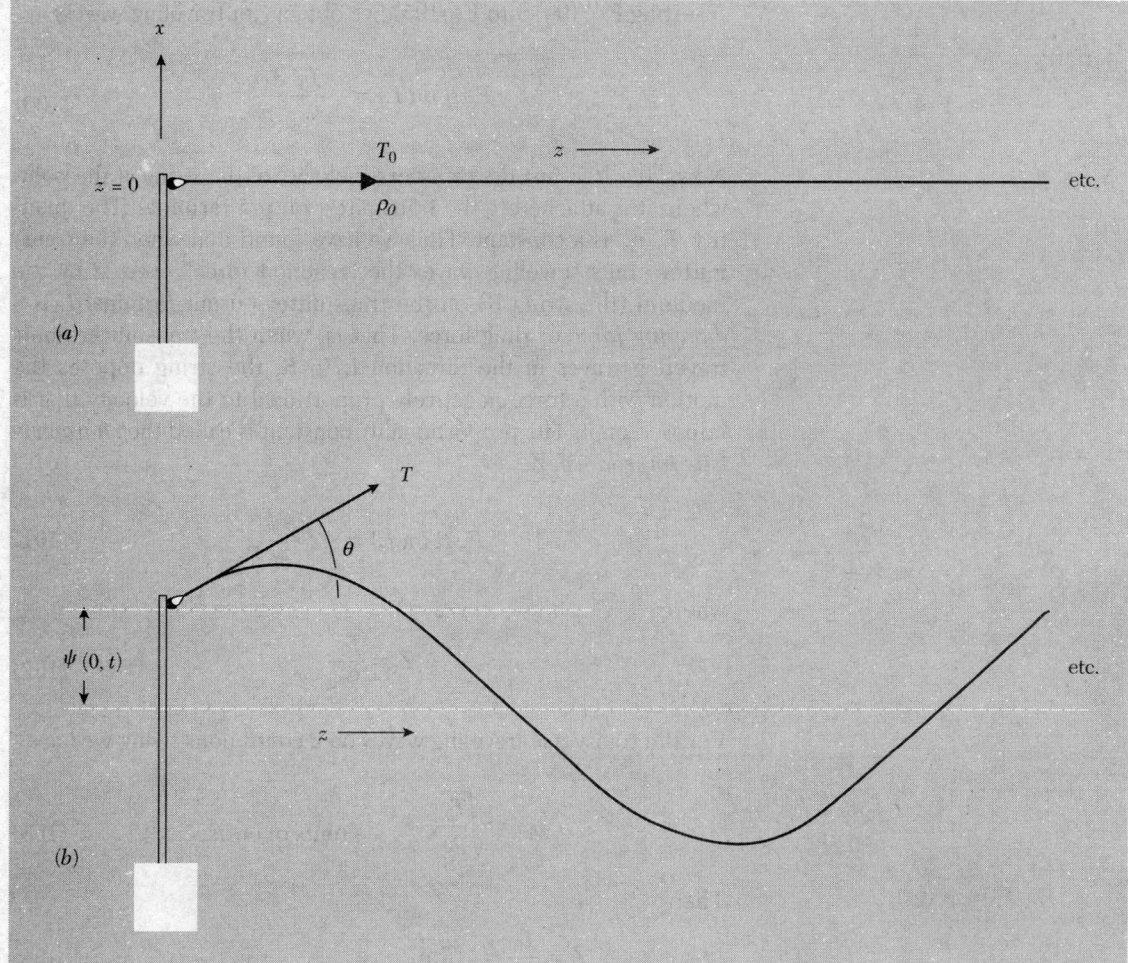

Fig. 4.8 Emission of transverse traveling waves. (a) Equilibrium. (b) General configuration.

Differentiation gives

$$\frac{\partial \psi}{\partial z} = kA \sin(\omega t - kz),$$ (97)

$$\frac{\partial \psi}{\partial t} = -\omega A \sin(\omega t - kz).$$ (98)

Comparing Eqs. (97) and (98) and using $v_\varphi = \omega / k$, we see that for a traveling wave traveling in the +z direction

$$\frac{\partial \psi}{\partial z} = -\frac{1}{v_\varphi} \frac{\partial \psi}{\partial t}.$$ (99)

Inserting Eq. (99) into Eq. (95), we obtain (for traveling waves)

$$F_x(R \text{ on } L) = -\frac{T_0}{v_\varphi} \frac{\partial \psi}{\partial t}. \tag{100}$$

Now, $\partial \psi / \partial t$ is just the transverse velocity of the string at the point where it is attached to the transmitter output terminal. The quantity T_0/v_φ is a constant. Thus we have found that when the transmitter emits traveling waves the "reaction force" exerted by the medium (the string R) on the transmitter output terminal L is a *damping force* or drag force. That is, when the transmitter emits traveling waves in the direction L to R, the string opposes the motion with a force negatively proportional to the velocity that is imposed on it. The proportionality constant is called the *characteristic impedance*, Z:

$$F_x(R \text{ on } L) = -Z \frac{\partial \psi}{\partial t}, \tag{101}$$

where

$$Z = \frac{T_0}{v_\varphi}. \tag{102}$$

For the transverse traveling waves on a continuous string we have

$$v_\varphi = \sqrt{\frac{T_0}{\rho_0}} \qquad \text{units of cm/sec.} \tag{103}$$

Then

$$Z = \frac{T_0}{v_\varphi} = \sqrt{T_0 \rho_0} \qquad \text{units of dyne/(cm/sec).} \tag{104}$$

Transmitter output power. The most characteristic thing about a damping force is that it "dissipates" or "absorbs" energy. In the present example, it is best to say that the string *absorbs* energy in the form of "radiation" by the transmitter output. The energy lost by the transmitter has not been dissipated, in the sense that it has not been "degraded" into "heat." Instead it has been radiated onto the string, which can transport it to a distant "receiver," where it can be completely recovered (as we shall learn later). The radiated output power is given by the product of the transverse force exerted by the transmitter on the string at $z = 0$ times the transverse velocity of the string at $z = 0$. Using the fact that $F_x(L \text{ on } R)$ is the negative of $F_x(R \text{ on } L)$ (which is Newton's third law) and using Eq. (101), we find the instantaneous output $P(t)$ to be given (in units of erg/sec) by

$$P(t) = F_x(L \text{ on } R) \frac{\partial \psi}{\partial t} \qquad \text{(general)}$$

$$P(t) = \left(Z \frac{\partial \psi}{\partial t} \right) \frac{\partial \psi}{\partial t} = Z \left(\frac{\partial \psi}{\partial t} \right)^2 \qquad \text{(traveling wave)}. \qquad (105)$$

The first equality in Eq. (105) is general. The second is not; it holds only for traveling waves.

In Eq. (105) we have expressed the output power in terms of the wave quantity $\partial \psi / \partial t$, which corresponds to the instantaneous transverse velocity of the string (at $z = 0$) in units of cm/sec. Another equally interesting and important wave quantity is the transverse force $F_x(R \text{ on } L)$ given (in units of dynes) by Eq. (95). The transmitter output power can be expressed in terms of this quantity by means of Eqs. (95) and (99):

$$P(t) = F_x(L \text{ on } R) \frac{\partial \psi}{\partial t} \qquad \text{(general)}$$

$$= \left[-T_0 \frac{\partial \psi}{\partial z} \right] \frac{\partial \psi}{\partial t} \qquad \text{(general)}$$

$$P(t) = \left[-T_0 \frac{\partial \psi}{\partial z} \right] \left[-v_\varphi \frac{\partial \psi}{\partial z} \right] \qquad \text{(traveling wave)} \qquad (106)$$

$$= \frac{v_\varphi}{T_0} \left[-T_0 \frac{\partial \psi}{\partial z} \right]^2$$

$$= \frac{1}{Z} \left[-T_0 \frac{\partial \psi}{\partial z} \right]^2.$$

The first and second equalities in Eq. (106) are general; the third is not; it holds only for traveling waves.

We take the trouble to express $P(t)$ in the different but equivalent ways of Eqs. (105) and (106) because we shall always find that there are two physically interesting wave quantities and in some systems we may wish to use one, while in other systems we may wish to use the other. For example, in the case of sound waves we shall find that the gauge pressure plays a role analogous to that of the transverse return force $- T_0 \, \partial \psi / \partial z$ for the string and that the longitudinal air velocity in the sound wave plays a role analogous to the transverse string velocity $\partial \psi / \partial t$. Similarly, in the case of electromagnetic radiation we shall find that the transverse magnetic field B_y plays a role analogous to that of the transverse string velocity $\partial \psi / \partial t$, while the transverse electric field E_x plays a role analogous to that of the return force $- T_0 \, \partial \psi / \partial z$ for the string.

Energy transport by a traveling wave. The radiated power $P(t)$ emitted at $z = 0$ by the transmitter in the form of traveling waves is equal to the amount of energy per unit time traveling in the $+z$ direction past any downstream point z. (We are neglecting damping.) In fact, when we derived our results for the energy flow from L to R (left to right) at the transmitter output terminal, we could have been speaking of a general downstream point z instead of the point $z = 0$. The only thing we required was that the medium be carrying traveling waves. If you will retrace the steps of the derivation with that consideration in mind, you will quickly see that for traveling waves the radiated power passing a given point z in the $+z$ direction is given by expressions analogous to Eqs. (105) and (106), except that the transverse velocity $\partial \psi/\partial t$ and the return force $-T_0 \, \partial \psi/\partial z$ are evaluated at the general point z instead of at $z = 0$. Thus for a traveling wave on a string, we have

$$P(z, t) = Z \left[\frac{\partial \psi(z, t)}{\partial t} \right]^2 \qquad (107)$$

or

$$P(z, t) = \frac{1}{Z} \left[-T_0 \, \frac{\partial \psi(z, t)}{\partial z} \right]^2. \qquad (108)$$

Example 9: Radiation of longitudinal waves on a spring

Next we consider the emission of longitudinal waves of compression and rarefaction on a spring. We shall be able to adapt these results to describe the radiation of sound waves, using the simple method of Newton (but correcting his famous oversight). The systems shown in Fig. 4.9.

In the equations of motion for longitudinal motion of a beaded spring, the quantity Ka enters in exactly the same way as does the equilibrium tension T_0 in the equations of motion for transverse oscillation of the beaded spring. (See Eq. (2.77), Sec. 2.4, and the discussion immediately following it.) That is why the phase velocities are obtained one from the other by interchanging T_0 and Ka [see Eq. (27), Sec. 4.2]. Similarly, we can find the characteristic impedance and the energy flux relations for longitudinal waves on a continuous spring by simply substituting Ka and T_0 in our results obtained for transverse oscillation. Thus, from Eqs. (103), (104), (107), and (108) we obtain for longitudinal waves the results

$$v_\varphi = \sqrt{\frac{Ka}{\rho_0}}, \qquad Z = \sqrt{Ka\rho_0}, \qquad (109)$$

and the power flow in a traveling wave (in erg/sec),

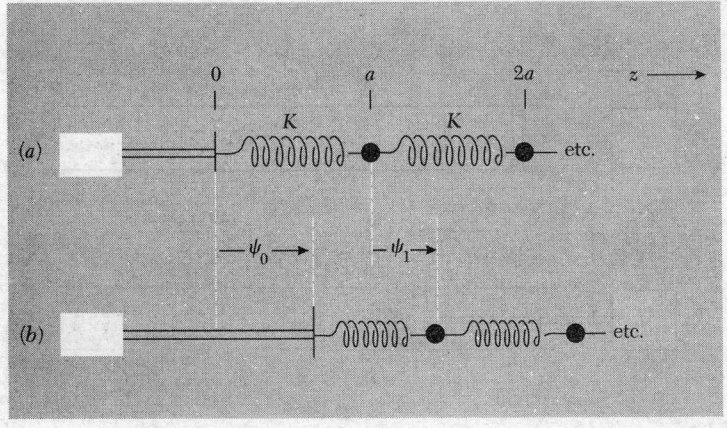

Fig. 4.9 Emission of longitudinal traveling waves. (a) Equilibrium. (b) General configuration.

$$P(z, t) = Z\left[\frac{\partial \psi(z, t)}{\partial t}\right]^2 = \frac{1}{Z}\left[-Ka\,\frac{\partial \psi(z, t)}{\partial z}\right]^2. \quad (110)$$

The quantity $\psi(z, t)$ is the displacement from its equilibrium position of that part of the spring having equilibrium position z; it is positive if the displacement is in the $+z$ direction. The corresponding velocity is $\partial \psi(z, t)/\partial t$. The quantity $-Ka\,\partial \psi(z, t)/\partial z$ turns out to be the force in the $+z$ direction exerted on that part of the spring with equilibrium position to the right of point z by that having equilibrium position to the left of point z, after the equilibrium value of that force, F_0, has been subtracted out (Prob. 4.29):

$$F_z(L \text{ on } R) = F_0 - Ka\,\frac{\partial \psi(z, t)}{\partial}. \quad (111)$$

The force F_0 in Eq. (111) is due to the stretching or compression of the springs in their equilibrium configuration. It makes no contribution to any waves. That is why it is only the excess above F_0, namely $-Ka\,\partial \psi/\partial z$, that appears in the second equality of Eq. (110).

Example 10: Sound waves

We will use Newton's model for sound waves as discussed in Sec. 4.2. The system is shown in Fig. 4.10.

Now, in Sec. 4.2 we found the phase velocity of sound by making use of Newton's analogy of sound waves with longitudinal waves on a continuous spring. We ended up by replacing the equilibrium linear mass density for the spring with the equilibrium volume mass density for the air and by replacing Ka for the spring by the equilibrium pressure p_0 for air times the famous factor γ. We can therefore easily guess the impedance and energy relations for sound waves.

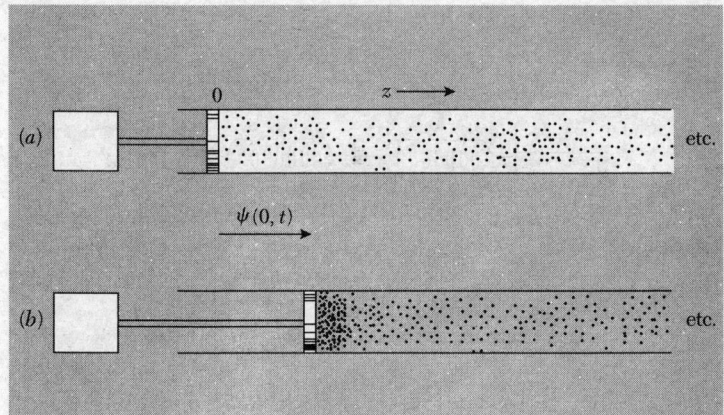

Fig. 4.10 Emission of longitudinal sound waves. (a) Equilibrium. (b) General configuration.

We simply replace Ka by γp_0 in the relations for longitudinal waves on a spring. Thus for sound waves we obtain [from Eqs. (109) and (110)] the results

$$v_\varphi = \sqrt{\frac{\gamma p_0}{\rho_0}}, \qquad Z = \sqrt{\gamma p_0 \rho_0}, \tag{112}$$

and the intensity of the energy flow in a traveling sound wave, in erg/cm² sec,

$$I(z, t) = Z \left[\frac{\partial \psi(z, t)}{\partial t} \right]^2 = \frac{1}{Z} \left[-\gamma p_0 \frac{\partial \psi(z, t)}{\partial z} \right]^2. \tag{113}$$

The quantity $\psi(z, t)$ is the displacement of a bit of air along the z direction from its equilibrium position, which is z. The quantity $\partial \psi(z, t)/\partial t$ is the corresponding velocity. The quantity $-\gamma p_0\ \partial \psi(z, t)/\partial z$ is equal to the force per unit area in the $+z$ direction exerted on the air to the right of z by that to the left of z (remember z is the equilibrium position, not the instantaneous position, of the air), after the equilibrium value of that force, per unit area, p_0, has been subtracted out:

$$\frac{F_z(L \text{ on } R)}{A} = p_0 - \gamma p_0 \frac{\partial \psi(z, t)}{\partial z}. \tag{114}$$

This follows from Eq. (111) for longitudinal waves on a spring, with the substitution of p_0 for F_0 and of γp_0 for Ka. The equilibrium pressure p_0 does not contribute any waves. We shall dignify $-\gamma p_0\ \partial \psi/\partial z$ by the name *gauge pressure*, p_g:

$$p_g = -\gamma p_0 \frac{\partial \psi(z, t)}{\partial z}. \tag{115}$$

For air at STP, we have $p_0 = 1$ atm $= 1.01 \times 10^5$ N/m^2 and $\rho_0 = 1.29 \times 10^{-3}$ gm/ cm^3. Thus Eqs. (112) give

$$v_\varphi = 3.32 \times 10^4 \text{ cm/sec}, \qquad (116)$$

$$Z = 42.8 \times 10 \frac{\left(\text{N/m}^2\right)}{(\text{m/s})}.$$

$$= 428 \frac{\left(\text{N/m}^2\right)}{(\text{m/s})} \qquad (117)$$

Standard sound intensities. The *intensity* of a traveling sound wave is defined as the energy propagating through a unit area per unit time. A commonly used standard of sound intensity is given by

$$\text{Standard intensity} = I_0 = 1 \text{ } \mu\text{w/cm}^2 = 10^{-2} \text{ J/m}^2 \text{ sec.} \qquad (118)$$

where we used the facts that $1 \text{ } \mu\text{w} = 10^{-6}$ watt and that 1 watt = 1 J/s. A person speaking in an average conversational tone emits about 10^{-5} J/sec of sound energy. The mouth aperture is about 10 cm^2 when speaking. Therefore if you speak into one end of a mailing tube, so that all the sound energy goes in the z direction (down the tube), the sound intensity is about $(10^{-5} \text{ J/s}) / 10^{-3} \text{ m}^2 = 10^{-2} \text{ J/s/m}^2 = I_0$. Thus you can get a feeling for the magnitude of I_0 by having someone talk to you through a (short) mailing tube. (A long tube attenuates the sound by friction on the rough cardboard walls and by radiation out the sides of the tube.) If the person shouts as loud as he can into the mailing tube, the intensity is about $100 I_0$. For intensities of 100 to 1000 times I_0, the listener feels pain.

The intensity of the faintest sound that can be heard depends on its frequency. At about A440 (i.e., 440 Hz or 440 cps), the average person's threshold of audibility is about $10^{-10} I_0$. Thus the human ear has the huge dynamic range of a factor of 10^{12} in intensity (from $100 I_0$ to $10^{-10} I_0$).

Nomenclature—decibel. Whenever the sound intensity increases by a factor of 10, it is said to have increased by 1 bel. Thus the dynamic range of the ear is about 12 bels. Whenever the intensity increases by a factor of $10^{0.1}$, it has increased by 0.1 bel or 1 decibel. Thus

> 1 db = 1 decibel = factor of $10^{0.1}$ = factor of 1.26 in intensity;
> 1 bel = 10 db = factor of 10 in intensity. $\qquad (119)$

A person with normal hearing can barely detect an increase of loudness of about 1 db.

The following applications involve calculations of sound impedance and flux.

Application: RMS gauge pressure for painful sound intensity

For painful sound intensity, what is the root-mean-square gauge pressure (in atmospheres)? We want the answer in atmospheres be cause we are interested in the question of whether the pain has the same cause as that which you feel if you swim down 4.67 m or so beneath the surface of water (without pumping air into your inner ear by swallowing). We know that ≈ 10 m of water gives 1 atmosphere of pressure, so at a 4.6 m depth the gauge pressure is about $\frac{1}{2}$ atmosphere of pressure. Is this the gauge pressure that a painful sound wave has?

Solution: Take $I = 1000I_0$ as the painful intensity. Then according to Eq. (113), we have

$$\langle p_g^2 \rangle^{1/2} = (ZI)^{1/2}$$
$$= (1000\, ZI_0)^{1/2}$$
$$= [(1000)(42.8)(10)]^{1/2} = 65 \text{ N/m}^2.$$

This is tiny compared with 1 atm $= 1.01 \times 10^5$ N/m^2. Thus we have the interesting result that the pain is not due simply to the time-averaged pressure's being too high, because 60 N/m^2 is 6×10^{-4} atm, which is like swimming under only $\frac{1}{2}$ cm of water.

Application: Amplitude for painfully loud sound

What is the amplitude A of the excursions of the air molecules for a painfully loud sound? Take $\psi(z, t) = A \cos(\omega t - kz)$. Then $\partial\psi/\partial t$, squared and averaged over one cycle at fixed z, equals $\frac{1}{2}\omega^2 A^2$. Then using Eq. (113) and assuming the frequency is 440 cps we find

$$A = \frac{(2I/Z)^{1/2}}{\omega}$$

$$= \frac{(2 \cdot 1000 \cdot 10/42.8)^{1/2}}{(6.28)(440)}$$

$$= 2.5 \times 10^{-2}\, \text{cm} = \tfrac{1}{4}\, \text{mm}.$$

Application: Amplitude for barely audible sound

What is the amplitude of the air excursions for barely audible sound? Suppose the intensity is $10^{-10}I_0$. The amplitude is proportional to the square root of I. Thus for frequency A440, the result is the square root of 10^{-13} times the result found in the application above, where we took $I = 1000I_0$. Thus

$$A = 10^{-6.5}(2.5 \times 10^{-2})$$

$$= \frac{2.5 \times 10^{-8}}{\sqrt{10}} \approx 10^{-8}\,\text{cm}.$$

This is about the diameter of an average atom. Thus your ear is so acutely sensitive that it is capable of detecting motions of the eardrum equaling about one atomic diameter!

Application: Audio output from typical hi-fi speaker

What is the approximate audio (sound) output (in watts) that you would expect to find from a typical hi-fi speaker? Assume that an average hi-fi enthusiast wishes to fill a long room that has reflecting side walls but a sound-absorbing rear wall with painfully loud traveling waves of intensity $1000I_0$. Suppose the room has cross-sectional area $3\,\text{m} \times 3\,\text{m} \approx 10\,\text{m}^2 \approx 10^5\,\text{cm}^2$. At the end where the speaker is radiating, the enthusiast can let the speaker drive the entire wall as a sounding board, or he can use the first part of the room to provide a gradually tapered "horn" so as to "match impedances" between the speaker and the room. (Impedance matching will be discussed in Chap. 5.) In any case, the audio output is given by

$$P = I \cdot (\text{area}) = (100)\, I_0 \cdot 10^5 = 10^7\, \mu\text{w}.$$
$$= 10 \text{ watts.}$$

Thus 10 watts of audio output is common in hi-fi sets.

Application: Sum of two nearly painful sounds

Suppose a person can barely stand the pain of intensity $100I_0$ at A440, but he cannot bear the pain of $200I_0$ at that same frequency. Suppose that this is also true for C512—he can stand $100I_0$ but cannot stand $200I_0$. What happens if both notes are sounded at once, each with intensity $100I_0$? The total intensity is now $200I_0$. Can he stand it? I don't know. (I have a guess.)

We hope we have convinced you that you are now equipped to answer some interesting questions about sound. We have not yet discussed standing waves of sound, but they behave just like standing longitudinal waves on a slinky. Therefore you should have no difficulty in understanding the home experiments on sound if you look at them now.

Example 11: Traveling waves on low-pass transmission line

The system for this important example is shown in Fig. 4.11. The driving force is the voltage $V(t)$ applied at $z = 0$. We shall only

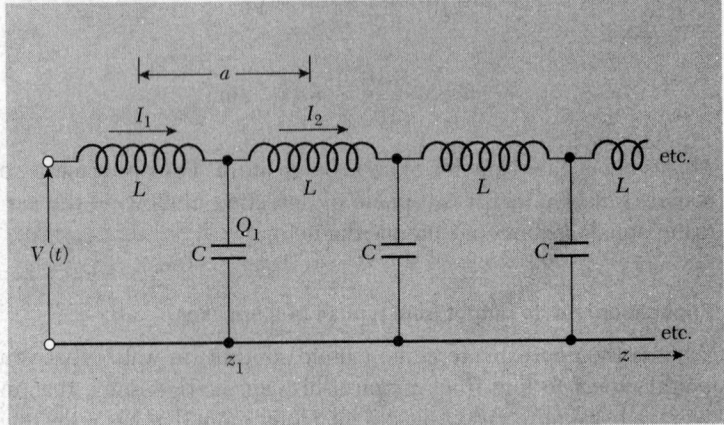

Fig. 4.11 Emission of traveling waves on a transmission line.

consider the long wavelength limit (i.e., the continuous limit), where $V(z, t)$ and $I(z, t)$ are continuous functions of z. If the transmission line is infinitely long (or is terminated in perfectly absorbing material), we have an open system carrying traveling waves of voltage $V(z, t)$ and current $I(z, t)$. If the driving voltage $V(t)$ at the input end has the form

$$V(t) = V_0 \cos \omega t, \qquad (120)$$

then the voltage wave $V(z, t)$ must equal $V_0 \cos \omega t$ at $z = 0$, and hence it is given by

$$V(z, t) = V_0 \cos (\omega t - kz). \qquad (121)$$

We wish to find the relation between $V(z, t)$ and $I(z, t)$. It will turn out that (for a traveling wave) they are proportional to one another (and not, for example, ±90 deg out of phase). We shall anticipate that result by writing

$$I(z, t) = I_0 \cos (\omega t - kz) + J_0 \sin (\omega t - kz) \qquad (122)$$

where, as we shall find, the constant J_0 has value zero.

Consider the first capacitor in Fig. 4.11. It has charge $Q_1(t)$, which corresponds to the potential difference $V_1(t)$, where

$$Q_1(t) = CV_1(t) = CV(z_1, t). \qquad (123)$$

Then

$$C \frac{\partial V(z_1, t)}{\partial t} = \frac{dQ_1}{dt}$$

$$= I_1 - I_2$$

$$= -(I_2 - I_1)$$

$$= -a \frac{\partial I(z_1, t)}{\partial z},$$

where in the last equality we used the continuous approximation. Thus

$$\frac{\partial V(z_1, t)}{\partial t} = -\left(\frac{C}{a}\right)^{-1} \frac{\partial I(z_1, t)}{\partial z}. \tag{124}$$

Inserting Eqs. (121) and (122) into Eq. (124), we see that the constant J_0 in Eq. (122) must indeed be zero. The remaining terms give

$$-\omega V_0 \sin(\omega t - kz) = -\left(\frac{C}{a}\right)^{-1} I_0 k \sin(\omega t - kz),$$

i.e.,

$$V_0 = \frac{(C/a)^{-1}}{v_\varphi} I_0, \tag{125}$$

where

$$V(z, t) = \frac{(C/a)^{-1}}{v_\varphi} I(z, t) \equiv ZI(z, t), \tag{126}$$

by definition of Z. Thus the phase velocity [from Sec. 4.2, Eq. (48)] and characteristic impedance are given (in the long-wavelength or continuous or "distributed-parameter" limit) by

$$v_\varphi = \sqrt{\frac{(C/a)^{-1}}{(L/a)}}, \tag{127}$$

$$Z = \frac{(C/a)^{-1}}{v_\varphi} = \sqrt{\left(\frac{L}{a}\right)\left(\frac{C}{a}\right)^{-1}} \tag{128}$$

The instantaneous power output of the transmitter at $z = 0$ is given by

$$P(t) = V(t)I(t) = V(0, t)I(0, t) = ZI^2(0, t). \tag{129}$$

Alternatively, $P(t)$ is given by

$$P(t) = V(0, t)I(0, t) = \frac{V^2(0, t)}{Z}. \tag{130}$$

Notice that we could have obtained Z by merely substituting C^{-1} for K and L for M in the results for longitudinal oscillations of masses on springs. Because of the importance of this example, we went through the details instead.

Example 12: Parallel-plate transmission line

This important example will lead us to results of great generality. According to Eqs. (55) and (59), Sec. 4.2, the capacitance per unit length and inductance per unit length for the parallel-plate transmission line (with vacuum between the plates) are

$$\frac{C}{a} = \frac{w\epsilon_0}{g}, \qquad \frac{L}{a} = \frac{\mu_0}{\epsilon_0}\frac{g}{w}, \tag{131}$$

where w is the width and g is the gap distance. Thus the characteristic impedance is [see Eq. (128)]

$$Z = \sqrt{\frac{\mu_0}{\epsilon_0}}\frac{g}{w}. \tag{132}$$

(Here the units of Z are volts/amps, i.e., ohms.) The radiated power $P(t)$ is given by Eq. (130) to be

$$P(t) = \frac{V^2}{Z} = \sqrt{\frac{\epsilon_0}{\mu_0}}\frac{w}{g}V^2(0, t). \tag{133}$$

Let us express the radiated power in terms of the only nonzero electric field component E_x, which is defined at every point between the plates, rather than in terms of the potential difference $V(0, t)$, which is the integral of the (uniform) electric field across the gap g:

$$V(0, t) = gE_x(0, t). \tag{134}$$

Then Eq. (133) becomes

$$P(t) = \sqrt{\frac{\epsilon_0}{\mu_0}}wgE_x{}^2(0, t). \tag{135}$$

Notice that wg is the cross-sectional area of the end of the transmission line. If we divide the radiated power by wg, we have the radiated intensity (in J/m^2). We don't want to designate this intensity by the symbol I, because the letter I is reserved at the moment for current. Instead, we shall use the symbol commonly used for electromagnetic waves, S. From our experience with strings and with sound waves, we know that we can write the intensity in the wave by simply replacing $z = 0$ with a general position z. Thus for a traveling plane wave of electromagnetic radiation propagating in the $+z$ direction in a parallel-plate transmission line, the energy per square centimeter per unit time passing a point z is given by the intensity

$$S(z, t) = \sqrt{\frac{\epsilon_0}{\mu_0}}E_x{}^2(z, t). \tag{136}$$

Let us now find the ratio of the only nonzero magnetic field component $B_y(z, t)$ to the electric field component $E_x(z, t)$. We can find that ratio because we have found that the ratio of $V(z, t)$ to $E(z, t)$ is the constant Z, and because we know how to relate V to E_x and I to B_y. Thus we have

$$V = ZI$$

i.e.,

$$gE_x = \sqrt{\frac{\mu_0}{\epsilon_0}} \frac{g}{w} I. \tag{137}$$

But according to Eq. (57), Sec.4.2, we have

$$wB_y = \mu_0 I. \tag{138}$$

By comparison of Eqs. (137) and (138), we find that, for a traveling plane wave of electromagnetic radiation propagating in the $+z$ direction in a parallel-plate transmission line, the electric and magnetic fields at every z and t are perpendicular to one another and to the direction of propagation, are equal in magnitude, and have algebraic signs such that the cross product $\mathbf{E} \times \mathbf{B}$ is in the direction of propagation. More briefly,

$$\boxed{E_x = \frac{1}{\sqrt{\mu_0 \epsilon_0}} B_y(z, t).} \tag{139}$$

Plane electromagnetic waves in transparent media. Suppose the transmission line is filled with material having ϵ_r and μ_r. The applied potential is $V(t)$. Then we can write the radiated power

$$P(t) = \frac{V^2}{Z},$$

where

$$V = gE_x,$$

and

$$Z = \sqrt{\frac{L/a}{C/a}} = \sqrt{\frac{\mu_r}{\epsilon_r}} \, Z_{\text{vacuum}},$$

i.e.,

$$Z = \sqrt{\frac{\mu}{\epsilon_0}} \frac{g}{w}. \tag{140}$$

These three equations give the intensity $S = P/gw$:

$$\boxed{S(z, t) = \sqrt{\frac{\epsilon}{\mu}} E_x^2 \, (z, t).} \tag{141}$$

Let us also find the ratio of B_y to E_x. For a given current I, the magnetic field is larger by a factor μ than it would be without the transparent material present. Thus

$$wB_y = \mu I.$$

But

$$V = ZI,$$

i.e.,

$$gE_x = \sqrt{\frac{\mu}{\epsilon}}\,\frac{g}{w}\,I.$$

Comparison of these expressions for E_x and B_y gives

$$\frac{B_y}{E_x} = \sqrt{\epsilon\mu}$$

Thus

$$\boxed{B_y = n\sqrt{\epsilon_0 \mu_0}\;E_x.} \tag{142}$$

Plane electromagnetic waves in unbounded vacuum. The results given for vacuum by Eqs. (136) and (139),

$$S(z, t) = \sqrt{\frac{\epsilon_0}{\mu_0}}E_x^{\,2}(z, t), \qquad B_y(z, t) = \sqrt{\mu_0 \epsilon_0}E_x, \tag{143}$$

were derived for the electromagnetic waves (waves of electric and magnetic field variation) that result from current and potential waves on a straight and parallel transmission line. Now, not only is a parallel-plate transmission line straight and parallel, but it is also uniform (assuming there are no edge effects). The electric and magnetic fields $E_x(z, t)$ and $B_y(z, t)$ are also uniform: as long as one stays between the plates and as long as the width w is so great that edge effects are negligible, E_x has the same value for all positions x and y (for a fixed position z and time t); B_y is similarly independent of x and y. Such waves are called *plane waves*. Any plane perpendicular to the z axis (the propagation axis of the waves) is a plane of constant phase, i.e., a plane of constant value of $\omega t - kz$. These planes are called *wave fronts*.

Now, there is more than one way to obtain traveling electromagnetic plane waves. One way is the way we have just studied, using a parallel plate transmission line. Another way to obtain traveling waves that are approximately plane waves is to go very far from a "point source" of electromagnetic radiation, such as a candle, or a street light, or a star. (In a later chapter we shall find how small the source must be before it is a sufficiently good approximation to call it a "point.") In this situation, all the radiation in the neighborhood of the observer is traveling in essentially the same direction, as long as one does not take too large a neighborhood transverse to the (approximate) propagation direction. (We shall learn later how large the "neighborhood" can be. It depends on the type of experiment you have in mind, as usual.) It turns out (as should by now seem plausible to you, and as we shall prove using Maxwell's equations in a later chapter) that the results in Eq. (143) are "local" properties of electromagnetic plane waves and do not depend on the boundary conditions, i.e., upon the current and charge configurations that radiated the waves. Of course, the fact that **E** happens to be along $\hat{\mathbf{x}}$ depends on the boundary conditions we specified by our arrangement of the parallel-plate transmission line. Therefore we should express these very important and general results in a more general way than Eq. (143). We shall do so now:

A traveling electromagnetic plane wave traveling in the +z direction in vacuum has the following properties (not all of which are independent):

1. $\mathbf{E}(z, t)$ and $\mathbf{B}(z, t)$ are perpendicular to $\hat{\mathbf{z}}$ and to each other.
2. The magnitude of $\mathbf{E}(z, t)$ equals the magnitude of $\mathbf{B}(z, t)$.
3. The directions of $\mathbf{E}(z, t)$ and $\mathbf{B}(z, t)$ are such that $\mathbf{E}(z, t) \times \mathbf{B}(z, t)$ is along $+\hat{\mathbf{z}}$.
4. The first three properties imply that $\mathbf{B}(z, t) = \hat{\mathbf{z}} \times \mathbf{E}(z, t)$, which is equivalent to the relations $B_y(z, t) = E_x(z, t)$ and $B_x(z, t) = -E_y(z, t)$.
5. The phase velocity is c, independent of frequency; i.e., electromagnetic waves in vacuum are nondispersive.
6. The instantaneous intensity (in units of erg/cm² sec) is given by

$$S(z, t) = \sqrt{\mu_0 \epsilon_0} \left[E_x^2 + E_y^2 \right]. \tag{144}$$

Some commonly used synonyms for this quantity are in tensity, flux, and energy flux.

The above relationships are very important and, as far as is known, completely general. They hold for all frequencies, from (say) $\nu = 1$ cycle per 100,000 years (corresponding to a wavelength of 100,000 light-years, about the diameter of our galaxy) to frequencies of (say) $\nu \approx 3 \times 10^{25}$ Hz, corresponding to a wavelength c/ν of 10^{-15} cm or a "photon energy" $h\nu$ of about 100 BeV (billion electron volts). (You must get used to the fact that different units are used in different frequency ranges.)

Application: Finding the solar constant   **Home experiment**

This is a numerical example illustrating the energy flux. (It is a combination home experiment and calculation. We hope you will do the experiment.)

Problem: Determine the root-mean-square electric field in traveling waves of ordinary sunlight at the earth's surface.

Solution: (Since this is a real experiment, we shall make various approximations and assumptions as we go along. The answer will therefore be full of qualifications, as are most experimental results.) Get a 200- or 300-watt incandescent light bulb having a clear glass envelope (i.e., not frosted glass) and a filament an inch or less in length. Turn it on. Close your eyes. Bring the glowing bulb up near your face. Use your eyes as a detector *with closed eyelids as a filter*. Your eyelids detect some of the invisible infrared—they feel warm. Your eyes, covered by the eyelid filters, see a "redness," due to light that penetrates the filter. Now turn off the light and go outside (assuming it is a sunny day). "Look" *with closed eyes* at the sun. Get the feel of the warmth on your eyelids and the "redness" as "seen" through your eyelids Go back to the electric light. Measure the distance R from the eyelids to the filament for which the intensity, as judged by your detectors, is the same as that from the sun.

The experiment is over and the rest is mostly arithmetic. Use the rated power P of the bulb and the distance R to calculate the flux at your eyelids, assuming the filament radiates equally strongly in all directions. Then the answer (in mixed units) is that the time-averaged intensity at your lids is

$$\langle S(z, t) \rangle \equiv S = \frac{P}{4\pi R^2}. \tag{145}$$

Then that is also the time-averaged intensity of sunlight at your lids, at least in the range of colors that you can detect (including some infrared detected by your lids). *Assuming* that the "spectra" of colors from the light bulb and the sun are not too different, we can assume that the *total* flux from the sun, including the ultraviolet which we are presumably not detecting by this technique, is given by Eq. (145). S is called the "solar constant" and is indexed in the *Handbook of Physics and Chemistry*. There, you will find that S equals 1.94 "small calories" per square centimeter per minute. To convert the units, we use the fact that 1 small calorie = 4.18 joule and that 1 joule/sec = 1 watt. According to the *Handbook*, the solar constant *at the top of the earth's atmosphere* is

$$S = \frac{(1.94)(4.18)\text{ joule}}{60\text{ sec}} = 135 \text{ milliwatt/cm}^2. \tag{146}$$

Assuming the *Handbook* value, Eq. (146), what is the rms electric field in volts/cm?

$$S = 1.35 \times 10^3 \, \text{J/m}^2 \, \text{sec} = \frac{c}{4\pi} \langle E^2 \rangle,$$

$$\langle E^2 \rangle = \frac{(12.57)(0.135 \times 10^7)}{3 \times 10^{10}} = 5.6 \times 10^{-4},$$

thus $\qquad\qquad\qquad E_{\text{rms}} = 7.2 \, \text{volt/cm}.$ $\qquad\qquad$ (147)

Measurement of energy flux of electromagnetic radiation. In the example just given, your eyes and eyelids were used to determine the solar constant at the earth's surface. Your eyes and the heat sensors in your lids are typical of many radiation detector, in that they are square-law detectors—they respond to the incident intensity but are insensitive to phase information. (That is also the case for sound detection by your ears.) Then the appropriate quantity to describe the incident flux is not the instantaneous value of $S(z, t)$, but rather its time-averaged value, averaged over one oscillation cycle,

$$S \equiv \langle S(z, t) \rangle = \sqrt{\mu_0 \epsilon_0} \langle \mathbf{E}^2(z, t) \rangle. \qquad (148)$$

(For a plane wave, this time-averaged intensity is independent of position z.)

A typical square-law detector consists of a band-pass filter (used to pass radiation of the desired frequency and exclude other "background" radiation) followed by a "sensitive element" that absorbs all the incident flux without loss (by reflection) and gives an "output signal" proportional to (or at least dependent upon) the amount of absorbed energy. One broad class of such detectors uses a sensitive *calorimeter* as the energy-absorbing sensitive element. The amount of energy absorbed per unit time is determined by measuring the rate of increase of temperature in some absorbing material or by measuring the equilibrium excess of the temperature of the sensitive element over that of a standard environment (which may be something very reproducible and very cold, like liquid helium), with the equilibrium maintained by a constant heat leak between the sensitive element and the environment. Such a detector might have a self-contained calibration device, where one would (for example) temporarily exclude the external radiation and instead turn on a current through a standard resistor contained in the sensitive element. The power dissipated by the resistor is easily measured by measuring the current and potential drop and must equal the absorbed power from radiation which gives the same temperature excess. There are many ingenious refinements of this method.

Another class of detectors consists of *photon counters*. A photomultiplier tube is a photon counter. Whenever the "photocathode" of a photomultiplier absorbs a photon, one "photoelectron" is produced. This photoelectron is then accelerated through a potential of about 100 volts to a "multiplying dynode," where one photoelectron produces 3 or 4 secondary electrons. These are accelerated to a second dynode, where each produces 3 or 4 more electrons, etc. Finally, after perhaps 10 such stages of amplification, i.e., after 10 dynodes, one has about $(3.5)^{10}$ electrons from each incident photon, which are then collected on a "collector plate" or "anode." They are passed through a resistor, giving a pulse of voltage. The pulses are recorded and can be counted. Each pulse corresponds to the absorption of *exactly one photon having electromagnetic energy hv*, where v is the oscillation frequency, and h is Planck's constant. For a given frequency v, the photomultiplier's *detection efficiency* $e(v)$ can be determined by using some standard source of radiation. Then the *average counting rate R* (in counts per second), averaged, over sometime interval to, is given by the observed number of counts N divided by the time t_0:

$$R = \frac{N \pm \sqrt{N}}{t_0}, \tag{149}$$

where the "standard deviation" of the number of counts, a conventional estimate of the statistical uncertainty in the measurement, is taken to be the square root of the number of counts. The measured quantity R is used to determine the energy flux in J/m² sec by the relation

$$R = \frac{1}{hv} \sqrt{\mu_0 \epsilon_0} \langle \mathbf{E}^2(z,t) \rangle A e(v), \tag{150}$$

where A is the area of the photocathode (in units of cm²), S is the time-averaged intensity, i.e., energy flux (in J/m² sec), S/hv is the time-averaged *photon flux* in photons/cm² sec, and $e(v)$ is the detection efficiency. The detection efficiency is the probability that a photon incident on the photocathode will be absorbed, producing a photoelectron. Typical detection efficiencies for photomultipliers range between 1 and 20%.

An example of a detector that is *not* a square-law detector consists of a receiving antenna, a tuned resonant circuit driven by the voltage induced in the antenna (by the traveling waves from a distant transmitter), an amplifier, and an oscilloscope. The oscilloscope signal shows the instantaneous phase of the incident radiation as well as its intensity; i.e., it gives a signal proportional to the instantaneous electric field at the antenna, rather than the time-averaged square of the electric field. It is possible to measure the phase in an electromagnetic wave with unlimited accuracy only if you have such a huge number of photons present that you cannot distinguish the individual photon counts. Then you can "sample" the electric field as a function of time by absorbing a large number of photons during each "instant." It is not possible to determine the phase constant φ of a single photon in a light wave described by $E_x = A \cos (\omega t - kz + \varphi)$.

Standard intensities for visible light—candlepower.   The Bureau of Standards keeps something called a "standard candle." Its brightness is about that of an ordinary candle. A *standard candle*, by definition, has a total output power emitted in all directions of about 20.3 milliwatts of visible light (taken to be at a frequency at the peak of visibility, about 5560 Å):

$$1 \text{ cd} = 1 \text{ candle} \approx 20 \text{ milliwatts of visible light.} \quad (151)$$

Surface brightness.   Each part of the radiating surface of a candle flame radiates light in all directions. When you look at a candle flame, it looks uniformly bright over its surface. It also looks just as "bright" when you are near it as when you are far from it. That is also true for the moon or a piece of white paper. It is approximately true for the surface of a frosted incandescent light bulb. The *surface brightness* is defined as the outgoing energy (of visible light) per unit area of surface per unit time. It can be measured in watts of visible light per unit area or in candles per unit area. An ordinary candle flame has a total surface area of about 2 cm^2 and a total output of about one candle. Thus the surface brightness of a candle is given by

$$\text{Surface brightness of candle} \approx \frac{1 \text{ cd}}{2 \text{ cm}^2} = 0.5 \text{ cd/cm}^2. \quad (152)$$

An ordinary 40-watt, 115-volt, incandescent "Mazda" lamp with a tungsten filament has an *absolute luminous efficiency* of about 1.8% (according to the *Handbook of Physics and Chemistry*, under "photometric quantities"; for comparison, a 100-watt lamp has about 2.5% efficiency). That means that about 1.8% of the 40 watts dissipated as "I^2R loss" in the filament emerges as visible radiation. Most of the rest goes into invisible radiation. (A small amount is also lost by conduction to the base of the bulb through the input leads to the filament. Some of the infrared is absorbed in the glass envelope, as is shown by the fact that the glass envelope gets very hot—even a clear glass envelope that is almost perfectly transparent to visible radiation.)

Let us estimate the surface brightness of a 40-watt bulb. (We can compare our result with that listed in the *Handbook*, 2.5 cd/cm^2.) My bulb has a diameter of about 6 cm. If I turn on the lamp and look at it, I see that, unlike the moon, it is not uniformly bright over the projected area. It is almost uniformly bright near the center but suffers a sudden decrease in brightness at a radius corresponding to a "full width at half-maximum brightness" of diameter about 2 cm; i.e., it has the appearance of a nearly uniformly bright projected surface of a sphere of 2-cm diameter. Therefore we shall approximate the light by a uniformly bright sphere of diameter 2 cm. The surface brightness of this "effective" sphere is the visible power divided by the surface area. The area is $4\pi r^2 = 4\pi = 12.6$ cm^2. The visible power is 40 watts times the efficiency, 0.0176. Expressing the answer in candles per square centimeter, cd/cm^2, where 1 cd/cm^2 is 20 mw/cm^2, we find

$$\text{Surface brightness of 40-watt bulb} = \frac{(40)(0.0176)}{(12.6)(20 \times 10^{-3})} \quad (153)$$

$$= 2.8 \text{ cd/cm}^2.$$

The value given in the *Handbook* is 2.5 cd/cm^2.

The "frosting" of an ordinary frosted light bulb (the kind considered above) consists of a roughening of the inner surface. Another common type of bulb is labeled "soft white." Unlike an ordinary frosted bulb, a "soft white" bulb gives an almost uniformly bright projection. It looks like the moon, but brighter.

Why the moon doesn't look brighter when it's closer.   Let us see why the apparent surface brightness of something that emits light in all directions, like a piece of white paper (or the moon, or sun, or blue sky) does not depend on how far you are from the surface. Suppose you are looking at a wall that is completely covered with incandescent light bulbs having "soft white" envelopes. Let D be the surface density of bulbs, measured in bulbs per unit area of wall. By definition, the surface brightness of the wall is the same as that of a single bulb. Now, the visual sensation of brightness depends on the amount of light energy entering the eye (from the source) within a "standard cone" with its apex at the eye and with a certain angular aperture. Thus you "look at" only a small part of a bright surface at any one time, and your brightness sensation depends on the energy entering your eye from the part of the surface intercepted by the standard cone. Suppose the distance from your eye to the wall is R, and suppose that you "look at" an area ΔA on the wall. The *solid angle* $\Delta\Omega$ subtended at your eye by the area ΔA is defined as

$$\Delta\Omega = \frac{\Delta A}{R^2}, \quad (154)$$

where the area ΔA is taken as the projected area perpendicular to your line of sight, and where ΔA is assumed to be small, so that any lateral dimension of the region ΔA is very small compared with R. A given constant solid angle $\Delta\Omega$ corresponds to a cone of given apex angle. The *brightness sensation* is *proportional to the energy entering your eye from some small constant solid angle* (i.e., a given cone) subtended at your eye by a part of the surface. The number N of light bulbs within the constant cone of solid angle $\Delta\Omega$ is the bulb density D times the area ΔA :

$$N = D\Delta A = D\Delta\Omega \cdot R^2. \tag{155}$$

Now suppose you move farther from the wall of light bulbs. Since D and $\Delta\Omega$ are constant, the number N of light bulbs you look at goes up in proportion to R^2. However, the intensity contributed to your sensation of brightness from a single light bulb falls off as $1/R^2$, because the power P of each bulb (Joule/sec) is distributed uniformly over an area $4\pi R^2$. These two tendencies "cancel" one another. The number N of contributing bulbs times $1/R^2$ is constant. Thus the light intensity S (J/cm² sec) coming to your eye from a cone of fixed solid angle $\Delta\Omega$ is constant:

$$S \text{ (at eye)} = \frac{NP}{4\pi R^2} = D\frac{\Delta\Omega}{4\pi}P. \tag{156}$$

Thus the wall of light bulbs looks equally "bright" whether you are near it or far from it, as does a piece of white paper.

In the above discussion we assumed that your line of sight was perpendicular to the wall of light bulbs. Suppose instead that the wall of bulbs is tilted at some fairly large angle to your line of sight. You might then argue that, since more bulbs are intercepted by the cone of constant solid angle, the surface should look brighter if it is tilted. However, that is not the case. The light bulbs are three-dimensional objects—spheres. When you look at the tilted wall, the bulbs partially obscure one another. If you take two shining frosted (soft white) bulbs and partially (or wholly) obscure one by the other, there is no light contribution from the obscured bulb. The projected area of "overlap" is no brighter than the projection of a single bulb.

When a sheet of white paper, or a surface sprinkled with salt, or sugar, or the surface of the moon is lit by illumination from the room or the sun, the material is illuminated to some considerable depth. Light emerging from the surface has been scattered many times. The net effect is a surface which reemits light something like a multilayered wall of soft-white frosted light bulbs. In order to see that much of the emerging light comes from a considerable depth, you can lay one sheet of white tissue on top of a dark surface. Then add a second sheet, a third, etc. The tissue gets more and more "white" as more layers are added.

Illumination—foot-candle. The total light intensity (in J/cm² sec) received at a given location is so metimes called the *illumination*. The illumination is proportional to the surface brightness of the source and to the total solid angle subtended at that location by the source. For example, if the moon were twice as large in diameter, its surface brightness would be unchanged (since that is due to its illumination by the sun). However, it would subtend four times the (former) solid angle, and the light flux S at the earth would be four times as great. The illumination provided by a standard candle at a distance of one foot is called a *foot-candle*. It is easy to show from Eq. (151) that.

$$1 \text{ foot-candle} \approx 1.8 \ \mu\text{w/cm}^2 \quad \text{(of visible light)}. \qquad (157)$$

According to Table 4.3, which gives some typical values of surface brightness for various interesting surfaces, a candle is about as bright as the sky. This means that if you hold a candle and look at it with the sky as a background, the candle flame should be difficult to see. Of course the color makes a difference; the candle is yellow, the sky is blue.

Application: Comparing 40-watt bulb and the moon

This is a numerical example: How far away should a 40-watt frosted bulb of "effective" diameter 2 cm be to provide the same illumination as the full moon? According to Table 4.3, the bulb has 10 times the surface brightness of the full moon. To provide the same illumination, it should therefore subtend $\frac{1}{10}$ the solid angle subtended by the moon, i.e., it should subtend $1/\sqrt{10} = (1/3.2)$ times the ordinary angular diameter of the moon. The angular diameter of the moon is about $\frac{1}{2}$ cm at an arm's length of 50 cm, which is $\frac{1}{100}$ radian. Therefore, we want the bulb to subtend an angular diameter of $\frac{1}{320}$ radian. Thus $\left(\frac{1}{320}\right) = 2 \text{ cm}/R; R = 2(320) = 640 \text{ cm} = 6.4$ meters. Of course 6.4 meters must be the distance for "full moonlight" for *any* 40-watt bulb, whether frosted or not. An unfrosted bulb will look brighter but will provide the same illumination.

Application: Satellite moon mirror

Suppose that farmers in Kansas and part of Nebraska living in a circular farm region with a diameter of 330 kilometers (the east-west length of Kansas) would like to plow their fields at night by the light of the full moon *at all times during the month*. The Department of Agriculture comes up with the solution: an earth satellite made of an inflated plastic bag with the shape of a circular disk and with a highly

Table 4.3   Surface brightness

Surface	Surface brightness, candle/cm²
Candle	0.5
40-watt frosted bulb	2.5
Clear sky	0.4
Moon	0.25
Sun	160,000
40-watt clear bulb (at filament)	200

SOURCE: *Handbook of Chemistry and Physics*, 48th ed. (The Chemical Rubber Co., Cleveland, ann. Pub.), indexed under "photometric quantities."

reflective surface. If the farmers wanted light equivalent to full sun-light, the smallest satellite mirror to do the job would be a plane mirror having the size of the farm area in Kansas and Nebraska. That would be impossible with present satellite technology. But these farmers only want full moonlight. According to Table 4.3, the moon is 640,000 times less bright than the sun. It subtends about the same solid angle as the sun. Thus the farmers want 64×10^4 times less intensity than the sun gives. Therefore the satellite mirror can have an area 64×10^4 times smaller than the farm region and still intercept enough sunlight to satisfy the farmers. (Instead of being a plane mirror, the mirror should be slightly diverging so as to spread the sunlight over the farm region.) Thus the diameter of the mirror can be 8×10^2 times smaller than the diameter of the region. The mirror diameter required is thus 330 km/800 = 410 meters. That *is* feasible!

Problems and Home Experiments

4.1   The end of a string at $z = 0$ is driven harmonically at frequency 10 cps and with amplitude 1 cm. The far end of the string is infinitely far away (or else the string is "terminated" so that there are no reflections). The phase velocity is 5 meter/sec. Describe (precisely) the motion of a point on the string located 325 cm downstream from the driving terminal. What is the motion of a second point located 350 cm downstream?

4.2   The phase velocity v_φ was introduced in describing traveling waves. It satisfies $v_\varphi = \lambda v$. We also know what λ and v mean for standing waves; therefore we can find v_φ by studying standing waves instead of traveling waves.

(*a*) Given a piano string of length 1 meter and with frequency A440 (440 cps) for its lowest mode, find the phase velocity.

(*b*) Show that, for a violin or piano string fixed at both ends, the period T of the *lowest mode* is given by the "down and back" time required for a pulse to travel from one end of the string to the other and then back to the first end, traveling always at the phase velocity. What are the periods of the higher modes?

(*c*) Explain the result of part (*b*) by thinking of a blow from the piano hammer near one end of the string as generating a "wave packet" or "pulse" which propagates back and forth at the phase velocity. Think of the Fourier analysis of the *time dependence* for the motion of any fixed point on the string. You need only the kind of Fourier analysis studied in Chap. 2.

(*d*) Consider a string fixed at $z = 0$ and free at $z = L$. Show that the period of the lowest mode equals the time it takes a pulse to go down and back and down and back, i.e., down and back *twice* at the phase velocity. Can you explain in a simple way why this result is so different from that in part (*b*)? Why does the pulse have to make two trips?

4.3 Assume that the piano string studied in part (*a*) of Prob. 4.2 has diameter 1 mm and is made of steel having volume density 7.9 gm/cm³. Find the tension in dynes and in newtons. (980 dyne = 1 gm-wt; 1 kgm weighs 2.2 lb.)

Ans. 485/n.

Home experiment **4.4 Phase velocity for waves on a slinky.**   (*a*) Measure the phase velocity by the method described in Prob. 4.2, i.e., using standing waves.

(*b*) *Calculation:* Show "theoretically" that the phase velocity of a slinky (consisting of a fixed number of turns, i.e., a fixed amount of actual material) is proportional to the length of the slinky. Thus if you double the length by stretching it out farther, the phase velocity increases by a factor of 2.

(*c*) Verify this experimentally, using standing waves as in part (*a*).

(*d*) Send a short "pulse" or "wave packet" along the slinky. At the same time, release the entire slink y from rest in a configuration that will give oscillation in the lowest transverse mode. Is the "down and back" time for the pulse equal to the period for the lowest mode?

Home experiment **4.5   Damping in rubber bands.**   Make a "rubber rope" two or three feet long by tying together some rubber bands cut open to form a single strand. Try to verify that the phase velocity for longitudinal waves is greater (if it is) than that for transverse waves. You will find the longitudinal modes are highly damped. Hold one of the bands against your moistened lips. Suddenly stretch it. Wait. Suddenly relax it. What, if anything, do the results of this experiment tell you about damping? Why are the longitudinal modes damped so much more than the transverse ones? To put it differently, how can you get decent transverse oscillations with such great damping?

Home experiment **4.6   Measuring the velocity of sound with wave packets.**   Here are two methods:

(*a*) Have a helper set off a firecracker a half-mile or so away. Start a stopwatch when you see the flash of light from the explosion. Stop it when you hear the sound. Pace off the distance. Perform the experiment at two different distances differing by a factor of order 2. Plot time delay versus distance for these two points. Does a plot of a straight line passing through the two points intersect the origin? If it doesn't, why doesn't it? If it does not, can you nevertheless determine the velocity? How?

(*b*) Find a school yard or playground with a broad flat space terminated at one side by a building, so that you get a clear echo when you clap your hands while standing 50 yards or so from the wall. The "down and back" time is then of order two-or three-tenths of a second. This is difficult to measure accurately, even with a stopwatch. Here is a method that requires only an ordinary watch (with a second hand). Lay the watch on the ground where y ou can watch it while you clap. Start clapping rhythmically, slowly at first. Listen both to your clap and the echo. Increase the tempo until the echoes come exactly on the "off beat." This can turn out to be an easy tempo like twice per second. Maintain this tempo for 10 seconds or so, looking at your watch and simultaneously counting the claps. (It may take a few minutes practice.) Pace off the distance to the flat wall that is doing the echoing. The rest is arithmetic.

4.7 Coaxial transmission line. Show that the capacitance per unit length, C/a, for a coaxial transmission line with inner conductor of radius r_1 and outer conductor of radius r_2 and with vacuum between the inner and outer cylindrical conductors is given (in esu, i.e., cm of capacitance per cm of length along the axis) by

$$\frac{C}{a} = \frac{\epsilon_0}{2\ln(r_2/r_1)}.$$

Show that the self-inductance per unit length, L/a, is given (in esu) by

$$\frac{L}{a} = \frac{\mu_0}{2\pi}\ln\frac{r_2}{r_1}.$$

To obtain C/a, use $Q = CV$ and Gauss's law (Vol. II, Sec. 3.5). To obtain L/a, use $L = \dfrac{\Phi}{I}$, where Φ is the magnetic flux produced by a current I [Vol. II, Sec. 7.8, Eqs. (7.53) and (7.54)].

4.8 Parallel-wire transmission line. First work Prob. 4.7, in which you can make use of the symmetry. This problem lacks that symmetry but can be easily worked by using the superposition principle: Calculate the contribution from the field of a single wire; then multiply by 2. Show that C/a and L/a are given by

$$\frac{C}{a} = \frac{1}{4\ln[(r+D)/r]},$$

$$\frac{L}{a} = \frac{\mu_0}{\pi}\ln\left(\frac{r+D}{r}\right),$$

where r is the radius of either wire and $r + D$ is the distance from the axis of one wire to the surface of the other. Note that the calculation is very similar to that for Prob. 4.7, except that one finds an interesting factor of 2. Explain this factor.

4.9 Show (for example, by a simple argument based on symmetry) that the electric and magnetic fields are zero outside the outer conductor of a coaxial transmission line and inside the inner conductor. Show that the electric and magnetic fields are zero outside of the region between the plates of a parallel-plate transmission line.

4.10   Show that the self-inductance of a parallel-plate transmission line is given by Eq. (59), Sec. 4.2, for alternating current as well as for steady current, as long as the wavelength is long compared with the plate thickness d_0. See the discussion in Sec. 4.2, including Eq. (60). Use Eq. (60) as your starting point.

4.11   According to Table 9.1, Vol. II, Sec. 9.1 , the dielectric constant of air at STP is 1.00059. (Assume its magnetic permeability is unity.) Thus according to Eq. (63), Sec. 4.3, the index of refraction of air at STP should be $\sqrt{1.00059} = 1.00029$. That agrees very well with the experimental value given in Table 4.1, Sec. 4.3. On the other hand, the dielectric constant of water is 80. Its index of refraction is *not* $\sqrt{80} \approx 9$ but is instead about 1.33. Why is there this huge discrepancy?

Home experiment

4.12   Water prism—dispersion of water.   Make a water prism as follows. Tape two microscope slides together to form a V-shaped "trough." Seal the ends of the trough with putty or clay or tape or something. Fill the trough with water. (Now fix the leaks!) Hold the prism up close to your eye and look at things through it. The colored edges you see on white objects are called "chromatic aberrations" when they are found in a lens—where they are undesirable. Now look at a point or line source of white light. (The best point source of light for this and other home experiments is made from a simple flashlight. Remove the glass "lens." Cover the aluminum reflector with a piece of black (or dark) cloth with a hole through which you can push the light bulb. Of course, this cannot be done with a "sealed-beam" flashlight. The best line source of light is a simple 25- or 40-watt "showcase lamp" with a *clear* glass envelope and a straight filament about 3 inches long, available for about 50 cents wherever light bulbs are sold.) Now put the *purple filter* from your optics kit between your eye and the light source. (Don't get the filter wet. It is gelatin and will disolve!) You will see two "virtual sources," one red, the other blue. (In order to understand the filter, look at the white light source with the filter both in and out, but using your diffraction grating instead of the prism. You can see that green is absorbed, while red and blue are transmitted.) Assume that the blue passed by the filter has wavelength averaging about 4500 Å and that the red has wavelength averaging 6500 Å. (After we have studied diffraction gratings, you will be asked to measure these wavelengths more accurately.) Measure the angular separation at your eye between the red and blue virtual sources of light. An easy method is to put a piece of paper with marked lines next to the light source. Walk toward the source. The angular separation of the lines changes and you can make the marked lines "coincide" with the two virtual sources. Then you can decide how many centimeters apart the colored virtual sources are. The angular separtion is that distance divided by the distance from your eye to the source. Tilt the prism to see whether the angular separation of the virtual sources depends sensitively on the angle of incidence of the beam on the first slide. Derive a formula for the bending of the light beam as a function of the prism angle and the index of refraction. (*Hint:* Normal incidence on the first slide is easiest for the derivation. Therefore do the experiment that way, or at least see if it matters.) Measure the prism angle. Do the parallel-sided microscope slides contribute to the angular separation or bending? How can you find out experimentally? Finally, determine a value for the rate of change of the index of refraction of water per thousand angstrom units. How does it compare with glass? (See Table 4.2, Sec. 4.3.) (It is conceivable that, although water has smaller index, it might have greater dispersion than glass. Is that the case?) For a pleasant surprise, repeat the experiment using heavy mineral oil. Try other clear liquids.

4.13 An infinite string with linear mass density 0.1 gm/cm and tension 450 N (1 Ib = 354 gm-wt; 1 gm-wt = 980 dyne) is driven at $z = 0$ in harmonic motion of amplitude 1 cm and frequency 100 cps. What is the time-averaged energy flux in watts?

Ans. About 40 watts. (Your answer should be slightly more accurate than this; i.e., is it 35 watts? 44 watts?)

4.14 One of the best wave demonstrators is a torsional wave machine. It consists of a long "backbone" along z with transverse "ribs" spaced about $a = 1$ cm apart. The backbone is a steel wire of square cross section with transverse dimensions about 2 mm by 2 mm. Each rib is an iron rod about 0.5 cm in diameter and 30 cm in length, fastened at its middle to the steel backbone. Let K denote the angular spring constant of the wire. That is, the return torque is K times the angle of twist (in radians). Let I denote the moment of inertia of 1 rod.

(*a*) Derive a formula for the wave velocity and the impedance for torsional waves (waves of twist of the wire). Define the impedance Z by "torque = Z times angular velocity." Assume the wavelength is long compared with the rib spacing a.

(*b*) Show that the exact dispersion law is given by $\omega^2 = 4\omega_1^2 \sin^2\left(\frac{1}{2}ka\right)$, and find an expression for ω_1.

(*c*) So far, we have neglected any return force due to gravity. Suppose now that when all the rods oscillate together (so that the wire backbone is never twisted, they oscillate with angular frequency ω_0 about their horizontal equilibrium position. What is the dispersion law? For the answer and some experimental results, see B. A. Burgel, *Am. J. Phys.* **35**, 913 (1967).

4.15 Whiskey-bottle resonator (Helmholtz resonator). If you **Home experiment** blow across the mouth of a jug or bottle, you get a tone because you have excited the lowest mode. If you try to estimate what frequency to expect, assuming that the bottle acts like a uniform tube closed at one end so that the length from bottom to top is $\frac{1}{4}\lambda$, you will get a surprise. The tone is much lower than you guessed. Here is Helmholtz's approximate derivation. It gives fairly good results. (Using an empty Jim Beam bottle, I predicted 110 cps and found 130 cps, according to my piano.) Assume that the air in

Problem 4.15

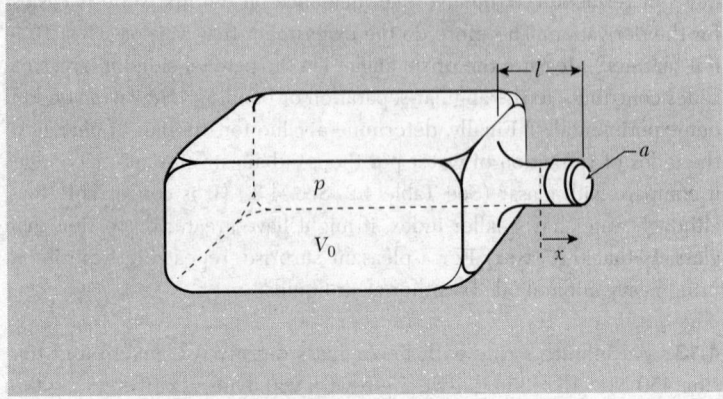

the large volume V_0 acts like a spring. It is attached to a mass which is the air in the neck of the bottle. This mass is $\rho_0 al$, where l is the length of the neck, a is its area, and ρ_0 is the density of the air. Helmholtz's approximation is to assume that all the motion is in the neck and that all the return force comes from pressure changes in V_0.

(a) Show that if x is the outward displacement of fluid along the neck, and if all the return force F_x is due to the pressure differential $p - p_0$, where p is the pressure in V_0 and p_0 is the equilibrium pressure, then

$$F_x = -\frac{\gamma p_0 a^2 x}{V_0},$$

where γ is the "ratio of specific heats," equal to about 1.4 for air.

(b) Show that this gives the single mode an angular frequency ω given by

$$\omega^2 = v^2 \frac{a}{V_0 l},$$

where v is the velocity of sound. In using this result, one must replace l by the "effective" length of the neck, which is the actual length plus about 0.6 of a neck radius at each end. If the actual neck length is zero, the formula still works fairly well (the length l then being completely due to "end corrections"). That case corresponds to something like one of the rectangular cans that paint thinner comes in.

If you blow very hard across the mouth of your bottle, you can excite the higher modes. Once you have heard them by blowing loudly, you can usually hear their faint presence even when you are blowing softly so as to excite mainly the lowest mode. There is no easy "one-dimensional" way to calculate the expected higher resonance frequencies. You will find that two bottles with different shapes have quite different frequency ratios for their first and second or third modes, even though you can calculate the lowest mode frequencies for each one fairly well using Helmholtz's approximation.

4.16 Sound velocity in air, helium, and natural gas. Get an ordinary whistle. Blow it and remember the pitch. Now connect the whistle to a helium bottle (available in any laboratory or physics department) and blow the whistle using helium. What is the pitch now? Measure experimentally the pitch ratio for helium as compared to air. The easiest method is to recognize the pitches and then use a table of frequencies versus pitch (found in the *Handbook of Physics and Chemistry*, or see Home Exp. 2.6). Show that the theoretically expected pitch ratio is about 3 to 1. Experimentally you may only get about 2.5. Why is that? Can you improve the experiment? How does the sound wavelength *in the whistle* compare in the two cases of helium and air? Instead of helium or air, use natural gas, connecting the whistle through a hose to the gas outlet for a gas stove or Bunsen burner. What is the pitch ratio? What can you learn about the molecular properties of the gas by measuring the pitch ratio for the gas and air?

Home experiment

4.17 Find the rms (root-mean-square) electric field (averaged over all frequencies) at a point in space 1 meter from a 40-watt light bulb.

4.18 Measuring the solar constant at the earth's surface. This experiment is described in Example 17, Sec. 4.4. Do the experiment, and give the result in watt/cm^2. You might try using several layers of glass, or perhaps just the glass of a window, to reduce the infrared "heat" detected by your eyelids from the light bulb, under the assumption that the earth's atmosphere has already done that for the sunlight to a great extent. Then, by limiting oneself to visible light (as seen through *closed* eyelids), perhaps one can come close to the solar constant outside the earth's atmosphere. The temperature of the tungsten filament is less than that of the sun, and the wavelength spectrum depends on color. Find a graph of emitted energy versus wavelength for the sun's surface temperature, which is about 5000°K, and one for the tungsten's temperature, which is about 3000°K. By roughly estimating the fraction of the total flux which is visible for each of these, see if you are underestimating or overestimating the sun's total flux (including the invisible) by comparing it with the light bulb only for visible frequencies.

Home experiment

4.19 Photomultiplier counting rate. Suppose you have a photomultiplier with the following properties: photocathode area = 1 cm^2; photocathode efficiency = 5%, averaged over the visible spectrum. Suppose you have a candle that emits 1 candlepower of visible light. How far must the candle be placed from the photomultiplier for the ouput counting rate to be as low as 10 counts per second? (We want it low so we can hear the individual counts.) Alternatively, if the candle is at 1 meter, how large a pinhole in a lighttight shield over the photomultiplier will give the same result? *Units:* a photon having 1 electron volt of energy has wavelength about 12,345 Å. (The last two digits are wrong; however, this is an easy and famous mnemonic.) Thus, if it has 2 eV, it has 6170 Å. Take all the photons at green, i.e., 5500 Å. Also, 1 eV = 1.6×10^{-12} erg.

4.20 Candlelight and romance. If a candle had the same intrinsic brightness as the moon, it would provide the same illumination as the moon when placed at a distance at which it subtended the same solid angle at your eye as the moon does. According to Table 4.3, Sec. 4.4, a candle has twice the brightness of the moon. Assume that full moonlight provides "perfect illumination for romance." Measure the approximate horizontally projected area of a candle flame. Calculate how far away the candle should be to provide "perfect illumination" as defined above.

4.21 Moonlight. According to the example following Table 4.3, Sec. 4.4, a 40-watt bulb of effective diameter 2 cm provides the same illumination as the moon when the bulb is placed at a distance of 6.4 meters. That calcula tion started with the surface brightness, as given in Table 4.3, and the effective diameter, as previously calculated. Of course, a 40-watt bulb with a clear glass envelope will provide the same illumination. (The brightness of the tungsten filament is much greater than that of the frosted surface, but the total power output is the same.) Use these facts to calculate the illumination provided by full moonlight in units of microwatts of *visible* light per square centimeter of illuminated surface. (Take into account the efficiency of the light bulb.)

Ans. Between 0.1 and 0.2 μw/cm^2. (Your answer should be to 2 significant figures; e.g., 0.13, or 0.18, or whatever it is.)

4.22 Sunlight. The sun subtends about the same solid angle as the moon. (If you are going to check that statement by holding a ruler at arm's length, be sure to use an adequate filter. Almost-crossed polaroids from your optics kit will work.) Use Table 4.3, Sec. 4.4, and the results of Prob. 4.21 to find the illumination provided by the sun.

Ans. About 90 mw/cm^2 of visible light.

4.23 Luminous efficiency of the sun. Assume that all the visible light of the sun penetrates the earth's atmosphere with negligible attenuation. Use the results of Prob. 4.22 and the *Handbook* value of the solar constant outside the earth's atmosphere to calculate the luminous efficiency of the sun. How does it compare with that of light bulbs? (A 5000-watt 1I5-volt bulb has 4.7% efficiency, according to the *Handbook*.)

Home experiment **4.24 Measuring the candlepower and luminous efficiency of a light bulb.** For this experiment you need an incandescent lamp (either clear or frosted glass envelope), a candle, two slabs of paraffin wax ("household wax" used to seal glasses of jellies, jams, and preserves), and a piece of aluminum foil. The candle is your standard. We *assume* that your candle is close to a standard candle and thus emits 1 candlepower, about 20 milliwatts of visible radiation. The lamp is the unknown. Its total power is known (marked on the bulb). Measure its visible output by comparing it with the candle as follows. Sandwich the aluminum foil between the two slabs of paraffin. Hold the paraffin sandwich near the candle; notice the brightness of the paraffin slab nearer the candle and the darkness of the slab farther from the candle. Hold it near the lamp. Now (at night, with only the candle and lamp on) hold the paraffin detector between the lamp and the candle with one slab facing each light source. Vary the location until the two slabs appear equally bright. Measure the distances. The rest is arithmetic. (Use the inverse-square law.) Give the visible output of the lamp in candle-power and in visible watts, assuming your candle is "standard." Give the efficiency of the bulb.

With a slightly more sophisticated arrangement, you should be able to measure the illumination of the sun in visible watts/cm^2 or (taking 1 cd/cm^2 = 20 mw/ cm^2) in candles per square cm. You may have a problem with background light coming from neither the sun nor from your standard light bulb. It helps to have a strong light bulb, perhaps 200 watts. Perhaps you need a mailing tube collimator, or a cardboard carton with suitable holes, or a dark cloth somewhere. Calibrate the light bulb in candles as described above. Find the distance of the bulb from the paraffin that gives equal illumination to that of the sun. From the geometry, you can calculate the flux from the bulb in cd/cm^2 and thus calculate the illumination of the sun in cd/cm^2.

4.25 Frosted light bulbs. Get an ordinary frosted bulb (of any wattage) and also a "soft white" bulb of the same wattage and bulb diameter. Turn them on and look. Notice the "bright core" of projected light for the frosted bulb. The soft white bulb is much more uniform. The frosted bulb therefore has a smaller "effective" surface than the soft white bulb. Since their power is the same, they presumably have the same total light output. The frosted bulb must therefore be brighter over the smaller surface of its bright core. Look at the central region of each lit bulb through a hole between your curled fingers, with your hand held at constant distance from your eye so that a constant solid angle is admitted. Which bulb is brighter (at the center)? Measure the diameter of the projected area of the bright core of the frosted bulb by holding a ruler against it. (You may need to reduce the brightness by looking through partially crossed polaroids from your optics kit.) Calculate a predicted ratio of surface brightnesses for the two bulbs, taking as a model an "effective sphere" for the frosted bulb with diameter equal to the diameter of the projected bright core. Now measure the ratio of surface brightnesses as follows. Put each bulb behind a cardboard carton (or something) with a hole in the cardboard that reveals only the central region of each bulb. Make the two holes the same size. Locate the bulbs so that they are several feet apart and shining at each other through the holes. Use the paraffin-sandwich technique described in Home Exp. 4.24 to measure the ratio of surface

Home experiment

brightnesses of the central regions of the two bulbs. How does your result compare with your calculation based on the "effective sphere" model?

Finally, break the bulbs to see how their frostings differ. (Wrap a bulb in a towel before br eaking it!) If this seems too wasteful of bulbs and money, you may take our word for it that the ordinary frosted bulbs we have broken open are slightly rough on the inner surface, which could be achieved by etching with acid or by sandblasting. The soft-white bulbs are coated with white powder (undoubtedly magnesium oxide). The powder wipes off at the touch of a finger, leaving a clear envelope. If you do break the bulb, *save* the largest hemispherical piece. It makes a very good "plano-convex" lens when partially filled with liquid. (To get a large piece, break the bulb by tapping at the *neck*.) You can use it to measure the indexes of refraction of water and mineral oil.

Home experiment   **4.26   Sound impedance.**   Sing a steady note into a cardboard mailing tube, with the tube held tight against your mouth so no air leaks around the edges. Vary the pitch to find the resonances. (They are not exactly where the free modes are that you hear when you tap the tube on your head. Your mouth and throat change the effective length at that end.) Sing a steady note *not* at a resonance. Suddenly remove the tube while continuing to sing. The change of impedance should be noticeable. Now sing at a resonance. Notice the markedly different feeling in your throat. The load is *not* a purely resistive load at resonance but rather is largely reactive. Now find a large jar or vase or bucket. (A large glass vase or large styrofoam or plastic jar or bucket will work very well.) Find a strong resonance by searching with your singing voice. Sing as loudly as you can at the resonance, with your mouth and throat closely coupled to the resonating system. If there were no radiation loss or other resistive loss, the load on your singing apparatus would be a purely reactive load. That is, as much energy would flow back into your throat on any one cycle as left it (during another part of the cycle). As it is, the feeling in your throat is remarkably different from that when you sing into an open medium. You will find that you have difficulty in controlling the pitch. The pitch "wobbles" because you are used to a resistive load and are experiencing a reactive one.

4.27   Suppose two traveling waves on an elastic string with 10^{-5} Newtons, $\rho = 1$ gm/cm, and $\omega = 10^3$ rad/sec are given by

$$\psi_1 = A \cos(\omega t - kz + \pi),$$

$$\psi_2 = A \cos\left(\omega t - kz + \frac{\pi}{4}\right).$$

Find the time-averaged intensity of the superposition of Ψ_1 and Ψ_2.

4.28   Three plane electromagnetic waves, given by

$$E_{1x} = E_0 \cos (kz - \omega t - \delta_1) = B_{1y},$$

$$E_{2x} = E_0 \cos (kz - \omega t - \delta_2) = B_{2y},$$

and

$$E_{3x} = E_0 \cos (kz - \omega t - \delta_3) = B_{3y},$$

travel over the same space. What are the maximum and minimum amplitudes and energy fluxes that can be produced by adjusting the values of the constants δ_1, δ_2, and δ_3?

4.29   "Gauge pressure" for longitudinal waves on a spring. Derive Eq. (111), Sec. 4.4, which is

$$F_z(L \text{ on } R) = F_0 - Ka \, \frac{\partial \psi(z, t)}{\partial z}.$$

Start with a lumped-parameter beaded spring. At equilibrium each spring is compressed so that it exerts a force F_0. The spring constant is K. The bead separation is a. Find the force on a given bead exerted in the $+z$ direction by the spring just to the left of that bead. Go to the continuous limit, and thus derive the desired relation. Note that in the continuous limit the product Ka is a property of the continuous spring and is independent of the length a.

4.30   Rubber ropes and slinkies.   For an ordinary rubber rope (or for the kind of spring that closes doors) the unstretched length is not negligible compared with the stretched length. Show that therefore the phase velocity is slower for transverse waves than for longitudinal waves. Show, for example, that if the stretched length is $\frac{4}{3}$ times the unstretched length, the longitudinal waves travel at twice the speed of the transverse waves. For a slinky, the unstretched length is about 3 in. and the stretched length can be 15 ft or so. What is the ratio of velocities in that case?

4.31   Are sound waves perfectly nondispersive?   We found in Sec. 4.2 that the phase velocity of sound is constant, independent of frequency. The dispersion law that gave that result was

$$\omega^2 = \frac{\gamma p_0}{\rho_0} k^2,$$

which is similar to the dispersion law for longitudinal oscillations on a continuous spring,

$$\omega^2 = \frac{K}{M} k^2.$$

For a lumped-parameter beaded spring, the dispersion law is

$$\omega^2 = \frac{K}{M} \frac{\sin^2 \frac{1}{2}ka}{\left(\frac{1}{2}a\right)^2},$$

which leads to a high-frequency cutoff. By analogy and by physical reasoning, guess a value for a high-frequency cutoff for sound in air at STP (standard temperature and pressure). Would you expect ultrasonic waves of frequency $\nu \approx 100$ Mc to travel at
ordinary sound velocity?

Ans. You expect a high-frequency cutoff $\nu_0 \approx 10^{10}$ Hz.

Chapter 5

Reflection

Chapter 5   Reflection

5.1   Introduction

In this chapter we shall apply the concept of impedance to find what happens when a traveling wave encounters a discontinuity in the medium. In Sec. 5.2 we consider a lumped resistive load that "matches" the characteristic impedance of the wave medium. This leads us to discover how to make "spacecloth" that can terminate electromagnetic waves without reflection. In Sec. 5.3 we consider reflections induced by "mismatched" impedances. By generalizing the results obtained for a transmission line, we learn how light waves are reflected at a discontinuity in index of refraction. Through studying multiple reflections in Sec. 5.5 we learn how to use a pane of glass to tell us something about the mean decay lifetime of excited neon atoms.

5.2   Perfect Termination

If a transmitter is coupled to a completely open medium and drives the medium at a frequency in its dispersive range, the transmitter emits traveling waves. The transmitter output terminal experiences a purely resistive drag force proportional to the characteristic impedance. The characteristic impedance depends on the medium and also on the geometrical configuration of the waves. (For example, a parallel-plate transmission line has a different impedance from that of a parallel-wire transmission line.)

As far as the transmitter is concerned, it cannot tell whether it is actually radiating traveling waves into an open medium or merely driving a lumped resistive load. If you were to disconnect the antenna of your local radio station and replace it by an equivalent resistor, the oscillator could not tell the difference. (This is slightly oversimplified, in that a radio antenna also has inductance and capacitance. Thus to "fool" the driving oscillator completely, you should replace the antenna with a suitable *LRC* circuit. The resistance *R* in this circuit is the "radiation resistance," and that is the characteristic impedance we are talking about.) Let us begin with a simpler example than a radio antenna.

Example 1: Continuous string

If you were to replace the string (which is being shaken by the oscillating output terminal of the transmitter) with a suitable "dashpot," the transmitter would experience the same drag force as it does when it is emitting traveling waves into an infinite string. By a *dashpot* we mean a device (which we shall call *R* for "right") with the following property: If its input terminal is forced to undergo a velocity $u(t)$, the dashpot responds to the driving force (which we shall call L for "left") by reacting with an opposing force *proportional to the velocity*.

$$F(R \text{ on } L) = -Z_R u(t), \tag{1}$$

where Z_R is a positive constant called the *impedance* of the dashpot. The impedance of a dashpot is said to be "purely resistive," because $F(R \text{ on } L)$ is proportional to the velocity. (A device involving either inertial mass or a spring would react with a force proportional respectively to the acceleration or to the displacement. In either case such a device would offer a "reactive" rather than a resistive load.) Now, when the transmitter is emitting traveling waves into an open system with characteristic impedance Z, the output terminal of the transmitter experiences a drag force

$$F(R \text{ on } L) = -Z \left(\frac{\partial \psi}{\partial t} \right)_{z=0}, \tag{2}$$

where $\partial \psi / \partial t$ is the velocity of the string at $z = 0$ and hence also of the output terminal. Thus, we see that if Z_R equals Z then the transmitter experiences the same "purely resistive" reaction when it drives the input terminal of the dashpot as when it drives an infinitely long string. One characteristic of a traveling wave is that every downstream point in the medium has the same "experience" as the output terminal of the transmitter, but at a later time. Thus, for any downstream point z of a system carrying traveling waves, the point L just to the left of z cannot tell whether the point R just to the right of z is actually the first point of a continuation of the string to infinity or is merely the input terminal of a dashpot having impedance $Z_R = Z$.

Impedance matching. This shows us that the way to obtain perfect termination of a continuous string, so that there are no reflections of transverse traveling waves incident on the termination, is to attach it to a resistive load consisting of a perfect dashpot having impedance

$$Z_R = Z = \sqrt{T_0 \rho_0}. \tag{3}$$

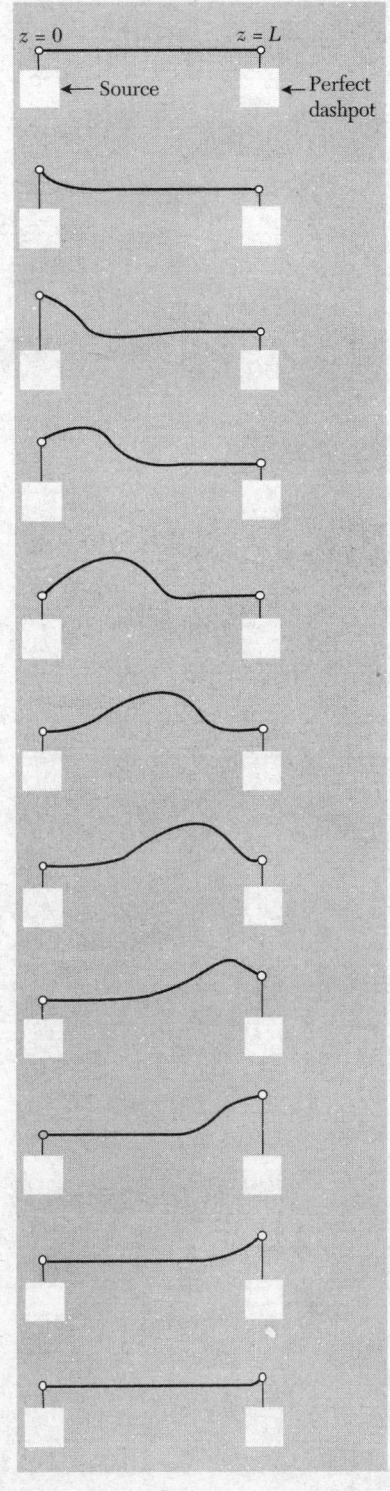

Fig. 5.1 *The source cannot tell the length of the string, if the string is perfectly terminated. As far as the source can tell, it may be connected to an infinitely long string, or it may be connected directly to the input terminal of the dashpot. Here it is shown connected to a finite string that is perfectly terminated.*

When Eq. (3) is satisfied, we say that the load impedance "matches" the characteristic impedance of the string. An example of perfect termination is shown in Fig. 5.1.

Distributed load.   A dashpot is a "lumped" resistive load, i.e., it occupies a region small compared with one wavelength. A way to achieve effective perfect termination without the stringent design requirement needed to satisfy the impedance-matching condition [Eq. (3)] is to use a very long "distributed" load that provides a small drag force, starting at the point $z = L$ where you wish to begin absorbing the wave energy. This drag force is then applied continuously and uniformly along the string at all z greater than L. If the drag force absorbs only a small fraction of the energy of the wave in a distance of one wavelength, then it turns out that it will not introduce a significant reflection, and will gradually absorb all the energy of the wave.

Example 2: Parallel-plate transmission line

This example will lead to a very general result. The input end of the line is shown in Fig. 5.2, along with a slab of resistive material that could be used either to replace the transmission line as the load on the transmitter or to terminate the transmission line without reflection. For the direction of current indicated, the resistance of the slab is the resistivity ρ times the length divided by the cross-sectional area (Vol. II, See, 4.7):

$$R = \rho \cdot \frac{(\text{length})}{(\text{area})} = \frac{\rho g}{dw}. \tag{4}$$

But the characteristic impedance Z of a parallel-plate transmission line is [see Eq. (4.132), Sec. 4.4]

$$Z = \sqrt{\frac{\mu_0}{4\pi\epsilon_0}} \frac{g}{w}. \tag{5}$$

If R is to be a perfect termination, we need $R = Z$. By equating Eqs. (4) and (5), we find

$$\frac{\rho}{d} = \sqrt{\frac{\mu_0}{4\pi\epsilon_0}}, \tag{6}$$

where ρ is in ohm-m and d is in cm. The ratio ρ/d is in ohm.

Resistance per square.   This slab can be characterized in a useful way which is independent of the thickness d as follows. Let us cut a square of length L and width L from this sheet of material of thickness d. Apply a voltage V between two opposite ends of the square. This gives a current flowing parallel to the surface of the sheet. The resistance of the square is equal to the resistivity times the length along the current (i.e., L) divided by

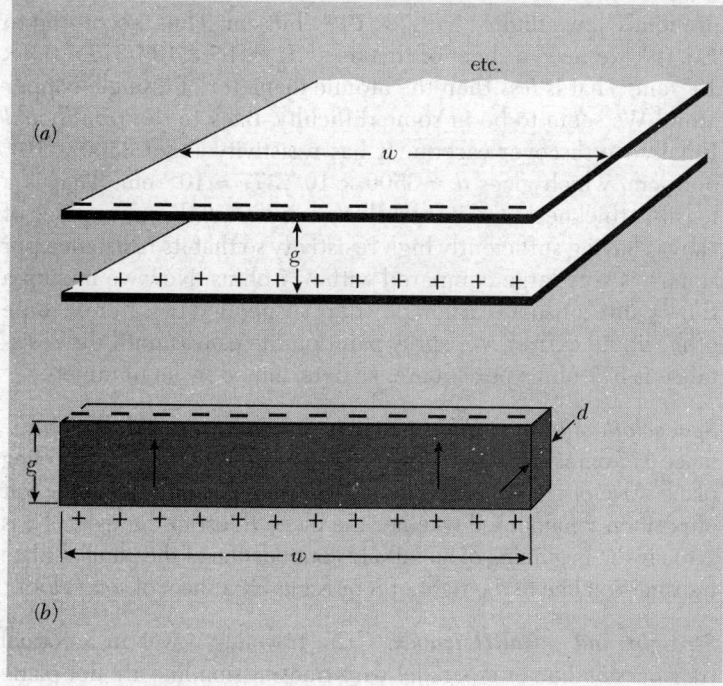

etc.

(a)

(b)

Fig. 5.2 *Terminating a parallel-plate transmission line. (a) Transmission line. (b) Resistive slab. When the potential difference across the slab has the sign indicated by the plus and minus signs, current flows in the direction of the arrows.*

the area perpendicular to the current (i.e., Ld). Thus, the resistance of the square is

$$R = \frac{\rho L}{Ld} = \frac{\rho}{d}. \tag{7}$$

Note that this is independent of the edge length L of the square. Consequently, ρ/d for a sheet of material can be characterized as its *resistance per square* (of any size) for current flowing from one edge to the opposite edge of the square. Thus, Eq. (6) tells us that *for perfect termination of a parallel-plate transmission line, the resistance per square should be*

Perfectly terminating sheet has $120\pi = 377$ ohms per square. (8)

Let us see how one could go about achieving perfect termination in practice. Let us design a terminating slab for a parallel-plate transmission line. We want a sheet of material having resistance 377 ohms per square, i.e., $\rho/d = 377$ ohms, so that

$$d \text{ (in cm)} = \frac{\rho \text{(in ohm-cm)}}{377 \text{ ohms}}. \tag{9}$$

Let us try a sheet of copper. How thick should it be? In the *Handbook of Physics and Chemistry* (indexed under "resistivity of metals"), we find $\rho_{Cu} \approx 1.7 \times 10^{-6}$ ohm-cm. Thus, according to Eq. (9), we need a sheet of thickness $d_{Cu} \approx 1.7 \times 10^{-6}/377 \approx 0.5 \times 10^{-8}$ cm. That is *less* than the atomic diameter of a single copper atom! We seem to be in some difficulty. Back to the *Handbook!* Finally we discover carbon. It has resistivity about 3500×10^{-6} ohm-cm, which gives $d \approx 3500 \times 10^{-6}/377 \approx 10^{-5}$ cm. That is a feasible thickness, and it works! We can start with (say) a piece of canvas having sufficiently high resistivity so that its resistance per square is very large compared with 377 ohms. Now we mix up a thin "paint" of soot (carbon powder) suspended in water or some other liquid carrier. We spray paint on the canvas until the resistance is 377 ohms per square, as determined by an ohmmeter.

Spacecloth.   In microwave jargon, a piece of material having resistance 377 ohms per square is called *spacecloth*. An incident traveling plane wave of electromagnetic radiation in a plane L just to the left of position z cannot tell whether the plane R just to the right of z is actually the beginning of an infinite continuation of the parallel-plate transmission line to the right of R or is merely a sheet of spacecloth.

Straight and parallel waves.   The traveling waves in a coaxial transmission line or a parallel-wire transmission line are not plane waves, because by definition a plane wave is composed of electric and magnetic fields whose values at a given instant t do not depend on x and y, but only on z (the propagation direction). They are instead members of a more general class, that of *straight and parallel* waves, which includes plane waves. Straight and parallel waves are those in which the fields **E** and **B** may depend on x and y, but in which the x and y dependence does not vary with z, the propagation direction. Thus, the waves in straight and parallel transmission lines (i.e., those that can be constructed from pairs of identical, straight, parallel wires) are straight and parallel waves.

A sheet of spacecloth provides perfect termination for *any* straight and parallel transmission line, and it does so for the following reason: It turns out that, in any neighborhood of a given point which is sufficiently small in the transverse direction, an incident straight and parallel traveling wave cannot be distinguished from a plane wave. That is, over a region of sufficiently small dimensions Δx and Δy transverse to the propagation direction z, the fields $\mathbf{E}(x,y,z,t)$ and $\mathbf{B}(x,y,z,t)$ may be regarded as constants, independent of x and y. Furthermore, it can be shown (using Maxwell's equations for a general demonstration) that, at a given x and y, straight and parallel waves satisfy relations analogous to those given for plane waves in transparent media, Sec. 4.4. Thus, for straight and parallel traveling waves at fixed x and y, $\mathbf{E}(x,y,z,t)$ and $\mathbf{B}(x,y,z,t)$ are perpendicular to each other and to $\hat{\mathbf{z}}$, have equal magnitudes, and have signs such that $\mathbf{E} \times \mathbf{B}$ is along $+\hat{\mathbf{z}}$; i.e.,

$\mathbf{B} = \hat{\mathbf{z}} \times \mathbf{E}$. Furthermore, the "local" energy flux (in the neighborhood $\Delta x \, \Delta y$) is given by an expression analogous to that which holds for plane waves. Thus, for straight and parallel traveling waves in vacuum, we have

$$S(x,y,z,t) = \frac{c}{4\pi} \, \mathbf{E}^2(x,y,z,t), \tag{10}$$

where S is the intensity in erg/cm²sec. Since this relation between the energy flux and the fields is locally the same as for plane waves, the "I^2R losses" from the currents induced in spacecloth by straight and parallel waves will exactly balance the incident energy flux, just as for plane waves. Therefore, the spacecloth in any given neighborhood $\Delta x \, \Delta y$ will absorb incident straight and parallel radiation without reflection, as long as the cloth has resistance 377 ohms per square.

Terminating a plane wave in free space.   After the discussion above you might guess that a sheet of spacecloth provides perfect termination not only for plane waves in a parallel-plate transmission line but also for plane waves in free space. This is a reasonable guess, but it is wrong. A plane wave incident on a sheet of spacecloth feels an impedance that is just *half* of the perfect termination value of 377 ohms per square!

We can easily understand that factor of one-half by considering the parallel-plate transmission line. If we have a parallel-plate transmission line extending from $z = -\infty$ to $+\infty$, we can terminate a wave incident from tile left by connecting a sheet of spacecloth across the face of the line at $z = 0$ *provided* we also disconnect the rest of the line extending from $z = 0$ to $+\infty$. If we fail to disconnect the line to the right of $z = 0$, then the voltage in the incident wave at $z = 0$ is applied across two equal impedances that are connected in parallel—one being the spacecloth, the other being the infinite continuation of the line. Thus the wave feels an effective impedance that is the impedance of two equal resistances in parallel, i.e., half of the impedance of either resistance. This is also what happens when a plane wave in empty space is incident on a sheet of spacedoth. The voltage being applied across the sheet of spacecloth at any instant is also being applied across the infinite continuation of empty space to the right of the spacecloth. The resultant impedance is just half that of the spacecloth or of the empty space. The wave is not completely absorbed; it is partly reflected, partly absorbed, and partly transmitted.

How can one "disconnect" the empty space to the right of a sheet of spacecloth? In the case of the transmission line it is easy. You can cut through the transmission line with a saw. The sawcut provides an infinite impedance that disconnects the transmission line to the right of the cut. The incident wave is then applied across the spacecloth in parallel with an infinite impedance. The resultant impedance is that of the spacecloth. Now, in the case of empty space, there is no way to make a "sawcut" to provide infinite impedance. Nevertheless there is an ingenious trick whereby one may effectively "disconnect" the space to the right of $z = 0$, for a harmonic wave with a single definite wavelength. The trick works for empty space and also for the transmission line. Consider the transmission line: Instead of trying to "disconnect" with a sawcut at $z = 0$, the trick is to "short-circuit" by means of a sheet of perfect conductor having zero resistivity. The short-circuit is located not at $z = 0$ but at $z = \frac{1}{4}\lambda$. The voltage is then always zero at $z = \frac{1}{4}\lambda$. To the left of the short-circuit the voltage has the form of a standing wave. (The spacecloth has not yet been inserted.) The current also has the form of a standing wave. It turns out that the places where the current is zero are one quarter-wavelength from the places where the voltage is zero. Thus the current is always zero at $z = 0$. This is like having infinite impedance due to a sawcut at $z = 0$. (Infinite impedance means the current is zero.) Thus the transmission line is effectively "disconnected" at $z = 0$ by a short-circuit at $z = \frac{1}{4}\lambda$.

The same trick works for empty space. A plane wave is perfectly terminated by a sheet of spacecloth at $z = 0$ followed by a sheet of perfect conductor (a "mirror") at $z = \frac{1}{4}\lambda$. All of the wave energy is dissipated in the spacecloth.

In the case of waves on a string, our "perfect dashpot" had its input connected to the string. Its other moving part (which moves relative to the input to provide the frictional damping) was anchored firmly to a rigid support. This is like its being connected to another string of infinite mass density extending from $z = 0$ to $+\infty$. Such a string would have infinite impedance. This is analogous to the sawcut in the transmission line and is why the dashpot provided perfect termination. If, instead, the other moving part of the dashpot were connected to a string of impedance Z_2 extending from $z = 0$ to $+\infty$, then the incident wave at $z = 0$ would feel an impedance that turns out to be equivalent to the impedance of the dashpot connected in parallel with the impedance Z_2 of the continuation of the string. Just as for the transmission line and for empty space, we can have perfect termination for waves on a string, provided the second connection of the dashpot is made either to solid support or to a quarter-wavelength string segment that is connected to a frictionless ring sliding on a rod, so as to give zero impedance at the rod and infinite impedance at the dashpot. This ensures that the dashpot output connection will not move. See Prob. 5.32.

Other methods of perfect termination. It is not always easy to make "perfect spacecloth." If you are satisfied to absorb the radiation with a distributed load that takes up a lot of space, you can do so with negligible reflection without having to satisfy the impedance-matching requirement of the lumped-parameter perfect termination, i.e., spacecloth. For example, if you want to absorb a flashlight beam with a negligible reflection, you can let the beam enter a hole in the side of a large lighttight cardboard box. Line the box with black (i.e., absorbing) material and provide some baffles so that the light has to bounce several times to get back out. If you look at such a hole in broad daylight, it looks much blacker than an ordinary black object like candle soot. Such a "black surface" is indistinguishable from perfect spacecloth, because essentially none of the radiation that enters the hole gets back out; it is as if the transparent medium (air) extended to infinity.

5.3   Reflection and Transmission

Continuous string.   Suppose we have a semi-infinite string with characteristic impedance Z_1 extending from $z = -\infty$ to $z = 0$. At $z = 0$ it is attached to a load consisting of the input terminal of a dashpot having an impedance Z_2 which is not equal to Z_1. At $z = -\infty$ there is a transmitter emitting traveling waves in the $+z$ direction. Thus there is an incident traveling wave given by

$$\psi_{\text{inc}}(z, t) = A \cos (\omega t - kz). \tag{11}$$

At $z = 0$ the incident wave is given by [setting $z = 0$ in Eq. (11)]

$$\psi_{\text{inc}}(0, t) = A \cos \omega t. \tag{12}$$

How mismatched load impedance generates reflection.   Let us call the last point on the string L (for "left") and the input terminal to the dashpot R (for "right"). If the dashpot had impedance Z_1, it would "match" the impedance of the string and terminate the incident wave without reflection. In this case, the "terminating force" exerted by the dashpot on the string, F_{term}, would be given by

$$F_{\text{term}}(R \text{ on } L) = -Z_1 \frac{\partial \psi_{\text{inc}}(0, t)}{\partial t}. \tag{13}$$

The actual force $F(R \text{ on } L)$ may be regarded as a superposition of this terminating force plus a *force excess*, F_{exc}, above (or below) that which is responsible for absorbing the incident wave. This excess force generates a traveling wave traveling in the $-z$ direction, just as if R were the output terminal of a transmitter. This wave is the reflected wave $\psi_{\text{ref}}(z, t)$. At $z = 0$ it satisfies the relation that always holds when a traveling wave is emitted, namely that the driving force is the impedance times the velocity:

$$Z_1 \frac{\partial \psi_{\text{ref}}(0, t)}{\partial t} = F_{\text{exc}}(R \text{ on } L), \tag{14}$$

where $F_{\text{exc}}(R \text{ on } L)$ is the force exerted as if by a transmitter. The total force $F(R \text{ on } L)$ is the superposition of the terminating force and the force excess, each acting "independently":

$$F(R \text{ on } L) = F_{\text{term}}(R \text{ on } L) + F_{\text{exc}}(R \text{ on } L). \qquad (15)$$

Combining Eqs. (13), (14), and (15) we get

$$F(R \text{ on } L) = -Z_1 \frac{\partial \psi_{\text{inc}}(0, t)}{\partial t} + Z_1 \frac{\partial \psi_{\text{ref}}(0, t)}{\partial t}. \qquad (16)$$

Now, the total force $F(R \text{ on } L)$ is provided by the drag force of the dashpot, given by $-Z_2$ times the velocity of point L. This velocity is the superposition of contributions from the incident and reflected waves:

$$\frac{\partial \psi(0, t)}{\partial t} = \frac{\partial \psi_{\text{inc}}(0, t)}{\partial t} + \frac{\partial \psi_{\text{ref}}(0, t)}{\partial t}. \qquad (17)$$

Thus, $F(R \text{ on } L)$, the drag exerted by the dashpot, is given by

$$F(R \text{ on } L) = -Z_2 \frac{\partial \psi(0, t)}{\partial t}$$

$$= -Z_2 \frac{\partial \psi_{\text{inc}}(0, t)}{\partial t} - Z_2 \frac{\partial \psi_{\text{ref}}(0, t)}{\partial t}. \qquad (18)$$

Equating the right-hand sides of Eqs. (16) and (18), we find (at $z = 0$)

$$-Z_1 \frac{\partial \psi_{\text{inc}}}{\partial t} + Z_1 \frac{\partial \psi_{\text{ref}}}{\partial t} = -Z_2 \frac{\partial \psi_{\text{inc}}}{\partial t} - Z_2 \frac{\partial \psi_{\text{ref}}}{\partial t},$$

i.e.,

$$\frac{\partial \psi_{\text{ref}}(0, t)}{\partial t} = \left[\frac{Z_1 - Z_2}{Z_1 + Z_2} \right] \frac{\partial \psi_{\text{inc}}(0, t)}{\partial t}. \qquad (19)$$

Reflection coefficient. Then we have (by integrating both sides of Eq. (19) and assuming there is no integration constant)

$$\psi_{\text{ref}}(0, t) = R_{12}\psi_{\text{inc}}(0, t) = R_{12}A \cos \omega t, \qquad (20)$$

where the quantity R_{12}, called the *reflection coefficient* for the displacement ψ, is given by

$$R_{12} = \boxed{\frac{Z_1 - Z_2}{Z_1 + Z_2}}. \qquad (21)$$

Since the reflected wave is a sinusoidal wave traveling in the $-z$ direction, its form for $z < 0$ is obtained from its form at $z = 0$ by replacing the variables $z = 0, t$ by $z, t + z/v_\varphi$, where v_φ is the magnitude of the phase velocity. Thus

$$\psi_{\text{ref}}(z, t) = R_{12}A \cos \left[\omega \left(t + \frac{z}{v_\varphi} \right) \right] = R_{12}A \cos(\omega t + kz). \qquad (22)$$

The total displacement $\psi(z, t)$ is given by the superposition

$$\psi(z, t) = \psi_{\mathrm{inc}}(z, t) + \psi_{\mathrm{ref}}(z, t),$$

i.e.,

$$\psi(z, t) = A \cos(\omega t - kz) + R_{12} A \cos(\omega t + kz). \tag{23}$$

Return force and displacement reflect with opposite signs. The physically interesting wave quantities for transverse waves on a string are not only the displacement, $\psi(z,t)$, but also the transverse velocity, $\partial\psi(z, t)/\partial t$, and the transverse projection of the tension, $-T_0 \partial\psi(z, t)/\partial z$, which gives the return force exerted by the string to the left of z on that to the right of z. We see from Eqs. (19) and (20) that the velocity wave $\partial\psi(z, t)/\partial t$ has the same reflection coefficient as the displacement wave $\psi(z,t)$. However, the "return-force" wave $- T_0\partial\psi(z,t)/\partial z$ has a reflection coefficient that is the same in magnitude but has the *opposite sign* to the reflection coefficient for $\partial\psi/\partial t$. Thus we have

$$\psi_{\mathrm{inc}} = A \cos(\omega t - kz), \qquad \psi_{\mathrm{ref}} = R_{12} A \cos(\omega t + kz); \tag{24}$$

$$\frac{\partial \psi_{\mathrm{inc}}}{\partial t} = -\omega A \sin(\omega t - kz), \qquad \frac{\partial \psi_{\mathrm{ref}}}{\partial t} = R_{12}[-\omega A \sin(\omega t + kz)]; \tag{25}$$

$$\frac{\partial \psi_{\mathrm{inc}}}{\partial z} = kA \sin(\omega t - kz), \qquad \frac{\partial \psi_{\mathrm{ref}}}{\partial z} = -R_{12}[kA \sin(\omega t + kz)]. \tag{26}$$

We see from Eqs. (25) that at $z = 0$ the part of the velocity contributed by the reflected wave is R_{12} times that contributed by the incident wave. We see from Eqs. (26) that at $z = 0$ the part of the return force contributed by the reflected wave is $-R_{12}$ times that contributed by the incident wave. Thus, we can summarize Eqs. (24), (25), and (26) by defining *reflection coefficients* for ψ, $\partial\psi/\partial t$, and $\partial\psi/\partial z$:

$$R_\psi = R_{\partial\psi/\partial t} = R_{12} = \frac{Z_1 - Z_2}{Z_1 + Z_2}, \tag{27}$$

$$R_{\partial\psi/\partial z} = -R_{12}. \tag{28}$$

Notice that R_{12} must lie between -1 and $+1$.

Reflection at boundary between dispersive media. Suppose that the string with impedance Z_1 extending from $z = -\infty$ to $z = 0$ is joined there to a string with impedance Z_2 which extends from $z = 0$ to $+\infty$. The point L just to the left of $z = 0$ cannot tell whether the point R just to the right of $z = 0$ is the beginning of an infinite string of impedance Z_2 or is merely the input terminal of a dashpot of impedance Z_2. Therefore, the reflection coefficients of

Eqs. (27) and (28) again give us the reflected wave in medium 1. Notice that $R_{21} = -R_{12}$. Therefore the sign of the reflection coefficient is reversed if the properties of the two media are interchanged. For example: R_ψ is negative for a wave incident from a light string to a heavy string (taking the tension to be the same in each string); R_ψ is positive for a wave incident from a heavy string to a light string.

Transmission at boundary between dispersive media. The point at $z = 0$ undergoes oscillations under the combined driving force of the incident and reflected waves in medium 1. It then acts as a source of traveling waves going in the $+z$ direction in medium 2. We wish to find the transmitted waves of displacement ψ_2, transverse velocity $\partial\psi_2/\partial t$, and return force $-T_2\,\partial\psi_2/\partial z$, where subscript 2 designates the transmitted wave in medium 2. We shall make use of the *boundary conditions*.

Boundary conditions end continuity. These boundary conditions are that $\psi(z, t)$ is the same just a little to the left of the boundary as it is a little to the right, i.e., *the displacement $\psi(z, t)$ is continuous*. Therefore, *the velocity $\partial\psi(z, t)/\partial t$ is continuous*. Also, *the return force $-T_0\,\partial\psi(z, t)/\partial z$ is continuous*. The boundary conditions of continuity for the displacement and velocity of a point on the string are obvious and need no comment. The boundary condition on return force is not so obvious. (For example, you might have thought that the slope $\partial\psi(z, t)/\partial z$, rather than the tension times the slope, would be continuous. However, if the tension changes at the boundary, the string will exhibit a "kink" at the boundary. The slope will not be continuous, but the tension times the slope will be.) To see that the return force is continuous, imagine that there is an infinitesimal element of mass at $z = 0$. This mass has transverse force $-T_1\,\partial\psi_1/\partial z$ exerted on it by the string on its left. It has transverse force $+T_2\,\partial\psi_2/\partial z$ exerted on it by the string to its right. The superposition of these two forces gives the mass of the infinitesimal element times its acceleration. But its mass is zero. Therefore the superposition gives zero:

$$-T_1\,\frac{\partial\psi_1}{\partial z} + T_2\,\frac{\partial\psi_2}{\partial z} = 0 \text{ at } z = 0,$$

which says that $T_0\,\partial\psi/\partial z$ is continuous. (*Note:* We are using T_0 to denote the equilibrium string tension in general, and using T_1 and T_2 to denote the equilibrium string tension in medium 1 and medium 2.)

Amplitude transmission coefficient. Let $\varphi(z, t)$ represent any one of the three wave quantities, displacement, velocity, or return force. In medium 1 the wave function $\varphi_1(z, t)$ is the superposition

$$\varphi_1(z, t) = \varphi_0 \cos(\omega t - k_1 z) + R\varphi_0 \cos(\omega t + k_1 z), \qquad (29)$$

where according to Eqs. (27) and (28) the reflection coefficient R is equal to $R_{12} \equiv (Z_1 - Z_2)/(Z_1 + Z_2)$ if $\varphi(z, t)$ represents either the displacement or velocity and is equal to $- R_{12}$ if $\varphi(z, t)$ represents the return force. In medium 2 this same wave quantity is a traveling wave going only in the $+z$ direction. (That is because, by hypothesis, the only external driving force is located at $z = -\infty$. It produces the incident wave. The discontinuity produces reflected and transmitted waves. However there is nothing to produce a wave traveling in the $-z$ direction in medium 2.) Thus we can write down the form of $\varphi_2(z, t)$ and at the same time define the *amplitude transmission coefficient T*:

$$\varphi_2(z, t) = T\varphi_0 \cos(\omega t - k_2 z). \tag{30}$$

The condition that $\varphi(z, t)$ be continuous at the boundary at $z = 0$ gives

$$\varphi_2(0, t) = \varphi_1(0, t),$$

i.e.,

$$T\varphi_0 \cos \omega t = \varphi_0 (1 + R) \cos \omega t,$$

i.e.,

$$\boxed{T = 1 + R,} \tag{31}$$

where R is R_{12} for ψ and $\partial\psi/\partial t$, and is $-R_{12}$ for the return force, $- T \, \partial\psi/\partial z$.

(*Note:* We use capital T for both string tension and transmission coefficient. In examples other than the string this will not cause any confusion.)

Notice that since R must lie between -1 and $+1$, T must lie between zero and $+2$. Thus *the transmission coefficient is always positive*.

Here are some interesting limiting cases:

Case 1: Perfect impedance matching

If $Z_2 = Z_1$ there is no reflected wave; i.e., R_{12} is zero. The transmission coefficient is unity. Notice that the condition $Z_2 = Z_1$ does not necessarily imply that the two media are identical. If the string density and tension *both* change at a junction in such a way that their product remains constant, then the impedances $Z_1 = \sqrt{T_1 \rho_1}$ and $Z_2 = \sqrt{T_2 \rho_2}$ are the same. However, the phase velocities $v_1 = \sqrt{T_1/\rho_1}$ and $v_2 = \sqrt{T_2/\rho_2}$ will not then be the same in the two media.

Case 2: Infinite drag

If Z_2/Z_1 is "infinite," R_{12} is -1. *Then the point $z = 0$ remains fixed.* The displacement and velocity have reflection coefficients -1, so

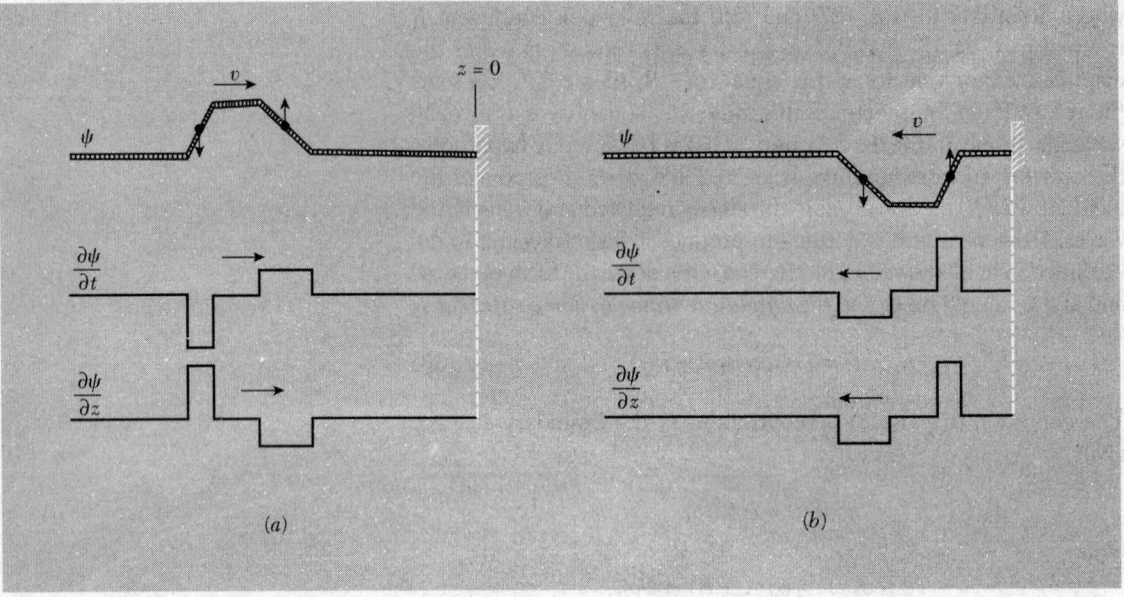

Fig. 5.3 *Reflection of an incident pulse at a fixed end of the string: (a) before reflection; (b) after reflection. (The string is joined at z = 0 to a string with infinite mass density.) The small vertical arrows on the three points indicate the instantaneous velocity of those string at those points. (The center point represents a zero-length arrow.)*

that the incident and reflected waves at $z = 0$ superpose to give zero displacement and velocity. An incident wave that is a positive displacement ("up") pulse becomes a negative ("down") pulse upon reflection. The transverse force has reflection coefficient +1, so that the force exerted on the string at $z = 0$ is the same direction as for perfect termination (i.e., down) but is twice as great as that necessary to produce perfect termination. The excess force is thus downward, and it generates a reflected wave with negative amplitude and equal magnitude to that of the incident wave.

Case 3: Zero drag

If Z_2/Z_1 is zero, the end of the string at $z = 0$ is a *free end.* Then the slope of the string remains zero there. Thus, the reflection coefficient for the return force is −1. An incident wave that is a positive return-force pulse becomes a negative pulse upon reflection. The reflection coefficients for the displacement and velocity are +1, and the string has twice the velocity at $z = 0$ that it would have if there were perfect impedance matching. An incident wave that is a positive displacement pulse is still a positive pulse after reflection. The limiting cases in which Z_2/Z_1 is infinite and zero are illustrated in Figs. 5.3 and 5.4.

General form of a sinusoidal wave. When we have an incident and a reflected wave in medium 1 we have

$$\psi(z, t) = A \cos (\omega t - kz) + RA \cos (\omega t + kz), \qquad (32)$$

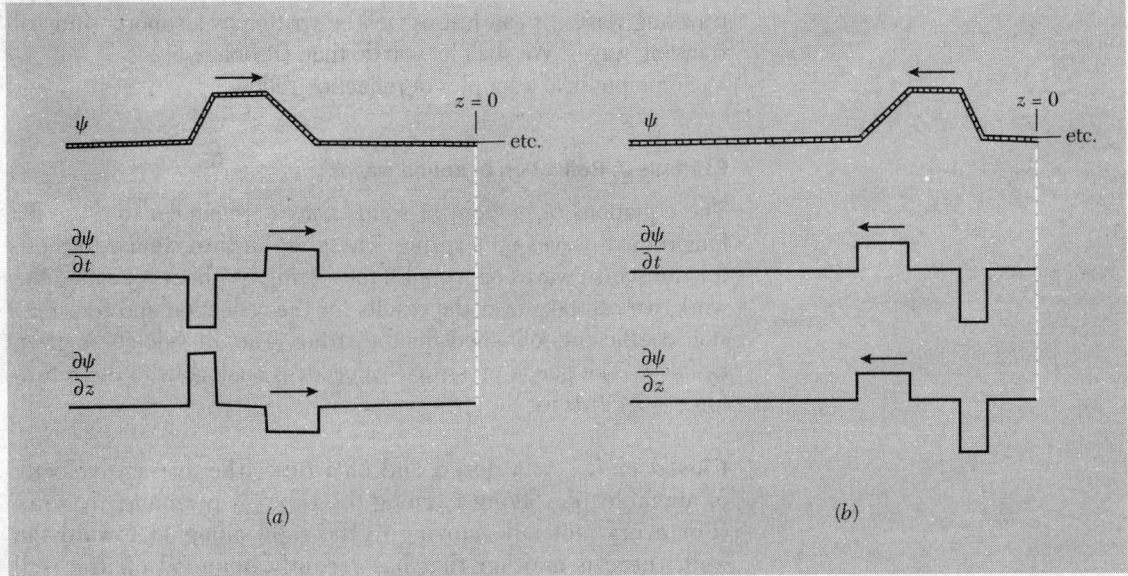

where R is the reflection coefficient and lies between -1 and $+1$. When R is zero, we have perfect termination. Then $\psi(z, t)$ is a "pure" traveling wave, i.e., a wave traveling in only the $+z$ direction. When R is -1, $\psi(z, t)$ is a "pure" standing wave, i.e., a wave with permanent nodes (zeros). There is a node at $z = 0$. When R is $+1$, $\psi(z, t)$ is again a pure standing wave, with a permanent antinode (maximum magnitude) at $z = 0$, i.e., with a permanent node one quarter-wavelength from $z = 0$. When R is neither zero nor ± 1, $\psi(z, t)$ is neither a pure standing wave nor a pure traveling wave; it is a more general sinusoidal wave. The most general sinusoidal wave (for given frequency ω) can be written *either as* a superposition of standing waves *or* as a superposition of traveling waves (or as some combination of both). Thus any sinusoidal wave $\psi(z, t)$ can be written in the form

$$\psi(z, t) = A \cos (\omega t + a) \sin kz + B \cos (\omega t + \beta) \cos kz, \quad (33)$$

which is a superposition of *two standing waves with nodes displaced by one quarter-wavelength* and with different amplitudes and phase constants. Alternatively, the *same* wave $\psi(z, t)$ can be written in the form

$$\psi(z, t) = C \cos (\omega t - kz + \gamma) + D \cos (\omega t + kz + \delta), \quad (34)$$

which is a superposition of *two traveling waves traveling in opposite directions* and having different amplitudes and phase constants. For example, the wave given by Eq. (32) is written as a superposition of

Fig. 5.4 Reflection of a pulse from a free end: (a) before reflection; (b) after reflection. (The string is joined at z = 0 to a string with negligible mass density.)

traveling waves; it can just as well be written as a superposition of standing waves. We shall let you do that. (Problem 5.20).

Some physical examples of reflection follow.

Example 3: Reflection of sound waves

The equations of motion for sound waves are similar to those for longitudinal waves on a spring. These are in turn similar to those for transverse waves on a continuous string. Without repeating the work, we can take over the results for the reflection and transmission coefficients obtained for the string. The air velocity is given by $\partial\psi/\partial t$. The gauge pressure $-\gamma p_0 \partial\psi/\partial z$ is analogous to the return force $-T_0 \partial\psi/\partial z$ for the string.

Closed end.   At a dosed end of a tube, the average velocity of air molecules along z (along the tube) is permanently zero. (For every molecule moving to the right along $+z$ toward the wall, there is another that has recently bounced off the wall and is moving to the left.) Thus the wave of the velocity $\partial\psi/\partial t$ must have reflection coefficient -1 at a closed end, so that the superposition of the incident and reflected velocity waves gives zero. The other physically interesting wave is the gauge pressure "returnforce" wave, $-\gamma p_0 \partial\psi/\partial z$. According to the mathematics we have just been through for the string, the gauge pressure wave must have a reflection coefficient equal in magnitude but opposite in sign to that of the velocity wave. Therefore, the gauge pressure must reflect with coefficient $+1$ at a closed end. The gauge pressure, therefore, has the same sign at a closed end as for a perfectly terminated sound wave but has twice the magnitude. On a "microscopic" level, we can see why the pressure at a closed end is double what it would be if the tube were continued there: Pressure is force per unit area. Force is momentum transfer per unit time. A molecule that bounces elastically from a wall reverses its z component of momentum. (If the wall is rough, this statement does not hold for each molecular collision, but it holds on the average; that is all we require.) Thus it transfers twice the momentum it would deliver if it were absorbed without bouncing or if it simply continued down the tube.

Open end.   At an open end of a tube, we have an experimental problem: We don't want to let the air escape into vacuum. Let us ask what happens if we let the tube terminate by opening into a large room full of air at the same pressure p_0 as the air in the tube (as it does in your home experiments with mailing tubes). At the open end of the tube, air can rush in and out freely. Thus the velocity wave is not constrained to be zero there (a sit is at a closed end). At locations in the room that are sufficiently far from the open end of the tube, the pressure p is permanently equal to the equilibrium pressure p_0. Right at the end of the tube, the pressure is not exactly equal to p_0, because the pressure waves emerging from the tube are still felt at the entrance to the tube. As soon as a region of compression (for example) reaches the open end of the tube, the air has a chance to spread out sideways, whereas inside the tube the air motion in the sound wave is purely along z. Thus the compression is "relieved" rapidly with increasing distance from the end of the tube, until at a certain distance into the room (which turns out to be of the order of a tube radius) the pressure is essentially equal to p_0. Thus, at an open end of a tube that ends in a large room, the gauge pressure is (approximately) permanently zero at a location just outside the tube. Let us call that (approximate) location the "effective" open end of the tube. Since the gauge pressure is permanently zero there, the reflection coefficient for gauge pressure must be −1 at an open end. The reflection coefficient for velocity is therefore +1. The impedance Z_2 offered by the room is effectively zero. (The fact that the air flows freely sideways is what gives zero impedance. Our formula $Z = \sqrt{\gamma p_0 \rho_0}$ for impedance cannot be used for the impedance offered by the room, because that formula is based on the assumption of strictly longitudinal motion.)

Here is a "microscopic" picture of what happens at an effective open end. Consider what happens when a compression reaches an open end. Before the compression reaches the open end, it is propagating by pushing on downstream air and transferring longitudinal momentum to it and by being pushed on the upstream side and receiving momentum. Suddenly the compression reaches an open end. There is no longer any impedance on the downstream side. The air rushes out into the room without having to transfer momentum downstream. This "excessive" outflow of air creates a deficit of pressure at the open end (a rarefaction). Air on the upstream side of this rarefaction now experiences less impedance than "usual" and rushes in to fill the rarefaction. This moves the rarefaction farther upstream. Molecules farther upstream rush in, etc. We notice that a *compression* traveling in the +z direction has produced a *rarefaction* traveling in the −z direction. A velocity wave consisting of a pulse of positive z component of velocity and traveling in the +z direction has produced a velocity wave traveling in the −z direction and consisting of a z component of *same* sign (the molecules always rush in the +z direction to fill the rarefaction). Thus we see that the reflection coefficient for the velocity wave is positive at an open end and that that for a pressure wave is negative.

Effective length of open-ended tube. The effective distance
beyond an open end of the tube at which the gauge pressure is zero
can be defined experimentally as follows. Consider a mailing tube
open at both ends. The lowest mode of free oscillation is one in which
the effective length of the tube is one half-wavelength. (At a given
instant when the air is rushing to the right at the right end of the tube,
it is rushing to the left at the left end. At the middle of the tube, there
is a permanent zero of air velocity, i.e., a node of the velocity standing wave. There is also an antinode of the pressure standing wave
there.) To find the effective length of the tube, tap it on something to
hear the pitch. (The lowest mode is the most easily excited, and that
is what you will hear.) Determine the pitch somehow. Calculate the
half-wavelength of sound at that pitch (frequency). It will be slightly
longer than the tube and can be considered the effective length of the
tube. An easier method is to use a standard driving frequency, i.e., use
a tuning fork. Then vary the length of the tube by gradually cutting off
pieces of a tube that is too long or by sliding a "trombone" extension
of the tube so as to vary the length. Tune for resonance. Notice that
when you have achieved resonance (i.e., a maximum in the loudness
of the radiation escaping from the tube when driven by the fork), the
pitch you hear for free oscillations excited by tapping the tube is the
same as the driving frequency of the fork. At "off-resonance" lengths,
the natural frequency is not the same as the tuning-fork frequency.
(What frequency do you hear when you drive the "off-resonance"
tube with the fork, the natural oscillation frequency or the frequency
of the fork? See the home experiments.)

Example 4: Reflection in transmission lines

A driving voltage $V(t)$ from a transmitter at the left end L of an
infinite transmission line of impedance Z_1 produces (in the continuous approximation) a traveling current wave $I(z, t)$ such that, at the
transmitter (at $z = 0$),we have

$$V_0 \cos \omega t = V(t) = Z_1 I(0, t). \tag{35}$$

The current and voltage traveling waves are given by

$$V(z, t) = V_0 \cos (\omega t - k_1 z), \quad I = I_0 \cos (\omega t - k_1 z), \quad V_0 = Z_1 I_0. \tag{36}$$

At a boundary where the characteristic impedance changes suddenly from Z_1 to Z_2, a reflected wave and a transmitted wave are
generated. There is no need to repeat the steps that we used for
the string. The reflection and transmission coefficients have forms
analogous to those that hold for waves on a string and sound waves.
Before writing down these formulas, let us consider the physical
situation in the limiting cases of zero impedance Z_2 (Example 5) and
infinite impedance Z_2 (Example 6) at the end of the line.

Example 5: Short-circuited end–zero impedance

If the right end of the line is short-circuited by being connected across a resistor having negligible resistance, the voltage across that end is permanently zero. The voltage reflection coefficient at a shorted end is therefore −1. The current, on the other hand,. has reflection coefficient +1 and has twice the value (at the end of the line) that it would have if the line were perfectly terminated. A wavefront of positive voltage propagating in the +z direction is reflected as a wavefront of negative voltage. A positive current wave is reflected as a positive current wave.

Example 6: Open-circuited end–infinite impedance

If the right-hand end is connected across an infinite resistance (or left "open" with no resistor at all), no current can flow across from one conductor to the other. Thus, the current is permanently zero at an open circuited end, and the current reflection coefficient must be −1. The voltage reflection coefficient is then +1.

From the above physical considerations we can deduce that the reflection coefficients for potential V and current I are given by

$$R_V = \frac{Z_2 - Z_1}{Z_2 + Z_1} \equiv -R_{12}, \qquad R_I = -R_V. \qquad (37)$$

As a check we notice that for $Z_2 = 0$ (shorted end) Eq. (37) gives $R_V = -1$ as it should; for $Z_2 = \infty$ (open end), $R_I = -1$.

Parallel-plate transmission line. The impedance (in statohm) is given by [Eq. (4.140), Sec. 4.4]

$$Z = \sqrt{\frac{\mu}{4\pi\epsilon}} \frac{g}{w}. \qquad (38)$$

Thus, if (for example) the gap g is doubled in going from line 1 to line 2, the impedance is then doubled.

Reflection coefficients for the fields. Instead of considering the potential and the current, we can focus our attention on the electric field E_x and the magnetic field B_y. In a given transmission line, the electric field is proportional to V and the magnetic field is proportional to I. Thus, we have

$$gE_x = V$$

$$wB_y = \mu I. \qquad (39)$$

Since a reflected wave in line 1 is in the same transmission line as the incident wave, i.e., with same gap g, width w, and permeability μ, we see that the reflection coefficient for E_x is the same as that for V and that the reflection coefficient for B_y is the same as that for I.

(To convince yourself that there is no "extra" minus sign relating the reflected current wave and the reflected magnetic field wave, make a sketch and use your right hand.) On the other hand, we see that the *transmission* coefficient for gE_x is the same as that for V. Similarly, the transmission coefficient for wB_y/μ is equal to that for I. We shall consider only the reflection coefficients.

Then we have

$$\text{Electlic field: } R_E = \frac{Z_2 - Z_1}{Z_2 + Z_1}. \tag{40}$$

The magnetic field B_y has a reflection coefficient equal in magnitude and opposite in sign to that of the electric field E_x.

In Example 7 we consider an important special case.

Example 7: Reflection in transmission line having discontinuity in ϵ

We assume that the cross-sectional geometry (i.e., width w wand gap g) does not change at the boundary and that the magnetic permeability μ does not change. (Many important media like glass, water, air, and the ionosphere have $\mu = 1$ to a high degree of accuracy.) Then, according to Eq. (38), the only quantity in the impedance Z that changes at the boundary is the dielectric constant ϵ. Then Z is proportional to $\dfrac{1}{\sqrt{\epsilon_r}}$, which is equal to $1/n$, where $n = \sqrt{\epsilon_r}$ is the index of refraction (for $\mu = 1$). Thus we find (after substituting $Z_1 = 1/n_1$ and $Z_2 = 1/n_2$ in Eq. (40) and multiplying by $n_1 n_2$ to clear fractions)

$$\boxed{R_E = \frac{n_1 - n_2}{n_1 + n_2}.} \tag{41}$$

Now we shall generalize the application of this result. Let the gaps of the transmission lines go to infinity and let the widths increase proportionally. The reflection coefficients for the local fields cannot depend on the boundary conditions. Therefore, Eq. (41) should hold even if the incident wave is emitted by a distant street light or a distant TV antenna. *The coefficients given by Eq. (41) hold for any straight and parallel electromagnetic waves normally incident on a surface where the dielectric constant changes suddenly (within less than a wavelength).*

We can immediately apply this result to the interesting case of visible light:

Example 8: Reflection of visible light

The reflection coefficient given by Eq. (41) holds for any electromagnetic plane wave reflecting at normal incidence at a boundary between two transparent media (if both have $\mu. = 1$). Thus, taking the index of refraction of air to be 1.00 and the index of refraction of glass to be 1.50 for visible light, we have, in going *from air to glass*,

$$R_E = \frac{1-n}{1+n} = \frac{1-1.5}{1+1.5} = -\frac{1}{5}. \tag{42}$$

Thus, the electric field reverses sign and is reduced in magnitude by a factor of 5. (In going from glass to air, the reflection coefficient has the opposite sign and is thus $+\frac{1}{5}$.) The reflected energy flux is proportional to the square of the electric field. Therefore the fraction of light intensity reflected at a single air-glass surface is $\frac{1}{25}$; i.e., 4 percent of the incident light intensity is reflected at normal incidence. See Home Exp. 5.1.

5.4 *Impedance Matching between Two Transparent Media*

Suppose that we wish to transmit traveling waves from one medium into another medium without generating a reflected wave. For example, we may wish to transfer sound energy from the air in a loudspeaker to the air of the room without generating reflections. (The reflections are undesirable because they make the effective load impedance felt by the driving mechanism a partly reactive load, with an impedance that varies with frequency, perhaps giving undesired resonances at some frequencies.) As another example, we may wish to transfer traveling waves of visible light from air into a glass lens or slab without generating a reflection. (The reflection may be undesirable both because of the loss of light intensity from the beam and also because we don't want reflected light to get into some other part of the apparatus.) In still another example, we may wish to devise a method by which two skindivers, equipped with underwater breathing apparatus, can talk with one another while under water. Each diver can talk out loud into a face mask (one that covers the mouth as well as eyes and nose), but very little of the sound wave is transmitted through the glass of the face mask into the water because T_{12}, the transmission coefficient, is very small. That is because the sound impedance of water differs greatly from that of air.

Solving the problem of transferring traveling waves from one medium to another without reflection is called *impedance matching*. We shall discuss two methods: one involves a "nonreflecting layer," the other involves "tapering." (It turns out that neither of these methods provides a solution to the communication problem of the skin divers. That solution was achieved by converting the audio frequencies of the voice to supersonic frequencies before attempting to radiate them into the water; it is easier to match impedances at these frequencies. Each skin diver is equipped with a supersonic transmitter and receiver and a frequency converter.)

Nonreflecting layer. Suppose medium 1 extends from $z = -\infty$ to $z = 0$. An impedance-matching device (medium 2) extends from $z = 0$ to $z = L$. Medium 3 extends from $z = L$ to $z = +\infty$. We wish to match impedances between media 1 and 3 for waves of angular frequency ω. That is, we want no reflected wave when a wave is incident from medium 1 traveling in the $+z$ direction. Now, there is no way that we can "turn off" the reflection generated at a discontinuity in impedance. The ingenious impedance-matching trick consists in making use of the fact that we can generate *two* reflected waves, one due to a discontinuity at $z = 0$ and the other due to a discontinuity at $z = L$. If we are clever, we can arrange things so that the *superposition* of these two waves gives a total reflected wave in medium 1 with zero amplitude.

Let us fill the region from $z = 0$ to L with a dispersive medium having characteristic impedance Z_2. It seems reasonable that, if we solve the impedance-matching problem, we will find that Z_2 lies between Z_1 and Z_3. Let us assume that is the case. According to our formula for reflection coefficients, we have

$$R_{12} = \frac{Z_1 - Z_2}{Z_1 + Z_2} = \frac{1 - (Z_2/Z_1)}{1 + (Z_2/Z_1)}, \quad R_{23} = \frac{Z_2 - Z_3}{Z_2 + Z_3} = \frac{1 - (Z_3/Z_2)}{1 + (Z_3/Z_2)}. \quad (43)$$

Therefore, (assuming $Z_1 < Z_2 < Z_3$) the two reflection coefficients R_{12} and R_{23} have the same sign. At first that sounds discouraging, because we want the two reflected waves to cancel one another to zero; but we have not yet taken into account the fact that the two reflected waves are generated at two different places, namely $z = 0$ and $z = L$. Let us now follow a given incident wave crest: At $z = 0$ the incident wave is partially reflected (with coefficient R_{12}) and partially transmitted, with a transmission coefficient T_{12} that is always positive. The transmitted wave propagates to $z = L$, where it is partially reflected with coefficient R_{23} and partially transmitted. The reflected wave travels back to $z = 0$, where it is partially transmitted with coefficient T_{12}. Thus, it emerges into medium 1 at $z = 0$, traveling in the $-z$ direction, with amplitude given by the amplitude of the incident wave times $T_{12}R_{23}T_{21}$, and with a phase constant that differs from that of the wave reflected at the first surface because of the time delay taken to go "down and back" a total distance of $2L$ in medium 2. Thus, we have in medium 1

$$\psi_{\text{inc}} = A \cos(\omega t - k_1 z), \quad (44)$$

$$\psi(\text{ref at } z = 0) = R_{12} A \cos(\omega t + k_1 z), \quad (45)$$

$$\psi(\text{ref at } z = L) = T_{12}R_{23}T_{21} A \cos(\omega t + k_1 z - 2k_2 L), \quad (46)$$

where $-2k_2 L$ is the phase (in radians) corresponding to propagating down and back a distance of $2L$ with angular wavenumber k_2. (The minus sign is due to the fact that we have a phase *lag*, i.e., a delay.) The incident wave and the two reflected waves given by Eqs. (45) and (46) are indicated in Fig. 5.5.

Small-reflection approximation. Aside from the two reflected waves shown, there are an infinite number of additional reflected waves, indicated by the ray labeled "etc." in Fig. 5.5. Now, in all our applications, Z_1, Z_2, and Z_3 do not differ by very much, and therefore the reflection coefficients are small compared with unity. In that case it turns out that the first two reflected waves (i.e., those shown) dominate, and we can, as a good approximation, neglect the additional contributions due to multiple internal reflection. For example, the next reflected ray to be added to the two shown is smaller in amplitude by a factor $R_{21} R_{23}$ than the second reflected ray. We can neglect this factor compared with unity if, for example, R_{21} and R_{23} are of order 0.1. To the same degree of approximation, we can replace $T_{12} T_{21}$ by unity in Eq. (46):

$$T_{12} T_{21} = (1 + R_{12})(1 - R_{12}) = 1 - R_{12}^2 \approx 1. \qquad (47)$$

Thus in this *small-reflection approximation* the total reflected wave is the sum of the two contributions from $z = 0$ and $z = L$ given by [using Eq. (47) in Eq. (46)]

$$\psi_{\text{ref}} \approx R_{12} A \cos (\omega t + k_1 z) + R_{23} A \cos (\omega t + k_1 z - 2 k_2 L), \quad (48)$$

where the quantity $2 k_2 L$ gives the "down and back" phase shift.

Solution for nonreflecting layer. The solution to the impedance-matching problem is now at hand: First, choose Z_2 so that $R_{12} = R_{23}$, i.e., so that [according to Eq. (43)]

$$\frac{Z_1}{Z_2} = \frac{Z_2}{Z_3}, \qquad Z_2 = \sqrt{Z_1 Z_3}. \qquad (49)$$

Then Eq. (48) becomes

$$\psi_{\text{ref}} \approx R_{12} A \left[\cos (\omega t + k_1 z) + \cos (\omega t + k_1 z - 2 k_2 L) \right], \qquad (50)$$

Now choose the length L so that the two contributions to this superposition cancel one another to zero, i.e., so that we have "completely destructive interference." That will be the case if $2 k_2 L$ is π, i.e., if the down and back distance, $2L$, is one half-wavelength in medium 2. *The total reflected wave is zero if Z_2 is the geometric mean of Z_1 and Z_3 and if the thickness L of the nonreflective layer is one quarter-wavelength of the wave in the layer.*

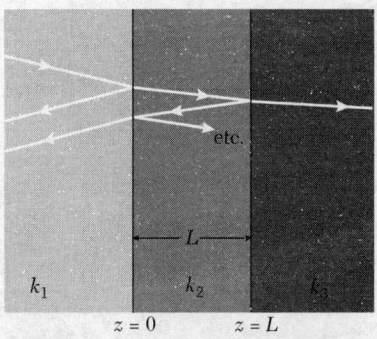

Fig. 5.5 *Incident wave and first two reflected waves. The rays are drawn at nonnormal incidence to prevent overlap on the sketch.*

Example 9: Optical impedance matching

When a beam of visible light passes through a slab of glass, it.
Passes through two surfaces. At each surface, the *intensity* suffers
a reflection given by the square of the amplitude reflection coefficient (since the intensity is proportional to the time-averaged
square of the electric field). Thus, according to Eq. (42), Sec. 5.3,
there is a loss of $(\frac{1}{5})^2 = \frac{1}{25} = 4\%$ in intensity at each surface. For
transmission through the two surfaces of a slab of glass, there is
thus an 8% loss. (We are neglecting "interference" due to superposition of waves reflected from the two surfaces. For ordinary
"white" light, these interference effects are zero when we average over a wide band of colors. However, see Home Exp. 5.10.)
This loss is intolerable in an optical instrument having many
glass-air interfaces. Consequently, it is common practice to "coat"
the lenses with a nonreflecting layer. According to our results of
Eq. (49), the impedance of the coating should be the geometric
mean of the impedance of air and glass. Thus, the index of refraction of the coating should be the square root of 1 times *n*, i.e.,
it should, for glass, have index $\sqrt{1.50} \approx 1.22$. It should also be
$\frac{1}{4}\lambda_2$ thick, where λ_2 is the wavelength of light in the coating. For
light of vacuum wavelength 5500Å, the wavelength in the coating
is 5500/1.22 =4500 Å. Thus, the thickness of the coating should
be 4500 Å/4 = 1120 Å = 1.12×10^{-5} cm. This can be accomplished
by putting the piece of glass to be coated, say a lens, in a vacuum
chamber containing a small crucible in which the coating material
is heated until it evaporates. Molecules from the evaporated material fly in straight lines in all directions and evenly coat the lens on
the side facing the crucible.

Here is an interesting question: Suppose the glass lens has been
coated with a noneflecting coating that is $\frac{1}{4}\lambda_2$ in thickness for
green light of vacuum wavelength 5500 Å. Then there is zero reflection of green light. What is the reflected intensity for *other* colors?
See Prob. 5.21.

Tapered section.   One possibly troublesome feature of the
quarter wavelength nonreflecting layer is that it only works well
at certain frequencies. If we have enough space available, we can
do much better than that. Suppose *L* is long compared with any
of the wavelengths that we would like to transmit without reflection. Let the impedance change gradually over the distance *L*. In
any given quarter-wavelength, the impedance changes very little.
For simplicity let us think of the impedance as increasing by a
series of tiny discrete steps, a new step being taken whenever the
distance *z* increases by $\frac{1}{4}\lambda$ for a wavelength somewhere among
those we are interested in transmitting. We will get rid of all
reflected waves if the amplitude reflected from one tiny step at

position z is canceled by that from the next tiny step, which is at a position downstream a distance $\Delta z = \frac{1}{4}\lambda$. (We are neglecting multiple reflections.) Now, the reflection from a tiny step where the impedance goes from Z_1 to $Z_2 = Z_1 + \Delta Z$ is given by the infinitesimal reflection coefficient ΔR, where

$$\Delta R = \frac{Z_1 - Z_2}{Z_1 + Z_2} \approx \frac{-\Delta Z}{2Z} \approx \frac{-1}{2Z}\left[\frac{dZ(z)}{dz}\right]\left(\frac{1}{4}\lambda\right). \qquad (51)$$

If the reflection from one little step is to be canceled by that from the next step, one quarter-wavelength downstream, ΔR must be a constant, independent of z. Call this constant by the name a. Then setting $\Delta R = a$ in Eq. (51) gives

$$\frac{dZ}{Z} = -\frac{8a}{\lambda}\, dz. \qquad (52)$$

Example 10: Exponential horn

If we take λ to be constant, independent of z, as for example for sound waves in air in a tube whose impedance is changing because the diameter is changing, then integration of Eq. (52) gives an exponential dependence of the impedance Z on distance, as you can easily show.

An impedance-matching exponential horn is commonly used in high fidelity audio loudspeakers so that a vibrating speaker piston of area A_1 can transfer sound energy to the room without reflection. Then the impedance offered to the mechanism that drives A_1 can be chosen to suit the properties of that mechanism. If A_1 were instead the area of a cylindrical tube which was driven by the speaker's driving mechanism at one end, and if the tube, without any flaring, suddenly ended in the room, then the tube would resonate at every wavelength for which the open end and the driven end were velocity antinodes. That would mess up the music.

Example 11: Tapered index of refraction

Similarly, optical impedance matching by the method of *tapering the index of refraction* can be accomplished by coating the optical element in question with successive thin layers of a variable mixture of substances having index varying from n_1 to n_2. Thus, the index can be made to change gradually from n_1 to n_2. This is a better method than a single nonreflective coating but technically more demanding. In this case the desired z dependence is *not* exponential. What is it? (Prob. 5.22.)

5.5 Reflection in Thin Films

Interference fringes. Any box of a dozen microscope slides will usually furnish several examples of two slides stuck closely together and exhibiting beautifully colored "interference fringes." Similarly, a drop of light machine oil placed on a surface of hot water will spread out and, when thin enough, exhibit the same sort of colored interference fringes. These fringes are due to interference between light reflected from both the front and the back surfaces of the thin film. Suppose, for example, that we have a thin film of air between two microscope slides. Where the glass surfaces touch, the film of air has zero thickness, and there is, of course, no reflection. That can be "accounted for" by noticing that, since $R_{21} = -R_{12}$, the reflection from the first surface has opposite sign to that from the second, and, since there is no phase shift introduced by a "down and back" distance of zero, the two contributions. cancel one another to zero. (If they did not, we would have a paradox and the whole theory of optics would collapse!) If the spacing now is increased from zero to $\frac{1}{2}\lambda$, then the "down and back" distance is λ. The relative phase of the two contributions is thus increased by 2π, and again one has zero net reflection. Halfway between these successive zeros of reflectivity the reflection is maximum. Thus for a given color the reflection maxima occur when the thickness is $\frac{1}{4}\lambda, \frac{3}{4}\lambda, \frac{5}{4}\lambda, \ldots$.

Example 12: Why the first fringe is white

Home experiment

Take two clean microscope slides. By a simultaneous combination of pressing the slides together and sliding them over one another, you can get the slides stuck together in intimate contact. (Don't press too hard—glass breaks!) Hold the pair of slides so that you can see the reflection of a broad source of light like the sky or an incandescent bulb with frosted glass envelope. Put a dark cloth or something beneath the slides, so as to reduce the background light. You should now see concentric "contours" consisting of colored fringes. The center of the pattern is "black." This is the region between zero thickness and the first maximum of reflected light. The first "fringe" (maximum of reflected skylight or bulblight) is essentially "white" in color. Let us see why this is so. Green is at the center of the visible spectrum and corresponds to $\lambda \approx 5500$ Å $= 5.5 \times 10^{-5}$ cm. The thickness of the air film between the slides is thus about $\frac{1}{4}(5.5) \times 10^{-5} = 1.37 \times 10^{-5}$ cm at the center of the first fringe of green. Blue has $\lambda \approx 4.5 \times 10^{-5}$ cm; therefore this same thickness corresponds to a wavelength fraction $(1.37/4.5)\lambda = 0.30\lambda$ for blue. Similarly, for red ($\lambda \approx 6.5 \times 10^{-5}$) this thickness is a fraction $(1.37/6.5)\lambda = 0.21\lambda$. Thus, blue and red are *also* at nearly their maximum reflectivity (corresponding to $\frac{1}{4}\lambda$) in the first fringe of green. That is why the first fringe is white.

The prettiest fringe. Successive fringes show more and more color. The most nearly monochromatic fringe should be the one where the thickness is some odd number N of quarter-wavelengths for green $\left(\frac{3}{4}, \frac{5}{4}, ..., N/4\right)$ and where blue is about $N + 1$ quarter-wavelengths and red is about $N-1$ quarter-wavelengths. Then blue and red are at reflection minima, and the fringe should be as green as possible. Thus the number N for the *prettiest fringe* is a constant of nature which we should perhaps know. (See Prob. 5.23.)

Expression for fringe intensity. We can obtain an expression for the intensity of a given color reflected from a film of air between two slabs of glass (or for a thin piece of glass in air) by suitable adaptation of our previous results for reflection from two discontinuities separating media 1, 2, and 3. In the present example, medium 3 is the same kind of medium as medium 1. Thus $R_{23} = R_{21} = -R_{12}$. Then you can show that the fractional reflected time-averaged intensity is given by (Prob. 5.24)

$$\frac{I_{\text{ref}}}{I_0} = R_{12}{}^2 \sin^2 k_2 L. \tag{53}$$

For glass to air or air to glass, we have $R_{12}{}^2 = 0.04$. Thus we have

$$\frac{I_{\text{ref}}}{I_0} \approx 0.16 \sin^2 k_2 L, \tag{54}$$

which is zero for $L = 0$ and for $L = \frac{1}{2}\lambda_2$ and reaches its first maximum at $L = \frac{1}{4}\lambda_2$. Notice that the maximum fractional intensity reflected from the film is 0.16; that is four times the fractional intensity that is reflected from a single air-glass interface.

One plus one equals four? How can we add the intensity from one surface to an equal intensity from the other and get a total of four times the intensity? The same way we can add them and get zero: $(1 + 1)^2 = 4$; $(1 - 1)^2 = 0$. *First* superpose the waves, *then* square and time-average to get the intensity.

Notice that if you look for colored fringes from the air film between two microscope slides the colored maxima have fractional intensity 0.16. The fractional intensity of the background light from the outer surface of the top slide is 0.04; that from the bottom is 0.04. [The interference from the top and bottom surfaces of the two slides is not noticed because it is of such high order (i.e., the interference is at such a large number of quarter wavelengths) that the colors overlap completely.] Thus, the colored fringes are twice as intense as the background light and are easy to see (especially if you put dark cloth under the slide so that there is no additional background.)

Home experiment   **Example 13: Fabry-Perot fringes in a microscope slide**

If you use a sufficiently monochromatic light source, you can easily see interference fringes due to superposition of the amplitudes of light reflected from the two surfaces of an ordinary microscope slide or a pane of window glass. A complete description of these fringes requires calculation of the reflection coefficient for nonnormal incidence as well as for normal incidence. That is easily done, but we shall not do it here. We shall only consider here the central fringe, i.e., the one corresponding to normal incidence, and ask the question, "How monochromatic must the light source be?" The answer can be obtained from Eq. (53). Suppose $L = 1$ mm $= 0.1$ cm. If a single wavenumber k_2 is present, then this central fringe is a maximum or a minimum depending on whether $\sin^2 k_2 L$ is 1.0 or 0.0. If a band of wavenumbers Δk_2 is present, then if the band is too broad, some wavenumbers will correspond to a maximum and some to a minimum, and the fringe will be "washed out." How narrow must the band be in order to get a good visible central fringe? (We can assume the fringes for nonnormal incidence will also be easily visible if the central fringe is.) Successive maxima of Eq. (54) are separated by an increase in $k_2 L$ of π. As a crude criterion, we can say that if $(\Delta k_2)L$ is less than π, we should get good fringes. You may then show that the required bandwidth is (Prob. 5.25)

$$\Delta(\lambda^{-1}) \approx 3.3 \text{ cm}^{-1}, \tag{55}$$

i.e.,

$$\Delta\nu = c\Delta (\lambda^{-1}) \approx 10^{11} \text{ Hz};$$

i.e., if we take $\lambda \approx 5.5 \times 10^{-5}$ cm (green light),

$$\lambda \approx 1.0 \text{ Å}.$$

Thus the bandwidth of the light is required to be less than 3.3 inverse centimeters (these are the units commonly used in spectroscopy). As we shall learn in Chap. 6, the bandwidth $\Delta\nu \approx 10^9$ Hz is approximately the "natural linewidth" for a freely decaying atom. It is hard to do better than that (except with a laser). Thus, to see interference fringes in a window pane, we can use a good light source of freely decaying atoms. A neon lamp works beautifully. (See Home Exp. 5.10.) So does a burning wad of toilet paper! (See Home Exp. 9.27.)

Problems and Home Experiments

Home experiment   **5.1   Reflection from glass.** A flat slab of glass reflects about 8% of incident light intensity for normal incidence, 4% in intensity being reflected from each surface. An ordinary silvered mirror reflects more than 90% of the visible light. Take a mirror and a single clean piece of glass (a microscope slide, for example). Compare the reflection from the mirror and from the slide with the two held close together so you can see both reflections at once. Look at the reflection of a broad light source like an incandescent bulb, or a piece of white paper, or a patch of sky. Compare the reflectivity of the slide and that of the mirror at near-normal incidence. *Now do the same at near-grazing incidence*. At near-grazing incidence, the source, mirror reflection, and slide reflection should be nearly indistinguishable; i.e., you get nearly 100% reflection at near-grazing incidence. At near-normal incidence, the glass should be noticeably dimmer than the mirror.

Next take four clean microscope slides. Lay them on top of one another in a series of "steps," with the first slide giving a "floor," the second slide giving the first "step,"and the remaining two slides giving a second "step" of double height. Thus, you can compare at the same time reflection at near-normal incidence from one slide, two slides, and four slides. Look at a broad source (the sky) reflected at near-normal incidence. Neglecting the complications of internal reflections, you should *transmit* about 0.92 through each slide. Thus, four slides should transmit $(0.92)^4 = 0.72$ and reflect $1 - (0.92)^4 \approx 0.28$.

Now make a *pile* of about a dozen clean slides; they should reflect $1 - (0.92)^{12} = 0.64$. Compare with the mirror. Suppose the formula keeps working (and the slides are clean). How many slides will equal one good mirror if the mirror reflects 93% of the intensity? Try it—compare the stack of slides with the mirror at near-normal incidence. Also look directly at the source through the stack to seethe transmitted light. (It takes about 32 slides, according to the formula. Needless to say, they should be free of fingerprints.) (Microscope slides cost about 35 cents per dozen. Three dozen slides is a good investment for home experiments.)

5.2   Interference in thin films.   (See Sec. 5.5.) Fill a pan with hot water. Put one drop of light penetrating oil on the water and watch the oil spread. (Use a light oil; salad oil, for example, is too heavy—it does not spread.) Look at the sky (or another broad source of light) reflected in the oil film as the film spreads out. (A black cloth or paper on the bottom of the pan helps by giving a dark background and eliminating undesired reflections from the bottom of the pan.) Notice that no colored fringes appear until the film has spread to 10 cm by 10 cm or so. Why is that? Watch the colored fringes as the film continues to spread. When it gets thin enough, you will get no more fringes; it will be "black" where the oil is thinnest. That is the region where the film is less than a quarter-wavelength thick. *Use this fact to estimate crudely the wavelength of visible light*. Take the "black" region to be (for a crude estimate) about an eighth of a wavelength thick; estimate the area of the film, and use that and the original size of the drop to find the film thickness, which then gives the wavelength.

Home experiment

5.3   Transitory standing waves on a slinky.   Attach one end of a slinky to a telephone pole or something. Hold the other end. Stretch the slinky out to 10 m or so. Shake the end of the slinky about 3 or 4 times as

Home experiment

rapidly as you can. A "wave packet" is thus propagated down the slinky. After you have sufficiently enjoyed following packets back and forth, try something new: This time, keep your attention fixed on a region near the fixed end of the slinky. As the packet comes in, reflects, and returns, you should see *transitory standing waves* during the time interval in which the incident and reflected wave packets overlap. (It may help to fix *both* ends of the slinky so that you can watch the process at close range at your end of the slinky.) That should help to convince you that a standing wave can always be regarded as the superposition of two traveling waves traveling in opposite directions.

Home experiment   **5.4 Multiple internal reflection in a microscope slide.**   Make a sketch showing a ray coming in from the left and hitting a slab of glass tilted at some angle, Show the first transmitted ray, the second (i.e., that transmitted after two internal reflections), the third, ... Now look at a point or line source through a microscope slide. Hold the slide close to your eye. Starting at normal incidence, gradually tilt the slide. Look for the "virtual sources" due to multiple reflections, (The effect is greater near grazing incidence.) Look also for the light that emerges, not by transmission out of the surface of the slide, but from the end. This is the "internally trapped" light, which finally escapes when it reaches the end surface at near-normal incidence rather than at the near-grazing incidence at which it encounters the sides of the slide.

Home experiment   **5.5 Reflections in transmission lines.**   Suppose a coaxial transmission line having 50 ohms characteristic impedance is joined to one having 100 ohms characteristic impedance.

(*a*) A voltage pulse of + 10 volts (maximum value) is incident from the 50 Ω line to the 100 Ω line. What is the "height" (in volts, including the sign) of the reflected pulse? Of the transmitted pulse?

(*b*) A + volt pulse is incident from the 100 Ω to the 50 Ω line. What are the reflected and transmitted pulse heights?

5.6 Irreversible impedance matching. Consider the transmission lines of Prob. 5.5.

(*a*) How can you insert an ordinary resistor so that an incident pulse traveling from the 50 Ω to the 100 Ω line is transmitted without generating any reflected pulse? We want to know how many ohms the resistance has, and we want a schematic sketch showing the center conductor and outer conductor of each of the lines at the place where they join and showing the resistor connected. (Do not worry about "distributing" the resistor. If the wavelengths are long compared with the diameter of the cable, there is no need to distribute the resistance.)

(*b*) What is the size of the transmitted pulse? (Suppose a + 10-volt pulse is incident.)

(*c*) Now suppose a + 1-volt pulse is sent down this line *in the "wrong" direction*, i.e., from the 100 Ω line to the 50 Ω line. What happens? Find the reflected and transmitted pulse heights.

(*d*) Next consider the problem of transmitting a pulse from the 100 Ω line to the 50 Ω line without generating any reflection. What should be the resistance value and how should it be connected at the place where the lines are joined? What is the pulse height transmitted if + 10 volts is incident? What happens now when a + 10 volt pulse is incident from the 50 Ω to the 100 Ω line, i.e., in the "wrong" direction?

5.7   Light of wavelength $\lambda = 5000$ Å is incident normally on a series of two transparent plastic disks separated by a distance large compared with the wavelength. If the index of refraction of the disks is $n = 1.5$, what fraction of the light is transmitted? Neglect absorption, internal multiple reflections, and interference effects.

Ans. $I_t/I_0 = 0.85$.

5.8   Compare the amplitude and intensity reflection coefficients for light normally incident on a smooth water surface (index $n = 1.33$) for the two cases of incidence from air to water and from water to air.

5.9   Reflections in a thin film of air.   Suppose you have two optically flat slabs of glass touching at one edge and spaced apart by a sheet of paper at the other edge, which is a distance L from the edge where they touch. Assume the paper has the thickness of one page of this book. (How can you measure that without a micrometer?) Suppose you want successive fringes of green light to be separated by 1 mm so that you can see them easily. How long must the length L of the "wedge" of air be?

5.10   Fabry-Perot fringes in a window pane.   For this experi-
ment you need a broad, almost monochromatic source of light. The best cheap source I know of is a General Electric lamp NE-40 (available, for example, at Brill Electronics, 610 E. 10th St., Oakland, Calif. 94604; list price $2.56). This provides a circular disk of glowing neon of diameter about 2.5 cm. It screws directly into a standard 60-cps 115-volt lamp socket. (Any neon lamp will work almost as well; for example a "circuit continuity tester" is available in most hardware stores for about $1.00.) Turn on the lamp and look at it with the diffraction grating (hold the grating up close to your eye). You can see (in the first-order spectrum, which occurs about 15 or 20 deg to the side of the central orange light) at least three clearly defined virtual sources. The brightest three are a yellow, an orange, and a red. (Actually there are about a dozen bright "lines" in the yellow, orange, and red.) The fact that the virtual light sources are clearly defined and not smeared out in angle shows that each individual color is a monochromatic light source (to the limits of your resolution). Each corresponds to a different atomic transition of excited neon atoms. Here is the experiment: Get an ordinary piece of glass. A microscope slide or a piece of window glass or the window pane in your room will do. Holding the neon lamp close to your nose, sight over the top of the lamp and look at the reflection of the lamp in the piece

Home experiment

of glass. If you see *two* reflections, get another piece of glass. (Window panes become wedges after many years of slow viscous flow of the glass.) Look for "fringes," i.e., "contour lines" of alternating bright and dark regions in the image of the lamp. After a minute of searching, you will see them. Once you find them, they are easy to see. (The glass should be roughly two feet from you.) They are due to interference between the front and back surfaces of the glass. To prove that, stick a piece of transparent tape on one surface and hold the glass first with this surface nearer you and then with it farther from you. Look for fringes in the image when the tape is in the region of the reflected image. The sticky side of the tape is "optically rough," i.e., there are irregularities smaller than a wavelength of light and finer on a transverse scale than the fringe spacing. In some tiny regions, the light goes from glass to tape without reflection (the index of the tape is nearly that of the glass) and does not reflect until it reaches the smooth outer surface of tape. In other tiny regions, the sticky surface does not touch the glass, and the reflection occurs at the glass-to-air surface (the air between the glass and the sticky side of the tape). You can now use the viewing of fringes as a way to find out if a slab of plastic or glass or cellophane is "optically smooth" on a transverse scale of the order of the fringe width. Scotch brand transparent tape is not; glass is. Try the pieces of polaroid and the quarter-wave and half-wave plates from your optics kit. Are they smooth to less than a wavelength? Try your red gelatin filter. Is it optically smooth? (You may have trouble finding a good flat spot to get a decent reflected image.)

An ordinary fluorescent lamp also works, although not as well as a neon lamp; it is perhaps more readily available. (The famous "green line of mercury" is the almost monochromatic light that gives the fringes.)

Here is an experiment that works with the neon lamp (I couldn't make it work with a fluorescent lamp). Look at the neon fringes from a piece of polaroid. Use a piece of polaroid in front of your eyes(or polaroid sunglasses). Try both orientations of the target polaroid. *Now turn the polaroid over and repeat the experiment.* Thus, there are four orientations: polaroid axes parallel and perpendicular and polaroid flipped over. Notice the size of the friuges. (Wider fringes mean a thinner film.) Polaroid consists of a sandwich of three layers, with two outer clear layers (the "bread") and the central layer of absorbing "ham." Here is the question: *Is the "ham" optically smooth on both sides?*

Here is another interesting experiment (or demonstration) with the neon Fabry-Perot fringes. At night, with no other illumination, illuminate your face with the neon lamp. Look at your image in a piece of glass held a foot or two away. Your face is now a "broad source of monochromatic light." Look for concentric circular fringes centered on the image of each eye. (The fringes are only circles if the glass is reasonably flat.) The effect is eerie.

Home experiment **5.11 Neon stroboscope.** If you have the bulb NE-40 described in Home Exp. 5.10, you can do other interesting things with it. With the lamp a foot from your eye, look in a direction such that your line of sight makes about a 45-deg angle with the line from your eye to the bulb. Notice the flicker! Now look at the bulb directly. The flicker disappears! Apparently, evolution has developed our peripheral vision so that it is sensitive to very rapid changes of light intensity. This seems wise. (You may also try this with a TV picture. Compare looking directly at it to looking peripherally.) Each plate of the NE-40 turns on and off at 60 cps. But they are 180 deg out of phase! When one is bright, the other is dark. Thus, you can use this bulb as a 60- or 120-cps stroboscope, depending on how you use it to illuminate an object. You can prove that the two plates are out of phase. Screw the bulb into a lamp that is not too heavy, so that you can easily shake it. Turn the lamp so that the two plates are seen edgewise. Now shake the lamp sideways vigorously at about 4 cycles per second (faster if you can) with as large an amplitude as possible (say 10 or 20 cm). Look at the orange streaks made by the plates. Do they occur together, or do they alternate? You can also use this shaking technique, plus an ordinary watch, to estimate the frequency. Assume the motion is sinusoidal. Measure the frequency and amplitude necessary to make the two red streaks look like an "alternating square wave." Since you know the light frequency must be related to 60 cps as some multiple that is an integer, you can easily determine the strobe frequency by this crude measurement.

Note: Rather than shake the lamp, it is easier to look at the reflection of the lamp in a mirror and shake the mirror. You can easily get a nice "alternating square wave" this way with the neon lamp. The same technique can be applied to examining the time structure of the illumination of the TV tube. Mask the TV so as to allow only a vertical strip to be seen. Shake the mirror about a vertical axis. The "sawtooth" that you will see will show you that *some* part of the TV tube is giving off light at any instant. Therefore, to have a good TV stroboscope, you should use a horizontal slit.

5.12 Continuity of a wave at a boundary.   For light (or other electromagnetic radiation) incident from medium 1 to medium 2, we found that, provided the magnetic permeability of the medium is unity (or does not change at the discontinuity) and provided the "geometry" is constant (parallel-plate transmission line of constant cross-sectional shape or slab of material in free space), then the reflection and transmission coefficients for the electric field E_x and magnetic field B_y are given by

$$R_E = \frac{k_1 - k_2}{k_1 + k_2}, \quad T_E = 1 + R_E = \frac{2k_1}{k_1 + k_2},$$

$$R_B = \frac{k_2 - k_1}{k_2 + k_1}, \quad T_E = 1 + R_B = \frac{2k_2}{k_2 + k_1},$$

where $k = n\omega/c$ and n is the index of refraction. Show that the reflection and transmission coefficients for E_x imply that E_x and $\partial E_x /\partial z$ are both continuous at the discontinuity, i.e., that they have the same instantaneous values on either side of the discontinuity. (By the field on the left side (medium 1) we mean, of course, the *superposition* of the incident and reflected waves.) Similarly; show that the reflection and transmission coefficients for the magnetic field B_y imply that B_y is continuous at the boundary but that $\partial B_y/\partial z$ is not continuous. Show that $\partial B_y/\partial z$ increases by a factor $(k_2/k_1)^2 = (n_2/n_1)^2$ in crossing from medium 1 to medium 2. It is important to notice that we mean the total field; not just the part traveling in a particular direction.

5.13 Show that for waves on a string the boundary condition that is analogous to constant magnetic permeability (across a discontinuity) for light is that the mass density of the string be constant. Show that an increase in dielectric constant for light in crossing the boundary is analogous to a decrease in string tension. Show that the transverse string velocity behaves like the magnetic field in a light wave, in the sense that it is continuous but that its z derivative increases by a factor $(k_2/k_1)^2$ in going from medium 1 to 2. Show that the transverse tension $- T_0 \partial \psi/\partial z$ behaves like the electric field, in that both it and its z derivative are continuous at the boundary. (In all cases, we are referring to the total field, not to components traveling in a particular direction.)

5.14 Suppose you have a coaxial cable with vacuum between the conductors and having (for example) 50 ohms characteristic impedance. Now suppose one end of this cable is pressed against a piece of spacecloth having resistance 377 ohms per square. The inner and outer conductors of the cable are thus connected by the spacecloth. At the other end of this cable, measure the DC resistance between the inner and outer conductors, using an ordinary ohmmeter. Neglect the resistance of the conductors themselves(the piece of cable is as short as you please). The resistance is entirely due to the terminating spacecloth. What will the ohmmeter read? (*a*) Guess. (*b*) Prove it.

Home experiment   **5.15   Effective length of open-ended tube for standing waves.**   Use a cardboard tube from a roll of paper towels or wax paper (or a mailing tube). Use a C523.3 tuning fork as a pitch standard. Tap the open-ended tube against your head and listen. Cut off some of the tube (if necessary) so that the pitch is slightly sharp (higher pitch than 523.3 cps). Then insert a slightly smaller tube in the end to act as a tuning element "trombone." (For example, using the cardboard tube from a roll of toilet paper, first cut along the length of the tube and then cut off some material so as to make a smaller tube. Next seal the crack along the seam with tape so that the smaller inner tube has no air outlets along its sides.) The mode you hear is the lowest mode of an open-ended tube. The tube contains one half-wavelength of the oscillation. Sound velocity is 332 meters/sec. Therefore, you "expect" the tube length to be

$$L = \frac{1}{2}\frac{v}{\nu} = \frac{1}{2}\frac{3.32 \times 10^4}{523.3} = 31.7 \text{ cm}$$

Instead you will find that the actual length L_0 is less than 31.7 cm by about 0.6 of a tube diameter. This can be interpreted as an "end effect" of about 0.6 of a tube radius at each end. To check that this is an end effect and not just a wrong value for the velocity of sound, try fat tubes and skinny tubes.

5.16 Resonance in cardboard tubes. Take the tube from Home Exp. 5.15. Hold the vibrating tuning fork near one end of the tube. If the tube's "trombone" extension is tuned for 523.3 Hz (cps), you will hear a nice loud sound. If not, vary the extension and tune for a resonance. *Question*: When the tube's natural vibration pitch differs from that of the tuning fork, what pitch do you hear when you drive the tube with the fork? (First, guess the answer from your knowledge of driven oscillators; then try it.) **Home experiment**

Here is a way to get a nice sharp resonance. Hold the tube vertically and dip the lower end into a sufficiently deep container of water. Hold the vibrating tuning fork at the open end. Raise and lower the tube and fork to tune for resonance. The tube plus one end correction should be $\lambda/4$.

Here is another good resonance. Fill a pop bottle about two-thirds full, so that when you blow across it you get something a little higher than C523.3. Put a straw in the bottle. Hold the vibrating fork at the mouth of the bottle. Tune for resonance by sucking water out with the straw.

You can search for resonances in mailing tubes, jars, rooms, and tunnels by singing a slow ascending "siren." You will both hear and "feel" the strong resonances. The change of impedance may actually "turn you off" or cause you to slide to a neighboring pitch.

5.17 Is your sound-detecting system (eardrum, nerves, brain) a phase-sensitive detector? Let us find out! Some people have said that for high- frequency sound you detect the direction by noticing a time delay between a crest at one ear and a crest at the other; i.e., you detect a phase shift for the vibration of one eardrwn relative to that of the other. The question of whether this is true boils down to this: Can you tell the difference between "both drums in, both drums out, both drums in, both drums out, ..." and "left drum in while right drum out, left drum out while right drum in, ... etc." **Home experiment**

First let us consider an open-ended mailing tube (tuned to C523 to get nice loud sounds). The tube is $\lambda/2$ in length. This means that when air is rushing to the right at the right-hand end, it is simultaneously rushing to the left at the left-hand end; i.e., the velocities at the two ends are 180 deg out of phase when the tube is resonating. To put it differently, the air is rushing *out* of both ends at the same time and *into* both ends at the same time. Now strike two C523 tuning forks on each other at equal distances from the ends of the prongs and put one fork at each end of the tube, so as to get beats. At a maximum in intensity, each fork is helping the air go where it wants to go to resonate; i.e., at the instant when the fork at one end is pushing air into the tube, the fork at the other end is also pushing air into the tube. Half of a fast cycle (at 523 Hz) later, each fork is sucking air out of its end of the tube. Half of a beat cycle later, we have a minimum in intensity of the sound coming out of the tube. (The minimum is zero if you struck the forks at equal distances from the ends.) This is because one fork is pushing air in at one end while the other is sucking it out at the other, which is the opposite of what is desired to sustain the resonance. It destroys the resonance. To put it differently, motion induced by the two forks is composed of two resonance vibrations superposed 180 deg out of phase, giving zero.

The point of all this is that the tube *can* tell whether the forks are vibrating "both in, both out ..." or instead "one in, the other out; one out, the other in;[bu8]etc." In one case you are at a beat maximum, in the other at a minimum. The question about your hearing apparatus is this: If you hold one fork at one ear and the other at the other ear, will you hear beats? Or anything having the mathematical structure of beats? For example, perhaps your system will tell you, "It's coming from the left side of the room; it's coming from the right side of the room; etc.," as you go through what would correspond to beat maxima and minima. That is, if the people who say that sound direction is determined by phase differences are right, the brain might say that if one drum is ahead of the other in phase by say 90 deg, then the sound is coming from the direction of the ear drum that leads by 90 deg. This direction would reverse at the beat frequency. To answer the question, do the experiment.

Another way to phrase the question (with the cardboard tube as an example) is this: Do you have a hole in your head?

Home experiment **5.18 Measuring the relative phase at the two ends of an open tube.** Suppose someone has taken a long hoselike tube, coiled it up in a box, and let one open end stick out one sipe of the box and the other out the other. You are not allowed to see how much of the tube is coiled inside the box. By adding a small tuning trombone to a protruding end, you find that you get a resonance at 523.3 cps from your tuning fork. That means that the total length is either $\frac{1}{2}\lambda$, or λ, or $\frac{3}{2}\lambda$, or .... How can you find out whether the tube is an odd or even number of half-wavelengths? Hold two vibrating forks at one end of the tube and listen to the beats. Get the rhythm in your head so that if you remove one fork momentarily and then replace it (without disturbing the continued vibrations of both forks), you can tell that the beat maximum comes "on the beat" (in musical jargon) just where it should be. Practice several times so that you can skip a beat, count beats in your head, and come back in step when you replace the fork. (You can adjust the rubber-band loading to get a convenient beat frequency. If you find all this difficult, you can use a metronome.) Now! This time, instead of replacing the (momentarily) removed fork at the Same end of the tube, carry it to the other end. Again listen for the beats. (Both forks have continued vibrating all this time.) Do they come back "on the beat," or do they come back "on the off-beat"? Depending on the experimental result, you should be able to decide whether the tube is an odd or even number of half-wavelengths. Predict the answer; then try the experiment with your half-wavelength tube. (Make another tube one wavelength long to get the opposite result.)

5.19   Overtones in tuning fork.   Does your C523.3 tuning fork emit nothing but a 523-cps sound? Strike the fork against something hard. You should hear a faint high tone in addition to the strong 523-cps tone. The high tone dies away in two or three seconds. It is a higher mode of the fork and is strongly damped because it involves greater bending of the prongs. What about the note an octave higher, C1046? This is difficult to listen for because of the presence of the fundamental, C523. To search for it, use a resonating tube. Tune a tube to C1046 by tapping it on your head and listening for the octave above C523. (Or simply cut it by "theory," subtracting 0.6 of a radius R for each end from $\lambda/2$ to get the length.) Hold the C523 fork at the end of the C1046 tube and listen. (Use a tube tuned at C523 as a control. Move the fork back and forth between the C523 and C1046 tubes.)

Home experiment

5.20   General sinusoidal wave.   Write the traveling wave $\psi(z, t) = A \cos(\omega t - kz)$ as a superposition of two standing waves. Write the standing wave $\psi(z, t) = A \cos \omega t \, \cos kz$ as a superposition of two traveling waves traveling in opposite directions. Consider the following superposition of traveling waves:

$$\psi(z, t) = A \cos(\omega t - kz) + RA \cos(\omega t + kz).$$

Show that this sinusoidal wave can be written as a superposition of standing waves given by

$$\psi(z, t) = A(1 + R) \cos \omega t \cos kz + A(1 - R) \sin \omega t \sin kz.$$

Thus the samewave can be thought of as a superposition either of standing waves or of traveling waves.

5.21   Nonreflecting coating.   A glass lens has been coated with a nonreflecting coating that is one quarter-wavelength in thickness *in the coating* for light of *vacuum* wavelength λ_0. The index of refraction of the coating is $\sqrt{n}$; that of the glass is n. Take the index of refraction to be constant, independent of frequency, over the visible frequency spectrum. Let I_{ref} denote the time-averaged reflected intensity and I_0 the incident intensity, for light at normal incidence. Show that the fractional reflected intensity has the following dependence on the wavelength of the incident light:

$$\frac{I_{\text{ref}}}{I_0} = 4\left[\frac{1-\sqrt{n}}{1-\sqrt{n}}\right]^2 \sin^2 \frac{1}{2}\pi\left(\frac{\lambda_0}{\lambda} - 1\right),$$

where λ is the vacuum wavelength of the incident light. Take $n = 1.5$ for glass. Suppose $\lambda_0 = 5500$ Å (green light). Then I_{ref} is zero for green. What is I_{ref}/I_0 for blue light of vacuum wavelength 4500 Å? What is it for red light of vacuum wavelength 6500 Å?

Ans. The red reflected fractional intensity is about 2×10^{-3}; the blue reflected intensity is about twice that for red. (*Your* answers should contain two significant figures.)

5.22 Impedance matching by "tapered" index of refraction.
Suppose you want to match optical impedances between a region of index n_1 and a region of index n_2, and you want to expend a total distance L in the impedance-matching transition region. What is the optimum z dependence of the index n between the two regions? Is it exponential? Why not?

Ans. The wavelength $\lambda = (c/v)/n$ should vary linearly with z, *i.e.*, if the transition region extends from $z = 0$ to $z = L$, we want $\lambda(z) = \lambda_1 + (z/L)$ $(\lambda_2 - \lambda)$.

5.23 The prettiest white-light fringe.
Look at the concentric interference fringes of two microscope slides pressed together. The center of the pattern is black (i.e., shows no reflected sky). The first fringe is white. Then the fringes get colorful. After a dozen fringes, they get all mixed up and overlapping and are again white. Which fringe (roughly) is the most monochromatic looking? To be more precise: define the "prettiest" fringe to be one that is "not red and not blue," where red has wavelength 0.65 μ (μ designates micron; one micron = 10^{-6} meter), blue has wavelength 0.45 μ, and "not present" means that, for the prettiest fringe, both red and blue give zero by destructive interference. Give the prettiest fringe number to the nearest integer, the wavelength for which the interference in this fringe is completely constructive, and the approximate color.

5.24 Interference in thin films.
Show that for monochromatic light at normal incidence the reflected intensity from a layer of air of thickness L between two glass microscope slides is given in the small-reflection approximation by

$$\frac{I_{\text{ref}}}{I_0} \approx 4R_{12}^2 \sin^2 k_2 L.$$

(Neglect the interference effects from the two outer surfaces of the two slides. Those fringes would be washed out by the spread in colors of any but very monochromatic light, as discussed under Home Exp. 5.10, "Fabry-Perot hinges in a window pane.")

5.25 Fabry-Perot fringes in a 1-mm glass slide. Derive the result that, for light to produce Fabry-Perot interference fringes in a glass slide 1 mm thick, the "line width" (i.e., bandwidth) of the light must be less than about 3 inverse centimeters in order for the fringes not to be washed out.

5.26 Multiple reflection. In the following derivations you are to use complex numbers. Suppose ψ_{inc} is the real part of $Ae^{i(\omega t - kz)}$, where A is real. Thus $\psi_{inc} = A \cos(\omega t - kz)$. At $z = 0$ the impedance suffers a sudden change from Z_1 to Z_2. At $z = L$ the impedance changes again from Z_2 to Z_3. Let $R_{12} = (Z_1 - Z_2)/(Z_1 + Z_2) = -R_{21}$, $R_{23} = (Z_2 - Z_3)/(Z_2 + Z_3)$. Assume that in medium 1 there is a reflected wave that is the real part of $RAe^{i(\omega t - kz)}$, where R is complex, and may be written $R = |R| e^{-i\delta}$.

(*a*) Show that if we neglect all contributions except the reflection from $z = 0$ and the first reflection from $z = L$, we obtain

$$R = R_{12} + T_{12}R_{23}T_{21}e^{-2ik_2L},$$

were $T_{12} = 1 + R_{12}$ and $T_{21} = 1 + R_{21} = 1 - R_{12}$.

(*b*) Show by explicit summation of the infinite series corresponding to an infinite number of multiple reflections that the exact solution for R is

$$R = R_{12} + \frac{\left(1 - R_{12}^2\right)R_{23}e^{-2ik_2L}}{1 - R_{23}R_{21}e^{-2ik_2L}},$$

where the first term, R_{12}, is due to the reflection at the first discontinuity at $z = 0$, and the rest is due to one or more reflections at $z = L$. Show that in the small reflection approximation this result reduces to that of part (*a*). Show that the exact result can be written in the form

$$R = \frac{R_{12} + R_{23}e^{-2ik_2L}}{1 + R_{12}R_{23}e^{-2ik_2L}}.$$

Show that this exact expression for R vanishes for the same combinations of R_{23}/R_{12} and k_2L as does the approximate expression for R obtained in the "small-reflection approximation" used in Sec. 5.5. Thus the approximate expression gives the zeroes correctly, but is inexact as to the intensity at the maxima.

5.27 Boundary-condition method for reflection and transmission coefficients. The physical situation is exactly as in Prob. 5.26. The method of solution will be completely different. Instead of summing an infinite series of multiply-reflected rays we take the following approach: Each "ray" of the superposition of multiply-reflected rays is continuous. Therefore the superposition itself is continuous. That is the basis of the method. Thus we do not bother with summing over multiple reflections. Instead we write $\psi(z, t)$ in the three regions: 1 ($z < 0$), 2 ($z = 0$ to L), and 3 ($z > -L$) as the real part of

$$\psi_1(z, t) = e^{i(\omega t - k_1 z)} + Re^{i(\omega t + k_1 z)},$$

$$\psi_2(z, t) = Fe^{i(\omega t - k_2 z)} + Be^{i(\omega t + k_2 z)},$$

and

$$\psi_2(z, t) = Te^{i[\omega t - k_3(z-L)]},$$

where R (reflected), F (forward), B (backward), and T (transmitted) are unknown complex numbers to be determined. (For simplicity we have taken the amplitude of the incident wave to be unity.) Notice that the term with complex amplitude F is the superposition of all of the multiply-reflected rays between $z = 0$ and L that are going in the forward direction at time t. Similarly the term with complex amplitude B is the superposition of all the backward-going rays. At the two discontinuities, $z = 0$ and $z = L$, you are to apply the boundary conditions of continuity. Assume $\psi(z, t)$ is continuous, and assume $\partial\psi(z, t)/\partial z$ is continuous. (This means that the string tension is constant, if we have a string; or that the equilibrium pressure p_0 times the factor γ constant, for sound waves; or that the magnetic permeability μ is constant, for electromagnetic waves.) These two boundary conditions at each of the two places give four linear equations in the four complex numbers T, F, B, and R. That is sufficient to determine T, F, B, and R uniquely. Justify that statement. Find T, F, B, and R. Show that your result for R is identical with that obtained by the method of multiple reflection, in Prob. 5.26.

5.28 Transmission resonance. (*a*) Show that for reflection due to two discontinuities (Prob. 5.26 and 5.27) the fractional time-averaged energy flux that is not reflected (and hence by energy conservation must be transmitted) is given by

$$1 - |R|^2 = \frac{\left(1 - R_{12}^2 - R_{23}^2 + R_{12}^2 R_{23}^2\right)}{1 + 2R_{12}R_{23}\cos 2k_2 L + R_{12}^2 R_{23}^2}.$$

(*b*) Show that if medium 3 has the same impedance as medium 1 this becomes

$$1 - |R|^2 = \frac{\left(1 - R_{12}^2\right)^2}{1 - 2R_{12}^2 \cos 2k_2 L + R_{12}^4}.$$

(*c*) Show that at certain values of $k_2 L$ the fractional time-averaged energy flux not reflected is unity, i.e., for those values, all the energy is transmitted and none reflected. Call anyone of these "resonance values" of k_2 by the name k_0. Show that the resonance values are given by $k_0 L = \pi, 2\pi, 3\pi$, etc.

(*d*) Show that for k_2 sufficiently near a resonance value k_0 the transmitted (time averaged) energy flux is given by

$$1 - |R|^2 \approx \frac{\left(1 - R_{12}^2\right)^2}{\left(1 - R_{12}^2\right)^2 + R_{12}^2 \left[2L\left(k_2 - k_0\right)\right]^2}.$$

Show that this form is that of a "Breit-Wigner resonance shape," as discussed in Sec.3.2, and has a full width at half-maximum transmitted intensity Δk_2, given by

$$\left(\Delta k_2\right) L \approx \frac{\left(1 - R_{12}{}^2\right)}{|R_{12}|},$$

provided that $|R_{12}|$ is not too much less than unity. (Show that for $|R_{12}| \ll 1$, the Breit-Wigner approximation is useless, because it does not hold except very near k_0, i.e., it doesn't hold out even to the "half-maximum transmitted power" points. In that case one should use the exact result.) Show that for $|R_{12}| \approx 1$, when the Breit-Wigner shape holds for many resonance widths away from k_0, the resonance full width is given by

$$\left(\Delta k_2\right) L \approx 2\left(1 - |R_{12}|\right).$$

5.29 Suppose that instead of tying the semi-infinite string directly to the output of the driving mechanism, it is coupled to the transmitter through a spring as follows:

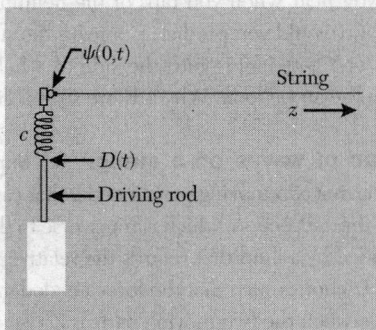

$\psi(0,t)$

String

$z \longrightarrow$

c

$D(t)$

Driving rod

Problem 5.29

The string tension is T, the mass density of the string is ρ, and the spring constant is K. The length of the spring is such that, if the displacement of the driving rod $D(t)$ is zero and the spring is relaxed, then $\psi(0, t)$ is zero. The motion of the rod is given by $D(t) = A \cos \omega t$. Assume there is a traveling wave of the form $\psi(z, t) = B \cos(\omega t - kz + \varphi)$. The problem is to figure out what the "boundary condition" is at $z = 0$ and then to apply it to find B/A and φ. (*Hint:* The algebra is easier if you use complex numbers.)

Ans. $\quad \tan \varphi = -\omega(T\rho)^{1/2}/K, B/A = \left[1 + \left(\omega^2 T\rho/K^2\right)\right]^{-1/2} = \cos \varphi.$

Notice that for K huge we get $\psi(0, t) = D(t)$ as we should expect. Why?

5.30 Suppose that a point a on a string at $z_a = 10$ cm oscillates in harmonic motion at frequency 10 cps with amplitude 1 cm. Its phase is such that at

$t = 0$ the point on the string is passing through its equilibrium position with upward velocity (positive displacement is upward).

(*a*) What is the magnitude and direction of the velocity of point *a* at $t = 0.05$ sec? Suppose the string parameters (mass per unit length and tension) are such that the wave velocity is 100 cm/sec.

(*b*) What is the wavelength of a traveling wave? What is the wavelength of a standing wave?

(*c*) Another point *b* at $z_b = 15$ cm oscillates with the same amplitude as that at $z_a = 10$ cm, but with a relative phase of 180 deg with respect to the oscillation at z_a. Can you tell whether we have here a pure traveling wave, a pure standing wave, or a combination?

(*d*) A third point *c* at 12.5 cm also oscillates with the same amplitude as that at *za* but 180 deg out of phase with point *a*. Point *b* oscillates as given above. Now tell us whether the wave is a traveling or a standing wave (or a combination).

Home experiment   **5.31 Resonances in toy balloons.** Get a helium-filled balloon. Hold it near your ear and tap it. Sing into the side of it and search for resonant pitches. Blow up another balloon with air to the same diameter as the helium balloon. Tap it. Estimate the ratio of frequencies of the lowest modes (the ones you hear when you tap) of the helium and air balloons. What frequency ratio would you predict: Compare the strength (loudness) of the resonances you get singing into the side of a helium balloon with those you get from the air balloon. Why is there such a difference?

5.32 Termination of waves on a string. (*a*) Suppose you have a massless dashpot having two moving parts 1 and 2 that can move relative to one another along the *x* direction, which is transverse to the string direction *z*. Friction is provided by a fluid that retards the relative motion of the two moving parts. The friction is such that the force needed to maintain relative velocity $\dot{x}_1 - \dot{x}_2$ between the two moving parts is $Z_d(\dot{x}_1 - \dot{x}_2)$, where Z_d is the impedance of the dashpot. The input (part 1) is connected to the end of a string of impedance Z_1 stretching from $z = -\infty$ to $z = 0$. The output (part 2) is connected to a string of impedance Z_2 that extends to $z = +\infty$. Show that a wave incident from the left experiences an impedance at $z = 0$ which is the same as that it would experience if connected to a "load" consisting of a string stretching from $z = 0$ to $+\infty$ and having impedance Z_L given by

$$Z_L = \frac{Z_d Z_2}{Z_d + Z_2}, \quad \text{that is,} \quad \frac{1}{Z_L} = \frac{1}{Z_d} + \frac{1}{Z_2}.$$

Thus it is as if the dashpot and string 2 were impedances connected "in parallel" and driven by the incident wave.

(*b*) Show that if string Z_2 extends only to $z = \frac{1}{4}\lambda_2$, where λ_2 is the wavelength in medium 2 (assuming we have a harmonic wave with a single frequency), and there is terminated by a dashpot of zero impedance (frictionless), the wave incident at $z = 0$ is perfectly terminated. Show that the output connection of the dashpot at $z = 0$ cannot tell whether it is connected to a string of infinite impedance or is instead connected to a quarter-wavelength string that is "short-circuited" by a frictionless dashpot at $z = \frac{1}{4}\lambda_2$. In either case the output connection remains at rest.

5.33 Acoustical properties of rooms. The acoustical properties of a room are determined mainly by the "reverberation time" as a function of frequency. Suppose the room is driven at steady state at a given frequency. Then the driving force (which may be an electrically driven organ pipe) is suddenly turned off. The stored sound energy will decay approximately exponentially with mean lifetime τ given by

$$\frac{1}{\tau} = \frac{1}{E_{\text{stored}}}\frac{dE_{\text{lost}}}{dt}.$$

At least that is the way a one-dimensional harmonic oscillator behaves, and we can assume that the room behaves that way also. Let ρ_E denote the sound energy density, and V denote the volume of the room. What is the stored energy? For a plane traveling wave, the energy flux (J/cm² sec) is the energy density times the velocity of sound, $v = 332$ m/sec. The sound waves in the room are not traveling waves, but they can be thought of as a superposition of traveling waves traveling in all directions. Approximately one-sixth of the energy can be thought of as traveling in each of the six directions, i.e., along the plus and minus x, y, and z axes.

Energy flux traveling in the $+x$ direction and encountering an open window travels out the window and is lost. An open window is said to have an absorption coefficient $a = 1.0$. The walls (and ceiling and floor) have a total area A that can be thought of as a sum of areas A_1, A_2, etc., with absorption coefficients a_1, a_2, etc. Derive the following approximate expression for the mean decay time τ:

$$\tau \sum (A_i a_i) \approx \frac{6V}{v}$$

where the sum covers all the surfaces of the room. See the accompanying table of absorption coefficients (Table 5.1).

In 1895 Wallace Sabine was asked to "do something" about the terrible acoustical properties of the lecture room in Harvard's new Fogg Art Museum, which had just been completed. We now ask you to estimate how bad it was (i.e., the duration of the residual sound), given the following information (from W. C. Sabine, *Collected Papers on Acoustics*, p. 30, Dover, 1964): $V = 2740$ m³; shape approximately a cube; plaster walls and ceiling, wood floor. Also take the "audible time duration" to be about four times τ. Sabine's experiments used human listeners as detectors. His experimental result for the audible time duration was 5.61 sec (which he reduced to 0.75 sec by adding various absorbing materials).

Table 5.1 Absorption coefficients, a_i, for $v = 512$ cps

Open window	1.00
Carpet rugs	0.20
Linoleum	0.12
Hair felt, 1 in. thick	0.78
Audience, per person (taking each person to have effective floor area 1 m²)	0.44
Wood	0.061
Plaster	0.033
Glass	0.027

SOURCE: Wallace C. Sabine, *Collected Papers on Acoustics*, pp. 223f. (Dover Publications, New York, 1964). A classic work.

Chapter 6

Modulations, Pulses, and Wave Packets

Chapter 6   Modulations, Pulses, and Wave Packets

6.1 Introduction

Up until now we have mainly studied waves and oscillations having a harmonic time dependence of $\cos(\omega t + \varphi)$ with a single frequency ω present. An exception was our study of beats in Sec. 1.5. There we learned that the superposition of two harmonic oscillations with nearly, but not quite, the same frequency leads to the very interesting phenomenon of beats. Chapter 6 is an extension of our study of beats. We shall study beats in space as well as in time, and beats resulting from the superposition of many frequency components as well as from just two components. We shall also study how the beats (or more generally "modulations," for more than two frequency components) propagate as traveling waves. It turns out that the modulations, called *wave groups* or *wave packets*, carry energy as they propagate and travel at the *group velocity*.

The best way for you to obtain personal experience with wave packets is to pitch pebbles into ponds and watch the expanding circular wave packets. (Dropping water droplets into dishes also works very well.) It is obvious that these expanding circular wave packets carry energy—they can set a distant cork to bobbing when the wave packet arrives. If you look closely, you will see that the little wavelets that make up the wave packet do not maintain constant positions relative to the packet. For water wave packets with wavelet wavelengths of more than a few centimeters, the wavelets travel almost twice as fast as the packet. They are "born" at the rear of the packet, travel to the front, and dwindle away. The wavelets travel at the phase velocity. The wave packet as a whole travels .at the group velocity.

We urge the reader to fill a bowl or tub with water and make some wave packets. (At first, you may not be able to seethe relative motion of wavelets .and wave packet. This is most easily seen by throwing pebbles or water drops in a large pone where you can follow the wave progress for a number of seconds. There is too little time in a small bowl.)

6.2 Group Velocity

We encountered several examples in Chap. 4 which showed that the phase velocity of a sinusoidal traveling wave is not necessarily the velocity at whi.ch energy or information is transported. For example, we found that the phase velocity of light in the ionosphere is greater than c. If signals could be propagated at a speed greater than c, the theory of relativity would be wrong.

Modulations carry signals. A signal cannot be sent with a harmonic traveling wave involving only a single frequency. That is because a harmonic traveling wave goes on and on forever, each cycle like the preceding cycle. It conveys no information except that it is there, so to speak. If you want to send a message, you must *modulate* the wave, which means to change something about it in a way that can be decoded at a distant "receiver." You may change the amplitude; that is called *amplitude modulation.* For example, you may modulate the amplitude so as to send a series of dots and dashes in Morse code, each pattern of dots and dashes representing a letter of the alphabet. Alternatively, you may vary the frequency or the phase constant in some way that can be decoded; those ways are called frequency modulation and phase modulation, respectively. In any of these cases, the driving force is not given by a simple harmonic force.

In order to discover how signals propagate, we must study the traveling waves emitted into an open medium by a transmitter at $z = 0$ whose displacement $D(t)$ does not have the simple harmonic time dependence $D(t) = A \cos \omega t$, but rather a more complicated dependence, $D(t) = f(t)$. It turns out that a wide class of functions $f(t)$ can be expressed as linear superpositions (sums) of harmonic functions of the form $A(\omega) \cos [\omega t + \varphi(\omega)]$, where the amplitude $A(\omega)$ and the phase constant $\varphi(\omega)$ are different for each frequency ω present and are determined by the function $f(t)$ that is to be expressed as a superposition. We will later study how to determine the amplitudes $A(\omega)$ and phase constants $\varphi(\omega)$ by the methods of Fourier analysis. For the present, let us consider a superposition containing only two terms. That will be sufficient to give us some very interesting results, which will eventually lead us to an understanding of how a pulse or wave group propagates in a dispersive medium (one where the phase velocity depends on wavelength).

Superposition of two harmonic oscillations to give amplitude-modulated oscillation. Let us assume that a transmitter at $z = 0$ drives a string stretching from $z = 0$ to $+ \infty$. The transmitter oscillates in a superposition of two harmonic motions with angular frequencies ω_1 and ω_2. We shall not miss any interesting results if we let the amplitudes and phase constants of the two contributions be the same. Therefore we assume that for the displacement of the oscillating output terminal of the transmitter we have

$$D(t) = A \cos \omega_1 t + A \cos \omega_2 t. \qquad (1)$$

From our previous study of beats [see Sec. 1.5, Eqs. (1.80) through (1.85)], we know that the superposition given by Eq. (1) can be written in the form of an *amplitude-modulated oscillation,*

$$D(t) = A_{\text{mod}}(t) \cos \omega_{\text{av}} t, \tag{2}$$

where

$$A_{\text{mod}}(t) = 2A \cos \omega_{\text{mod}} t, \tag{3}$$

with

$$\omega_{\text{mod}} = \tfrac{1}{2}(\omega_1 - \omega_2),$$

$$\omega_{\text{av}} = \tfrac{1}{2}(\omega_1 + \omega_2). \tag{4}$$

If ω_1 and ω_2 are of comparable magnitude, the modulation frequency ω_{mod} is small compared with the average frequency ω_{av}. Then the form of Eq. (2) can be thought of as an *almost harmonic oscillation* at frequency ω_{av} with an amplitude that is almost, but not quite, constant; it is harmonically *amplitude-modulated* at the relatively slow modulation frequency ω_{mod}.

We have in Eqs. (2) and (3) the simplest possible amplitude-modulated oscillation, in that it involves only a single modulation frequency ω_{mod}. A more general amplitude-modulated oscillation would have the form of Eq. (2) but with $A_{\text{mod}}(t)$ given by a superposition of many different terms similar in form to Eq. (3), each term having its own modulation frequency and its own amplitude and phase constant. For example, in AM radio ν_{av} would be the "carrier frequency" of perhaps 1000 kc (kilocycles per second). The modulation frequencies would be audible frequencies in the range from 20 cps to 20 kc.

Superposition of two sinusoidal traveling waves to give amplitude-modulated traveling wave. Let us examine the traveling waves radiated by a transmitter whose output terminal oscillates with the time dependence given by Eqs. (1) or (2). The medium is coupled to the transmitter in such a way that, at $z = 0$, $\psi(z, t)$ is given by

$$\psi(0, t) = D(t) = A \cos \omega_1 t + A \cos \omega_2 t. \tag{5}$$

Because the waves satisfy the superposition principle, the two contributions to the transmitter displacement given by the linear superposition Eq. (5) will give two "independent" traveling waves. Thus, the traveling wave $\psi(z, t)$ will be the superposition of the two sinusoidal traveling waves $\psi_1(z, t)$ and $\psi_2(z, t)$ that would be present if one or the other of the transmitter oscillations $A \cos \omega_1 t$ or $A \cos \omega_2 t$ were present by itself. We know that $\psi_1(z, t)$ is obtained from $\psi_1(0, t)$ by replacing $\omega_1 t$ by $\omega_1 t - k_1 z$. This just expresses the fact that the phase velocity is ω_1/k_1. Similarly, $\psi_2(z, t)$ is obtained by replacing $\omega_2 t$ by $\omega_2 t - k_2 z$. Thus the traveling wave $\psi(z, t)$ is given by making both of those replacements in Eq. (5):

$$\psi(z, t) = A \cos(\omega_1 t - k_1 z) + A \cos(\omega_2 t - k_2 z). \qquad (6)$$

Of course, we can make these same substitutions of $\omega_1 t - k_1 z$ for $\omega_1 t$ and $\omega_2 t - k_2 z$ for $\omega_2 t$ in Eqs. (2), (3), and (4) to find the form of the traveling wave analogous to the amplitude-modulated almost harmonic oscillation given by Eqs. (2), (3), and (4). In this way, we obtain the *amplitude-modulated, almost sinusoidal traveling wave*

$$\psi(z, t) = A_{\mathrm{mod}}(z, t) \cos(\omega_{av} t - k_{av} z), \qquad (7)$$

where (as you can easily show)

$$A_{\mathrm{mod}}(z, t) = 2A \cos(\omega_{\mathrm{mod}} t - k_{\mathrm{mod}} z), \qquad (8)$$

with

$$\omega_{\mathrm{mod}} = \tfrac{1}{2}(\omega_1 - \omega_2),$$
$$k_{\mathrm{mod}} = \tfrac{1}{2}(k_1 - k_2), \qquad (9)$$
$$\omega_{av} = \tfrac{1}{2}(\omega_1 + \omega_2),$$
$$k_{av} = \tfrac{1}{2}(k_1 + k_2). \qquad (10)$$

Notice that $\omega_{av} t - k_{av} z$ is obtained where we had $\omega_{av} t$ by replacing $\omega_1 t$ with $\omega_1 t - k_1 z$ and by replacing $\omega_2 t$ with $\omega_2 t - k_2 z$. Similarly, $\omega_{\mathrm{mod}} t - k_{\mathrm{mod}} z$ is obtained where we had $\omega_{\mathrm{mod}} t$ by making these same replacements.

Modulation velocity. Now we ask the very interesting question: at what velocity do the modulations propagate? Suppose that ω_{mod} is small compared with ω_{av}. Then the transmitter output at $z = 0$ has the form of amplitude-modulated oscillations (shown in Fig. 1.13, Sec. 1.5). The question is at what velocity a given modulation wave crest [i.e., a place where $A_{\mathrm{mod}}(z, t) = +1$] propagates. The answer is given by inspection of Eq. (8); we see that to follow a given constant value (such as a crest) of the modulation amplitude $A_{\mathrm{mod}}(z, t)$, we need to maintain a constant value for its argument $\omega_{\mathrm{mod}} t - k_{\mathrm{mod}} z$. Thus when t increases by dt, z must increase by dz in such a way that the increment of $\omega_{\mathrm{mod}} t - k_{\mathrm{mod}} z$, namely $\omega_{\mathrm{mod}} dt - k_{\mathrm{mod}} dz$, remains zero:

$$\omega_{\mathrm{mod}} dt - k_{\mathrm{mod}} dz = 0. \qquad (11)$$

To satisfy this condition, we must travel at the *modulation velocity*,

$$\boxed{\frac{dz}{dt} = v_{\mathrm{mod}} = \frac{\omega_{\mathrm{mod}}}{k_{\mathrm{mod}}} = \frac{\omega_1 - \omega_2}{k_1 - k_2}.} \qquad (12)$$

Now, ω and k are related by a dispersion relation:

$$\omega = \omega(k). \tag{13}$$

This dispersion relation gives ω_1, once k_1 is specified, and it gives ω_2, once k_2 is specified:

$$\omega_1 = \omega(k_1), \quad \omega_2 = \omega(k_2). \tag{14}$$

Therefore, the modulation velocity given by Eq. (12) can be expressed (by using a Taylor's series expansion of $\omega(k)$ at $k = k_{av}$)

$$v_{mod} = \frac{\omega(k_1) - \omega(k_2)}{k_1 - k_2} = \frac{d\omega}{dk} + \cdots, \tag{15}$$

where the derivatives of the function $\omega(k)$ are evaluated at the average wavenumber k_{av}.

Group velocity. In most of the interesting applications of Eq. (12), ω_1 and ω_2 differ by only a small fraction of their average value. Then we can neglect all terms except the first one in Eq. (15). The quantity $d\omega/dk$, evaluated at a suitable average value of k, is called the *group velocity*:

$$\boxed{\text{Group velocity} \equiv v_g = \frac{d\omega}{dk}.} \tag{16}$$

Thus we see that a "signal" consisting of a wave crest of the modulation amplitude propagates, not at the average phase velocity $v_{av} = \omega_{av}/k_{av}$, but at the group velocity $v_g = d\omega/dk$.

In Fig. 6.1, we show the propagation of the traveling wave $\psi(z, t)$ given by Eq. (7) or Eq. (6), with the specifications that the average frequency be 8 times the modulation frequency and that the group velocity $d\omega/dk$ (evaluated at the average frequency) be equal to half of the phasevelocity ω_{av}/k_{av}.

Here is a briefer derivation of the modulation velocity. The difference in phase between waves 1 and 2 of the superposition Eq. (6) is given by

$$\varphi_1(z, t) - \varphi_2(z, t) = (\omega_1 t - k_1 z + \varphi_1) - (\omega_2 t - k_2 z + \varphi_2)$$
$$= (\omega_1 - \omega_2)t - (k_1 - k_2)z + (\varphi_1 - \varphi_2).$$

Now, at certain values of the phase difference $\varphi_1(z, t) - \varphi_2(z, t)$, the two components are in phase, producing constructive interference and a maximum of the magnitude of the modulation amplitude. At other values of the phase difference $\varphi_1(z, t) - \varphi_2(z, t)$, they are out of phase, producing destructive interference and zeros of the

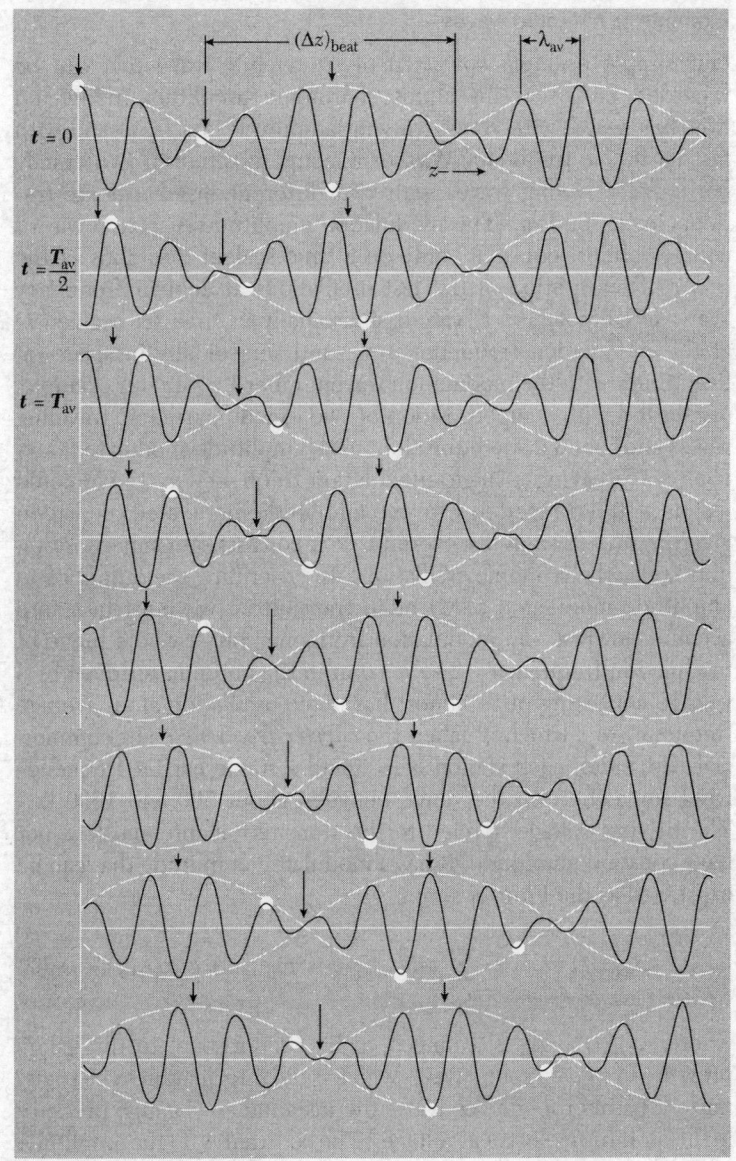

Fig. 6.1 *Group velocity. The arrows follow the beats, which travel at the group velocity v_g. The white circles follow individual wave crests, which travel at the average phase velocity v_{av}.*

modulation amplitude. Thus, in order to travel at the speed of a modulation, we should travel at the speed that corresponds to maintaining constant phase difference $\varphi_1(z,t) - \varphi_2(z,t)$. Therefore we take the total differential of the above expression and set it equal to zero:

$$(\omega_1 - \omega_2)dt - (k_1 - k_2)dz = 0.$$

The modulation velocity is dz/dt; that gives Eq. (12).

Example 1: AM radio waves

Our simple example consisted of a traveling wave that can be regarded *either* as an almost harmonic amplitude-modulated traveling wave, with slowly varying amplitude $A_{mod}(z, t)$ and with fast harmonic frequency Way, or asa superposition of two exactly harmonic traveling waves with two different fast harmonic frequencies ω_1 and ω_2. The modulation amplitude $A_{mod}(z, t)$ was of course "almost constant" only on a time scaleof durations of the order of one period of the fast oscillations at angular frequency Way. Actually $A_{mod}(z, t)$ varied sinusoidally in time (at a given z) at the modulation frequency ω_{mod}, and sinusoidally in space (at fixed time) with the modulation wavenumber k_{mod}. In our example, we started with a superposition of two exactly harmonic traveling waves and found it was equivalent to an amplitude-modulated traveling wave having a single modulation frequency ω_{mod}. We could just as well have started with the amplitude modulated oscillation given by Eq. (2) and discovered that it consists of a superposition of two exactly harmonic oscillations. In describing the output of an amplitude-modulated (AM) radio transmitter, we must take into account not just one modulation frequency but a whole range of modulation frequencies. The current in the antenna is driven by a voltage consisting of an almost harmonic oscillation at an average frequency ω_{av}, which is called the *carrier frequency*. (In commercial AM radio, each station is assigned a single carrier frequency lying somewhere in the range between about 500 and 1600 kc.) The driving voltage applied to the transmitter antenna does not have constant amplitude. It has a modulated amplitude that can be expressed as the Fourier series

$$A_{mod}(t) = A_0 + \sum_{\omega_{mod}} A(\omega_{mod}) \cos\left[\omega_{mod}t + \varphi(\omega_{mod})\right], \quad (17)$$

where $A_{mod}(t) - A_0$ is arranged to be proportional to the gauge pressure in a given sound wave, which is the information to be transmitted. (A microphone converts the instantaneous gauge pressure in the air into an electrical voltage.) The constant A_0 in the amplitude of the driving voltage gives a contribution that is present whether or not anyone is talking or singing into the microphone. The remaining terms are due to the sound waves picked up by the microphone. The modulation frequencies in Eq. (17) are thus the frequencies of the sound waves. They lie in the audible range from 20 to 20,000 cps and are called "audio" frequencies. The audio frequencies are small compared with the carrier frequency. The driving voltage $V(t)$ is given by an almost harmonic oscillation at frequency ω_{av}:

$$V(t) = A_{mod}(t) \cos \omega_{av}t$$

$$= A_0 \cos \omega_{av}t + \sum_{\omega_{mod}} A(\omega_{mod}) \tag{18}$$
$$\cos[\omega_{mod}t + \varphi(\omega_{mod})] \cos \omega_{av}t.$$

This expression may be written as a *superposition of exactly harmonic oscillations,*

$$V(t) = A_0 \cos \omega_{av}t$$

$$+\Sigma \tfrac{1}{2} A(\omega_{mod}) \cos[(\omega_{av} + \omega_{mod})t + \varphi(\omega_{mod})]$$

$$+\Sigma \tfrac{1}{2} A(\omega_{mod}) \cos[(\omega_{av} - \omega_{mod})t - \varphi(\omega_{mod})]. \tag{19}$$

Sidebands. The amplitude-modulated voltage $V(t)$ is thus a superposition of harmonic oscillations consisting of a single term with frequency ω_{av}, called the *carrier oscillation*, a sum of many harmonic oscillations with frequencies $\omega_{av} + \omega_{mod}$, called the *upper sideband*, and a sum of many harmonic oscillations with frequencies $\omega_{av} - \omega_{mod}$, called the *lower sideband*. In order to radiate traveling waves that carry all the information in sounds in the audio frequency range from zero to 20 kc, the voltage $V(t)$ must consist of a superposition of harmonic components with angular frequencies ω in a frequency range from the lowest frequency of the lower sideband to the highest frequency of the upper sideband. Thus the radiated frequencies occupy the *frequency band*

$$\omega_{av} - \omega_{mod}(\max) \le \omega \le \omega_{av} + \omega_{mod}(\max), \tag{20}$$

i.e.,

$$v_{av} - v_{mod}(\max) \le v \le v_{av} + v_{mod}(\max). \tag{21}$$

Bandwidth. The maximum frequency minus the minimum frequency is called the *bandwidth*:

$$\text{Bandwidth} \equiv \Delta v = v(\max) - v(\min) = 2v_{mod}(\max). \tag{22}$$

Thus to transmit the carrier and two sidebands due to amplitude modulations occupying the complete audible frequency range requires a bandwidth of twice 20 kc or 40 kc. (Actually, commercial AM radio stations are only allowed to broadcast a bandwidth of 10 kc. Thus they can only carry audible information in the range zero to 5 kc. This is completely adequate for ordinary speech and fairly adequate for music; the highest note on the piano has frequency about 4.2 kc.)

The "music" travels at the group velocity. The driving force $V(t)$ given by Eqs. (18) or (19) leads to radiation of electromagnetic traveling waves. These waves, may be regarded as a superposition of harmonic components occupying a certain frequency band $\Delta\omega$ centered at ω_{av}. Alternatively, they may be regarded as a single almost harmonic traveling wave having a "fast" oscillation frequency Wayequal to the carrier frequency and having an "almost constant," slowly varying amplitude $A_{mod}(z, t)$, consisting of a superposition of terms like that in Eq. (8). (In that example, where there are only two harmonic components, the upper sideband consists of the single frequency $\omega_1 = \omega_{av} + \omega_{mod}$; the lower sideband consists of the single frequency $\omega_2 = \omega_{av} - \omega_{mod}$.) The modulations propagate through the medium (air, ionosphere,...) at the modulation velocity. In the case of an AM radio station with (for example) a carrier frequency of 1000 kc and a bandwidth of 10 kc, the frequency band extends from 995 kc to 1005 kc. Since the bandwidth is small compared with the average frequency, we expect that the neglected higher terms in the Taylor's series expansion [Eq. (15)] are indeed negligible, and the group velocity as given by Eq. (16) is completely adequate to describe the propagation of the modulations.

Frequency modulation, phase modulation, and related topics are discussed in Probs. 6.27 to 6.32. (There is another important modulation technique, called *pulse-code modulation.*†

Now we shall consider some physical examples of group velocities. In the cases that involve electromagnetic traveling waves, we shall not limit ourselves to frequencies of AM radio ($\nu \sim 10^3$ cps) but shall also include visible light ($\nu \sim 10^{15}$ cps), microwaves ($\sim 10^{10}$ cps), and other frequencies.

Example 2: Electromagnetic radiation in vacuum

The dispersion relation is given by

$$\omega = ck. \tag{23}$$

The phase and group velocities are given by

$$v_\varphi = \frac{\omega}{k} = c, \qquad v_g = \frac{d\omega}{dk} = c. \tag{24}$$

Thus the phase and group velocities are both equal to c for light (or other electromagnetic radiation) in vacuum. Modulations propagate at velocity c.

---

† J. S. Mayo, "Pulse-Code Modulation," *Scientific American*, p. 102 (March 1968).

Example 3: Other nondispersive waves

Light waves in vacuum are nondispersive; i.e., the phase velocity does not depend on frequency (or on wavenumber). Whenever that is the case, the group velocity equals the phase velocity, since in general we have

$$\omega = v_\varphi k, \tag{25}$$

$$v_g = \frac{d\omega}{dk} = v_\varphi + k\frac{dv_\varphi}{dk}. \tag{26}$$

Thus the group and phase velocities are equal if dv_φ/dk is zero. Other examples of nondispersive waves are audible *sound waves*, where we have

$$\omega = \sqrt{\frac{\gamma p_0}{\rho_0}}\, k, \tag{27}$$

and *transverse waves on a continous string*, where we have

$$\omega = \sqrt{\frac{T_0}{\rho_0}}k. \tag{28}$$

Example 4: Electromagnetic waves in the ionosphere

The dispersion relation for sinusoidal waves is

$$\omega^2 = \omega_p{}^2 + c^2 k^2 \tag{29}$$

for frequencies exceeding the cutoff frequency, $v_p \approx 20\,\mathrm{Mc}$. Differentiating Eq. (29) with respect to k gives

$$2\omega\frac{d\omega}{dk} = 2c^2 k, \tag{30}$$

i.e.,

$$\left(\frac{\omega}{k}\right)\left(\frac{d\omega}{dk}\right) = v_\varphi v_g = c^2. \tag{31}$$

Thus the phase velocity and group velocity are given by

$$v_\varphi = \sqrt{c^2 + \frac{\omega_p{}^2}{k^2}} \geq c,$$

$$v_g = c\left(\frac{c}{v_\varphi}\right) \leq c. \tag{32}$$

We see that although the phase velocity always exceeds c, the group velocity is always less than c. Thus a *signal* cannot be transmitted at a velocity greater than c.

Example 5: Surface waves on water

At equilibrium, the surface of a body of water is flat and horizontal. When a wave is present, there are two kinds of restoring forces that tend to flatten the wave crests and restore equilibrium: one is gravity, the other is surface tension. For wavelengths of more than a few centimeters, gravity dominates. For millimeter wavelengths, surface tension dominates.

Because of the great incompressibility of water, the excess of water that appears in a wave crest must flow in from the neighboring trough regions. Individual water drops in a water wave therefore undergo a motion that is a combination of longitudinal motion (forward and backward) and transverse (up and down) motion. If the wavelength is small compared to the equilibrium depth of water, we have what are called deep-water waves. Then the individual water droplets in a traveling wave move in circles. A floating duck or a droplet at the surface) undergoes a uniform circular motion with radius equal to the amplitude of the harmonic wave and with period equal to that of the wave. On the crest of a traveling wave, the duck has its maximum forward velocity; in a trough, it has its greatest backward velocity. Water droplets below the surface travel in smaller circles; it turns out that the radius of gyration decreases exponentially with depth. The motion is negligibly small a few wavelengths below the surface. (These properties of water waves are derived in Sec. 7.3.)

The dispersion relation for deep-water waves is given approximately by

$$\omega^2 = gk + \frac{T}{\rho} k^3, \tag{33}$$

where $\rho \approx 1.0$ gm/cm^3 and $T \approx 72 \times 10^{-3}$N/m (surface tension) for water; $g = 980$ cm/sec^2.

We shall let you show that when g and $(T/\rho)k^2$ are equal so that gravity and surface tension make equal contributions to the return force per unit displacement per unit mass (i.e., to ω^2), then the phase and group velocities are equal. You can show that this occurs at a wavelength $\lambda = 1.70$ cm. The phase and group velocities are then both 23 cm/sec. For wavelengths much less than 1.7 cm, surface tension dominates; then the group velocity is 1.5 times the phase velocity. For wavelengths much greater than 1.7 cm, gravity dominates; then the group velocity is half the phase velocity. (See Prob. 6.19.)

In Table 6.1 we give wave parameters for wavelengths ranging from 1 mm (such as can be excited by a tuning fork driving a styrofoam cup full of water) up to 64 meters (very long ocean waves).

Application

Here is an example that makes use of Table 6.1. Suppose you are having a picnic at the beach. Someone wonders about the wavelength of waves in the open ocean twenty or thirty miles out from the coast. You tell them to wait a minute you'll tell them the wavelength. You take out your watch and time the waves breaking on your beach. You find an average of 12 waves per minute, i.e., one

Table 6.1 *Deep-water waves*

λ, cm	v, cps	v_ϕ, cm/sec	v_g, cm/sec	$\dfrac{v_g}{v_\phi}$
0.10	675	67.5	101.4	1.50
0.25	172	43.0	63.7	1.48
0.50	62.5	31.2	44.4	1.42
1.0	24.7	24.7	30.7	1.24
1.7	13.6	23.1	23.1	1.00
2	11.6	23.2	21.4	0.92
4	6.80	27.2	17.8	0.65
8	4.52	36.2	19.6	0.54
16	3.14	50.3	25.8	0.51
32	2.22	71	35.8	0.50
100	1.25	125	62.5	0.50
200	0.884	177	88.5	0.50
400	0.625	250	125	0.50
800	0.442	354	177	0.50
1600	0.313	500	250	0.50
3200	0.221	708	354	0.50
6400	0.156	1000	500	0.50

per five seconds: $v = 0.2$ cps. The weather has been constant for several days, so you can assume that the waves are at steady state (aside from local winds that do not affect the big ocean swells). The frequency is thus 0.2 cps at sea, as well as at your beach. (Of course the wavelength is different, because the waves breaking on your beach are not deep-water waves. *The wavelength depends on the water depth at your local beach. The steady-state driving frequency does not.*)

According to the table, the wavelength of the waves in the open ocean should be about 40 meters.

How far have the wave crests now breaking on your beach traveled in the last hour? If most of the time was spent traveling in deep water, then according to Table 6.1 the phase velocity has been about 8 meters/sec, i.e., about 29,000 meters per hour. Thus the waves have traveled about 30 kilometers (20 miles) in the last hour, and since the weather has been constant for many hours you should feel confident that your estimate of wavelength in the open ocean is a good one.

If you are not at the beach but are at a seismograph within ten or twenty miles of the beach, you can answer the same question.

6.3   *Pulses*

We wish to consider a situation where a transmitter at $z = 0$ describes a motion which is a superposition of many harmonic oscillations, all with equal amplitudes and with closely neighboring frequencies lying in a narrow band between the lowest, ω_1, and the highest, ω_2. We have already considered the situation where there are only two frequencies. In that case we get modulations which propagate with the group velocity.

Rotating-vector diagram.   In preparation for the more complicated case involving many harmonic components with slightly differing frequencies, let us reexamine the case of just two frequencies using the technique of the *rotating-vector diagram*. (See Vol. I, p. 125.) The harmonic oscillation

$$\psi(t) = A \cos \omega t \tag{34}$$

is the real part of the complex harmonic oscillation

$$\psi_c(t) = Ae^{i\omega t}, \tag{35}$$

where the subscript c stands for complex. A graphical representation of $\psi_c(t)$ is given by a vector of length A in the complex plane rotating counterclockwise at angular frequency ω. [The projection of this rotating vector on the horizontal axis (i.e., the real axis) gives the harmonic motion Eq. (34).] Instead of visualizing this rotating vector throughout one cycle, we shall imagine that we take "stroboscopic snapshots." Then, if the stroboscope has the same frequency as the rotating vector, the vector will appear to stand still, i.e., every snapshot will catch the vector in the same position. (See Fig. 6.2*a*.) If the angular frequency ω of the rotating vector is slightly greater than that of the stroboscope, ω_s, the vector will appear to rotate slowly forward (counterclockwise) at the *difference* angular frequency $\omega - \omega_s$ (see Fig. 6.2*b*); if, instead, $\omega - \omega_s$ is negative, the vector will appear to rotate slowly in the "retrograde" (clockwise) direction (see Fig. 6.2*c*). The subscript s stands for stroboscope.

Now let us consider a superposition of two harmonic waves having the same amplitude but slightly different frequencies,

$$\psi(t) = A \cos \omega_1 t + A \cos \omega_2 t. \tag{36}$$

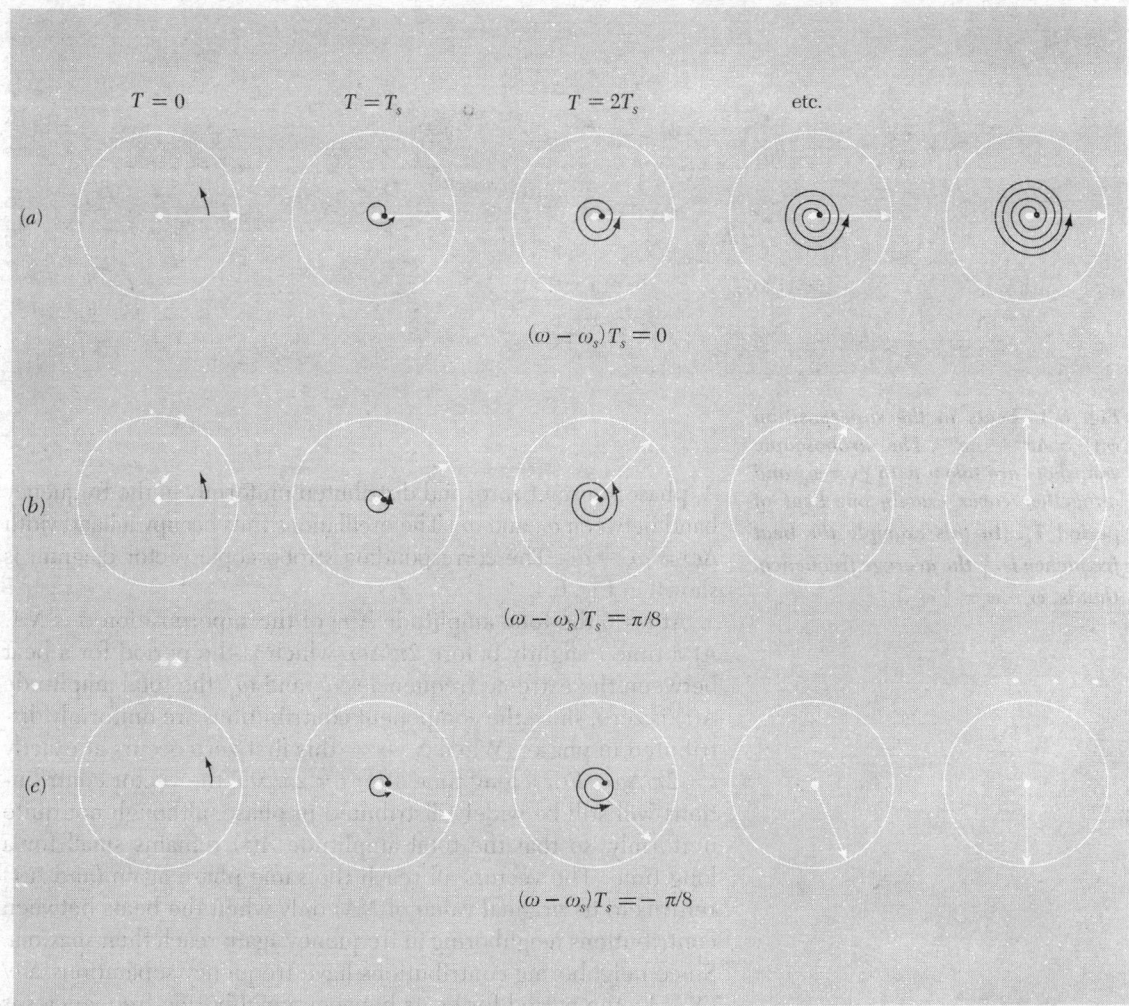

$$(\omega - \omega_s)T_s = 0$$

$$(\omega - \omega_s)T_s = \pi/8$$

$$(\omega - \omega_s)T_s = -\pi/8$$

Fig. 6.2 *Stroboscopic snapshots of rotating complex vector $e^{i\omega t}$. The spirals are to help you keep track of the number of rotations of the vector. The time interval between snapshots is $T_s = 2\pi/\omega_s$.*

We "strobe" (take stroboscopic snapshots of) the rotating vectors $Ae^{i\omega_1 t}$ and $Ae^{i\omega_2 t}$ at the frequency

$$\omega_s = \omega_{av} = \tfrac{1}{2}(\omega_1 + \omega_2). \tag{37}$$

Thus (taking $\omega_2 - \omega_1$ to be positive) $\omega_2 - \omega_{av}$ is positive and $\omega_1 - \omega_{av}$ is negative. Recall that $\psi(t)$ may be written [as in Eq. (2), Sec. 6.2] as the product of a slowly varying amplitude $A(t)$ times a fast oscillation of frequency ω_{av}. Our strobe frequency of ω_{av} will cause the fast oscillation to "stand still," and only $A(t)$ will change between snapshots. Thus we obtain the snapshots shown in Fig. 6.3.

Constructing a pulse. Now let us consider the situation where $\psi(t)$ is a superposition of very many oscillations, all of equal amplitude

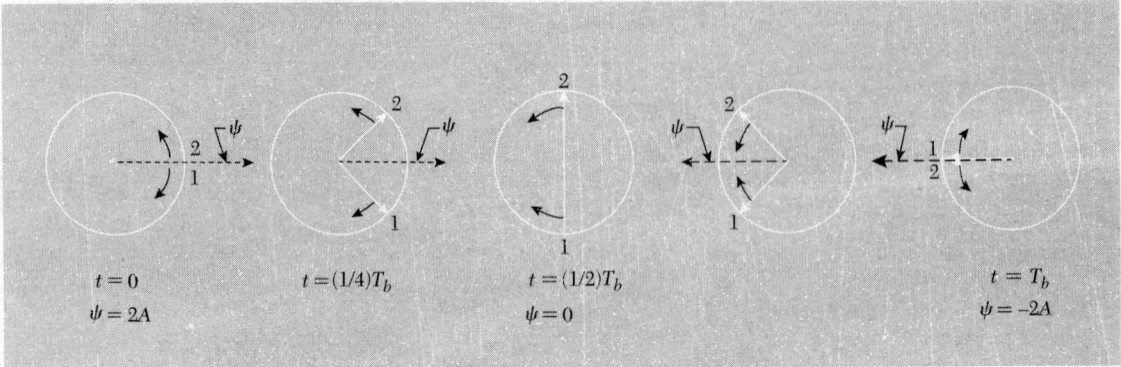

$t=0$ $t=(1/4)T_b$ $t=(1/2)T_b$ $t=T_b$

$\psi=2A$ $\psi=0$ $\psi=-2A$

Fig. 6.3 Beats in the superposition $\psi(t)=Ae^{i\omega_1 t}+Ae^{i\omega_2 t}$. *The stroboscopic snapshots are taken with* $\omega_s=\omega_{av}$ *and altogether cover exactly one beat of period* T_b. [*In this example the beat frequency is* $\frac{1}{4}$ *the average frequency; that is,* $\omega_2-\omega_1=\frac{1}{4}\omega_{av}$.]

A, phase constant zero, and distribu⁺ed uniformly in the frequency band between ω_1 and ω_2. The oscillations thus occupy a bandwidth $\Delta\omega=\omega_2-\omega_1$. The corresponding stroboscopic vector diagram is shown in Fig. 6.4.

At $t=0$ the total amplitude $A(t)$ of the superposition ψ is NA. At a time t slightly before $2\pi/\Delta\omega$, which is the period for a beat between the extreme frequencies ω_2 and ω_1, the total amplitude $A(t)$ is zero, since the component contributions are uniformly distributed in phase. (When $N\to\infty$, this first zero occurs at exactly $t=2\pi/\Delta\omega$.) For a long time after $t=2\pi/\Delta\omega$, the vector contributions will still be widely distributed in phase, although not quite uniformly, so that the total amplitude $A(t)$ remains small for a long time. The vectors all reach the same phase again (and $A(t)$ returns to its original value of NA) only when the beats between contributions neighboring in frequency again reach their maxima. Since neighboring contributions have frequency separations $\Delta\omega/(N-1)$, the period for beats between neighboring frequencies is $(N-1)$ times the beat period corresponding to the frequency separation $\Delta\omega$. Thus if $N\to\infty$, the total amplitude $A(t)$ remains small "forever," never returning to its original value. We then have what is called a *pulse*, i.e., a function of time that is significantly different from zero only during a limited time interval.

Time duration of pulse. Let us denote the *duration* of the pulse, i.e., the time interval during which $\psi(t)$ is "substantial," by the symbol Δt. That interval is given approximately by the time interval from $t=0$, when all the frequency components between ω_1 and ω_2 are in phase, to time t_1, when all the frequency components are distributed uniformly in phase over a total phase interval of 2π radians:

$$\Delta t\approx t_1,\qquad\qquad(38)$$

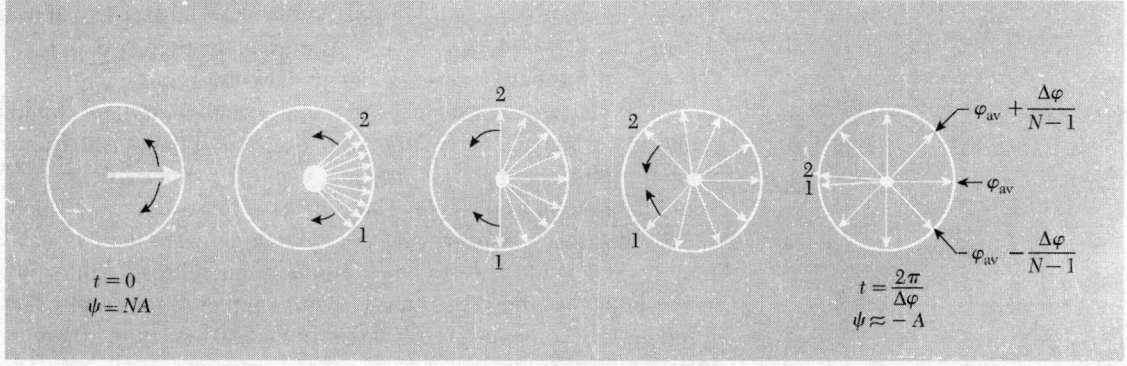

where

$$(\omega_2 - \omega_1)\, t_1 = 2\pi. \tag{39}$$

Thus the bandwidth $\Delta\omega = \omega_2 = \omega_1$ and the time interval Δt satisfy the relation

$$\Delta\omega\, \Delta t \approx 2\pi, \tag{40}$$

i.e.,

$$\boxed{\Delta\nu\Delta t \approx 1.} \tag{41}$$

Equation (41) is a specific example of a very general (and very important) mathematical relationship between the time duration Δt of a pulse $\psi(t)$ and the bandwidth $\Delta\nu$ of the frequency spectrum of harmonic components which superpose to form the pulse. This relation is of broad applicability in all of physics, wherever there are phenomena with the form of a pulse in time or some other variable. The general relationship is independent of the detailed specific shape of $\psi(t)$, as long as $\psi(t)$ has the characteristic that defines a pulse, namely that $\psi(t)$ be substantially different from zero only during a single limited time interval of duration Δt.

Bandwidth-time interval product. The *general* relationship that holds between the frequency bandwidth $\Delta\nu$ and the time duration Δt describing a pulse is given by

$$\boxed{\Delta\nu\Delta t \geq 1.} \tag{42}$$

The inequality sign in Eq. (42) results from the fact that if we superpose a number of harmonic oscillations occupying a frequency

Fig. 6.4 Stroboscopic snapshots of N oscillations ($N = 9$ here) distributed uniformly within the frequency interval $\Delta\omega = \omega_2 - \omega_1$. The strobe frequency is ω_{av}. The oscillation having $\omega = \omega_{av}$ appears to "stand still."

band Δv, we only get a pulse of duration as short as $\Delta t \approx 1/\Delta v$ if we choose the relative phase constants appropriately. In the example of Fig. 6.4, all the harmonic components have the same phase constant. If their phase constants were not all equal, then there would be no time when all components were exactly in phase (as they are at $t = 0$ in the example), i.e., no time when the superposition $\psi(t)$ would be as large as possible. Then the time interval during which $\psi(t)$ is substantially nonzero (i.e., not too much smaller than its maximum value) would have to be chosen as a much broader interval. In the limit that the phases are chosen completely randomly, the duration Δt becomes arbitrarily large. In that extreme limit, there is no recognizable pulse.

Pounding the piano.   Suppose that you devise a method to strike all the keys on a piano at once. The bandwidth of the resulting sound is about 4000 cps (the range of the piano). Thus *if* all the strings were excited exactly in phase at $t = 0$, you would thereby make a sound that would be very loud for a time Δt given by $\Delta t \approx \frac{1}{4000} \approx 0.2$ millisec and would be relatively weak after that. If your method of striking all the keys at once merely makes use of your arms or a long board or something, then it is imposible for you to excite every string at the same instant to within a small fraction of its period, i.e., a small fraction of about 10^{-3} sec. Instead, the phase constants are more likely to be essentially random. The sound then does not have the character of a pulse but just sounds like a steady noise.

Harmonic oscillation of limited duration.   Here is another illustration of Eq. (41). Suppose an oscillator is turned on, builds up rapidly (in a few cycles) to constant amplitude A, oscillates for about n cycles according to the harmonic oscillation $A \cos \omega_0 t$, is then turned off, and dies out in a few cycles, as shown in Fig. 6.5. *Since the oscillation does not continue forever, it is not a pure harmonic oscillation of frequency ω_0.* Certainly the (angular) frequency $\omega = \omega_0$ dominates, but, according to what we have said, it cannot be the only frequency ω present. There must be a band of frequencies centered at $\omega \approx \omega_0$. Here is a very simple way to estimate (crudely) the bandwidth $\Delta\omega$. Using the definition of frequency as cycles per second, we simply count the total number of cycles in the time interval Δt during which the oscillator is turned on and divide by Δt. Thus, counting n cycles,
we find

$$\nu \,(\text{dominant}) = \frac{n}{\Delta t}, \qquad\qquad (43)$$

which is bound to be about equal to $\nu_0 = T_0^{-1}$, according to Fig. 6.5. However, we see by inspection of Fig. 6.5 that it is not possible

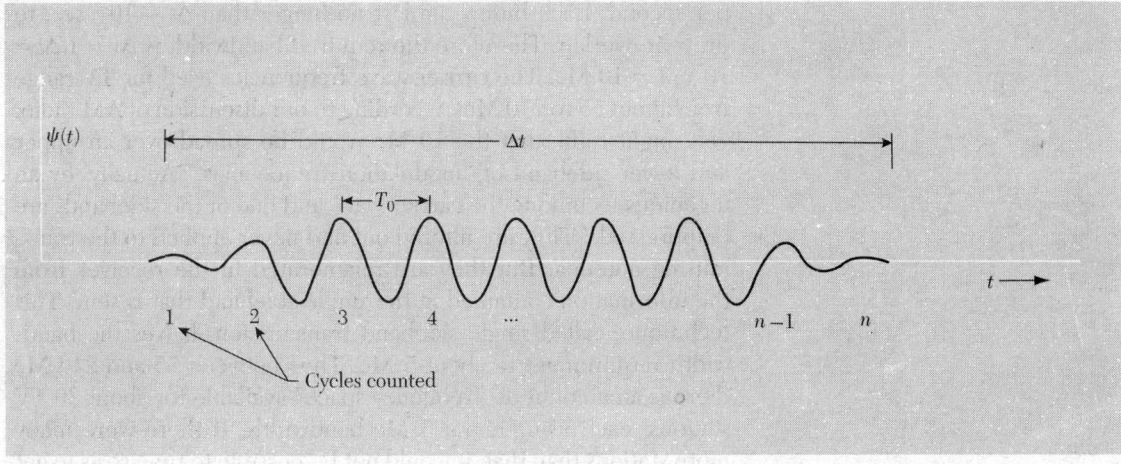

Fig. 6.5 *Harmonic oscillation of limited duration.*

to specify n *exactly*. There is an uncertainty of order $\pm \frac{1}{2}$ cycle at each end of the pulse, where we have to decide "Shall we count one more, or is it dead?" You may remark, "It doesn't matter very much, especially if n is large—the error is small compared with n." Yes, but it is exactly *that* error that we are concerned with. According to Eq. (43), an *uncertainty bandwidth* Δn of width approximately 1 in the number of cycles n leads to a *fractional frequency bandwidth* $\Delta \nu / \nu$ given by

$$\frac{\Delta \nu}{\nu} = \frac{\Delta n}{n} \approx \frac{1}{n}. \tag{44}$$

Taking the product of Eqs. (43) and (44) gives $\Delta \nu \approx 1/\Delta t$.

Example 6: Television bandwidth

The picture on the TV screen consists of a rectangular grid pattern of black and white spots. A given spot is "white" if the phosphorescent TV screen was recently (within $\approx \frac{1}{50}$ sec) struck by the electron beam at that spot. The spot separation is about 1 mm. A typical screen 50 cm × 50 cm thus has 500 lines with 500 spots per line, or 25×10^4 spots. Each spot is renewed every $\frac{1}{30}$ sec. (Every other horizontal line is skipped during a given traversal of the electron beam over the screen. The skipped lines are renewed on the next traversal. Thus a given region of the screen that includes many horizontal lines has a flicker rate of 60 cps. That is the "TV strobe" rate.) Thus the rate at which the instructions "turn off, turn on,..." must be sent to the electron beam is about $30 \times 25 \times 10^4$ or 8×10^6 times per second. The transmitted and received antenna voltages must therefore have about 10^7 little voltage on-off bumps per second. Each bump can last no longer than $\Delta t \sim 10^{-7}$ sec, to prevent overlap. Therefore the required bandwidth is $\Delta \nu \approx 1/\Delta t \approx 10^7$ cps = 10 Mc. The carrier wave frequencies used for TV range from about 55 to 210 Mc. According to our discussion of AM radio, you might think that the 10 Mc would be spread over an upper and lower sideband of "modulation frequencies." Actually, by an ingenious technique the carrier wave and one of the sidebands are "suppressed." They are filtered out and never applied to the transmitting antenna. But they are regenerated in the receiver from the information contained in the single sideband that is sent. This technique, called single sideband transmission, halves the bandwidth requirement to about 5 Mc. Thus between 55 and 210 Mc there is an amount of "frequency space" available for about 30 TV stations, each using about 5 Mc bandwidth. If there were many more stations than that, it would not be possible to tune so as to get a single station.

Example 7: Broadcasting with visible light

A *laser* is a device that holds promise of eventually providing as much control over electromagnetic radiation at frequencies of visible light as one has now at radio and microwave frequencies. Many individuals are hard at work developing techniques to modulate the light output in a manner analogous to the way a radio or TV transmitter modulates its carrier wave. Assume that suitable modulation techniques will be developed over most of the visible frequency range. Then we can consider how many TV channels can be accommodated within the "frequency space" consisting of the visible frequency range of light. The light is to be used as the carrier wave. The required bandwidth is 10 Mc per channel. Visible light has wavelengths from about 6500 Å (red) to 4500 Å (blue), i.e., frequencies from $\nu = c/\lambda = 3 \times 10^{10}/6.5 \times 10^{-5} \approx 4.6 \times 10^{14}$ Hz $= 4.6 \times 10^8$ Mc to $\nu = 3 \times 10^{10}/4.5 \times 10^{-5} \approx 6.6 \times 10^{14}$ Hz $= 6.6 \times 10^8$ Mc. Thus the total frequency band that might be available is 4.6 to 6.6×10^8 Mc, i.e., a bandwidth of 2×10^8 Mc. That would allow 2×10^7 nonoverlapping TV channels, each having 10 Mc bandwidth. (We could perhaps demand that at least one in a million of these channels be allocated to educational TV.)

Exact solution for the pulse $\psi(t)$ produced by a "square" frequency spectrum. Now we shall find an explicit expression for the pulse $\psi(t)$ formed by superposing N different harmonic components having equal amplitudes A, equal phase constants (zero), and frequencies distributed uniformly between the lowest frequency, ω_1, and the highest frequency, ω_2. This is the superposition that is illustrated in the "stroboscopic snapshots" of Fig. 6.4. It is given by

$$\psi(t) = A \cos \omega_1 t + A \cos (\omega_1 + \delta\omega)t + A \cos (\omega_1 + 2\delta\omega)t$$
$$+ \cdots + A \cos \omega_2 t, \qquad (45)$$

where $\delta\omega$ is the frequency spacing between neighboring contributions, i.e.,

$$\delta\omega \equiv \frac{\omega_2 - \omega_1}{N-1} = \frac{\Delta\omega}{N-1}. \qquad (46)$$

Equation (45) expresses $\psi(t)$ as a *linear superposition over many exactly harmonic components*. We wish now to find an alternative expression for $\psi(t)$ in the form of an *almost harmonic oscillation with a single "fast" oscillation frequency*, ω_{av}, given by

$$\omega_{av} = \tfrac{1}{2}(\omega_1 + \omega_2), \qquad (47)$$

and having an amplitude $A(t)$ that is "almost constant" on the time scale of the fast oscillations. That is, from our experience with a superposition of just two harmonic oscillations (Sec. 5.2), we hope to find an expression of the form

$$\psi(t) = A(t) \cos \omega_{av} t. \qquad (48)$$

We shall indeed find such an expression. It will turn out that if the bandwidth $\Delta\omega$ is small compared with ω_{av}, then $A(t)$ is slowly varying on the time scale of the fast oscillations. (However, our answer will be exact regardless of this condition.) Then we shall have written $\psi(t)$ as an amplitude-modulated almost harmonic oscillation. We shall find that $\psi(t)$ has the form of a pulse, as we have already shown qualitatively in the discussion following Fig. 6.4. From the exact expression, we will be able to see just what we mean by the statement that the bandwidth–time duration product is approximately unity.

In order to simplify the algebra, we shall use complex numbers. The superposition Eq. (45) is the constant A times the real part of the complex function $f(t)$, where

$$f(t) = e^{i\omega_1 t} + e^{i(\omega_1 + \delta\omega)t} + e^{i(\omega_1 + 2\delta\omega)t} + \cdots + e^{i(\omega_1 + \Delta\omega)t} \qquad (49)$$
$$\equiv e^{i\omega_1 t} S,$$

where [letting $a = e^{i\delta\omega t}$ and using $\Delta\omega = (N-1)\,\delta\omega$] the sum S is the geometric series

$$S = 1 + a + a^2 + \cdots + a^{N-1}.$$

Then

$$aS = a + a^2 + \cdots + a^{N-1} + a^N,$$
$$(a-1)S = a^N - 1,$$

$$S = \frac{a^N - 1}{a - 1} = \frac{e^{iN\delta\omega t} - 1}{e^{i\,\delta\omega t} - 1}$$

$$= \frac{e^{(1/2)(iN\delta\omega t)}}{e^{(1/2)(i\,\delta\omega t)}} \cdot \left[\frac{e^{(1/2)(iN\delta\omega t)} - e^{-(1/2)(iN\delta\omega t)}}{e^{(1/2)(i\,\delta\omega t)} - e^{-(1/2)(i\,\delta\omega t)}} \right]$$

$$= e^{(1/2)i(N-1)\delta\omega t}\, \frac{\sin \frac{1}{2}N\delta\omega t}{\sin \frac{1}{2}\,\delta\omega t}$$

$$= e^{(1/2)(i\,\Delta\omega t)}\, \frac{\sin \frac{1}{2}N\delta\omega t}{\sin \frac{1}{2}\,\delta\omega t}.$$

Thus

$$f(t) = e^{i\omega_1 t}S = e^{i[\omega_1 + (1/2)\,\Delta\omega]t}\, \frac{\sin \frac{1}{2}N\delta\omega t}{\sin \frac{1}{2}\,\delta\omega t}$$

$$= e^{i\omega_{av}t}\, \frac{\sin \frac{1}{2}N\delta\omega t}{\sin \frac{1}{2}\,\delta\omega t}.$$

Finally, $\psi(t)$ is the constant A times the real part of $f(t)$:

$$\psi(t) = A \cos \omega_{av}t\, \frac{\sin \frac{1}{2}N\delta\omega t}{\sin \frac{1}{2}\,\delta\omega t},$$

i.e.,

$$\psi(t) = A(t) \cos \omega_{av}t, \tag{50}$$

where

$$A(t) = A\, \frac{\sin \frac{1}{2}N\delta\omega t}{\sin \frac{1}{2}\,\delta\omega t}. \tag{51}$$

Equation (51) is exact. Let us check to see whether it reduces to the familiar form for beats when there are only two terms present: Setting $N = 2$ in Eq. (51) and using the identity $\sin 2x = 2 \sin x \cos x$, with $x = \frac{1}{2}\delta\omega t$, we obtain

$$N = 2: \quad \psi(t) = [2A \cos \tfrac{1}{2}\,\delta\omega t]\cos \omega_{av}t$$

$$= 2A \cos \tfrac{1}{2}(\omega_1 - \omega_2)t \cos \omega_{av}t.$$

This is the same expression we found for beats in Sec. 1.5.

A more convenient form for Eq. (51) is obtained by eliminating the constant A in favor of $A(0)$, the value of $A(t)$ at time $t = 0$. Inspection of Eq. (51) shows us that we must exercise care in evaluating $A(t)$ at $t = 0$, because both the numerator and denominator of Eq. (51) vanish there. That problem is easily resolved by expanding the numerator and denominator in a Taylor's series at $t = 0$. Letting $\theta \equiv \frac{1}{2}\delta\omega t$, we have

$$\frac{\sin N\theta}{\sin \theta} = \frac{N\theta - \frac{1}{6}(N\theta)^3 + \cdots}{\theta - \frac{1}{6}\theta^3 + \cdots}. \tag{52}$$

For sufficiently small θ, we can neglect all terms except the first term in the numerator and the first term in the denominator. Thus we find

$$\lim_{\theta \to 0} \left\{ \frac{\sin N\theta}{\sin \theta} \right\} = N. \tag{53}$$

Then Eq. (51) gives

$$A(0) = NA, \qquad A = \frac{A(0)}{N}, \tag{54}$$

i.e.,

$$A(t) = A(0)\frac{\sin \frac{1}{2}N\delta\omega t}{N\sin \frac{1}{2}\delta\omega t}. \tag{55}$$

Now let us go to the interesting limit where N is huge. When N gets large enough, the frequency spacing $\delta\omega$ between neighboring harmonic components will become small enough so that it cannot be resolved by whatever experimental apparatus we have in mind. (This is physics, not mathematics; we must *always* have some apparatus in mind, eventually.) Then we can think of the frequency components as being essentially continuously distributed. Such a sufficiently large N is given the nickname "infinity." For huge N we can neglect the difference between N and $N - 1$. Then

$$N = \text{huge:} \qquad N\delta\omega \approx (N - 1)\,\delta\omega = \Delta\omega. \tag{56}$$

Thus we let N go to "infinity" and $\delta\omega$ go to "zero." Their product is always the bandwidth $\Delta\omega$. In the denominator term $\sin \frac{1}{2}\,\delta\omega t$ in Eq. (55), we assume that $\delta\omega$ goes to zero but that t does not go to infinity (the experiment must end sometime). Then we can neglect all but the first term in the Taylor's series for $\sin \frac{1}{2}\,\delta\omega t$. Thus we obtain

$$\begin{aligned} A(t) &= A(0)\frac{\sin \frac{1}{2}N\delta\omega t}{N\sin \frac{1}{2}\delta\omega t} \\ &= A(0)\frac{\sin \frac{1}{2}\Delta\omega t}{N\cdot \frac{1}{2}\delta\omega t}, \\ &= A(0)\frac{\sin \frac{1}{2}\Delta\omega t}{\frac{1}{2}\Delta\omega t}, \end{aligned} \tag{57}$$

and

$$\psi(t) = A(t)\cos\omega_{\text{av}}t. \tag{58}$$

Now let us go back to the expression for $\psi(t)$ as a superposition, Eq. (45), and express it in an appropriate way to correspond to the limit $\delta\omega \to 0$. We can use Eqs. (54) and (56) to write

$$A = \frac{A(0)}{N} = \frac{A(0)}{\Delta\omega} \, \delta\omega. \tag{59}$$

Then the superposition Eq. (45) can be expressed as

$$\psi(t) = \frac{A(0)}{\Delta\omega} [\delta\omega \cos \omega_1 t + \delta\omega \cos (\omega_1 + \delta\omega)t + \dots$$
$$+ \delta\omega \cos \omega_2 t]. \tag{60}$$

But, in the limit $\delta\omega \to 0$, the expression in brackets is just the integral of $\cos \omega t$ times $d\omega$ (we replace letter δ by letter d), integrated from $\omega = \omega_1$ to ω_2. Thus Eq. (60) becomes

$$\psi(t) = \frac{A(0)}{\Delta\omega} \int_{\omega_1}^{\omega_2} \cos \omega t \, d\omega. \tag{61}$$

Fourier integral.   Equation (61) is an example of a *continuous harmonic superposition* or *Fourier integral*. It turns out that any (reasonable) nonperiodic function $\psi(t)$ can be expressed as a continuous Fourier superposition of the general form

$$\psi(t) = \int_0^\infty A(\omega) \sin \omega t \, d\omega + \int_0^\infty B(\omega) \cos \omega t \, d\omega. \tag{62}$$

The continuous functions $A(\omega)$ and $B(\omega)$ are called *Fourier coefficients* of $\psi(t)$ by analogy with the same name given to the constants in a Fourier series consisting of discrete frequencies.

By comparison of Eq. (61) and (62), we see that the function $\psi(t)$ given by Eqs. (57) and (58) has Fourier coefficients

$$A(\omega) = 0 \qquad \text{for all } \omega,$$

$$B(\omega) = 0 \qquad \text{for } \omega \text{ not between } \omega_1 \text{ and}$$
ω_2,

$$B(\omega) = \frac{A(0)}{\Delta\omega} \qquad \text{for } \omega \text{ between } \omega_1 \text{ and } \omega_2. \tag{63}$$

Fourier frequency spectrum.   A plot of the Fourier coefficients versus ω is called the *frequency spectrum* of the continuous Fourier superposition. The spectrum given by Eq. (63) is as simple as a spectrum can be. It is "flat" [i.e., $B(\omega)$ is constant] over a certain limited frequency band of width $\Delta\omega$, and it is zero elsewhere. Such a spectrum is sometimes called a "square" spectrum because of the appearance of its plot. [In general, we have to give two plots, one of $A(\omega)$ and one of $B(\omega)$.]

In Fig. 6.6, we plot the pulse $\psi(t)$ and its Fourier coefficient $B(\omega)$. Notice that $A(t)$ has its first zero (for positive t) at a time

t_1 which satisfies $t_1 = 2\pi/\Delta\omega$. That is how long it takes for all the frequency components to become uniformly distributed in relative phase over an interval of 2π radians, as we had already concluded from the "stroboscopic snapshots" of Fig. 6.4. For the time duration Δt during which the amplitude $A(t)$ of $\psi(t)$ is relatively large, we could take the interval between the two zeros of $A(t)$ at $t = -t_1$ and $t = +t_1$. However, that is too large. It is more reasonable to take Δt to be that interval outside of which $\psi(t)$ "never recovers" (its lost amplitude). A convenient definition of the full width Δt (for this particular pulse) is to take half of the time interval between the two zeros at $t = \pm t_1$. Thus we can *define* the duration of this pulse to be

$$\Delta t = t_1 = \frac{2\pi}{\Delta\omega},$$

$$\Delta\nu\,\Delta t = 1. \tag{64}$$

Equation (64) has an "equals" sign rather than an "approximately equals" sign because we have defined precisely what we mean by the duration Δt for this pulse. According to our definition, $A(t)$ at the ends of the interval Δt is given by

$$A\left(\frac{t_1}{2}\right) = A(0)\frac{\sin(\pi/2)}{\pi/2} = \frac{2}{\pi}A(0). \tag{65}$$

Thus at the beginning and end of the interval Δt the amplitude $A(t)$ is down by a factor $2/\pi$ from its maximum value.

An "almost harmonic oscillator" with displacement $\psi(t) = A(t)\cos\omega_{av}t$ has stored energy proportional to $A^2(t)$. Thus the energy is maximum at the center of the pulse (at $t = 0$) and is down to a fraction $(2/\pi)^2 = 0.406$ at the beginning and end of the interval Δt. Thus our definition of the duration Δt corresponds to the interval during which the oscillator has 40 per cent or more of its maximum stored energy.

In Sec. 6.4, we shall study further examples of pulses and their corresponding continuous Fourier superpositions.

Traveling wave packet.   Suppose a transmitter at $z = 0$ describes a motion in the form of a pulse similar to that of Fig. 6.6. Since the transmitter radiates waves into the medium for a time of limited duration and since the waves propagate away from the transmitter, they will form a pulse of waves with a limited extent in space. Such a pulse is called a *wave packet* or *wave group*. The wave packet propagates with the group velocity. Because k and ω are related through the dispersion relation $k(\omega)$, the existence of a band $\Delta\omega$ of frequencies emitted by the transmitter implies a corresponding

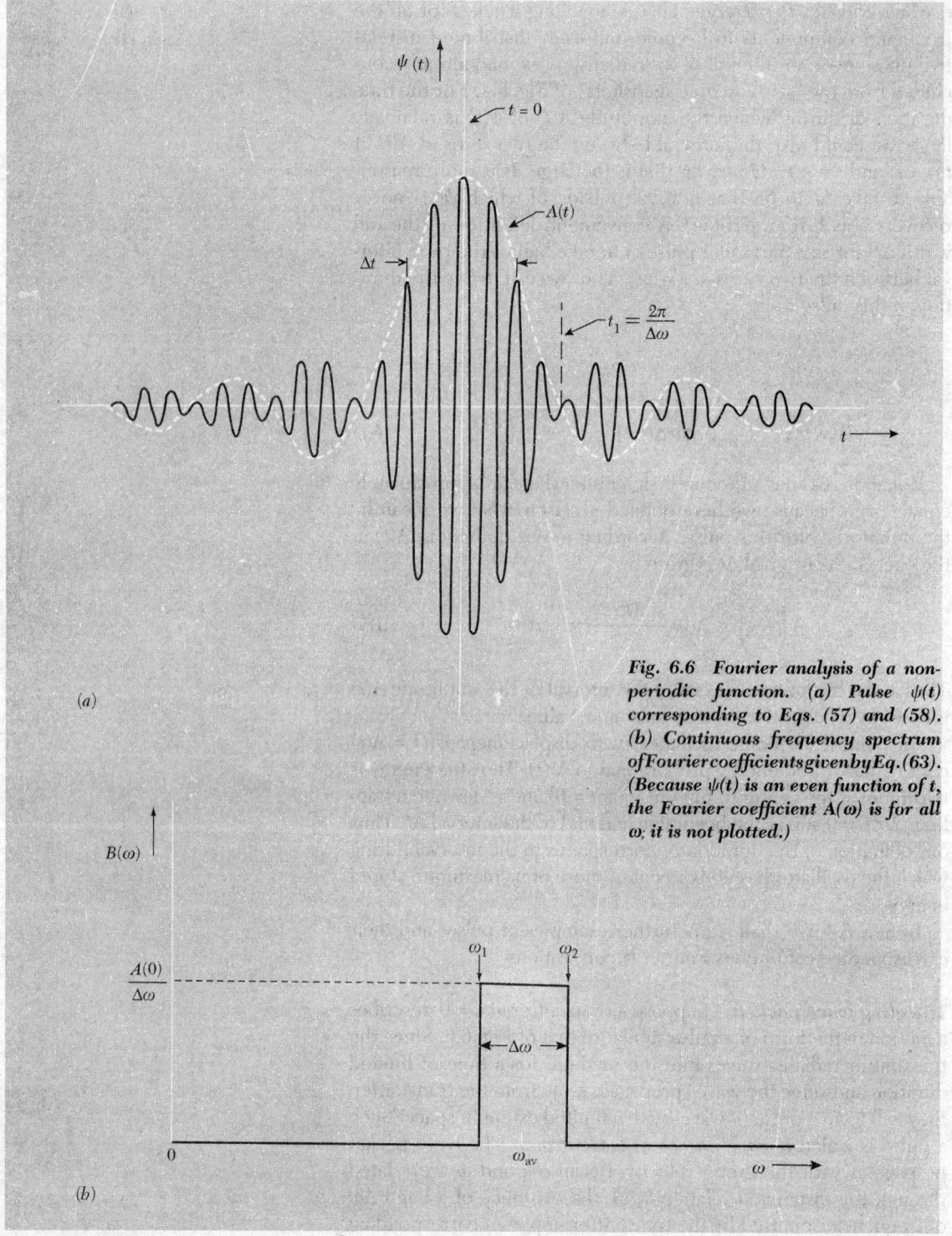

Fig. 6.6 *Fourier analysis of a non-periodic function. (a) Pulse* $\psi(t)$ *corresponding to Eqs. (57) and (58). (b) Continuous frequency spectrum of Fourier coefficients given by Eq.(63). (Because* $\psi(t)$ *is an even function of* t, *the Fourier coefficient* $A(\omega)$ *is for all* ω; *it is not plotted.)*

band Δk of wavenumbers (and corresponding wavelengths)in the wave packet. Related to the dominant frequency ω_0 there will be a dominant wavenumber $k_0 = k(\omega_0)$ [i.e., k_0 is obtained by substituting $\omega = \omega_0$ in the functional relation $k(\omega)$]. The band Δk is centered at k_0 and is obtained by differentiating the dispersion relation and setting $\omega = \omega_0$:

$$\Delta k = \left(\frac{dk}{d\omega} \right)_0 \cdot \Delta\omega = \frac{\Delta\omega}{v_g}, \tag{66}$$

where we used $v_g = (d\omega/dk)_0$. [The subscript zero means that the derivative is evaluated at the center of the band. Also, we neglect higher terms in the Taylor's series expansion of the dispersion relation, for which Eq. (66) may be considered the first term.]

Product of length times the wavenumber bandwidth. A packet of length Δz traveling at group velocity v_g passes a given fixed point z in a time interval Δt given by

$$\Delta z \approx v_g \, \Delta t. \tag{67}$$

Taking the product of Eqs. (66) and (67), we obtain

$$\boxed{\Delta k \, \Delta z \approx \Delta\omega \, \Delta t.} \tag{68}$$

Thus since $\Delta\omega \, \Delta t \geq 2\pi$ we have $\Delta k \, \Delta z \geq 2\pi$, i.e., using the wavenumber $\sigma \equiv k/2\pi = \lambda^{-1}$, we have

$$\boxed{\Delta\sigma \, \Delta z \gtrsim 1.} \tag{69}$$

This relation is completely analogous to the general relation $\Delta v \, \Delta t \geq 1$, but it applies to a pulse in space instead of time.

Another simple way to derive Eq. (69) is to consider the "uncertainty bandwidth" in the number of cycles contained in Δz. Thus σ (in cycles per unit length) is given by

$$\sigma \approx \frac{\text{cycles} \pm \frac{1}{2}}{\Delta z}, \tag{70}$$

so that the wavenumber bandwidth $\Delta\sigma$ is approximately $1/\Delta z$. This derivation is the analog in space of the derivation of $\Delta v \Delta t \approx 1$ that we gave following Eq. (44).

Spread of wave packet with time. Finally we should point out that the length Δz of a packet does not remain constant as the packet propagates in a dispersive medium; the packet spreads as it progresses. That is because the group velocity $v_g = d\omega/dk$ depends on k (or ω). Therefore the band Δk contains a band of group velocities Δv_g, given approximately by,

$$\Delta v_g = \left(\frac{dv_g}{dk}\right)_0 \Delta k = \left(\frac{d^2\omega}{dk^2}\right)_0 \Delta k. \tag{71}$$

A group which starts out at $t = 0$ with width $(\Delta z)_0$ will have at time t a width $(\Delta z)_t$ given approximately by

$$(\Delta z)_t \approx (\Delta z)_0 + (\Delta v_g)t. \tag{72}$$

The time Δt that the packet takes to pass a fixed z will increase accordingly. [Equation (68) holds at all times and, of course, $\Delta \omega$ and Δk are constant.]

Because of this spreading of the packet, the relations $\Delta\sigma\,\Delta z \approx 1$ and $\Delta v\Delta t \approx 1$ cannot hold except at $t = 0$. In order to have a transmitter output that satisfies $\Delta v\Delta t \approx 1$ we must take all harmonic components to be in the *correct* phase at $t = 0$. However, once we let the group propagate a sufficient distance in the medium, we *cannot* have the entire band Δk in phase at a downstream point, because some parts of the group arrive early and some late due to the variation in group velocity. Thus the phases of different frequency components of the wave are different from one another at a downstream point, unlike the situation at $t = 0$. Then we get $\Delta\sigma\,\Delta z \approx \Delta v\Delta t > 1$.

Of course, if the medium is "nondispersive," then the packet does *not* spread and the relation $\Delta\sigma\,\Delta z \approx \Delta v\Delta t \approx 1$ will be preserved.

Wave packets in water. You can make a nice expanding circular wave packet by tossing a pebble in a pond. With some practice you can follow a packet with your eyes and watch individual wavelets grow at the back, pass through the packet, and "disappear" at the front. (The phase velocity is greater than the group velocity for water wavelengths greater than 1.7 cm, as is usually the case for waves from a moderate-sized pebble. A picture of a wave group with phase velocity twice the group velocity is shown in Fig. 6.7.) I strongly urge the student to study water wave packets in sink, bathtub, and pond. Because they move rather rapidly (see Table 5.1, Sec. 5.2), it takes some practice but the effort is amply repaid. (See the Home Experiments.)

6.4   *Fourier Analysis of Pulses*

In Sec.6.3 we encountered our first example of a function of time $\psi(t)$ expanded in a continuous Fourier superposition (Fourier

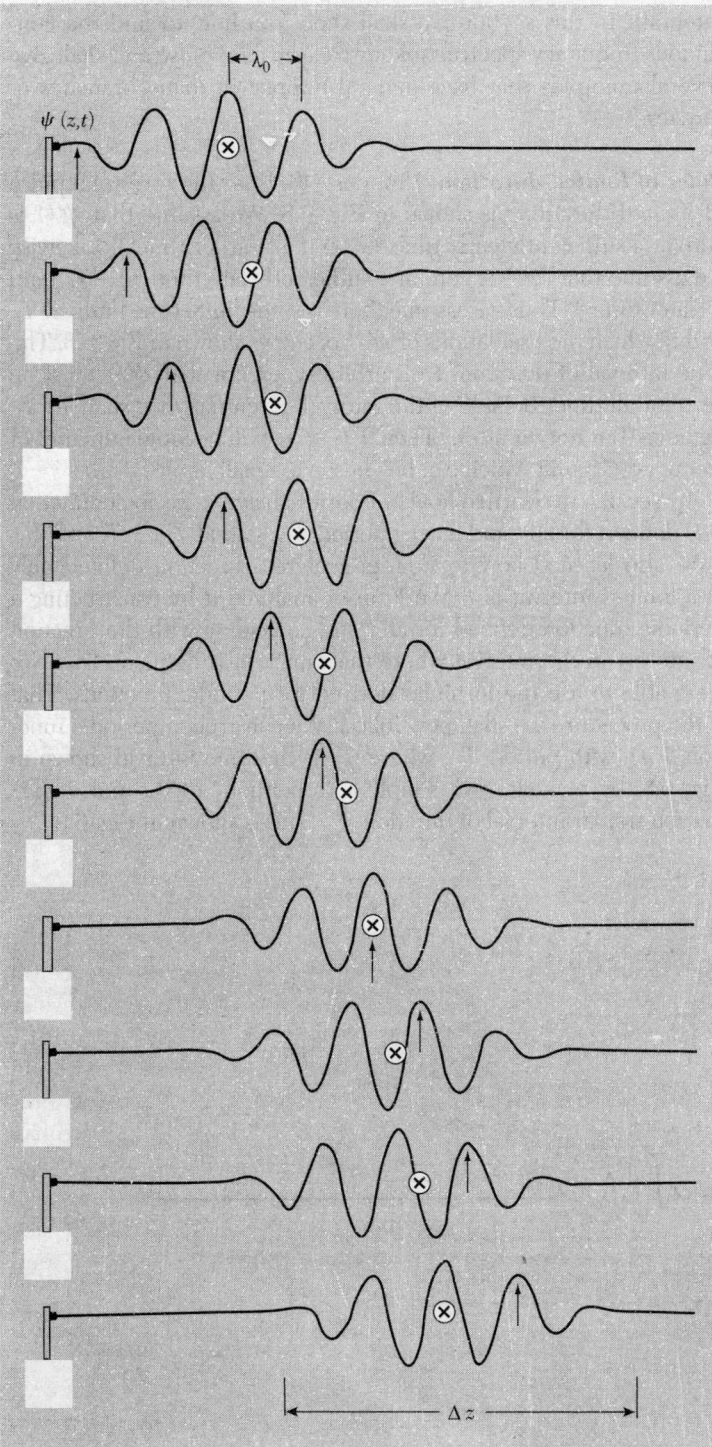

Fig. 6.7 Wave packet with phase
velocity twice the group velocity. The
arrow travels at the phase velocity,
following a point of constant phase
for the dominant wavelength. The
cross travels at the group velocity
with the packet as a whole.

integral). In this section we shall show you how to find the continuous frequency spectrum of any (reasonable) pulse and shall give several examples that have general interest in many branches of physics.

Pulse of limited duration. Suppose $\psi(t)$ has the form of a pulse of limited duration, as shown in Fig. 6.8. We assume that $\psi(t)$ is zero at a sufficiently early time t_0 (and all earlier times). Likewise we assume that $\psi(t)$ is zero at a sufficiently late time $t_0 + T_1$ (and all later times). Thus we assume there is some finite time interval T_1 within which the oscillations of $\psi(t)$ occur, as shown in Fig. 6.8. The time interval of duration T_1 is arbitrary, except that $\psi(t)$ must be zero for all times outside of the interval. Eventually we shall let T_1 be huge (but not infinite). (Then $1/T_1 \equiv \nu_1$ will become our "unit of frequency," a unit which can be chosen as small as we wish.)

In Sec. 2.3 we learned how to Fourier-analyze a periodic function $F(t)$ defined for all t and having period T_1, so that $F(t + T_1) = F(t)$. We also learned how to Fourier-analyze a function defined only in a limited interval of t. We Fourier-analyzed it by constructing a periodic function defined for all t and coinciding with the function of interest in the interval where that function is defined. Then we were able to use the formulas derived for periodic functions. That is the procedure we shall now follow. We construct a periodic function $F(t)$ with period T_1, where T_1 is the time interval shown in Fig. 6.8, by making $F(t)$ simply a "repetition" of the pulse $\psi(t)$ in each similar interval of duration T_1. This is shown in Fig. 6.9.

Fig. 6.8 A pulse $\psi(t)$. For times earlier than t_0 or later than $t_0 + T_1$, the function $\psi(t)$ is zero.

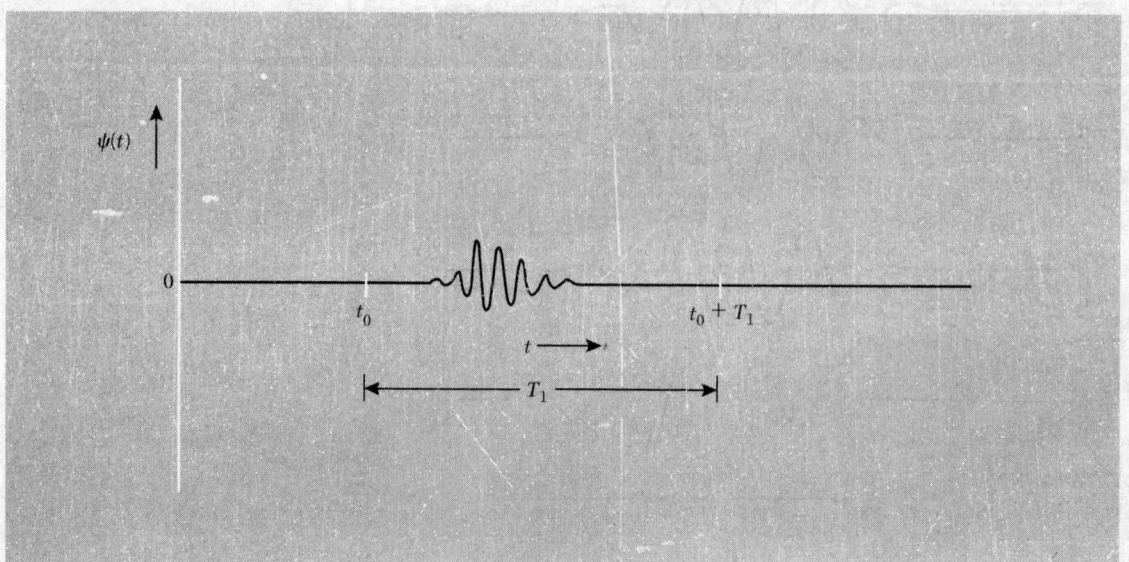

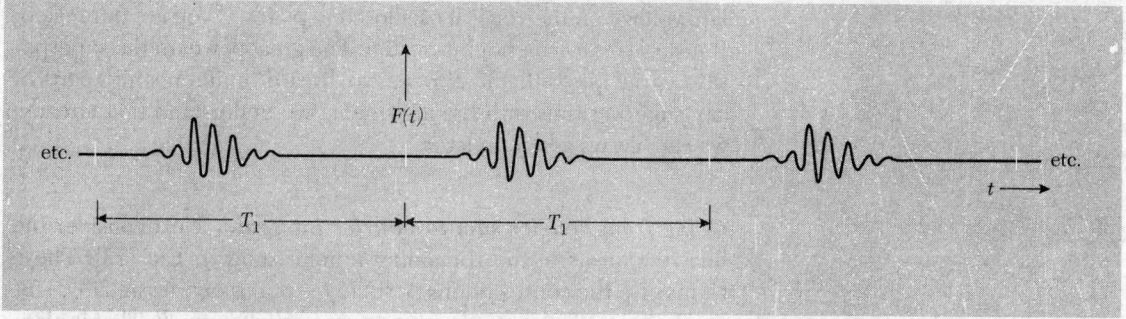

Fig. 6.9 Periodic function F(t) with period T_1 constructed by "repeating" the pulse $\psi(t)$ in successive time intervals of duration T_1.

The Fourier series for the periodic function $F(t)$ is given in Eqs. (2.49) through (2.52), Sec. 2.3. We recopy here those results we need:

$$F(t) = B_0 + \sum_{n=1}^{\infty} A_n \sin n\omega_1 t + \sum_{n=1}^{\infty} B_n \cos n\omega_1 t \qquad (73)$$

with

$$\omega_1 = 2\pi\nu_1 = \frac{2\pi}{T_1}. \qquad (74)$$

Then

$$B_0 = \frac{1}{T_1} \int_{t_0}^{t_0+T_1} F(t) \, dt, \qquad (75)$$

$$B_n = \frac{2}{T_1} \int_{t_0}^{t_0+T_1} F(t) \cos n\omega_1 t \, dt, \qquad (76)$$

$$A_n = \frac{2}{T_1} \int_{t_0}^{t_0+T_1} F(t) \sin n\omega_1 t \, dt, \qquad (77)$$

where

$$n=1, 2, 3,....$$

We will now adapt Eqs. (73) through (77) to our present problem, which is to express the pulse $\psi(t)$ as a superposition of harmonic oscillations.

First we note that the constant term B_0 given in Eq. (73) must be absent (i.e., B_0 is zero). That is because we assumed $\psi(t)$ to be zero for sufficiently early and sufficiently late times. There is no "constant displacement," or constant voltage, or whatever it is, included in our $\psi(t)$. (That does not mean that we cannot tell you (for example) the

DC voltage on the vertical oscilloscope plates, if you are interested. It just means *we* are not interested. The great power of the superposition principle is that it allows us to discard "uninteresting" parts of any superposition with the argument "we understand that already; we can always add it on later.")

Going from Fourier sum to Fourier integral. Next consider the first few terms in the remaining infinite sums in Eq. (73). These terms give the contributions $A_1 \sin \omega_1 t + B_1 \cos \omega_1 t$, $A_2 \sin 2\omega_1 t + B_2 \cos 2\omega_1 t$, etc. *These first few terms are negligibly small.* That is obvious from our sketch of Fig. 6.8. We see that there is no component of $\psi(t)$ varying as slowly as an oscillation with period T_1. The artificially constructed function $F(t)$ indeed has a frequency component with period T_1. But since T_1 is arbitrary (except for the properties we have specified), we can double it, i.e., replace it by a new T_1 twice as large. Then double that, etc., etc. We see that since T_1 can be made as large as we please, the angular frequency $\omega_1 = 2\pi/T_1$ can correspondingly be made as small as we please. Thus the artificially introduced constants A_1 and B_1, although not strictly zero, are (like B_0) strictly uninteresting. The constants A_2 and B_2 are essentially zero (for T_1 large enough). In fact, we can take T_1 so large that all the first few constants A_n and B_n are negligible, where "first few" can mean, for example, the first ten thousand or so. Now let us consider some n so large that A_n and B_n are not completely negligible. Consider two successive terms in Eq. (73), designated by n and $n + 1$:

$$F(t) = \cdots + A_n \sin n\omega_1 t + A_{n+1} \sin (n\omega_1 + \omega_1)t + \cdots \qquad (78)$$

If T_1 is sufficiently large, we can assume that ω_1 is so small and n so large (the first few n having been passed—they had negligible coefficients) that A_{n+1} differs only infinitesimally from A_n. Then we may regard $n\omega_1$ as the continuous variable ω and A_n as a continuous function of ω:

$$\omega = n\omega_1. \qquad (79)$$

Let $\delta\omega$ be the increment of ω when n increases by δn in going from n to $n + \delta n$:

$$\delta\omega = \omega_1\,\delta n, \quad \delta n = \frac{\delta\omega}{\omega_1}. \qquad (80)$$

Now let δn be sufficiently small so that all the coefficients A_n in the band from n to $n + \delta n$ are essentially equal to one another. We may then group together all the terms in Eq. (78) that correspond to the band δn, taking them all to have the samefrequency ω (the averagevalue of ω in the band $\delta\omega$). Since all the terms are equal (in a band), and since there are δn terms, we may write the infinite series Eq. (78) in the form [using Eqs. (79) and (80)]

$$F(t) = \cdots + \delta n A_n \, \sin \, n\omega_1 t + \cdots$$

$$= \cdots + \delta\omega \, \frac{A_n}{\omega_1} \, \sin \, \omega t + \cdots$$

$$\equiv \cdots + \delta\omega \, A(\omega) \, \sin \, \omega t + \cdots$$

$$= \int_0^\infty A(\omega) \, \sin \, \omega t \, \, d\omega + \cdots \qquad (81)$$

To reach the last equation, we merely recognized that the sum over successive bands of width $\delta\omega$ can be written as an integral, with $\delta\omega$ replaced by the more comon symbol $d\omega$. The ellipsis ($\cdots$) represents the remaining terms in Eq. (73), which originate in the sum $\Sigma B_n \cos n\omega_1 t$. This sum also becomes an integral. Thus we obtain the complete expression

$$F(t) = \int_0^\infty A(\omega) \sin \, \omega t \, \, d\omega + \int_0^\infty B(\omega) \cos \, \omega t \, \, d\omega, \qquad (82)$$

$$A(\omega) = A(n\omega_1) = \frac{A_n}{\omega_1},$$

$$B(\omega) = B(n\omega_1) = \frac{B_n}{\omega_1}. \qquad (83)$$

Notice that we have let the continuous variable ω start at zero. We can do that because we know A_n and B_n are zero near $n = 0$, so that $A(\omega)$ and $B(\omega)$ must be zero near $\omega = 0$.

According to Eqs. (83) and (77), $A(\omega)$ is given by

$$A(\omega) = \frac{2}{\omega_1 T_1} \int_{t_0}^{t_0 + T_1} F(t) \sin \, \omega t \, \, dt,$$

i.e., since $\omega_1 T_1 = 2\pi$,

$$A(\omega) = \frac{1}{\pi} \int_{-\infty}^\infty \psi(t) \sin \, \omega t \, \, dt,$$

where we used the fact that the integral of the artificially constructed periodic function $F(t)$ over one of its periods equals the integral of the nonperiodic pulse $\psi(t)$ over all time.

Fourier integral. Finally we can discard the periodic function $F(t)$ given in Eq. (82) and write the *Fourier integral*,

$$\psi(t) = \int_0^\infty A(\omega)\sin \omega t \ d\omega + \int_0^\infty B(\omega) \cos \omega t \ d\omega, \qquad (84)$$

$$A(\omega) = \frac{1}{\pi}\int_{-\infty}^\infty \psi(t)\sin \omega t \ dt, \qquad (85)$$

$$B(\omega) = \frac{1}{\pi}\int_{-\infty}^\infty \psi(t)\cos \omega t \ dt. \qquad (86)$$

We can now apply these formulas to interesting examples.

Application: Square frequency spectrum

Suppose that $A(\omega)$ is zero for all ω, and suppose that $B(\omega)$ is a constant for values of ω in the interval from ω_1 to ω_2 and is zero for all other ω. Let us choose the constant value of B in this interval so that the "area" of a plot of $B(\omega)$ versus ω is unity; i.e.,

$$B(\omega) = \frac{1}{\Delta\omega} \quad \text{for } \omega_1 \leq \omega \leq \omega_2 = \omega_1 + \Delta \omega; \qquad (87)$$

$$B(\omega) = 0 \qquad \text{elsewhere.}$$

(Notice that since $B(\omega)$ has been chosen to have dimensions of inverse frequency, $\psi(t)$ should turn out to be dimensionless.) Here is the solution for $\psi(t)$:

$$\psi(t) = \int_0^\infty A(\omega)\sin \omega t \ d\omega + \int_0^\infty B(\omega)\cos \omega t \ d\omega$$

$$= 0 + \int_{\omega_1}^{\omega_2}\frac{1}{\Delta\omega}\cos \omega t \ d\omega = \frac{1}{\Delta\omega}\cdot\frac{\sin \omega t}{t}\Big|_{\omega=\omega_1}^{\omega=\omega_2}$$

$$\psi(t) = \frac{\sin \omega_2 t - \sin \omega_1 t}{\Delta\omega t} = \frac{\sin \omega_2 t - \sin \omega_1 t}{(\omega_2 - \omega_1)t}. \qquad (88)$$

The numerator of Eq. (88) is a superposition of a type we have encountered before, which gives modulations at the modulation frequency $\frac{1}{2}(\omega_2 - \omega_1)$. The denominator contains a factor t that makes $\psi(t)$ largest at $t = 0$.

age frequency ω_0 and with a slowly varying amplitude:

$$\omega_0 = \frac{1}{2}(\omega_2 + \omega_1), \qquad \frac{1}{2}\Delta\omega = \frac{1}{2}(\omega_2 - \omega_1);$$

$$\omega_2 = \omega_0 + \frac{1}{2}\Delta\omega, \qquad \omega_1 = \omega_0 - \frac{1}{2}\Delta\omega. \qquad (89)$$

$$\psi(t) = \frac{\sin(\omega_0 + \frac{1}{2}\Delta\omega)t - \sin(\omega_0 - \frac{1}{2}\Delta\omega)t}{\Delta\omega t}$$

$$= \left[\frac{\sin\frac{1}{2}\Delta\omega t}{\frac{1}{2}\Delta\omega t}\right]\cos\omega_0 t. \tag{90}$$

Thus $\psi(t)$ is a fast oscillation with slowly varying amplitude $A(t)$:

$$\psi(t) = A(t)\cos\omega_0 t,$$

$$A(t) = \frac{\sin\frac{1}{2}\Delta\omega t}{\frac{1}{2}\Delta\omega t}. \tag{91}$$

The result, Eq. (91), is identical with that which we obtained in Sec. 6.3, where we studied a superposition of N harmonic oscillations having N different discrete frequencies distributed uniformly between ω_1 and ω_2. When we took the limit $N \to \infty$, we obtained Eq. (91). [See Eqs. (57) and (58), Sec. 6.3.] The pulse $\psi(t)$ and its Fourier coefficient $B(\omega)$ are plotted in Fig. 6.6.

Application: Square pulse in time

Suppose that $\psi(t)$ is zero for all time except for an interval of duration Δt centered at t_0 and extending between t_1 and t_2. In that interval $\psi(t)$ is constant; we set the constant so that the integral of $\psi(t)$ over the interval Δt is unity:

$$\psi(t) = \frac{1}{\Delta t}, \quad t_1 \leqq t \leqq t_2 = t_1 + \Delta t. \tag{92}$$

Let us find the Fourier coefficients $A(\omega)$ and $B(\omega)$.

Notice that if t_0 is zero, then $\psi(t)$ is an even function of t, and thus $A(\omega)$ must be zero (because $\sin\omega t$ is an odd function). For an arbitrary t_0, we need both $A(\omega)$ and $B(\omega)$, i.e., we have both the odd function $\sin\omega t$ and the even function $\cos\omega t$. By a trick we can save half the work. Let us simply replace t by $t - t_0$ in our general results. Then, since $\psi(t)$ is an even function of $t - t_0$, we have

$$\psi(t) = \int_0^\infty B(\omega)\cos\omega(t - t_0)d\omega, \tag{93}$$

with

$$B(\omega) = \frac{1}{\pi}\int_{-\infty}^\infty \psi(t)\cos\omega(t - t_0)dt. \tag{94}$$

We shall let you perform an easy integration to find (Prob. 6.20)

$$B(\omega) = \frac{1}{\pi}\frac{\sin\frac{1}{2}\Delta t\omega}{\frac{1}{2}\Delta t\omega}. \tag{95}$$

The square pulse of Eq. (92) and its Fourier coefficient $B(\omega)$ are plotted in Fig. 6.10. Notice that if we define $\Delta\omega$ to be the interval from the minimum frequency, which is zero, to the first zero of the Fourier coefficient $B(\omega)$, then we have

$$\Delta\omega \, \Delta t = 2\pi, \; \Delta\nu \, \Delta t = 1. \tag{96}$$

Use of piano to Fourier-analyze a handclap. Here is an application of the Fourier spectrum plotted in Fig. 6.10. Supposethat you would like to know the approximate duration of the loud sound that you hear when you clap your hands. Suppose that you don't happen to have a microphone, audio amplifier, and oscilloscope, but you do have a piano. Hold down the damper pedal so that all the strings can vibrate, hold your hands near the sounding board, and clap. The piano Fourier-analyzes the clap and preserves the analysis on the vibrating strings. If you can estimate the highest pitch at which the sound intensity from the strings is large, then that frequency must be roughly $\nu \approx 1/\Delta t$. This physical example gives us some additional insight into the meaning of Fourier analysis, as follows:

To a certain approximation, all the strings are pushed in the same direction by an air pressure wave of duration Δt. They start to oscillate at their natural frequency in a beginning transient oscillation. Those strings with frequency small compared with $1/\Delta t$ do not get through an appreciable fraction of one natural oscillation cycle before the force ends. These strings are accelerated for the entire interval Δt. The string whose period is exactly Δt is accelerated by the pressure wave for the first halfcycle of duration $\frac{1}{2} \Delta t$ and is decelerated on the next half-cycle. It ends up decelerated as much as accelerated, and thus it does not vibrate at all after the force ceases. Thus the strings with natural frequencies from zero to a value somewhat less than $1/\Delta t$ are excited with positive amplitude. The string with frequency $1/\Delta t$ has zero amplitude: that frequency is also the first zero of the Fourier coefficient $B(\omega)$ given by Eq. (95). Strings with frequency between $1/\Delta t$ and $2/\Delta t$ make between one and two complete cycles during Δt. The first cycle is wasted, in the sense that no net impulse is picked up from the pressure pulse. The string with frequency $2/\Delta t$ goes through two complete cycles and picks up no impulse. Thus $B(\omega)$ has its second zero at frequency $2/\Delta t$. The string with frequency $1.5/\Delta t$ does fairly well; the first cycle is wasted, but the force pushes in the same direction for the first half of the second cycle. Then the force ends. This string gets "$\frac{1}{3}$ value" out of the pulse, in the sense that it goes through three half-cycles of its natural oscillation, two of which get canceling contributions. By contrast, a string with frequency $\frac{1}{2}(1/\Delta t)$ goes through one half-cycle during Δt and should have three times the final vibration amplitude of the string that has $v = (\frac{3}{2})(1/\Delta t)$. Indeed, we see from Eq. (95) that $B(\omega)$ is three times larger in magnitude for $\omega \, \Delta t = \pi$ than for $\omega \, \Delta t = 3\pi$.

This example shows how a piano or other similar apparatus can be used as a Fourier-analyzing device. (We have neglected the fact that the coupling from the air to the strings may not be Uniformly good.) Notice that it would be very difficult to extract the *phase* information from the piano Fourier-analyzer. But your ear is not interested in phase. That is a common situation: often we are not interested in knowing $A(\omega)$ and $B(\omega)$ separately. It is then enough to know the *Fourier intensity*, $I(\omega)$, which can be defined by

$$I(\omega) = A^2(\omega) + B^2(\omega). \tag{97}$$

Delta function of time. If the duration of the square pulse Δt is much shorter than the period of that oscillation with the highest frequency we can detect (i.e., the shortest period), then the Fourier coefficient $B(\omega)$ is constant over the entire detected frequency spectrum. That is obvious from Fig. 6.10. If we let Δt go to zero, then the first zero of $B(\omega)$ moves to $+ \infty$, and any finite frequency has $B(\omega) = 1/\pi$, independent of frequency. The pulse defined in Eq. (92) is called a delta function of time, when Δt is sufficiently small. For example, since the highest note on the piano has $v \approx 5000$ cps, any short sound that is shorter than a millisec or so should excite all the strings about equally well. The piano analyzer would not distinguish such a sound from another sound ten times larger in amplitude and ten times shorter in duration; the strings would have the same final motion in either case.

Application: Damped harmonic oscillator—natural linewidth

We want to find the frequency spectrum, "line shape," of visible light emitted by an atom that has a mean decay lifetime $\tau \approx 10^{-8}$ sec. If we want only the bandwidth, we can quit now: the bandwidth Δv must be of order 10^8 cps, since the duration of the pulse is about 10^{-8} sec. We want to do better than that, however. We want to find the detailed shape of the spectrum, *assuming the decay has the time dependence of a damped harmonic oscillator.* Therefore we assume that $\psi(t)$ is zero for all time less than $t = 0$, that at $t = 0$ it is suddenly given an excitation, and that it thereafter describes the damped harmonic oscillation

Fig. 6.10 Square pulse ψ(t) and its Fourier coefficient B(ω).

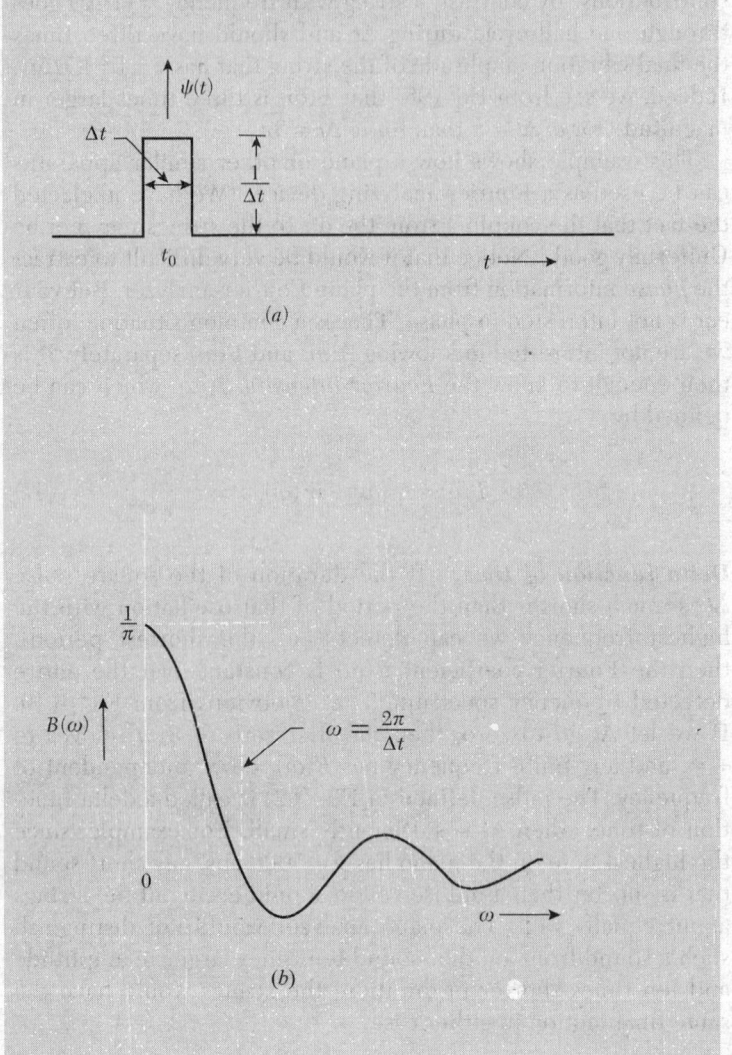

Fig. 6.10 Square pulse ψ(t) and its Fourier coefficient B(ω).

$$\psi(t) = e^{-(1/2)\Gamma t} \cos \omega_1 t. \tag{98}$$

(We take the amplitude constant to be unity in order to save writing in what follows.) The damping constant is the inverse of the mean decay lifetime:

$$\Gamma = \frac{1}{\tau}. \tag{99}$$

The spring constant is related to the mass M and the undamped natural frequency ω_0 by [see Eq. (3.5), Sec. 3.2]

$$K = M\omega_0^2. \tag{100}$$

The almost harmonic damped oscillation frequency ω_1 is related to ω_0 and Γ by

$$\omega_1^2 = \omega_0^2 - \tfrac{1}{4}\Gamma^2. \tag{101}$$

Let us expand Eq. (98) in a continuous Fourier superposition:

$$\psi(t) = \int_0^\infty A(\omega)\sin \omega t \ d\omega + \int_0^\infty B(\omega)\cos \omega t \ d\omega. \tag{102}$$

Then

$$2\pi A(\omega) = 2\int_{-\infty}^\infty \psi(t)\sin \omega t \ dt = \int_0^\infty e^{-(1/2)\Gamma t} 2 \cos \omega_1 t \ \sin \omega t \ dt$$

$$= \int_0^\infty e^{-(1/2)\Gamma t}\left[\sin(\omega + \omega_1)t + \sin(\omega - \omega_1)t\right]dt, \tag{103}$$

$$2\pi B(\omega) = 2\int_{-\infty}^\infty \psi(t)\cos \omega t \ dt = \int_0^\infty e^{-(1/2)\Gamma t} 2 \cos \omega_1 t \ \cos \omega t \ dt$$

$$= \int_0^\infty e^{-(1/2)\Gamma t}\left[\cos(\omega + \omega_1)t + \cos(\omega - \omega_1)t\right]dt, \tag{104}$$

Any table of definite integrals gives

$$\int_0^\infty e^{-ax}\sin bx \ dx = \frac{b}{b^2 + a^2},$$

$$\int_0^\infty e^{-ax}\cos bx \ dx = \frac{a}{b^2 + a^2}. \tag{105}$$

Thus Eqs. (103) and (104) give

$$2\pi A(\omega) = \frac{(\omega + \omega_1)}{(\omega + \omega_1)^2 + (\tfrac{1}{2}\Gamma)^2} + \frac{(\omega - \omega_1)}{(\omega - \omega_1)^2 + (\tfrac{1}{2}\Gamma)^2}, \tag{106}$$

$$2\pi B(\omega) = \frac{\tfrac{1}{2}\Gamma}{(\omega + \omega_1)^2 + (\tfrac{1}{2}\Gamma)^2} + \frac{\tfrac{1}{2}\Gamma}{(\omega - \omega_1)^2 + (\tfrac{1}{2}\Gamma)^2}. \tag{107}$$

We can make use of Eq. (101) to eliminate ω_1^2 in favor of ω_0^2. After some algebra we get

$$2\pi A(\omega) = \frac{2\omega(\omega^2 - \omega_0^2) + \omega\Gamma^2}{(\omega_0^2 - \omega^2)^2 + \Gamma^2\omega^2}, \tag{108}$$

$$2\pi B(\omega) = \frac{\Gamma(\omega^2 + \omega_0^2)}{(\omega_0^2 - \omega^2)^2 + \Gamma^2\omega^2}, \sim \tag{109}$$

$$I(\omega) \equiv \left[2\pi A(\omega)\right]^2 + \left[2\pi B(\omega)\right]^2 = \frac{4\omega^2 + \Gamma^2}{(\omega_0^2 - \omega^2)^2 + \Gamma^2\omega^2}. \tag{110}$$

Comparison of free decay with forced oscillation. It is interesting to compare these Fourier components of the freely decaying damped harmonic oscillator with amplitudes and intensities obtained when the same system undergoes steady-state forced oscillations at frequency ω. We recopy here the results of Eqs. (3.17) and (3.32) through (3.35) Sec. 3.2:

$$A_{el}(\omega) = \frac{F_0}{M} \frac{(\omega_0^2 - \omega^2)}{(\omega_0^2 - \omega^2)^2 + \Gamma^2\omega^2}, \tag{111}$$

$$A_{ab}(\omega) = \frac{F_0}{M} \frac{\Gamma\omega}{(\omega_0^2 - \omega^2)^2 + \Gamma^2\omega^2}, \tag{112}$$

$$|A|^2 = [A_{el}(\omega)]^2 + [A_{ab}(\omega)]^2 = \frac{F_0^2}{M^2} \frac{1}{(\omega_0^2 - \omega^2)^2 + \Gamma^2\omega^2}, \tag{113}$$

$$P(\omega) = \frac{1}{2}\frac{F_0^2}{M^2} \frac{\Gamma\omega^2}{(\omega_0^2 - \omega^2)^2 + \Gamma^2\omega^2}, \tag{114}$$

$$E(\omega) = \frac{1}{2}\frac{F_0^2}{M^2} \frac{\frac{1}{2}(\omega^2 + \omega_0^2)}{(\omega_0^2 - \omega^2)^2 + \Gamma^2\omega^2}. \tag{115}$$

We see that the Fourier amplitude $B(\omega)$ for free decay is proportional to the stored energy $E(\omega)$ for forced oscillation. We see that $A(\omega)$ for free decay has one contribution proportional to $\omega A_{el}(\omega)$ for forced oscillation and another contribution proportional to $A_{ab}(\omega)$. For reasonably weak damping, the contribution proportional to A_{ab} is negligible except when ω is very near the resonance frequency ω_0; thus $A(\omega)$ is essentially proportional to $\omega A_{el}(\omega)$. The Fourier intensity $I(\omega)$ has one contribution proportional to the power absorption $P(\omega)$ for forced oscillation and another contribution that is negligible for reasonably weak damping, i.e., for is $\Gamma^2 \ll \omega^2$. Thus $I(\omega)$ for free decay is essentially proportional to the power $P(\omega)$ for forced oscillation.

Lorentz line shape—relation to resonance curve. For weak damping and for ω not too far from ω_0, the Fourier amplitude

$B(\omega)$ and the Fourier intensity $I(\omega)$ are each proportional to the "Lorentz line-shape curve," $L(\omega)$, given by

$$L(\omega) = \frac{(\frac{1}{2}\,\Gamma)^2}{(\omega_0 - \omega)^2 + (\frac{1}{2}\,\Gamma)^2}. \tag{116}$$

The damping constant Γ, which is equal to the full frequency width at halfmaximum of the Lorentz line-shape curve, is called the *linewidth* $\Delta\omega$ of the frequency spectrum of the Fourier superposition that describes the free decay:

$$(\Delta\omega)_{\text{f.d.}} = \Gamma. \tag{117}$$

The Lorentz line shape, Eq. (116), has exactly the same form as the Breit-Wigner resonance response curve $R(\omega)$ which gives (for weak damping) the frequency dependence of $A_{ab}(\omega)$, $|A|^2$, $E(\omega)$, and $P(\omega)$ for forced oscillations [Eq. (3.36), Sec. 3.2]:

$$R(\omega) = \frac{(\frac{1}{2}\,\Gamma)^2}{(\omega_0 - \omega)^2 + (\frac{1}{2}\,\Gamma)^2}. \tag{118}$$

The full resonance width at half-maximum is given by

$$(\Delta\omega)_{\text{res}} = \Gamma. \tag{119}$$

Thus we have the remarkable result that for a weakly damped harmonic oscillator *the Fourier spectrum for free decay has the same frequency dependence as does the resonance response for driven oscillations*. We may summarize by writing the equalities

$$\boxed{(\Delta\omega)_{\text{f.d.}} = (\Delta\omega)_{\text{res}} = \frac{1}{\tau_{\text{f.d}}}.} \tag{120}$$

Measurement of natural frequency and frequency width. The close relationship between the Fourier components for free decay and the resonance response for steady-state forced oscillations has important experimental consequences. Suppose we want to study (*a*) the lowest mode of a piano string and (*b*) the first excited state of an atom. Here are three methods that we may use:

1. *Time dependence of free oscillation.* Excite the system suddenly at $t = 0$, using either a hammer or a collision with another atom. Then take high-speed photographs of the motion of the damped decaying oscillator, and plot the displacement versus

time. This can be done for the piano string. It cannot be done for the atom, even in principle, as you will learn in Vol. IV (*Quantum Physics*).

2. *Resonant response to forced oscillation.* Drive the system at steady state with a harmonic force $F_0 \cos \omega t$. Vary the driving frequency. Measure the absorbed power $P(\omega)$ as a function of frequency. This can be done with the piano string. It can also be done with some excited states of atoms by driving them with steady-state electromagnetic radiation and observing the absorbed power P as a function of ω to obtain ω_0 and Γ.

3. *Fourier analysis of emission spectrum.* Excite the system suddenly. Fourier-analyze the radiation that it emits. This can be done with the piano string. It can also be done with some excited states of atoms by looking at the frequency spectrum of the emitted light. The easiest thing to measure is the emitted intensity of radiation as a function of frequency. That is proportional in turn to the Fourier intensity $I(\omega)$. Determination of $I(\omega)$ gives the mode frequency ω_0 and width Γ.

In Fig. 6.11 we plot a damped harmonic oscillation and the Fourier coefficients $A(\omega)$ and $B(\omega)$. In order to obtain as an exact equality the relation $\Delta\omega \, \Delta t = 2\pi$ for the bandwidth–time interval product, we can define the time duration Δt to be 2π times the mean decay time τ. Then Eq. (120) gives $\Delta\omega \, \Delta t = 2\pi$.

6.5 *Fourier Analysis of a Traveling Wave Packet*

Suppose a transmitter at $z = 0$ drives a continuous homogeneous onedimensional open system in such a way that the wave function $\psi(z, t)$ of the traveling waves has time dependence at $z = 0$ given by a known function of time $f(t)$:

$$\psi(0, t) = f(t) \tag{121}$$

Any reasonable function $f(t)$ can be expanded in a superposition of harmonic oscillations. If $f(t)$ is not a periodic function of time, the superposition is continuous (in frequency) and is given by the Fourier integral

$$f(t) = \int_0^\infty \left[A(\omega) \sin \omega t + B(\omega) \cos \omega t \right] d\omega. \tag{122}$$

Traveling waves in homogeneous dispersive medium. Each harmonic component of the superposition Eq. (122) gives rise to its own harmonic traveling wave, with its angular wavenumber k given by the dispersion relation.

$$k = k(\omega). \tag{123}$$

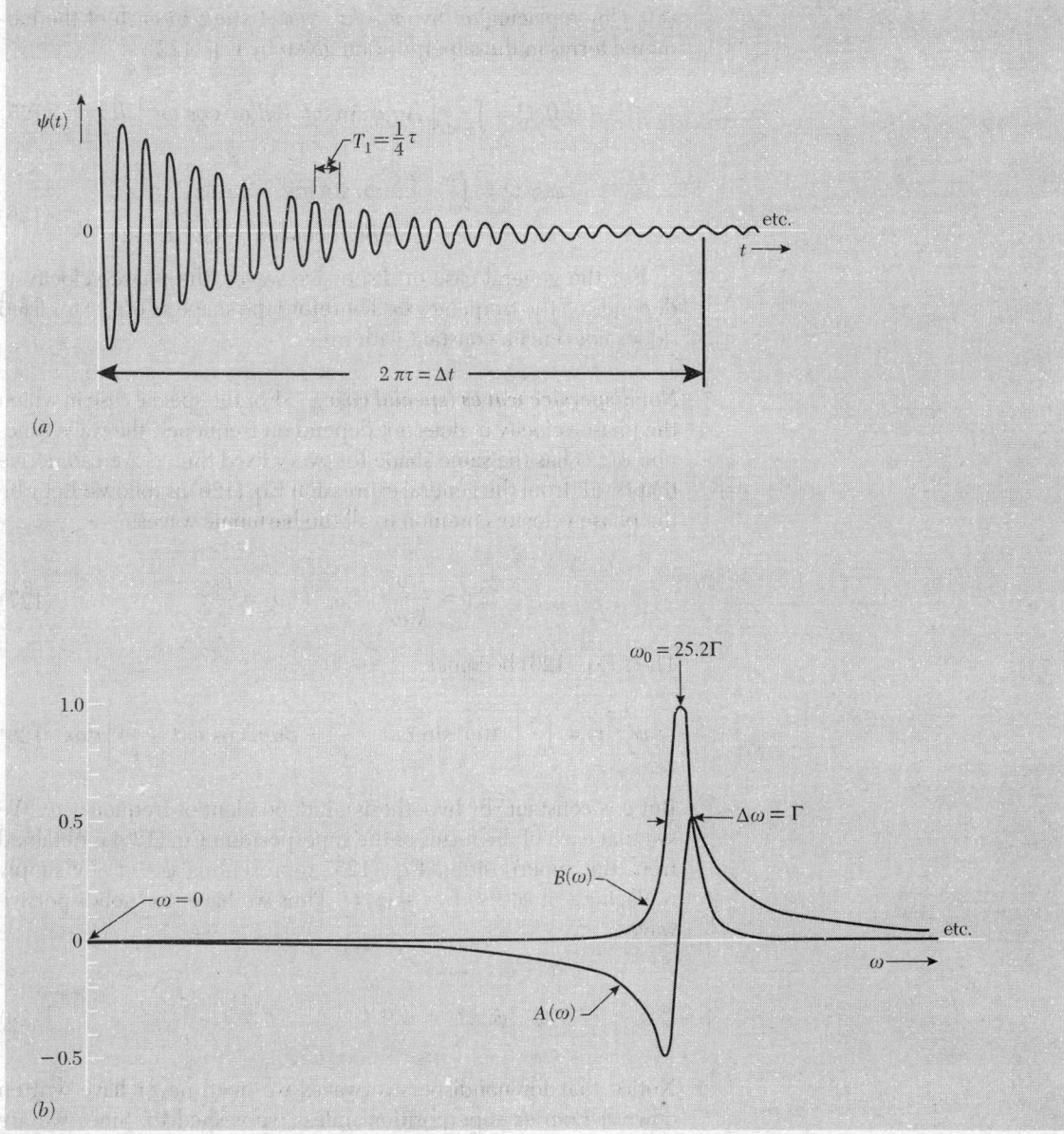

Fig. 6.11 *Weakly damped harmonic oscillator. (a) Pulse* $\psi(t) = e^{-(1/2)t/\tau}\cos\omega_1 t$, *with the choice* $\omega_1 = 8\pi\Gamma$, *that is,* $\tau = 4T_1$. *(b) Fourier coefficients in the continuous superposition of harmonic terms* $\int_0^\infty [A(\omega)\sin\omega t + B(\omega)\cos\omega t]\,d\omega$.

Each frequency component of the harmonic traveling wave travels at its own phase velocity

$$v_\varphi = \frac{\omega}{k(\omega)}. \qquad (124)$$

The total traveling wave $\psi(z, t)$ is just the superposition of all these harmonic traveling waves. That means that we obtain $\psi(z, t)$ from

$\psi(0, t)$ by replacing ωt by $\omega t - kz = \omega t - k(\omega)z$ in each of the harmonic terms in the superposition given by Eq. (122):

$$\psi(0, t) = \int_{\omega=0}^{\infty} \Big[A(\omega) \sin \omega t + B(\omega) \cos \omega t \Big] d\omega, \qquad (125)$$

$$\psi(z, t) = \int_{\omega=0}^{\infty} \Big\{ A(\omega) \sin [\omega t - k(\omega)z] \\ + B(\omega) \cos [\omega t - k(\omega)z] \Big\} d\omega. \qquad (126)$$

For the general case of dispersive waves, the phase velocity v_φ depends on the frequency ω. Therefore the shape of $\psi(z, t)$ for fixed t does not remain constant with time.

Nondispersive waves (special case). For the special case in which the phase velocity v_φ does not depend on frequency, the wave function $\psi(z,t)$ has the same shape for every fixed time t. We can derive that result from the general expression Eq. (126) as follows: Let v be the phase velocity common to all the harmonic waves:

$$v = \frac{\omega}{k(\omega)}; \text{ i.e., } k(\omega) = \frac{\omega}{v}. \qquad (127)$$

Then Eq. (126) becomes

$$\psi(z, t) = \int_{0}^{\infty} \Big[A(\omega) \sin \omega(t - \frac{z}{v}) + B(\omega) \cos \omega(t - \frac{z}{v}) \Big] d\omega. \qquad (128)$$

But v is constant (by hypothesis), independent of frequency ω. We see that each of the terms of the superposition Eq. (128) is obtained from the superposition Eq. (125) representing $\psi(0, t)$ by simply replacing t in $\psi(0, t)$ by $t - (z, t)$. Thus we have, for nondispersive waves,

$$\psi(z, t) = \psi(0, t'), \qquad t' \equiv t - \frac{z}{v}. \qquad (129)$$

Notice that for nondispersive waves we need never have written down a Fourier superposition unless we wished to. Since we are given $\psi(0, t)$, we could obtain $\psi(z, t)$ immediately from Eq. (129) without the necessity for the intermediate steps of Fourier analysis. Equation (129) says that a traveling wave in a nondispersive medium travels without changing its shape. That is, the displacement (or electric field, or whatever it is) at the downstream point z has the same value at time t as the displacement at $z = 0$ at the earlier time $t - (z/v)$.

Here is an example of a nondispersive wave in which we make no use of Fourier analysis or of harmonic functions. Suppose that we have nondispersive waves (for example, audible sound waves or light in vacuum). Suppose that at $z = 0$ the displacement satisfies

$$\psi(0, t) = Ae^{-(1/2)t^2/\tau^2}. \tag{130}$$

Equation (140) is a *Gaussian-shaped pulse*. It is maximum at $t = 0$ and becomes very small for times much earlier than $t = 0$ or much later than $t = 0$ (in units of τ). We can Fourier-analyze Eq. (140), but we do not need to, since the medium is by hypothesis nondispersive. We can immediately write down the form of the traveling wave:

$$\psi(z, t) = \psi(0, t') = Ae^{-(1/2)(t')^2/\tau^2}$$

$$= Ae^{-(1/2\tau^2)[t-(z/v)]2}. \tag{131}$$

Home experiment

Nondispersive waves and classical wave equation. Every harmonic traveling wave of the form

$$\psi(z, t) = A \cos\,[\omega t - k(\omega)z] \tag{132}$$

satisfies (as you can easily show) the differential equation

$$\frac{\partial^2 \psi(z, t)}{\partial t^2} = \frac{\omega^2}{\kappa^2} \frac{\partial^2 \psi(z, t)}{\partial z^2} = v_\varphi^2(\omega) \frac{\partial^2 \psi(z, t)}{\partial z^2}. \tag{133}$$

For the special case in which the waves are nondispersive, we have $v_\varphi = v$, a constant velocity independent of ω. In that case, every term in a superposition of traveling harmonic waves [like Eq. (128)] satisfies the same differential equation, namely

$$\boxed{\frac{\partial^2 \psi(z, t)}{\partial t^2} = v^2 \frac{\pi^2 \psi(z, t)}{\partial z^2},} \tag{134}$$

where $\psi(z, t)$ is supposed to represent any one of the harmonic traveling waves in the superposition. But since each term satisfies Eq. (134), so does the entire superposition; i.e., the total wave function $\psi(z, t)$ satisfies Eq. (134). This partial differential equation is called the *classical wave equation for nondispersive waves*, or simply the classicalwave equation.

Waves that preserve their shape satisfy classical wave equation. We made use of the harmonic traveling waves of Eq. (132)

in obtaining Eq. (134). That was not necessary. Any traveling wave which preserves its shape as it travels must satisfy Eq. (134). Thus suppose we are given $\psi(0, t) = f(t)$, and we are given that the wave travels without change of shape, i.e.,

$$\psi(z, t) = f(t'), \qquad t' \equiv t - \frac{z}{v}. \tag{135}$$

You can easily see that $\psi(z, t)$ as given by Eq. (135) satisfies the classical wave equation. (Prob. 6.26.) Similarly, any nondispersive traveling wave traveling in the $-z$ direction also satisfiesthe classical wave equation, as you can see by replacing v by $-v$ in your delivation. Also, any superposition of nondispersive traveling waves traveling in both directions satisfies the classical wave equation, since all the terms in the superposition satisfy it.

A harmonic standing wave of the form

$$\psi(z,t) = A \cos k(z - z_0) \cos \omega(t - t_0)$$

Satisfies Eq. (133), as you can easily show. If the medium is nondispersive, then all harmonic standing waves satisfy the classical wave equation, Eq. (134). This follows from Eq. (135) with $v_\varphi = v$ for all frequencies. (For a standing wave, v_φ means ω/k, even though the concept of phase velocity is not a natural one to use in describing standing waves.) It also follows from the fact that a standing wave can be regarded as a superposition of traveling waves traveling in opposite directions. In fact, our first encounter with the classical wave equation was in studying standing waves on a continuous string in Sec. 2.2.

Problems and Home Experiments

6.1 Show that the sum of two traveling harmonic waves $A_1 \cos (\omega t - kz + \varphi_1)$ and $A_2 \cos (\omega t - kz + \varphi_2)$ traveling in the $+z$ direction and *having the same frequency* ω is itself a harmonic traveling wave of the same kind. That is, the sum can be written in the form $A \cos (\omega t - kz + \varphi)$. Find out how A and φ are related to A_1, A_2, φ_1, and φ_2. (*Hint:* The use of complex numbers or a rotating vector diagram helps immensely.)

6.2 Consider electromagnetic radiation in a medium with dielectric constant $\epsilon(\omega)$. Suppose the magnetic permeability μ is 1. Then $n(\omega) = [\epsilon(\omega)]^{1/2}$. According to the theory of relativity, no signal can propagate faster than $c = 3.0 \times 10^{10}$ cm/sec. What limitation does that put on the possible variation of $\epsilon(\omega)$ with ω? [Assume $\epsilon(\omega)$ is positive for all ω.]

Ans. $\omega (dn/d\omega) + (n - 1) \geqq 0.$

Home experiment   **6.3** Measure the approximate bandwidth broadcast band received in yom AM radio by twisting the tuning dial and seeing what the extremes are for receiving a given station. (The highest reading on the AM dial is usually 130. This means 1300kc.) How does your result compare with the result that one needs $\Delta \nu \approx 40$ kc to cover both sidebands for very high fidelity sound reproduction?

6.4   Tubas can play very low, for example C_1 at 32.7 cps (the lowest C on the piano, called C_1). Flutes can play very high, their highest note normally being C_7 at 2093 cps (one octave below the highest note on the piano). Each note on the equal tempered scale differs from its neighbor by a factor of about 1.06. Flutes can play very fast; tubas don't go so fast. Is that the fault of the tuba player? Or the tuba? Could the tuba be redesigned so that tuba players could play as fast as flutists? What would calculate as reasonable top speed for tubists playing scales near C32.7? For flutists playing near C2093? First you have to decide on a reasonable musical criterion, then do physics.

Ans.   2 notes/sec for tuba; 120 per sec for flute (wow!).

6.5   A man brought his AM radio into a repair shop and complained that the tuning wasn't fine enough. He wanted a given station to be very sharply defined on the dial. So they fixed it according to his desires. Then he brought it back again. What was his complaint this time?

6.6   (*a*) One way to measure the velocity of sound in air is to clap your hands and determine the time delay between the clap and an echo from a known reflector. Another way is to measure the length of a mailing tube that resonates at a known frequency (and correct for end effects). Do these methods detelmine phase velocity? Group velocity?

(*b*) One way to measure the velocity of light is to send a chopped light beam through the air from Mt. Wilson to Mt. Palomar, reflect it from a mirror, and time the round trip. Another way is to find the length of a resonant cavity oscillating in a known mode at a known frequency. Do these methods determine phase velocity? Group velocity?

6.7   Show that for light of index $n(\lambda)$,

$$\frac{1}{v_g} = \frac{1}{v_\varphi} - \frac{1}{c} \lambda \frac{dn(\lambda)}{d\lambda},$$

where λ is the vacuum wavelength of the light.

6.8   The velocity of light in vacuum is quoted in tables as $c = 2.997925 \times 10^{10}$ cm/sec. It is pretty well known. Suppose you measured the velocity of light by reflecting a chopped light beam between Mt. Wilson and Mt. Palomar and timing the round trip. Suppose that at first you neglect to take into account the fact that the path is in air rather than vacuum. Estimate the correction that must be added to or subtracted from your measured value to obtain the velocity in vacuum, assuming that the light travels at the phase velocity in air. Repeat your estimate of the correction, assuming that the light travels at the group velocity in air. (For the index of refraction of air, use $n = 1 + 0.3 \times 10^{-3}$.) In order to estimate the correction using the group velocity, use the result of Prob. 6.7. Also, assume that an air molecule is indistinguishable from a glass molecule. Therefore, *if* there were as many air molecules per unit volume in air at STP as there are glass molecules in glass, you could obtain $dn/d\lambda$ directly from Table 4.2, Sec. 4.3. But there aren't. For air, $N \approx 2.7 \times 10^{19}$ molecules/cm^3. For glass, $N \approx 2.6 \times 10^{22}$ molecules/cm^3. Find $dn/d\lambda$ for air (for average visible light), using Table 4.2 and a suitable correction for number density. Finally, does it matter which correction you use (assuming you are aiming for the kind of accuracy quoted above)? Which correction *should* you use?

6.9 Show that for a damped harmonic oscillator the decay lifetime τ is given by

$$\frac{1}{\tau} = \frac{1}{E_{stored}} \frac{dE_{lost}}{dt}.$$

6.10 Suppose you tap a mailing tube on your head. You hear the pitch of the lowest mode for a short time. Assume the oscillation is damped harmonic motion. Thus there is a certain decay time τ. Now suppose you double the length of the tube; the lowest mode frequency is halved. But suppose you some how excite the tube so it vibrates at the original frequency (which is now the second mode of the longer tube). The excitation is sudden, and the air then oscillates freely and undergoes damped oscillations.

(*a*) Assume that all the energy loss is due to radiation out the ends of the tube. Compare the new decay time with the old one.

(*b*) Assume that the tube diameter is so small that the energy loss at the ends of the tube is negligible compared with frictional loss along the walls of the tube and radiation out the sides of the tube. Compare the new and old decay times again.

(*c*) Suppose you measure the resonant full widths of the new and old tubes by driving each of them with the same tuning fork (oscillating at the lowest mode frequency of the original short tube) and varying the length of the tube with a paper "trombone." Compare the "length full width" ΔL for the two cases mentioned above. Be careful—relate ΔL to the frequency full width. Make use of the results of Prob. 6.9.

Home experiment **6.11 Water wave packets.** The best way to understand the difference between phase and group velocities is to make water wave packets. To make expanding circular wave packets having dominant wavelength 3 or 4 cm or longer, throw a big rock in a pond or pool. To make straight waves (the two-dimensional analog of three-dimensional plane waves) with wavelengths of several centimeters, float a stick across the end of a bathtub or a large pan of water. Give the stick about two swift vertical pushes with your hand. After some practice, you should see that for these packets the phase velocity is greater than the group velocity. (See Table 6.1, Sec. 6.2.) You will see little wavelets grow from zero at the rear end of the packet, travel through the packet, and disappear at the front. (It takes practice; the waves travel rather fast.) Another good method is to put a board at the end of a bathtub and tap the board.

To make millimeter-wavelength waves (surface tension waves), use an eyedropper full of water. Squeeze out one drop and let it fall on your pan or tub of water. First let the drop fall from a height of only a few millimeters. This gives dominant wavelengths of only a few millimeters. To see that these waves really are due to surface tension, add some soap to the water and repeat the experiment. You should notice a decrease in the group velocity when you add the soap. (To see that the longer wavelength waves are *not* due to surface tension, you can repeat the experiment at long wavelengths.) To lengthen the dominant wavelength of the group, let the water drop fall from a greater height.

Here is a way to see (without doing a difficult measurement) that millimeter waves have a faster group velocity than waves of a centimeter or so. Generate a packet that has both millimeter and centimeter waves by dropping a water drop from a height of a foot or so into a circular pan filled to the brim. (A coffee can works very well.) Drop the drop near the center of the circular pan. Notice that after reflection from the rim the group comes to a focus at a point that is conjugate to the point where the drop hit. (By two conjugate points we mean points located on a line through the center of the circle and lying at equal distances on either side of the center.) When the packet is passing through the conjugate focus, there is a transitory standing wave there (similar to the transitory standing wave you get when you shake a wave packet onto a slinky tied to a wall). This enables you to judge the average arrival time of the packet. Look to see if there is a difference in arrival times for short-wavelength contributions to the packet as compared with longer wavelength contributions. It is difficult to measure, but you can *see* the effect fairly easily.

An experiment I have not yet tried is to find a smooth running stream with flow velocity roughly equal to the group velocity for reasonable wavelengths. One should be able to make wave packets that travel upstream at about the flow velocity, so that the packet remains nearly at rest in your reference frame (assuming you are wading, not floating with the current). Surely that would be a most pleasant way to study wave packets.

6.12 Shallow-water wave packets—tidal waves. In Prob. 2.31 you derived the dispersion law for sawtooth shallow-water standing waves, obtaining the result $v_\varphi \approx 1.1\sqrt{gh}$. For sinusoidal shallow-water waves the result turns out to be $v_\varphi = \sqrt{gh}$. Thus shallow-water waves are nondispersive. (The phase velocity does not depend on the wavelength.) Instead of standing waves we now consider shallow-water traveling wave packets. Since the waves are nondispersive, a single "solitary wave" or "tidal wave" will propagate without changing its shape (approximately). Such waves,

Home experiment

called *tsunami*, can be excited by undersea earthquakes in the ocean. The average water depth in the deep ocean is about 5 kilometers: $h = 5 \times 10^5$ cm. Tidal waves of horizontal length much longer than 5 km are therefore "shallow-water" waves. Tsunami waves propagate in the deep ocean at a velocity

$$v = \sqrt{gh} = \sqrt{(980)5 \times 10^5} = 2.2 \times 10^4 = 220 \text{ meter/sec}$$

$$= 495 \text{ miles/hour,}$$

which is somewhat slower than a typical jet airplane. How long does it take such a tidal wave to propagate from Alaska to Hawaii?

In 1883 the volcano Krakatoa blew up, creating the world's biggest explosion. (Krakatoa is located in Sunda Strait, between Sumatra and Java. An account of the explosion can be found in any encyclopedia.) Huge tidal waves and atmospheric waves were created. Recently it has been discovered that there are air traveling waves with velocity about 220 m/sec. (Recall that ordinary sound velocity at $0°C$ is 332 m/sec. On the average the air is colder than that, so the velocity is less than that.) The existence of these air waves probably explains how the tidal water waves from Krakatoa appeared on the far sides of land masses that should have blocked the water waves. Apparently the tidal waves "jumped over" the land masses by coupling to the air waves having the same velocity (and same excitation time). See the article by F. Press and D. Harkrider, "Air-Sea Waves from the Explosion of Krakatoa." *Science* **154,** 1325 (Dec. 9, 1966).

In the experiment make your own shallow-water tidal waves as follows: Take a square pan a foot or two long. Fill it with water to a depth of about $\frac{1}{2}$ or 1 cm. Give the pan a quick nudge (or lift one end and drop it suddenly). You will create two traveling wave packets, one at the near end and one at the far end, traveling in opposite directions. Follow the bigger of the packets. Measure the velocity by timing the wave for as many pan lengths as you can (probably about four). A stopwatch helps. Alternatively, you can count out loud as the packet hits the walls, memorize the "musical tempo," and finally measure the tempo with an ordinary watch. How well do your results agree with $v = \sqrt{gh}$? As the depth of the water increases, you will finally get to the point where the waves are not shallow-water waves. Then the dispersion relation gradually goes over to the deep-water gravitational-wave dispersion relation $\omega_2 = gk$, i.e.,

$$v_\varphi = \lambda v = \sqrt{\frac{g\lambda}{2\pi}}$$

(We shall derive this relation in Chap. 7.) Thus the wave packet will spread out and not maintain its shape. For sufficiently shallow water (less than 1cm, roughly), the shape is maintained fairly well for several feet.

Finally, make a traveling tidal wave in your bathtub by suddenly pushing the entire end of the tub water with a board. Measure the down and back time and thus measure the velocity. Is it $\sqrt{gh}$? Notice the breakers!

6.13 Musical trills and bandwidth. This experiment requires a piano. Trill two adjacent notes (a halftone apart). First pick two notes near the top of the keyboard. Trill slowly, then as fast as you can. Estimate the trill frequency. Can you still easily make out the two notes of the trill? Now trill two adjacent notes near the bottom of the keyboard, first very slowly, then gradually more rapidly. Is there a speed at which the two notes blend into a messy, indistinguishable mixture? Estimate the frequency where things get messy. Then do arithmetic and decide how good your ear and brain are at recognizing two separate maxima in the Fourier analysis, even when the frequency widths of the peaks (at half-maximum intensity) are not small compared to the frequency spacing between the maxima.

6.14 Group velocity at cutoff. Show that for a system of coupled pendulums the group velocity is zero at both the lower and upper cutoff frequencies (minimum and maximum frequencies for sinusoidal waves). What is the phase velocity at these two frequencies? Make a sketch of the dispersion relation, i.e., a plot of ω versus k. Show how one can read at a glance the group and phase velocities from such a diagram.

6.15 Fourier analysis of exponential function. Consider a function $f(t)$ that is zero for negative t and equals exp $(-t/2\tau)$ for $t \geq 0$. Find its Fourier coefficients $A(\omega)$ and $B(\omega)$ in the continuous superposition

$$f(t) = \int_0^\infty [A(\omega) \sin \omega t + B(\omega) \cos \omega t] \, d\omega.$$

6.16 Truncated sine wave with one oscillation. Suppose $f(t)$ is zero except in the interval from $t = t_1$ to $t = t_2$ of duration $\Delta t = t_2 - t_1$ and centered at $t_0 = \frac{1}{2}(t_1 + t_2)$. Suppose that in this interval $f(t)$ makes exactly one sinusoidal oscillation at angular frequency ω_0, starting and ending with value zero at t_1 and t_2 (i.e., $\Delta t = T_0 = 2\pi/\omega_0$). Find the Fourier coefficients $A(\omega)$ and $B(\omega)$ in the continuous superposition

$$f(t) = \int_0^\infty [A(\omega) \sin \omega(t - t_0) + B(\omega) \cos \omega(t - t_0)] \, d\omega.$$

Make a rough plot of the Fourier coefficients versus ω and a sketch of $f(t)$.

6.17 Beaded string. Derive an expression for the group velocity of traveling waves on a beaded string. Plot (roughly) the dispersion relation for the beaded string from $k = 0$ to the maximum value. Plot (roughly) the group velocity versus k and the phase velocity versus k from $k = 0$ to k_{max}.

6.18 Phase and group velocities for light in glass. Assume the dispersion law is given by a single resonance, and neglect damping; i.e., assume

$$c^2 k^2 = \omega^2 \left(1 + \frac{\omega_p^2}{\omega_0^2 - \omega^2} \right), \qquad \omega_p^2 = \frac{4\pi N e^2}{m},$$

where N is the number of resonating electrons per unit volume.

(a) Sketch the square of the index of refraction, n^2, versus ω, for $0 \leq \omega < \infty$. The important features are the value and slope at $\omega = 0$, at ω slightly less than ω_0 and slightly greater than ω_0, at $\omega = \sqrt{\omega_0^2 + \omega_p^2}$, and at infinity. How do you interpret the region where n^2 is negative? The region near ω_0?

(b) Derive the following formula for the square of the group velocity:

$$\left(\frac{v_g}{c}\right)^2 = \frac{1 + \dfrac{\omega_p^2}{\omega_0^2 - \omega^2}}{\left[1 + \dfrac{\omega_p^2 \omega_0^2}{(\omega_0^2 - \omega^2)^2}\right]^2}$$

Sketch $(v_g/c)^2$ versus ω. Show that $(v_g/c)^2$ is always less than unity, as required by the theory of relativity. Show that v_g^2 is negative in the same frequency region where n^2 is negative. For what frequency is the group velocity greatest? What is the group velocity at that frequency?

6.19 Phase and group velocities for deep-water waves.   The dispersion law is

$$\omega^2 = gk + \frac{Tk^3}{\rho},$$

where $g = 980$, $T = 72$, and $\rho = 1.0$ (all in CGS units). Derive formulas for the group velocity and for the phase velocity. Show that the group velocity equals the phase velocity when gk and Tk^3/ρ are equal and that this occurs for wavelength 1.7 cm and velocity 23.1 cm/sec. Show that for *surface-tension waves*, i.e., waves with wavelength very short compared with 1.7 cm, the group velocity is 1.5 times the phase velocity. Show that for *gravity waves*, i.e., waves with wavelength long compared with 1.7 cm, the group velocity is half the phase velocity. Extend Table 6.1, Sec. 6.2 to include wavelength 128 meters and 256 meters. Give the wave velocities in km/hr as well as cm/sec. (To observe frequencies as low as four or five per minute, go to a beach on a sheltered ocean bay on a day when there is no strong offshore wind. Then the only waves are those coming from far out at sea.)

6.20 Fourier analysis of a single square pulse in time.   Consider a square pulse $\psi(t)$ which is zero for all t not in the interval t_1 to t_2. Within that interval, $\psi(t)$ has the constant value $1/\Delta t$, where $\Delta t = t_2 - t_1$. Let t_0 be the time at the center of the interval. Show that $\psi(t)$ can be Fourier-analyzed as follows:

$$\psi(t) = \int_0^\infty A(\omega) \sin\, \omega(t - t_0)d\omega + \int_0^\infty B(\omega) \cos\, \omega(t - t_0)d\omega,$$

with the solution

$$A(\omega) = 0, \qquad B(\omega) = \frac{1}{\pi} \frac{\sin \frac{1}{2} \Delta t \omega}{\frac{1}{2} \Delta t \omega}$$

Sketch $B(\omega)$ versus ω. In the limit where Δt goes to zero, $\psi(t)$ is called a "delta function" of time, written $\rho(t - t_o)$. What is $B(\omega)$ for this delta function of time?

6.21 Fourier analysis of a truncated harmonic oscillation. Suppose $\psi(t)$ is zero outside of the interval from t_1 to t_2, which has duration $t_2 - t_1 = \Delta t$ and central value $\frac{1}{2}(t_1 + t_2) = t_0$. Suppose $\psi(t)$ is equal to $\cos \omega_0(t - t_0)$ within that interval.

(*a*) Show that $\psi(t)$ can be Fourier-analyzed as follows:

$$\psi(t) = \int_0^\infty B(\omega) \, \cos \, \omega(t - t_0),$$

$$\pi B(\omega) = \frac{\sin \left[(\omega_0 + \omega) \frac{1}{2} \Delta t \right]}{\omega_0 + \omega} + \frac{\sin \left[(\omega_0 - \omega) \frac{1}{2} \Delta t \right]}{\omega_0 - \omega}$$

(*b*) Show that if Δt is much shorter than the period of any frequency that we can measure or are interested in, then $\pi B(\omega)$ has the constant value Δt.

(*c*) Show that if $\Delta(t)$ contains many oscillations, i.e., if $\omega_0 \Delta t \gg 1$, then, for ω sufficiently near ω_0, $B(\omega)$ is given essentially by the second term only:

$$\pi B(\omega) \approx \frac{\sin \left[(\omega_0 - \omega) \frac{1}{2} \Delta t \right]}{\omega_0 - \omega}, \qquad |\omega_0 - \omega| \ll |\omega_0 + \omega|.$$

(*d*) Sketch $\psi(t)$ and $B(\omega)$ for part (*c*).

This problem can help us to understand *collision broadening* of spectral lines. An undisturbed atom emitting almost monochromatic visible light has a mean decay time of about 10^{-8} sec, and thus the Fourier spectrum of its radiation has a bandwidth $\Delta\nu$ of about 10^8 cps. If the atoms are in a gas-discharge-tube light source, then it turns out that the bandwidth of the emitted light (called "linewidth" in optics) is about 10^9 cps, rather than 10^8 cps. Part of the reason for this "line broadening" is the fact that the atoms do not radiate in a free and undisturbed manner; they *collide*. A collision results in a sudden change in amplitude or phase constant or both. That is similar to the situation illustrated by the truncated harmonic oscillator. A given atom may spend most of its time "unexcited." Occasionally it is excited into an oscillatory motion of the optical (valence) electrons (we are speaking classically; a more accurate picture requires quantum mechanics). The atom begins to oscillate as a damped harmonic oscillation with decay time of order 10^{-8} sec. However, within a time Δt of about 10^{-9} sec (in a typical gas-tube light source), it has a collision that truncates the oscillation in some random way. If one adds the light from many such sources, the bandwidth Δv will be given by $\Delta v \approx (1/\Delta t) \approx 10^9$ cps.

6.22 Fourier analysis of almost periodically repeated square pulse. A single square pulse of duration Δt in time gives a continuous frequency spectrum that has its most important contributions between zero and $v_{max} = \Delta v$, with $\Delta v \approx 1/\Delta t$. (See Prob. 6.20.) A periodically repeated square pulse of duration Δt, repeated at time intervals T_1 (with $T_1 > \Delta t$), gives a *discrete* frequency spectrum consisting of *harmonics* (integral multiples) of $v_1 = 1/T_1$, with the most important contributions extending from zero to $v_{max} = \Delta v$, with $\Delta v \approx 1/\Delta t$. (See Prob. 2.30). Now consider an "almost periodic" repeated square pulse of duration Δt, repeated at time intervals T_1, for a total time T_{long}, where the time T_{long} is long compared to the period T_1. If T_{long} were infinite, we would have an exactly periodically repeated square wave, as described above. In that case each of the discrete harmonics would be "infinitely narrow."

(*a*) Show that for a finite value of T_{long} the Fourier analysis of this almost periodically repeated square pulse consists of a superposition of *almost discrete* harmonics of the fundamental frequency $v_1 = 1/T_1$, each harmonic being actually a coutinuum of frequencies extending over a narrow frequency band of width $\delta v \approx 1/T_{long}$. The most important harmonics lie between zero and $v_{max} \approx 1/\Delta t$. You need not perform any integrations. Use qualitative arguments.

(*b*) Sketch qualitatively the shapes of $\psi(t)$ and of the Fourier coefficients $A(\omega)$ or $B(\omega)$ that you would expect, without worrying about the distinction between $A(\omega)$ and $B(\omega)$.

6.23 Mode-locking of a laser to achieve narrow pulses of visible light. (First work Prob. 6.22.) A laser consists (crudely speaking) of a region of length L with mirrors at each end to reflect light back and forth. Under proper conditions, when the space is filled with suitable excited atoms, the radiation from each atom stimulates other excited atoms to radiate with phase relations that give constructive interference among all the radiating atoms, for radiation along the length of the laser (back and forth between the mirrors). Then all the atoms oscillate in phase, and the system of atoms plus radiation oscillates in a *normal mode*. The frequencies of the possible normal modes of free oscillation are harmonics of a fundamental frequency v_1. The period $T_1 = 1/v_1$ is just the "down and back" time required for light to propagate down and back between the mirrors. Thus $T_1 = 2L/(c/n)$, where n is the index of refraction. Then $v_1 = 1/T_1$, and the possible modes have frequencies $v = mv_1$, where $m = 1, 2, 3$, etc. Now, if there were no mirrors, the excited atoms would independently radiate their usual light. For a helium-neon gas laser, this is the red neon light of wavelength 6328 Å. The damping time τ for a single atom in that case would be about 10^{-9} sec, giving a frequency bandwidth Δv of about 10^9 cps. When, instead, one has a normal mode of the entire system (of atoms plus radiation), the damping time for the mode of the entire system is very much longer than the free-decay time τ for a single atom. The damping of the mode is caused by leakage of light out through the end mirrors, imperfectly parallel light "walking" sideways off the mirrors, and other factors. The damping time T_{long} can be hundreds or thousands of times the free-decay time. This means that each mode has a frequency width $\delta v \approx 1/T_{long}$ which is hundreds or thousands of times narrower than the natural linewidth Δv. The natural linewidth Δv does play an important role, however. Since it takes initially freely decaying atoms to get the whole system excited into a mode, the only modes appreciably excited are those for which the mode frequency mv_1 lies somewhere in the band Δv of the freely decaying atoms. For visible light and with a length L on the order of 1 meter, it is easy to see that the harmonic number m is a very large integer.

(*a*) What is the order of magnitude of the mode integer m?

(*b*) Sketch the shape of the frequency spectrum of the important modes of a laser.

In other words, put into graphical form what has been said so far. Label the frequency separation v_1 between "adjacent" mode frequencies, the frequency width δv of each mode, and the frequency width Δv of the most easily excited modes.

Now we continue: When any complicated system is given some excitation and then allowed to oscillate, it oscillates in some more or less complicated superposition of its normal modes. If it is excited in a rather "brutal" fashion, there may be many modes present, with no particularly simple phase relationship between the different modes. We could call such a superposition an "incoherent" superposition of modes. That is what you get ordinarily if you excite a laser in such a way that several of its modes are excited. For example, it is not hard to excite a laser in such a way that practically all the modes in the band Δv are excited. The phase relation between the different modes is "random," in the following sense: If you look at the system at one time and determine the relative phases of the modes, and then look at a later time, much later than the decay time T_{long}, the relative phases among the modes will be unpredictably different. That is because during a time of order T_{long} the energy has all leaked out of a given mode and has meanwhile been replenished by newly excited atoms. The mode is thus effectively "turned on again" about once every interval T_{long}. The "turn-on time" is random. Therefore the phase has changed unpredictably in a time of order T_{long}. Now, the frequency spectrum of the important modes that you sketched in part (*b*) is quite similar to the frequency spectrum of the Fourier analysis of an almost periodically repeated square pulse, as given in Prob. 6.22. There is one extremely important difference, however. In the Fourier analysis of the almost periodic square wave, there is a very definite and completely specified phase relationship between each of the frequency components that make up the superposition. This is *not* the case for an incoherent mixture of laser modes.

(*c*) Show that a superposition of an incoherent mixture of laser modes, each of bandwidth $\delta v \approx 1/T_{\text{long}}$ and occupying a total frequency region of width Δv, gives a time dependence $\psi(t)$ that is an almost periodic function of t with period T_1. Show that this almost periodic function will only maintain a recognizable similarity to itself during successive periods T_1 contained in time intervals of order T_{long}. Show that, although it may happen by chance that during a given time interval of order T_{long} the almost periodic function $\psi(t)$ happens to look like a periodically repeated square wave of duration $\Delta t \approx 1/\Delta v$, this would only happen by rare accident. Ordinarily we would expect $\psi(t)$ to be significantly different from zero during the entire period T_1. Thus we would have $\Delta t \gg 1/\Delta v$. Now we are ready to understand the effect of the beautiful invention of *mode-locking*. Suppose that we can somehow get all the important laser modes "locked" in phase with one another; never mind how, yet. Then we may expect that this *coherent* superposition of modes, all with the same phase constant, will give an almost periodic function $\psi(t)$ consisting of repeated pulses of duration $\Delta t \approx 1/\Delta v$, repeated at intervals T_1, with a pulse shape that remains roughly constant over times of order T_{long}. That expectation has been realized experimentally. Here is the ingenious mode-locking trick: Turn on the laser. Some mode near the center of the band Δv will ordinarily be the first to start oscillating. Call this mode v_0. Now arrange it so that (for example) the transparency of the medium, (or of the mirrors, or of some object through which the light must pass) is varied or *modulated* sinusoidally about some average value, the modulation frequency being chosen to equal the fundamental frequency $v_1 = 1/T_1$ that corresponds to the "down and back" time, T_1. Then the first mode to oscillate will have an amplitude that is not constant but is modulated at the modulation frequency v_1:

$$\psi_{\text{1st mode}} = [A_0 + A_{\text{mod}} \cos \omega_1 t] \cos \omega_0 t,$$

where the modulated amplitude is $A_0 + A_{\text{mod}} \cos \omega_1 t$. This "almost harmonic" oscillation can be written as a superposition of exactly harmonic oscillations at frequencies $\omega_0, \omega_0 + \omega_1$, and $\omega_0 - \omega_1$:

$$\psi_{\text{1st mode}} = A_0 \cos \omega_0 t + \tfrac{1}{2} A_{\text{mod}} \cos (\omega_0 + \omega_1) t + \tfrac{1}{2} A_{\text{mod}} \cos (\omega_0 - \omega_1) t.$$

The terms in $\cos (\omega_0 + \omega_1) t$ and $\cos (\omega_0 - \omega_1) t$ now act as driving forces. They help turn on the modes $\omega_0 + \omega_1$ and $\omega_0 - \omega_1$. Thus these modes are not turned on randomly but are *driven* into oscillation. They therefore have a definite phase relation (that given above) to the central mode ω_0. Once the modes $\omega_0 + \omega_1$ and $\omega_0 - \omega_1$ are turned on, their amplitude is modulated by the same physical effect that is modulating ω_0, and with the same phase. Therefore these modes in turn contain components which act as driving forces to turn on their neighbors (one of which is already on, the other is not). In this way the modes $\omega_0 + 2\omega_1$ and $\omega_0 - 2\omega_1$ are turned on. As modes with frequencies farther and farther from ω_0 are turned on, they start up with definite phase relationships. That is how it works.

For a gas laser, the natural decay time τ is of order 10^{-9} sec; therefore the natural linewidth $\Delta \nu$ is of order 10^9 Hz. Hence one can generate pulses of width $\Delta t \approx 10^{-9}$ sec by mode-locking a gas laser. For a solid laser, for example one made of a polished ruby, the natural damping time for individual atoms is of order 10^{-11} sec or 10^{-12} sec. (The atomic oscillations are rapidly damped because of collisions with neighboring atoms in the solid.) Therefore the bandwidth of radiation from atoms emitting red ruby light is about 10^{12} sec^{-1}. This is also the bandwidth of the easily excited laser modes. Using a solid laser, therefore, one can generate ultrashort light pulses of duration $\Delta t \approx 1/\Delta \nu \approx 10^{-11}$ or 10^{-12} sec. Of course this is merely the duration of the light pulse from a single atom decaying in the solid, according to classical mechanics. Then why should we be so enthusiastic about the result? For one thing, a single atom doesn't give much light, whereas we have here a huge number of atoms all emitting at once so as to get an extremely *powerful* light pulse of short duration. Even more important than this is the fact that according to quantum mechanics (and experiment) a single atom does not emit light in a continuous stream as described by our classical model. Instead the light "photon" comes out in a discrete "bundle." For a single atom there is no way of telling exactly when this bundle of energy is going to be emitted. Only the probability versus time is known. Thus one cannot actually obtain synchronized short light pulses using single atoms.

These ultrashort light pulses can be used for many interesting experiments. See A. de Maria, D. Stetser, and W. Glenn, Jr., "Ultrashort Light Pulses," *Science* **156**, 1557 (June 23, 1967).

6.24 Frequency delta function. In Sec. 6.4 we considered the superposition

$$\psi(t) = \int_0^\infty B(\omega) \cos \omega t \, d\omega$$

of a "square" frequency spectrum given by setting $B(\omega) = 1/\Delta \omega$ for ω in the interval from ω_1 to $\omega_2 = \omega_1 + \Delta \omega$ and setting $B(\omega) = 0$ elsewhere. We found that superposition to be

$$\psi(t) = \left[\frac{\sin \frac{1}{2} \Delta \omega t}{\frac{1}{2} \Delta \omega t} \right] \cos \omega_0 t,$$

where ω_0 is the frequency at the center of the band $\Delta \omega$. Let t_{max} be a time longer than the duration of whatever experiment you have in mind. Show that if $\Delta \omega$ is sufficiently small so that $\Delta \omega t_{max} \ll 1$, then, as far as you can tell (in your experiment of duration t_{max}), $\psi(t)$ is an exactly harmonic oscillation of constant amplitude and phase constant. The Fourier coefficient $B(\omega)$ is then called a "delta function of frequency." A delta function of frequency has the properties that it is zero everywhere except in a tiny region, $\Delta\omega$, and that its integral over all ω gives unity. Show that $B(\omega)$ given above has these properties in the limit $\Delta\omega \ll 1/t_{max}$ and hence is an example of a delta function of frequency.

6.25 Resonance in tidal waves. Take the ocean to have uniform depth of 5 km. (That is about the average depth.) Show that a tidal wave generated by (for example) an earthquake travels at about 220 m/sec. Suppose there were no continents. Suppose that the water were confined in "canals" running along lines of constant latitude, so that the water could not move north and south but only east and west. At what latitude would a traveling tidal wave (generated by an earthquake) take 25 hours to encircle the globe? Call this latitude θ_0. (At the equator, θ_0 is zero. At the poles, it is 90 deg.)

The sun and moon provide gravitational driving forces that drive the tides. Consider the moon. (The sun provides half as large a driving force as does the moon.) A moon "day" (the time between successive transits of the moon) lasts about 25 hours. It turns out that if the earth did not turn on its axis the tidal bulges of high water due to the moon would be directly under the moon and also at the diametrically opposite point. At new moon and at full moon, the sun and moon cooperate to give very high tides. Therefore at these times of the month you would expect high water to occur exactly at noon and midnight, with low water at sunrise and sunset (according to the "static model" of a nonrotating earth). At least that is what you would expect out on an island in the ocean. (In a harbor you must wait for the water to flow in and out.) Now consider the "canal model," and a rotating earth. At new moon and full moon, when would you expect high water to occur in the canal at the equator? When would you expect it to occur for a canal at greater latitude than θ_0? (*Hint:* Consider a driven oscillator.)

For further reading on tidal waves, seiches in Lake Geneva, the possible evolution of the Earth-Moon system, and other fascinating subjects, see that popular classic *The Tides*, by George H. Darwin (Charles Darwin's son), written in 1898, available from W. H. Freeman and Company, San Francisco (1962), paperback $2.75. Fourier analysis was just starting to be used in those days, and Darwin describes some simple, ingenious Fourier-analyzing machines, among other things.

6.26 Nondispersive waves. Show that any differentiable function $f(t')$ of $t' = t - (z/v)$ satisfies the classical wave equation, i.e., show

$$\frac{\partial^2 f(t')}{\partial t^2} = v^2 \frac{\partial^2 f(t')}{\partial z^2}$$

Show also that any differentiable function $g(t'')$ of $t'' = t + (z/v)$ satisfies the classical wave equation. Make up an example of a function $f(t')$ and show explicitly that it satisfies the classical wave equation.

6.27 Amplitude modulation and nonlinearity. (*a*) One way to produce an amplitude-modulated carrier wave is to pass a current $I = I_0 \cos \omega_0 t$ oscillating at the carrier frequency ω_0 through a resistance R which is not constant but has a component that varies at the modulation frequency ω_{mod}, that is, $R = R_0(1 + a_m \cos \omega_{mod}t)$. (In a "carbon-granule" microphone the resistance is modulated by the motion of a diaphragm, which compresses the carbon granules that provide the resistance.) The voltage $V = IR$ across the resistor is an amplitude-modulated carrier wave. Find the expression for V in terms of a *superposition* of carrier (frequency ω_0), upper sideband (frequency $\omega_0 + \omega_{mod}$), and lower sideband (frequency $\omega_0 - \omega_{mod}$).

(*b*) Alternatively, suppose we happen to start with two voltages, one oscillating at the carrier frequency, the other at the modulation frequency. The problem is this: How can you physically combine these two voltages, $V_0 = A_0 \cos \omega_0 t$, and $V_m = A_m \cos \omega_{mod}t$ in such a way as to produce an amplitude-modulated carrier wave? First, suppose you merely superpose the two voltages, i.e., you put them both on the broadcasting antenna. Will this work?

(*c*) Next, suppose that the voltages in part (*b*), after being superposed, are then applied to the input of a voltage amplifier. (For example they may be applied between control grid and cathode of a radio tube.) Suppose that the amplifier is a linear amplifier, i.e., its output (for example the plate-to-cathode voltage of the tube) is proportional to the input. Will this work?

(*d*) Finally, suppose that the amplifier output has both a linear and a quadratic component, as follows:

$$V_{out} = A_1 V_{in} + A_2 (V_{in})^2.$$

Let $V_{in} = V_0 + V_m$ as defined in part (*b*). Show that because of the nonlinear quadratic term $A_2(V_{in})^2$ the amplifier output includes, among other things, an amplitude modulated carrier wave, with modulation amplitude proportional to A_m.

(*e*) The amplitude-modulated carrier wave in (*d*) contributes Fourier components with frequencies $\omega_0, \omega_0 + \omega_{mod}$, and $\omega_0 - \omega_{mod}$. What *other* frequency components are there in V_{out}? Make a diagram showing a complete frequency spectrum of the amplifier output. Describe how you could get rid of these other (undesired) components, using bandpass filters. Suppose that ω_{mod} is small compared with ω_0. How selective do the filters have to be?

6.28 Amplitude demodulation and nonlinearity. Suppose that your receiving antenna picks up an amplitude-modulated carrier wave with voltage given by

$$V = V_0 (\cos \omega_0 t)(1 + a_m \cos \omega_{mod}t).$$

How can you recover the modulation voltage, $a_m \cos \omega_{mod}t$? Assume that you have at your disposal whatever bandpass filters you wish, and that you also have at your disposal a nonlinear amplifier of the type described in Prob. 6.27 such that

$$V_{\text{out}} = A_1 V_{\text{in}} + A_2 (V_{\text{in}})^2.$$

(Hint: Express the amplitude-modulated carrier wave as a superposition, pass it through the nonlinear amplifier, and then filter it.)

6.29 Frequency modulation (FM).   A frequency-modulated voltage can be written in the form (for example)

$$V = V_0 \cos\left[\omega_0 (1 + a_m \cos \omega_{\text{mod}} t) t\right] = V_0 \cos \omega t,$$

with

$$\omega = \omega_0 + \omega_0 a_m \cos \omega_{\text{mod}} t.$$

One way of producing a frequency-modulated carrier wave to transmit music is by use of a "capacitative microphone." The sound waves move a diaphragm which moves one plate of a capacitor. The capacitor then has capacitance (for example)

$$C = C_0 (1 + c_m \cos \omega_{\text{mod}} t).$$

Suppose this capacitance is part of an LC circuit with natural oscillation frequency $\omega = \sqrt{1/LC}$. The voltage across the capacitor is, for example, $V = V_0 \cos \omega t$. Show that for c_m small in magnitude compared with unity, one obtains a frequency-modulated voltage with amplitude a_m proportional to c_m. Find the proportionality constant between c_m and a_m.

6.30 Phase modulation (PM).   A phase-modulated voltage can have the form (for example)

$$V = V_0 \cos (\omega_0 t + a_m \sin \omega_{\text{mod}} t) = V_0 \cos (\omega_0 t + \varphi),$$

with

$$\varphi = a_m \sin \omega_{\text{mod}} t.$$

The "instantaneous frequency" is obtained by differentiating the quantity in parentheses with respect to time:

$$\omega = \omega_0 + \frac{d\varphi}{dt} = \omega_0 + a_m \omega_{\text{mod}} \cos \omega_{\text{mod}} t.$$

By comparison with Prob. 6.29 we see that phase modulation and frequency modulation are closely related. (Sometimes both are loosely called FM.)

(*a*) Show that the phase-modulated voltage can be written as a superposition of harmonic oscillations having frequencies ω_0, $\omega_0 \pm \omega_{\text{mod}}$, $\omega_0 \pm 2\omega_{\text{mod}}$, $\omega_0 \pm 3\omega_{\text{mod}}$, etc. [*Hint:* First expand $\cos (\omega_0 t + \varphi)$. Next expand $\sin \varphi$ and $\cos \varphi$ in their infinite Taylor's series. Then use the trigonometric relations developed in Prob. 1.13.]

(*b*) Show that if the modulation amplitude a_m is small compared with unity we can reasonably neglect all terms in the superposition except those with frequencies ω_0 and $\omega_0 \pm \omega_{\text{mod}}$. Thus we see that for small phase-modulation amplitude we have just the carrier and essentially only one upper and one lower sideband. Hence for small a_m the required bandwidth is the same as for AM (amplitude-modulated) transmission. For large a_m the required bandwidth is larger because of the additional sidebands at $\omega_0 \pm 2\omega_{mod}$, etc.

(*c*) Compare the relative phase of carrier and two neighboring sidebands in PM (phase modulation) with those of the carrier and two sidebands for AM found in Prob. 6.27. The phase relations are (as you will find) different. That is one way to distinguish PM and also FM) from AM.

(*d*) Suppose you wish to convert an AM voltage to a PM voltage. You are given whatever bandpass filters you wish, and also are given a circuit that will perform a desired arbitrary phase shift. After you have tried to invent a method, turn to Prob. 9.58 where you will be led by the hand. (This problem is given in Chap. 9 because there is a very beautiful analogy with the phase-contrast microscope (Prob. 9.59).]

6.31 Single sideband transmission.

If information to be transmitted occupies a band of modulation frequencies from $\omega_{mod}(min)$ to $\omega_{mod}(max)$ then the AM or FM broadcast band extends from $\omega_0 - \omega_{mod}(max)$ to $\omega_0 + \omega_{mod}(max)$, where ω_0 is the carrier frequency. The bandwidth is thus $2\omega_{mod}(max)$. Bandwidth is precious, because each station in a region must occupy a different band to prevent mutual overlap and interference of the signals.

(*a*) Suppose you are broadcasting AM radio waves, and you use a bandpass filter to separate out the carrier and upper sideband, discarding the lower sideband. You broadcast just the carrier and upper sideband. Invent a way to re-create the lower sideband *in the receiver* by putting the received signal (carrier and upper sideband) through a nonlinear amplifier of the type described in Probs. 6.27 and 6.28. Discuss the necessary amplitude and phase relations so that you will end up with a signal proportional to the original AM signal.

(*b*) You can decrease the transmitted bandwidth even further if you suppress not only the lower sideband but also the carrier. Suppose you transmit only the upper sideband. Suppose that the receiver has its own "local oscillator" that puts out a signal $V = A \cos \omega_0't$, where ω_0' is as close as possible to being equal to ω_0. (It will never be exactly equal to ω_0, because of unavoidable drifts due to various causes.) Invent a method by which you can combine the signal from the local oscillator with that received from the transmitter (the upper sideband) so as to re-create the lower sideband. Use nonlinear amplifiers, filters, phase shifters—whatever you need.

(*c*) Suppose that the carrier frequency is $\omega_0 = 100$ Mc (1 Mc is 10^6 cps) and that the local oscillator frequency) ω_0' used for single sideband transmission (with carrier also suppressed) exceeds ω_0 by (for example) 30 cps. That is an error of only one part in three million. Suppose that the music consists of a flute playing the note A440 (at 440 cps). What note will come out of your loudspeaker after you have finally recreated the sidebands and demodulated? That result will tell you why at present (1968) single sideband transmission includes the carrier as well as one sideband, for commercial

TV. For voice communication the carrier can be suppressed, since nobody cares if your voice pitch is not precisely reproduced.

6.32 Frequency multiplexing. It often happens that we want to transmit two or more completely independent "channels" of information using the same carrier frequency ω_0. These channels may carry information in the form of bands of modulation frequencies $\omega_{mod}(1)$, $\omega_{mod}(2)$, etc., for channels 1, 2, etc. If the modulation frequency bands do not overlap, one can simply modulate the carrier with all the modulation channels at once. For example you could superpose the carrier and the modulation voltages from all the channels at the input to a nonlinear amplifier, just as you did with a single modulation voltage (single channel) in Prob. 6.27. The amplifier output would then consist (among other things) of an amplitude-modulated carrier that is equivalent to a superposition containing frequencies ω_0, $\omega_0 \pm \omega_{mod}(1)$, $\omega_0 \pm \omega_{mod}(2)$, etc.

(*a*) Justify the preceding statement.

In the receiver you would demodulate, for example as in Prob. 6.28, in order to recover the modulation bands $\omega_{mod}(1)$, $\omega_{mod}(2)$, etc. These bands could then be separated by bandpass filters, provided that the modulation frequency bands do not overlap. Finally we would have separate output information for channels 1, 2, etc., with no "crosstalk" or "overlap," i.e., without the channel 1 output giving us false signals derived from channel 2, etc.

Since in most cases of interest, the modulation frequencies carried by the separate channels occupy overlapping frequency bands, the above method will not work. For example, in FM stereo broadcasting there are two channels, one of which is to give (eventually) a loudspeaker output derived entirely from one input microphone (near "the woodwinds"), the other of which is to give an output from the other input microphone (near "the brass instruments"). The modulation frequencies for the two channels are those of the music, and they overlap.

As another example, in long distance telephone transmission using a single wire, or a single carrier frequency for radio transmission, the various channels consist of various simultaneous telephone conversations. The modulation frequencies are those of the human voice. Similarly in "telemetering" the readings of instruments in an earth satellite back to the ground station, each instrument has a separate channel. The modulation frequencies depend on how the instrument is designed. (For example, a thermometer may consist of a capacitor whose capacitance changes with temperature. This capacitance may then determine the frequency ω_{mod} in an LC circuit in an oscillator.) The modulation frequencies may largely overlap.

What is needed, then, is a means to "label" each channel, so that the channels can be kept separate. One method would be to use a different carrier frequency for each channel. That is what one does with separate radio or TV stations. But there is a much more convenient method, called *frequency multiplexing*. In frequency multiplexing each channel is "labeled" with its own "subcarrier" frequency, as follows: Call the subcarrier frequencies ω_1, ω_2, etc., for channels 1, 2, etc. (The subcarrier frequencies are large compared with the modulation frequencies. The main carrier frequency ω_0 in turn is large compared with any of the subcarriers.) Subcarrier ω_1 is amplitude modulated (or frequency modulated) at modulation frequency $\omega_{mod}(1)$ by channel 1. This gives a channel 1 amplitude-modulated output that consists of a superposition with frequencies ω_1, $\omega_1 + \omega_{mod}(1)$, and $\omega_1 - \omega_{mod}(1)$. Similarly channel 2 has an output with frequencies ω_2, $\omega_2 + \omega_{mod}(2)$, and $\omega_2 - \omega_{mod}(2)$. The subcarrier frequencies ω_1 and ω_2 are chosen to be sufficiently far apart so that there is no overlap in the two bands that surround the two carriers, i.e., taking ω_1 less than ω_2, the highest fequency in the upper sideband $\omega_1 + \omega_{mod}(1)$ is less than the smallest frequency in the lower sideband $\omega_2 - \omega_{mod}(2)$. For example, for FM stereo broadcasting, typical subcarrier frequencies are $\nu_1 = 20$ kc, $\nu_2 = 40$ kc. If the modulation frequencies (the music) extend from zero to 10 kc, then channel 1 will include a band from 10 to 30 kc and channel 2 a band from 30 to 50 kc. So far we seem to have two carriers (for two channels). But we have not yet gotten to the output antenna! We now superpose the outputs of all the channels and regard this multi channel multiband superposition as one big band of modulation frequencies extending from the lower end of the lower sideband of channel 1 to the upper end of the upper sideband of the highest channel. We use this complete band to modulate the main carrier ω_0, by (for example)superposing this multichannel band on the carrier voltage and applying the resultant to the input of a nonlinear amplifier, as in Prob. 6.27.

(*b*) If you use the nonlinear amplifier of Prob. 6.27, what will the amplifier output consist of? Rather than use formulas you can make a qualitative sketch of intensity versus frequency. Show the frequency bands near ω_0 (the main carrier) that you will apply to the transmitting antenna. Show also the other frequencies coming out of the amplifier that you will filter out and discard.

(*c*) At the receiver you can "demultiplex" as follows: Apply the signal consisting of the carrier ω_0 and its multiband upper and lower sidebands to the input of a nonlinear amplifier, as in Prob. 6.28. The amplifier output will include among other things, the subcarrier ω_1 and its sidebands $\omega_1 \pm \omega_{mod}(1)$, and similarly for the other channels. Justify that statement. The various subcarriers and their sidebands are nonoverlapping and can now be separated by bandpass filters. Each channel then provides its own output, with no "crosstalk."

6.33 Multiplex Interferometric Fourier Spectroscopy (MIFS). In 1967, the technique of infrared astronomy was revolutionized by a new technique called Multiplex Interferometric Fourier Spectroscopy, or MIFS for short. The new technique provides a factor of 100 improvement in frequency resolution over old techniques, and a factor of 60,000 decrease in the time spent collecting light in order to determine a frequency

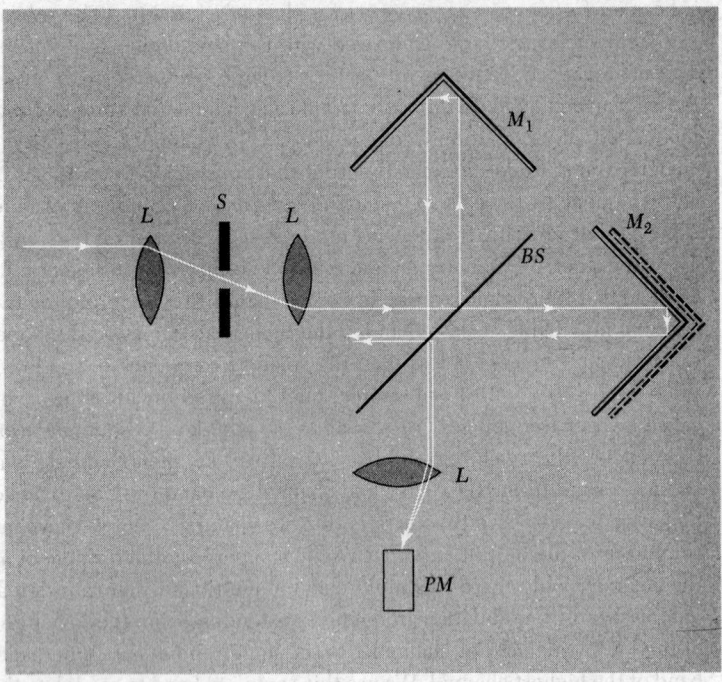

spectrum. The MIFS technique is an ingenious application of the concept of frequency multiplexing discussed in Prob. 6.32.

The frequency spectrum of a star emitting *visible* light can be obtained with a diffraction grating followed at a suitable distance downstream by a photographic emulsion. The entire spectrum is obtained at once because different wavelengths are diffracted in different directions and hence to different parts of the film. The blackness of the film at a given diffraction angle gives the intensity of that wavelength component.

For infrared light (i.e., wavelengths of order 10^{-4} cm) there is no photographic film that is adequate. The diffraction grating still works and can be used. In place of the film one can use a photomultiplier tube with a movable slit. The slit location gives the diffraction angle and hence gives the wavelength. The photomultiplier current gives the intensity. If you want narrow resolution (in frequency or wavelength), you must use a narrow slit, so as to have narrow angular resolution. If you want a complete frequency spectrum, you must count for sufficient time at one slit location to measure the intensity at the corresponding wavelength, then move the slit by one slit width and count for sufficient time at the new position, etc. To obtain a complete spectrum in frequency range ν_1 to ν_2 with every part of the range measured with bandwidth $\Delta\nu$, it takes $(\nu_2 - \nu_1)/\Delta\nu$ separate intensity measurements. For a range of wavelengths from 1 to 3 microns (1 micron $= 10^{-4}$ cm), we have a range of wave numbers from 1 to $\frac{1}{3}$ times 10^4 cm^{-1} (inverse centimeters), that is, $\lambda_1^{-1} - \lambda_2^{-1} = \frac{2}{3} \times 10^4$ cm^{-1}. For a typical good resolving power of $\Delta(\lambda^{-1}) = \Delta(v/c) \approx 0.1 \text{ cm}^{-1}$, we would need about $\frac{2}{3}$ $\times 10^5 \approx 60,000$ separate measurements to cover the complete spectrum. Since each measurement may occupy one night, that would take us several hundred years!

Of course, if you had 60,000 photomultipliers you could measure the entire spectrum at once, but that is obviously impractical. If one photomultiplier extended over the entire diffraction pattern from the diffraction grating, you would measure all wavelengths at once. The photomultiplier output would be proportional to the total intensity averaged over the entire spectrum, however, and you could never tell what part came from what wavelength. It would be like having all the telephone conversations from San Francisco to New York coming over one line without any way of separating them. The problem of sending separate telephone conversations over one line has been solved by "labeling" each conversation with its own "subcarrier," and then "multiplexing" all the subcarriers together, as discussed in Prob. 6.32. If only there were a way to *label* each separate infrared wavelength somehow with a "subcarrier" frequency so as to identify the wavelength! Then all the infrared light could be focused at the same time onto one photomultiplier. The photomultiplier output could be Fourier-analyzed so as to resolve it into separate subcarrier bands. The intensity for each subcarrier would then give the intensity for the corresponding infrared wavelength.

(*a*) Invent a way of labeling each wavelength with a subcarrier by using a mechanical "chopper," consisting of a rotating wheel with holes or slits that pass light incident on a slit and block it otherwise. Your main problem is to devise a way to make the chopping frequency depend on the infrared wavelength.

The elegant method used in the MIFS technique is as follows. No diffraction grating or mechanical chopper is used. Instead a Michelson interferometer with one movable mirror is used. (This type of interferometer, used in the Michelson-Morley experiment, is shown in the accompanying sketch above.) Light from the star is incident in (for example) the x direction on a half-silvered mirror "beam splitter" oriented at 45 deg to the incident beam. The beam splitter reflects half of the light in the y direction and transmits half in the x direction. Mirrors then redirect the two beams back to the beam splitter, which reflects half of the recombined light in the $-y$ direction to a photomultiplier. (The other half is transmitted in the $-x$ direction back toward the star and is lost.) For a given wavelength λ the photomultiplier current is a maximum or a minimum depending on whether the two recombined beams are in phase or 180 deg out of phase. This in turn depends on whether their path lengths (from beam splitter to mirror to beam splitter to photomultiplier) differ by an even number of half-wavelengths (to give zero relative phase) or an odd number of half-wavelengths (to give 180 deg relative phase).

(*b*) Now suppose that one of the mirrors is moved at a perfectly well-known uniform velocity v. Show that infrared light of frequency ν gives a photomultiplier output having a time dependence which includes a component that oscillates harmonically as $\cos \omega_{mod} t$ with modulation frequency $v_{mod} = 2(v/c)v$. Alternatively, show that if the position of the mirror is varied in some arbitrary way (as far as time-dependence is concerned) and the photomultiplier output is measured as a function of x, then the photomultiplier output has an x dependence which includes a component that varies as $\cos k_{mod} x$, with modulation wavenumber given by $k_{mod} = 4\pi/\lambda$. If there are many wavelengths present, then the photomultiplier output is a superposition consisting of a constant (the average over the whole spectrum) plus one Fourier component for each modulation wavenumber k_{mod}. Thus if we Fourier-analyze the output, the intensity at each modulation wavenumber k_{mod} gives us the corresponding intensity at the infrared wavelength λ. The important thing is that while the data (a recording of photomultiplier output versus x) is being taken, all infrared wavelengths are being measured at once. Each wavelength is "labeled" by the modulation frequency (or wavenumber) that it produces in the photomultiplier output. Thus the modulation frequency acts like a "subcarrier" in enabling the various simultaneously recorded wavelengths to be separated by Fourier analysis of the photomultiplier output.

This technique may offer the best method of detecting life on Mars without going there. Analysis of the infrared spectrum of the atmosphere of Mars would indicate its composition and a search could be made for components that are the result of life processes. The MIFS technique is so sensitive that, with telescopes now being planned, not only the major components but also trace components of gases down to perhaps one part in 10^9 may be determined. These prospects, as well as a more detailed description of MIFS, are given in five related articles in the British magazine *Science Jot/mal* for April, 1967: "Detecting Planetary Life from Earth," by J. Lovelock, D. Hitchcock, P. Fellgett, J. and P. Connes, L. Kaplan, and J. Ring, p. 56, and in *The Physics Teacher* for April, 1968: "Remote Sensing of Planetary Atmospheres by Fourier Spectroscopy," by Reinhard Beer, p. 151.

Chapter 7

Waves in Two and Three Dimensions

7.1 Introduction

Practically all the waves that we have considered so far have been "one-dimensional." That is, they have been waves propagating along a straight line, which we usually call the z axis. In Sec. 7.2 we shall introduce three-dimensional waves by a rotation of the coordinate system used to describe a one-dimensional plane traveling wave. We thus obtain the three-dimensional form of plane harmonic traveling waves.

We shall see that there is something more to having extra dimensions than is implied by a mere change of variables. One has qualitatively new features because the extra dimensions give extra degrees of freedom. For example, in three dimensions and in vacuum one can have an electromagnetic wave that is a pure traveling wave in one direction, a pure standing wave in another, and an exponential wave in still another direction! In one dimension it is not possible to have exponential electromagnetic waves in vacuum, because the dispersion relation $\omega^2 = c^2 k^2$ cannot become $\omega^2 = -c^2 \kappa^2$ for some frequency ranges. In order to have exponential waves in one dimension, one needs a cutoff frequency, i.e., one needs a dispersion relation like that of the ionosphere, $\omega^2 = \omega_0^2 + c^2 k^2$, which can become $\omega^2 = \omega_0^2 - c^2 \kappa^2$ for sufficiently low frequency. In three dimensions we shall find that k is the magnitude of a vector, called the propagation vector. Thus the dispersion relation for electromagnetic waves in vacuum becomes $\omega^2 = c^2(k_x^2 + k_y^2 + k_z^2)$. Under certain circumstances one can have one or two of the components k_x^2, etc., replaced by $-\kappa_x^2$, etc., and still have the return force per unit displacement per unit inertia, ω^2, be positive, as it must be. We shall examine electromagnetic waves in waveguides and total reflection of light as examples. In Sec. 7.3 we shall study water waves (for ideal water) and find their space dependence and dispersion law. (There are several home experiments by which you can easily verify the dispersion law for water waves.) Sec. 7.4 is devoted to demonstrating through Maxwell's equations those things we learned in Chap. 4 when we studied waves in parallel-plate transmission lines. In Sec. 7.5 we shall derive the radiation from an oscillating point charge. We use it to find the "natural linewidth" of visible light and the reason why the sky is blue.

7.2 Harmonic Plane Waves and the Propagation Vector

Suppose we have a *harmonic traveling plane wave* propagating in a homogeneous dispersive medium in the direction of the unit vector

$\hat{\mathbf{z}}'$, along the $+z'$ axis. Suppose that at the plane $z' = 0$ the wave function has the time dependence

$$\psi(z', t) = A \cos \omega t. \tag{1}$$

Then at the plane given by a fixed value of z' the wave function is given by

$$\psi(z', t) = A \cos(\omega t - kz'). \tag{2}$$

We wish to express this wave function in terms of a general Cartesian coordinate system x, y, z, instead of by the coordinate z' along the propagation direction . Let the origin of the x, y, z system be in th plane $z' = 0$. Let $\mathbf{r} = x\hat{\mathbf{x}} + y\hat{\mathbf{y}} + z\hat{\mathbf{z}}$ denote a point in space as measured from the origin of the x, y, z system. The plane $z' = $ constant is described in the x, y, z system by the plane $z' = \mathbf{r} \cdot \hat{\mathbf{z}}' = $ constant. Thus the quantity kz' in Eq. (2) can be written

$$kz' = k(\hat{\mathbf{z}}' \cdot \mathbf{r}) = (k\hat{\mathbf{z}}') \cdot \mathbf{r} \equiv \mathbf{k} \cdot \mathbf{r} \tag{3}$$

Propagation vector. The quantity $k\hat{\mathbf{z}}'$ is called the *propagation vector* $\mathbf{k}$:

$$\mathbf{k} \equiv k\,\hat{\mathbf{z}}'. \tag{4}$$

The magnitude of $\mathbf{k}$ is k; the direction of $\mathbf{k}$ is $\hat{\mathbf{z}}'$, along the wave propagation direction. Equation (3) becomes

$$kz' = \mathbf{k} \cdot \mathbf{r} = k_x x + k_y y + k_z z. \tag{5}$$

The physical meaning of the wavenumber k is the number of radians of phase per unit displacement along the propagation direction $\hat{\mathbf{z}}'$, so that kz' is the phase accumulated in distance z'. (We are temporarily reversing our usual sign convention for phase; we usually think of phase as increasing positively when ωt increases for fixed z'.) The meaning of k_x is the number of *radians of phase per unit displacement along the $+x$ axis*, i.e., along $\hat{\mathbf{x}}$, with similar meanings for k_y and k_z. For example, suppose $\hat{\mathbf{x}}$ makes an angle θ with $\hat{\mathbf{z}}'$. Suppose that the wavelength is λ. Then if one advances along the direction $\hat{\mathbf{z}}'$ by a distance λ (at fixed time), the phase increases by 2π. If instead one advances along $\hat{\mathbf{x}}$, one must travel a distance $\lambda / \cos \theta$ before z' increases by one wavelength. Then the phase has increased by 2π in a distance along $\hat{\mathbf{x}}$ that is larger than λ by a factor $(\cos \theta)^{-1}$, i.e., the increase in phase per unit distance along $\hat{\mathbf{x}}$ is smaller than k by a factor $\cos \theta$. That is the way a vector is supposed to behave: if you take a projection $\mathbf{k} \cdot \hat{\mathbf{x}} = k_x$ of the vector along some direction $\hat{\mathbf{x}}$, you get a number less than the magnitude of the vector by a factor equal to the cosine of the appropriate angle. This condition ensures that the sum of the squares of the components equals the square of the magnitude. Thus we see that k_x has the right relation to k to be the x component of a vector $\mathbf{k}$ whose magnitude is k.

Why not a wavelength vector? That last sentence sounds so obvious as not to be worth stating. Here is a counter example to show that it is worth checking such an obvious requirement. Consider the following plausible (but wrong) argument: The phase velocity of a traveling wave is given by $v_\varphi = \lambda v$. When we want to describe a wave propagating in the direction $\hat{\mathbf{z}}'$ in three dimensions, it is supposedly a good idea to define a "wavelength vector" $\boldsymbol{\lambda}$ as follows:

$$\mathbf{v}_\varphi = \lambda v \,\hat{\mathbf{z}}' = (\lambda \,\hat{\mathbf{z}}')v = \lambda v?$$

The wavelength λ is defined as the distance between wave crests, for displacement along z', and is naturally the magnitude of the "vector" $\boldsymbol{\lambda}$. Similarly λ_x is the distance between wave crests for displacement along x. But notice the following horrible property of λ_x: it is longer than λ! Thus if $\hat{\mathbf{x}}$ is perpendicular to $\hat{\mathbf{z}}$, the quantity λ_x is infinite, whereas it would be zero if it were the x component of an ordinary vector directed along $\hat{\mathbf{z}}'$. The conclusion is that *there is no such vector* $\boldsymbol{\lambda}$ that can be defined in any sensible fashion, since nothing should be called a vector that has "components" larger than the magnitude of the vector.

Plane of constant phase. The traveling wave given by Eq. (2) can now be written in the equivalent forms

$$\psi(x, y, z, t) = A \cos (\omega t - kz')$$
$$= A \cos (\omega t - k_x x - k_y y - k_z z)$$
$$= A \cos (\omega t - \mathbf{k} \cdot \mathbf{r}). \qquad (6)$$

The argument of the sinusoidal wave function is called the *phase* $\varphi(x,y,z,t)$:

$$\varphi(x, y, z, t) = \omega t - kz'$$
$$= \omega t - k_x x - k_y y - k_z z.$$
$$= \omega t - \mathbf{k} \cdot \mathbf{r}. \qquad (7)$$

At fixed time t, the places with equal φ define a plane called a *wave-front*:

$$d\varphi = \omega\, dt - \mathbf{k} \cdot d\mathbf{r}$$

$$= 0 - \mathbf{k} \cdot d\mathbf{r} \qquad \text{at fixed time}$$

$$= 0 \qquad \text{only if } d\mathbf{r} \text{ is perpendicular to } \mathbf{k}. \qquad (8)$$

Then at fixed t the phase will have the same value at all places reached by adding up vectors $d\mathbf{r}$ that are perpendicular to the propagation direction $\hat{\mathbf{k}}$, i.e., $d\varphi = 0$ in going from one such place to another, which means going about in a plane. That is why such a wave is called a plane wave.

Phase velocity.   The phase velocity is equal to dz'/dt for fixed φ:

$$\varphi = \omega\, dt - k\, dz' = 0,$$

$$v_\varphi = \frac{dz'}{dt} = \frac{\omega}{k}. \qquad (9)$$

Three-dimensional dispersion relations.   Here are the three–dimensional forms of some of the dispersion relations you have be come familiar with:

Case 1: Electromagnetic waves in vacuum

$$\omega^2 = c^2 k^2 = c^2\,(k_x^{\,2} + k_y^{\,2} + k_z^{\,2}). \qquad (10)$$

Case 2: Electromagnetic waves in a dispersive medium

$$\omega^2 = \frac{c^2}{n^2} k^2 = \frac{c^2}{n^2}\,(k_x^{\,2} + k_y^{\,2} + k_z^{\,2}). \qquad (11)$$

Case 3: Electromagnetic waves in the ionosphere

$$\omega^2 = \omega_p^{\,2} + c^2 k^2 = \omega_p^{\,2} + c^2\,(k_x^{\,2} + k_y^{\,2} + k_z^{\,2}). \qquad (12)$$

The dispersion relations are always independent of the boundary conditions. The boundary conditions are of course the determining factor in deciding whether one has (for example) standing waves, or traveling waves, or (as we shall see) a mixed type.

Standing waves.   Two traveling plane waves traveling in opposite directions and having the same amplitude (and frequency) may be superposed to form a standing plane wave of the form

$$\psi(x, y, z, t) = A \cos\,(\omega t + \varphi) \cos\,(\mathbf{k} \cdot \mathbf{r} + a). \qquad (13)$$

Writing out $\mathbf{k} \cdot \mathbf{r} = k_x x + k_y y + k_z z$ and using trigonometric identities we may write this standing wave as a superposition of terms each of which has the general form

$$\psi(x, y, z, t) = A \cos (\omega t + \varphi) \cos (k_x x + a_1) \cos (k_y y + a_2) \cos (k_z z + a_3).$$

(14)

When we express a harmonic wave in terms of standing waves of the form of Eq. (14) we can define k_x, k_y, and k_z to be positive quantities. The physical reason is that in a standing wave the waves are not propagating in a definite direction, as they are in a traveling wave, but are "going in both directions at once." Algebraically we see that if (for example) k_x in Eq. (14) is negative we can replace k_x by $-k_x$ and replace a_1 by $-a_1$ without affecting ψ (x, y, z, t). Thus we can let all three of k_x, k_y, and k_z be positive and make compensating changes (if necessary) in the phase constants a_1, a_2, and a_3.

Mixed traveling and standing wave. In one dimension (the dimension z', for example) we can have a pure traveling wave, which we can write as a superposition of two standing waves. Similarly we can have a pure standing wave, which we can write as a superposition of traveling waves. Alternatively, we can have a wave which is a superposition of a more general type, neither pure traveling wave nor pure standing wave. The same situation holds in three dimensions, but with the added freedom that each of the three dimensions is "independent," in the sense that one can have a wave that is (for example) a constant along x, a pure standing wave along y, and a pure traveling wave along z:

$$\psi(x, y, z, t) = \psi(y, z, t) = A \sin (k_y y) \cos (k_z z - \omega t).$$

(15)

Later we shall meet several examples of mixed types of waves similar to Eq. (15).

Three-dimensional wave equations and the classical wave equation. Any three-dimensional sinusoidal harmonic wave, whether a standing, traveling, or mixed type, satisfies the following relations (as you can easily show):

$$\frac{\partial^2 \psi(x, y, z, t)}{\partial t^2} = -\omega^2 \psi(x, y, z, t),$$

$$\frac{\partial^2 \psi}{\partial x^2} = -k_x^2 \psi, \qquad \frac{\partial^2 \psi}{\partial y^2} = -k_y^2 \psi, \qquad \frac{\partial^2 \psi}{\partial z^2} = -k_z^2 \psi.$$

(16)

We thus find the following wave equations corresponding respectively to the dispersion relations given by Eqs . (10), (11), and (12):

Case 1: Electromagnetic waves in vacuum

Using Eqs. (16) and (10), we find that for a single harmonic component having frequency ω and wavenumber k the wave function satisfies the differential equation

$$\frac{\partial^2 \psi}{\partial t^2} = c^2 \left\{ \frac{\partial^2 \psi}{\partial x^2} + \frac{\partial^2 \psi}{\partial y^2} + \frac{\partial^2 \psi}{\partial z^2} \right\}. \tag{17}$$

Since c is independent of frequency, the wave Eq. (17) is satisfied by every harmonic component and thus by an arbitrary superposition of standing and traveling electromagnetic waves in vacuum. Equation (17) is the three-dimensional form of the *classical wave equation for nondispersive waves*. A similar equation holds for any other three-dimensional nondispersive waves—for example, for ordinary sound waves in air. In vector notation, the right-hand side of Eq. (18) is c^2 times the divergence of the gradient of ψ, written div grad ψ or $\nabla \cdot \nabla \psi$, sometimes called $\nabla^2 \psi$, "del-squared psi":

$$\frac{\partial^2 \psi}{\partial t^2} = c^2 \nabla^2 \psi. \tag{18}$$

Case 2: Electromagnetic waves in a homogeneous dispersive medium

The dispersion relation Eq. (11) gives for a harmonic wave of frequency ω the wave equation

$$\frac{\partial^2 \psi}{\partial t^2} = \frac{c^2}{n^2(\omega)} \nabla^2 \psi. \tag{19}$$

Since n depends on the frequency ω, there is not much gained by writing down this wave equation. To solve it, we usually need to go to Fourier superpositions and consider one frequency at a time, so that we might as well just use the dispersion relation. The classical wave Eq. (18) is different in that respect; i.e., one may work with pulses or other nonharmonic waves without ever using Fourier analysis.

Case 3: Electromagnetic waves in the ionosphere

Using the dispersion relation Eq. (12) and using Eq. (16), we find the three-dimensional Klein-Gordon wave equation,

$$\frac{\partial^2 \psi}{\partial t^2} = -\omega_p^2 \psi + c^2 \nabla^2 \psi. \tag{20}$$

Some physical examples of two-dimensional sinusoidal harmonic waves follow:

Example 1: Electromagnetic waves in a rectangular waveguide

A rectangular waveguide can be made by adding conducting side plates to a parallel-plate transmission line, as shown in Fig. 7.1. The space inside the waveguide is vacuum. We shall only consider wave

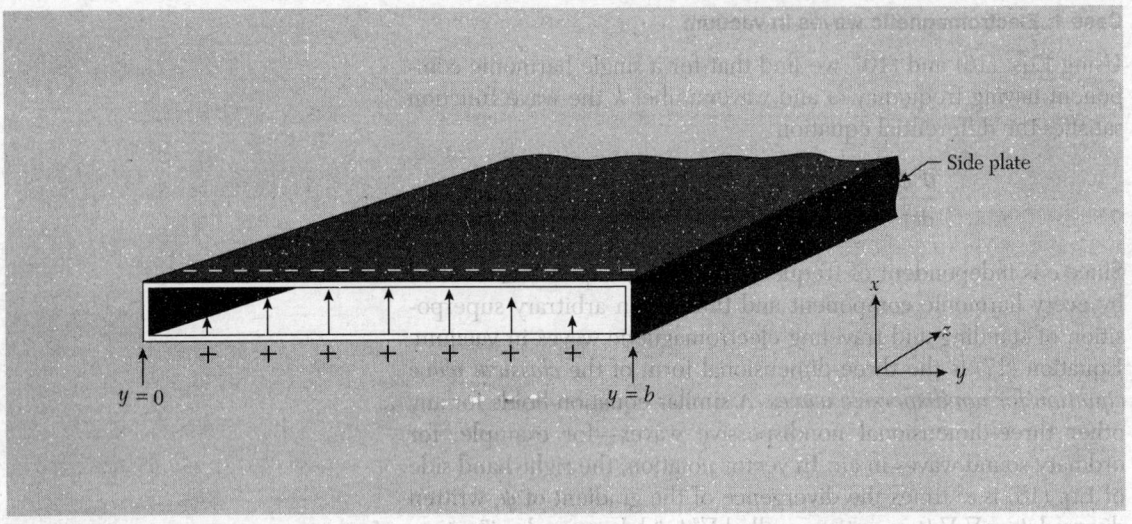

Fig. 7.1 Rectangular waveguide made by short-circuiting a parallel-plate transmission line by adding conducting side plates at $y = 0$ and $y = b$. The arrows represent the instantaneous electric field at the input end of the guide.

modes in which the electric and magnetic fields are both independent of x (for fixed y and z, and for x inside the guide). The appropriate wave equation is the two-dimensional version of the classical wave equation, Eq. (17). Letting ψ stand for the electric field E_x, we have

$$\frac{\partial^2 \psi}{\partial t^2} = c^2 \frac{\partial^2 \psi}{\partial y^2} + c^2 \frac{\partial^2 \psi}{\partial z^2}. \tag{21}$$

We choose a definite frequency ω, so that Eq. (2 1) becomes

$$-\omega^2 \psi = c^2 \frac{\partial^2 \psi}{\partial y^2} + c^2 \frac{\partial^2 \psi}{\partial z^2}. \tag{22}$$

The conducting sideplates force the electric field E_x to be zero at $y = 0$ and $y = b$. Therefore $\psi(y, z, t)$ must be a standing wave with respect to the y direction with permanent nodes at $y = 0$ and b. We assume there is a driving voltage at $z = 0$. Therefore electromagnetic waves propagate in the $+z$ direction down the guide. The waves must therefore be traveling waves with respect to the z direction. Equation (22) is satisfied by the mixed standing and traveling wave

$$\psi(y, z, t) = A \sin k_y y \cos (k_z z - \omega t), \tag{23}$$

provided we have the dispersion relation

$$\omega^2 = c^2 k_y^{\,2} + c^2 k_z^{\,2}. \tag{24}$$

By our choice of $\sin k_y y$, we satisfied the condition $E_x = 0$ at $y = 0$. However, we also need $\sin k_y y$ to be zero at $y = b$:

$$k_y b = \pi, 2\pi, \ldots, m\pi, \ldots. \tag{25}$$

These waves are called TE modes (transverse electric field modes). It is not necessary for us to study the magnetic field separately, since it is determined by the electric field.

Low-frequency cutoff frequency.   Let us consider the lowest mode, i.e., that having $m = 1$ in Eq. (25). That is the mode sketched in Fig. 7.1, since the figure shows one half-wavelength from $y = 0$ to b. Inserting Eq. (25) into Eq. (24) for $m = 1$, we obtain

$$\omega^2 = \frac{c^2\pi^2}{b^2} + c^2 k_z^2. \tag{26}$$

Thus the dispersion relation between ω and k_z (for this mode with $k_y b = \pi$) is similar in appearance to the dispersion relation for plane waves traveling in the z direction in the ionosphere, namely

$$\omega^2 = \omega_p^2 + c^2 k^2, \tag{27}$$

or to the dispersion relation for coupled pendulums (in the long-wavelength limit), namely

$$\omega^2 = \frac{g}{l} + \frac{Ka^2}{M} k^2. \tag{28}$$

Therefore we expect that the quantity $c^2\pi^2/b^2$ acts as (the square of) a low-cutoff frequency and that for driving frequency ω below this cutoff, the dispersion relation Eq. (26) goes over to the dispersion relation

$$\omega^2 = \frac{c^2\pi^2}{b^2} - c^2 \kappa_z^2. \tag{29}$$

That surmise is correct. For frequency $\omega < \pi c/b$, the wave equation Eq. (21) has the solution

$$\psi(y, z, t) = A \sin k_y y \cos \omega t \, e^{-\kappa_z z} \tag{30}$$

provided that ω, k_y, and κ_z are related by

$$\omega^2 = c^2 k_y^2 - c^2 \kappa_z^2, \tag{31}$$

that Eq . (25) is satisfied, and that ω^2 is less than $c^2\pi^2/b^2$ (taking $m = 1$), so that Eq. (29) is satisfied with κ_z^2 positive. (Note that we could have included a term with exp $(+\kappa_z z)$ in Eq. (30). However, the boundary condition that the waveguide extend to $z = +\infty$ requires zero for the coefficient of such a term.)

Physical origin of waveguide cutoff frequency.   Let us think of the frequency as fixed and the width b as variable. According to Eq. (26), if b is infinitely large, the dispersion relation is that for electromagnetic plane waxes in vacuum propagating along the z direction; the waves think they are in a parallel-plate transmission line. For finite b, k_y (which is π/b) is not zero. Thus if we want to think of the wave function as a superposition of plane traveling waves (which we are always free to do—even when we have a pure standing wave), we see that a decrease of b from infinity to something finite changes the wave from a pure traveling wave traveling along $+\hat{z}$ to a superposition with a nonzero component of propagation vector along $\hat{y}$. Actually, we must have traveling waves going simultaneously in the $+y$ and $-y$ directions and superposing, to give a standing wave along $\hat{y}$. The components of $\mathbf{k}$ along plus and minus $\hat{y}$ are necessary to satisfy the boundary condition introduced by the conducting side plates. The *magnitude* of $\mathbf{k}$ is always given by the vacuum dispersion relation,

$$k^2 = \frac{\omega^2}{c^2} = k_z{}^2 + k_y{}^2. \qquad (32)$$

Therefore the increase of the y components from zero to something finite necessarily leads to a decrease in the z component of $\mathbf{k}$. Thus as b decreases further, the y components increase further, and the z component decreases further. At any fixed b, the wave function can be thought of as superposition of plane waves crisscrossing down the guide and superposing so as to satisfy the boundary condition at the side plates. (More physically, we can say that the currents generated in a side plate by an incident "crissing" plane wave generate a specularly reflected wave that "crosses" back in the opposite y direction.) With this picture, we see that when b gets sufficiently small the z component of $\mathbf{k}$ will be zero. Then the wave is bouncing back and forth between the side plates, and there is no flow of waves down the tube. This says that the cutoff period $T_{\text{c.o.}}$ must be the time for a plane wave to travel from one side of the guide to the other and back in vacuum at velocity c:

$$T_{\text{c.o.}} = \frac{2b}{c}.$$

Then

$$\omega_{\text{c.o.}} = 2\pi\nu_{\text{c.o.}} = \frac{2\pi}{T_{\text{c.o.}}} = \frac{2\pi}{2b/c} = \frac{c\pi}{b}. \qquad (33)$$

By comparison of Eqs. (33) and (26), we see that Eq. (33) does in fact give the cutoff frequency.

For a frequency below cutoff, the wave amplitude decreases exponentially with increasing z, even though the waves are in vacuum. The physical reason for the decrease in the electric field is this: With the conducting side plates in place, the charges on the top and bottom plates can run around through the side plates and neutralize one another. In the region at $z = 0$, the driving source of voltage furnishes new charge and maintains the electric field.

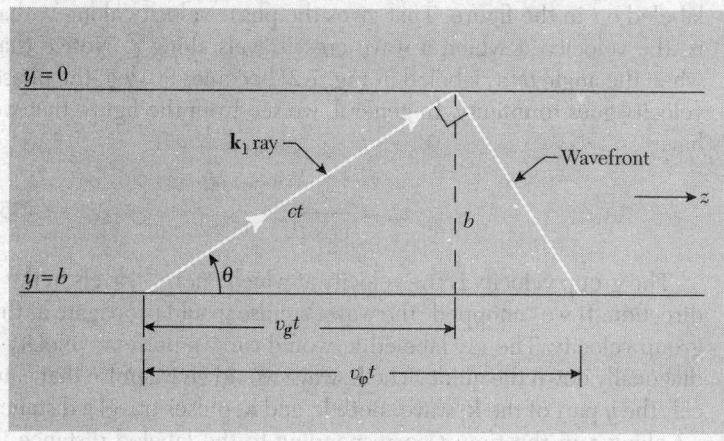

*Fig. 7.2   One of the Crisscross waves
in a waveguide.*

Farther downstream, the driving force has less influence, and when
the frequency is too slow, the charges have time to neutralize one
another.

Crisscross traveling waves.   The mixed standing wave and traveling wave of Eq. (23) is equivalent to a superposition of traveling
plane waves crisscrossing down the guide. You can see that algebraically by showing (Prob. 7.1) the identity

$$\psi = A \sin k_y y \cos (k_z z - \omega t)$$

$$= \tfrac{1}{2} A \sin (\mathbf{k}_1 \cdot \mathbf{r} - \omega t) - \tfrac{1}{2} A \sin (\mathbf{k}_2 \cdot \mathbf{r} - \omega t), \qquad (34)$$

where

$$\mathbf{k}_1 = \hat{\mathbf{z}} k_z + \hat{\mathbf{y}} k_y, \quad \mathbf{k}_2 = \hat{\mathbf{z}} k_z - \hat{\mathbf{y}} k_y.$$

The crisscrossing lies in the fact that $\mathbf{k}_1$ and $\mathbf{k}_2$ have opposite y components.

Phase velocity, group velocity, and c.   The crisscross-traveling-
wave picture gives   very simple way to see the relations between
the phase and group velocities. Consider just one of the two traveling waves superposed in Eq. (34), as shown in Fig. 7.2. Consider a
little bit of wavefront (called a "ray" in optics) that travels diagonally
across the guide a distance ct in time t, as shown by the labeled "$\mathbf{k}_1$
ray" in Fig. 7.2. We are interested in the phase velocity and the
group velocity *in the z direction*. (We know that there is a traveling
wave only in that direction. The k_2 partner to the $\mathbf{k}_1$ wave shown
cancels the y part of the traveling wave $\mathbf{k}_1$ but has the same z part.)
While the ray travels a distance ct, the intersection of the wavefront
with any fixed value of y (for example $y = b$) travels the distance

labeled $v_\varphi t$ in the figure. That gives the phase velocity along z, that is, the velocity at which a wave crest travels along z. Notice that when the angle θ (as labeled in Fig. 7.2) becomes 90 deg, the phase velocity goes to infinity. In general, we see from the figure that we have

$$v_\varphi = \frac{c}{\cos\theta}. \tag{35}$$

The group velocity is the velocity at which energy travels in the z direction. If we "chopped" the wave, a pulse would propagate at the group velocity. The ray labeled $\mathbf{k}_1$ would carry a pulse at velocity c diagonally down the guide. The $\mathbf{k}_2$ wave would give a pulse that cancels the y part of the $\mathbf{k}_1$ wave. Both $\mathbf{k}_1$ and $\mathbf{k}_2$ pulses travel a distance $v_g t$ along z in the time t corresponding to the labeled distance in Fig. 7.2. We see that we have

$$v_g = c\cos\theta. \tag{36}$$

We could now verify that v_φ and v_g as given by Eqs. (35) and (36) are correct by using the dispersion relation. Instead, we shall turn the problem around and derive the dispersion relation, given Eqs. (35) and (36):

$$v_\varphi = \frac{\omega}{\kappa_z} = \frac{c}{\cos\theta},$$

$$v_g = \frac{d\omega}{d\kappa_z} = c\cos\theta.$$

Then

$$v_\varphi v_g = \frac{\omega}{k_z}\frac{d\omega}{dk_z} = c^2, \tag{37}$$

i.e.,

$$\frac{d(\omega^2)}{d(k_z{}^2)} = c^2,$$

i.e.,

$$d(\omega^2) = c^2\, d(k_z{}^2).$$

Integration gives

$$\omega^2 = c^2 k_z{}^2 + \text{constant}. \tag{38}$$

The constant can be determined by setting $k_z = 0$, so that $\omega = \omega_{c.o.}$, and requiring that the "over and back" time $T_{c.o.}$ be $2b/c$. Thus we get the dispersion law Eq. (26). The higher modes are obtained by letting the cutoff frequency be a harmonic of the lowest possible cutoff frequency. That gives the more general case [Eqs. (24) and (25)]

$$\omega^2 = c^2 k_z^2 + \frac{c^2 \pi^2 m^2}{b^2}. \quad (39)$$

Example 2: Reflection and transmission of light incident from glass to air

This is another two-dimensional wave example. Suppose we have a lot of glass from $z = -\infty$ to $z = 0$. At the plane $z = 0$ the glass ends; the vacuum begins and extends to $z = +\infty$. You might think that the vacuum would always act as a dispersive medium, as it does for plane waves. But, as you have seen in Example 1 (the rectangular waveguide), when we do not have plane waves (i.e., when E_x varies along the y axis as well as along z, the axis of propagation), the waveguide becomes reactive under certain conditions (width too narrow, or what amounts to the same thing, frequency too low), even though it still has nothing in it but vacuum. Something similar happens for light incident from glass to air, if the angle of incidence becomes too large—i.e., as the light gets too near to grazing incidence. This is of great practical importance in the design of many optical instruments that make use of total internal reflection in order to get 100 percent reflection. An example is shown in Fig. 7.3.

Now let us see how it works. The light waves satisfy the wave equation in each medium, glass and vacuum. (We are considering a single frequency ω.) The boundary between the glass and vacuum is at $z = 0$. The propagation vector $\mathbf{k}_1$ of the incident wave has a component k_z along $\hat{\mathbf{z}}$ and a component k_y along $\hat{\mathbf{y}}$. Thus we have a two-dimensional problem (somewhat as we did for the waveguide in the TE mode). The geometry is shown in Fig. 7.4.

In glass, the magnitude k_1 of the propagation vector $\mathbf{k}_1$ is equal to the product of the index of refraction n times the vacuum magnitude ω/c of the propagation vector. The magnitude k_2 of $\mathbf{k}_2$ is just ω/c:

$$k_2 = \frac{\omega}{c}, \qquad k_1 = n\frac{\omega}{c}. \quad (40)$$

The dispersion relation in medium 2 (the vacuum to the right of $z = 0$) is thus

$$\frac{\omega^2}{c^2} = k_2{}^2 = k_{2y}{}^2 + k_{2z}{}^2. \quad (41)$$

Next, we claim that k_{2y} must equal k_{1y}. This is because the meaning of k_{1y} is 2π times the number of wave crests per unit length along $\hat{\mathbf{y}}$ in medium 1. Similarly, k_{2y} is 2π times the number of wave

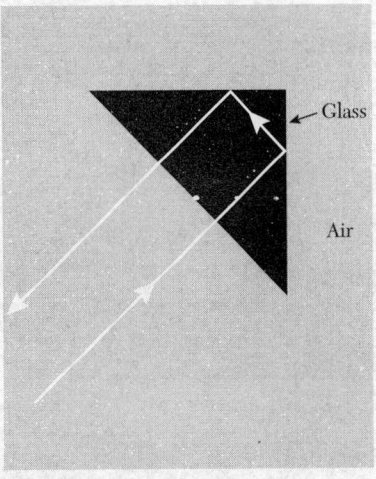

Fig. 7.3 *Retrodirective prism used to deflect light through 180 deg without loss of intensity.*

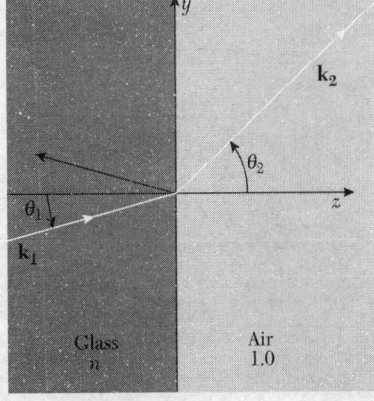

Fig. 7.4 *Reflection and transmission of ray incident from glass to vacuum.*

crests per unit length along $\hat{\mathbf{y}}$ in medium 2. But as you travel along the y axis at $z = 0$, the number of crests that you pass just inside the glass must be the same as the number just outside the glass in the vacuum. You cannot "lose" any crests per unit length along $\hat{\mathbf{y}}$ in going from the glass to the vacuum. Thus

$$k_{2y} = k_{1y}$$

$$= k_1 \sin \theta_1$$

$$= n \frac{\omega}{c} \sin \theta_1, \qquad (42)$$

where the second equality in Eq. (42) is obvious from Fig. 7.4 and the third equality uses Eq. (40). Inserting Eq. (42) into Eq. (41) gives

$$\frac{\omega^2}{c^2} = \frac{n^2 \omega^2}{c^2} \sin^2 \theta_1 + k_{2z}{}^2, \qquad (43)$$

i.e., we have the dispersion relation

$$k_{2z}{}^2 = \frac{\omega^2}{c^2} (1 - n^2 \sin^2 \theta_1). \qquad (44)$$

Critical angle for total internal reflection. As we increase the angle of incidence θ_1, the z component of the propagation vector $\mathbf{k}_2$ becomes smaller and smaller. Finally we reach an angle of incidence for which k_{2z} is zero (we are assuming n is greater than unity, as for example for visible light in glass or water). That gives us the *cutoff angle*, otherwise called the *critical angle of incidence for total internal reflection* or simply the critical angle θ_{crit}. According to Eq. (44), the critical angle is given by

$$\boxed{n \sin \theta_{\text{crit}} = 1.} \qquad (45)$$

(For glass of index $n = 1.52$ this gives $\theta_{\text{crit}} = 41.2°$.) At the critical angle of incidence, the beam emerging in vacuum is tangential to the surface of the glass.

Snell's law. For angles θ_1 between zero and θ_{crit}, the beam is partly reflected and partly refracted into the vacuum. Then there exists an angle θ_2 as drawn in Fig. 7.4, and the relation $k_{2y} = k_{1y}$ is equivalent to *Snell's law* (derived by a different method in Sec. 4.3):

$$k_{2y} = k_2 \sin \theta_2 = n_2 \frac{\omega}{c} \sin \theta_2,$$

$$k_{1y} = k_1 \sin \theta_1 = n_1 \frac{\omega}{c} \sin \theta_1;$$

then $k_{2y} = k_{1y}$ gives

$$\boxed{n_1 \sin \theta_1 = n_2 \sin \theta_2.} \tag{46}$$

Total internal reflection. For angles of incidence greater than the critical angle, the dispersion relation is obtained from Eq. (44) by replacing k_{2z}^2 with $-\kappa_{2z}^2 \equiv -\kappa^2$:

$$\kappa^2 = \frac{\omega^2}{c^2} [n^2 \sin^2 \theta_1 - 1], \tag{47}$$

with

$$n \sin \theta_1 > 1.$$

Then the wave function (electric field or magnetic field) is given in medium 2 (vacuum) by a wave that is a traveling wave along y, but an exponential wave along z:

$$\psi(y, z, t) = A \cos (\omega t - k_y y) e^{-\kappa z}, \tag{48}$$

where k is given by Eq. (47) and k_y is $k_1 \sin \theta_1 = n(\omega/c) \sin \theta_1$. The timeaveraged energy density is proportional to the time-averaged square of $\psi(y, z, t)$ that is,

$$\text{Energy density} \propto e^{-2\kappa z}. \tag{49}$$

As an application of Eq. (47), consider the retrodirective prism shown in Fig. 7.3. The light is incident internally from glass to air at angle of incidence $\theta_1 = 45°$. This angle exceeds the crictical angle $\theta_{crit} = 41.2°$ (for glass of index $n = 1.52$). Therefore the ray is totally reflected. The mean exponential decay distance (of the fields penetrating into the vacuum) is given (for $\theta_1 = 45°$) by

$$\delta = \kappa^{-1} = \frac{c}{\omega} [n^2 \sin^2 \theta_1 - 1]^{-1/2}$$

$$= \frac{\lambda}{2\pi} \left[\frac{(1.52)^2}{2} - 1 \right]^{-1/2} = 0.4\lambda$$

Thus at a distance of several wavelengths into the "forbidden" region (the vacuum) the fields are negligible.

A beautiful demonstration of total internal reflection is provided when you swim with a faceplate (so your eyes can see well under water). With your eyes several inches under the water surface look ahead at the "underside" of the surface. It looks "shiny," somewhat like liquid mercury. That is because your line of sight exceeds the critical angle. Then the surface acts like a perfect mirror for light reflected into your eyes.

A more convenient way to observe total internal reflection from the underside of a water surface is to look up at the underside of the surface while looking through the flat vertical side of a transparent glass or plastic container.

Barrier penetration of light.   If the vacuum does not extend to infinity but is terminated by another slab of glass, then we should include a second term in Eq. (48) with a positive exponential, exp (+κz). Then one has a typical barrier-penetration problem.

A beautiful and ingenious experiment which verifies the exponential attenuation of the energy density has been performed by D. D. Coon, a graduate student (as of spring 1965) at Princeton.† Although his experiment is inherently quantum-mechanical, it verifies this result of classical optics. This is but one of many results of classical optics which are retained in quantum mechanics. (Classical optics is what we have been studying whenever we dealt with electromagnetic waves of those wavelengths which we call "light," instead of, say, "microwaves.") Coon set two prisms with a variable air gap so that light (from the green line of mercury) was incident through one prism on the gap at an angle beyond the critical angle. The light energy transmitted through the air gap into the second prism is proportional to the energy density at the surface of the second prism. Now, quantum mechanics tells us that light of frequency ω comes in indivisible units called photons, each with energy exactly $\hbar\omega$. Thus for a given ω the energy density is proportional to the number of photons. Coon measured the energy density by counting transmitted photons as a function of the air gap and verified the predicted exponential dependence given by Eq. (49). His experiment was the first to verify it at wavelengths of less than 1 cm and the first to verify it at any wavelength by detecting individual photons.

A qualitative demonstration of barrier penetration and of the rapid decrease of the fields in a light wave with distance from the glass into the "forbidden" vacuum (or air) region is easily performed with a glass prism or cube. While looking at a spot on the surface that is totally reflecting along your line of sight, touch that spot *lightly* with one finger, from the far side of the surface. The finger is then invisible. It lies in the "forbidden" region. Now press the finger *tightly* against the surface. Then you will see your "fingerprint." The ridges of the whorls on your finger make intimate contact with the totally

---

†D. D. Coon, *Am. J. Phys.* 34, 240 (1966).

reflecting glass surface and spoil the total reflection. The valleys of the whorls do not quite touch the glass and thus do not spoil the total reflection. They look like silvery whorls separating the ridges. The depth of these valleys must be several wavelengths, i.e., several times the mean penetration depth $\delta = k^{-1}$. If the valleys were shallower than δ, the fields would pentrate the "barrier" between the glass and skin to a significant extent, and would then interact with the skin so as to spoil the total reflection.

The above demonstration of barrier penetration may be performed using, in place of the glass prism or cube, a transparent rectangular container filled with water.

7.3 Water Waves

Water waves are easily observed. Since your early childhood you have seen them in bathtub, lake, and sea. Undoubtedly you have experienced great aesthetic pleasure in watching them in all their beauty and complexity. Now we wish to enjoy the intellectual pleasure of understanding them. That understanding requires simplicity. We will therefore neglect some properties of real water. For example, we will neglect viscocity, which is the result of internal friction. (Professor Richard P. Feynman gives such idealized water the wonderful name "dry water.") We will also confine ourselves to gentle waves of small amplitude. No breakers for us!

In spite of our simplifications we will learn the geometrical structure and the dispersion relation $\omega(k)$ for gentle water waves. You can verify all the results by easy home experiments using a shoe box or fish tank. (See Home Exp. 7.11.)

At equilibrium (i.e., when there are no waves) the surface of a body of water is flat and horizontal. When a wave is present there are two restoring forces that tend to flatten the wave crests: *gravity* and *surface tension*.

Because of the great incompressibility of water, the excess of water that appears in a wave crest must flow in from the neighboring trough regions.

Individual water drops in a water wave therefore undergo a motion which is some combination of longitudinal motion (along the wave propagation direction and transverse (up and down) motion.

If the equilibrium water depth is small compared to the wavelength (of harmonic waves) the waves are called *shallow-water waves*, or *tidal* waves. It turns out that these waves have a propagation velocity that is independent of wavelength but depends on the *depth*.

If the wavelength is small compared to the equilibrium water depth we have what are called *deep-water waves*. The individual water droplets in a traveling harmonic deep-water wave do not have any *average* translation. They move in circles! For example, a floating cork (or a water droplet at the surface) undergoes a uniform circular motion with radius equal to the amplitude of the harmonic wave, and period equal to that of the wave. In a trough the cork has its maximum backward velocity; on a crest it has an equally large velocity forward (with respect to the direction of wave propagation). Water droplets below the surface travel in smaller circles; the circular radius decreases exponentially with increasing depth, and is negligibly small a few wavelengths below the surface.

Straight waves.   Let us consider water waves having a single wavelength λ and having long straight parallel crests and troughs. Such waves are called *straight waves*. They are the two-dimensional analog of three-dimensional plane waves.

   Suppose we have an infinite lake of uniform equilibrium depth h. When there are no waves, the surface of the water is a plane, which we call the plane $y = 0$. Positive y is measured vertically upward. We shall take the wave propagation direction to be along the horizontal direction $\hat{\mathbf{x}}$. Thus the wave crests and troughs are along lines perpendicular to $\hat{\mathbf{x}}$.

   Let x and y designate the *equilibrium position* of a given drop of water. (No matter where this drop goes in a wave motion, its equilibrium position is still x, y. The equilibrium position labels a given drop and does not tell us where the drop is when there is a wave present.) The variable x goes from $x = -\infty$ to $+\infty$. The variable y goes from $y = -h$ (the bottom of the lake) to $y = 0$ (the surface of the lake).

   When a wave is present, a given drop undergoes a motion that combines up-and-down motion (along y) and forward-and-backward motion (along x). Let $\psi(x, y, t)$ denote the instantaneous vector displacement from its equilibrium position of the water drop with equilibrium position x, y. The displacement vector in a straight water wave has only an x component and a y component:

$$\psi(x, y, t) = \hat{\mathbf{x}}\,\psi_x(x, y, t) + \hat{\mathbf{y}}\psi_y(x, y, t). \tag{50}$$

The instantaneous velocity $\mathbf{v}$ of the water droplet with equilibrium coordinates x, y is the partial derivative of ψ with respect to t:

$$\mathbf{v}(x, y, t) = \frac{\partial \psi(x, y, t)}{\partial t} = \hat{\mathbf{x}}\,\frac{\partial \psi_x}{\partial t} + \hat{\mathbf{y}}\,\frac{\partial \psi_y}{\partial t}. \tag{51}$$

Properties of ideal water.   In the following paragraphs we shall examine some of the properties of ideal water.

1. *Conservation of mass.* When you studied electric currents (Vol. II , Sec. 4.2), you found that the *conservation of electric charge* is expressed by the *equation of continuity:*

$$\nabla \cdot (\rho \mathbf{v}) = -\frac{\partial \rho}{\partial t}. \tag{52}$$

Equation (52) merely says that the reason the charge density ρ in an infinitesimal volume changes with time is that a current $\rho \mathbf{v}$ flows out the surface of the volume. In the present case we let ρ designate the water's mass density; then Eq. (52) expresses conservation of mass. Now, to a good approximation, *water is incompressible.* Therefore the mass density ρ is a constant, independent of time and position, and hence, the right side of Eq. (52) is zero. We can also factor ρ from the left side of Eq. (52) and discard it. Then we use Eq. (51) to express $\mathbf{v}$:

$$0 = -\frac{\partial \rho}{\partial t} = \nabla \cdot (\rho \mathbf{v}) = \rho \nabla \cdot \mathbf{v},$$

i.e.,

$$0 = \nabla \cdot \mathbf{v} = \nabla \cdot \left(\frac{\partial \boldsymbol{\psi}}{\partial t} \right) = \frac{\partial}{\partial t} (\nabla \cdot \boldsymbol{\psi}),$$

i.e.,

$$\nabla \cdot \boldsymbol{\psi} = \text{constant}. \tag{53}$$

2. *Absence of bubbles.* The constant in Eq. (53) can only be zero. Otherwise, according to Gauss's theorem, the surface integral of ψ over the surface of a little sphere would not be zero, which could only mean that we have bubbles. We assume there are no bubbles. Thus we have found that conserved, incompressible, bubbleless water satisfies

$$\nabla \cdot \boldsymbol{\psi} = \frac{\partial \psi_x(x, y, t)}{\partial x} + \frac{\partial \psi_y(x, y, t)}{\partial y} = 0. \tag{54}$$

3. *Absence of whirlpools.* In a whirlpool, the line integral of the velocity $\mathbf{v}$ around a circular path enclosing the vortex is not zero. On an infinitesimal scale, the presence of little vortices or whirlpools would imply (by Stokes' law) that the curl of $\mathbf{v}$ is not zero. (See Vol. II, Secs. 2.15 through 2.18 for a review of the meaning of the curl of a vector.) We assume there are no vortices. Thus we assume

$$0 = \nabla \times \mathbf{v} = \nabla \times \frac{\partial \boldsymbol{\psi}}{\partial t}$$
$$= \frac{\partial}{\partial t} (\nabla \times \boldsymbol{\psi}),$$

i.e.,

$$\nabla \times \boldsymbol{\psi} = \hat{\mathbf{z}} \left(\frac{\partial}{\partial x} \psi_y - \frac{\partial}{\partial y} \psi_x \right) = 0. \tag{55}$$

Standing water waves. We want to use our intuition to help us find the form of water waves without too much algebra. You should get a rectangular aquarium or container of some sort. (An ordinary cardboard carton lined with a plastic bag of the sort used to line garbage cans will work fairly well. Any cardboard carton will work for about ten minutes before falling apart. If it is painted on the inside, it will last indefinitely.) Fill it with 6 or 8 inches of water. Shake it gently along x and try to find sinusoidal-appearing modes. You will find that the lowest mode looks something like Fig. 7.5.

If you stir some coffee grounds into the water, you can see the motion of the water. You will notice that all the coffee grounds are motionless at the same time, and that the x and y displacements are zero at the same time. That is what we expect for a normal mode, i.e., for a standing wave: all degrees of freedom ("moving parts") oscillate in phase. Therefore we can assume that for sufficiently small oscillations the time dependence of both ψ_x and ψ_y is given by a harmonic oscillation with the same phase constant, i.e., that the time dependence is given by a common factor $\cos \omega t$.

Next let us assume that the x dependence of the vertical displacement ψ_y is that of a sinusoidal standing wave. If the mode looks like that shown in Fig. 7.5, then ψ_y has a node at $x = 0$. Therefore ψ_y contains the factor $\sin kx$ (rather than $\cos k_x$). We can thus write

$$\psi_y(x, y, t) = \cos \omega t \sin kx \, f(y), \qquad (56)$$

where $f(y)$ is an as-yet-unknown function of y.

Fig. 7.5 *Lowest sinusoidal mode in a rectangular fish tank.*

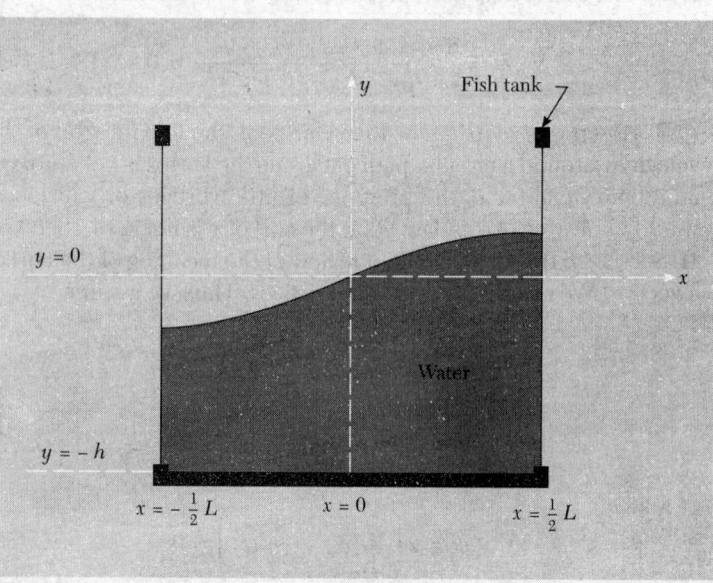

Boundary conditions at the walls. What is the x dependence of ψ_x? At the ends of the tank, a water droplet can only go up and down. It cannot leave the wall. Thus the places where ψ_y has its maxima (at the walls) are places where ψ_x has nodes. Thus we must have $\cos kx$ for ψ_x where we had $\sin kx$ for ψ_y:

$$\psi_x(x, y, t) = \cos \omega t \cos kx \, g(y), \tag{57}$$

where $g(y)$ is an as-yet-unknown function of y.

Relation between horizontal and vertical motions. Now let us make use of the facts that the div and curl of ψ are zero. Then you can easily show that Eqs. (56) and (57) give

$$\nabla \cdot \psi = 0: \quad -kg(y) + \frac{df(y)}{dy} = 0; \tag{58}$$

$$\nabla \times \psi = 0: \quad \frac{dg(y)}{dy} - kf(y) = 0. \tag{59}$$

We can eliminate $g(y)$ from Eqs. (58) and (59) by differentiating Eq. (58) with respect to y and then using (59) to eliminate dg/dy. That gives

$$\frac{d^2 f}{dy^2} = k^2 f, \tag{60}$$

which has the general solution

$$f(y) = Ae^{ky} + Be^{-ky}. \tag{61}$$

Then we obtain $g(y)$ from Eq. (58):

$$g(y) = Ae^{ky} - Be^{-ky}. \tag{62}$$

Boundary condition at the bottom. Finally we put in the boundary condition that at the bottom of the lake there is no vertical motion of water droplets—they cannot leave the bottom. The condition $\psi_y = 0$ at $y = -h$ is equivalent to $f(y) = 0$ at $y = -h$. Then Eq. (61) gives $B = -Ae^{-2kh}$.

Our final result for a standing sinusoidal water wave in a lake of equilibrium depth h is then

$$\psi_y = A \cos \omega t \sin kx (e^{ky} - e^{-2kh}e^{-ky}), \tag{63}$$

$$\psi_x = A \cos \omega t \cos kx (e^{ky} + e^{-2kh}e^{-ky}). \tag{64}$$

Equations (63) and (64) give the instantaneous displacement of a water droplet with *equilibrium* position x, y. As you can easily show from these equations, the motion of a given droplet (or coffee ground) in a standing water wave consists of harmonic oscillation along a straight line in the xy plane. That can also be seen by watching the coffee grounds in your tank.

Deep-water waves. If the depth h is huge compared with the wavelength, then the factor e^{-2kh} is essentially zero, and we can neglect the second terms in the y dependences $f(y)$ and $g(y)$. In that case Eqs. (63) and (64) become

$$\psi_y = A \cos \omega t \sin kx \, e^{ky}, \qquad (65)$$

$$\psi_x = A \cos \omega t \cos kx \, e^{ky}. \qquad (66)$$

We see that the waves are sinusoidal in the x direction and exponential in the y direction. The amplitude attenuation length δ is $1/k$, which equals $\lambda/2\pi$. The quantity $\lambda/2\pi$, which is called the *reduced wavelength*, is designated by the symbol $\lambdabar$ (pronounced "lambda bar"). Thus we have for deep-water waves

$$f(y) \; = \; e^{ky} \; = \; e^{-k|y|} \; = \; e^{-|y|/\lambdabar}; \qquad (67)$$

the amplitude attenuation length for deep-water waves equals the reduced wavelength. Therefore the oscillation amplitude of a water drop whose equilibrium position is one wavelength beneath the surface is less than that of a drop at the surface by a factor of $e^{-2\pi} \approx 1/500$. We see that the equilibrium water depth need only be of order one wavelength for the wave motion at the bottom to be essentially negligible and thus for the "deep-water wave" approximation to be an excellent one.

Shallow-water waves. By a shallow-water wave we mean one for which the equilibrium depth h is small compared to the attenuation depth $\lambdabar$. In that case, we can approximate the y dependence of ψ_x and ψ_y by retaining only the first interesting terms in the Taylor's series expansion of $f(y)$ and $g(y)$. Thus we can easily show that, for $h \ll \lambdabar$, Eqs. (63) and (64) become

$$\psi_y = 2A \cos \omega t \sin kx[k(y + h)], \qquad (68)$$

$$\psi_x = 2A \cos \omega t \cos kx. \qquad (69)$$

We see that for a shallow-water wave the horizontal motion ψ_x is independent of the equilibrium vertical position y of the water drop. The vertical motion ψ_y varies linearly with the depth of the droplet, being zero at the bottom and maximum at the surface. At the surface the maximum vertical motion is less than the maximum horizontal motion by a factor $h/\lambdabar \ll 1$.

In our "idealized water" model we have neglected friction of the water rubbing on the rough bottom. For deep-water waves this omission is not important. For shallow-water waves the friction is important, as you can easily see if you excite shallow-water standing waves

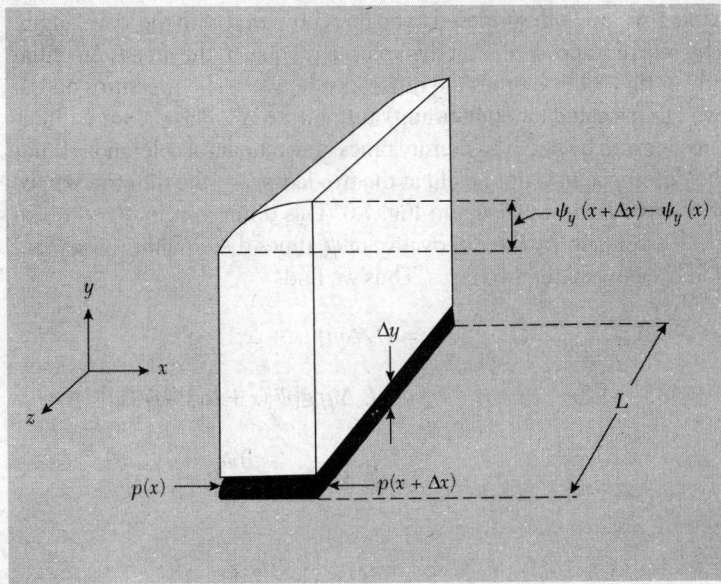

Fig. 7.6 Gravitational return force along x on a volume element of water. The shaded volume experiences a force that is proportional to the pressure difference $p(x + \Delta x) - p(x)$. This pressure difference is proportional to the difference in water height $\psi_y(x + \Delta x) - \psi_y(x)$.

in a rectangular pan (as in Home Exp. 7.11). You will notice that the coffee grounds get swept away from the regions of maximum horizontal velocity and collect at the regions where the horizontal velocity is always zero, i.e., at the maxima for vertical motion. Another approximation was in our neglect of the "internal" friction in the liquid, the viscosity. If you wish to see the effect of viscosity, try repeating any of the home experiments using mineral oil instead of water.

Dispersion relation for gravitational water waves. We have learned the geometrical structure of (ideal) water waves, but we still know nothing about the relation between the "shape" (the wavelength and depth) and the frequency. This is because we have not said anything about the return forces that act on the water in the waves. (Recall that the return force per unit displacement per unit mass is ω^2. This is a very general result and holds for harmonic water waves as well as for any other harmonic waves.)

In studying modes in Chap. 1 we learned that, since in a mode all moving parts have the same value of ω^2, we can find the relation between the mode frequency and mode shape by considering the motion of a single degree of freedom of one moving part, once the shape of the mode is known. In the present problem that shape is given by Eqs. (63) and (64). Therefore we need consider only the x (or y) motion of a single drop of water. We choose to consider the x motion of an infinitesimal volume of water that lies very near the surface.

Consider a tiny volume which at equilibrium extends a small distance Δx along the propagation direction x, a distance L along the "uninteresting" direction z, and a small vertical distance Δy. The dimensions

Δx and Δy are both supposed to be tiny compared with the wavelength. The return force along x on this volume is equal to the area $L \Delta y$ of the side of this volume times the difference between the pressures on the two faces located (at equilibrium) at x and $x + \Delta x$. This pressure difference is given by ρg (mass density times gravitational acceleration) times the difference in water height at the two faces, i.e., the difference in ψ_y at the two faces, as shown in Fig. 7.6. This difference in ψ_y is in turn given essentially by the x derivative of ψ_y times the equilibrium separation Δ_x between the two faces. Thus we find

$$F_x = -L \, \Delta y[p(x + \Delta x) - p(x)]$$

$$= -L \, \Delta y \rho g[\psi_y(x + \Delta x) - \psi_y(x)]$$

$$= -L \, \Delta y \, \Delta x \rho g \frac{\partial \psi_y}{\partial x}$$

$$= -(\Delta M)g \left[\frac{\partial \psi_y}{\partial x} \right]_{y=0} \tag{70}$$

where $\Delta M \equiv \rho L \, \Delta y \, \Delta x$ is the mass of the water in the volume element. This force produces an acceleration along x. The acceleration along x is $\partial^2 \psi_x / \partial t^2$, which is equal to $-\omega^2 \psi_x$, since we have harmonic motion. Thus Newton's second law for the acceleration of the mass ΔM is

$$F_x = (\Delta M) \frac{\partial^2 \psi_x}{\partial t^2},$$

which gives [using Eq. (70) for F_x]

$$(\Delta M)g \left[\frac{\partial \psi_y}{\partial x} \right]_{y=0} = (\Delta M)\omega^2 [\psi_x]_{y=0}. \tag{71}$$

Now use ψ_y and ψ_x as given by Eqs. (63) and (64). Then Eq. (71) gives

$$\omega^2 = gk \frac{(1 - e^{-2kh})}{(1 + e^{-2kh})}. \tag{72}$$

Equation (72) is the desired dispersion relation. In the interesting limiting cases of deep- and shallow-water gravity waves the dispersion relation and corresponding phase velocities are easily obtained from Eq. (72); they are

$$\text{Deep water:} \qquad \omega^2 = gk, \qquad v_\varphi = \sqrt{g\lambda}, \tag{73}$$

$$\text{Shallow water:} \qquad \omega^2 = gk(h/\lambda), \qquad v_\varphi = \sqrt{gh}. \tag{74}$$

Thus shallow-water gravity waves are nondispersive. Deep-water gravity waves are dispersive: the phase velocity doubles when the wavelength quadruples.

Surface tension waves. In deriving the dispersion law Eq. (72) we neglected the return force contributed by surface tension. For a given volume element of displaced water the surface tension's contribution to the return force is proportional to T (surface tension constant) times the curvature of the surface. The curvature is proportional to k^2. Thus the surface tension contribution is proportional to Tk^2. The gravitational contribution is proportional to the weight Mg, i.e., to ρg. Thus we might guess that the relative contribution to ω^2 of surface tension and gravity is proportional to the dimensionless ratio $Tk^2/\rho g$. That guess is correct. (See Prob. 7.33.)

Traveling water waves. We shall let you show (Prob. 7.31) that the form of traveling water waves is

$$\psi_y = A \cos(\omega t - kx)(e^{ky} - e^{-2kh}e^{-ky}), \tag{75}$$

$$\psi_x = A \sin(\omega t - kx)(e^{ky} + e^{-2kh}e^{-ky}). \tag{76}$$

From Eqs. (75) and (76) you can easily show that for deep-water traveling waves a given water droplet travels in a circle in the xy plane, traveling forward when on a crest and backward when in a trough. For a general water depth h, the water droplet travels in an ellipse. This elliptical motion is similar to the circular motion found in a deep-water traveling wave except that the circle is "squashed flat" between the top surface and the bottom of the pan (or lake or ocean). At least that is the case if the friction at the bottom is negligible. If it is not negligible, then the water travels relatively easily forward (on the crests) but rubs on the bottom when trying to go backward in the troughs. The result is that water is carried forward more on the crests than backward in the troughs, and there is a net translation of water. When that is the case, the waves are near to (or are) "breaking." Thus the breakers at the beach carry water with them. (The return flow is the "undertow.") A skin diver swimming at what he regards as a safe distance offshore from a rocky beach (onto which he would not like to be translated) may be in trouble (at least I was) when a wave of exceptionally long wavelength comes in.

7.4 Electromagnetic Waves

In this section we shall use Maxwell's equations to give general proofs of several things that we already know from having studied parallel-plate transmission lines. We will thus "strengthen our foundations" as well as prepare for a better understanding of electromagnetic waves in three-dimensional space.

Maxwell's equations for vacuum. These are given by (see Vol. II, p. 264)

$$\frac{\partial \mathbf{E}}{\partial t} = + \frac{1}{\mu_0 \epsilon_0} \vec{\nabla} \times \mathbf{B} = C^2 \vec{\nabla} \times \mathbf{B} \quad (77a)$$

$$\frac{\partial \vec{\mathbf{B}}}{\partial t} = -\vec{\nabla} \times \vec{\mathbf{E}} \quad (77b)$$

$$\nabla \cdot \mathbf{E} = 0 \quad (77c)$$

$$\nabla \cdot \mathbf{B} = 0. \quad (77d)$$

Classical wave equation for electromagnetic waves in vacuum. We shall find a partial differential equation for $\mathbf{E}$ by eliminating $\mathbf{B}$ from Eqs. $(77a)$ through $(77d)$. We start by differentiating Eq. $(77a)$ with respect to t. Then we use Eq. $(77b)$:

$$\frac{\partial \mathbf{E}}{\partial t} = c^2 \vec{\nabla} \times \vec{\mathbf{B}},$$

$$\frac{\partial^2 \mathbf{E}}{\partial t^2} = c^2 \frac{\partial}{\partial t} (\vec{\nabla} \times \vec{\mathbf{B}})$$

$$= c^2 \vec{\nabla} \times \frac{\partial \vec{\mathbf{B}}}{\partial t}$$

$$= c^2 \left[\vec{\nabla} \times \left(-\vec{\nabla} \times \vec{\mathbf{E}} \right) \right]$$

$$= -c^2 \left[\vec{\nabla} \times \left(\vec{\nabla} \times \vec{\mathbf{E}} \right) \right] \quad (77e)$$

It can be shown [Appendix, Eq. (39)] that for any vector $\mathbf{C}$

$$\nabla \times (\nabla \times \mathbf{C}) = \nabla(\nabla \cdot \mathbf{C}) - (\nabla \cdot \nabla)\mathbf{C} \cdot \quad (78)$$

Subsituting $\mathbf{E}$ for $\mathbf{C}$ in Eq. (78) and using the fact that $\nabla \cdot \mathbf{E} = 0$ [Eq. $(77c)$], we obtain from Eq. $(77e)$

$$\frac{\partial^2 \mathbf{E}(x, y, z, t)}{\partial t^2} = c^2 \ \nabla^2 \mathbf{E}(x, y, z, t). \quad (79a)$$

This vector equation consists of three separate partial differential equations:

$$\frac{\partial^2 E_x}{\partial t^2} = c^2 \ \nabla^2 E_x; \quad \frac{\partial^2 E_y}{\partial t^2} = c^2 \ \nabla^2 E_y; \quad \frac{\partial^2 E_z}{\partial t^2} = c^2 \ \nabla^2 E_z. \quad (79b)$$

Thus E_x, E_y, and E_z each satisfy the classical wave equation for non-dispersive waves [see Eq. (18), Sec. 7.2]. Similarly, one can eliminate **E** from Maxwell's equations and obtain the classical wave equation for the three components of **B** (Prob. 7.12).

Electromagnetic plane waves in vacuum. An electromagnetic *plane wave* consists of space- and time-dependent electric and magnetic fields $\mathbf{E}(x, y, z, t)$ and $\mathbf{B}(x, y, z, t)$ having the following properties:

1. There is a unique propagation direction, which we take to be along $\hat{\mathbf{z}}$. (The waves can be any combination of traveling or standing waves.)
2. None of the components of **E** or **B** depends on either of the transverse coordinates x and y.

Thus we have

$$\mathbf{E} = \hat{\mathbf{x}}E_x(z, t) + \hat{\mathbf{y}}E_y(z, t) + \hat{\mathbf{z}}E_z(z, t) \tag{80}$$

$$\mathbf{B} = \hat{\mathbf{x}}B_x(z, t) + \hat{\mathbf{y}}B_y(z, t) + \hat{\mathbf{z}}B_z(z, t). \tag{81}$$

Of course the fact that we *have* plane waves [waves of the form of Eqs. (80) and (81)] has something to do with where the waves came from, how they were produced, etc. We are not interested in the sources now. We just assume that the waves came from somewhere and have the form of Eqs. (80) and (81).

Electromagnetic plane waves are transverse. Now let us apply Maxwell's equations to Eqs. (80) and (81). First we use Gauss's law, which says that div **E** is $4\pi\rho$. In vacuum, ρ is zero. Also, since there are no x or y dependences of any of the components, the partial derivatives with respect to x and y give zero. Thus we have

$$\nabla \cdot \mathbf{E} = \frac{\partial E_z(z, t)}{\partial z} = 0, \tag{82}$$

which says that E_z is independent of z. That E_z is also independent of t can be seen by considering Maxwell's "displacement current" equation,

$$\frac{\partial \vec{\mathbf{E}}}{\partial t} = c^2 \vec{\nabla} \times \vec{\mathbf{B}}. \tag{83}$$

Take the z component of Eq. (83). The right-hand side involves $\partial B_y/\partial x$ and $\partial B_x/\partial y$, both of which are zero. Thus $\partial E_z/\partial t$ is zero. We conclude that E_z is a constant. For simplicity, we take the constant to be zero. (We are not thereby losing generality. We are merely using the superposition principle to "turn off" any constant field that we already understand. We can always add it back if occasion calls for it.)

Similarly, the fact that we have $\nabla \cdot \mathbf{B} = 0$ tells us that $B_z(z, t)$ has no z dependence. That it also has no time dependence is seen by considering the z component of Faraday's law,

$$\frac{\partial \vec{B}}{\partial t} = -\vec{\nabla} \times \vec{E}, \qquad (84)$$

which gives $\partial B_z / \partial t$ to be zero. Thus, although there may be some steady magnetic fields due to big steady currents somewhere, they have no space or time dependence and are not of present interest to us. We therefore take B_z to be zero (again using the superposition principle).

So far, we have concluded that (aside from nonwavelike constant fields) *electromagnetic plane waves are transverse waves*. That is, the electric and magnetic fields are perpendicular to the direction of propagation $\hat{z}$.

Coupling of E_x and B_y. We are left with E_x, E_y, B_x, and B_y, and the as-yet-unused x and y components of Eqs. (83) and (84). The x component of Eq. (83) and the y component of Eq. (84) give

$$\frac{\partial E_x}{\partial t} = -c^2 \frac{\partial B_y}{\partial z}, \qquad \frac{\partial B_y}{\partial t} = -\frac{\partial E_x}{\partial z}. \qquad (85)$$

Similarly the y component of Eq. (83) and the x component of Eq. (84) give

$$\frac{\partial E_y}{\partial t} = c^2 \frac{\partial B_x}{\partial z}, \qquad \frac{\partial B_x}{\partial t} = \frac{\partial E_y}{\partial z}. \qquad (86)$$

According to Eqs. (85), E_x and B_y are not independent. They are "coupled" by two first-order linear partial differential equations, Eqs. (85). Thus if, for example, E_x is constant in both space and time, then so is B_y. On the other hand, if E_x is completely known as a function of both z and t, then, as we shall show, B_y is also completely determined (aside from constant fields in which we are not interested). Similarly, according to Eqs. (86), E_y and B_x are coupled. If E_y is known, then B_x is determined: if E_y is zero, B_x is zero (or constant).

Linear and elliptical polarization. The fields E_x and E_y are not coupled by Maxwell's equations (for the plane waves we are considering). They are "independent." That means that it is possible to produce (by a suitable radiating source) electromagnetic plane waves with E_x different from zero but with E_y equal to zero for all z and t. In that case, the waves are said to be *linearly polarized along* $\hat{x}$. The electric field E_x and magnetic field B_y are then the only nonzero (or rather nonconstant) fields. Similarly we can have electromagnetic waves that are linearly polarized along $\hat{y}$, where E_y and B_x are the only nonzero fields. We can also have any combination of E_x and E_y with (in the case of a single frequency) an arbitrary relative phase between E_x and E_y. Then we have a general state of polarization, called *elliptical polarization*. We will study polarization in Chap. 8.

You may have noticed that Eqs. (86) relate E_y and *minus* B_x in the same way that Eqs. (85) relate E_x and B_y. The minus sign may be puzzling at first. However, as you can easily show, if you had linearly polarized waves with E_x and B_y both positive (at a given instant) and if you rotated the coordinate axes by 90 deg so as to put the new y axis along the electric field, then the new x axis would be along the *negative* of the magnetic field. (Prob. 7.34). Therefore Eqs. (86) are physically equivalent to Eqs. (85). That means that we shall not be missing anything if we confine ourselves to studying the consequences of Eqs. (85).

From now on we shall assume that we have only the single linear polarization state corresponding to nonzero E_x and B_y, i.e., corresponding to Eqs. (85). It will be simplest if we consider first a pure harmonic traveling wave propagating in the +z direction. Then we will immediately see how to get the equivalent result for a pure harmonic wave propagating in the −z direction. A superposition of these with arbitrary amplitudes and phase constants is then the most general solution for a given frequency and includes pure standing waves as a special case.

Traveling harmonic wave.   Assume that E_x is given by

$$E_x = A \cos (\omega t - kz). \tag{87}$$

Then Eqs. (85) and the relation $\omega = ck$ give

$$\frac{\partial B_y}{\partial z} = -\frac{1}{c^2} \frac{\partial E_x}{\partial t}$$
$$= \frac{\omega}{c^2} \frac{\partial E_x}{\partial t} = \frac{1}{c} \frac{\partial E_x}{\partial z} \tag{88}$$

$$\frac{\partial B_y}{\partial t} = -\frac{\partial E_x}{\partial z}$$
$$= -kA \sin (\omega t - kz)$$
$$= \frac{1}{c} \frac{\partial E_x}{\partial t} \tag{89}$$

According to Eqs. (88) and (89) the variation of B_y with respect to z and t is the same as that of E_x. Thus we see that in a traveling harmonic plane wave propagating in the +z direction B_y and E_x are equal, aside from uninteresting additive constants, which we "superpose to zero."

If we consider a harmonic traveling wave propagating in the −z direction, we find that B_y is the negative of E_x, as you can easily see by replacing k with −k in the above equations. Both directions of propagation are included in the summarizing statements

$$\text{Traveling wave:} \begin{cases} \left| \mathbf{E}(z, t) \right| = \left| \mathbf{B}(z, t) \right|, \\ \quad\quad \mathbf{E} \cdot \mathbf{B} = 0, \\ \quad\quad \hat{\mathbf{E}} \times \hat{\mathbf{B}} = \hat{\mathbf{v}}. \end{cases} \tag{90}$$

Standing harmonic wave.   Assume that E_x is given by

$$E_x(z, t) = A \cos \omega t \cos kz. \tag{91}$$

Then we shall let you show (Prob. 7.36)

$$B_y(z, t) = A \sin \omega t \sin kz = E_x(z - \tfrac{1}{4}\lambda, \ t - \tfrac{1}{4}T). \tag{92}$$

We see from Eqs. (91) and (92) that in an electromagnetic stand-ing plane wave in vacuum $\mathbf{E}$ and $\mathbf{B}$ arc perpendicular to one another and to $\hat{\mathbf{z}}$, have the same amplitude, and are 90 deg out of phase both in space and in time. (This behavior is similar to that of the pressure and velocity in a standing sound wave or that of the tranverse tension and velocity in a standing wave on a string.)

Energy flux in a plane wave.   The energy density of electromag-netic fields in vacuum is given by

$$\text{Energy density} = \frac{1}{2} \left(\epsilon_0 \mathbf{E}^2 + \frac{\mathbf{B}^2}{\mu_0} \right). \tag{93}$$

(This expression is given for static fields in Vol. II, pp. 102 and 256, and can be shown to hold generally.) We are interested in the energy in any linear combination of traveling and stand-ing plane waves. In particular, we are interested in the flow of energy. Let us therefore obtain an expression for the energy in an infinitesimal volume element having area A perpendicular to the z axis and infinitesimal thickness Δz along the z axis. (We shall then examine the rate of change of this energy with time.) The energy $W(z, t)$ in this volume element is the energy density times the volume $A \, \Delta z$:

$$W(z, t) = \frac{A\Delta z}{2} \left(\epsilon_0 E_x^{\ 2} + \frac{B_y^{\ 2}}{\mu_0} \right). \tag{94}$$

Differentiating the energy $W(z,t)$ with respect to t gives

$$\frac{\partial W(z, t)}{\partial t} = A\Delta z \left(\epsilon_0 E_x \frac{\partial E_x}{\partial t} + \frac{B_y}{\mu_0} \frac{\partial B_y}{\partial t} \right). \tag{95}$$

Now use Maxwell's equations, Eqs. (85), to eliminate $\partial E_x / \partial t$ and $\partial B_y / \partial t$:

$$\frac{\partial W(z,t)}{\partial t} = A\Delta z\left[\epsilon_0 E_x \frac{\partial E_x}{\partial t} + \frac{B_y}{\mu_0}\frac{\partial B_y}{\partial t}\right]$$

$$= A\Delta z\left[\epsilon_0 E_x\left(-c^2\frac{\partial B_y}{\partial z}\right) + \frac{B_y}{\mu_0}\left(-\frac{\partial E_x}{\partial z}\right)\right]$$

$$= -\frac{A\Delta z}{\mu_0}\left[E_x\frac{\partial B_y}{\partial z} + B_y\frac{\partial E_x}{\partial z}\right]$$

$$= -\frac{A\Delta z}{\mu_0}\frac{\partial}{\partial z}(E_x B_y)$$

$$= -\frac{A\Delta z}{\mu_0}\left[\frac{(E_x B_y)_{z+\Delta z} - (E_x B_y)_z}{\Delta z}\right] \tag{96}$$

The last step corresponds, in the limit of infinitesimal Δz, to the definition of the partial derivative of $E_x B_y$ with respect to z (at fixed time); i.e., we evaluate the quantity $E_x B_y$ at positions z and $z + \Delta z$, subtract one result from the other, divide by Δz, and take the limit as Δz goes to zero. Thus we have found that the rate of change of energy in the volume $A\,\Delta z$ is given by

$$\frac{1}{A}\frac{\partial W(z,t)}{\partial t} = \frac{1}{\mu_0}\left[\left(E_x B_y\right)_z - \left(E_x B_y\right)_{z+\Delta z}\right]$$

$$= S_z(z,t) - S_z(z + \Delta z, t), \tag{97}$$

where

$$S_z(z,t) \equiv \frac{1}{\mu_0}E_x(z,t)B_y(z,t)$$

$$= \frac{1}{\mu_0}(\vec{\mathbf{E}} \times \vec{\mathbf{B}})_z. \tag{98}$$

Thus the rate of change of energy in the volume element $A\,\Delta z$ is the value of a quantity $AS_z\,(z,t)$ evaluated at z, the left edge of the interval, minus the value of this same quantity at $z + \Delta z$, the right edge of the interval. The quantity $S_z(z,t)$ must therefore be the instantaneous rate of flow of energy per unit area in the $+z$ direction at point z. The increase of energy in the volume element (if there is an increase) results from the difference of the inflow (from the left) minus the outflow (to the right). The z component $S_z(z,t)$ of the *flux vector* S is defined as the rate of energy flow in the $+z$ direction per unit area (in erg/cm^2 sec) at z, t. (Of course, that is the only direction of energy flux in our problem, since we chose $\hat{\mathbf{z}}$ for the propagation direction.)

Poynting vector.   The general form of the flux vector is

$$\boxed{\vec{\mathbf{S}} = \frac{1}{\mu_0}(\vec{\mathbf{E}} \times \vec{\mathbf{B}}),} \tag{99}$$

which is independent of the choice of coordinates. The flux vector **S** is also called the *Poynting vector*.

Energy density and flux in traveling wave. For a linearly polarized wave traveling in the $+z$ direction, we can take $\mathbf{E} = \hat{\mathbf{x}} E_x$ and $\mathbf{B} = \hat{\mathbf{y}} B_y$, with $B_y = E_x$ for every z, t. Thus (with E_0 in statvolt/cm)

$$E_x = E_0 \cos(\omega t - kz),$$

$$B_y = \frac{E_0}{c} \cos(\omega t - kx) \qquad (100)$$

$$\text{Energy density} = \frac{1}{2}(\epsilon_0 E_x{}^2 + \frac{1}{\mu_0} B_y{}^2) = -\frac{\epsilon_0}{2} E_0{}^2 \cos^2(\omega t - kx), \quad (101)$$

$$\begin{array}{l}\text{Energy} \\ \text{flux}\end{array} = S_z = \frac{1}{\mu_0} \frac{E_0{}^2}{c^2} \cos^2(\omega t - kz) = \frac{\epsilon_0}{2} E_0{}^2 \cos^2(\omega t - kx). \quad (102)$$

Notice that the energy flux S_z (in erg /cm^2 sec) for a traveling wave is simply the energy density (in erg /cm^3) times the velocity of light (in cm/sec).

The time-averaged energy flux (at fixed z) equals the space-averaged energy flux (at fixed t). Both are independent of z and t and are obtained from Eq. (102) by replacing $\cos^2(\omega t - kz)$ by its average value of $\frac{1}{2}$.

Energy density and flux in standing wave. For a standing wave we have

$$E_x = E_0 \cos \omega t \cos kz,$$

$$B_y = \frac{E_0}{c} \sin \omega t \sin kz. \qquad (103)$$

The electric energy density and magnetic energy density have their maxima at times separated by $\frac{1}{4}$ period and at positions separated by $\frac{1}{4}$ wavelength. We shall let you show (Prob. 7.36) that in any region of length $\frac{1}{4}\lambda$ the total energy is constant. The energy in the electric field oscillates harmonically about its average value at a frequency 2ω, between the limiting values of zero and twice the average value. So does the energy in the magnetic field. Thus the energy oscillates back and forth from being purely electric, with maximum energy density at one place, to being purely magnetic, with maximum energy density at another place $\frac{1}{4}\lambda$ away. This is somewhat like the behavior of a harmonic oscillator: The total energy of the oscillator is constant but oscillates back and forth between being purely potential energy with the mass at one place and being purely kinetic energy with the mass at another place. The potential and kinetic energies each oscillate harmonically about their average values at frequency 2ω, where the factor of two arises from the fact that the potential energy is large and positive twice per oscillation cycle (as is the kinetic energy). The electric field E_x in a standing wave is somewhat analogous to the displacement from equilibrium of the mass of a harmonic oscillator, while the magnetic field B_y is somewhat analogous to the velocity of the mass.

Flux of linear momentum in traveling plane wave—radiation pressure. If a beam of electromagnetic radiation is absorbed without reflection (by a perfect termination, for example), thereby giving an amount of energy W to the absorber, it also gives momentum (along the propagation direction) to the absorber, as we shall show. The amount of momentum turns out to be W/c. If the beam is reflected through 180 deg by a mirror (without any absorption), then it gives twice this amount of momentum to the mirror; i.e., if an amount of energy W is reflected without absorption, then the mirror acquires momentum $2W/c$ along the propagation direction. Thus the radiation pushes on things that absorb it or reflect it. This push is called *radiation pressure*. Every amount of energy W in a *traveling* electromagnetic plane wave has an amount of momentum **P** given by

$$\vec{\mathbf{P}} = \frac{W\hat{\mathbf{z}}}{c^2}, \tag{104}$$

where $\hat{\mathbf{z}}$ is along the direction of propagation.

A simple derivation of Eq. (104) is obtained by using the idea that light in a traveling wave is bundled into packages called *photons*. A photon is like a "particle" having zero rest mass. A relativistic particle with rest mass M and momentum P has energy W given by

$$W = [(cP)^2 + (Mc^2)^2]^{1/2}. \tag{105}$$

If M is zero, we get Eq. (104).

The above derivation is short, and perhaps misleading. The fact that electromagnetic radiation turns out to be "quantized," in the sense that it can only deliver energy in quantized "bits" of magnitude $\hbar\omega$, really has nothing to do with radiation pressure, i.e., has nothing to do with Eq. (104). Therefore we should be able to give a purely classical derivation of Eq. (104) without using the idea of photons or "particles," and we shall now do that. (You will study the quantum aspects of light in Vol. IV.)

Consider a particle of charge q acted on by a traveling plane wave. Take the charge q to be positive, and suppose the particle is released from rest at $t = 0$. The force **F** on the particle is given by the Lorentz force,

$$\mathbf{F} = q\vec{\mathbf{E}} + q\vec{\mathbf{v}} \times \vec{\mathbf{B}} \tag{106}$$

At first (during the first few oscillations, for example), the magnitude of the velocity **v** is small. Therefore the motion of the charge is mainly due to **E**. Thus **v** is along **E** and reverses direction at the same rate that **E** reverses direction. But **B** reverses whenever **E** reverses. Thus **v** × **B** always has the same sign. The force on q due to **B** is therefore always in the propagation direction, the direction of **E** × **B**. Thus the charge q undergoes a motion that is a superposition of a transverse oscillation at the frequency of the fields plus a slowly increasing velocity along the propagation direction. We shall now show that the time-averaged rate at which the charge acquires momentum along z is $1/c$ times the time-averaged rate at which the charge absorbs energy from the traveling wave. (The charge does not keep the energy that it absorbs. If it is a charge in a piece of perfectly terminating spacecloth, then it is continually transferring energy to the material through the resistive drag acting on the charge. If it is a charge in free space, then it turns out to be continually radiating energy in all directions. The amount of energy radiated in the direction of the incident traveling wave is negligible, so that a negligible amount of the absorbed energy is returned to the traveling wave.)

Here is the derivation. Our "standard traveling wave" has $\mathbf{E} = \hat{\mathbf{x}} \, E_x$, $\mathbf{B} = \hat{\mathbf{y}} \, B_y$, and $B_y = E_x$. The velocity **v** of the charged particle is given by $\mathbf{v} = \hat{\mathbf{x}}\dot{x} + \hat{\mathbf{y}}\dot{y} + \hat{\mathbf{z}}\dot{z}$. Inserting these values into Eq. (106) and using $\hat{\mathbf{x}} \times \hat{\mathbf{y}} = \hat{\mathbf{z}}$, $\hat{\mathbf{y}} \times \hat{\mathbf{y}} = 0$, and $\hat{\mathbf{z}} \times \hat{\mathbf{y}} = -\hat{\mathbf{x}}$, we get

$$\mathbf{F} = \hat{\mathbf{x}} q E_x + q\dot{x} B_y \hat{\mathbf{z}} - q\dot{z} B_y \hat{\mathbf{x}}. \tag{107}$$

Now we take the time average of Eq. (107) over one cycle. The first term, $\hat{\mathbf{x}} \, qE_x$, averages to zero. So does the last term, containing $\dot{z}B_y$. That is because we can assume that the increment of velocity along z during one cycle is negligible, i.e., w can take the slowly increasing velocity $\dot{z}$ to be constant during one cycle. The field B_y then averages to zero over the cycle. The remaining term $\dfrac{q}{c}\dot{x}B_y\hat{\mathbf{z}}$ does not average to zero, because the transverse velocity $\dot{x}$ is oscillating at the same rate as B_y. Thus we have (recalling that force is time rate of change of momentum) for the time average (denoted by brackets $\langle \ \rangle$)

$$\langle \mathbf{F} \rangle = \left\langle \frac{d\mathbf{P}}{dt} \right\rangle = \hat{\mathbf{z}} q \left\langle \dot{x} B_y \right\rangle. \tag{108}$$

Now let us consider the rate at which work is done on the charge by the traveling wave. The instantaneous rate of doing work on q is given by

$$\frac{dW}{dt} = \vec{\mathbf{v}} \cdot \mathbf{F} = \vec{\mathbf{v}} \cdot \left(q\vec{\mathbf{E}} + q\vec{\mathbf{v}} \times \vec{\mathbf{B}} \right)$$

$$= q\vec{\mathbf{v}} \cdot \vec{\mathbf{E}}$$

$$= q\dot{x}E_x.$$

Averaging over one cycle gives

$$\left\langle \frac{dW}{dt} \right\rangle = q \left\langle \dot{x}E_x \right\rangle. \tag{109}$$

By comparing Eqs. (108) and (109), and using the fact that $B_y = \dfrac{E_x}{c^2}$ (for a traveling wave), we see that

$$\left\langle \frac{d\mathbf{P}}{dt} \right\rangle = \frac{\hat{\mathbf{z}}}{c^2} \left\langle \frac{dW}{dt} \right\rangle. \tag{110}$$

Thus, during a time interval in which the electron removes energy W from the traveling wave, it also removes momentum $\hat{\mathbf{z}}\,(W/c)$ from the wave. It is not possible to remove the energy W without removing the momentum $\hat{\mathbf{z}}\,(W/c)$. This is the same as saying that the radiation has momentum given by Eq. (104). Radiation pressure from the sun is discussed in Probs. 7.13, 7.14, and 7.15.

Angular momentum in traveling plane wave. We shall show that a traveling plane wave can transfer not only energy and linear momentum to the charge q, but also angular momentum. In order to do this, it has to drive the charge in a circular motion. Obviously, this will not happen in the case of the "linearly polarized" field that we have been considering. It does happen if the fields are "circularly polarized." Let us consider a traveling wave propagating in the $+\hat{\mathbf{z}}$ direction with an electric field $\mathbf{E}$ that has constant magnitude and rotates (at fixed z) with angular velocity ω about the z axis, the sense of rotation being along $+\hat{\mathbf{z}}$ as given by a right-hand rule for rotation. Thus E_x and E_y are harmonic functions of time (at fixed z), and E_x leads E_y by 90 deg in phase. The magnetic field $\mathbf{B}$ is given (as always for a traveling wave) by $\mathbf{B} = \hat{\mathbf{z}} \times \mathbf{E}$. Since the electric field drives the charge q (and the magnetic field bends it), we can assume that at steady state, q is traveling in a circle at angular velocity ω with the same sense of rotation as the fields. (The charge q is also slowly drifting in the $+z$ direction because of the radiation pressure exerted on it by the traveling wave. This we can neglect.) Thus the configurations of the fields and of the position $\mathbf{r}$ and velocity $\mathbf{v}$ of the charge are as shown in Fig. 7.7. Notice that $\omega\mathbf{r}$ has the same

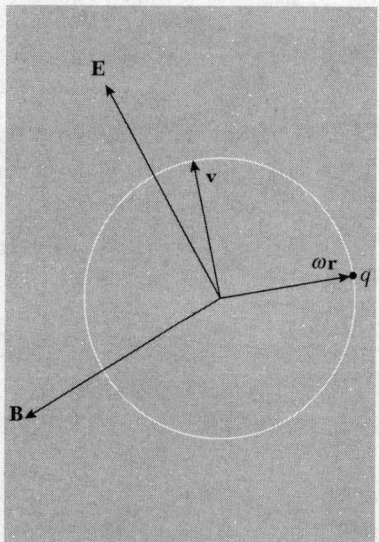

Fig. 7.7 *Circularly polarized light drives charge q in circular path. $\hat{\mathbf{z}}$ is out of the paper.*

magnitude as **v** and that the relative directions of $\omega\mathbf{r}$ and **v** are as indicated.

The torque τ on the charge q is equal to $\mathbf{r} \times \mathbf{F}$. Thus (multiplying by ω) we see that

$$\omega\tau = \omega\mathbf{r} \times \mathbf{F}$$

$$= \omega\vec{\mathbf{r}} \times q\vec{\mathbf{E}} + \omega\vec{\mathbf{r}} \times q(\vec{\mathbf{v}} \times \vec{\mathbf{B}}). \tag{111}$$

We average this torque over one cycle. From Fig. 7.7 we see that $\mathbf{v} \times \mathbf{B}$ is along $\hat{\mathbf{z}}$, and thus $\mathbf{r} \times (\mathbf{v} \times \mathbf{B})$ is along $-\mathbf{v}$. Since each component of **v** averages to zero over one cycle, we see that the magnetic field makes no net contribution to the time-averaged torque. From Fig. 7.7 we also see that $\omega\mathbf{r} \times \mathbf{E}$ is along $\hat{\mathbf{z}}$, and has the same algebraic magnitude as $\mathbf{v} \cdot \mathbf{E}$. Therefore it is given by

$$\omega\mathbf{r} \times \mathbf{E} = \hat{\mathbf{z}}\,\mathbf{v} \cdot \mathbf{E}. \tag{112}$$

Thus the time-averaged torque on q given by Eq. (111) is

$$\langle \tau \rangle = \left\langle \frac{d\mathbf{J}}{dt} \right\rangle = \frac{\hat{\mathbf{z}}}{\omega} \langle q\mathbf{v} \cdot \mathbf{E} \rangle = \frac{\hat{\mathbf{z}}}{\omega} \left\langle \frac{dW}{dt} \right\rangle, \tag{113}$$

where we used the facts that the torque is the time rate of change of angular momentum **J** and that $q\mathbf{v} \cdot \mathbf{E}$ is the rate of doing work on q. According to Eq. (113), a charge q that absorbs an amount of energy W from a circularly polarized traveling plane wave in which the sense of rotation is along $+\hat{\mathbf{z}}$ also absorbs an angular momentum **J** given by

$$\mathbf{J} = \hat{\mathbf{z}}\frac{W}{\omega}.$$

A better way to express this result is to use the unit vector $\hat{u}$ for the rotation direction, which can be along either $+\hat{\mathbf{z}}$ or $-\hat{\mathbf{z}}$. Thus the result is that a circularly polarized plane traveling wave carries angular momentum

$$\boxed{\mathbf{J} = \hat{\omega}\frac{W}{\omega},} \tag{114}$$

where $\hat{\omega}$ is either along, or opposite to, the propagation direction.

As we shall learn in Chap. 8, a linearly polarized traveling plane wave of amplitude A can be regarded as a superposition of two circularly polarized traveling plane waves, each having amplitude

$\frac{1}{2}$ A, but with opposite senses of rotation. Thus it carries no angular momentum.

As you will study in Vol. IV, electromagnetic plane traveling waves only transfer energy in "quantized" bits of energy $\Delta W = \hbar\omega$. According to Eq. (114) such a wave must transfer a correspondingly quantized amount of angular momentum $\Delta J = \hbar\omega$ when it is absorbed (or emitted). It is important to realize that Eq. (114) only holds for plane traveling waves. Thus it holds at sufficiently large distance from a radiating "point source."

It turns out that if you send "right-handed" circularly polarized light through a transparent "half-wave retardation plate," you will reverse the handedness. That gives a recoil torque to the plate, since the plate must provide (by recoil) *twice* the angular momentum given by Eq. (114). This is discussed in Prob. 8.19.

Electromagnetic waves in a homogeneous medium. We have used Maxwell's equations to study electromagnetic plane waves in vacuum. In Supplementary Topic 9 we use Maxwell's equations to study electromagnetic waves in a homogeneous medium that is not vacuum. We then obtain the result

$$k^2 = \frac{\omega^2}{c^2}\,\mu_r\epsilon_r, \tag{115}$$

where ϵ is the dielectric constant and μ is the magnetic permeability. This result is the same as that we obtained in Sec. 4.3 by considering electromagnetic waves in a parallel-plate transmission line [Eq. (4.66)].

7.5 Radiation from a Point Charge

In this section we shall find the electric and magnetic fields in the *outgoing spherical traveling waves* emitted by an oscillating point charge. The results will help us to understand the properties of electromagnetic radiation emitted by atoms, radio stations, and stars, and will also tell us why the sky is blue.

Maxwell's equations with source terms. We must use Maxwell's complete equations, including the "source terms" that give the contributions of charges and currents:

$$\vec{\nabla}\cdot\vec{E} = \frac{\rho}{\epsilon_0} \tag{116}$$

$$\vec{\nabla}\times\vec{E} = \frac{-\partial\vec{B}}{\partial t} \tag{117}$$

$$\nabla\cdot\mathbf{B} = 0 \tag{118}$$

$$\vec{\nabla} \times \mathbf{B} = \frac{1}{c^2} \frac{\partial \vec{\mathbf{E}}}{\partial t} + \mu_0 \vec{\mathbf{J}}. \qquad (119)$$

We have already made use of all four of these equations for vacuum (where ρ and $\mathbf{J}$ are zero). We thus found (in Sec. 7.4) that $\mathbf{E}$ and $\mathbf{B}$ obey the classical wave equation for non-dispersive waves that propagate with velocity c. Futhermore, we have already found the relations between $\mathbf{E}$ and $\mathbf{B}$ at large distances from the source, because we can assume that in regions sufficiently far from the source the waves are indistinguishable from plane waves (if we do not try to correlate the fields at one place with those at another too far away). It only remains to make use of the source terms in Maxwell's equations to find how the radiated waves depend on the motion of the source. Now, there are two "sources" in Maxwell's equations. One is the charge density ρ; the other is the current density $\mathbf{J}$. These two sources are not independent; they are related by *conservation of charge*:

$$\frac{\partial \rho}{\partial t} + \nabla \cdot \mathbf{J} = 0. \qquad (120)$$

[You can easily verify Eq. (120) by using Eqs. (116) and (119) and the fact that $\nabla \cdot \nabla \times \nabla = 0$. See Vol. II, Eq. (4.9).] Therefore we shall not need to use $\mathbf{J}$ explicitly, because we shall automatically impose charge conservation as we follow the motion of the point charge q. Thus the current will be implicitly present but need not concern us. We can concentrate on the effect of the charge as given by Eq. (116).

Gauss's law and conservation of flux of E. Equation (116) is equivalent to *Gauss's law*. (See Vol. II , Secs. 1.10 and 2.10) For a motionless point charge, Gauss's law [or Eq. (116)] gives the familiar inverse-square field (Vol. II , Sec. 1.11),

$$\mathbf{E} = q \frac{\hat{\mathbf{r}}}{r^2}, \qquad (121)$$

where $\mathbf{r} = r\,\hat{\mathbf{r}}$ is the vector displacement of a given observation point from the charge q. For a moving charge, we can make use of the concept of lines of force and the *conservation of flux* of $\mathbf{E}$ (which is equivalent to charge conservation). [See Vol. II, Secs. 5.3 and 5.4.]

Motion of the charge. We shall now use Gauss's law to find the radiation field emitted by a point charge that undergoes the following motion: The positive charge q is at rest at the origin of an inertial frame from time $t = -\infty$ until $t = 0$. At $t = 0$, it accelerates in the $+x$ direction with a constant acceleration, a, for a short

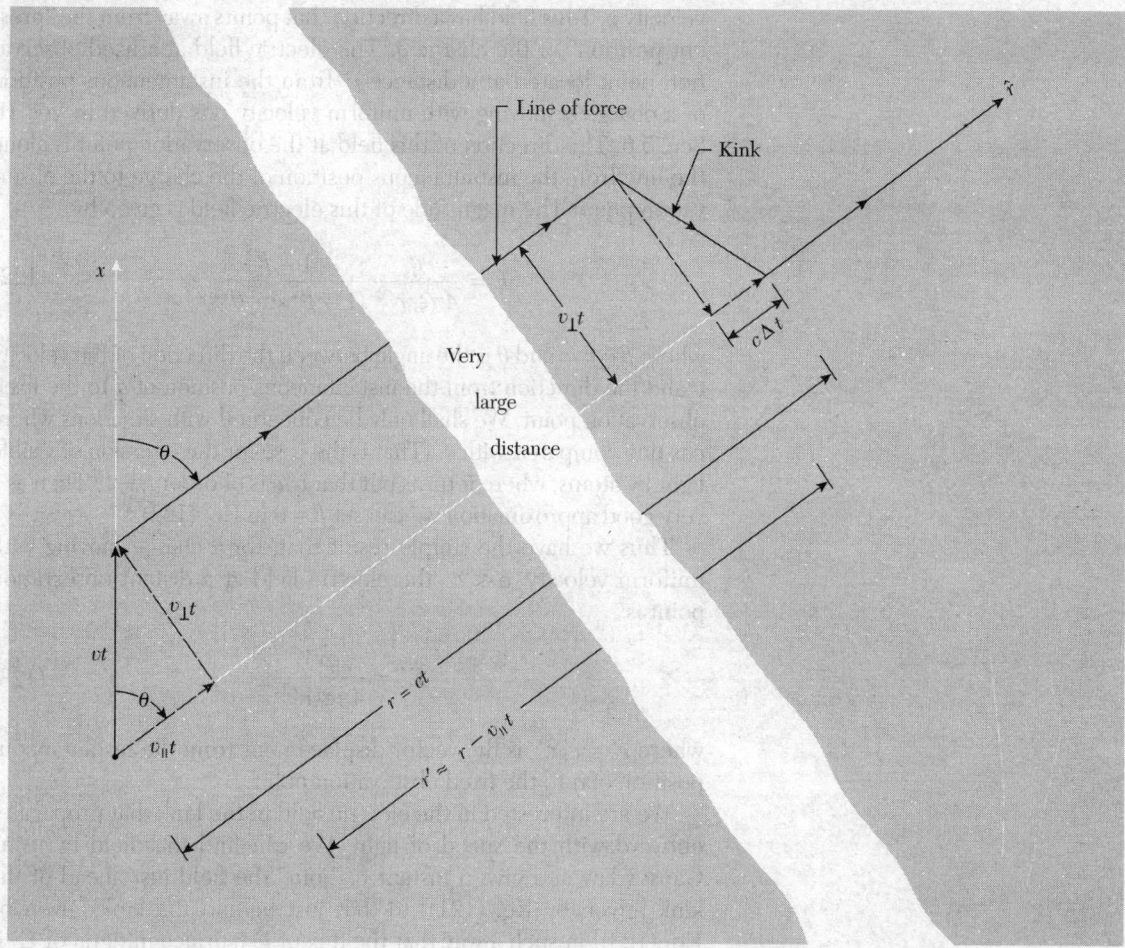

Line of force

Kink

$\hat{r}$

x

$v_\perp t$

$c\Delta t$

Very

large

distance

θ

$v_\perp t$

vt

θ

$v_\parallel t$

$r = ct$

$r' = r - v_\parallel t$

Fig. 7.8 Radiation from an acceler-ated point charge. The kink in the line of force of E propagates at the veloc-ity c. The figure is drawn for the case $t \gg t$ and $v(= a\,\Delta t) \ll c$. The compo-nents of v perpendicular to and paral-lel to the direction $\hat{r}$ from q to the observation point are labeled $v_\perp$ and $v_\parallel$, respectively.

time Δt. After that, it coasts at constant velocity $v = a\,\Delta t$. Before $t = 0$, the electric field is given everywhere in the inertial frame by Eq. (121); the magnetic field is zero. The lines of force of **E** are directed away from the position of q during all that time. The sudden acceleration at $t = 0$ creates "kinks" in the lines of force of **E** and creates lines of **B**. These propagate outward from the source at the velocity c. (We used all of Maxwell's equations in that sentence!) We are only going to find the fields at a large distance; therefore we need only find **E**. (Our results for plane waves will then give **B**.)

Consider a time t that is large compared with Δt. At positions whose distance r from the origin are greater than ct, the "news" of the acceleration has not yet arrived (i.e., the kink has not arrived). At positions with r less than $ct - \Delta t$, the kink has already passed by, and the electric field is that due to a charge moving at steady velocity v. This field has a direction that points away from the "present position" of the charge q. The electric field at a fixed observation point located at a distance r' from the instantaneous position of a charge q moving with uniform velocity v is derived in Vol. II, Sec. 5.6. The direction of this field at the observation point is along the line from the instantaneous position of the charge to the observation point. The magnitude of this electric field is given by

$$E = \frac{q}{4\pi\epsilon_0 r'^2} \frac{1 - \beta^2}{(1 - \beta^2 \sin^2\theta)^{3/2}}, \tag{122}$$

where $\beta = v/c$ and θ is the angle between the direction of the velocity v and the direction from the instantaneous position of q to the fixed observation point. We shall only be concerned with situations where v is tiny compared with c. (That is the case for the emission of visible light by atoms, where it turns out that v/c is of order $\frac{1}{137}$.) Then as a very good approximation we can set $\beta = 0$ in Eq. (122).†

Thus we have the simple result that, for a charge moving with uniform velocity $v \ll c$, the electric field at a distant observation point is

$$\mathbf{E}' = \frac{q\hat{\mathbf{r}}'}{4\pi\epsilon_0 r'^2}, \tag{123}$$

where $\mathbf{r}' = r'\,\hat{\mathbf{r}}'$ is the vector displacement from the instantaneous position of q to the fixed observation point.

We are interested in the electric field in the kink that propagates outward with the speed of light. We can find that field by using Gauss's law at a given instant to "join" the field just ahead of the kink [given by Eq. (121)] to that just behind the kink [given by Eq. (123)] in such a way that the flux of $\mathbf{E}$ (surface integral of $\mathbf{E}$) is conserved. (See Vol. II, Sec. 5.7.)

We now consider a time t that is large compared with the time duration Δt of the acceleration. Then we can neglect the distance $\frac{1}{2}a(\Delta t)^2$ that the charge traveled in time Δt compared with the much larger distance ut that it has traveled at constant velocity. We consider an observation point whose vector displacement $\mathbf{r}$ from the origin makes an angle θ with the velocity $\mathbf{v}$. The time t is chosen so that the kink is starting to sweep past the observation point at time t. Thus $r = ct$. Now consider $\mathbf{r}'$ at the rear of the kink. Since $v \ll c$, the charge has traveled a distance vt that is very small compared to $r = ct$. Therefore the direction of $\hat{\mathbf{r}}'$ is essentially parallel to the direction $\hat{\mathbf{r}}'$ The distance r' is therefore given essentially by

---

†The general case for arbitrary v ($v \leq c$) is given by J. R. Tessman and J. T. Finnell, Jr., *Am. J. Phys.* **35**, 523 (1967).

$$r' = r - vt\,\cos\theta = r\left(1 - \frac{v}{c}\cos\theta\right) \approx r,\ \text{since}\ \frac{v}{c} \ll 1. \quad (124)$$

The geometry is shown in Fig. 7.8.

Let $E_\perp$ and $E_\parallel$ denote the magnitudes of the components of **E** that are respectively perpendicular to and parallel to the propagation direction $\hat{\mathbf{r}}$, where **E** is the electric field in the space occupied by the kink. Conservation of electric flux implies continuity of the lines of force. Therefore the ratio of the transverse (perpendicular) component $E_\perp$ to the longitudinal (parallel) component $E_\parallel$ is obtained by simple inspection of Fig. 7.8. The right triangle whose hypotenuse is the line of force of **E** in the kink and whose sides are $E_\perp$ and $E_\parallel$ is similar to the right triangle with sides of length $v_\perp t$ and $c\,\Delta t$. Thus by inspection of Fig. 7.8 we see that

$$\frac{E_\perp}{E_\parallel} = \frac{v_\perp t}{c\Delta t}, \quad (125)$$

or, since $v_\perp$ is $a_\perp \Delta t$ and t is r/c,

$$\frac{E_\perp}{E_\parallel} = \frac{(a_\perp \Delta t)(r/c)}{c\Delta t} = a_\perp \frac{r}{c^2}, \quad (126)$$

where $a_\perp$ is the magnitude of the transverse component of the acceleration a.

We still need to know $E_\parallel$, the longitudinal component of **E** inside the kink. We find it by applying Gauss's law to the little pillbox-shaped volume shown in Fig. 7.9. There is no charge inside the "pillbox," so the electric flux entering it must be the same as that leaving it. We chose the pillbox in such a way that the entering flux is $E_\parallel$ times the area of the entrance surface of the pillbox and the flux that leaves is the radial field E_r just ahead of the kink times an equal area. We conclude from Fig. 7.9 that $E_\parallel$ and E_r are equal. But E_r is given by the inverse-square field, Eq. (121). Thus we have

$$E_\parallel = E_r = \frac{q}{4\pi\epsilon_0 r^2}. \quad (127)$$

[If this pillbox argument is used at the rear of the kink, one obtains the result that $E_\parallel$ must equal E_r' as given by Eq. (123). But E_r' is equal to E_r because r' and r are essentially equal, according to Eq. (124). Thus we obtain Eq. (127). Equation (125) may also be obtained by a pillbox argument. Our simpler method, "inspection" of the direction of **E** in the kink, is equivalent to the pillbox argument, as you can easily show (Prob. 7.16).]

Radiation field.   Combining Eqs. (126) and (127), we find for the magnitude of the transverse field in the kink

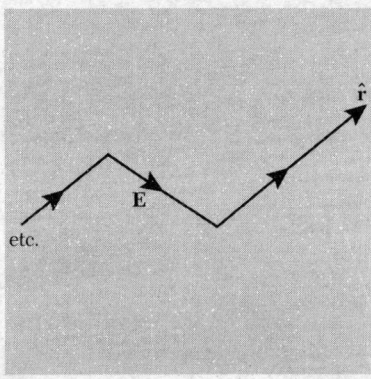

Fig. 7.9 Electric field E in the kink. The dotted lines show an imagined Surface used for application of Gauss's law.

$$E_\perp = \left(a_\perp \frac{r}{c^2} \right) E_\parallel = a_\perp \frac{r}{c^2} \frac{q}{4\pi\epsilon_0 r^2} = \frac{qa_\perp}{r 4\pi\epsilon_0 c^2} . \quad (128)$$

We now include the direction of $\mathbf{E}_\perp$ by noticing from Fig. 7.8 that $\mathbf{E}_\perp$ at point $\mathbf{r}$ at time t is along the negative of the direction of $\mathbf{a}_\perp$ at the earlier t', where $t' = t - (r/c)$. We also dignify $\mathbf{E}_\perp$ by a name, the *radiation field* $\mathbf{E}_{rad}$:

$$\mathbf{E}_{rad}(\mathbf{r}, t) = -\frac{qa_\perp(t')}{4\pi\epsilon_0 rc^2},$$
$$t' = t - \frac{r}{c}. \quad (129)$$

Note that since the radial component of $\mathbf{E}$ in the kink is the same as the radial field ahead of an d behind the kink, it carries no "news;" it is not "radiation;" it is not part of a traveling wave. A detector that could only detect radial electric field would not notice the kink at all. That is why we only include the transverse field in the kink under the name "radiation field." This result is to be expected from our results for plane waves in Sec. 7.4, where we learned that the longitudinal components of $\mathbf{E}$ and $\mathbf{B}$ are constant in space and time for a plane wave and are therefore not to be called part of the wave. (In the present example of radiation from a point charge, the fields at a distant point $\mathbf{r}$ are expected to be similar to those for a plane wave over a limited region transverse to $\hat{\mathbf{r}}$.)We shall boldly assume that we can take over other results for traveling plane waves, namely that $\mathbf{B}$ and $\mathbf{E}$ are perpendicular to one another and to the propagation direction $\hat{\mathbf{r}}$ and that the magnitudes of $\mathbf{B}$ and $\mathbf{E}$ are equal at every instant and place.

Generalization to arbitrary (nonrelativistic) point charge. Suppose we have a point charge q that is undergoing some complicated three-dimensional motion. We shall call this an "arbitrary" motion, but it must always satisfy our assumption $v \ll c$. Furthermore we assume for simplicity that q remains in the neighborhood of the origin of coordinates. Thus q may be one of the electrons in a distant radio antenna or in a distant atom. What we mean by "neighborhood" and "distant" is that the vector displacement $\mathbf{r}'$ from the instantaneous position of q to the fixed observation point can be taken as nearly constant in direction and length. Thus a "distant" atom may be 10^{-5} cm away from the observation point, since the "neighborhood" occupied by an atom is only about 10^{-8} cm in radius. For a radio antenna of 10-meter length to be just as "distant," it should be about 10,000 meters away.

What is the form of the radiation field due to this "arbitrarily" moving charge at a distant observation point? Equation (129) was derived for an especially simple motion, consisting of a constant acceleration for a short time Δt followed by motion at constant velocity. We have discovered that the resulting radiation field at the observation point at time t is entirely due to the transverse acceleration $\mathbf{a}_\perp(t')$ at the earlier "retarded time" $t' = t - (r/c)$. Now, for an arbitrary motion in which $\mathbf{a}(t')$ is a continually (but continuously) varying quantity, we can regard $\mathbf{a}(t')$ as constant in magnitude and direction for a sufficiently short time interval $\Delta t'$. Therefore the acceleration $\mathbf{a}(t')$ produces during $\Delta t'$ a radiation field at the distant observation point that is given by Eq. (129) and that sweeps past the observation point in a time interval Δt. Now we come to a complication. The retarded time t' at which the acceleration occurs is given by

$$t' = t - \frac{r'}{c}. \tag{130}$$

The radiation emitted by q during time interval $\Delta t'$ sweeps past the observation point in a time interval Δt given by

$$\Delta t = \Delta\left(t' + \frac{r'}{c}\right) = \Delta t' + \frac{\Delta r'}{c}, \tag{131}$$

where $\Delta r'$ is the change of the distance between the charge q and the observation point during the interval $\Delta t'$. We see that in general Δt does not equal $\Delta t'$. Therefore, at a given instant t at the observation point, there is an "overlap" between contributions to the radiation field emitted at different retarded times t'.

Avoiding "overlap." We do not wish to study this general case. We notice that $\Delta r'$ is equal to the longitudinal component of the velocity times $\Delta t'$. Therefore, for $v \ll c$, we can to a good approximation neglect $\Delta r'$ in Eq. (131):

$$\Delta t = \Delta t' + \frac{(\Delta r')}{c}$$

$$= \Delta t' + \frac{(v_\parallel \Delta t')}{c}$$

$$\approx \Delta t' \quad \text{for } \frac{v_\parallel}{c} \ll 1. \tag{132}$$

Thus for $v \ll c$ there is negligible overlap between Δt and $\Delta t'$. Then there is a one-to one correspondence between the radiation detected at time t and the transverse acceleration at a *single retarded time t'*. In that case, the radiation field $\mathbf{E}_{rad}(\mathbf{r}, t)$ is given by Eq., (129) for all t.

Once E is known, so is B.   From now on we assume that Eq. (129) holds for a distant observation point, with **r** essentially constant. We also assume that $\mathbf{B}_{\text{rad}}$ is given by the relationship that holds for a plane wave. Thus (dropping the subscript "rad" from the radiation fields) we have

$$\mathbf{B}(\mathbf{r}, t) = \hat{\mathbf{r}} \times \mathbf{E}(\mathbf{r}, t). \tag{133}$$

Energy radiated by a point charge.   For a distant observation point, the energy flux vector $\mathbf{S}(\mathbf{r}, t)$ is given by

$$\mathbf{S}(\mathbf{r}, t) = \frac{1}{\mu_0} \left(\vec{\mathbf{E}} \times \vec{\mathbf{B}} \right)$$

$$= \frac{1}{c\mu_0} [\mathbf{E}(\mathbf{r}, t)]^2 \hat{\mathbf{r}}$$

$$= \frac{1}{c\mu_0} \left[\frac{-q\mathbf{a}_\perp(t')}{4\pi\epsilon_0 rc^2} \right]^2 \hat{\mathbf{r}}$$

$$= \frac{q^2 \left[\mathbf{a}_\perp(t') \right]^2}{4\pi\epsilon_0 r^2 c^3} \frac{\hat{\mathbf{r}}}{4\pi}, \tag{134}$$

where the units of **S** are erg/cm^2 sec. The energy flux in erg/sec passing through the infinitesimal area dA located at the observation point **r** (and oriented perpendicular to **r**) is given by the magnitude of the flux vector **S** times the area dA. Let us call this energy flux dP (P is for power in erg/sec; the letter d indicates that we are considering the infinitesimal power passing through dA):

$$dP(\mathbf{r}, t) = \left| \mathbf{S}(\mathbf{r}, t) \right| dA$$

$$= \frac{q^2}{4\pi\epsilon_0 r^2 c^3} \mathbf{a}_\perp{}^2(t') \frac{dA}{4\pi}. \tag{135}$$

Let $\theta(t')$ be the angle that the instantaneous retarded acceleration $\mathbf{a}(t')$ makes with the constant direction **r** from the neighborhood of q to the observation point. Then, according to Fig. 7.8, we see

$$\mathbf{a}_\perp{}^2(t') = \mathbf{a}^2(t') \sin^2 \theta(t'). \tag{136}$$

Then Eq. (135) can be written

$$dP(\mathbf{r}, t) = \frac{q^2}{4\pi\epsilon_0 r^2 c^3} \mathbf{a}^2(t') \sin^2 \theta(t') \frac{dA}{4\pi}. \tag{137}$$

Total instantaneous power radiated in all directions.   Let us hold both t' and r fixed and integrate dP over all directions $\hat{\mathbf{r}}$ (i.e.,

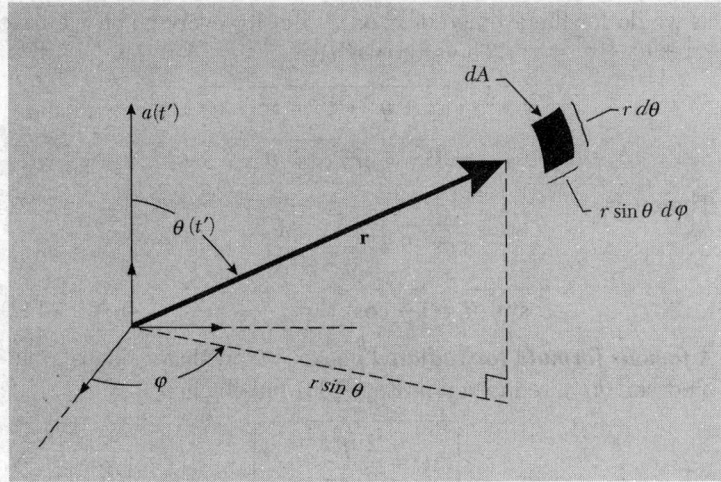

Fig. 7.10 Spherical polar coordinates. The infinitesimal area dA at the end of the radius vector r and oriented perpendicular to r has magnitude $r^2\, d\varphi\, \sin\theta\, d\theta$.

over the surface of a sphere of radius r). If it were not for the factor $\sin^2\theta(t')$, we could perform the integration trivially by simply replacing the infinitesimal area dA by the total area of the sphere $4\pi r^2$. As it is, we must include the variation of $\sin^2\theta(t')$ as we integrate over different infinitesimal areas dA at different observation points distributed over the sphere. Thus we can write

$$P(t) = \frac{q^2}{4\pi\epsilon_0 c^3}\,\mathbf{a}^2(t')\overline{\sin^2\theta(t')} = \frac{q^2\mathbf{a}^2}{4\pi\epsilon_0 c^3}\,(t')\overline{\sin^2\theta(t')}$$

$$t' = t - \frac{r}{c}, \tag{138}$$

where

$$\overline{\sin^2\theta(t')} \equiv \int \sin^2\theta(t')\,\frac{dA}{4\pi r^2}. \tag{139}$$

In order to evaluate this integral, one can use *spherical polar coordinates* as shown in Fig. 7.10. The infinitesimal area dA is the area of a tiny rectangle having edge lengths $r\,d\theta$ and $r\sin\theta\,d\varphi$. Thus

$$\frac{dA}{r^2} = \frac{(r\,d\theta)(r\sin\theta\,d\varphi)}{r^2} = d\theta\,\sin\theta\,d\varphi. \tag{140}$$

Then you can easily show (Prob. 7.40)

$$\overline{\sin^2\theta(t')} = \tfrac{2}{3}. \tag{141}$$

Here is a short derivation of Eq. (141). The vector $\mathbf{r}$ has the component $r\cos\theta$ along the polar axis. Let us call this axis the z axis. Then $z = r\cos\theta$. When we average z^2 over all directions θ (holding r fixed on a sphere), we must get the same answer as we do for the average of x^2 or y^2. But for every point, we have $x^2 + y^2 + z^2 = r^2$. Therefore we have

$$r^2 = \overline{r^2} = \overline{x^2 + y^2 + z^2} = \overline{x^2} + \overline{y^2} + \overline{z^2}$$

$$= 3\overline{z^2} = 3r^2 \overline{\cos^2 \theta}.$$

Therefore

$$\overline{\cos^2 \theta} = \frac{r^2}{3r^2} = \frac{1}{3};$$

$$\overline{\sin^2 \theta} = \overline{1 - \cos^2 \theta} = 1 - \tfrac{1}{3} = \tfrac{2}{3}. \tag{142}$$

A famous formula for radiated power.   Now that we have evaluated $\sin^2 \theta(t')$, we insert it in Eq., (138) and obtain

$$\boxed{\begin{aligned} P(t) &= \frac{2}{3} \frac{q^2 \mathbf{a}^2(t')}{4\pi\epsilon_0 c^3}, \\ t' &= t - \frac{r}{c}. \end{aligned}} \tag{143}$$

According to Eq. (143), the radiated power passing outward across a sphere of radius r_1 at time t_1 has the same value as at any other radius r_2 and time t_2 which correspond to the same retarded time t', That just means that energy is conserved and that the energy is traveling outward at the velocity of light. Notice that this result depends on the fact that the radiation field varies as the inverse first power of r. Therefore the outgoing energy flux $|\mathbf{S}|$ in erg/cm² sec falls off as the inverse square of r. The energy is distributed over a sphere of area proportional to r^2. These two factors r^{-2} times r^2, nullify one another, so that the total outgoing energy per unit time is constant on a sphere whose radius is expanding at the velocity of light.

Radiation and "near-zone" fields.   It turns out that the exact solution for the time-dependent electric and magnetic fields produced by a moving charge contains fields that vary in proportion to r^{-2} and to r^{-3} as well as the "radiation" fields that vary in proportion to r^{-1}. At sufficiently small distances, these time-dependent inverse-square and inverse-cube fields dominate. They are sometimes called "near-zone" fields. If one is in the "near zone" of a radio antenna or an atom, these fields are important. At sufficiently great distances r, they become negligible compared to the inverse-first-power field, i.e., the radiation field. Thus, for example, they do not contribute any net outgoing energy flux at large distances. In the near zone they do contribute to the energy flux vector $\mathbf{S}(r, t)$. Their contributions give an energy flux that travels outward part of the time and inward part of the time, somewhat as in a standing wave. Thus an oscillating point charge does not produce a "pure" outgoing spherical traveling wave, but rather a

combination of both traveling and standing waves, with the standing waves dominant at small distances and the traveling waves
dominant at large distances. A distant detector is only influenced
by the traveling waves. A close detector is influenced by both the
standing and traveling waves.

Definition of solid angle.   Let dA be an infinitesimal area located
at the observation point **r** and oriented perpendicular to **r**. *The differential solid angle $d\Omega$ subtended by dA at the origin is defined by*

$$d\Omega = \frac{dA}{r^2},\qquad(144)$$

in dimensionless units called steradians, abbreviated ster. Now consider a sphere of radius r with center at the origin. The surface area
of this sphere is composed of many infinitesimal areas dA, each of
which is oriented perpendicular to the radius vector connecting it
to the origin. Therefore a differential element of solid angle can be
assigned to each infinitesimal area. The tot al solid angle subtended
by the sphere is obtained by adding up all the infinitesimal solid
angles and is thus given by the total area of the sphere divided by r^2:

$$\Omega = \int d\Omega = \int \frac{dA}{r^2} = \frac{4\pi r^2}{r^2} = 4\pi \text{ ster.}\qquad(145)$$

Here is another derivation of Eq. (145). According to Eq. (140),
the differential solid angle $d\Omega$ is given in spherical polar coordinates by

$$d\Omega = d\varphi \sin\theta\, d\theta,\qquad(146)$$

with $d\varphi$ and $d\theta$ positive, or by

$$d\Omega = d\varphi\, d(\cos\theta),\qquad(147)$$

with $d\varphi$ and $d(\cos\theta)$ positive. The variable φ goes from zero to
2π. The variable θ goes from 0 to π. The variable $\cos\theta$ goes from
-1 to $+1$. The total solid angle subtended by any closed surface surrounding the origin is

$$\Omega = \int d\Omega = \int_0^{2\pi} d\varphi \int_{-1}^{+1} d(\cos\theta) = (2\pi)\cdot 2 = 4\pi \text{ ster.}\quad(148)$$

Power radiated into a differential solid angle $d\Omega$.   We can
use the definition of solid angle to write Eq. (1.37) in the simpler
form

$$dP(\mathbf{r},t) = \frac{q^2}{4\pi\epsilon_0 c^3} \mathbf{a}^2(t') \sin^2 \theta(t') \frac{d\Omega}{4\pi}. \tag{149}$$

Electric dipole radiation. If the motion of q is a harmonic motion along a fixed direction $\hat{\mathbf{x}}$, the resulting radiation is called electric dipole radiation. Then we have

$$x(t') = x_0 \cos\omega t'$$

$$\mathbf{a}(t') = \hat{\mathbf{x}}\ddot{x}(t') = -\omega^2 \hat{\mathbf{x}} x(t'). \tag{150}$$

The power radiated into a solid angle $d\Omega$ and averaged over one cycle of oscillation is

$$dP(\mathbf{r}) = \frac{q^2}{4\pi\epsilon_0 c^3} \sin^2 \theta \frac{d\Omega}{4\pi}$$

$$= \frac{q^2}{4\pi\epsilon_0 c^3} \omega^4 \langle x^2(t')\rangle \sin^2 \theta \frac{d\Omega}{4\pi}. \tag{151}$$

The total time-averaged power radiated in all directions is obtained by integrating over the total solid angle. Then we simply replace $d\Omega$ by $\Omega = 4\pi$ and $\sin^2 \theta$ by its average value of $\frac{2}{3}$ in Eq. (151):

$$P = \frac{2}{3}\frac{q^2}{(4\pi\epsilon_0)c^3} \omega^4 \langle x^2(t')\rangle. \tag{152}$$

Natural linewidth for atom emitting light. We can use Eq. (152) to obtain a simple classical estimate of the lifetime for free decay of an excited atom emitting electric dipole radiation. Remarkably enough, the result will turn out to agree with the experimentally observed values, even though we shall use no explicit quantum theory.

We consider a simple classical model of an atom. The atom consists of an "electron" with charge $q = -e$ and mass m which is bound to a heavy "nucleus" by a spring with spring constant $m\omega_0^2$. If the atom is given an excitation energy E_0 at time zero, it oscillates with weakly damped harmonic motion at frequency ω_0. (We neglect the slight change of the frequency ω_0 caused by the damping; i.e., we do not bother to use $\omega_1^2 = \omega_0^2 - \frac{1}{4}\Gamma^2$ instead of ω_0^2.) The energy of the atom is given by

$$E(t) = E_0 e^{-t/\tau}. \tag{153}$$

The inverse mean life, $1/\tau$ is equal to the fractional decrease of energy per unit time:

$$\frac{1}{E}\left[-\frac{dE}{dt}\right] = \frac{1}{\tau}. \tag{154}$$

The energy $E(\tau)$ is given by

$$E(t) = \tfrac{1}{2}m\omega_0^2 x^2(t) + \tfrac{1}{2}m\dot{x}^2(t).\qquad(155)$$

We can neglect the change of $E(t)$ during one cycle and replace the instantaneous quantities on the right-hand side of Eq. (155) by the time-averaged values:

$$E(t) = \tfrac{1}{2}m\omega_0^2\langle x^2(t)\rangle + \tfrac{1}{2}m\langle\dot{x}^2(t)\rangle$$

$$= \tfrac{1}{2}m\omega_0^2\langle x^2\rangle + \tfrac{1}{2}m\omega_0^2\langle x^2\rangle,$$

i.e.,

$$E(t) = m\omega_0^2\langle x^2\rangle.\qquad(156)$$

We now assume that the damping is entirely due to the loss of energy by radiation of electromagnetic radiation. The radiation is electric dipole radiation, with the radiated power given by Eq. (152):

$$-\frac{dE}{dt} = P = \frac{2}{3}\frac{e^2}{(4\pi\epsilon_0)c^3}\omega_0^4\langle x^2\rangle.\qquad(157)$$

Combining Eqs. (154), (156), and (157), we obtain the *natural line-width* for an atom emitting light,

$$\Delta\omega = \frac{1}{\tau} = \frac{P}{E} = \frac{2}{3}\frac{e^2}{\epsilon_0 c^3 m}\frac{\omega_0^2}{4\pi} = \frac{2}{3}\frac{e^2}{4\pi\epsilon_0}\frac{\omega_0^2}{mc^3},\qquad(158)$$

where we have used the fact that for a damped oscillator the full frequency width at half maximum power (in the Fourier spectrum of the radiation) equals the inverse of the mean lifetime. Equation (158) applies to any damped electric dipole radiation for which the damping is entirely due to radiation. For an atom emitting visible light, we can take $\lambda_0 = 5000$ Å $= 5 \times 10^{-5}$ cm, $v_0 = c/\lambda_0 = 3 \times 10^{10}/5 \times 10^{-5} = 6 \times 10^{14}$ Hz. For e and m we take the electron charge $e = 1.6 \times 10^{-19}$ C esu and mass $m = 9.1 \times 10^{-31}$ kg. Thus we get

$$\tau = \frac{3}{2}\frac{\epsilon_0 c^3}{e^2}\frac{m\,(4\pi)}{\omega_0^2} = \frac{3}{2}\frac{4\pi\epsilon_0}{e^2}\frac{mc^3}{\omega_0^2}$$

$$= \frac{3}{2}\times\frac{4\pi\times 8.85}{(1.6\times 10^{-19})^2}\times\frac{10^{-12}\times 9.1\times 10^{-31}\times(3\times 10^8)^3}{(2\pi)^2(6\times 10^{14})^2}$$

$$= 1.127\times 10^{-8}\text{ sec.}\qquad(159)$$

It is important to remember $\tau \sim 10^{-8}$ sec for freely-decaying atoms emitting visible light.

Here is a question to which we can apply our result s for dipole radiation:

Why is the sky blue? We shall look at the frequency dependence of the scattering of sunlight into our eyes by a single atom of air. We shall find that blue is scattered more strongly than red. That is why the sky is blue. (Sunsets are red because the blue has largely been removed, leaving red.) You may demonstrate this color effect for your self very easily as follows: Get a glass bowl or jar of water and a flashlight. Add a few drops of milk to the water and stir it up. Shine the flashlight beam through the water in such a way that you can either look at the light beam from the side by virtue of the scattering from suspended milk molecules or instead look straight through the water at the flashlight bulb. Notice the blue tinge to the scattered light (that's the blue sky). Notice the reddish tinge to the directly observed flashlight bulb (that's the sunset). Keep adding milk gradually, a few drops at a time, to simulate the effect of gradually increasing smog.

Consider an electron in a "classical milk molecule" driven at steady state by the electric field of the traveling electromagnetic wave produced by the flashlight. If the flashlight beam is directed along $\hat{\mathbf{z}}$, the electric field in the traveling wave has only x and y components. Let us consider only the x component of the electric field in the flashlight beam. (The y component gives similar results.) Furthermore, let us consider a single color, i.e., a single Fourier component of the "white" light (which consists of frequencies covering the visible range, as well as others that we do not detect with our eyes). Then the electric field $E_x(t)$ at the location of the milk molecule is given by

$$E_x = E_0 \cos \omega t. \tag{160}$$

Suppose an "electron" of the milk molecule is bound to the milk nucleus with spring constant $m\omega_0^2$. Let us neglect damping (i.e., assume the driving frequency ω is not near the resonant frequency ω_0). Then the equation of motion of the electron is

$$m\ddot{x} = -m\omega_0^2 x + qE_x. \tag{161}$$

At steady state, $x(t)$ is a harmonic oscillation at frequency ω. Therefore $\ddot{x}(t)$ is $-\omega^2 x(t)$. Then Eq. (161) gives

$$-m\omega^2 x(t) = -m\omega_0^2 x(t) + qE_x$$

$$x(t) = \frac{qE_x(t)}{m(\omega_0^2 - \omega^2)}. \tag{162}$$

The harmonic oscillation $x(t)$ emits dipole radiation. The total radiated power is given by Eq. (152):

$$P = \frac{2}{3} \frac{e^2}{4\pi\epsilon_0 c^3} \omega^4 \langle x^2 \rangle$$

$$= \frac{2}{3} \frac{e^2}{4\pi\epsilon_0 c^3} \omega^4 \left[\frac{-e}{m(\omega_0^2 - \omega^2)} \right]^2 \langle E^2 \rangle. \tag{163}$$

Now, in studying the index of refraction of a classical glass molecule (Sec. 4.3), we found that the effective angular frequency ω_0 is large compared with ω for frequencies ω corresponding to visible light. Therefore we can take $\omega_0 \gg \omega$ in Eq. (163). Thus we see that the scattered power is proportional to the fourth power of the driving frequency ω, i.e., to the inverse fourth power of the driving wavelength:

$$P \propto \omega^4 \propto \frac{1}{\lambda^4}. \tag{164}$$

Blue-sky law. Equation (164) is called "Lord Rayleigh's blue-sky law." The wavelength ratio of red light of wavelength 6500 Å to blue light of wavelength 4500 Å is $\frac{6}{4}\frac{5}{5} = 1.44$. The fourth power of 1.44 is 4.3. Thus, according to Eq. (164), blue light is scattered about 4 times as effectively as red light. That is why the sky is blue. Why is it so bright? See Supplementary Topic 8.

Integrated cross section for scattering. Suppose you had a billiard ball of radius R sitting in a broad uniform beam of steel BB's traveling in the $\hat{\mathbf{z}}$ direction with velocity v. Those BB's that hit the ball are elastically scattered out of the beam. The energy that they carry is removed from the beam and directed in other directions. The total number of BB's scattered per unit time is the product of the incident *number flux*, in BB's per square centimeter per second, times the *integrated cross section* $\sigma = \pi R^2$ of the ball:

$$\text{Scattered BB's per second} = \sigma \times (\text{incident BB flux}). \tag{165}$$

Since the BB's are assumed to be scattered elastically, each scattered BB has the same energy as an incident BB. Thus we can multiply both sides of Eq. (165) by the energy of one BB. Then Eq. (165) becomes

$$\text{Scattered energy per second} = \sigma \times (\text{incident energy flux}). \tag{166}$$

We can now *define* by a suitable interpretation of Eq. (166) the integrated cross section for elastic scattering of light by a classical milk molecule: The "scattered" energy per unit time is *defined* as the radiated power P of the driven electron, and the incident energy flux is the electromagnetic energy flux S_z. Thus, by analogy with Eq. (166), we define σ_{sc} by

$$P = \sigma_{sc} \cdot \frac{1}{\mu_0 c} \langle E_x^2 \rangle. \tag{167}$$

By comparison of Eqs. (167) and (163), we obtain

$$\sigma_{sc} = \frac{\mu_0 c P}{\langle E_x^2 \rangle} = \frac{2}{3} \left(\frac{e^2}{4\pi} \right) \mu_0^2 \omega^4 \left[\frac{-e}{m(\omega_0^2 - \omega^2)^2} \right]^2.$$

$$= \frac{2}{3} \frac{e^4}{4\pi\epsilon_0^2 m^2 c^4} \tag{168}$$

Waves in Two and Three Dimensions

Thus our result Eq. (164), which says that for that $\omega_0 \gg \omega$ the scattered intensity is proportional to ω^4, is expressed more precisely by Eq. (168), which gives the frequency dependence of the *integrated cross section for elastic scattering of light by an atom* (for this classical model). The quantity e^2/mc^2 has the dimensions of length. (It must have, since σ has dimensions of length squared and the frequency dependence of σ occurs in a dimensionless ratio.) It is called for historical reasons *the classical radius of the electron*, r_0, or the *Lorentz radius of the electron:*

$$r_0 = \frac{e^2}{4\pi\epsilon_0 mc^2} = \frac{(1.6\times10^{-19})^2}{4\times3.14\times8.85\times10^{-12}\times9.1\times10^{-31}\times(3\times10^8)^2}$$
$$= 2.812\times10^{-15}\,\text{m.} \tag{169}$$

Classical Thomson scattering cross section. If the electron is bound to the nucleus with a spring of spring constant zero, it is not bound at all; it is free. If the spring constant is zero, ω_0 is zero. Thus the elastic scattering cross section for scattering of light by a free electron, otherwise called the *classical Thomson scattering cross section*, is obtained by setting $\omega_0 = 0$ in Eq. (168):

$$\sigma_{\text{Thomson}} = \tfrac{8}{3}\pi r_0^2 = \tfrac{8}{3}(3.14)(2.82\times10^{-13})^2 = 0.67\times10^{-24}\,\text{cm}^2. \tag{170}$$

Now, a cross section of 10^{-24} cm^2 may not seem large to you, but in some branches of physics (namely nuclear physics) and at a certain time in history it seemed as big as the side of a barn. Therefore it is called a barn:

$$1\,\text{barn} = 10^{-24}\,\text{cm}^2. \tag{171}$$

(Nuclear cross sections are ordinarily given in millibarns, abbreviated mb.) Thus the Thomson cross section given by Eq. (170) is easily remembered: the Thomson cross section is very big; it is two-thirds of a barn.

Problems and Home Experiments

7.1 Show that the identity given in Eq. (34), Sec. 7.2, holds. This identity is the basis of the "crisscross traveling wave" description of waves in a waveguide. It is an illustration of the fact that three-dimensional traveling harmonic waves form a "complete set" of functions for describing three-dimensional waves. Of course three dimensional standing waves also form a complete set.

7.2 (*a*) Show that for glass of index 1.52 the critical angle for internal reflection is about 41.2 deg.

(b) What is the critical angle for water of index 1.33? Will a water prism in the shape of an isosceles right triangle (as shown in Fig. 7.3) give retrodirection of light without any loss (by refraction into air). First assume that the water extends right up to the air. Then worry about the glass microscope slides that form the sides of your water prism.

7.3 Retrodirective water prism. Make a water prism from two microscope slides and some putty or tape. Check the results of Prob. 7.2b by shining a flashlight beam down onto the surface of the water.

Home experiment

7.4 Show that the retrodirective glass prism shown in Fig. 7.3 works at other angles of incidence than the normal incidence shown, in the sense that it directs the light back in the opposite direction from the incident direction.

7.5 Calculate the mean penetration distance (the mean amplitude attenuation distance $\kappa^{-1} = \delta$) for visible light of wavelength 5500 Å retrodirected by the glass prism of Fig. 7.3.(We mean the distance normal to the rear glass-to-air surface.) Assume the incident light beam is at normal incidence as shown in the figure. Take the index of refraction tobe 1.52   *Ans.* δ $= 2.2 \times 10^{-5}$ cm.

7.6 Light in vacuum. For light (or microwaves) in a waveguide we found that, if the frequency is below cutoff, the z direction (along the guide) is "reactive." The other two directions were not reactive. Is it possible in principle by some ingenious method to construct a "generalized waveguide" in which the waves will be reactive in all three directions x, y, and z?

7.7 Fiber optics. It is possible to "pipe" light around in waveguides made of glass fibers. The light stays in the glass because it makes glancing collisions at the glass-to-air surface and is thus incident at angles greater than the critical angle. However, if the fiber is too small in diameter, the fiber becomes a waveguide in which the frequency of the light is below cutoff. Assume the fibers have a square cross section (like a rectangular waveguide). Estimate the minimum edge length for a fiber if the fiber is to be dispersive, i.e., if it is to carry traveling visible light waves.

Ans. Edge $> 1.7 \times 10^{-5}$ cm for $\lambda = 5000$ Å.

7.8 Critical angle for reflection from the ionosphere. Replace the glass to the left of $z = 0$ in Fig. 7.4 by vacuum. Replace the air to the right of $z = 0$ by a plasma—the ionosphere, idealized so as to have a sharp boundary (and a uniform composition). Show that for every angle of incidence θ_1 there is a cutoff frequency $\omega_{c.o.}$ which depends on θ_1 and that at normal incidence this cutoff frequency is the plasma oscillation frequency ω_p. Show that for every frequency ω above the plasma oscillation frequency ω_p there is a critical angle for total reflection such that for angles of incidence greater than the critical angle the wave is exponential in the ionosphere. As an example, take the plasma oscillation frequency to be $\nu_p = 25\,\text{MHZ}$ (megahertz) and find the critical angle for microwaves of frequency $\nu = 100\,\text{MHZ}$.

Ans. For fixed θ_1, $\omega_{\text{c.o.}} = \omega_p / \cos\theta_1$. For fixed frequency ω above ω_p, $\cos\theta_{\text{crit}} = \omega_p / \omega$.

Home experiment

7.9   Fish's view of the world above the water.   For this experiment you need a quiet pond somewhere or a home with a swimming pool. Otherwise (in a public pool), you must be the first one in the pool, so that the water surface is still smooth. Use a face plate (a skindiver's glass viewing mask). Swim down, turn on your back and look up. As a problem, predict (now) what you will see.

7.10   Phase velocity of water waves versus water depth.   Suppose you have a rectangular aquarium (or a cardboard carton painted on the inside, or something) of length 25 cm along x. You fill it to equilibrium height and excite the lowest sinusoidal mode (shown in Fig. 7.5).

(*a*) What is the phase velocity (in cm/sec) for deep-water waves? (Remember that you can define the phase velocity even though the waves are standing waves.)

(*b*) Make a plot of phase velocity (in cm/sec) versus water depth, h (in cm), for this mode and this aquarium, using the exact dispersion relation (for small-amplitude waves) given by Eq. (72), Sec. 7.3. On the plot show the "deep-water limit.". Also plot the expression for the shallow-water phase velocity on the same plot, plotting it as if it held for all h independent of wavelength. Your exact plot should thus show the "transition" between shallow- and deep-water phase velocities.

Home experiment

7.11   Dispersion law for water waves.   Get a rectangular tank measuring about a foot along x. (Anything from 15 cm to 60 cm will do.) It should be at least $\frac{2}{3}$ as deep as it is long (so that you can reach the deep-water limit). The best tank is an aquarium. (They cost about \$5 in a typical pet shop.) The cheapest tank is a cardboard carton (for example a shoe box, hat box, or grocery carton). If you spray water-sealing paint on the inside of the carton, it will last longer, and you will not get so wet. Damping caused by the flexing of the cardboard reduces the lifetimes of the modes and makes the cardboard less desirable than a glass (or stiff plastic) tank. Also it is nice to be able to see in through the sides of a glass tank. But cardboard is adequate.

(*a*) *Lowest mode.* This is shown in Fig. 7.5. Calculate λ for your tank for this mode. Calculate δ for your tan k for this mode. Plot the theoretical expression for the phase velocity $v_\varphi = \lambda\nu$ for your tank for this mode as a function of water depth h, as discussed in Problem 7.10. (Use the "exact" dispersion relation of Eq. (72), Sec. 7.3.) Now fill the tank to some arbitrary height h. Stir in some coffee grounds so that you can see the motion throughout the water. Excite the lowest mode by gently pushing the tank back and forth. When you see that you have it, let go. Measure the frequency. (An ordinary watch is adequate) Calculate your experimental result for v_φ and put an experimental point on your graph of the theoretical expression for the phase velocity. Repeat the experiment for different values of h. You should have at least one "shallow-water" experimental point, at least one "deep-water" point, and at least one point in the transition region $h \approx \delta$.

(b) *Next highest mode.* If you have a cardboard tank, you can excite the mode where the center of the tank at $x = 0$ (in Fig. 7.5) is at an antinode in ψ_y and the tank length is one wave length. How can you excite that mode? If your tank is rigid, you cannot excite that mode (at least not easily). Why not? In that case the next easily excited mode has L equal to three half-wavelengths and has a node in ψ_y at $x = 0$ (Fig. 7.5). Calculate δ for this tank and this mode. Calculate the expected frequency. Now try shaking the tank at that frequency to excite the mode. Measure the frequency of the free oscillations in this mode, once you have learned how to excite it.

(c) *Transient beats.* You need a metronome for this experiment. Borrow one, or make one by hanging a can of soup or something by a string of variable length, with a piece of paper for the bob to hit to make a noise. With the metronome ticking, shake the tank gently and uniformly at the metronome tempo. Vary the length of the string (or the metronome frequency) in small steps so as to sweep through the resonance frequency for the second mode described in part (b) above. You should notice transient beats at the beat frequency between the driving force and the natural oscillation frequency. It will be obvious when you reach the resonance frequency. (You will also notice in these experiments many things that are not accounted for by the "small-oscillations" theory!) It may be possible for you to estimate the width $\Delta\omega$ of the resonance. (I have not tried that.) In any case, *calculate* the width of the resonance by measuring (crudely) the mean decay time for the mode and then using the famous relation between bandwidth and decay time for a damped mode $\Delta\nu\Delta t \approx 1$.

7.12 Obtain the classical wave equation for **B**, as suggested following Eq. (79b), Sec. 7.4.

7.13 Radiation pressure of the sun.   Given that the solar constant (outside the earth's atmosphere) is 1.94 small calories per square centimeter per minute (which is 1.35×10^3 J/m² sec), calculate in N/m² the radiation pressure on the earth (at normal incidence) under the two assumptions (a) and (b). Then compare the result to atmospheric pressure of air at sea level.

(a) The earth is "black" and absorbs all the light.
(b) The earth is a perfect mirror and reflects all the light.

$\quad\quad\quad\quad$ *Ans.* (a) About 5×10^{-11} atmosphere (1 atm $\approx 10^5$ N/m²).

7.14 Radiation pressure. (First work Prob. 7.1.3.) The radiation pressure of the sun on the earth gives an effective repulsive force between the sun and the earth.

(*a*) Show that this effective repulsive force satisfies the inverse-square law. Thus, if the earth were twice as far away, the net force on the earth would be four times smaller, *as would the gravitational force.*

(*b*) Review Kepler's law. Show that it can be written (for circular orbits) in the form $\omega^2 R^3 = MG$, where ω is the angular frequency of a planet going around the sun, R is the distance from the sun to the planet, M is the mass of the sun, and G is the gravitational constant.

(*c*) Show that for a spherical "black" object having mass density ρ and radius r and traveling in a circular equilibrium orbit about the sun, Kepler's law must be "corrected" to read $\dfrac{4\pi^2}{7^2} R^3 = MG - \dfrac{P}{4\pi c}\dfrac{3}{4pr}$ where P is the total electromagnetic output power of the sun.

(*d*) Given the solar constant (Prob. 7.13) and given that the sun is 93 million miles from earth, calculate P in J/sec(watt).

(*e*) Suppose you have a "dust particle" in a circular orbit about the sun. Take the mass density to be the same as that of water (1.0 gm/cm^3). For what particle radius r does the outward radiation pressure equal the inward gravitational attraction? What happens to such dust particles (and smaller ones)?

(*f*) Suppose you have a "comet" consisting of small particles of dust or ice or something, all having the same mass density and same radius. Will such a "comet" change its "shape" as it passes the sun? (We are no longer talking about circular orbits; we are talking about elliptical orbits. But you should still be able to guess the answer.)

(*g*) It is said that the long tail of a comet (extending away from the sun) is due to radiation pressure. Suppose you have a comet (a cloud of dust particles) in a circular equilibrium orbit. The comet has an angular frequency common to all of the particles. But the particles do not all have the same equilibrium radius; the comet extends from R_1 to R_2, where R_1 is closest to the sun and R_2 is farthest from the sun. Suppose you can measure R_1 and R_2 (merely by looking at the extended size of the comet with a telescope). Show how you can use that information and other easily obtained information to discover the distribution (or limits of the distribution) of radius sizes r, assuming all the particles are "black" and have the density of water. Of course all of this does not prove that radiation pressure is more important than, for example, the "solar wind" of protons emitted by the sun in determining the out-ward pressure on dust particles and comet tails.

7.15 Sailing by sunlight. Suppose you would like to design a solar sail that can "hover" in space, the gravitational pull of the sun exactly canceled by the radiation pressure. Suppose the sail consists of aluminized plastic. Take the mass density of the sail as 2.0 (Aluminum has density 2.7 gm/cm^3; plastic has density about 1.) Suppose there is no "pay load," so that the sail need support only its own weight. Suppose the sun light is completely reflected. Show that for the sail to hover at rest (in the inertial frame) its thickness d must be given by

$$\rho d = \frac{2P/c}{MG},$$

where the symbols are as defined in Problem 7.14. Show (from Prob. 7.13) that $P = 3.8 \times 10^{26}$ W. Show (using Kepler's law for the earth, with $R = 149$ million kilometers and $\nu =$ once per year) that $MG = 1.3 \times 10^{26}$ cm³/sec². Show that for $\rho = 2$ the thickness d required is about 10^{-4} cm (unless I made a mistake). This is 1 micron and is thinner by a factor of 10 or 100 than we would like. We would also like to support a pay load. It looks as though we need some orbital motion to keep from falling into the sun. Show that the result of this problem gives the size of a "shiny cubical dust particle" of density 2.0 that will hover about the sun, if it is oriented with one face toward the sun.

7.16 Radiation from a point charge.   Use a "pillbox integration" to obtain Eq. (125), Sec. 7.5, which says $E_\perp/E_\parallel = v_\perp t/c\Delta t$. See the discussion following Eq. (127), Sec. 7.5.

7.17 Electric dipole radiation from a "dipole doublet" radio antenna.   Consider the radio transmitter and antenna shown in the diagram. The current I is assumed to be uniform over the entire length l of the antenna. The leads from the oscillator to the antenna are very close together or are twisted around one another. Therefore the *net* current in the outgoing and incoming leads is effectively zero, and the leads do not radiate appreciably compared with the antenna. The little balls on the ends of the antenna are capacitors to collect the charge accumulated from the current I. These balls are not necessary—charge accumulates at the ends of the conductors and tends to make the current not perfectly uniform, but we can neglect that. *The antenna length l is very small compared with the wavelength λ of the electromagnetic radiation.*

(*a*) Show that at a distant observation point **r** the radiated electric field **E** is given by

$$\mathbf{E}_{\text{rad}}(\mathbf{r}, t) = -\frac{l\,\dot{\mathbf{I}}_\perp(t')}{rc^2}, \quad t' = t - \frac{r}{c},$$

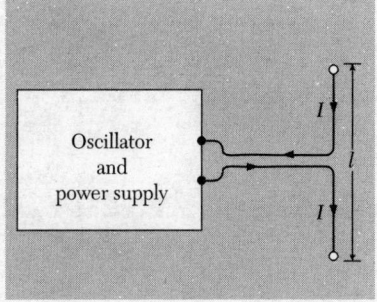

Problem 7.17

where **I** is a vector whose direction and magnitude are those of the current in the antenna, and $\mathbf{I}_\perp$ is the projection of **I** transverse to the line of sight $\hat{\mathbf{r}}$ from the antenna to the distant observation point. [*Hint*: You can derive this formula by inventing an "equivalent point charge q moving with equivalent velocity $\mathbf{v}(t')$" that will give results indistinguishable from those of **I**.]

(*b*) Show that for wavenumber k the characteristic impedance Z experienced by the oscillator (i.e., the resistive load it thinks it is hooked up to) is given by

$$Z = (kl)^2 \cdot 20 \text{ ohms},$$

where, we recall c^{-1} statohm = 30 ohm.

7.18   Scattering of light by milk molecules.   Fill a glass jar with water. Shine a flashlight beam from the side into the water. Look at light scattered through an angle of about 90 deg, and also look through the water at the flashlight bulb. Add a few drops of milk to the water and stir it up. Keep looking and adding more milk. Notice the blue tinge to the scattered light and the yellowish or reddish tinge to the remaining transmitted light. Explain. Notice that when you get enough (or too much) milk in the water the scattered light no longer looks bluish. It looks whitish, like fog or smog. The "sunset" keeps getting more and more red, however. Explain. Finally, it gets so thick that you cannot see the flashlight bulb at all, and the scattered light is white. You also cannot "see" the flashlight "beam" in the liquid. The "air" has become a "white cloud." Explain. Look at the scattered light with your polaroid. (We will explain *that* in Chap 8!)

7.19   Radiation by a thin sheet of charge.   Suppose the xy plane at $z = 0$ is filled with a very thin sheet of positive charge of uniform charge density σ. All the charges oscillate along the x direction with the same amplitude and frequency.

(*a*) Show by Gauss's law that, for z positive, $E_z(z, t) = 2\pi\sigma$, whether or not the charges are all oscillating or are all at rest. (This is like a stretched spring in the slinky approximation, where the z component of tension is constant, independent of the motion).

(*b*) Show by drawing a sketch of a line of force that the radiation field is given by

$$\frac{E_x(z, t)}{E_z(z, t)} = -\frac{\dot{x}(t')}{c},$$

where $\dot{x}(t')$ is the *velocity* of anyone of the charges at the retarded time $t' = t - (z/c)$. Thus the radiation field at positive z is

$$E_x(z, t) = -2\pi\sigma\,\frac{\dot{x}(t')}{c}.$$

(Instead of a sketch you may wish to use a Gaussian pillbox argument.) Notice the peculiar fact that, in contrast to the case of radiation from a single point charge, where it is the (retarded) *acceleration* that is proportional to the radiation field, here it is the earlier (retarded) *velocity* that is proportional to the radiation field. Can you give a qualitative explanation for "what has happened?" (*Hint:* Consider the contributions from various point charges distributed over the plane.)

7.20   Radiation by a thin sheet of charge.   Derive the result of Prob. 7.19 by adding up (integrating) all the point charge contributions from the plane. In order to get your integral to converge, assume that the sheet is not exactly of zero thickness but has thickness d (where d is very small compared with the wavelength λ). Assume that the sheet absorbs or scatters radiation (as it must) and that the mean amplitude attenuation constant is κ. Show that this gives an exponential attenuation factor (e.a.f.)

$$\text{e.a.f.} = e^{-ar}, \quad a \equiv \frac{\kappa d}{z},$$

where k is the wave number and r is the distance from the point charge contribution to the observation point located a perpendicular distance z from the sheet. Define $\varphi = kr - kz$. Notice that $\varphi = 0$ for the point charge at $x = y = z = 0$, that is, the point charge closest to the observation point at $x = y = 0, z = z$. Show that if $x(t')$ is given by the real part of

$$x(t') = x_0 e^{i\omega t'},$$

then the contribution to E_x from an annular ring in the plane with radius p and radial thickness $d\rho$ can be juggled until it has the form

$$dE_x = 2\pi k x_0 e^{i(\omega t - kz)} e^{-i\varphi} e^{-\beta\varphi} \, d\varphi,$$

where

$$\beta \equiv \frac{a}{k} = \frac{\kappa d}{kz},$$

and where the fact that the projection of the acceleration perpendicular to the line of sight must be used has been neglected. (This is based on the assumption that, since this projection factor is 1 for small φ, we can assume that it is slowly decreasing with increasing φ and can be lumped into our "experimental" attenuation factor $e^{-\beta\varphi}$. All of this is possible because of the amazing fact that when we are done we can take $\beta = 0$ and find an answer that is independent of β. The factor $e^{-\beta\varphi}$ is called a *convergence factor*. It is necessary in order to get an answer, but it doesn't matter what value you use for β as long as β is small in magnitude compared with unity.) Next show that E_x is the real part of the integral of dE_x from $\varphi = 0$ to $\varphi = \infty$. Show that

$$\int_0^\infty e^{-i\varphi} e^{-\beta\varphi} \, d\varphi = \frac{1}{i + \beta} \approx -i \quad \text{for} \quad \beta \ll 1.$$

Finally, take the real part and show that you get the same result as in Prob. 7.19. Thus you can now *explain the physical origin of the "effective 90-deg phase shift"* that makes the *total* field 90 deg retarded in phase compared with the contribution from the nearest charge, the one at $x = y = z = 0$. *The "average" charge is effectively one quarter-wavelength farther away than the nearest charge.*

7.21 Approximate expression for index of refraction: Consider a plane wave incident on a thin sheet of charge. The charges are in a thin plane, the xy plane at $z = 0$. The thickness of the plane is Δz. The number density of charges is N (in units of particles per cm³). Each charge has the same charge q, and mass m, and each is bound by a spring of spring constant $m\omega_0^2$. Assume that each charge experiences forces due to its spring and due to the incident driving plane wave. Neglect the contribution from the other charges (i.e., neglect the polarization contribution to the field).

Take the incident electric field (at $z = 0$) to be the real part of $E_0 e^{i\omega t}$. Find the radiated field in the forward direction. Superpose that with the incident field. Show that the total field is then given at $z = 0$ (for these approximations) by the real part of

$$E_{\text{tot}} = E_0 e^{i\omega t}\left\{1 - \frac{i\omega 2\pi N q^2 \Delta z}{mc(\omega_0^2 - \omega^2)}\right\}.$$

Show that if you think of the slab of charge as a slab with thickness Δz and index of refraction n, then insertion of the slab gives a phase shift corresponding to a time delay t_0, i.e., instead of $E_0 e^{i\omega t}$ at the rear of the slab (at $z = 0$), one has a field $E_0 e^{i\omega(t - t_0)}$, where

$$\omega t_0 = \frac{\Delta z}{\lambda} 2\pi(n - 1) = k\Delta z(n - 1).$$

Show that for $\omega t_0 \ll 1$ this gives

$$n - 1 = \frac{2\pi N q^2}{m(\omega_0^2 - \omega^2)}.$$

Show that this is what one finds (approximately for n near unity, in the more exact result derived in Sec. 4.3.

7.22 Angular momentum of circularly polarized plane traveling wave. Derive simply the famous result $J = W/\omega$ follows: Assume the plane wave is produced by a sheet of charge with all the charges going in similar circles. Each charge is constrained to move in a circle of fixed radius r by a frictionless tube. The charges slow down as they lose energy. Thus their angular velocity ω decreases, their energy decreases, and their angular momentum decreases, all because of the loss to radiation. (They always have $v \ll c$, however.) Show that the loss of angular momentum of the charges moving in circles is ω^{-1} times the loss of energy. Q.E.D.

7.23 What are the time-averaged energy flux, energy density, and linear momentum per unit volume in a uniform monochromatic light beam of intensity 1000 watts/ cm²?

7.24 An electron oscillates harmonically with an amplitude of 10^{-8} cm and a frequency of 10^{14} Hz; what is the total average power radiated?
                                          Ans. Approximately $\frac{1}{3} \times 10^{-17}$ W.

7.25 How can an object absorb light energy with out absorbing linear momentum? How can it absorb linear momentum with negligible absorption of energy? How can it absorb angular momentum with negligible absorption of energy?

7.26 Suppose you had a superconducting oscillator and antenna emitting microwave radiation of wave length 100 cm. At $t = 0$ you remove the source of power that replenishes the energy lost by radiation. There is no ordinary resistance anywhere in the circuit. Find the mean decay time of the damped harmonic oscillations of the electrons in the antenna. Use the results of Prob. 7.17.

Ans. Let L be the inductance in the oscillating LC circuit that gives the oscillation frequency. Let l be the length of the antenna ($l \ll \lambda$). Then

$$\frac{1}{\tau} = \frac{2}{3} \frac{l^2}{L} \frac{\omega^2}{c^3}.$$

This can be compared with the expression for the inverse decay lifetime for a single charge e having mass m:

$$\frac{1}{\tau} = \frac{2}{3} \frac{e^2}{4\pi\epsilon_0} \frac{\omega^2}{mc^3}.$$

7.27 A radio station 10 miles away radiates 50 watts of vertically polarized radio waves. What is the maximum instantaneous voltage driving the electrons in your receiving antenna if the antenna is 20 cm long and is oriented vertically? Neglect all reflections of the waves from the ground, buildings, etc.

7.28 Smith-Purcell light source. A narrow beam of electrons of kinetic energy 300 kev travels at grazing incidence parallel to the surface of a metallic diffraction grating that has scratches separated by $d = 1.67$ microns. The beam travels perpendicular to the scratches. The "mirror-image" induced charge that travels with a given electron suffers a sudden deflection whenever it encounters a scratch, since the induced charge must follow the surface. Thus a "radiation kink" is propagated out from each scratch when the electron passes it. Suppose the observe r is at angle θ to the electron beam, where $\theta = 0$ is along the beam.

(*a*) Show that the observer receives radiation pulses with period T between pulses, where $T = (d/v) - (d \cos \theta)/c$; Show that the wavelength is then equal to $d(\beta^{-1} - \cos \theta)$.

(*b*) Would you expect this to be the only wavelength observed at a given θ? (Think about the Fourier analysis of the time dependence of the radiation impulses that arrive at time intervals T.)

(*c*) Put in numbers for 300 keV electrons observed at $\theta = 15°$. What colors would you expect to see?

(*d*) Would you expect the light to be polarized?

Now read about the lovely experiment of S. J. Smith and E. M. Purcell (author of Vol. II), *Phys. Rev.* **92,** 1069 (1953).

7.29 Form of standing water waves. In the text we used an intuitive argument to show that if the vertical displacement in a standing wave has x dependence $\sin kx$ then the horizontal displacement must have the dependence $\cos kx$.

(*a*) Obtain this same result algebraically. Assume

$$\psi_y = \cos \omega t \sin kx f(y),$$

$$\psi_x = [\cos \omega t (\cos kx \, g(y) + \sin kx \, h(y)].$$

Then show that $h(y)$ must be zero.

(*b*) Show that the results obtained for the motion of a water droplet in a standing wave correspond to harmonic oscillation back and forth along a straight line.

7.30 Suppose that at the surface of the ocean there are traveling waves with 3 m amplitude and wavelength 9 m. If yon were a fish (or a Scuba diver), how far beneath the surface should you swim if you wished the amplitude of your motion to be 15 cm?

Ans. About 4.5 m.

7.31 Form of traveling water waves. Assume that ψ_y has the form

$$\psi_y = A\cos(\omega t - kx)f(y),$$

where $f(y)$ is an unknown function of y. Now assume that water is conserved, is incompressible, and has no bubbles to show that ψ_y and ψ_x are given by Eqs. (75) and (76), Sec. 7.3.

7.32 The dispersion law for water waves, Eq. (72), Sec. 7.3, was derived when we were considering standing waves. What is the dispersion law for traveling waves?

7.33 Dispersion law for surface-tension waves. The surface of the water acts like stretched membrane. At equilibrium the tension along x is the surface tension constant, $T = 72 \times 10^{-4}$ N/m, times the length L along the "uninteresting" z direction. If the surface has a convex curvature, the surface tension contributes a downward pressure. Show that for a sinusoidal wave the downward pressure is given by

$$p = Tk^2\psi_y.$$

Show that the gravitational weight of the water gives a pressure that is a constant (the value at equilibrium) plus a contribution

$$p = pg\psi_y.$$

Show that the contribution to the re turn force per unit mass per unit displacement, ω^2, due to surface tens ion can be obtained from the result for the gravitational return force by replacing pg by Tk^2. Thus show that the complete dispersion law is given by

$$\omega^2 = \left(gk + \frac{T}{p}k^3\right)\left[\frac{1 - e^{-2kh}}{1 + e^{-2kj}}\right].$$

7.34 Plane electromagnetic waves. Show that, for electromagnetic plane waves in vacuum, those Maxwell's equations that give the relation between E_y and B_x are "equivalent" to the Maxwell equations relating E_x and B_y, in the sense that one set of equations can be obtain ed from the other merely by rotating the coordinate system by 90 deg about the z axis (which is the propagation axis). Make a sketch showing the orientations of **E**, **B**, and the x and y axes.

7.35 Standing electromagnetic waves in vacuum. Show that if $E_x(z, t)$ is the standing wave $E_x = A \cos \omega t \cos kz$, then $B_y(z, t)$ is the standing wave $A \sin \omega t \sin kz$.

7.36 Energy relations in electromagnetic standing waves. Assume a standing wave of the form given in Prob. 7.35. Find the electric and magnetic energy densities and the Poynting vector as functions of space and time. Consider a region of length $\frac{1}{4}\lambda$ extending from a node in E_x to an antinode in E_x. Sketch a plot of E_x and B_y versus z over that region at the times $t = 0$, $T/8$, and $T/4$. Sketch a plot of the electric energy density, the magnetic energy density, and the total energy density over that region for the same times. Give the direction and magnitude of the Poynting vector S_z for those same times.

7.37 First-order coupled linear differential equations for waves on a string. Consider a continuous homogeneous string of linear mass density ρ_0 and equilibrium tension T_0. As you know, such a string can carry nondispersive waves with velocity $= \sqrt{T_0 / \rho_0}$. Define the wave quantities $F_1(z, t)$ and $F_2(z, t)$ as follows:

$$F_1(z, t) \equiv -\frac{T_0}{v}\frac{\partial \psi_x}{\partial z}, \qquad F_2(z, t) \equiv \rho_0 \frac{\partial \psi_x}{\partial t}.$$

Thus F_1 is $1/v$ times the transverse return force exerted on the port ion of the string to the right of z by that to the left of z, and F_2 is the transverse momentum per unit length. Show that F_1 and F_2 satisfy the first-order coupled equations

$$\frac{1}{v}\frac{\partial F_1}{\partial t} = -\frac{\partial F_2}{\partial z}, \qquad \frac{1}{v}\frac{\partial F_2}{\partial t} = -\frac{\partial F_1}{\partial z}.$$

Show that one of these equations is "trivial," i.e., is essentially an identity. Show that the other is equivalent to Newton's second law. Notice that these equations are similar in form to Maxwell's two equations relating E_x and B_y, with E_x analogous to F_1 and B_y to F_2. Similarly, one of the two Maxwell equations can be regarded as a "trivial identity," if one knows the special theory of relativity.

7.38 Find suitable wave quantities $F_1 (z, t)$ and $F_2(z, t)$ for longitudinal waves on a beaded string such that F_1 and F_2 satisfy first-order coupled equations of the same form as those in Prob. 7.37. Do the same for sound waves. Do the same for electromagnetic waves in a transmission line. (In this last case, the coupled equations are not merely "similar in form" to Maxwell's equations; they *are* Maxwell's equations, expressed in terms of current and voltage instead of the fields E_x and B_y.)

7.39 Show by direct integration that the average value of $\sin^2 \theta$, averaged over all directions, is $\frac{2}{3}$, where θ is the angle between a given direction and a fixed axis, the "polar" axis, and where each infinitesimal solid angle carries a "weight" (in the averaging) proportional to the solid angle. Use spherical polar coordinates in performing the integration .

7.40 Mirages on the highway.   Driving on a hot summer day you will often see far ahead what appear to he pools of water reflecting the sky or the headlights of an approaching car. As you draw nearer, the reflections suddenly disappear when the angle of reflection (measured from the highway surface) becomes greater than a certain critical angle. These reflections or "mirages" are due to total internal reflection of light incident from cooler air (the denser medium) to hotter air near the hot pavement. The hotter air is less dense and has smaller index. (Recall that $n^2 - 1$ is proportional to the air density.) Suppose the air near the pavement is hotter by an amount ΔT than that several inches above the pavement. Assume as an approximation that the temperature change is sudden. Take the cool air temperature to be $T = 300°\text{K}$ (degrees Kelvin), and the temperature increment ΔT to be l0°C near the pavement. The index of refraction n of the air is about 1.0003. Let φ be the angle of incidence of a ray at the critical angle for total internal reflection, with φ measured from the pavement, that is, φ is 90° minus the angle of incidence as measured from the normal to the pavement . Assuming $n - 1 \ll 1$, derive the formula $\varphi \approx [2(n-1)$ $\Delta T/T]^{1/2}$ for $\varphi \ll 1$. If your eyes are 1.2 m above the pavement , how far ahead of you will you see the near edge of the apparent "pool of water "?

Ans. About 1000 ft.

7.41 Waveguide.   A rectangular waveguide has internal transverse dimensions of 5×10 cm.

   (*a*) What is the frequency in megacycles per sec of the lowest frequency electromagnetic wave that will pass down the guide without being attenuated?

   (*b*) Show by a sketch the direction and variation with position of the electric field for this wave.

   (*c*) Find the phase and group velocity (expressed as a multiple of *c*) for a wave with frequency equal to $\frac{5}{4}$ of the lowest frequency that is passed without attenuation.

   (*d*) Find the mean attenuation length for a wave with frequency $\frac{4}{5}$ of the lowest frequency not attenuated.

7.42 Reflection coefficient for electric field.   Given the analogy that inductance per unit length in a transmission line is equivalent to mass per unit length for a stretched string, and inverse capacitance per unit length is like string tension, and given that $C = \epsilon_r C_{\text{vac}}$ and $L = \mu_r L_{\text{vac}}$, and that the phase velocity is *c* for vacuum.

   (*a*) Show by analogy with the string that the index of refraction *n* is $(\epsilon_r \mu_r)^{\frac{1}{2}}$ and that the characteristic impedance Z is $\left(\dfrac{\mu_r}{\epsilon_r}\right)^{\frac{1}{2}}$ times its value for vacuum in the transmission line. Then show that the reflection coefficient for the electric field is $R = [1 - (n/\mu)] / [1 + (n/\mu)]$ in going from vacuum to the medium. This is also the reflection coefficient for the electric field for plane waves normally incident from vacuum to a surface where the medium begins.

(b) We shall now give (or ask you to give) a more rigorous derivation of the reflection coefficient, using Maxwell's equations. Show by using Maxwell's equations and a suitable line integration that the tangential electric field is continuous at the boundary, pro vided $\partial \mathbf{B}/\partial t$ is not infinite at the boundary. (It isn't.) Thus, assuming the incident electromagnetic wave is linearly polarized with electric field along x, show $E_{x(\text{ine})} + E_{x(\text{ref})} = E_{x(\text{tr})}$.

(c) Use Maxwell's equations for a medium as given in Supplementary Topic 9. Consider the field $\dfrac{1}{\mu_0} B - M$. This field is equal to $\mathbf{B}/\mu$, by definition of μ, and is also called $\mathbf{H}$. Show that the tangential component of $\mathbf{H}$ is continuous, provided that the partial time derivative of $\epsilon_0 E + P$ is not infinite. (It isn't.) Then show that for a wave incident from vacuum $B_{y(\text{ine})} + B_{y(\text{ref})} = (1/\mu)B_{y(\text{tr})}$. Now use the fact that in the medium B_y is n times E_x, and use the relation between B_y and E_x in the incident and reflected waves to obtain the reflection coefficient $R = E_{x(\text{ref})}/E_{x(\text{ine})}$. Show that $R = [1 - (n/\mu)] / [1 + (n/\mu)]$.

Chapter 8

Polarization

Chapter 8   *Polarization*

8.1   *Introduction*

We learned in Chap. 7 that the electric and magnetic fields in electromagnetic plane waves are transverse to the direction of propagation $\hat{\mathbf{z}}$. There are two transverse directions, $\hat{\mathbf{x}}$ and $\hat{\mathbf{y}}$, and the fields with one orientation with respect to $\hat{\mathbf{x}}$ and $\hat{\mathbf{y}}$ are independent of those with an orientation differing by 90 deg. One may therefore have various amounts (amplitudes) of the fields in each of the two transverse directions and various possible relative phases. A specific relation of the amplitudes and phases of the two independent transverse fields is called a *state of polarization*.

When electromagnetic waves impinge on (and interact with) matter, it often happens that different polarization states of the incident radiation do not interact with the material in the same way. For example, we might find a material in which charged particles are free to move along $\hat{\mathbf{x}}$ but cannot move at all along $\hat{\mathbf{y}}$. In that case, E_x can do work on the charged particles, but E_y cannot. Then the electromagnetic wave energy associated with E_x may be reduced by being converted into charged-particle kinetic energy and thence by particle-particle collisions into heat energy, whereas the amplitude of E_y is not affected. Or instead it may only happen that the phase of E_x is shifted relative to that of E_y, without any diminution of energy (i.e., without decrease in the amplitude of E_x). In all such cases of *asymmetrical interactions*, the polarization state of the electromagnetic radiation is modified or changed by the interaction. This fact has many important consequences. By studying the effect of well-understood materials on an incident beam of unknown polarization state, the polarization state can be determined. Conversely, by measuring the modification of a known polarization state by a material, one can learn something about the material. For example, the direction of magnetic field in "our" spiral arm of our galaxy is now being mapped out by measuring the polarization direction of radio waves from extragalactic sources as a function of the direction of the source and of the wavelength of the radiation [G. L. Berge and G. A. Seielstad, *Scientific American* p. 46, (June 1965).

It is important to recognize that the concept of polarization applies only to waves that have at least two independent "polarization directions." Consider, for example, a sound wave propagating in air along $\hat{\mathbf{z}}$. Once one knows the frequency, amplitude, and phase constant of such a wave, there is nothing left to specify. We know that the displacements of the air in a sound wave are along the propagation direction—the waves are longitudinal waves. However, we do not ordinarily say that these waves are "longitudinally polarized"; that would be poor terminology. We reserve the term *polarization state* to describe waves for which there are at least two alternative polarization directions. In the cases of sound waves in a *solid* or of waves on a slinky, there are three possible polarization states—one longitudinal and two transverse polarization directions are available. In such a case, one can have longitudinally polarized waves or two different transversely polarized waves (or a general superposition of all three polarizations).

8.2 *Description of Polarization States*

All the waves that we study consist of some physical quantity whose displacement from its equilibrium value varies with position and time. The displacement can be described by a vector $\psi\,(x, y, z, t)$. We usually study plane waves for which ψ has the form $\psi(z, t)$, where z is measured along the propagation direction. (Here we include both standing waves and traveling waves.) The quantities $\partial\psi(z, t)/\partial t$ and $\partial\psi/(z, t)/\partial z$ are often the quantities with the most interesting physical properties. We have seen this to be the case for waves on a string and for sound waves, where in each case $\psi(z, t)$ designates the displacement of the particles of the medium away from their equilibrium positions.

For plane waves propagating along $\hat{\mathbf{z}}$, we can write the displacement as

$$\psi(z, t) = \hat{\mathbf{x}}\psi_x(z, t) + \hat{\mathbf{y}}\psi_y(z, t) + \hat{\mathbf{z}}\psi_z(z, t). \tag{1}$$

In the case of transverse waves on a string, ψ has only x and y components. Such waves are said to have transverse polarization. (Actually, one can also have longitudinal waves on a string consisting of variations of the tension and of the longitudinal velocity of the particles of the string.) For sound waves in air, the displacement ψ is along the propagation direction $\hat{\mathbf{z}}$. These are called longitudinal waves but are not usually termed longitudinally polarized. (Actually, it is possible to have transverse sound waves in a tube. These transverse waves can be thought of as longitudinal waves that do not head down the tube but rather bounce from one side of the tube to the other. The net propagation direction is down the tube, but the air oscillations have transverse components as well as longitudinal components.) In the case of electromagnetic plane waves, the displacement ψ is transverse to $\hat{\mathbf{z}}$, as we saw in Sec. 7.5. We found there that **E** and **B** are always transverse to $\hat{\mathbf{z}}$ for plane waves in vacuum. (It is possible to have longitudinal components of **E** and **B** if, for example, the waves are enclosed in a waveguide or cavity.)

Polarization of transverse waves. From now on we shall only consider transverse waves of the form

$$\psi(z, t) = \hat{\mathbf{x}}\psi_x(z, t) + \hat{\mathbf{y}}\psi_y(z, t). \tag{2}$$

We shall have two physical examples in mind in the discussion that follows: one is transverse waves on a stretched string or slinky; the other is electromagnetic plane waves in vacuum. For waves on a string, $\psi(z, t)$ will denote the instantaneous transverse displacement of the string from its equilibrium position. The other physically interesting quantities are the transverse velocity $\partial\psi/\partial t$ and the transverse force $-T_0\,\partial\psi/\partial z$ exerted by the string to the left of a position z on that to the right of z. These are known if $\psi(z, t)$ is known. For electromagnetic plane waves, $\psi(z, t)$ will denote the transverse electric field $\mathbf{E}(z, t)$. The other physically interesting quantity is the transverse magnetic field $\mathbf{B}(z, t)$, which is known if $\mathbf{E}(z, t)$ is known. For example, we can always resolve a general $\mathbf{E}(z, t)$ into a superposition of traveling waves traveling in both $+z$ and $-z$ directions. Letting $\mathbf{E}^+$ denote the part of $\mathbf{E}$ contributed by wave traveling in the $+z$ direction and $\mathbf{E}^-$ the part contributed by waves traveling in the $-z$ direction, we can write

$$\mathbf{E}(z, t) = \mathbf{E}^+(z, t) + \mathbf{E}^-(z, t). \tag{3}$$

Then from our study of traveling waves (Sec. 7.4) we know that the magnetic field $\mathbf{B}^+$ corresponding to $\mathbf{E}^+$ is equal to $\hat{\mathbf{z}} \times \mathbf{E}^+$, and that the magnetic field $\mathbf{B}^-$ corresponding to $\mathbf{E}^-$ is equal to $-\hat{\mathbf{z}} \times \mathbf{E}^-$. Thus the magnetic field corresponding to the superposition Eq. (3) is

$$\mathbf{B}(z, t) = \hat{\mathbf{z}} \times \left[\mathbf{E}^+(z, t) - \mathbf{E}^-(z, t) \right]. \tag{4}$$

We shall not make explicit use of Eq. (4). We only wished to show you (or remind you) that $\mathbf{B}$ is known "automatically" (in the sense that Maxwell's equations are "automatic") once $\mathbf{E}$ is known (assuming we know that we have plane waves in vacuum).

Effective point charge. Another physical picture which is very helpful in the case of electromagnetic plane waves is that obtained if we think of the plane waves as having been emitted by a harmonically oscillating point charge at the origin of coordinates, which is assumed to be far enough away so that the radiated waves are plane waves to a sufficiently good approximation. If the instantaneous transverse displacement of the charge q is denoted by

$$\psi(t) = \hat{\mathbf{x}}x(t) + \hat{\mathbf{y}}y(t) = \hat{\mathbf{x}}x_0 \cos(\omega t + \varphi_1) + \hat{\mathbf{y}}y_0 \cos(\omega t + \varphi_2), \tag{5}$$

then we know from our discussion of the radiation from a point charge (Sec. 7.5) hat the electric field $\mathbf{E}(z, t)$ is given by

$$\mathbf{E}(z, t) = -\frac{q\mathbf{a}_\perp(t')}{rc^2}$$

$$= -\frac{q\ddot{\boldsymbol{\psi}}(t')}{zc^2}.$$

Thus, since $\ddot{\boldsymbol{\psi}} = -\omega^2\boldsymbol{\psi}$, we have

$$\mathbf{E}(z, t) = \frac{q\omega^2\boldsymbol{\psi}(t')}{zc^2} = \frac{q\omega^2\boldsymbol{\psi}(t - \frac{z}{c})}{zc^2}. \tag{6}$$

Thus, when we are considering traveling plane electromagnetic waves, we may think of $\boldsymbol{\psi}(z, t)$ as denoting the electric field $\mathbf{E}(z, t)$; alternatively, we may think of it as denoting (aside from a known proportionality constant $q\omega^2/zc^2$) the displacement of a positive charge q at the earlier retarded time $t' = t - z/c$. Even if $\mathbf{E}(z, t)$ is not really produced by a single charge q, we can "invent" the charge q, defining it by Eq. (6). (Lacking any explicit knowledge of the source of radiation, we cannot tell that the radiation is *not* produced by the effective point charge q.)

Linear polarization. In the case of transverse waves (electromagnetic plane waves and transverse waves on a string), if the displacement is an oscillation back and forth along a fixed line transverse to $\hat{\mathbf{z}}$, the waves are said to be *linearly polarized*. There are two independent transverse directions. These may be taken to be $\hat{\mathbf{x}}$ and $\hat{\mathbf{y}}$. Let us consider a fixed value of z. Then we need not specify whether we are dealing with standing waves or traveling waves or both; i.e., we need not specify the phase relations between oscillations at different values of z, since we are now considering only a single value of z. Then the oscillations corresponding to a linearly polarized plane wave can have one or the other of the forms

$$\boldsymbol{\psi}(t) = \hat{\mathbf{x}}A_1 \cos \omega t, \tag{7}$$

$$\boldsymbol{\psi}(t) = \hat{\mathbf{y}}A_2 \cos \omega t, \tag{8}$$

where we have suppressed z from the notation and set the phase constant to zero. More generally, we can have a linearly polarized oscillation along a line that is neither $\hat{\mathbf{x}}$ nor $\hat{\mathbf{y}}$. Such an oscillation can always be written as a superposition of the two independent linearly polarized oscillations given by Eqs. (7) and (8), where the

x and y components of the superposition have the *same phase constant* (or else phase constants that differ by π):

$$\psi(t) = \hat{\mathbf{x}}A_1 \cos \omega t + \hat{\mathbf{y}}A_2 \cos \omega t, \tag{9}$$

i.e.,

$$\psi(t) = \left(\hat{\mathbf{x}}A_1 + \hat{\mathbf{y}}A_2\right) \cos \omega t. \tag{10}$$

The vector $\hat{\mathbf{x}}A_1 + \hat{\mathbf{y}}A_2$ has magnitude and direction that are independent of time. Therefore $\psi(t)$ as given by Eq. (10) consists of an oscillation along a fixed line. The amplitude A of the oscillation is given by

$$A = \sqrt{A_1^2 + A_2^2}. \tag{11}$$

The direction of $\psi(t)$ is (for linear polarization) always either along $+\hat{\mathbf{e}}$ or (half a cycle later) along $-\hat{\mathbf{e}}$, where $\hat{\mathbf{e}}$ is the unit vector

$$\hat{\mathbf{e}} = \frac{A_1}{A}\hat{\mathbf{x}} + \frac{A_2}{A}\hat{\mathbf{y}}. \tag{12}$$

We can see that $\hat{\mathbf{e}}$ is a unit vector:

$$\hat{\mathbf{e}} \cdot \hat{\mathbf{e}} = \frac{(A_1\hat{\mathbf{x}} + A_2\hat{\mathbf{y}})^2}{A^2}$$

$$= \frac{A_1^2\hat{\mathbf{x}} \cdot \hat{\mathbf{x}} + A_2^2\hat{\mathbf{y}} \cdot \hat{\mathbf{y}} + 2A_1A_2\hat{\mathbf{x}} \cdot \hat{\mathbf{y}}}{A^2}$$

$$= \frac{A_1^2 + A_2^2}{A^2} = 1. \tag{13}$$

The displacement $\psi(t)$ for a linearly polarized wave (at fixed z) is shown in Fig. 8.1.

Linearly polarized standing wave. Suppose we wish to describe a linearly polarized "pure" standing wave having a node in ψ at (for example) $z = 0$. Then we simply multiply the linearly polarized displacement for fixed z [given by Eq. (10)] by $\sin kz$:

$$\psi(z, t) = (\hat{\mathbf{x}}A_1 + \hat{\mathbf{y}}A_2) \sin kz \cos \omega t. \tag{14}$$

Linearly polarized traveling wave. In order to describe a traveling wave propagating (for example) in the $+z$ direction, we simply replace ωt by $\omega t - kz$ in the linearly polarized displacement for fixed z:

$$\psi(z, t) = (\hat{\mathbf{x}}A_1 + \hat{\mathbf{y}}A_2) \cos (\omega t - kz). \tag{15}$$

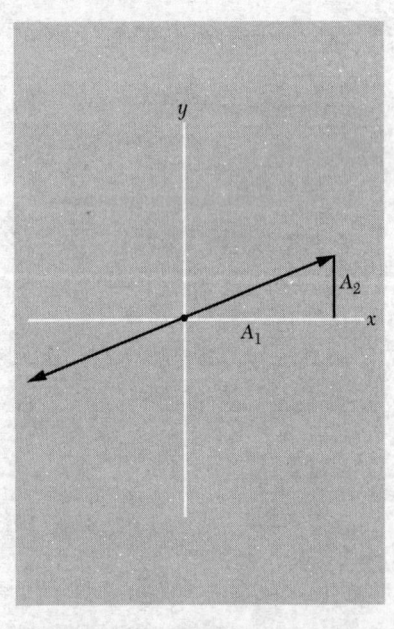

Fig. 8.1 Linear polarization. The displacement $\psi(t)$ for fixed z, given by Eqs. (9) and (10), oscillates harmonically along the line indicated by the double-headed arrow.

Circular polarization. If the displacement in a transverse wave is a motion in a circle, the waves are said to be circularly polarized. We consider at first, a fixed value of z. We do not (yet) specify whether the waves are propagating along $+\hat{\mathbf{z}}$ or along $-\hat{\mathbf{z}}$ (or even whether they are traveling waves). If the thumb of your right hand points along $+\hat{\mathbf{z}}$ when the fingers curl in the direction of rotation, then the oscillation is said to be circularly polarized along $+\hat{\mathbf{z}}$. (Similarly, we use the right-hand rule to define circular polarization along $-\hat{\mathbf{z}}$.) An oscillation circularly polarized along $+\hat{\mathbf{z}}$ can be expressed as a superposition of a linearly polarized oscillation along $\hat{\mathbf{x}}$ and a linearly polarized oscillation along $\hat{\mathbf{y}}$ with the same amplitude as the oscillation along $\hat{\mathbf{x}}$. Taking the x, y, z axes to be (as usual) a right-handed set of axes, so that $\hat{\mathbf{x}} \times \hat{\mathbf{y}} = \hat{\mathbf{z}}$, we see that, for circular polarization along $+\hat{\mathbf{z}}$, , the $\hat{\mathbf{x}}$ oscillation *leads* the $\hat{\mathbf{y}}$ oscillation by 90 deg:

$$\psi(t) = \hat{\mathbf{x}}A \cos \omega t + \hat{\mathbf{y}}A \cos\left(\omega t - \frac{\pi}{2}\right)$$

$$= \hat{\mathbf{x}}A \cos \omega t + \hat{\mathbf{y}}A \sin \omega t. \qquad (16)$$

Similarly, for circular polarization along $-\hat{\mathbf{z}}$, the $\hat{\mathbf{x}}$ oscillation *lags* the $\hat{\mathbf{y}}$ oscillation by 90 deg:

$$\psi(t) = \hat{\mathbf{x}}A \cos \omega t + \hat{\mathbf{y}}A \cos\left(\omega t + \frac{\pi}{2}\right)$$

$$= \hat{\mathbf{x}}A \cos \omega t - \hat{\mathbf{y}}A \sin \omega t. \qquad (17)$$

According to our discussion of electromagnetic plane waves (Sec. 7.4), circularly polarized plane waves carry angular momentum $\mathbf{J} = \pm(W/\omega)\hat{\mathbf{z}}$, where W is the energy and ω is the angular frequency. The sign of the angular momentum is the same as that of the sense of rotation of the fields. Thus the angular momentum is along $+\hat{\mathbf{z}}$ for circular polarization along $+\hat{\mathbf{z}}$ and is along $-\hat{\mathbf{z}}$ for circular polarization along $-\hat{\mathbf{z}}$. (In our discussion so far, $\hat{\mathbf{z}}$ is a direction fixed in space. The above discussion holds for either direction of propagation of traveling waves; it also holds for standing waves.) Circularly polarized waves on a string or slinky also carry angular momentum, of course.

The displacement $\psi(t)$ for circularly polarized oscillation at fixed z is shown in Fig. 8.2.

Circularly polarized standing wave. A circularly polarized standing wave with polarization (and angular momentum) along $+\hat{\mathbf{z}}$ is obtained by multiplying the appropriate circularly polarized oscillation for fixed z [given by Eq. (16)] by a sinusoidal function of z.

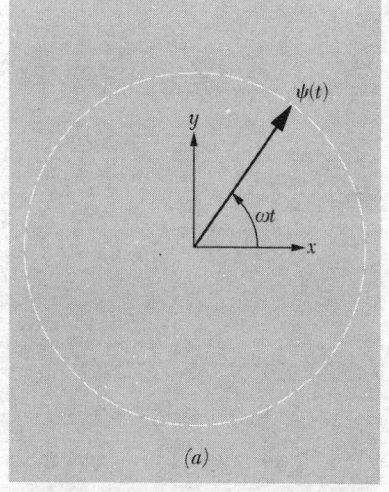

(a)

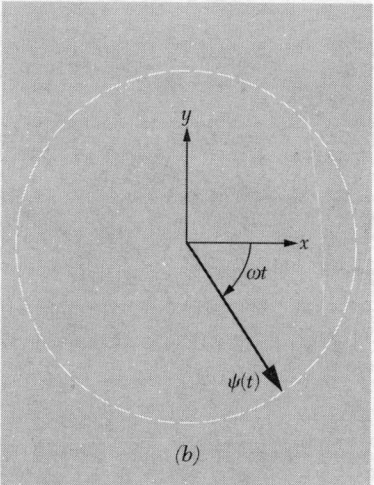

(b)

Fig. 8.2 *Circular polarization. (a) Circular polarization and angular momentum along $+\hat{\mathbf{z}}$, where $\hat{\mathbf{z}}$ is fixed in space and is independent of the propagation direction. (b) Circular polarization and angular momentum along $-\hat{\mathbf{z}}$.*

Thus, for a standing wave with a node at $z = 0$ (for example) and circular polarization along $+\hat{\mathbf{z}}$, we have

$$\psi(z, t) = \left[\hat{\mathbf{x}} \cos \omega t + \hat{\mathbf{y}} \cos \left(\omega t - \frac{\pi}{2} \right) \right] A \sin kz. \quad (18)$$

Circularly polarized traveling wave. A circularly polarized traveling wave with circular polarization (and angular momentum) along $+\hat{\mathbf{z}}$ is most easily obtained by replacing ωt by $\omega t - kz$ (for propagation along $+\hat{\mathbf{z}}$) in the circularly polarized oscillation given by Eq. (16):

$$\psi(z, t) = A \left\{ \hat{\mathbf{x}} \cos \left[\omega t - kz \right] + \hat{\mathbf{y}} \cos \left[\left(\omega t - \frac{\pi}{2} \right) - kz \right] \right\}. \quad (19)$$

Similarly, if we want a wave propagating along $-\hat{\mathbf{z}}$, we replace ωt by $\omega t + kz$; if we want a wave with angular momentum along $-\hat{\mathbf{z}}$, we start with the circularly polarized oscillation given by Eq. (17) and make similar replacements of ωt by $\omega t - kz$ or $\omega t + kz$.

Handedness conventions for circularly polarized traveling waves. Suppose we have a circularly polarized traveling wave propagating in the $+\hat{\mathbf{z}}$ direction. Suppose its angular momentum is also in the $+\hat{\mathbf{z}}$ direction, and therefore the sense of rotation of the fields (for electromagnetic waves) or of the displacements (for waves on a slinky) is along $+\hat{\mathbf{z}}$ as given by a right-hand rule. It is natural to call such a polarization "right-handed"; we shall call that convention the *angular-momentum convention.* According to the angular-momentum convention, a circularly polarized traveling wave is called right-handed if its angular momentum is along the propagation direction and is called left-handed if the angular momentum is opposite to the propagation direction. That convention is, however, opposite to the convention that is customarily used in optics. For example, it is opposite to the convention used to label the circular polarizer in your optics kit as "left-handed." The *optics convention* can be called the "screw-shape convention" or simply the *screwy convention.* Its justification can be obtained by considering the emission of a circularly polarized traveling wave on a slinky that you shake. Suppose you shake one end of a slinky in a rapid circular motion that is clockwise as seen from your viewpoint. A circularly polarized wave packet travels down the slinky away from you. Its sense of rotation is clockwise; the angular momentum is along the propagation direction. The wave is right-handed by the angular-momentum convention. Now "stop the motion" with a mental snapshot, and look at the instantaneous shape of the slinky. Is it a right-handed screw or a left-handed screw? The optics convention is to let the handedness of the screw be used to name the handedness of the polarization. Unfortunately the handedness is that of a left-handed screw! (To see that, think of the motion of your hand and the slinky as the wave is being emitted. Picture the present slinky configuration near to your hand. The slinky slightly downstream from your hand has a present angular position corresponding to the angular position of your hand at a slightly earlier time—it lags the present position of your hand. The slinky even farther downstream lags even more, since its wave was emitted even earlier. As you progress farther downstream at fixed time, you are tracing out a left-handed screw.) Thus the screwy convention assigns labels with opposite handedness to those assigned in the angular-momentum convention. The angular-momentum convention is easier on the brain. The optics convention is most easily remembered by remembering that it is screwy.

It is instructive to have a tangible experience of different transverse polarizations by playing with a slinky. To obtain standing waves, tie one end to a telephone pole and shake the other. To simulate a "free" end, tie one end of the slinky to about 10 m of string, and tie the other end of the string to the pole. Linearly or circularly polarized standing waves are easy to generate. Traveling harmonic waves are difficult to generate, because it is not easy to terminate a slinky with an impedance-matching resistive load. [I suspect that the combination of a long string (to "free" the end) with suitable styrofoam ("massless") pistons or paddles stirring buckets of water can be made to work.] However, one can easily send a pulse or wave packet down the slinky and follow it as it reflects at fixed or free ends.

Properties of transverse polarization states. By playing with a slinky or by studying the above equations, you can verify the following properties of transverse polarization states (which hold for electromagnetic plane waves as well):

1. In a linearly polarized wave, the displacement at fixed z goes through zero twice per cycle.

   In a standing wave, all points go through zero simultaneously.

   In a traveling wave, all points undergo the same motion as all others, but with a phase shift corresponding to the propagation time of the wave between the points.

2. In a circularly polarized standing or traveling wave, the displacement at fixed z has constant magnitude.

   If the slinky is carrying a circularly polarized traveling wave, a snapshot at a fixed time t would show the slinky to be in the shape of a corkscrew. If instead it carries a circularly polarized standing wave, the slinky always lies completely in a plane. The single snapshot could not distinguish this shape from either that of a linearly polarized standing wave or that of a linearly polarized traveling wave. (A second snapshot taken slightly later, combined with the first snapshot, would tell us which of these three possibilities held.)

3. The reflection at a termination of a traveling slinky wave packet with circular polarization along $+\hat{\mathbf{z}}$ (a direction fixed in space) gives a reflected wave with circular polarization along the same direction. This is true for reflection from either fixed or free ends (or any other kind of load). Thus the sense of the rotation with respect to the fixed direction $\hat{\mathbf{z}}$ is preserved under reflection. This follows directly from conservation of angular momentum. A fixed or free end of a slinky cannot exert any torque, and therefore the angular momentum with respect to the fixed $+\hat{\mathbf{z}}$ axis is preserved under reflection. Of course, the handedness is reversed because the propagation direction is reversed upon reflection. Electromagnetic radiation has the same behavior as the slinky. By this we mean that the sense of rotation with respect to a fixed direction $\hat{\mathbf{z}}$ of circularly polarized light or microwaves or any other electromagnetic radiation is not changed by a reflection through 180 deg, but the handedness, i.e., the sense of rotation with respect to the propagation direction, is reversed. The fact that the handedness of light is reversed under reflection is not new to you. You know that if you look at your right hand in a mirror it looks like a left hand. That may not seem to be obviously related to the preservation (under reflection from a mirror) of the sense of rotation of circularly polarized light with respect to a fixed direction $\hat{\mathbf{z}}$, but in fact it is related. Both can be thought of as being the result of conservation of momentum transverse to $\hat{\mathbf{z}}$ in the interaction of the wave with the reflecting medium, i.e., the wall in the case of the slinky or the electrons of the mirror in the case of electromagnetic radiation. (However, see Home Exp. 8.27.)

General transverse polarization—elliptical polarization. At a fixed z, a general transversely polarized oscillation has the form

$$\psi(t) = \hat{\mathbf{x}}A_1 \cos(\omega t + \varphi_1) + \hat{\mathbf{y}}A_2 \cos(\omega t + \varphi_2). \qquad (20)$$

If φ_2 equals either φ_1 or $\varphi_1 \pm \pi$, we have linear polarization. If φ_2 is $\varphi_1 - \frac{1}{2}\pi$ and A_2 equals A_1, we have circular polarization along $+\hat{\mathbf{z}}$. If φ_2 is $\varphi_1 + \frac{1}{2}\pi$ and A_2 equals A_1, we have circular polarization along $-\hat{\mathbf{z}}$. For the general case where A_2 and A_1 are unequal and φ_2 and φ_1 are arbitrary, the displacement ψ describes an elliptical path. We can see this as follows: Call ψ_x and ψ_y by the names x and y. Then x is $A_1 \cos(\omega t + \varphi_1)$ and y is $A_2 \cos(\omega t + \varphi_2)$. Expand each of these cosines, so that x is a certain linear combination of $\cos \omega t$ and $\sin \omega t$, and y is another linear combination. Now solve these two linear equations for $\sin \omega t$ and $\cos \omega t$. You will then find that $\sin \omega t$ and $\cos \omega t$ are each a (different) linear combination of x and y. Now add the squares of $\sin \omega t$ and $\cos \omega t$. The result (which equals 1) is a quadratic expression involving x^2, y^2, and xy. Such an expression is called a conic section. If the possible values of x and y have limited magnitude (as they do here), the conic section is an ellipse. (See Prob. 8.1.) In Fig. 8.3 we show what happens as we

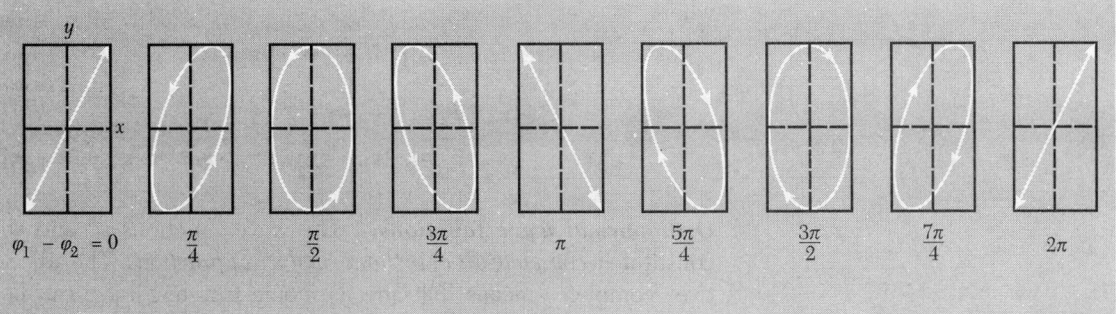

$$\varphi_1 - \varphi_2 = 0 \qquad \frac{\pi}{4} \qquad \frac{\pi}{2} \qquad \frac{3\pi}{4} \qquad \pi \qquad \frac{5\pi}{4} \qquad \frac{3\pi}{2} \qquad \frac{7\pi}{4} \qquad 2\pi$$

vary the relative phase $\varphi_1 - \varphi_2$ in Eq. (20). [You can demonstrate the effect of relative phase on polarization using clear cellophane tape. See Home Exp. 8.16.]

Fig. 8.3 General polarization. The amplitude for the y motion is taken to be twice that for the x motion. The y motion lags the x motion by the indicated phase constant, which is $\varphi_1 - \varphi_2$.

Complex notation. When there are several phase constants in a superposition of waves, it is sometimes a convenience to use complex numbers. We shall illustrate this by considering a traveling harmonic electromagnetic wave propagating in the $+\hat{\mathbf{z}}$ direction:

$$\mathbf{E}(z, t) = \hat{\mathbf{x}}E_x(z, t) + \hat{\mathbf{y}}E_y(z, t)$$

$$= \hat{\mathbf{x}}E_1 \cos(kz - \omega t - \varphi_1) + \hat{\mathbf{y}}E_2 \cos(kz - \omega t - \varphi_2). \quad (21)$$

The electric field given by Eq. (21) is easily seen to be the real part of the following *complex wave function:*

$$\mathbf{E}_c(z, t) = e^{i(kz - \omega t)}(\hat{\mathbf{x}}E_1 e^{-i\varphi_1} + \hat{\mathbf{y}}E_2 e^{-i\varphi_2}). \qquad (22)$$

The fact that $\exp i(kz - \omega t)$ can be factored out of the complete expression for $\mathbf{E}_c$ is sometimes a help in evaluating expressions involving superpositions of several different waves. We should always return to the real electric fields $\mathbf{E}$ before applying any results to a physical situation. (There is no $\sqrt{-1}$ in Maxwell's equations; there is no such thing as an electric field having strength $\sqrt{-1}$ volt/cm.)

Complex wave functions and complex amplitudes. The complex quantity $\mathbf{E}_c$, whose real part is the electric field $\mathbf{E}$, can be thought of as a superposition,

$$\mathbf{E}_c(z,t) = A_1 \psi_1(z, t) + A_2 \psi_2(z, t), \qquad (23)$$

where

$$\psi_1(z,t) = \hat{\mathbf{x}}e^{i(kz-\omega t)},$$

$$\psi_2(z,t) = \hat{\mathbf{y}}e^{i(kz-\omega t)}, \tag{24}$$

$$A_1 = E_1 e^{-\varphi_1}, \qquad A_2 = E_2 e^{-\varphi_2}, \tag{25}$$

Orthonormal wave functions. The wave functions ψ_1 and ψ_2 constitute a *complete set of orthonormal wave functions*. The adjective "complete" means that any harmonic traveling wave can be expanded in a superposition of ψ_1 and ψ_2 with suitable complex constant coefficients A_1 and A_2. The adjective "orthonormal" means that one has

$$\psi_1^* \cdot \psi_1 = \psi_2^* \cdot \psi_2 = 1, \quad \psi_1^* \cdot \psi_2 = \psi_2^* \cdot \psi_1 = 0, \tag{26}$$

where the asterisk indicates complex conjugation (i.e., replacing of i by $-i$). Thus we have

$$\psi_1^* \cdot \psi_1 = [\hat{\mathbf{x}}e^{-i(kz-\omega t)}] \cdot [\hat{\mathbf{x}}e^{i(kz-\omega t)}] = \hat{\mathbf{x}} \cdot \hat{\mathbf{x}} = 1.$$

$$\psi_1^* \cdot \psi_2 = [\hat{\mathbf{x}}e^{-i(kz-\omega t)}] \cdot [\hat{\mathbf{y}}e^{i(kz-\omega t)}] = \hat{\mathbf{x}} \cdot \hat{\mathbf{y}} = 0.$$

Because of the orthonormality conditions, Eqs. (26), the absolute magnitude squared of the complex vector $\mathbf{E}_c$ has a very simple expression:

$$\left|\mathbf{E}_c\right|^2 \equiv (\mathbf{E}_c^*) \cdot (\mathbf{E}_c)$$

$$= (A_1^* \psi_1^* + A_2^* \psi_2^*) \cdot (A_1 \psi_1 + A_2 \psi_2)$$

$$= \left|A_1\right|^2 + \left|A_2\right|^2$$

$$= E_1^2 + E_2^2. \tag{27}$$

Time-averaged energy flux in complex notation. The counting rate of a photomultiplier detector in a beam of traveling electromagnetic waves is proportional to the time-averaged energy flux of the beam. More precisely, for angular frequency ω a detector with area A and photocathode conversion efficiency ϵ will have an average counting rate R (in units of counts per second) given by

$$R = \frac{\langle S \rangle}{\hbar \omega} \cdot A \cdot \epsilon, \tag{28}$$

where the time-averaged energy flux (in erg/cm^2 sec) is

$$\langle S \rangle = \frac{4\pi\epsilon_0}{\mu_0} \langle \mathbf{E}^2 \rangle, \tag{29}$$

and

$$\left\langle \mathbf{E}^2 \right\rangle = \left\langle (\hat{\mathbf{x}} E_x + \hat{\mathbf{y}} E_y)^2 \right\rangle$$

$$= \left\langle E_x^{\ 2} \right\rangle + \left\langle E_y^{\ 2} \right\rangle$$

$$= \frac{1}{2} E_1^{\ 2} + \frac{1}{2} E_2^{\ 2}. \tag{30}$$

The factors of $\frac{1}{2}$ in the last line of Eq. (30) result from the time, average over the square of the harmonic oscillations given by Eq. (21).

By comparison of Eqs. (27) and (30) we see that if we wish to work with the complex quantity $\mathbf{E}_c$ whose real part is the electric field $\mathbf{E}$, we can obtain the correct expression for the time-averaged energy flux if we use half the absolute magnitude squared of $\mathbf{E}_c$ in place of the time-averaged square of $\mathbf{E}$:

$$\mathbf{E} = \operatorname{Re} \mathbf{E}_c \equiv \text{real part of } \mathbf{E}_c, \tag{31}$$

$$\left\langle \mathbf{E}^2 \right\rangle = \tfrac{1}{2} \left| \mathbf{E}_c \right|^2, \tag{32}$$

where

$$\left\langle \mathbf{E}^2 \right\rangle = \left\langle E_x^{\ 2} \right\rangle + \left\langle E_y^{\ 2} \right\rangle,$$

$$\left| \mathbf{E}_c \right|^2 = \left| E_{xc} \right|^2 + \left| E_{yc} \right|^2. \tag{33}$$

Other complete representations of polarized light. The most general polarization state can be represented as a superposition of waves with linear polarization along $\hat{\mathbf{x}}$ and along $\hat{\mathbf{y}}$. Of course there are an infinite number of directions (fixed in the inertial frame) that we could have chosen for $\hat{\mathbf{x}}$. Thus there are an infinite number of linear polarization representations that can be used. In complex notation, we can say (very fancily) that there are an infinite number of complete sets of orthonormal wave functions ψ_1 and ψ_2 that can be used to form the basis for a superposition (with complex coefficients) that gives $\mathbf{E}_c$. For example, suppose that the unit vectors $\hat{\mathbf{e}}_1$ and $\hat{\mathbf{e}}_2$ are obtained from our original $\hat{\mathbf{x}}$ and $\hat{\mathbf{y}}$ by rotating $\hat{\mathbf{x}}$ and $\hat{\mathbf{y}}$ through an angle φ (in the direction of rotation from $\hat{\mathbf{x}}$ to $\hat{\mathbf{y}}$). Then you can easily show that

$$\hat{\mathbf{e}}_1 = \hat{\mathbf{x}} \cos \varphi + \hat{\mathbf{y}} \sin \varphi, \quad \hat{\mathbf{e}}_2 = -\hat{\mathbf{x}} \sin \varphi + \hat{\mathbf{y}} \cos \varphi, \tag{34}$$

The complete set of orthonormal wave functions corresponding to the linear polarization representation with linear polarization along $\hat{\mathbf{e}}_1$ and $\hat{\mathbf{e}}_2$ is given by

$$\psi_1 = \hat{\mathbf{e}}_1 e^{i(kz - \omega t)}, \qquad \psi_2 = \hat{\mathbf{e}}_2 e^{i(kz - \omega t)}, \tag{35}$$

You can easily check to see that ψ_1 and ψ_2 satisfy the orthonormality conditions, Eqs. (26).

Circular polarization representation. A general polarization state of a harmonic traveling wave can also be represented as a superposition of right-handed and left-handed circularly polarized components with suitable amplitudes and phase constants. For example, a wave linearly polarized along $\hat{\mathbf{x}}$ can be written in either of the equivalent forms

$$\mathbf{E} = \hat{\mathbf{x}} A \cos(kz - \omega t) \tag{36}$$

or

$$\mathbf{E} = \frac{A}{2} \left\{ \hat{\mathbf{x}} \cos\left[\omega t - kz\right] + \hat{\mathbf{y}} \cos\left[\left(\omega t - \frac{\pi}{2}\right) - kz\right] \right\}$$

$$+ \frac{A}{2} \left\{ \hat{\mathbf{x}} \cos\left[\omega t - kz\right] + \hat{\mathbf{y}} \cos\left[\left(\omega t + \frac{\pi}{2}\right) - kz\right] \right\}. \tag{37}$$

(The terms in $\hat{\mathbf{y}}$ have equal amplitudes and 180-deg phase difference; they add up to zero.) The representation of $\mathbf{E}$ given by Eq. (36) is a linear polarization representation with amplitude A. The representation of $\mathbf{E}$ given by Eq. (37) is a superposition of circularly polarized components having angular momentum along $+\hat{\mathbf{z}}$ and $-\hat{\mathbf{z}}$ and each having amplitude $\frac{1}{2}A$. The complex expressions analogous to Eqs. (36) and (37) are as follows:

$$E_c = A\hat{\mathbf{x}}e^{i(kz - \omega t)} \tag{38}$$

and

$$\mathbf{E}_c = \frac{1}{2}A \left[\hat{\mathbf{x}}e^{i(kz-\omega t)} + \hat{\mathbf{y}}e^{i\{kz-[\omega t-(\pi/2)]\}} \right]$$

$$+ \frac{1}{2}A \left[\hat{\mathbf{x}}e^{i(kz-\omega t)} + \hat{\mathbf{y}}e^{i\{kz-[\omega t+(\pi/2)]\}} \right] \tag{39}$$

Now we use the facts

$$e^{i(\pi/2)} = \cos\frac{\pi}{2} + i\sin\frac{\pi}{2} = i, \tag{40}$$

$$e^{-i(\pi/2)} = \cos\frac{\pi}{2} - i\sin\frac{\pi}{2} = -i$$

to write Eq. (39) in the briefer form

$$\mathbf{E}_c = \tfrac{1}{2}A\left[\left\langle\hat{\mathbf{x}} + i\hat{\mathbf{y}}\right\rangle e^{i(kz-\omega t)}\right] + \tfrac{1}{2}A\left[\left\langle\hat{\mathbf{x}} - i\hat{\mathbf{y}}\right\rangle e^{i(kz-\omega t)}\right], \tag{41}$$

We can now define a complete set of orthonormal circularly polarized wave functions by

$$\psi_+ = \left(\frac{\hat{\mathbf{x}} + i\hat{\mathbf{y}}}{\sqrt{2}} \right) e^{i(kz - \omega t)} \tag{42}$$

$$\psi_- = \left(\frac{\hat{\mathbf{x}} - i\hat{\mathbf{y}}}{\sqrt{2}} \right) e^{i(kz - \omega t)}$$

You may easily check that ψ_+ and ψ_- are orthonormal, i.e., that we have

$$\psi_+^* \cdot \psi_+ = \psi_-^* \cdot \psi_- = 1; \quad \psi_+^* \cdot \psi_- = \psi_-^* \cdot \psi_+ = 0. \tag{43}$$

Then the most general polarization state for a harmonic traveling wave can be written in the form

$$\mathbf{E}_c(z, t) = A_+ \psi_+ + A_- \psi_-, \tag{44}$$

where A_+ and A_- are complex constants. For the special case of linear polarization corresponding to Eq. (38), we see that A_+ and A_- are

$$A_+ = A_- = \frac{1}{\sqrt{2}} A. \tag{45}$$

The time-averaged counting rate R for a photomultiplier in a beam of harmonic traveling waves can be expressed in terms of the complex coefficients of any complete set of wave functions. Thus instead of using the $\hat{\mathbf{x}}, \hat{\mathbf{y}}$ linear polarization representation [see Eqs. (28) through (33)], we can use the $+\hat{\mathbf{z}}, -\hat{\mathbf{z}}$ angular momentum representation:

$$R = \frac{\langle S \rangle}{\hbar \omega} A \cdot \boldsymbol{\epsilon}, \tag{46}$$

where A is the area (not the amplitude!), $\boldsymbol{\epsilon}$ is the efficiency, and

$$\langle S \rangle = \frac{4\pi\epsilon_0}{\mu_0} \langle \mathbf{E}^2 \rangle, \tag{47}$$

$$\langle \mathbf{E}^2 \rangle = \tfrac{1}{2} |\mathbf{E}_c|^2, \tag{48}$$

$$|\mathbf{E}_c|^2 = |A_+\psi_+ + A_-\psi_-|^2 = |A_+|^2 + |A_-|^2.$$

We shall seldom use complex wave functions. Our main purpose in introducing them here is to ease the mental readjustment required when you study quantum physics in Vol. IV. (The wave functions used in quantum physics are nearly always complex. The square root of −1 appears explicitly in the wave equations of quantum mechanics.)

8.3   *Production of Polarized Transverse Waves*

In this section we shall examine several methods for producing a desired polarization state. Control of polarization is easiest when you control the radiation process, shaking a slinky or broadcasting electromagnetic waves from an antenna of your own design. However, it may happen that you have no control over the radiation process. In that case, whether you start with light from an electric bulb or the sun, the problem is to select somehow a desired polarization state from the existing complicated superposition of different states. Perhaps the undesired polarization components can be absorbed by a sheet of polaroid. Or perhaps you can arrange to reflect light with negligible reflection of the undesired polarization components and then look only at the reflected radiation. This kind of selective reflection is what makes the blue sky polarized and what makes light reflected from water or glass or concrete or a knee polarized.

Polarization by selective emission.   When you shake a slinky, you control the polarization state of the waves by controlling the direction of the shaking. Similarly, the radio waves or microwaves emitted by an antenna have polarization which depends on the motion of the electrons in the antenna. If the antenna is a straight piece of wire normal to $\hat{z}$, the electrons oscillating along the wire shake the electric lines of force in that direction, and the electromagnetic waves propagated along $\hat{z}$ are linearly polarized with the electric field parallel to the antenna. Those radiated in other propagation directions are linearly polarized along the direction of the projection of the antenna perpendicular to the propagation direction. If there is one straight antenna along $\hat{x}$ and another along $\hat{y}$, and if they are driven by equal currents which have the same phase, the radiation propagated along $\pm\hat{z}$, will be linearly polarized along the 45-deg direction between $\hat{x}$ and $\hat{y}$. If the x current is equal in amplitude to the y current but leads it by 90 deg in phase, the electromagnetic radiation radiated along either the $+\hat{z}$ or $-\hat{z}$ axes will be circularly polarized with angular momentum along $+\hat{z}$ The radiation emitted along $+\hat{z}$ will be right-handed (by the angular momentum convention); that radiated along $-\hat{z}$ will left-handed. The radiation is indistinguishable (at a sufficiently large distance) from that which would be produced by a single oscillating "equivalent point charge" q describing a circular motion

$$\psi = A\left[\hat{x}\cos\omega t + \hat{y}\sin\omega t\right], \qquad (49)$$

where the amplitude A (and the phase constant) of the circular motion of q are related to the radiated circularly polarized electric field by Eq. (6), Sec. 8.2. The polarization of the radiation emitted in any direction from this system of two antennas is just what one would get from the motion of the equivalent point charge as given by Eq. (49). From a general observation point, the projected circular motion of the equivalent charge looks like (and is) an elliptical motion. Thus the polarization for a general emission direction is elliptical. For example, for emission in a direction perpendicular to $\hat{\mathbf{z}}$, the polarization is linear (a special case of a "degenerate ellipse"). All these results follow directly from our discussion of the radiation point charge (Sec. 7.5) with only two provisos: (i) we must be sufficiently far from the antenna so that we can neglect the "near-zone" fields, and (ii) the antenna must be short compared with a wavelength, so that it takes only one equivalent charge to represent the motion of all the electrons in the antenna. (For an antenna that is several wavelengths long electrons in different parts of the antenna contribute with different phases; then it takes more than one equivalent charge and we have what is called "multi pole" radiation, as contrasted with the "dipole" radiation obtained from a single harmonically oscillating charge.)

Polarization by selective absorption. If you start with a general state of polarization, one way to produce a given polarization is to get rid of the undesired components of the waves by arranging to have them do work on some "moving parts," while the desired component does not. As an example, consider standing waves on a slinky. Suppose $\hat{\mathbf{z}}$ is horizontal (along the slinky), $\hat{\mathbf{y}}$ is vertical, and $\hat{\mathbf{x}}$ is horizontal. A vertical massless (styrofoam) push rod is attached to a massless piston that stirs the water in a bucket. The piston will be driven by the y component of vibrations. If we start out with a standing wave that includes both x and y vibrations in equal amounts, the y vibrations will soon be damped out as their energy is converted into heat in the water bucket (provided we do not keep regenerating them by shaking).

Skein of wires. In the case of microwaves, we can accomplish selective absorp-tion by means of a skein of parallel conducting wires stretched along $\hat{\mathbf{y}}$, as shown in Fig. 8.4. Suppose that the electric field in the incident electromagnetic radiation (microwave radiation) has both x and y components. We may consider the effect of the wires on these two components separately. First consider the y component, along the wires. The electric field of the incident radiation drives the electrons along the wire. The wire (if it is made of copper or silver or any good metallic conductor)

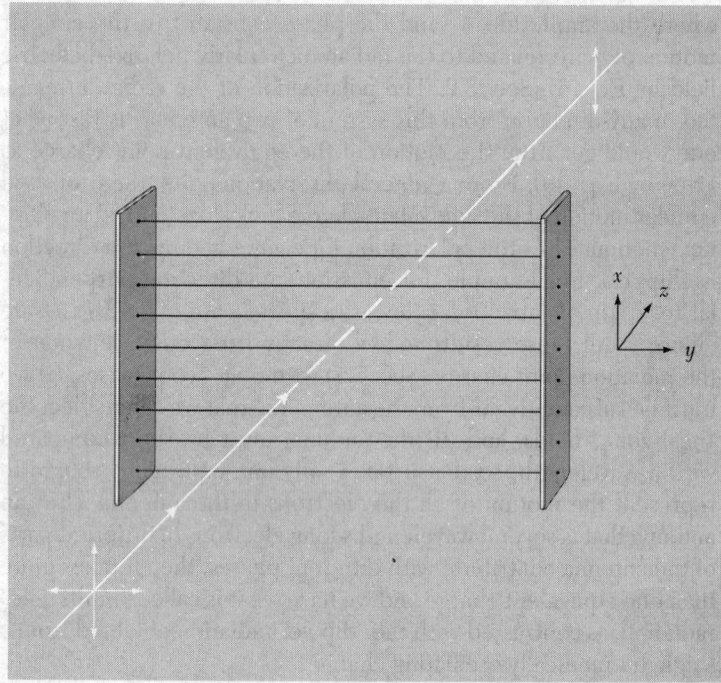

Fig. 8.4 Skein of wires absorbs those microwaves with E along ŷ.

acts as a resistive load. The conduction electrons reach terminal velocity in a time which is short compared with the period of the microwaves (which we may take as having a frequency of 1000 Mc, for example). The electric field does work on the electrons; they transfer some of their energy to the copper lattice through collisions. They also radiate. It turns out that their radiation in the forward direction interferes destructively with the incident radiation and cancels it essentially to zero. In the backward direction, the radiation due to motion of the electrons driven along $\hat{\mathbf{y}}$ gives a reflected wave. (Actually, only a small part of the energy of the incident radiation with **E** along $\hat{\mathbf{y}}$ is turned into heat in the wires. Most is reflected back in the $-\hat{\mathbf{z}}$ direction.) Thus the skein of wires eliminates the y component.

Now consider what happens along $\hat{\mathbf{x}}$. The electrons are not free to move along $\hat{\mathbf{x}}$ because they cannot leave the wire. Instead of reaching a steady terminal velocity (as they do for motion along $\hat{\mathbf{y}}$), the electrons soon build up a surface charge along the $+x$ and $-x$ edges of the wire. When the field due to the surface charge is sufficient to cancel the incident field (inside the wire), the electrons stop moving. This happens in a time which is short compared with the microwave period. Thus the electrons are always in a sort of static equilibrium (or nearly so) with no velocity or acceleration. They do not absorb energy, nor do they radiate. Consequently the x component of the radiation is unaffected.

It may have occurred to you that surface charge will also build up on the $+y$ *ends* of the wires. However, the resulting field from these end charges (which tends to cancel the incident field along $\hat{y}$ inside the wires) can be made as small as we wish in the region of interest (near the center of the skein) by making the wires sufficiently long in the y direction.

For visible light with $\lambda \sim 5 \times 10^{-5}$ cm, it is not easy to make parallel conducting "wires" spaced at less than λ. Nevertheless, it has been done!†

Polaroid. In 1938 Edwin H. Land invented Polaroid, which behaves somewhat like a wire grid. In the manufacturing process, a sheet of plastic consisting of long hydrocarbon chains is greatly stretched in one direction. This lines up the molecules. Next, the sheet is dipped in a solution containing iodine. The iodine attaches to the long hydrocarbon chains and provides conduction electrons that can move along the chains but not perpendicular to them. One thus has effective "wires" along the hydrocarbon chains. The electric field component along the wires is absorbed; that perpendicular to the wires is transmitted with very little attenuation. [A simple analogy of a rope and a picket fence is sometimes used as a reminder of the action of a skein of wires on incident electromagnetic waves. The rope passes between the slats of the fence. Waves are absorbed if the transverse *velocity* of the rope is transverse to the direction of the slats. The transverse velocity for waves on a rope is analogous to the *magnetic* field in an electromagnetic wave. Thus the mnemonic is that magnetic field transverse to the wires is absorbed; i.e., the electric field parallel to the wires is absorbed. It is not a very good mnemonic, because one must remember that it is the transverse velocity (rather than the transverse component of tension) that makes the rope hit the slats, and one must also remember that the analog of rope velocity is magnetic field rather than electric field (which is analogous to transverse tension). The mnemonic requires remembering more things than the simple correct explanation.]

A sheet of polaroid thus has an axis (lying in the sheet) called the axis of *easy transmission.* If **E** is along this axis, the light is transmitted with very little absorption. If **E** is perpendicular to the easy-transmission axis, the light is almost completely absorbed. The easy-transmission axis is perpendicular to the direction of stretching of the plastic, i.e., it is perpendicular to the "wires."

---

†G. R. Bird and M. Parrish, Jr., *J. Opt. Soc. Am.* **50,** 886 (1960), evaporated gold at a glancing angle onto a plastic diffraction grating having 50,000 parallel scratches per inch. The gold deposited on the downstream sides of the scratches to form parallel conducting "wires."

When you look at a white sheet of paper through a piece of polaroid, the paper looks gray. That is because half the light coming from the paper is absorbed by the polaroid, so the paper naturally looks darker. On the other hand, a piece of clear cellophane (or other clear plastic) transmits almost all the light incident on it.

Your optics kit has five gray-colored pieces of plastic. Take them out and look at them. One of them is a piece of circular polarizer (to be discussed later). The other four are pieces of Polaroid HN-32 linear polarizer, otherwise called simply polaroids (to be discussed now). The circular polarizer can be identified as follows: Put a dime or penny (or any shiny piece of metal) on a table. Lay one of the gray pieces of plastic on top of the dime. Look at the dime through the plastic. Now flip the plastic over and look again. Does the dime look the same? If so, it is not the circular polarizer. (The fascinating asymmetry exhibited by the circular polarizer will be discussed later.) Look at an incandescent light through the two pieces of polaroid, with the polaroids superposed face to face and held close in front of your eye. Rotate one polaroid relative to the other. When the light is essentially extinguished, the polaroids are said to be "crossed." Their easy-transmission axes are then at 90 deg to one another.

When the easy axes are parallel, most of the light that gets through the first polaroid also gets through the second. Ideally, one would get the following results: The light from the light bulb is "unpolarized"—that means that there is just as much intensity with linear polarization along a transverse direction $\hat{\mathbf{x}}$ as there is along the perpendicular transverse direction $\hat{\mathbf{y}}$ (where $\hat{\mathbf{x}}$ is *any* transverse direction—any $\hat{\mathbf{x}}$ and $\hat{\mathbf{y}}$ form a complete representation for description of polarized light). If the first polaroid had a perfect impedance-matching nonreflective coating on both surfaces, and if all the hydrocarbon chains were perfectly parallel, and if the thickness were sufficient to completely absorb the undesired polarization component, then 50% of the intensity from the light bulb would be transmitted. However, there are no nonreflective coatings on polaroids. Therefore about 4% in intensity is lost at each surface. [The index of refraction of the plastic is about the same as that of glass, i.e., about 1.5. Therefore the reflected intensity is $\left[(n-1)/(n+1)\right]^2 \approx 0.04$ from each surface. We can neglect interference effects between the two surfaces when we average over a reasonable band of colors. Thus there is a total loss of 8%.] If the hydrocarbon chains were perfectly aligned, there would be no further loss. Such Polaroid would be labeled HN-46, meaning that 46% of the incident unpolarized light is transmitted. Your polaroid is labeled HN-32, which means that about 32% of the intensity of the original 100% is transmitted through the first polaroid. That means that about 64% of the desired component of the unpolarized incident light is transmitted. (Less than 10^{-4} of the intensity of the other component is transmitted over most of the color spectrum.) If the second polaroid is parallel to the first, it will transmit about 64% of the intensity incident on it, since all the light has the correct polarization direction to be transmitted. Thus the intensity passed by two parallel HN-32 linear polarizers is about

$$I_{\text{out}} = I_{\text{in}} \times 0.32 \times 0.64 = 0.21\, I_{\text{in}}, \tag{50}$$

if I_{in} is the intensity of incident unpolarized light.

Perfect polarizer—Malu's law. By perfect polarizer we shall mean "HN-50" polaroid. (It doesn't exist, but it is easier to discuss than real polaroid.) We neglect all intensity loss due to reflections at surfaces. We assume the undesired component is completely absorbed and the desired component (that with **E** parallel to the easy axis, i.e., perpendicular to the hydrocarbon chains) is completely transmitted. If linearly polarized light is normally incident along $\hat{\mathbf{z}}$ with transverse electric field amplitude **E** and if $\hat{\mathbf{e}}$ is the direction of easy transmission of the perfect polarizer, then only the amplitude component $(\mathbf{E} \cdot \hat{\mathbf{e}})\hat{\mathbf{e}}$ is transmitted. The transmitted energy flux I_{out} is less than the incident flux I_{in} by a factor $(\mathbf{E} \cdot \hat{\mathbf{e}})^2/(\mathbf{E}^2)$:

$$I_{\text{out}} = I_{\text{in}}\, \cos^2 \theta \equiv I_{in}(\hat{\mathbf{E}} \cdot \hat{\mathbf{e}})^2, \tag{51}$$

where $\hat{\mathbf{E}} \equiv \mathbf{E}/|\mathbf{E}|$ is a unit vector along the direction of **E**. Equation (51) is often called *Malu's law*. See Fig. 8.5.

Two successive polaroids Nos. 1 and 2 with easy axes $\hat{\mathbf{e}}_1$ and $\hat{\mathbf{e}}_2$ at 90 degrees to one another are called "crossed" polaroids. The first Polaroid transmits **E** along $\hat{\mathbf{e}}_1$ and the second completely absorbs this field. No light is transmitted beyond the second polaroid. However, if a third polaroid is inserted between the crossed polaroids, the transmitted field is not zero, provided $\hat{\mathbf{e}}_3$ is along neither $\hat{\mathbf{e}}_1$ nor $\hat{\mathbf{e}}_2$. See Prob. 8.3. Also, you can prove this with the pieces of polaroid from your optics kit.

Polarization by single scattering. On a sunny day look at the blue sky through a piece of polaroid held close to one eye so that you can see a large area of sky. Rotate the polaroid and watch for a minimum that looks like a dark swath across the sky. The light coming from that part of the sky is strongly polarized. Measure (roughly) the angle between the line connecting your head and the sun and the line connecting your head and the region of greatest polarization of the blue sky. (You will find it is about 90°.) Measure the direction of polarization. (You can find the axis of easy transmission

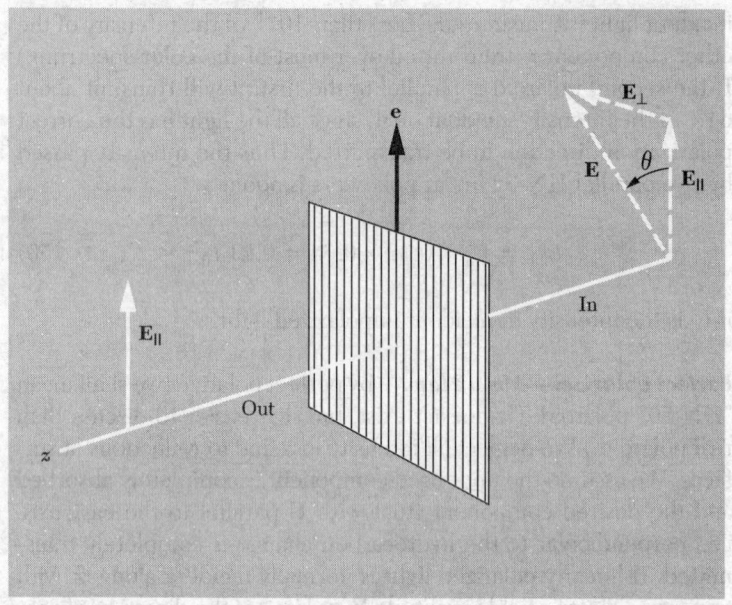

Fig. 8.5 Perfect polarizer. The axis for easy transmission of E is along ê. $E_{\parallel}$ the component of E that is parallel to ê, is transmitted. The other component, $E_{\perp}$, is completely absorbed.

of a polaroid by looking at a source having known polarization. For example, look at light reflected from a window or from a wooden or plastic floor. As we will show later in this section, the reflected light is polarized parallel to the flat surface, say the floor.)

The explanation for the polarization of the blue sky is as follows: Let $\hat{z}$ be the direction of propagation of light from the sun to a given air molecule (see Fig. 8.6). The electric field in the sunlight is unpolarized. (You may verify this by cutting a small hole in a piece of cardboard, holding the cardboard so that the sunlight makes a bright spot on the floor where it comes through the hole, covering the hole with a piece of polaroid, rotating the polaroid, and looking for a variation in the intensity of the spot on the floor. Don't look at the sun!) The electrons in an air molecule act somewhat like oscillators driven by the incident light. They therefore oscillate in a superposition of motions along $\hat{x}$ and $\hat{y}$ (the directions transverse to $\hat{z}$). The oscillating electrons radiate in all directions, but they do not radiate equally well in all directions. From our previous discussion in Sec. 7.5, we know that the amplitude and polarization direction of the electric field radiated by a single oscillating point charge are proportional to the "projected" amplitude of motion of the oscillating charge as seen by an observer who looks toward the radiating oscillating charge. By the projected amplitude of motion we mean the amplitude of that vector component of the electron's motion which is perpendicular to the propagation direction $\hat{r}$ from the oscillating charge to the observer. If $\hat{r}$ is along $\hat{y}$, then the observer sees

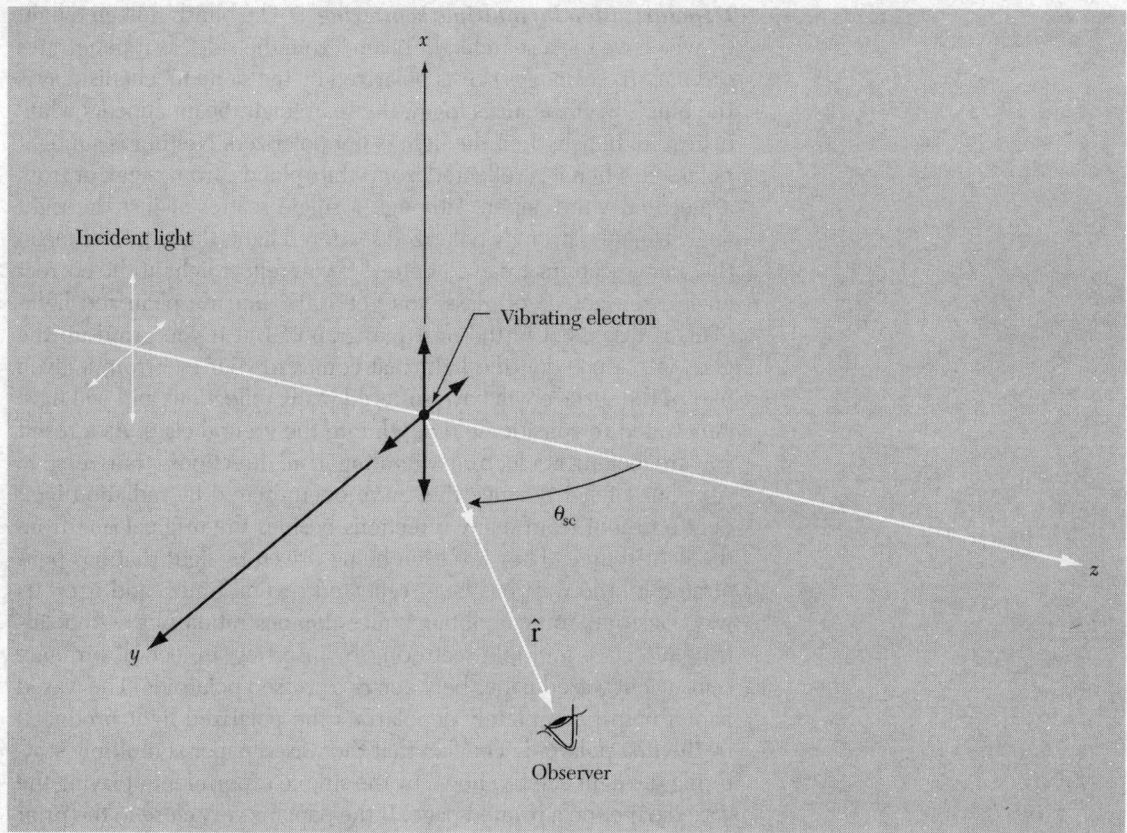

only the $\hat{\mathbf{x}}$ component of the electron's motion. He therefore observes radiation which is 100% linearly polarized along $\hat{\mathbf{x}}$. The intensity is only half what it would be if he were looking back along the $\hat{\mathbf{z}}$ axis and could see both the $\hat{\mathbf{x}}$ and $\hat{\mathbf{y}}$ motions of the electron. [In our example it is difficult to look back directly along $\hat{\mathbf{z}}$ because then we are blinded by the sun. But you can look at various angles and see that, if you look near the sun, the sky appears unpolarized. That is also the case if you look at light that required large scattering angles (as near 180 deg as you can manage) to get to your eye.] This polarization process is illustrated in Fig. 8.6.

Bees don't need polaroids to detect the polarization of the blue sky; they use it to navigate.† Some people (but not I) can also detect the polarization without polaroid; they see "Haidinger's brush."‡

Fig. 8.6 Polarization by single Scattering. The $\hat{\mathbf{y}}$ axis is chosen to lie in the plane of $\hat{\mathbf{z}}$ and $\hat{\mathbf{x}}$. The observer sees the full projected motion of the electron along $\hat{\mathbf{x}}$, but he sees a projected amplitude along $\hat{\mathbf{y}}$ that is only a fraction $\cos\theta_{\mathrm{sc}}$ of the actual $\hat{\mathbf{y}}$ motion. At $\theta_{\mathrm{sc}} = 90°$, the scattered radiation is 100% polarized along $\hat{\mathbf{x}}$.

---

†Karl von Frisch, *Bees, Their Vision, Chemical Sense, and Language* (Cornell University Press, Ithaca, N.Y., 1950).

‡M. Minnaert, *Light and Colour* (Dover Publications, Inc., New York, 1954). This is a wonderful book full of "outdoor experiments."

Depolarization by multiple scattering.   The bluish reflected light
by which we see a searchlight beam "from the side" as it penetrates
ordinary (i.e., smoggy) air is polarized by the same mechanism as is
the blue sky. If the air is foggy, the searchlight beam appears white
instead of bluish; then the light is not polarized. Neither is sunlight
polarized when it is reflected from white clouds, from sugar, or from
a piece of white paper. Although a single scatter at just the right
angle can give strongly polarized scattered light, this does not mean
that many scatters are even better! If you reflect light at the correct
angle from apiece of glass, you get 100% linearly polarized light.
(This is discussed in the next paragraph.) But if you grind up the
glass into a powder, the light that comes to your eye from a given
part of the surface will have suffered many reflections and will have
penetrated to considerable depth into the ground class. As a result
you are looking at electrons vibrating in all directions (transverse to
your line of sight), since they have been excited by radiation incident on them from many directions besides the original one from
the light source. They are even being driven by light that has penetrated a little way, has been reflected several times, and is on its
way back out. You can obtain a nice demonstration of the depolar-
izing effect of multiple scattering by inserting a piece of ordinary
translucent waxed paper between two crossed polaroids. The waxed
paper almost completely depolarizes the polarized light produced
by the first polaroid. The fact that the waxed paper is multiply scattering the light can be shown by the simple experiment of laying the
waxed paper on a printed page. If the paper is very close to the print
you can easily see the black letters. If the waxed paper is moved an
inch above the page, the letters become so blurred that they cannot be distinguished. To understand this, think of the "black light"
from a letter on the page to your eye as a little flashlight beam—this
beam gets diffused by the waxed paper. Another good experiment
is to take a flashlight beam and shine it on something through the
waxed paper. Look at the size of the transmitted spot as you move
the flashlight farther and farther behind the waxed paper. A clear
piece of glass or plastic does not multiply-scatter incident light (you
can read through it whether or not it is close to the print) and does
not depolarize.

Polarization by specular reflection—Brewster's angle.   Look at
the reflection of something in a piece of ordinary window glass or
from a smooth surface of water. Test the polarization of the reflected
light with a piece of polaroid. You will find that at an angle of incidence of about 56° (measured from the incident ray to the normal
to the surface) for glass of index $n = 1.5$, or at an angle of incidence of about 53° for water (index about 1.33), the reflected light is
100% linearly polarized parallel to the surface. This special angle of

incidence is called Brewster's angle. By rotating the polaroid to the proper position, you can completely extinguish the reflected light, if it is incident at Brewster's angle. (If you properly orient the polaroid and hold it close to one eye so as to see a broad range of angles, you will see a band of "extinction" centered at Brewster's angle.)

For any angle of incidence, the angles of the incident and refracted rays, namely θ_1 and θ_2, are related by Snell's law,

$$n_1 \sin \theta_1 = n_2 \sin \theta_2 \tag{52}$$

The incident and reflected rays make equal angles with the normal. (This is called the law of specular reflection.) Thus at the particular angle of incidence θ_1, for which $\theta_1 + \theta_2$ is 90 deg, the reflected ray makes an angle of 90 deg with the refracted (i.e., transmitted) ray, as shown in Fig. 8.7. The direction of oscillation of the electrons in the glass is transverse to the direction of the transmitted ray (since that is the direction of their driving force). For any angle of incidence, the component of electron motion perpendicular to the plane of incidence (perpendicular to the plane of the paper in Fig. 8.7) is completely "visible" to an observer looking at the reflected light (the

Fig. 8.7 Brewster's angle. The angles are drawn correctly for glass (n = 1.5). The reflected light is 100% polarized perpendicular to the plane of incidence, i.e., the plane of the incident ray and the normal. (The circled dot indicates electric field polarization into or out of the paper.)

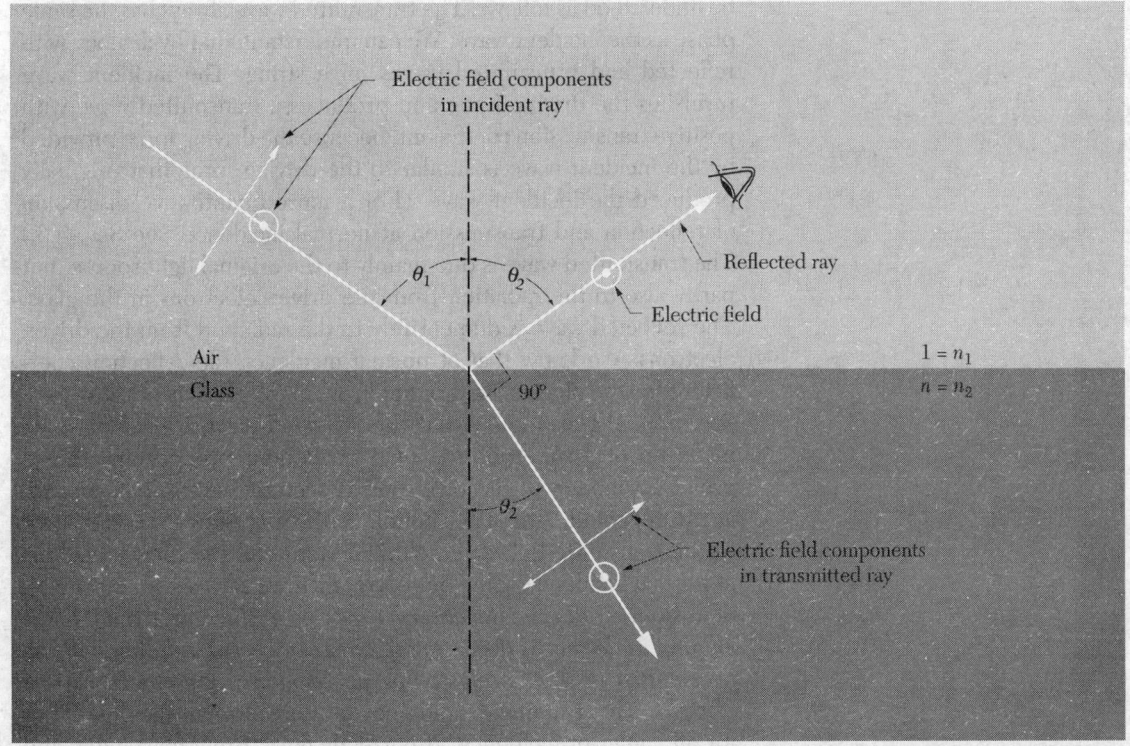

light radiated by the driven electrons in the glass), since this component of the motion is perpendicular to the direction of propagation from electron to observer (i.e., the direction of the reflected ray). However, the component of electron motion that lies in the plane of incidence is not perpendicular to the direction of the reflected ray. Only the component of motion projected perpendicular to the reflected ray contributes to the reflected radiation. At Brewster's angle of incidence, the component of the motion of the electrons in the plane of incidence is exactly along the line from the electron to the observer and contributes no reflected light. Thus the reflected light is completely polarized perpendicular to the plane of incidence. From Fig. 8.7 we see that this condition corresponds to $\theta_1 + \theta_2$ equal to 90 deg. Therefore Eq. (4) gives [using $n_1 = 1$, $n_2 = n$, and $\sin \theta_2$ equal to $\sin (90° - \theta_1)$, which is $\cos \theta_1$]

$$\tan \theta_1 = n, \quad \theta_1 = \text{Brewster's angle.} \tag{53}$$

Phase relations for specularly reflected light. The phase relations between the incident, transmitted, and reflected light are interesting. They are shown in Fig. 8.8. These phase relations can be understood as follows. The transmitted wave always has the same phase as the incident wave. We can understand that by analogy with reflected and transmitted waves on a string. The incident wave furnishes the driving force and produces a transmitted wave with positive transmission coefficient, because the driving force provided by the incident wave is similar to the driving force that originally produced the incident wave. (For a more quantitative discussion of reflection and transmission at normal incidence, see Sec. 5.3.) The transmitted wave is due mainly to the original light source, but partly also to the radiation from the driven electrons in the glass. The reflected wave is due entirely to the radiation from the driven electrons. We know that at normal incidence tlle reflection coefficient for the electric field (going from air to glass) is negative (see Sec. 5.3). We also know that the reflected electric field must be made up of a superposition of contributions proportional to the projected motion of the electrons as seen by an observer looking at the reflected light. The motion of the electrons is proportional to the transmitted electric field. Therefore all the phase relations at normal incidence come out correctly if we say that for light incident from air to glass *the observer looking at the reflected light sees an amplitude that is the negative of the projected amplitude of the transmitted field, as projected perpendicular to the observer's line of sight.* This statement holds not only for normal incidence but for all angles of incidence. It correctly gives Brewster's angle, and

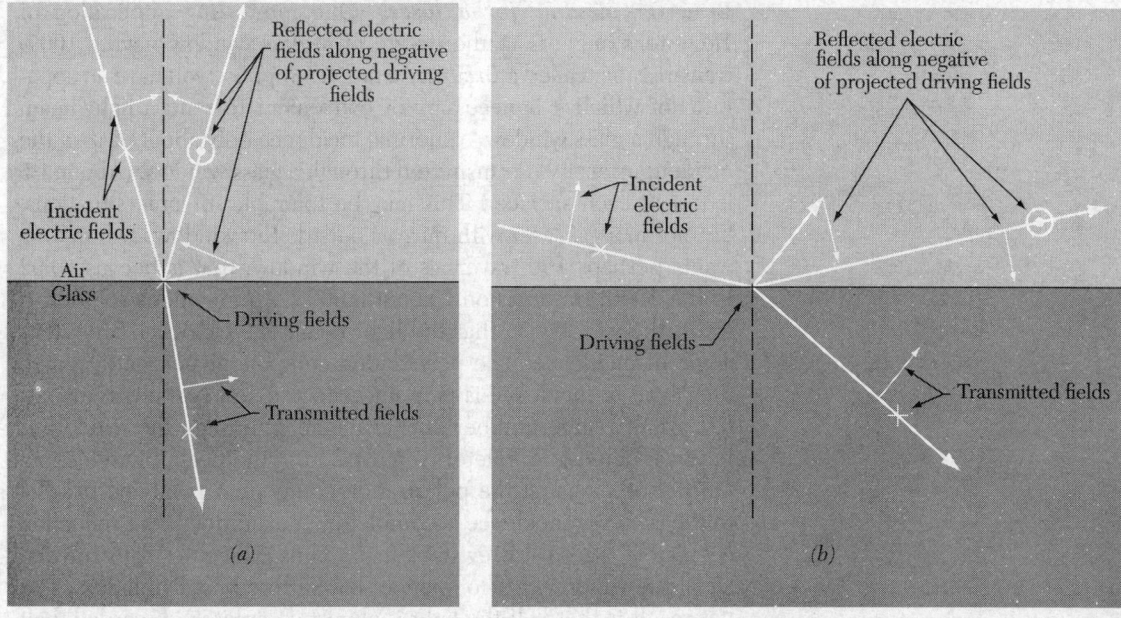

Fig. 8.8 Phase relations in light reflected from glass. (a) θ_1 less than Brewster's angle. (b) θ_1 greater than Brewster's angle. (A dot means E is Out of the paper, a cross means E is into the paper, and an arrow in the plane of the paper indicates E along the arrow.)

it gives the phase relationships for all other angles of incidence as well. (It gives the intensity only approximately.) You can easily verify the phase relations in Fig. 8.8 using two pieces of polaroid and one microscope slide. (See Home Exp. 8.26.)

Intensity relations for specularly reflected light. We shall not derive these relationships.† Using polaroids and a microscope slide, you can easily verify that the linear polarization component perpendicular to the plane of incidence is reflected with gradually increasing fractional intensity as the angle of incidence is increased from 0 deg (normal incidence) to 90 deg (grazing incidence). At normal incidence, about 4% of the intensity is reflected from a single surface; about twice that is reflected from a microscope slide having two surfaces. At grazing incidence, essentially 100% of the light is reflected. For the component polarized in the plane of incidence, the intensity reflected from the two surfaces of the slide decreases from about 8% at normal incidence to zero at Brewster's angle (56 deg) and then gradually increases to essentially 100% at grazing incidence. See Home Exp. 8.26.

---

† A beautiful derivation of these relations, called Fresnel's formulas, is given by R. Feynman, *The Feynman Lectures on Physics,* vol. I, chap. 33 (Addison Wesley, Reading, Mass., 1963).

Brewster window for a laser. One interesting application of
Brewster's angle is in the design of a glass window having 100%
transmission, called a *Brewster window.* Suppose you have an apparatus in which it is necessary or convenient to send a light beam
through a glass window. At normal incidence only about 92% of the
incident intensity is transmitted through a glass window. (About 4%
is lost at each surface.) This may be tolerable in some situations,
but not in a gas laser with mirrors outside the windows, where one
wants perhaps 100 traversals of the window; that is because 0.92
to the 100th power is only about 0.0003. An ingenious solution is
to tilt the window so that the light beam is incident at Brewster's
angle of incidence. The polarization component perpendicular to
the plane of incidence is partially reflected and partially transmitted. After a large number of transmissions through the window, it
has been almost completely removed from the beam by reflection.
On the other hand, the polarization component polarized parallel
to the plane of incidence is completely transmitted—the reflection
coefficient is zero at Brewster's angle. Thus even after many traversals of the window this component has suffered negligible loss. The
net result is that half the light is almost completely discarded, half
is almost completely retained, and the light emitted by the laser
is 100% linearly polarized. The inexpensive gas lasers that can be
found around any physics department usually have Brewster windows. Find such a laser. Test the polarization of the output with a
piece of polaroid. Turn it off and take off the lid to see the Brewster
windows. (Some lasers don't use Brewster windows—their output
is not linearly polarized.) The operation of a Brewster window is
illustrated in Fig. 8.9.

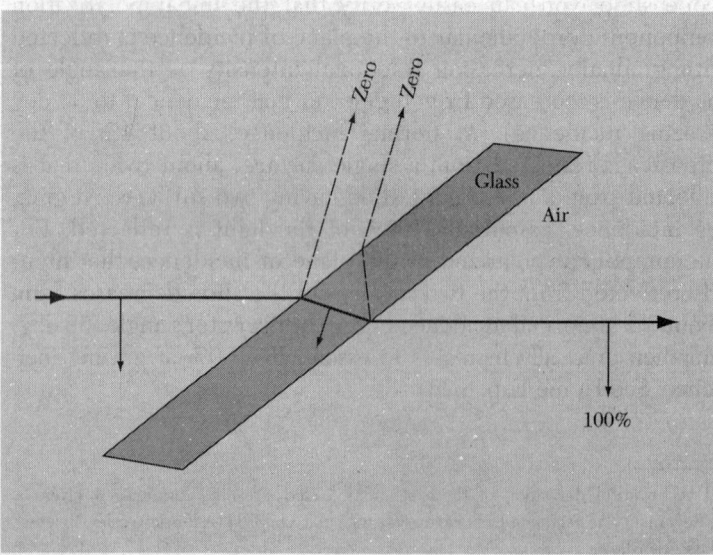

**Fig. 8.9 Brewster window. The fig-
ure is drawn for n = 1.5.**

Polarization of the rainbow.   Even more spectacular than the polarization of the blue sky is the polarization of the rainbow. It is an interesting exercise to try to predict whether the polarization is radial or tangential (with respect to the bow). If you cannot wait for rain in order to verify your prediction, get a garden hose and wait for sun (or use a source of light at night). For an explanation of how a rainbow works, see M. Minnaert, *Light and colour* (Dover Publications Inc., New York, 1954).

8.4   Double Refraction

In Sec. 8.3 we learned how to change the state of polarization of a beam of electromagnetic waves by means of selective absorption or reflection (selective in the sense that one polarization component is absorbed or reflected more than the other). In this section, we shall learn to change the state of polarization by changing the relative phase of the two components.

Cellophane.   Take two pieces of polaroid and cross them so that no light gets through. Now slip an ordinary piece of cellophane (from a candy wrapper or bread wrapper or something—almost any kind of clear plastic will do) between the crossed polaroids. Now light gets through! Since the cellophane is perfectly transparent, with none of the "dark" appearance of polaroid, we know that the cellophane cannot be absorbing light. The only way it can change the polarization of the light is to change the relative phases of different polarization components. (Then there will be no loss of intensity, as you can easily show.)

Now rotate the piece of cellophane between the polaroids, keeping the polaroids crossed. You should find two angles at 90° from one another (in 180° of rotation) where the cellophane has its greatest effect and two angles at 90° from one another where the cellophane has no effect. Thus cellophane has two special directions, oriented at 90° to one another and lying in the plane of the cellophane, which are related to the property of introducing relative phase shifts in different polarization components of light.

Just to show that not all transparent plastic has this peculiar property, find a piece of household plastic wrap (such as Saran Wrap) or a piece of the stretchy polyethylene used by dry cleaners to protect clothes. Try this between crossed polaroids. You will find that it has little effect. Not much light gets through. (There will be same effect, but it will be small compared with that of the cellophane.) Since (or if) you have now found a piece of plastic that has no "optic axes," i.e., no special directions in the plane of the sheet, try to create a special direction. Take a piece of this stretchy plastic (like Saran Wrap) and stretch it. Lay the plastic between the crossed polaroids with the stretch direction at 45° to the axes of the crossed polaroids. You should now get a huge effect.

Here is the explanation for the behavior of stretched Saran Wrap. Before the stretching, the long organic molecules of the plastic were coiled in a spaghettilike mess going in all directions. However, the molecules tend to get straightened out and aligned by the stretching. The electrons in a single, long chainlike organic molecule have different "effective spring constants" for vibrations along the hydrocarbon chain as compared with vibrations in the two directions perpendicular to it. Therefore the polarizability of the molecule is different for displacements along and perpendicular to the hydrocarbon chain. After the stretching operation, the direction of the long dimension of the molecules tends to lie along the stretching direction. One of the directions perpendicular to the stretching direction can be taken to lie in the plane of the sheet of plastic. (The other is perpendicular to the sheet and does not concern us.) The electric susceptibility (induced polarization per unit volume and per unit incident electric field) for electric fields along the stretch direction will thus be different from that for fields perpendicular to the stretch direction. Thus the dielectric constant is different for these two directions, and thus the *index of refraction is different for these two directions.*

Slow and fast axes of retardation plate. These two directions, the stretch direction and the direction perpendicular to it (and lying in the plane of the sheet), are called the *optic axes.* The optic axis yielding the larger of the two indices of refraction (for **E** directed along it) is called the *slow axis.* (Larger index of refraction means slower phase velocity.) The other optic axis is called the *fast axis.* We will call the two corresponding indices of refraction n_f and n_s (f stands for fast, s for slow), with $n_s > n_f$. A sheet of cellophane or plastic or other material having these properties is called a *retardation plate.*

Now let us examine the effect of a retardation plate on an incident traveling electromagnetic plane wave. First let us resolve the incident electric field into orthogonal components along the slow axis $\hat{\mathbf{e}}_s \equiv \hat{\mathbf{x}}$ and the fast axis $\hat{\mathbf{e}}_f \equiv \hat{\mathbf{y}}$. Suppose that there is vacuum for $z < 0$, and the retardation plate begins at $z = 0$ and extends to $z = \Delta z$, after which we again have vacuum. Suppose that the oscillations of the electric field of the incident wave at $z = 0$ are given by the real part of the complex quantity

$$\mathbf{E}_c(0, t) = e^{i\omega t}[\hat{\mathbf{x}} A_s e^{i\varphi s} + \hat{\mathbf{y}} A_f e^{i\varphi f}]. \qquad (54)$$

The amplitudes A_s and A_f and phase constants φ_s and φ_f are those we get when we resolve the incident electric field into linearly polarized components along $\hat{\mathbf{x}}$ and $\hat{\mathbf{y}}$. (Since these amplitudes and phase constants are arbitrary, Eq. (54) represents a general polarization.) Now consider the transmitted wave inside the retardation plate, between $z = 0$ and Δz. We neglect any loss due to reflection at the first surface and then merely replace ωt by $\omega t - kz$ in Eq. (54). But we must remember that k is not the same for $\mathbf{E}$ along $\hat{\mathbf{e}}_s$ as it is for $\mathbf{E}$ along $\hat{\mathbf{e}}_f$. Thus, recalling that k is proportional to the index of refraction and is given by $n\omega/c$, we have inside the retardation plate

$$\mathbf{E}_c(z, t) = e^{i\omega t} [\hat{\mathbf{x}} A_s e^{i\varphi_s} e^{-in_s\omega z/c} + \hat{\mathbf{y}} A_f e^{i\varphi} e^{-in_f\omega z/c}]. \tag{55}$$

Relative phase retardation. By the time the wave has reached the exit of the plate at $z = \Delta z$, each component has suffered a phase retardation relative to the phase it would have had if the plate had been vacuum (with $n = 1$). For the s component, this is given by $(n_s - 1)\omega \Delta z/c$:

$$\text{Phase retardation of } E_s \text{ relative to vacuum} = (n_s - 1)\frac{\omega\Delta z}{c}. \tag{56}$$

Similarly, we have

$$\text{Phase retardation of } E_f \text{ relative to vacuum} = (n_f - 1)\frac{\omega\Delta z}{c} \tag{57}$$

Subtracting Eq. (57) from Eq. (56), we find the retardation in phase of E_s relative to E_f:

$$\text{Phase retardation of } E_s \text{ relative to } E_f = (n_s - n_f)\frac{\omega\Delta z}{c}$$

$$= (n_s - n_f)2\pi \frac{\Delta z}{\lambda_{\text{vac}}}, \tag{58}$$

where λ_{vac} is the wavelength in vacuum.

Quarter-wave plate. Consider the following example, which should help you keep the signs straight. Suppose incident linearly polarized light has $\mathbf{E}$ along the 45-deg line between $\hat{\mathbf{e}}_s$ and $\hat{\mathbf{e}}_f$. Then A_s and A_f are equal, and φ_s and φ_f are equal. Suppose the thickness of the plate is such that the slow component suffers a retardation of $\frac{1}{4}$ cycle relative to the fast component, i.e., it suffers a phase retardation of $\pi/2$ relative to the fast component. Such a retardation plate is called a *quarter-wave plate*. The wave emerging at the rear of the plate has equal amplitudes for the slow and fast components,

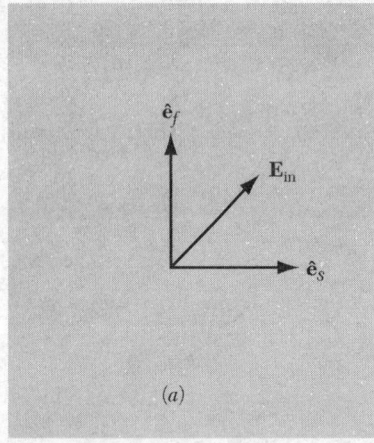

(a)

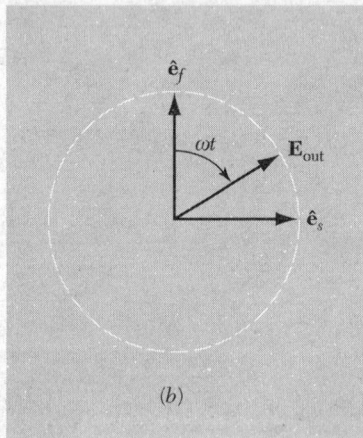

(b)

Fig. 8.10 Quarter-wave plate. Incident linearly polarized light with E at 45 deg to each optic axis. (a) Input. (b) Out-put. This result is obtained for propagation direction either into or out of the paper.

and the fast component leads the slow by 90 deg in phase. This means that we have circularly polarized light, with the sense of rotation from $\hat{\mathbf{e}}_f$ toward $\hat{\mathbf{e}}_s$. These results are implied by Eq. (55). They are shown in Fig. 8.10.

Getting the signs straight is crucial to your understanding of retardation plates. Here is another phrasing of the argument that may help convince you that the sense of rotation is as shown in Fig. 8.10 for a quarter-wave plate. If the two components of polarization are traveling in vacuum, then at any given location z and time t the oscillation along $\hat{\mathbf{x}}$ and the oscillation along $\hat{\mathbf{y}}$ both correspond to the same earlier retarded time of emission at the light source. These two polarization components are now passed through a slab with n_s greater than n_f. At the output of the slab, the instantaneous value of E_s must have been emitted at an earlier retarded time than the simultaneous instantaneous value of E_f at the same place (the back of the slab). This is because the traveling wave that carries E_s has traveled the same distance as that which carried E_f, but it has done so at slower phase velocity; therefore it must have started sooner. Thus E_f corresponds to a more recent time of emission and therefore to a greater advance in phase of the oscillation than E_s, and thus E_f *leads* E_s. These phase relations are illustrated in Fig. 8.11.

Properties of retardation plates. You should convince yourself of the truth of the following numbered statements and "rules." (You should not memorize them. You should understand them so well that you can forget the answers and figure them out whenever you want to.)

(i) A half-wave plate (one twice as thick as a quarter-wave plate) converts linearly polarized light into linearly polarized light, with the direction of polarization of the output obtained from that of the input by reflection in one of the optic axes. (We are almost never concerned with which axis, i.e., with the absolute phase. We don't care about a minus sign in the direction of the amplitude.) That is, a half-wave plate reverses the *relative* sign of the linear components of the incident amplitude.

(ii) A half-wave plate converts right-handed circularly polarized light into left-handed circularly polarized light and vice versa.

(iii) A quarter-wave plate converts linearly polarized light with polarization lying somewhere between $\hat{\mathbf{e}}_s$ and $\hat{\mathbf{e}}_f$ into elliptically polarized light with sense of rotation from $\hat{\mathbf{e}}_f$ to $\hat{\mathbf{e}}_s$. If the incident polarization is at 45 deg to $\hat{\mathbf{e}}_s$ and to $\hat{\mathbf{e}}_f$, the output polarization is circular. (*Note:* This means that if we rotate the linear polarization of $\mathbf{E}_{in}$ in Fig. 8.10 by 90 deg, the output will rotate at frequency ω in the opposite sense to that shown in Fig. 8.10. To use the "rule" to see this, simply reverse the sign convention on one or the other of

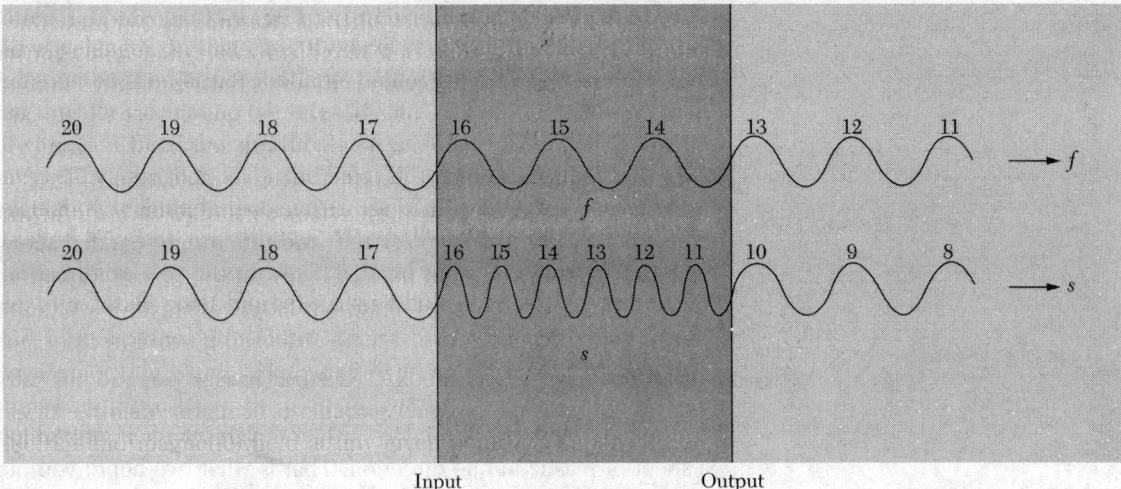

Input   Output

Fig. 8.11 Relative phase retardation of slow and fast polarization components. The integers give the emission time at the light source. At the input to the re-tardation plate, the two polarization components are shown as having the same emission time. At the output, the slow component emitted at cycle 10 is present at the same time as the fast component emitted at cycle 13. The fast component leads the slow component by three full cycles.

$\hat{\mathbf{e}}_f$ or $\hat{\mathbf{e}}_s$ in Fig. 8.10, so that E_{in} will again lie between the $\hat{\mathbf{e}}_s$ and $\hat{\mathbf{e}}_f$ unit vectors. Then the rule says that the rotation of the output is from $\hat{\mathbf{e}}_f$ to $\hat{\mathbf{e}}_s$.)

(iv) A quarter-wave plate converts circularly polarized light into linearly polarized light. To obtain a simple rule, label the signs of the slow and fast axes so that the sense of rotation of the incident circularly polarized light is from fast axis to slow axis. Then the quarter-wave plate converts the circularly polarized light into linear polarization at 90 deg to the direction lying midway between $\hat{\mathbf{e}}_s$ and $\hat{\mathbf{e}}_f$. (The f vibration already led in phase by $\frac{1}{4}$ cycle. After the quarter-wave plate, it leads by $\frac{1}{2}$ cycle.)

(v) A retardation plate has no influence on the state of polarization of linearly polarized incident light with **E** along either $\hat{\mathbf{e}}_s$ or $\hat{\mathbf{e}}_f$.

(vi) A retardation plate cannot convert "unpolarized" light (the kind you get directly from a light bulb or the sun) into polarized light. We shall study unpolarized light in Sec. 8.5. For now, we just say rather vaguely that for unpolarized light there is a "random" phase relation between the x and y components when you average over the observation time interval. The relative phase shift introduced by the retardation plate still leaves the relation of x and y phases as random as before, i.e., if φ_x and φ_y are randomly related, so are φ_x and $\varphi_y + \Delta\varphi$.

(vii) A circular polarizer is obtained if we make a "sandwich" by gluing together (face to face) a piece of polaroid and a piece of quarter-wave plate with its optic axes at 45 deg to the easy-transmission direction of the polaroid. The unpolarized light must be incident on the polaroid face of the sandwich.

(viii) A circular polarizer that produces right-handed circularly polarized light will transmit with 100% efficiency (neglecting small losses by reflection) right-handed circularly polarized light traveling in the reverse direction (i.e., incident on the quarter-wave plate face of the sandwich.) It will completely absorb left-handed circularly polarized light incident on the quarter-wave plate face. (This fact is easily remembered by analogy with a die and a screw. A die that converts a cylindrical "unpolarized" rod into a right-handed screw will also "transmit" 3 right-handed screw in the reverse direction, but it will completely annihilate a left-handed screw traveling in the reverse direction.) This fact has interesting consequences. See Home Exp. 8.18.

We have been considering retardation plates made by stretching a sheet of plastic in one direction. That is what (we hope) you did with a piece of Saran Wrap. That is the way Polaroid Corporation makes the quarterwave and half-wave plates that are included in your optics kit. That is the way cellophane acquires its properties (it comes out between rollers that squeeze it and line up the molecules). An ordinary piece of window glass is isotropic and exhibits no double refraction (i.e., has no optic axis). But if you look at a piece of stressed plate glass between crossed polaroids, you will see transmitted light in some places. Safety glasses are highly stressed and exhibit interesting patterns of double refraction. Plastic drawing triangles and dishes show beautiful colored stress patterns when put between crossed polaroids. The color effects are partly due to the variation of the index of refraction with color (i.e., with wavelength), but mostly they are due to the variation of the phase shift with wavelength.

Most crystalline materials exhibit double refraction If (like the stretched plastic) they have only one direction of anisotropy, they are called *uniaxial*. The direction along the axis of anisotropy is called the "extraordinary" direction The other two. directions perpendicular to the anisotropy axis are called "ordinary" directions. The corresponding indices of refraction are called n_e and n_o (e for extraordinary and o for ordinary) for electric field along the e and o directions. The anisotropy axis can be either a fast axis or a slow axis, depending on the crystal structure. Table 8.1 shows some examples, with indices of refraction for light of wavelength 5890 Å (yellow light emitted by excited sodium atoms).

Table 8.1 Some uniaxial crystals

Material	n_e	n_o	e axis
Quartz	1.553	1.554	Slow
Calcite	1.486	1.658	Fast
Ice	1.307	1.306	Slow

Optical activity. Here is an interesting home experiment. Fill a glass-bottomed jar or glass (*not* a plastic one) with about two inches of Karo corn syrup (obtained at any grocery store). Put a light source under the jar, a piece of polaroid under the jar, and one above it. Now look through the syrup. You will see beautiful color effects. Now investigate quantitatively: Use the red or green gelatin filter in your optics kit in order to work with a reasonably small band of wavelengths. (You may look at a light bulb using your Edmund diffraction grating from the optics kit with the filter either in or out to see what sort of band of colors you are using.) Vary the depth of the syrup. You should find that the linearly polarized light remains linearly polarized, but the direction of polarization is rotated by about 30° clockwise (as you face the light) per inch of syrup. This phenomenon is called *optical activity*.

Here is the explanation. The linearly polarized light produced by the first polaroid is a superposition of equal amounts of right-and left-handed circularly polarized light (see Fig. 8.12):

$$\mathbf{E}_c = E_0 \hat{\mathbf{x}} e^{i\omega t} = \frac{E_0}{2} \left[\hat{\mathbf{x}} e^{i\omega t} + \hat{\mathbf{y}} e^{i[\omega t - (1/2)\pi]} \right]$$
$$+ \frac{E_0}{2} \left[\hat{\mathbf{x}} e^{i\omega t} + \hat{\mathbf{y}} e^{i[\omega t + (1/2)\pi]} \right] \tag{59}$$

The sugar molecules have a helical structure. All sugar molecules made from corn have the same handedness. A helix has the same handedness whether viewed from one end or the other. Therefore a solution of randomly oriented sugar molecules has a net handedness like that of the individual molecules. Because of the helical structure of the molecules, the sugar solution has different indices of refraction for traveling waves with right-handed and with left-handed circular polarization. As the linearly polarized wave progresses through the sugar solution, one of the circularly polarized components gets ahead of the other in phase. A little thought and a sketch should convince you that the direction of rotation of the linear polarization is the same as the direction of rotation of the fast circular component (the fast one being the one with the smaller index of refraction). Here is food for thought: what happens if we

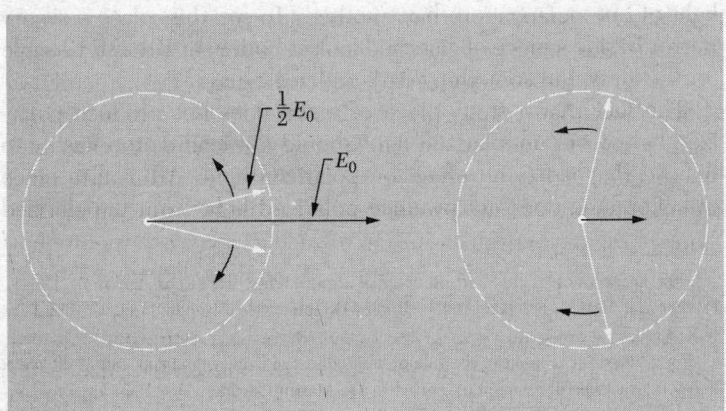

Fig. 8.12 *Linearly polarized oscillation of amplitude E_o is a superposition of left- and right-handed circularly polarized oscillations each having amplitude $\frac{1}{2} E_0$. The direction of the linearly polarized superposition depends on the relative phases of the circular components.*

send light through a sugar solution, reflect it from a mirror, and send it back through the solution in the opposite direction? Does the rotation get doubled? Or is it returned to zero?

Pasteur's first great discovery.   Louis Pasteur's first great discovery was that racemic acid, an optically inactive form of tartaric acid, is an equal mixture of left-handed and right-handed tartaric acid. He succeeded in recognizing under a microscope the right-handed and left-handed crystals in the racemic mixture, and he separated the crystals into two piles using a fine tweezer. when dissolved in water, one pile rotated the plane of polarized light in the same direction as does natural tartaric acid produced by grapes. The other pile rotated the polarization by an equal amount in the opposite direction. That kind of tartaric acid had never been seen in the world before! †

The observed single-handedness of helical organic molecules made by present-day living organisms is undoubtedly a fundamental clue for the unraveling of the history of the evolution of life on our planet. All present **DNA** molecules (the stuff of life) are right-handed helixes! why? Because of original chance? Did the oceans once contain equal amounts of right-handed and left-handed primitive **DNA?** Did the right-handed **DNA** learn how to eat up the left-handed kind? Nobody knows, yet. ‡

Metallic reflection.   After observing the strong polarization (which is 100% at Brewster's angle) obtained by specular reflection from a dielectric material like glass or water, one finds it something of a shock when looking for polarization upon reflection from ordinary aluminized or silvered mirrors (or any other such silvery-looking substances, as automobile chrome trim or a table knife) to discover essentially zero polarization. That is because a silvery-looking metal reflects both polarizations almost completely. That is why it is silvery looking; it would be darker if it were polarizing by reflecting less light of one polarization than another. (To see this, place a silvery mirror beside a piece of glass and look at both near Brewster's angle for the glass. Put something dark under the glass.)

The fact that a shiny piece of metal does not produce polarized light from unpolarized light should not lead us too hastily to believe that it has no effect on polarized light. After all a piece of cellophane does not produce polarized light from unpolarized

---

† For an account of this and other great experiments of Pasteur, see Rene Dubos, *Pasteur and Modern Science* (Anchor Books, Doubleday & Company, Inc., Garden City, N.Y., 1960).

‡For a beautiful account of the role of handedness in living organisms and in the weak decay interactions of elementary particles, see Martin Gardner, *The Ambidextrous Universe* (Basic Books, Inc., Publishers, New York, 1964).

incident light, but it can change the polarization state of incident polarized light. That is exactly what a piece of shiny metal does. You can verify this in an easy home experiment by which you convert linearly polarized light into circularly polarized light by metallic reflection. See Home Exp. 8.28.

8.5 *Bandwidth, Coherence Time, and Polarization*

In this section, we shall discuss the polarization of light emitted by atoms. We shall use a classical picture of an electron bound to a heavy nucleus. In this picture, the electron oscillates and emits classical electromagnetic waves, just as if the atom were a little radio antenna. This classical picture omits the "graininess" of the emission of light, i.e., it omits the fact that light is emitted and absorbed in "lumps" called photons. Aside from that, the classical picture gives many of the same results as does sophisticated quantum theory. The main difference is that in the classical theory we think of electromagnetic waves as carrying a continuous energy flux, whereas in quantum theory we learn that the energy flux is not continuous. However, Maxwell's equations (the equations of classical electromagnetic theory) do give the right predictions for the *average* energy flux. Classically we think of the electric and magnetic fields of the electromagnetic radiation as being quite "tangible," with their squares giving the "actual" energy density in the wave. Quantum theory reinterprets the classical energy density as being the average number of photons times the energy of one photon. (When the average number of photons in a given volume is less than unity, it is called the probability of finding one photon.) You will study quantum theory in Volume IV. We only give you these remarks so as to assure you that the results we shall obtain with a classical picture still hold in quantum theory, after suitable reinterpretation of energy flux as probability flux times photon energy.

Classical atom emitting polarized radiation. Let us consider a single classical atom situated at $x = y = z = 0$. The electron may be oscillating in a superposition of motions along $\hat{\mathbf{x}}$, $\hat{\mathbf{y}}$, and $\hat{\mathbf{z}}$. The observer of the emitted light is situated somewhere far away on the positive z axis. Only the x and y components of the electron's motion contribute to the electromagnetic waves observed (the light).

Suppose that at $t = 0$ the electron is excited into vibration, perhaps by a collision. After $t = 0$ the electron vibrates freely at a natural frequency ω_0. The polarization state of the emitted radiation depends on the amplitudes of the x and y components of the motion and on the relative phase of the x and y motions. The electron does not vibrate forever. It loses energy by radiation, with a *mean decay time* τ (the time for the energy to decrease by a factor e, also called *mean life*). After several mean lives, the electron has lost most of its energy, and its subsequent radiation is negligible. During the time (of order τ) that it radiates, its x and y motions retain constant relative phase. (They both oscillate with the same frequency, and we assume the atom is undisturbed during this interval.) Therefore the polarization of the emitted radiation is constant during this time interval.

At a later time, the atom may suffer a second collision, again exciting the electron into a motion that is a superposition of oscillations along $\hat{\mathbf{x}}$, $\hat{\mathbf{y}}$, and $\hat{\mathbf{z}}$, all with the same natural frequency ω_o, and with amplitudes and phase constants that depend on the circumstances of the collision. If the atom is in a gas and is being bombarded uniformly from all sides, we may assume that there is practically no correlation between the x and y amplitudes and phases in successive excitations. Therefore the polarization state of the radiation emitted during the second time interval (of order τ) following the second excitation bears no relation to that emitted following the first excitation.

Duration of polarization state. Now suppose that instead of one atom we have many atoms excited at any one instant. They are all located in a small region near $x = y = z = 0$, and the observer out on the z axis sees an electromagnetic wave that is a superposition of the waves emitted by the individual atoms. Let us call by the name "instant" a time interval which is short compared with the mean decay time τ but which nevertheless contains many oscillations at frequency ω_0. Suppose the observer describes the radiation in terms of the amplitudes for E_x and for E_y and the relative phase between E_x and E_y. At any "instant," E_x is a superposition of contributions from all the atoms radiating at that instant. The same is true of E_y. All atoms oscillate with the same dominant frequency ω_0, but with different amplitudes and phase constants. Therefore the superposition E_x has dominant frequency ω_0 and has amplitude and phase constant that depend on the amplitudes and phase constants of all the contributing atoms. (The same statement holds for E_y.) During any time interval that is short compared with τ, all the atoms that are vibrating lose only a small fraction of their energy, and they maintain the same phase constants. Therefore the amplitude and phase constant of the superposition that gives E_x (or E_y) do not change substantially during a time interval short compared with τ. *The polarization state of the total electromagnetic wave remains constant during a time interval short compared with τ.* In particular, the relative phase of E_x and E_y remains constant. Now suppose we wait for several mean lives τ, and then reexamine the state of polarization of the total wave. After a long interval of many mean lives, the atoms that were radiating (at the beginning of the interval) have decayed to zero and have been replaced by new atoms. (It does not matter what fraction of the "new" atoms are old atoms that have been reexcited.) The motion of the new atoms has no correlation with that of the old, except that we may assume for simplicity that the average excitation energy is about the same for the new atoms as for the old. When we add the x component of radiation from all the atoms, we get the x component E_x of the total wave. It should have about the same amplitude as the E_x obtained from the old set of atoms. But the phase constant of the new E_x is completely unpredictable from the phase constant of the old E_x. The same statement holds for E_y. Furthermore, since the relative phase of the x and y motions of the new set of atoms does not have any correlation with the relative phase of the x and y motions for the old set, we see that *the relative phase of E_x and E_y "drifts" in a completely unpredictable and "random" manner over time intervals long compared with τ.*

We have assumed that the electron in the atom oscillates freely during the decay time τ. We have also assumed that the atom is at rest. Then the Fourier spectrum of the radiation emitted by a single atom has a bandwidth $\Delta\omega$ that is about equal to τ^{-1}. Typical mean decay times are about 10^{-8} sec for atoms emitting visible light, giving a bandwidth about 10^8 rad / sec.) In a gas-discharge tube the atoms are not at rest but travel at velocities of order 10^5 cm/sec. This velocity gives a Doppler shift, whose sign depends on whether the atom moves toward or away from the observer The Doppler "broadening" gives a bandwidth that is greater than the "natural" width τ^{-1} by a factor of order 100. Furthermore, collisions often cut short the radiation process before the atom has had a chance to decay. In that case the bandwidth is further increased by "collision broadening."

Coherence time. When all factors of "frequency broadening" have been taken into account, we finally arrive at some bandwidth $\Delta\omega$, which may be much greater than τ^{-1}. In that case, the time during which the polarization state may be taken to be roughly constant is not the natural decay time τ but is what we shall call the *coherence time* t_{coh}, given by

$$t_{\text{coh}} \approx \frac{1}{\Delta\nu}. \qquad (60)$$

We can understand Eq. (60) as follows. The state of polarization remains essentially unchanged as long as the relative phase of E_x and E_y drifts by an amount that is small compared with 2π. Therefore the coherence time t_{coh} is given approximately by the time that it takes the maximum and minimum frequencies of the band to get out of phase by 2π:

$$\Delta\omega t_{\text{coh}} \approx 2\pi, \qquad (61)$$

which is the same as Eq. (60).

The fact that there is a finite bandwidth $\Delta\omega$ does not necessarily mean that the polarization will change after a time interval of order $(\Delta\nu)^{-1}$. For example, there may be a piece of Polaroid between the radiating atoms and the observer. In that case, the x and y components of the radiation seen by the observer maintain a constant phase relation, although the bandwidth is still $\Delta\nu$. That is because the x and y components are no longer "independent." The polaroid has, so to speak, "examined" the x and y components of the incident radiation and at any instant has "selected for transmission" only those parts of the incident x and y components which superposed with the proper phase relation to drive the electrons in the polaroid along the easy direction, transverse to the "wires." Those parts of the incident x and y radiation which had a relative phase that drove electrons "along the wires" were absorbed.

Here is another example. Suppose we have two identical gas-discharge light sources giving light of the same dominant frequency ω_0, same bandwidth, and same average intensity. With a suitable slab of glass or a "halfsilvered" mirror, we can make an arrangement such that the two sources appear to the observer to be superposed on top of one another (i.e., their images are superposed). Light from each source ends up going in the $+z$ direction to the observer. Now we put a piece of Polaroid over each source, so that one source gives radiation linearly polarized along $\hat{\mathbf{x}}$, and the other gives radiation linearly polarized along $\hat{\mathbf{y}}$ (after both light beams are finally going in the $+z$ direction). If the observer makes a measurement of polarization during a time interval short compared with the coherence time $(\Delta\nu)^{-1}$, he will find some definite polarization state. If he waits for a time interval long compared with $(\Delta\nu)^{-1}$ and makes another measurement of polarization, he will find a polarization state that is completely uncorrelated with the previous measurement. As a matter of fact, the observer will find it essentiany impossible to distinguish this radiation from that which he would get by removing one of the sources and at the same time removing the piece of polaroid from the remaining source.

Unpolarized radiation defined. We are now prepared to say what is meant by "un polarized" light. Unpolarized light is light whose two polarization components (x and y, or right- and left-handed) are emitted "independently" (i.e., are not locked in phase, as by a piece of polaroid) and whose amplitudes and relative phase for the two polarization components have been measured by a technique that averages over a time interval long compared with the coherence time $(\Delta\nu)^{-1}$. There is no such thing as "intrinsically" unpolarized light. All you have to do to change "unpolarized" light into "completely polarized" light is to invent a technique that allows you to measure the polarization before the phases have a chance to drift.

Measurement of polarization.   A quantitative description of the "amount of polarization," which means the amount of correlation between phases and amplitudes that is maintained during the measurement interval, can be given as follows: Suppose we represent the instantaneous state of polarization by E_1, E_2, φ_1, and φ_2 in the linear polarization representation, where **E** is the real part of

$$\mathbf{E}_c = e^{i(\omega_0 t - k_0 z)}(\hat{\mathbf{x}}E_1{}^{ei\varphi_1} + \hat{\mathbf{y}}E_2{}^{ei\varphi_2}). \tag{62}$$

We could write this as a continuous Fourier superposition of exactly harmonic waves occupying a small band of frequencies; but we can equally well think of Eq. (62) as an almost harmonic wave with dominant frequency ω_0 and with amplitudes and phase constants E_1, E_2, φ_1, and φ_2 that are not quite constant in time but vary slowly (in some unpredictable manner).

Now let us see how to express E_1, E_2, φ_1, and φ_2 in terms of measurements of intensity only. (We use the word *intensity* as a synonym for energy flux.) That is the kind of measurement that is easiest to make. We imagine that we have polaroids and quarter-wave plates and also a photomultiplier with which we can measure the photon flux (number of photons per unit area per unit time) for any experimental setup. The average photon flux is proportional to the average classical energy flux. The average classical energy flux is proportional to the average over 'one cycle of the square of the electric field. We assume we have a photomultiplier with a photocathode of known area and with known detection efficiency and that we can therefore determine the time-averaged square of the electric field in the light beam incident on the photocathode.

Measurement time.   Let T be the total time duration of all the measurements about to be described; we shall call it the *measurement time*. It is the time during which we shall determine *all* of the interesting constants E_1, E_2, φ_1, and φ_2. We want to complete the measurement before the state of polarization has a chance to change. Therefore we assume that the measurement time T is small compared with the coherence time $\Delta \nu^{-1}$. Our description will seem to be that of a rather leisurely experiment that takes "all day"; that is not necessarily the case. We should (by the exercise of sufficient ingenuity) be able to arrange things so that we measure everything simultaneously. In that case, the basic limitation on the measurement time T should be the instrumental resolving time. If the instrument is a typical photomultiplier, the resolution time is about 10^{-9} sec. Therefore we should be able to measure the "instantaneous" polarization state for radiation with coherence time longer than 10^{-9} sec, say 10^{-8} sec.

Measurements for four constants. We must leave something for the experimenter to do. Therefore we shall not specify exactly how he is to accomplish the measurements (now to be described) in a time T of order 10^{-8} sec. Here then is a (leisurely) procedure for finding the four constants that describe the light beam, given that we already know its frequency and direction. We assume that we have in addition to the calibrated photomultiplier a perfect polaroid (or a calibrated one) and a quarterwave plate. The procedure is as follows:

1. Put the polaroid in front of the photomultiplier. Choose arbitrary transverse axes $\hat{\mathbf{x}}$ and $\hat{\mathbf{y}}$. Rotate the axis of easy transmission to $\hat{\mathbf{x}}$. Measure the time-averaged photon counting rate. The result gives

$$\left\langle E_x^2 \right\rangle = \frac{1}{2} E_1^2. \tag{63}$$

2. Rotate the polaroid to $\hat{\mathbf{y}}$ and measure the counting rate. This gives

$$\left\langle E_y^2 \right\rangle = \frac{1}{2} E_2^2. \tag{64}$$

3. Rotate the polaroid to the 45-deg direction between $\hat{\mathbf{x}}$ and $\hat{\mathbf{y}}$. Call this direction $\hat{\mathbf{e}}$. Then we have that the unit vector $\hat{\mathbf{e}}$ is given by

$$\hat{\mathbf{e}} = \frac{\hat{\mathbf{x}} + \hat{\mathbf{y}}}{\sqrt{2}}. \tag{65}$$

The electric field component transmitted by the polaroid is the dot product of $\hat{\mathbf{e}}$ with the incident field $\mathbf{E}$. Using the complex $\mathbf{E}_c$, we have from Eqs. (65) and (62)

$$\hat{\mathbf{e}} \cdot \mathbf{E}_c(z, t) = e^{i(\omega_0 t - k_0 z)} \left(\frac{E_1}{\sqrt{2}} e^{i\varphi_1} + \frac{E_2}{\sqrt{2}} e^{i\varphi_2} \right). \tag{66}$$

The transmitted photon flux now gives us a measurement of

$$\left\langle (\hat{\mathbf{e}} \cdot \mathbf{E})^2 \right\rangle = \frac{1}{2} \left[\frac{1}{2} E_1^2 + \frac{1}{2} E_2^2 + E_1 E_2 \cos(\varphi_1 - \varphi_2) \right]. \tag{67}$$

Since we have already determined E_1^2 and E_2^2 (E_1 and E_2 are positive real numbers) from Eqs. (63) and (64), we see that Eq. (67) gives us $\cos(\varphi_1 - \varphi_2)$.

We still need $\sin(\varphi_1 - \varphi_2)$ to pin down the relative phase. (We are not usually interested in the absolute phases.) To get it we use the quarter-wave plate:

4. Leave the polaroid along $\hat{\mathbf{e}}$ at 45 deg to $\hat{\mathbf{x}}$ and $\hat{\mathbf{y}}$. Then Eq. (66) gives the transmitted field. Now insert a quarter-wave plate *ahead* of the polaroid with its slow axis along $\hat{\mathbf{x}}$ or $\hat{\mathbf{y}}$. For definiteness, suppose the slow axis is along $\hat{\mathbf{y}}$. Then $\mathbf{E}_c$ as given by Eq. (62)

has φ_2 replaced by $\varphi_2 - \frac{1}{2}\pi$. (Also, both φ_1 and φ_2 have acquired an uninteresting constant, which we drop.) Consequently, $\hat{\mathbf{e}} \cdot \mathbf{E}_c$ as given by Eq. (66) has φ_2 replaced by $\varphi_2 - \frac{1}{2}\pi$. Now measure the photon flux behind the system of quarter-wave plate plus polaroid. The flux is given by an expression like Eq. (67) with φ_2 replaced by $\varphi_2 - \frac{1}{2}\pi$. Thus we determine

$$\left\langle (\hat{\mathbf{e}} \cdot \mathbf{E})^2 \right\rangle = \frac{1}{2}\left[\frac{1}{2}E_1{}^2 + \frac{1}{2}E_2{}^2 - E_1 E_2 \sin(\varphi_1 - \varphi_2) \right]. \quad (68)$$

Thus we have completely determined E_1, E_2, and $\varphi_1 - \varphi_2$ by measurements represented by Eqs. (63), (64), (67), and (68). These are the results one gets if the measurement time T is short compared with the coherence time.

As we mentioned earlier, if the light beam is put through a polarizer (such as a piece of polaroid or a circular polarizer) before it reaches the detecting equipment, then the polarization coherence time is not so short as $\Delta \nu^{-1}$. Instead, the polarization coherence time is infinite (at least if no one removes the piece of polaroid). Then you can make the leisurely measurements implied by the above description; you can also use your eye instead of a photomultiplier. You should practice using your optics kit to determine the polarization state of a light source of unknown polarization. If the source is linearly polarized or circularly polarized or elliptically polarized and has coherence time longer than the few minutes it takes you to make the measurement, then you can completely determine the polarization state using your eye, a polaroid, and a quarter-wave plate. (You can also make use of the circular polarizer and half-wave plate.)

Our description of a general measurement of polarization was more general than is necessary in many practical cases. For example, if the light turns out to be linearly polarized along a certain transverse direction, then it is silly to use general Cartesian directions $\hat{\mathbf{x}}$ and $\hat{\mathbf{y}}$. As soon as you find that the light is linearly polarized, you will naturally mentally orient $\hat{\mathbf{x}}$ along the polarization direction. In that case, the relative phase $\varphi_1 - \varphi_2$ is irrelevant, since the amplitude along $\hat{\mathbf{y}}$ is zero. Similarly, if you find for example that the light is circularly polarized and is right-handed, then it is silly to use the linear polarization representation (used in the general description above) to describe the light.

Circular polarizer. The circular polarizer in your optics kit is a sandwich consisting of a piece of linear polarizer glued to a quarter-wave plate. The easy axis of the polaroid is at 45 deg to the optic axes of the quarter-wave plate. The "input" end is the linear-polarizer face of the sandwich. The ouput end is the quarter-wave face. If you shine unpolarized light from a light bulb at the input end, you get out *left-handed* light (by the screwy optics convention). Thus the polarizer has absorbed all of the light except that which corresponds to motion of electrons in circles that would look *counterclockwise* to you (if you could see them) as you look at the light bulb. If you add your half-wave plate behind the output end, you will convert this left-handed (screw) light into right-handed light. If instead you let the left-handed light reflect at near-normal incidence from a mirror, you will convert it to right-handed light.

You can use your circular polarizer "run backward" as an analyzer. It then "passes" light with the same handedness that it produces (when run "forward"); it absorbs light with the opposite handedness. We can understand this as follows. The phase retardation of the slow linear polarization component relative to the fast linear component is independent of the direction of traversal of the quarter-wave plate. When the circular polarizer is run forward, the linear polarizer followed by the quarter-wave plate produces circular polarization with $\mathbf{E}$ rotating from $\hat{\mathbf{f}}$ toward $\hat{\mathbf{s}}$. If this light is reflected from a mirror, it continues to rotate in the same direction with respect to an axis fixed in space (by angular momentum conservation). As it passes back through the quarter-wave plate, the phase retardation proceeds as before and thus gives an additional 90 deg of phase lag by the time the light again reaches the linear polarizer traveling in the reverse direction. That means the light is linearly polarized at 90 deg to its original direction along the easy axis of the linear polarizer, because one linear component has suffered a sign reversal relative to the other. The light is therefore absorbed. That explains why a mirror or shiny piece of metal looks "black" (actually it is dark blue) when you lay your circular polarizer on it with input end upward. The mirror reverses the handedness. Similarly, *any* right-handed (screw) light is absorbed by your polarizer [which produces left-handed (screw) light] when it is incident on the output face. On the other hand, if left-handed (screw) light is incident on the output face of your left-handed polarizer, the linear component along the $\hat{\mathbf{s}}$ axis of the quarter-wave plate *leads* the $\hat{\mathbf{f}}$ component. The quarter-wave plate reduces this lead from 90 deg to zero. Thus, when the light reaches the linear polarizer, the $\hat{\mathbf{s}}$ and $\hat{\mathbf{f}}$ linear components are in phase, and the light is completely transmitted through the linear polarizer. It emerges linearly polarized and with all of its intensity (neglecting the losses that we usually neglect).

Here is an example: Suppose you look at a light source through a polaroid; rotating the polaroid about the line of Sight produces no variation in intensity. Then suppose you look at it through your left-handed polarizer (run backward); the intensity is unchanged. (What can you conclude at this stage?) Then you put your half-wave plate between the source and your left-handed polarizer and repeat the last measurement; the light is completely absorbed. *Conclusion:* It is left-handed circularly polarized light.

Quarter-wave plate and half-wave plate.   Take one of the two pieces of clear plastic from your optics kit. Hold it face to face with a piece of polaroid, with its edge at 45 deg to the edge of the polaroid. Look through both pieces at a light bulb or the sky, with the polaroid toward the source. Place a second polaroid on the other side of the clear plastic. Rotate the second polaroid. Now repeat the experiment with the other piece of plastic. Which one is the quarter-wave plate and which is the half-wave plate? Repeat the experiment with the edge of the clear plastic parallel to the edge of the polaroids.

The label on the sheet of quarter-wave retardation plate from which your piece was cut does not say "quarter-wave retarder"; instead it says "retardation value is 140 ± 20 mμ." A mμ is a mil-limicron $= 10^{-3} \cdot 10^{-6}$ meter $= 10^{-7}$ cm $= 10$ Å. Thus the retardation is 1400 Å. That is one quarter-wavelength if the wavelength is 4 $\times$ 1400 Å = 5600 Å (which is green). It is some other fraction of a wavelength if the wavelength is other than 5600 Å. Let us try to understand what the label means. The relative phase retardation $\Delta\varphi$ between s and f components traversing a retardation plate of thickness Δz and with indices n_s and n_f is

$$\Delta\varphi = 2\pi \left(n_s - n_f \right) \frac{\Delta z}{\lambda}. \qquad (69)$$

For a quarter-wave plate, the phase retardation corresponds to $\frac{1}{4}$ cycle, i.e., to $\frac{1}{2}\pi$ radians. Thus a quarter-wave plate must have

$$\left(n_s - n_f \right) \Delta z = \tfrac{1}{4}\lambda. \qquad (70)$$

The labeling indicates that $(n_s - n_f)\,\Delta z$ is $\frac{1}{4}\lambda_0$, where λ_0 is 5600 Å. That is the "spatial retardation," *independent* of λ (over most of the visible range). This just means that to a good approximation $n_s - n_f$ is independent of wavelength. For an arbitrary (visible) wavelength, we thus have

$$\Delta\varphi = \frac{\pi}{2} \frac{5600 \text{ Å}}{\lambda}. \qquad (71)$$

Similarly your half-wave plate has retardation value 280 ± 20 mμ.

Unpolarized light.   If you use your optics kit to determine the polarization of light from a light bulb, you will find that the linear polarizer gives no variation in intensity for any angle about the line of sight; neither is there any variation in intensity when a quarter-wave plate is inserted between the source and the polarizer. In terms of the x, y linear polarization representation of Eqs. (63),

(64), (67), and (68), these facts imply (we use a "bar" over the measured quantities to remind us that the measurements take place during the "measurement time" T)

$$\frac{1}{2}\overline{E_1^2} = \frac{1}{2}\overline{E_2^2}$$

$$= \frac{1}{2}\left[\frac{1}{2}\overline{E_1^2} + \frac{1}{2}\overline{E_2^2} + \overline{E_1 E_2 \cos(\varphi_1 - \varphi_2)}\right]$$

$$= \frac{1}{2}\left[\frac{1}{2}\overline{E_1^2} + \frac{1}{2}\overline{E_2^2} - \overline{E_1 E_2 \sin(\varphi_1 - \varphi_2)}\right]. \qquad (72)$$

In other words, for *any* choice of axes x and y, the time average of E_x^2 equals the time average of E_y^2, and the time average of cos $(\varphi_1 - \varphi_2)$ and of sin $(\varphi_1 - \varphi_2)$ is zero. Now of course there is no angle $\varphi_1 - \varphi_2$ with the property that both its sign and cosine are zero! The essential thing about Eq. (72) is the bar, which indicates the time average over time interval T. The reason the relative phase has zero for its time-averaged cosine and for its time-averaged sine is that the relative phase has been wandering at random for the long time T during which we made the measurement. It has taken on all values between $-\pi$ and $+\pi$ (relative phase is only defined over an interval of 2π). Both the sine and cosine have been as often positive as negative and have averaged to zero.

If we could complete the measurement in less than 10^{-10} sec (for a typical Doppler-broadened gas-discharge source), we would get a very different result. We would find the light is completely polarized at any "instant," where an instant includes many oscillations but is small compared with the inverse bandwidth (the coherence time).

One can simulate unpolarized light with a slinky. Shake it one way for a while, then shake it another way for a while. Make a time exposure photograph over a time T. If T is short compared with the time of unchanged shaking, the photograph will indicate complete polarization. If T is long, then anyone examining the picture will say "The slinky is unpolarized"—that just means T was too long.

Partial polarization. If T is neither short nor long compared with the coherence time, the radiation is called *partially polarized*. In that case, there is some distinguishable difference between the results of the four intensity measurements that give $\overline{E_x^2}$, $\overline{E_y^2}$, $\overline{\cos(\varphi_1 - \varphi_2)}$, and $\overline{\sin(\varphi_1 - \varphi_2)}$. There are many different ways one can express the fact that the polarization has been "washed out" over the measurement time T. For example, one can define a "fractional polarization P" by

$$P^2 \equiv \left[\overline{\sin(\varphi_1 - \varphi_2)}\right]^2 + \left[\overline{\cos(\varphi_1 - \varphi_2)}\right]^2, \qquad (73)$$

where $\overline{\sin(\varphi_1 - \varphi_2)}$ and $\overline{\cos(\varphi_1 - \varphi_2)}$ are defined by the intensity measurements that give the results of Eqs. (63), (64), (67), and (68). If T is small compared with the coherence time, then P is 1. If T is large compared with the coherence time, then P is zero. For intermediate T, P lies between 0 and 1. Of course, P is not the whole story; the whole story requires all four measurements.

Problems and Home Experiments

8.1 Carry out explicitly the steps outlined following Eq. (20), Sec. 8.2, to show that Eq. (20) represents a displacement $\psi(t)$ that follows an elliptical path.

8.2 Unpolarized light from a mercury discharge tube is passed through your green gelatin filter, which isolates the green line. Slits and lenses form a parallel beam propagating in the $+z$ direction. The beam is well defined at $z = 0$. At $z = 100$ there is a photomultiplier counting photons from the beam. The average counting rate is 64 counts per minute: $R = 64$.

(*a*) We insert a quarter-wave plate with fast axis along $\hat{\mathbf{x}}$ at $z = 10$. What is R now? (Neglect small losses due to reflections, etc.)

(*b*) We insert a linear polarizer with easy transmission axis along ($\hat{\mathbf{x}}$ + $\hat{\mathbf{y}}$)/$\sqrt{2}$ at $z = 20$. Now what is R? (*Note:* As we keep inserting things in this problem, we leave all the old stuff in place. The z locations are just to help keep the order straight.)

(*c*) We add a half-wave plate with fast axis along $\hat{\mathbf{x}}$ at $z = 30$. What is R?

(*d*) Now we add a linear polarizer with easy axis along $\hat{\mathbf{x}}$ at $z = 40$. What is R?

(*e*) A left-handed circular polarizer is now added at $z = 50$. What is the maximum counting rate possible (with the polarizer run in the forward direction)? What is the minimum possible rate?

(*f*) With the left-handed polarizer of part (*e*) set for maximum counting rate, we insert a half-wave plate with fast axis along ($\hat{\mathbf{x}} + \hat{\mathbf{y}}$)/$\sqrt{2}$ at $z = 60$, followed by a linear polarizer at $z = 70$ with easy transmission axis along $\hat{\mathbf{y}}$. What is R?

8.3 Circularly polarized light of intensity I_0 (intensity means energy flux per unit area per unit time; this is proportional to a photomultiplier's output current, for light at a given frequency) is incident on a single polaroid. Show that the output intensity (intensity of the light emerging from the rear of the polaroid) is $\frac{1}{2} I_0$.

8.4 Linearly polarized light with polarization direction at angle θ from $\hat{\mathbf{x}}$ is incident on a polaroid with easy axis along $\hat{\mathbf{x}}$. The first polaroid is followed by a second polaroid with its easy axis along the direction of polarization of the original incident light. Show that if the input intensity is I_0, the output intensity is $I_0 \cos^4 \theta$.

8.5   Circularly polarized light of intensity I_0 is incident on a sandwich of three polaroids. The first and third polaroids are crossed, i.e., their easy axes are at 90 deg to one another. The middle polaroid makes an angle θ with the axis of the first polaroid. Show that the output intensity is $\frac{1}{2} I_0 \cos^2 \theta \sin^2 \theta$.

8.6   A very large number $N + 1$ of polaroids are arranged in a sandwich. The angle of the easy axis of each polaroid is a constant angle a greater than that of its immediate predecessor in the sandwich. Thus the last polaroid is at an angle $\theta = Na$ from the first. Neglecting any losses due to reflection at the many, surfaces, and supposing that linearly polarized light of intensity I_0 is incident on the first polaroid with its polarization along the easy axis of the first polaroid, find the output intensity. Take N to be very large and retain only the first interesting terms in an appropriate power series (Taylor's expansion).

$$\textit{Ans.} \quad I = I_0 (1 - \frac{\theta^2}{N} + \text{higher-order terms}).$$

That means that even if θ is 90 deg, so that the first and last polaroids are crossed, the output intensity is equal to the input intensity if we have enough intermediate polaroids. We can "gently" rotate the plane of polarization and lose nothing! Thus if we cemented a large number of polaroids together (using clear cement with the same index as the polaroids, so as to minimize reflections), we would have something like a gigantic sugar molecule, which rotates the plane of polarization without absorbing any energy.

Another way to get "macroscopic optical activity" is to twist pieces of tinfoil into corkscrews (all with the same handedness), embed them in styrofoam (a good approximation to a massless, rigid, electron-free supporting medium), and send linearly polarized microwaves through the stuff. The plane of polarization of the microwaves will be rotated.

8.7   Suppose you have linearly polarized incident light with polarization along $\hat{\mathbf{x}}$. You desire linearly polarized light with polarization at 30 deg to $\hat{\mathbf{x}}$, i.e., along

$$\hat{\mathbf{e}} = \hat{\mathbf{x}} \cos 30° + \hat{\mathbf{y}} \sin 30°.$$

How can you obtain this transmitted field (a) at the cost of some loss of intensity; (b) without loss of intensity and without using any polaroids?

8.8   What is the transmitted intensity for unpolarized light of intensity I_0 incident on crossed polaroids with a half-wave plate between them, (a) when the retardation plate's optic axis (say the slow axis) is parallel to the easy axis of one of the polaroids; (b) when the wave plate's optic axis is at 45 deg to one of the easy axes?

8.9   Answer the same questions as in Prob. 8.8, but use a quarter-wave plate.

8.10 Test various sheets of plastic (drawing instruments, cellophane, Scotch-brand cellophane tape, etc.) for double refraction by rotating them between crossed polaroids. How can you tell if you are lucky enough to find a quarter-wave plate or halfwave plate? Try the experiment of stretching Saran Wrap to make it doubly refracting. **Home experiment**

8.11 Saran Wrap quarter-wave plate. Get a roll of Saran Wrap or Handi-Wrap (clear stretchy plastic used to wrap sandwiches) at a grocery store. About six or seven parallel layers make a very good quarter-wave plate. (A quarter-wave plate of the same size—about 30 cm²—can be obtained from Polaroid Corporation, Cambridge, Mass. 02141, for about $l3.) It can be "tuned" for different colors by adding or subtracting one layer. For example, if seven layers is a perfect quarter-wave plate for 5600 Å (green), then eight layers should be perfect for wavelength $(\frac{8}{7})(5600) = 6400$ Å (red). To get rid of some of the wrinkles, you can tape the stuff on the side of a cardboard carton with a hole cut in the cardboard. **Home experiment**

8.12 Color dependence of retardation plates. A "half-wave plate" is truly a half-wave plate only at a particular wavelength. The half-wave plate in your optics kit is a half-wave plate for 5600 Å. Get a bright white line source. (Any incandescent bulb with a clear glass envelope will do, for example, a 150-watt bulb with a straight filament that is a helix about an 2.5 cm long and 1 mm in diameter.) Look at the white source with your diffraction grating. (Spread the colors out perpendicular to the line source to get the best resolution.) Now take two *parallel* polaroids. Put the half-wave plate between the polaroids at 45 deg. Then the color for which the half-wave plate is a half-wave plate will have its linear polarization flipped by 90 deg and will be absorbed. Look through the sandwich using your diffraction grating. (Hold everything close to one eye.) Do you see the dark band at green? *That* is the color of 5600 Å! (*Note:* Rotate the last polaroid slightly to tune for the maximum blackness in the absorption band.) **Home experiment**

8.13 Saran Wrap half-wave plate. Make a half-wave plate by the method described in Home Exp. 8.11. It should take about 12 to 15 layers (if your Saran Wrap is like my Handi-Wrap). Perhaps you can use the method of Home Exp. 8.12 to "tune" the plate, adding layers until the absorption band is at 5600 Å (as determined by comparison with the half-wave plate in your kit). That will tell you fairly accurately the value of $(n_s - n_f)\,\Delta z$ for one layer. **Home experiment**

8.14 Slinky polarization. Find a slinky and a partner. You and your partner hold opposite ends of the slinky. **Home experiment**

(*a*) Let each shake the slinky in a clockwise circular rotation (from his own point of view). If this doesn't convince you that linear polarization is a superposition of opposite circular polarizations, nothing else will.

(*b*) With each person using a book as a straightedge to guide his hand, let one partner shake linear polarization at 45 deg to the horizontal, and let the other shake linear polarization at 90 deg to the first. (The 45 deg angle is so as to prevent gravity from giving a big asymmetry.) One counts out loud, " 1, 2, 3, 4, 1, 2, 3, 4, . . ." four beats to a cycle, or perhaps four to a half-cycle), with "one" coming at a reproducible phase of the motion of his hand. The other shakes in phase, or 180 deg out of phase, or 90 deg out of phase. It takes some concentration not to be distracted by what you see.

(*c*) With the far end fixed to something (your partner can now go home), shake out a circularly polarized wave packet of one or two turns. Verify that it conserves angular momentum upon reflection. Verify that if the angular momentum is along the propagation direction, the shape is that of a left-handed screw, and that the handedness reverses upon reflection.

Home experiment **8.15 Clear cellophane tape half-wave plate.** Stick one layer of clear cellophane tape on a microscope slide (the slide is for mechanical support). Test it for being a half-wave plate by the method of Home Exp. 8.12. Estimate $(n_s - n_f) \Delta z$.

Home experiment **8.16 General polarization with clear cellophane tape.** Stick 16 layers of clear cellophane Scotch tape on a microscope slide. There will probably be bubbles and it will then be difficult to see through. Here is an improved technique: Lay a clean microscope slide on a table. Put a "bead" of oil (Three-in-One oil or mineral oil or something) on the center of the slide. Take a piece of tape as long as the slide plus about two or three inches extra at each end. Stick the tape on the slide, making a good optical contact by spreading the oil. Stick the tape to the table as you do so. Put a bead of oil in the center region of the tape. Now add another layer of tape. (The tape will stick to itself out near the ends where there is no oil.) Oil, tape, oil, tape, ... On the 16th tape layer, put one more bead of oil, then a final microscope slide. Add more tape to hold the slide—but don't obscure the clear sensitive region. Now you have a package with flat outer glass surfaces and with 16 layers of tape. It should be reasonably clear and easy to see through.

Here is the experiment: Tape a polaroid to one face of the package with the polaroid axis at 45 deg to the tape axis. Tape the diffraction grating on also. Hold this package so as to look at your bright white line source. With your other hand, hold a linear polarizer (polaroid) at the exit end of the package and parallel to the first polaroid.

(*a*) Notice the several black bands! Those black bands all have linear polarization that is absorbed by the linear polarizer. They are separated in relative phase (of linear polarization components along the fast and slow Scotch tape axes) by 2π. The "bright" region between two successive dark bands has relative phase that varies from zero to 2π and thus sweeps through all the varieties of polarization shown in Fig. 8.3, Sec. 8.2.

(*b*) Rotate the rear "analyzing" polaroid by 90 deg. The black regions become bright and the bright become black! Why?

(*c*) Replace the polaroid by your circular polarizer run backward as an analyzer, i.e., with the output end toward the source. (When you lay it on a dime and the dime turns dark blue, then the input end is upward.)

(*d*) Put both the linear and circular polarizers in place, splitting the field of view between the two. They should be parallel to the first polarizer (45 deg to the slide). Move the package so that you see through first the circular, then the linear analyzer. The bands move a quarter of the distance between black bands (i.e., $\pi/2$ radians of phase). Now rotate the linear polarizer and repeat. The direction of displacement of the bands from the circular to the linear should reverse. By such methods, you should be able to convince yourself that the polarization varies from (for example) linear upper right to right-handed circular to linear upper left to left-handed circular to linear upper right as you go through 2π. Make a sketch of polarization versus color (wavelength), indicating linear polarization by double-headed arrows and circular polarization by "circular arrows."

8.17 Clear cellophane tape–Polaroid-Lyot filter.

(The following experiment requires four pieces of polaroid.) Make a l6-layer Scotch tape retarder as in Home Exp. 8.16. Make an 8-layer retarder by the same technique, and also make a 4-layer retarder. Let us call these packages by the names l6LR (16-layer retarder), 8LR, and 4LR. Let us also call a Polaroid linear polarizer by the name P, and let P(45°) mean that the easy axis of the P is at 45 deg to the tape axis. Call the diffraction grating DG. Now do the following experiments:

(*a*) Make a sandwich consisting of DG : P(45°) : 16LR : P(45°). Look at your white line source. This is just Home Exp. 8.16. Now repeat the experiment using the 8LR in the place of the 16LR.

(*b*) Add the 8LR and another P at the exit end of the l6LR package, so that you have the package DG : P(45°) : l6LR : P(45°) : 8LR : P(45°). Look at the line source.

(*c*) Now add 4LR : P(45°) at the exit of the package of part (*b*) and look again. Notice that you are wiping out the "sidebands" by the successive filters! You finally end up with a bandpass filter. (You can clean things up with your gelatin filter if you wish.) The bandwidth is given by the l6LR package. If you want to halve the bandwidth, you need a 32LR. Even with mineral oil and care, my package starts to get a little difficult to see through then.) This kind of filter was invented by B. F. Lyot in 1932. Astronomers use a Lyot filter made with quartz retardation plates instead of Scotch-tape retardation plates. They typically achieve a bandwidth 1 Å, centered on (for example) the Hα spectral line of the Balmer series of hydrogen, with wavelength 6563 Å. It is used to photograph the sun. The total transmitted filtered intensity is the *product* of the intensity transmission curves for the

Home experiment

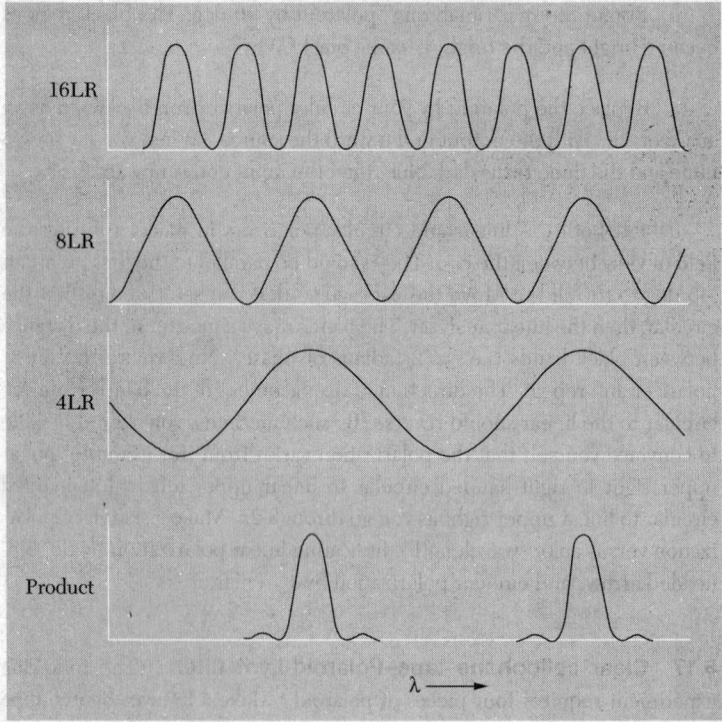

Problem 8.17 *Lyot filter. The curve16LR gives the transmitted intensity for the filter made of the package P(45°) : 16LR: P(45°). The curves 8LR and 4LR are the corresponding curves for individual filters using 8LR or 4LR in place of 16LR When the entire package is used, the transmission curve is the product of the three individual curves, as shown.*

individual 16LR, 8LR, and 4LR filters. (The order does not matter. You can run the filter forward or backward.) This is shown in the figure.

Question: Tell us why it is that the total transmitted intensity is obtained by taking the product of the intensity transmission curves for the individual filters. Why not (for example) add the three amplitudes for the three filters (when only one is present), then square, then time average?

Home experiment **8.18 Circular polarizer.** (*a*) Lay your circular polarizer on a piece of aluminum foil (available in any grocery store) or on an ordinary mirror or shiny knife. Turn it so that the foil looks "black" (or dark blue). Flip it over and look. (Do the same with a piece of polaroid.) Flip it back to "black" position. Now lift the polarizer slightly off the metal, so that light can get to the metal without going through the polarizer. Look at the "shadow" or "image" of the polarizer as you slowly lift it off the metal and put it back. Explain what you see.

(*b*) Take the aluminum foil and make a V-shaped crease in it. Use illumination such that most of the light comes from a definite direction (a lamp or a window). Put the circular polarizer on the foil, partly covering the crease and partly covering the uncreased foil. Notice that the crease now looks bright, while the rest of the foil is still dark. Explain! (*Hint:* When you look at your right hand in a single mirror, it looks like a left hand. What does it look like in a "double mirror" made by joining two mirrors at right angles to form a trough-shaped mirror?

(*c*) Now take the foil and crinkle it all up in a mess to make a "rough" surface. Lay the circular polarizer on it. Look closely. Explain the statement, "It is depolarized on a large scale but is completely polarized on a small scale." Explain the statement, "That is somewhat analogous to the depolarization of light on a large *time* scale and the complete polarization of light on a sufficiently small time scale."

(*d*) Lay the polarizer on a piece of ordinary white paper. Can you tell the circular polarizer from a piece of polaroid in this way? Explain.

(*e*) Lay the circular polarizer on the smooth metal surface again. Insert your half-wave plate between the polarizer and the metal. First predict what you will see, then do the experiment. Repeat, using your quarter-wave retarder. (*Note:* Every color is retarded by a different amount, as you have learned. The effects can be somewhat enhanced by using your green gelatin filter. That is not really necessary, as long as you understand that "black" is an approximate description.)

8.19 Angular momentum of light.

Suppose right-handed circularly polarized light (by the angular momentum convention) is incident on an absorbing slab. The slab is suspended by a vertical thread. The light is directed upward and hits the underside of the slab.

(*a*) If the circularly polarized light beam has 1 watt of visible light of average wavelength 5500 Å and if all of this light is absorbed by the slab, what is the torque exerted on the slab? (Give the answer in N-m.) Remember that torque is rate of change of angular momentum, and that the slab is of course absorbing the angular momentum of the radiation.

(*b*) Suppose that instead of an absorbing slab you use an ordinary silvered mirror surface, so that the light is reflected back at 180 deg to its original direction. What is the torque now?

(*c*) Suppose that the slab is a transparent half-wave plate. The light goes through the plate and doesn't hit anything else. What is the torque? (Neglect reflections at the surfaces of the slab.)

(*d*) Suppose the slab is a transparent half-wave plate with the top surface silvered so that the light goes through the half-wave plate, reflects from the mirror, and returns through the plate. What is the torque?

(*e*) The slab is a transparent half-wave plate. Above the slab is a fixed quarter-wave plate (ie., not attached to the slab) silvered on the top surface so as to reflect light back through the slab. What is the torque exerted on the slab?

(*f*) How can you get the biggest torque?

(*g*) Suppose that the thread on which the slab is suspended, together with the slab, has a natural period for torsional oscilliations of 10 min. How would you design an experiment so as to "magnify" the effect of the torque, so that you can finally get a decent measurement? (We want only the ingenious idea, not engineering details.) Now read how the experiment was actually done by R. A. Beth, *Physical Review* **50**, 115 (1936).

Home experiment **8.20 Polarization by scattering.** (*a*) Put a few drops of milk in a glass jar of water. Shine a flashlight beam through the liquid. Look at the bluish-tinged light scattered from the "milk molecules." Test the polarization with your linear polarizer. Do this for 90-deg scattering (scattering through an angle of deviation of 90 deg) and for scattering at small angles (near 0 deg) and large angles (near 180 deg). (*Note:* you should put a small piece of tape or something on your linear polarizer to label the easy transmission axis. This axis is found by looking at the specularly reflected light from glass, or a wood or plastic floor, or any painted surface, at angle of incidence near 45 deg (that is close enough to Brewster's angle).

(*b*) Linearly polarize the flashlight beam (collimated with a piece of cardboard so the beam is smaller than the polaroid) and look at the scattered light from different directions in the plane at 90 deg to the beam (or rotate the polaroid at the flashlight).

(*c*) Study Fig. 8.6 and its caption. Define the fractional polarization *P* by

$$P = \frac{I(\hat{\mathbf{x}}) - I(\hat{\mathbf{y}})}{I(\hat{\mathbf{x}}) + I(\hat{\mathbf{y}})},$$

where $I(\hat{\mathbf{x}})$ is the intensity of scattered light polarized along $\hat{\mathbf{x}}$ and $I(\hat{\mathbf{y}})$ is the intensity of scattered light polarized along the projection of $\hat{\mathbf{y}}$ seen by the observer. Show that the dependence of *P* on the scattering angle θ_{sc} of Fig. 8.6 is given by

$$P = \frac{1 - \cos^2\theta_{sc}}{1 + \cos^2\theta_{sc}}.$$

Notice that *P* is zero at scattering of 0 or 180 deg and is 1 at 90 deg.

(*d*) Add a little milk; the beam gets whitish. Look at the polarization at 90 deg, where it is maximum. Add more milk. Explain what is happening. Would you expect sunlight scattered from a white cloud to be polarized? Try the experiment and see.

Home experiment **8.21 Polarization of the rainbow.** Is it polarized? You can use the fine spray from a garden hose instead of rain.

Home experiment **8.22 Moonlight and earthlight.** When the moon appears half-full, the illuminated portion is scattering sunlight through about 90 deg to your eye. We know that for 90 deg scattering the blue sky is almost completely linearly polarized. Do you predict that half-moon-light is polarized? Do the experiment. Now think about how the earth looks from the moon at "half-earth." Is the earthlight polarized? (You can look for twenty-four hours while the earth turns.)

Ans. Sometimes; it depends on the time and the weather. Why?

8.23 Suppose that .a beam of linearly polarized light is incident on a half-wave plate which is rotating about the beam axis with angular velocity ω_0. Show that the output light is linearly polarized, with the polarization direction rotating at $2\omega_0$.

8.24 Look at a light bulb through a piece of polaroid. Is the light polarized? Now insert a piece of cellophane (or your quarter- or half-wave plate) between the light bulb and the polaroid. Now is the light polarized? Reflect the light from a silvery metal, like a table knife. Is the reflected light polarized?

8.25 Measuring the index of refraction by finding Brewster's angle. You need a light bulb (perhaps covered with a piece of cardboard with a hole to get a reasonably small source), a piece of glass, a table, a cardboard box or something to give a measurable location for your eye, and a single piece of polaroid. Lay the piece of glass flat on the table and look at the reflection of the bulb. (You will see two reflections, one from the front and one from the rear surface. If you wish to, you can eliminate the one from the rear surface by spraying the rear surface of the glass with black paint.) Vary the angles until the polaroid reveals that the reflected light is completely polarized. Measure the appropriate distances and obtain the index of refraction by the formula for Brewster's angle, $\tan \theta_B = n$. With this crude setup, you cannot measure to better than a few degrees, so that you probably cannot distinguish Brewster's angle for glass from that for a smooth surface of water.

Home experiment

8.26 Phase relations in specular reflection of light from glass. We are trying to verify the relationships shown in Fig. 8.8. The experimental setup is as in Home Exp. 8.25, except that between the piece of glass lying on the table and the light we place a piece of polaroid with easy axis rotated at 45 deg to the horizontal. (*Suggestion:* For a convenient mounting, put a little nonhardening putty or glazing compound on a microscope slide and stick the corner of the polaroid into the putty. The microscope slide can be the glass surface used for reflecting the light.)

Home experiment

Suppose that when your eye is at the microscope slide and you are looking back at the light bulb through the first polaroid, the easy direction of the polaroid is "upper right to lower left." Keep analyzing the polarization of the reflected light as you vary the angle of incidence by changing the position of the slide or light bulb. You will find that near normal incidence the polarization of the reflected light is along "upper left to lower right." As you move the slide and approach Brewster's angle, the polarization remains linear, but it rotates toward the horizontal. It becomes horizontal at Brewster's angle and then continues to rotate in the same direction as you go beyond Brewster's angle toward grazing incidence; i.e., it becomes "lower left to upper right." Thus in going from normal to grazing incidence, the polarization rotates by 90 deg, as predicted by Fig. 8.8. (At normal incidence, both components are equally well reflected, as they must be since they don't know which is which, so to speak. Therefore the polarization is at 45 deg. At grazing incidence, both components are almost completely reflected, so that they are equally well reflected. Therefore the polarization is at 45 deg.) It is interesting to notice that the polarization remains linear through all of this experiment. This means that there are no phase shifts other than 0 or

180 deg introduced between those components in the plane of incidence and those perpendicular to it. Thus the incident waves always experience a purely resistive load when they reflect. That is what we expect for reflection from a transparent medium.

Home experiment   **8.27 Conservation of angular momentum.** Reflection at normal incidence changes right-handed circularly polarized light into left-handed light. (When you look at your right hand in a mirror, it looks like a left hand.) What about grazing incidence? Is the handedness the same after reflection, or is it reversed? Predict the answer using Fig. 8.8.

(*a*) Now do the experiment. (Use your quarter-wave plate and one polarizer taped together to make a circular polarizer. Use your circular polarizer run backward as an analyzer. Make sure you know the handedness of the circular polarizer you make relative to that in the kit.) What does your hand look like at grazing incidence? What does the following "philosophical" statement mean: "You must be skeptical about using intensity measurements to make predictions that involve signs, i.e., phases?" What has this statement to do with the way your hand looks in a mirror?

(*b*) Do a similar experiment with linear polarization. With one polaroid, make "upper right" linearly polarized light. Look at its reflection at grazing incidence. Is the reflected light "upper right" or "lower right?" If you look at the image of a pencil held so it is aligned along "upper right," what does its image look like, "upper right" or "lower right?" What is the relevance of the "philosophical" statement above?

Home experiment   **8.28 Phase changes in metallic reflection.** The experimental setup is like that for Home Exp. 8.26. But instead of a piece of glass, use any shiny flat piece of metal, for example the stainless steel blade of a table or kitchen knife, or any chromium or silver-plated object. (Do *not* use an ordinary mirror, i.e., a rear-surfaced piece of glass; it won't work.) You will need two polaroids and one quarter-wave plate. First verify that ∥ (parallel) or ⊥ (perpendicular) linearly polarized light (i.e., respectively parallel or perpendicular to the plane of incidence) retains its polarization upon reflection. (That is analogous to the action of a retardation plate on linearly polarized light that is polarized parallel or perpendicular to the optic axis; nothing happens to the polarization.) Next linearly polarize the incident light at 45 deg to the plane of incidence. Set the angle of incidence so that if the light bulb is 0.3 m above the table, the knife is about 1.0 m from the light bulb. Now analyze the reflected light, using your polaroid and quarter-wave plate (or using your circular polarizer run backward as an analyzer, with or without a half-wave plate). You will find that the polarization is elliptical. By varying the angle of incidence, you can find a spot where the polarization of the reflected light is almost perfectly circular. If you now tilt the polarizing polaroid by five or ten degrees away from 45 deg, tilting the easy axis toward the vertical so as to increase the ∥ component slightly, you can get completely circularly polarized reflected light. (This slight tilt is necessary to compensate for the fact that the ∥ component is not as completely reflected as is the ⊥ component.)

Rotating the polarizing polaroid from an "upper right" to an "upper left" direction of the easy transmission axis reverses the handedness of the reflected light.

Here is a qualitative explanation of the phenomenon. The metal is a reactive medium. Both polarization components of the incident light are almost totally reflected. There is a phase shift corresponding to the time required for the fields to penetrate into the reactive medium for about one exponential attenuation distance (on the average) and then to turn around and come back out. The ∥ and ⊥ polarization components do not have the same phase shift for the following reason. The ⊥ component is parallel to the surface; the electrons are free to move parallel to the surface and move so as to cancel the incident radiation. The time delay and phase shift is due to the inertia of the electrons. The ⊥ component therefore has a certain phase retardation, due to this time delay. Now consider the ∥ component. At near-normal incidence; the ∥ component is nearly parallel to the surface and therefore behaves like the ⊥ component. The phase retardation is therefore the same as for the ⊥ component. Both components receive a minus sign upon reflection, in addition to the phase shift due to penetration. Therefore the incident polarization "upper right" (as seen looking back at the light bulb from the metallic reflector) becomes "upper left" after reflection. But now suppose we are not near normal incidence. Then the ∥ component of electric field is not parallel to the surface. We can resolve it into a component parallel to the surface and a component perpendicular to the surface. That component parallel to the surface continues to behave in the usual way and undergoes a phase retardation as before. But the component perpendicular to the surface behaves in a completely different manner; charges are not free to move perpendicular to the surface. The surface acquires a surface charge, and the charges come to rest very rapidly. The time delay due to the inertia of the electrons moving parallel to the surface is not present for the motion perpendicular to the surface, because the motion is so tiny. Therefore this part of the ∥ component reflects with negligible time delay.

In order to complete the explanation, we must be able to calculate the phase retardation of each component and to see how it depends on the angle of incidence. That is difficult.

8.29 Optical activity. Suppose you send linearly polarized light through a length L of Karo corn syrup and find that, for $L = 5$ cm, red light is rotated by 45 deg. Now reflect the light that has passed through the syrup from a mirror and send it back through the syrup, so that the total length is 10 cm. (If you do the experiment, make the angle of reflection not quite 180 deg; then look at the "image light bulb" through both the "real syrup" and the "image syrup." As a control experiment, you can look through the "image syrup" alone by moving your head.) *Question:* After the two traversals, is the linear polarization at 0 or 90 deg to the original direction?

Home experiment

8.30 Finding the fast axis of your quarter-wave plate. Given that your circular polarizer makes left-handed light (by the screwy optics convention—or right-handed light by the angular momentum convention). Find the fast axis of your quarter-wave plate. (Once you find it, put a notch or tape or something on it.)

Home experiment

8.31 Effective spring constants for Saran Wrap molecules. Stretch a piece of Saran Wrap and lay it behind a polaroid at 45 deg to the easy axis of the polaroid. Don't stretch too hard—we don't want to go beyond a relative phase shift of $\pi/2$. Now determine the handedness of the elliptically polarized light. You can do this with your circular polarizer and half-wave plate. Once you know the handedness, you know whether the axis of stretch is the slow or fast axis. Now assume that the stretching has lined up the molecules with their long dimension along the stretching direction. You now can figure out whether the index of refraction is larger or smaller along the long direction of the molecule. Large index of refraction means large dielectric constant, which means large molecular polarizability, which means weak effective spring constant (as long as the frequency of the light is less than the effective natural vibration frequency of the electrons in the molecule. That is the case for visible light in glass. We can assume it is also the case here.) Thus if (for example) the stretch axis turns out to be the slow axis, it means that the effective spring constant for vibration along the molecule is less than for vibration perpendicular to it. What is the experimental result?

Home experiment

8.32 Iceland spar (calcite crystal). Get a nice big (an inch or so thick) crystal of Iceland spar. (See any mineral or lapidary store, for example Frazier's Mineral & Lapidary, 1724 University Ave., Berkeley, Calif.; or see Central Scientific Co., 1700 W. Irving Park Rd., Chicago, Ill., 60613, Catalogue item 87595, $0.65.) Put a black pencil dot on a piece of paper, set the crystal on the paper, and look at the dot through the crystal; you will see two dots. Now look at these two dots with a linear polarizer. They are each 100% polarized! Rotate the crystal about a vertical axis as you look at the dots. One rotates, and the other doesn't! The extraordinary dot has its **E** vector along the optic axis. It is the one that moves. Now use your two eyes and your depth perception to decide which of the two dots is closer to you. Convince yourself with a slab of glass or with a sketch (or a slab of water, such as a fish aquarium) that things look closer when seen through material with index n bigger than 1. Is it the extraordinary or the ordinary dot that is closer and hence has the larger index? Does your experimental result agree with the indices in Table 8.1, Sec. 8.4? Show by using a pencil as a marker in space and looking down at normal incidence that the ordinary dot has no lateral displacement. Thus the ray which enters normal to the surface, continues normal to the surface, and emerges normal to the surface is the ordinary ray. The extraordinary ray does not travel normal to the surface! Show by a time-reversibility argument that an extraordinary ray that leaves the real dot and is normally incident on the surface must leave the exit surface at normal incidence, even if it travels obliquely inside the crystal. (The top and bottom surfaces are parallel for any orientation of the crystal on the piece of paper.) Does the extraordinary ray bend in an attempt to become more parallel to the optic axis, or instead does it try to become more perpendicular? (Think of the indices of refraction.) The physical explanation of the deflection of the extraordinary ray is as follows: Resolve **E** of the incident extraordinary ray into a component along the optic axis and a component perpendicular. The indices are different for **E** along these two directions, so the polarizability is different. Thus the amplitude of oscillation of the driven electrons is different, so they do not radiate the same amount (or one component. of motion does not radiate the same as the other). When you superpose the radiation fields due to these two components of electron motion, they give a wave traveling in a "skewed" direction. All you have to do now is get straight on which way the ray is skewed. Does the result agree with your experimental observation?

8.33 Navigation by the Vikings. At high latitudes (say above the Arctic Circle) the magnetic compass is unreliable. The sun is also difficult to use for navigation; it may be below the horizon even at noon. Airline navigators then sometimes use a "twilight compass" that locates the sun's position below the horizon by means of the variation with direction of the polarization of the blue sky. The compass contains a piece of polaroid. Some natural crystals have properties similar to polaroid—one such substance is tourmaline; another is cordierite. When linearly polarized light is viewed through a cordierite crystal, the crystal is clear (with a yellowish tinge) when the polarization is along the axis of easy transmission, and the crystal is dark blue when the polarization is 90 deg to this axis. Such substances are called "dichroic."

The Viking sailors of the ninth century navigated their ships without benefit of either magnetic compass or polaroid. At night they used the stars. In the day they used the sun, when it was not obscured by clouds. According to ancient Scandinavian sagas, the Viking navigators could always locate the sun, even when it was behind the clouds, by using magical "sun stones." It was long a mystery what these "sun stones" were. The mystery has probably been solved by a Danish archeologist, who knew about the Vikings, and a ten-year-old boy, who knew about the twilight compass (his father is chief navigator of the Scandinavian Airlines System). Archeologist Thorkild Ramskou had written in an archeology journal, ". . . but there seems to be a possibility that it was an instrument which in clouded weather could show where the sun was." The boy read this; to him it sounded like a twilight compass. The boy's father, Jorgen Jensen, passed on this observation to Ramskou. Ramskou and the jeweler of Denmark's royal court collected and tested various dichroic crystals found in Scandinavia. The best "sun stone" turned out to be cordierite. Ramskou found that he could locate the sun to $\pm 2\frac{1}{2}$ deg and track it until it was 7 deg below the horizon.

Here is the question: According to the account in *Time* magazine, July 14, 1967, p. 58, the ancient Scandinavian sagas said that the sun could always be located by the magical "sun stones" *no matter what the weather.* Do you believe that? Explain.

8.34 Polarization "projection operator." If a piece of linear polaroid with easy axis along $\hat{\mathbf{x}}$ is placed in a beam of light containing a mixture of all sorts of polarization, the polaroid absorbs all light that does not have linear polarization along $\hat{\mathbf{x}}$. It has an "output" at the rear of the polarizer consisting of light linearly polarized along $\hat{\mathbf{x}}$. We shall call this piece of polaroid a "projection operator." It "projects out" the $\hat{\mathbf{x}}$ polarization without loss (neglecting small reflections) and delivers it at its output end. Note that this "x projection operator" can be run either forward or backward; i.e., either face of the polaroid may be used as the input end. Now consider a piece of circular polarizer consisting of a piece of linear polarizer (input end) glued to a quarter-wave plate with optic axis at 45 deg to the easy axis of the polaroid. This polarizer puts out (for example) right-handed light. But it absorbs half of any right-handed light incident. If it is run backward, it passes incident right-handed light and absorbs left-handed light. But when it thus passes right-handed light incident on the quarter-wave-plate face, it delivers it out the polaroid face as linearly polarized light. Therefore it is *not* what we are calling a polarization projection operator. Here is the problem: Invent circular polarization projection operators, one for left-handed and one for right-handed light. The right-handed projection operator should transmit incident right-handed light with no loss (neglecting small reflections) and should deliver it as right-handed light. It should absorb left-handed light. *Question:* Is your circular polarization projection operator reversible? Can you use either face for the input end?

Home experiment **8.35 Glare elimination.** Suppose you wish to shine your flashlight through a glass window to illuminate something on the far side of the window. How can you get rid of the bothersome glare from light specularly reflected from the glass? Suppose instead that you are trying to look at something through rain, at night, using a flashlight beam for illumination. Will the trick used for eliminating window glare get rid of the light reflected from the raindrops? Suppose that instead of visible light you are using 10-cm microwaves emitted and received by the same antenna system, i.e., radar. How can you arrange the phase relations in two antennas oriented along x and y so as to eliminate glare from raindrops?

Home experiment **8.36 Colors in clear plastics.** Find a piece of clear plastic with a glossy finish on both sides,—for example, a plastic icebox dish or other container. Look at specular reflection of the sky at angle of incidence 45 deg or so. Do you see colors? (Put dark cloth or paper underneath to reduce background.) To enhance the effect, hold a piece of polaroid in front of your eye. Explain the origin of the colors.

Chapter 9

Interference and Diffraction

Chapter 9 Interference and Diffraction

9.1 Introduction

Most of our studies so far have been essentially one-dimensional, in the sense that there was only one path by which a wave emitted at one place could go to another place. Now we shall consider situations where there are different possible paths from an emitter to a detector. These lead to what are called *interference* or *diffraction phenomena*, resulting from constructive and destructive superposition of waves that have different phase shifts, depending on the path taken.

In Sec. 9.2 we consider the superposition at a detector of the waves emitted by two point sources having the same frequency and a constant phase relation. Examples are water waves emitted by two screwheads jiggling the surface of a pan of water or light emitted by the currents in the edges of two slits which are illuminated by a line or point source (Home Exp. 9.18) or sound waves emitted by two loudspeakers driven by the same audio oscillator.

In Sec. 9.3 we consider interference between two "independent" sources, i.e., sources whose phases are not constrained to maintain a definite relation. We find that the interference pattern remains constant only for time intervals of order $(\Delta \nu)^{-1}$, where $\Delta \nu$ is the frequency bandwidth of the sources. Nevertheless, by a sufficiently fast measurement one can determine the interference pattern.

In Sec. 9.4 we find how large a source can be and still behave like a point source, when the source consists of independently radiating parts and when the detector averages over long time intervals [i.e., long compared with $(\Delta \nu)^{-1}$]. The result can be verified in an easy home experiment (Home Exp. 9.20). Another home experiment (Home Exp. 9.21) demonstrates the coherence of a Lloyd's mirror.

In Sec. 9.5 we give a crude derivation of the result that a beam of spatial width D has an angular divergence ("width") of order $\Delta \theta \approx \lambda/D$ about the dominant direction of travel. This fact is mathematically related (by the theory of Fourier analysis) to the fact that a pulse of time width Δt has a frequency width of order $(\Delta t)^{-1}$.

In Sec. 9.6 we use Huygens' construction to find the interference patterns of single and multiple slits. The emphasis is on optical and electromagnetic phenomena. There are several home experiments involving diffraction gratings and various diffraction patterns. For these experiments we strongly advise the student to get a "display lamp"—a light bulb with a clear glass envelope and a single straight filament about 7.5 cm long (about 40 cents in most grocery or hardware stores). Most of the experiments use one of these as a line source.

In Sec. 9.7 we study so-called "geometrical" optics. We first derive the law of specular reflection and Snell's law of refraction from the wave properties of light. Then we consider various mirrors, prisms, and thin lenses.

9.2   *Interference between Two Coherent Point Sources*

Coherent sources.   The simplest situation involving interference is that in which there are two identical point sources at different locations, each emitting harmonic traveling waves of the same frequency into an open homogeneous medium. If each source has a perfectly definite frequency (rather than a dominant frequency and a finite frequency bandwidth), then the relative phase of the two sources (the difference between their phase constants) does not change with time and the two sources are said to be relatively coherent, or simply, *coherent.* (Even if they have different frequencies, they are "coherent" if each is monochromatic, since their relative phase is always completely determined.) If each source has the same dominant frequency and each has a finite bandwidth $\Delta \nu$, then, if the sources are "independent," the relative phase of the two sources will only remain constant over times of the order of $(\Delta \nu)^{-1}$. On the other hand, two sources may be "locked" in phase with one another because they are driven by a common driving force. In this case, even though the phase constant of each source will drift in an uncontrollable manner through a phase of order 2π in a time $(\Delta \nu)^{-1}$, where $\Delta \nu$ is the bandwidth of the common driving force, the relative phase will remain constant. The sources are then said to be coherent even though they are not monochromatic.

As an example of two coherent sources of waves, consider two rods which touch the surface of a body of water. If the rods are identically driven in vertical oscillations, they produce surface-tension waves on the water. The relative phase of the rods is constant because they are driven by a common source. As another example of two coherent sources, consider two identical radio antennas driven at constant relative phase by the same oscillator. Even if the oscillator is not perfectly monochromatic, the relative phase of the two antenna currents remains constant. As an example of two coherent sources of visible light, consider two small holes or parallel slits in an opaque screen which is illuminated on one side by a distant "point" source of light. Currents are induced in the edges of the slits by the electric field of the electromagnetic radiation (light) emitted by the point source. The two slits are then said to be coherent sources of light. See Fig. 9.1.

In all these examples we need a "detector" that is responsive to the waves. In the case of the surface-tension waves in water, we may

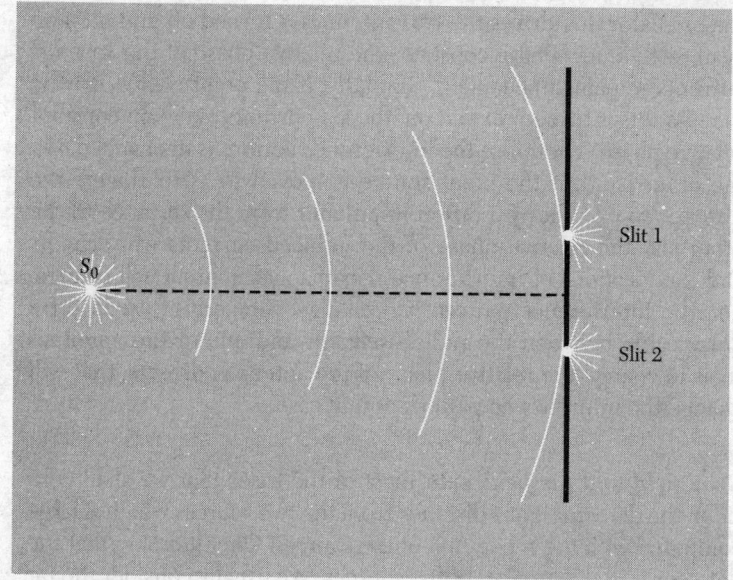

Fig. 9.1 *Two coherent sources of light. Currents in the edges of slits 1 and 2 are driven by incident waves emitted by point source* S_0. *The phase constant of* S_0 *may drift or change suddenly, but the relative phase of the slit currents remains constant.*

use a tiny piece of cork which floats on the surface and whose vertical displacement can be measured. In the case of the radio waves, we may use a detector consisting of a receiving antenna, a tuned resonant circuit, and an oscilloscope. In the case of the visible light, we may use our eyes, or a photographic emulsion, or a photomultiplier whose output current we can measure. In any case, the detector will experience a total wave that is the linear superposition of two contributions, one from each source.

Constructive and destructive interference. For some locations of the detector, the arrival of a wave crest (or trough) from one source is always accompanied by the simultaneous arrival of a crest (or trough) from the other source. Such a location is called a region of *constructive interference* or an *interference maximum*. At other locations the arrival of a crest from one source is always accompanied by the arrival of a trough from the other, and we then have a region of *destructive interference* or an *interference minimum*. Since (by hypothesis) the two sources maintain a constant relative phase, a region that is one of constructive interference at a given time will always be a region of constructive interference, and likewise a region of destructive interference at a given time will remain one for all time.

Interference pattern. The pattern formed by the various regions of interference maxima and minima is called an *interference pattern*. Even though the waves are traveling waves, the interference pattern is stationary in the sense just mentioned. Notice that even if the oscillator that drives the two antennas is turned off and then on again with a new phase constant, the relative phase of the antenna currents remains unchanged. Similarly, if the point source driving the two slits is turned off and on, the slit currents maintain constant relative phase. Therefore the interference pattern is unchanged. On the other hand, if the point source is moved so as to change the distance to one slit by a different amount from the distance to the other slit, the relative phase of the induced currents will change, and the locations of interference maxima and minima will change, i.e., the interference pattern will change. Similarly if we insert a delay cable between the radio oscillator and one of the antennas so as to change the relative phase of the antenna currents, that will change the interference pattern in that case.

Near field and far field. In most of the cases that we shall consider, the detector is at a distance from the two sources which is large compared with the separation of the sources. One then says that the detector is in the *far field* of the sources. We usually consider the far field because we can make simplifying geometrical approximations. As far as the effect of distance on wave amplitude is concerned, we can then say that the two identical sources are essentially at the same distance from the detector. In this case, each source will contribute a traveling wave having essentially the same amplitude as that contributed by the other (provided the sources are identical).

At a given position of the detector (often called the field point P), the time dependence of the total wave function is therefore given by superposition of two harmonic oscillations having the same frequency and amplitude but having (generally) different phase constants. The two phase constants (at a given field point) depend on the phase constants of the two oscillating sources and on the number of wavelengths between each source and the field point. If the distance from the field point P to one source equals that to the other source or if they differ by a whole number of wavelengths, and if the sources oscillate in phase, then P is at an interference maximum and the amplitude of its harmonic oscillation is twice the amplitude it would have if either source were present alone. (If the sources oscillate 180° out of phase, P is at an interference node and has zero amplitude.) If the distance from the field point P to one source exceeds that to the other by $\frac{1}{2}\lambda$ (plus any whole number of wavelengths) and if the sources oscillate in phase, then P is at an interference node and has zero amplitude. The approximation consists in taking the amplitudes of the individual contributions from the two sources to be exactly equal, in spite of the facts that they are in general at slightly different distances from the field point and that the amplitudes fall off with distance. Thus the amplitude at an interference minimum is generally not *exactly* zero.

A second important simplification that can be used in the far field is the approximation that the direction from source 1 to the field point P is parallel to the direction from source 2 to P. We shall utilize this approximation when we calculate (below) the interference pattern from two point sources. We now give an approximate criterion that is helpful in deciding in a given case whether use of the far-field approximation is justified. We consider a field point P at which the direction from source 1 to P is perpendicular to the line joining source 1 and source 2. (See Fig. 9.2.) The far-field approximation is justified provided we can take the direction from source 2 to P to be parallel to that from 1 to P. In this case one can assume that the relative phase of the two wave contributions at P is essentially the same as the relative phase of the two sources (for the geometry of Fig. 9.2). This approximation breaks down badly if the distance L_{2P} from source 2 to P exceeds the distance L_{1P} by one half-wavelength (or more), since then the two wave contributions at P differ in phase by 180 deg (or more) when the two sources are in phase.

"Boundary" between near and far.   Let us define a crude sort of "boundary distance" L_0 between sources and field point, such that when L_{1P} and L_{2P} are very large compared with L_0, the far-field approximation is a good one. Thus L_0 is a rough boundary between the far-field and the near-field regions. The natural choice for the boundary distance L_0 is a distance L_{1P} at which L_{2P} exceeds L_{1P} by exactly one half-wavelength. We obtain an approximate expression for this approximate boundary as follows: According to Fig. 9.2, we have (exactly)

$$L_{2P^2} = L_{1P^2} + d^2,$$

i.e.,

$$L_{2P^2} - L_{1P^2} = (L_{2P} - L_{1P})(L_{2P} + L_{1P}) + d^2.$$

But, for the case of interest, L_{2P} and L_{1P} are nearly equal to one another and both are essentially equal to L_0, since L_{2P} exceeds L_{2P} by $\frac{1}{2}\lambda$:

$$d^2 = (L_{2P} - L_{1P})(L_{2P} + L_{1P}) \approx \left(\tfrac{1}{2}\lambda\right)(L_0 + L_0).$$

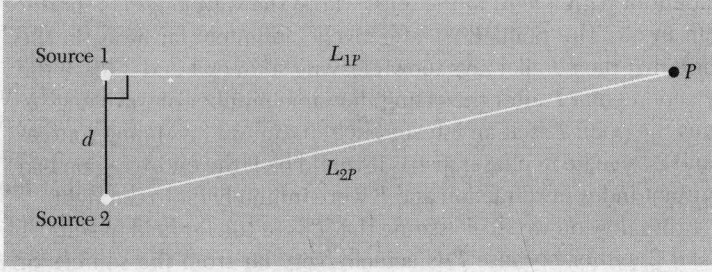

Fig. 9.2   *Far field. The detector at P is in the far field of the two sources provided L_{2P} exceeds L_{1P} by much less than one wavelength, for the Configuration shown.*

Thus for a rough criterion we can say that far-field approximations are justified for field points P much farther from the sources than a distance L_0 satisfying the relation

$$L_0\lambda \approx d^2. \tag{1}$$

Use of a converging lens to obtain far-field interference pattern. You will study experimentally the two-slit inteference pattern for visible light. (See Home Exp. 9.18.) The two coherent sources are produced as in Fig. 9.1. A typical slit separation is $\frac{1}{2}$ mm. Let us calculate how far downstream from the slits the field point must be in order to be in the far field of the double slit. Using Eq. (1) with $\lambda = 5000$ Å and $d = \frac{1}{2}$ mm, we get

$$L_0 \approx \frac{d^2}{\lambda} = \frac{(0.5 \times 10^{-1}\,\mathrm{cm})^2}{5.0 \times 10^{-5}\,\mathrm{cm}} = 50\ \mathrm{cm}.$$

Thus one should be perhaps $10L_0 \approx 5$ meters from the slit to be in the far field. That is inconvenient and unnecessary; here is how we can get a far field pattern with the double slit held right in front of your detector: The detector is your eye, which consists essentially of a photosensitive surface (the retina) and a lens. (We shall study lenses in Sec. 9.7.) The lens has a variable focal length that is varied by changing the tension in the accommodation muscles of the eye. When you look at a distant object, these muscles are relaxed (for a normal eye); the lens is then shaped so that rays from a distant point source striking different parts of the lens surface are brought to a "focus" at the retina. (If the refractive power of the lens is either too strong or too weak, the rays will not focus at the retina, and the distant object will appear blurred.) Since the source is distant, these rays are almost parallel. But this same lens (with accommodation muscles relaxed) will focus *any* parallel rays on the retina, whether or not they arise from a "distant point source." The focusing action of the lens is shown in Fig. 9.3. It turns out (as we shall show in Sec. 9.7) that although the actual distance from source 1 to P (in Fig. 9.3) is less than that from source 2 to P, *the number of wavelengths is the same.* That is possible because the path from S_1 to P has a larger amount of path length in the lens, where the wavelength is shorter than in air. The point P is "effectively" infinitely far away, in the sense that the parallel rays shown leaving sources 1 and 2 reach the detection point P after traversing the same number of wavelengths. Thus the point P is at an interference maximum (assuming sources 1 and 2 oscillate in phase) just as it would be if the entire region had constant index of refraction and P were infinitely far to the right.

From now on we shall assume that P is in the far field of sources 1 and 2, either because P is actually very far from the sources or

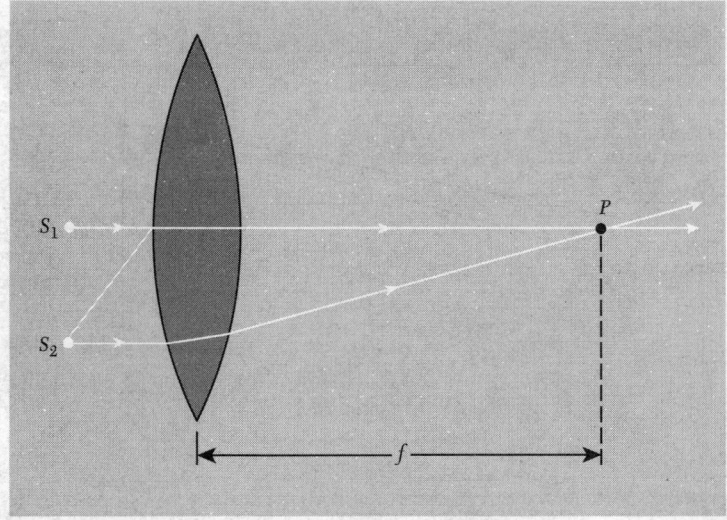

Fig. 9.3 Converging lens. Parallel rays from sources 1 and 2 are focused at point P provided the two sources oscillate with the same phase constant. The distance from the center of the lens to the focal point P is called the focal length f, for a lens whose thickness is small compared with f.

because we are using a lens and P is "effectively" very far from the sources.

Far-field interference pattern. In Fig. 9.4 we show two point sources emitting electromagnetic waves that are detected at a distant field point P. We are only going to look at the interference pattern in the plane containing the two sources and the field point P. Our results will also apply to two "line" sources (consisting of slits, in the case of light), or to two radio antennas, or to surface waves in water.

Principal maximum. When the distances r_1 and r_2 from sources 1 and 2 to the field point P are large compared with the separation d, then the two rays along the lines of sight from the two sources to point P are nearly parallel, both being at essentially the same angle θ to the z axis as shown in the figure. In that case, the path difference $r_2 - r_1$ is essentially equal to $d \sin\theta$. Therefore, if the two sources oscillate in phase, P lies in a region of constructive interference when $d \sin\theta = 0, \pm\lambda, \pm2\lambda$, etc. The interference maximum at $\theta = 0$ is called the *principal* or *zeroth-order maximum*. The first maximum on either side, where $d \sin\theta$ is $\pm\lambda$, is called the *first order maximum*, etc. The regions of destructive interference, where the total wave is always zero, are called nodes. They occur at angles where the path difference $d \sin\theta$ is $\pm\frac{1}{2}\lambda, \pm\frac{3}{2}\lambda$, etc.

We now derive an expression for the total electric field at P under the assumption that both sources undergo the same harmonic "motion," except that they may have different phase constants. We shall use a mental picture for the sources of two oscillating point

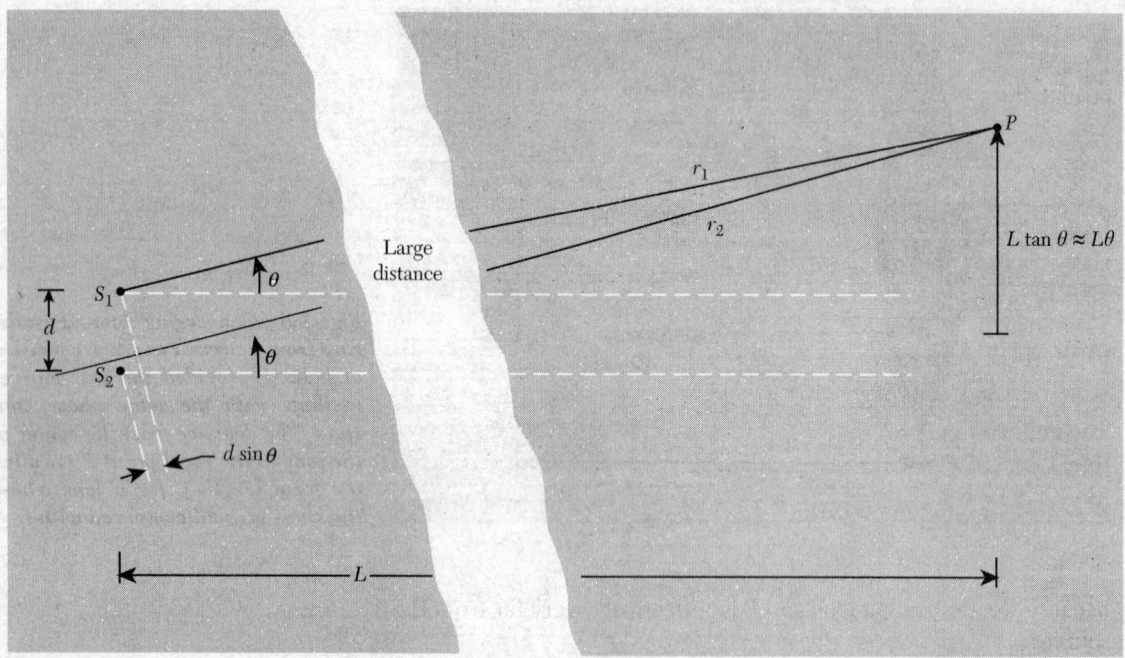

Fig. 9.4 Two point sources emitting waves which are detected at a distant field point P.

charges. We consider a single polarization component, which we can take to be one or the other of the two independent directions transverse to the line of sight from the sources to P. We need not specify the polarization, because the results obtained hold independently for either (or any other) polarization; for example, left-handed or right-handed circular polarization). However, for concreteness we consider the linear polarization component along $\hat{\mathbf{y}}$, where $\hat{\mathbf{y}}$ is perpendicular to the plane of Fig. 9.4. Then the motions of the point charges 1 and 2 have y components

$$y_1(t) = y_0 \cos(\omega t + \varphi_1),$$

$$y_2(t) = y_0 \cos(\omega t + \varphi_2). \tag{2}$$

The field point P is located at the angle θ given by Fig. 9.4 and at a distance r, where we take r to be the average of r_1 and r_2 (i.e., we put the origin of coordinates halfway between the two sources). The radiation field $E_1(t)$ at field point P due to the earlier retarded motion $y_1(t_1')$ is given by

$$E_1(t) = -\frac{q\ddot{y}_1(t_1')}{r_1 c^2}$$

$$= \frac{\omega^2 q y_0 \cos(\omega t_1' + \varphi_1)}{r_1 c^2}. \tag{3}$$

The radiation field $E_2(t)$ due to $y_2(t_2')$ is given by an analogous expression. In the far-field approximation, we take r_1 and r_2 both essentially equal to the average distance r:

$$r \equiv \tfrac{1}{2}(r_1 + r_2), \tag{4}$$

$$E_1(t) = A(r)\cos(\omega t_1' + \varphi_1),$$
$$E_2(t) = A(r)\cos(\omega t_2' + \varphi_2), \tag{5}$$

$$A(r) \equiv \frac{\omega^2 q y_0}{rc^2}. \tag{6}$$

The emission times t_1' and t_2' of the radiation detected at the later time t are given by

$$\omega t_1' = \omega\left(t - \frac{r_1}{c}\right) = \omega t - kr_1$$

$$\omega t_2' = \omega\left(t - \frac{r_2}{c}\right) = \omega t - kr_2. \tag{7}$$

Relative phase due to path difference.   Because of the fact that the path difference $r_2 - r_1$ depends on the angle θ, the relative phase of the two waves at P depends on θ. It is just this variation of the relative phase with angle that gives rise to the interference pattern. This *relative phase due to path difference* is important, so we give it a name, $\Delta\varphi$:

$$\Delta\varphi = \omega t_1' - \omega t_2'$$
$$= k(r_2 - r_1)$$
$$= k(d\sin\theta)$$
$$= 2\pi\frac{d\sin\theta}{\lambda}, \tag{8}$$

where $d\sin\theta$ is the path difference as indicated in Fig. 9.4. All the various lines of Eq. (8) are equivalent mathematically, but they correspond to different mental pictures, each of which should be learned independently. Thus, in the first line, we think about different emission times; in the last line, we think about the fact that the phase difference is 2π times the number of wavelengths of path difference; in the second and third lines; we think about the number of radians of phase per unit distance (the wave number k) times the path difference. In addition to $\Delta\varphi$ as given by Eq. (8), there is of course the phase difference $\varphi_1 - \varphi_2$ of the oscillations of the two sources.

The total field E at P is the superposition of E_1 and E_2:

$$E(r, \theta, t) = E_1 + E_2$$

$$= A(r) \cos(\omega t_1' + \varphi_1) + A(r) \cos(\omega t_2' + \varphi_2)$$

$$= A(r) \cos(\omega t + \varphi_1 - kr_1) + A(r) \cos(\omega t + \varphi_2 - kr_2). \quad (9)$$

"Average" traveling wave. Rather than express E as a superposition of two outgoing spherical traveling waves from sources 1 and 2, we can express it as a single "average" outgoing spherical traveling wave with an amplitude that is modulated as a function of the propagation direction θ and with a phase constant that is the average of the phase constants φ_1 and φ_2 of the two sources. To show this, we use the trigonometric identities

$$\cos a + \cos b = \cos\left[\tfrac{1}{2}(a+b) + \tfrac{1}{2}(a-b)\right]$$
$$+ \cos\left[\tfrac{1}{2}(a+b) - \tfrac{1}{2}(a-b)\right]$$
$$= 2\cos\tfrac{1}{2}(a+b)\ \cos\tfrac{1}{2}(a-b),$$

with

$$a = \omega t + \varphi_1 - kr_1,$$
$$b = \omega t + \varphi_2 - kr_2.$$

Then

$$\tfrac{1}{2}(a+b) = \omega t + \tfrac{1}{2}(\varphi_1 + \varphi_2) - k \cdot \tfrac{1}{2}(r_1 + r_2)$$
$$= \omega t + \varphi_{av} - kr, \quad (10)$$

$$\tfrac{1}{2}(a-b) = \tfrac{1}{2}(\varphi_1 - \varphi_2) - \tfrac{1}{2}k(r_1 - r_2)$$
$$= \tfrac{1}{2}(\varphi_1 - \varphi_2) + \tfrac{1}{2}\Delta\varphi. \quad (11)$$

Then Eq. (9) becomes

$$E(r, \theta, t) = \left\{ 2A(r) \cos\left[\tfrac{1}{2}(\varphi_1 - \varphi_2) + \tfrac{1}{2}\Delta\varphi\right]\right\}$$
$$\cos(\omega t + \varphi_{av} - kr)$$
$$= A(r, \theta) \cos(\omega t + \varphi_{av} - kr), \quad (12)$$

with amplitude $A(r, \theta)$ given by

$$A(r, \theta) = 2A(r) \cos\left[\tfrac{1}{2}(\varphi_1 - \varphi_2) + \tfrac{1}{2}\Delta\varphi\right],$$

$$\Delta\varphi = k(r_2 - r_1) = 2\pi \frac{d \sin\theta}{\lambda} \quad (13)$$

Photon flux. The photon flux at the field point P is proportional to the time-averaged energy flux $\langle S \rangle$. If we have only the single polarization component along $\hat{\mathbf{y}}$ that we have been considering, the energy flux is given by

$$\langle S \rangle = \frac{1}{\mu_0 c} \langle \mathbf{E}^2 \rangle, \quad (14)$$

with

$$\mathbf{E} = \hat{\mathbf{y}}E(r, \theta, t). \qquad (15)$$

Then

$$\left\langle \mathbf{E}^2 \right\rangle = \left\langle \left[A(r, \theta) \cos\left(\omega t + \varphi_{av} - kr\right) \right]^2 \right\rangle$$

$$= \tfrac{1}{2} A^2(r, \theta), \qquad (16)$$

with

$$A^2(r, \theta) = \left\{ 2A(r) \cos\left[\tfrac{1}{2}(\varphi_1 - \varphi_2) + \tfrac{1}{2}\Delta\varphi \right] \right\}^2. \qquad (17)$$

Two-slit interference pattern.   Let us hold r fixed and look at the variation of the photon flux with angle θ. According to Eqs. (14) through (17), we have [calling the photon flux $I(\theta)$]

$$I(\theta) = I_{max} \cos^2\left[\tfrac{1}{2}(\varphi_1 - \varphi_2) + \tfrac{1}{2}\Delta\varphi \right]. \qquad (18)$$

According to Eq. (18), *the intensity varies as the squared cosine of half the relative phase*, where the relative phase is partly that of the oscillating sources and partly that due to the dependence of the path difference on angle.

Sources oscillating in phase.   If φ_1 and φ_2 are equal, the angular dependence of the two-slit (or two-point-source) pattern is

$$I(\theta) = I_{max} \cos^2 \tfrac{1}{2}\Delta\varphi$$

$$= I_{max} \cos^2\left[\pi \frac{d \sin\theta}{\lambda} \right]. \qquad (19)$$

In Fig. 9.5 we plot this angular distribution in the region near $\theta = 0$ under the assumption that the sources are separated by many wavelengths ($d \gg \lambda$), so that $I(\theta)$ goes through many maxima and minima while θ is still rather small. This enables us to make a diagram in which we show several maxima and minima in the same small region (near $\theta = 0$).

Sources oscillating out of phase.   If φ_1 and φ_2 differ in phase by $\pm\pi$, then half their phase difference is $\pm \tfrac{1}{2}\pi$, so that Eq. (18) gives

$$I(\theta) = I_{max} \sin^2 \tfrac{1}{2}\Delta\varphi$$

$$= I_{max} \sin^2 \frac{\pi d \sin\theta}{\lambda}. \qquad (20)$$

In Fig. 9.6 we plot Eq. (20) near $\theta = 0$ for the case where d is many wavelengths, so that several maxima of $I(\theta)$ occur near $\theta = 0$.

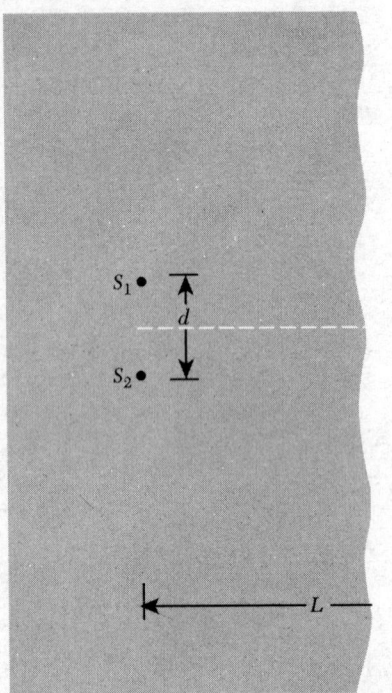

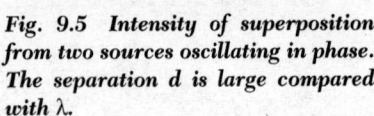

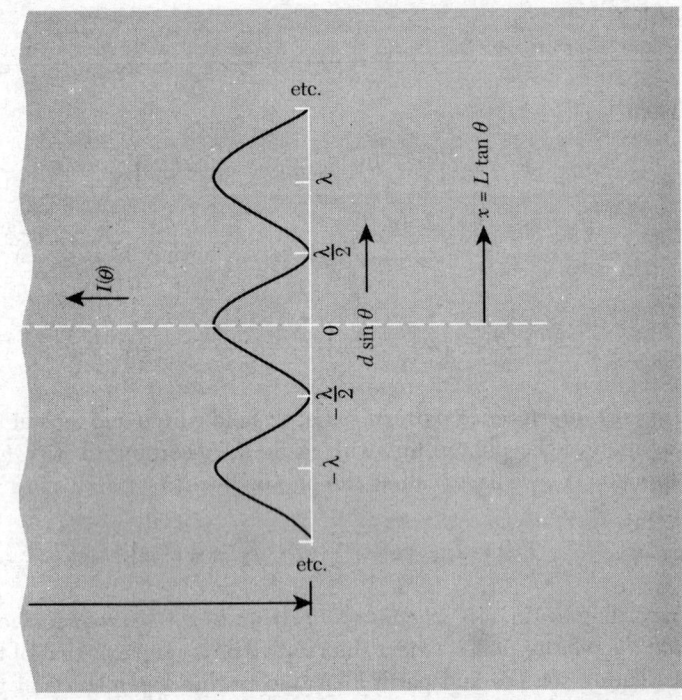

Fig. 9.5 *Intensity of superposition from two sources oscillating in phase. The separation d is large compared with λ.*

Interference pattern near θ = 0°. When you look at a line source of light with a double slit, you cannot usually tell exactly where θ = 0 occurs. Thus Figs. 9.5 and 9.6 contain more information than is usually available (at least in the home experiments). The important information is the angular interval between successive maxima or the corresponding spatial interval on a detecting screen (which may be your retina, for example). Successive maxima in Figs. 9.5 and 9.6 correspond to an increase in path difference of one wavelength, i.e., to an increase of $d \sin \theta$ by an amount λ. For θ near 0 deg, we can use the small-angle approximation $\sin \theta \approx \theta$. Then *the angular interval between successive maxima is λ/d radians.* Let us call this angular interval θ_0:

$$\theta_0 \approx \frac{\lambda}{d}. \tag{21}$$

Let us call the corresponding spatial separation between successive maxima by the name x_0. According to either Fig. 9.5 or 9.6, for θ near 0 deg, x_0 is the distance L times θ_0:

$$x_0 \approx L\theta_0 \approx \frac{L\lambda}{d}. \tag{22}$$

Energy conservation. If source 2 is turned off, the electric field at P is given by source 1 only:

$$E = E_1 = A(r) \cos(\omega t + \varphi_1 - kr_1). \tag{23}$$

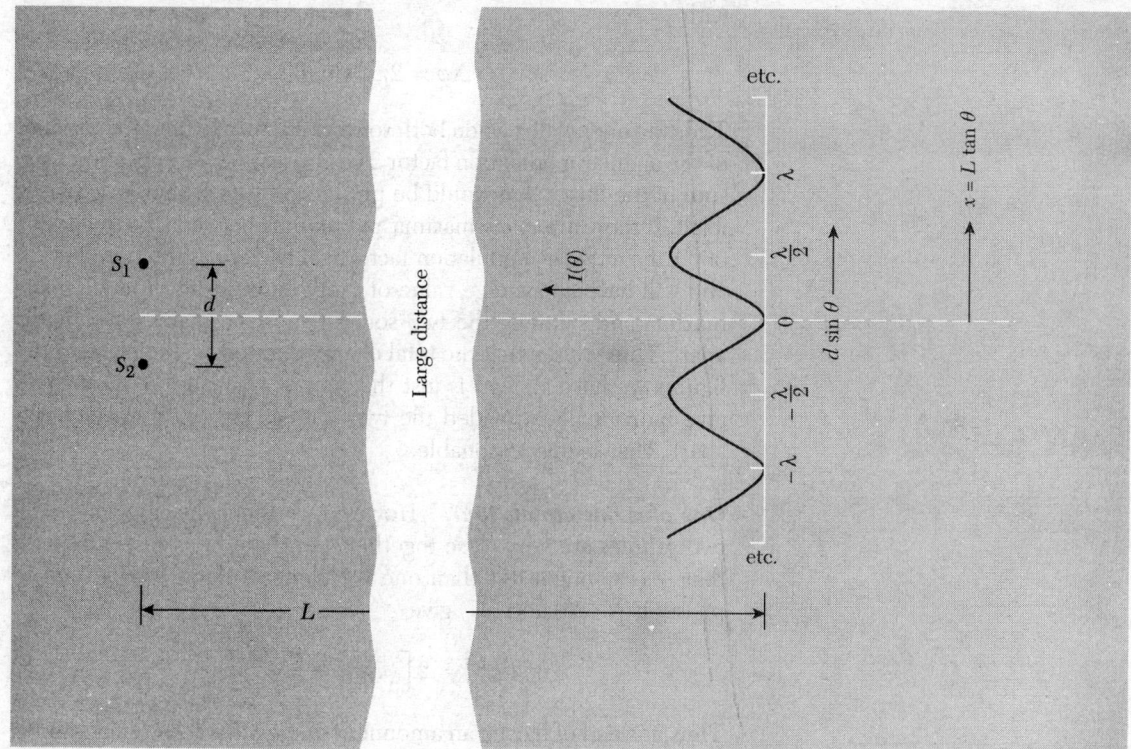

Fig. 9.6 *Intensity of superposition from two sources oscillating 180 deg out of phase.*

The photon flux is then proportional to

$$\left\langle E_1{}^2 \right\rangle = A^2(r)\left\langle \cos^2\left(\omega t + \varphi_1 - kr_1\right)\right\rangle$$
$$= \tfrac{1}{2}A^2(r), \tag{24}$$

which is independent of θ. Similarly, if only source 2 is turned on, the photon flux is proportional to

$$\left\langle E_2{}^2 \right\rangle = \tfrac{1}{2}A^2(r). \tag{25}$$

When both sources are turned on, the photon flux is proportional (with the same proportionality constant as above) to

$$\left\langle E^2 \right\rangle = \left\langle (E_1 + E_2)^2 \right\rangle$$
$$= \tfrac{1}{2}A^2(r, \theta)$$
$$= \tfrac{1}{2}\cdot\left\{ 2A(r)\cos\left[\tfrac{1}{2}(\varphi_1 - \varphi_2) + \tfrac{1}{2}\Delta\varphi\right]\right\}^2$$
$$= A^2(r)\cdot 2\cos^2\left[\tfrac{1}{2}(\varphi_1 - \varphi_2) + \tfrac{1}{2}\Delta\varphi\right].$$

Using Eqs. (24) and (25), we write this in the form

$$\left\langle E^2 \right\rangle = \left[\left\langle E_1{}^2 \right\rangle + \left\langle E_2{}^2 \right\rangle\right] 2\cos^2\left[\tfrac{1}{2}(\varphi_1 - \varphi_2) + \tfrac{1}{2}\Delta\varphi\right], \tag{26}$$

with

$$\Delta \varphi = 2\pi d \sin \theta / \lambda. \tag{27}$$

Thus the energy flux when both sources are turned on is the product of the angular modulation factor $2 \cos^2[\frac{1}{2}(\varphi_1 - \varphi_2) + \frac{1}{2}\Delta\varphi]$ times the sum of the fluxes that would be produced by each source acting by itself. If there are many maxima and minima between $\theta = 0°$ and $\theta = 360°$, the angular modulation factor will be zero as often as it is 2.0 and will have an average value of unity. In order to produce many maxima and minima, the two sources must be many wavelengths apart. Thus we see that the total energy emitted (in the plane of the figures we have shown) is just the sum of what the sources would give individually, provided the two sources are many wavelengths apart. That seems reasonable.

One plus one equals four. However, consider the case where the two sources are very close together. Let them be at a separation d that is very much less than one wavelength. If the sources are in phase, Eqs. (26) and (27) give

$$\langle E^2 \rangle \approx 2\left[\langle E_1{}^2 \rangle + \langle E_2{}^2 \rangle\right]. \tag{28}$$

Thus, instead of having an amount of energy that is the sum of what the two sources give individually, we get twice that amount. This may seem strange. Doesn't it violate energy conservation? No. The implication is that each source emits twice as much energy when the other source is sitting on top of it (and oscillating in phase) as it does when oscillating by itself. How can that be? We have prescribed the motion of each source by Eqs. (2), independent of the separation d. The energy output doubles *not* because the motion of either source changes but because the *impedance* experienced by each source has doubled! Why is that? It is because the resistive drag force exerted on the electrons in one antenna by the radiated field (taking the .two radio antennas as an example) is due not just to the field being emitted by that antenna; it is that force *plus* the force due to the field being emitted by the other antenna. Since the currents are in phase (by hypothesis) and since the antennas are in very close proximity, the net drag force exerted on the electrons in one antenna is twice what it would be if the other antenna were not present. The power supply must therefore push twice as hard to maintain the prescribed velocity, and thus we get twice the work done by the power supply. Since this holds for each antenna, we have accounted for the twofold increase in total energy emission.

One plus one equals zero. If the sources oscillate 180 deg out of phase and if you then superpose one antenna almost on top of the other, you get almost zero for the total wave amplitude. In the limit that the antennas are on top of one another, the output is zero, according to Eq. (20). The power supply does no work, and no energy is radiated. The field emitted by one antenna pushes on the electrons in the other antenna in such a way as to help the oscillator. In the limit of zero separation of the antennas, the electrons in the two antennas drive one another with no help from the oscillator. We then have a "closed" system with energy going out of one antenna into the other and back again. The antennas are then just part of the resonant circuit of the oscillator, and the power supply need only replenish the losses due to the resistance of the antennas. The radiation resistance—the characteristic impedance—has gone to zero.

9.3 *Interference between Two Independent Sources*

Independent sources and coherence time. Suppose that each of two sources has dominant angular frequency ω_0 and bandwidth $\Delta\omega$. Suppose further that the sources are independent. That is, they are not driven by a common driving force. Then there is nothing that keeps them exactly in phase. In the case of two radio antennas, that would mean that each antenna is driven by a separate oscillator and power supply. In the case of sources of visible light, it means we have two independent sources with different atoms contributing to each source. For example, we may have a mercury vapor lamp consisting of a gas discharge in a glass tube surrounded by an opaque jacket which is pierced by two pin holes or slits. Each pinhole is illuminated by different atoms of the gas. Alternatively, we may have two pinholes or slits in an opaque piece of material set in front of an ordinary light bulb. (In order to have a reasonably small band of frequencies, we could put a red gelatin filter over the slits.)

We shall suppose that the frequency bandwidth $\Delta\nu$ is small compared with the dominant frequency ν_0. Then there are many oscillations at frequency ν_0 during a time interval of length $(\Delta\nu)^{-1}$. The time interval $(\Delta\nu)^{-1}$ is the coherence time, t_{coh}; it is the time interval required for frequency components at extremes of the frequency band to get out of phase by about 2π. Thus if t_{coh} is defined by

$$\Delta\omega t_{\text{coh}} \approx 2\pi, \tag{29}$$

we see that t_{coh} is $2\pi/\Delta\omega$, i.e., t_{coh} is $(\Delta\nu)^{-1}$. For time intervals smaller than $(\Delta\nu)^{-1}$, we can think of the relative phase of the two sources as remaining essentially constant. (There can be many oscillations in such a time interval, because we assume $\nu_0(\Delta\nu)^{-1}$ is large.)

"Incoherence" and interference. Let us consider only the situation in which the separation d between the two sources is large compared with the wavelength λ. Then the interference pattern looks like Fig. 9.5 at times when the relative phase of the two sources happens to be zero. It looks like Fig. 9.6 at times when the relative phase happens to be 180 deg. For relative phases between 0 and 180 deg, the interference pattern lies between those shown in Figs. 9.5 and 9.6.

If the detector is one which requires a long time to detect the intensity at a given position, such as the eye (which has a resolution time of about $\frac{1}{20}$ sec), then the plot of time-averaged intensity versus θ will show no θ dependence, because during a time long compared with $(\Delta \nu)^{-1}$ the interference pattern will have taken on all appearances between the extremes given by Figs. 9.5 and 9.6, and every value of $d \sin \theta$ will have experienced the same time-averaged intensity. One then says that the two point sources are "incoherent." The time-averaged energy flux (the photon flux) is then just the sum of the fluxes one would get for either source by itself. The interference pattern is "washed out" because of the long time average during the measuring process. This fact is expressed algebraically by noting that Eq. (26), Sec. 9.2, gives $\langle E^2 \rangle \approx \langle E_1{}^2 \rangle + \langle E_2{}^2 \rangle$, independent of θ, provided that the relative phase $\varphi_1 - \varphi_2$ takes on all possible values with roughly equal amounts of time spent in each small interval of relative phase between zero and 2π. That follows from

$$\left\langle \cos^2 \left[\tfrac{1}{2}(\varphi_1 - \varphi_2) + \tfrac{1}{2} \Delta \varphi \right] \right\rangle = \tfrac{1}{2}, \tag{30}$$

for fixed $\Delta \varphi$ and for $\varphi_1 - \varphi_2$ uniformly distributed from zero to 2π.

It is clear that there are no "intrinsically" incoherent sources. "Incoherence" is merely the result of a measurement process which throws away information that is available in the interference pattern if one has the technique to look at times comparable to or shorter than $(\Delta \nu)^{-1}$. For visible light, the coherence times are of order 10^{-9} to 10^{-8} sec (for a source consisting of independently radiating atoms in a gas-discharge tube), so that it takes some experimental ingenuity to measure the interference pattern before it changes. Nevertheless, it has been done in a very beautiful experiment by R. Brown and R. Twiss.[†]

Brown and Twiss experiment. The method by which Brown and Twiss effectively "read" the interference pattern in a time less than 10^{-8} sec is as follows: One uses two photomultipliers at different values of x (as defined in Figs. 9.5 and 9.6) and with a variable

† R. Hanbury Brown and R. O. Twiss, "The Question of Correlation between Photons in Coherent Light Rays," *Nature* **178**, 1447 (1956). For a more recent experiment using lasers, see R. Pfleegor and L. Mandel, "Interference of Independent Photon Beams," *Phys. Rev.* **159**, 1084 (1967).

separation $x_1 - x_2$. The output current of one photomultiplier, I_1, is multiplied by that of the other, I_2, in a fast circuit which can follow current fluctuations that occur in times of order 10^{-8} sec. (In other words, the fast circuitry has 100-Mc bandwidth.) The product $I_1 I_2$ is determined "instantaneously," i.e., in a time interval at the 10^{-8} sec level, but then the average of this product, $\langle I_1 I_2 \rangle$, is taken over a long time interval of many minutes. The separation $x_1 - x_2$ of the two photomultipliers is varied, and the time average of the product of currents is taken at each separation. Finally, one plots the time-averaged product versus $x_1 - x_2$. Now, the instantaneous current in each photomultiplier is proportional to the flux of light energy, i.e., to $I(\theta)$ at that photomultiplier. First let us consider the case where the separation $x_1 - x_2$ is zero, so that each photomultiplier is subject to the same instantaneous light flux. Let us perform a very crude average of the product of the two currents. Let us say that $I(\theta)$ only takes on the four values indicated by a, b, c, and d in Fig. 9.7. Let us call the corresponding currents by the names a, b, c, and d and give them units in which we shall have $a = 0$, $b = \frac{1}{2}$, $c = 1$, and $d = \frac{1}{2}$. For one-fourth of the "instants" (of duration about 10^{-8} sec), PM 1 (photomultiplier 1) has current I_1 corresponding approximately to a. At the same times, I_2 is also equal to a, since PM 2 is at the same place as PM 1. One-fourth of the time, each has current corresponding to b, one-fourth of the time to c, and one-fourth of the time to d as the interference pattern shifts. Thus the time average of the product of the two currents for $x_2 = x_1$ is given (in our crude approximation) by

$$(I_1 I_2)_{\text{av}} = \tfrac{1}{4}(aa + bb + cc + dd)$$

$$= \tfrac{1}{4}\left(0 \cdot 0 + \tfrac{1}{2} \cdot \tfrac{1}{2} + 1 \cdot 1 + \tfrac{1}{2} \cdot \tfrac{1}{2}\right) = \frac{3}{8}. \tag{31}$$

Now let us find the average of $I_1 I_2$ when the separation $x_2 - x_1$ is that between an "instantaneous" interference maximum and the

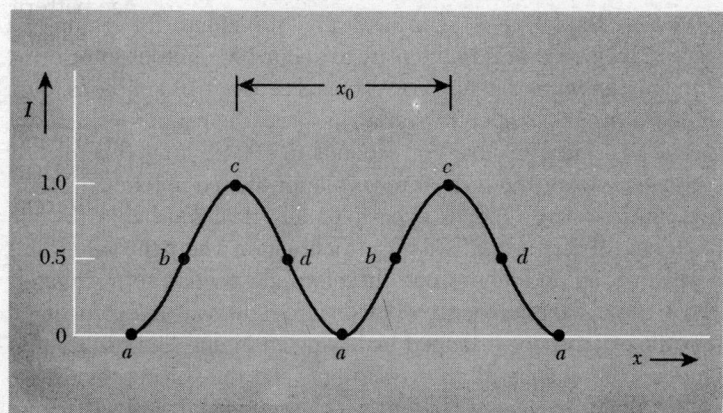

Fig. 9.7 *Graphical illustration of Brown and Twiss experiment.*

neighboring minimum, i.e., when it is half of x_0, where x_0 (as shown in Fig. 9.7) is the separation between successive maxima of the instantaneous two-slit interference pattern. (It is given by Eq. (22), Sec. 9.2.) If $x_2 - x_1 = \frac{1}{2} x_0$, then at an instant when PM 1 happens to have current a, PM 2 has (according to Fig. 9.7) current c. When PM 1 has b, PM 2 has d, etc. Thus, for the time average over the four representative currents a, b, c, d for PM 1, we have

$$
\begin{aligned}
(I_1 I_2)_{\text{av}} &= \tfrac{1}{4}(ac + bd + ca + db) \\
&= \tfrac{1}{4}\left(0 \cdot 1 + \tfrac{1}{2} \cdot \tfrac{1}{2} + 1 \cdot 0 + \tfrac{1}{2} \cdot \tfrac{1}{2}\right) \\
&= \tfrac{1}{8}.
\end{aligned}
\tag{32}
$$

We see that $(I_1 I_2)_{\text{av}}$ is three times larger when $x_2 - x_1$ is zero than when it is half of the separation between successive maxima of the instantaneous pattern. Thus we see that a plot of $(I_1 I_2)_{\text{av}}$ versus $x_2 - x_1$ will determine the relative phase $\Delta\varphi = 2\pi\, d \sin\theta/\lambda$.

The crux of the technique of Brown and Twiss is that in the product $I_1 I_2$ each current is only averaged over times of order 10^{-8} sec, and during this time the currents are essentially constant. The average $\langle I_1 I_2 \rangle$ over a time interval of minutes is just what they would get by averaging over several dozen coherence times, say over 10^{-6} sec. (They average over much longer times so as to average out photomultiplier noise and for other experimental reasons.) The product $\langle I_1 \rangle \langle I_2 \rangle$, on the other hand, is independent of $x_1 - x_2$, because each photomultiplier has sampled the entire interference pattern during the time of averaging. The essential thing is to find out which separations $x_1 - x_2$ correspond to situation that I_1 is large when I_2 is large and is small when I_2 is small (as when $x_1 - x_2$ is zero) and which separations correspond to I_1 small when I_2 is large and vice versa.

In terms of photons, one finds that the probability for detecting a photon in photomultiplier 2 is larger than average when photomultiplier 1 has "recently" (within 10^{-8} sec) detected a photon, provided one has $x_1 = x_2$; it is smaller than average if one has $x_1 - x_2 = \frac{1}{2} x_0$. To put it very crudely and "semiclassically," if one has (for example) a wave that corresponds in intensity to about 100 photons interfering with another wave that also corresponds to about 100 photons, then if the wavetrains happen to overlap in space, their superposition can give a total intensity that corresponds to 400 photons (completely constructive interference) or to zero intensity (completely destructive interference). This is experimentally distinguishable (by the technique of Brown and Twiss) from a situation where the wavetrains never overlap and where one therefore always has approximately $100 + 100 \approx 200$ photons. It is obvious from this way of expressing it that the experiment is helped by having an intense light source (to increase the chance of an overlap between the wavetrains or two photons) and by having photons with narrow bandwidths [because the length of the wavetrain is essentially c times the mean decay time τ (that is, $c/\Delta \nu$), and a long wavetrain means a higher chance for overlap].

9.4 How Large Can a "Point" Light Source Be?

In Fig. 9.1 we showed how one can obtain two coherent sources of light (two sources whose relative phase remains constant) by irradiating two slits in an opaque screen with radiation from a "point" source of light. On the other hand, if the source is so broad that one slit is illuminated mostly by one set of atoms and the other by another independent set, then the two slits are completely incoherent, i.e., their phases are uncorrelated [for measurement times that are long compared with $(\Delta \nu)^{-1}$]. These two extremes are illustrated in Fig. 9.8.

Classical point source.   The closest we can get to having a point source is to have a single atom. According to the classical picture, this atom emits electromagnetic waves in all directions and drives the slit currents in Fig. 9.8a with equal phase. (The quantum theory gives effectively the same result.) A practical source of light will have a huge number of radiating atoms. If they were all sitting at exactly the same point, we would have a point source. (It would be even more like a classical point source than is a single real atom.) But in any practical source the atoms are in a region of finite dimensions. How large can a light source be and still be an "effective" point source (meaning that the slit currents in the double slit irradiated by the "point" source maintain constant relative phase)?

Simple extended source.   Let us consider a very simple source that is not a point source. It consists of three independent point sources a, b, and c, each with the same dominant frequency, bandwidth, and average intensity, arrayed as shown in Fig. 9.9. Suppose we start out with only point source a turned on. Then slits 1 and 2 are driven at constant relative phase (which happens to be zero for our figure) and are coherent over any time interval. Next turn on both sources a and c. Source c is a light source with the same frequency and bandwidth as source a but not correlated in phase with source a. Thus c and a do not maintain constant relative phase over times long compared with $(\Delta \nu)^{-1}$. Nevertheless, the relative phase of slits 1 and 2 remains zero for all time, because source c drives the slit currents with zero relative phase just as does source a. The slit currents may be regarded as a superposition of the currents induced by the two sources, and if each source contribution gives zero relative phase between the slit currents, so does the superposition. Thus we conclude that we can extend the point source along the line connecting a and c without spoiling the coherence of slits 1 and 2.

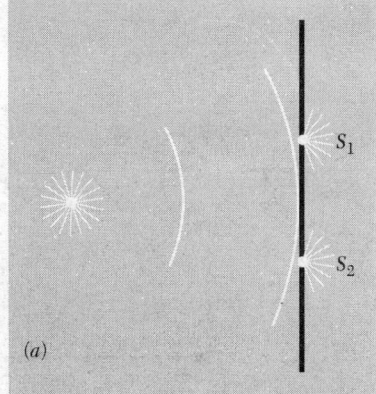

(a)

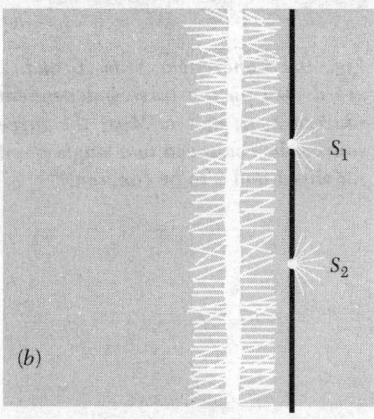

(b)

Fig. 9.8 (a) Sources 1 and 2 are driven by a single point source and maintain constant relative phase. They are coherent. (b) Sources 1 and 2 are driven by different sets of independently radiating atoms. For measurement times long compared with $(\Delta \nu)^{-1}$, they are incoherent.

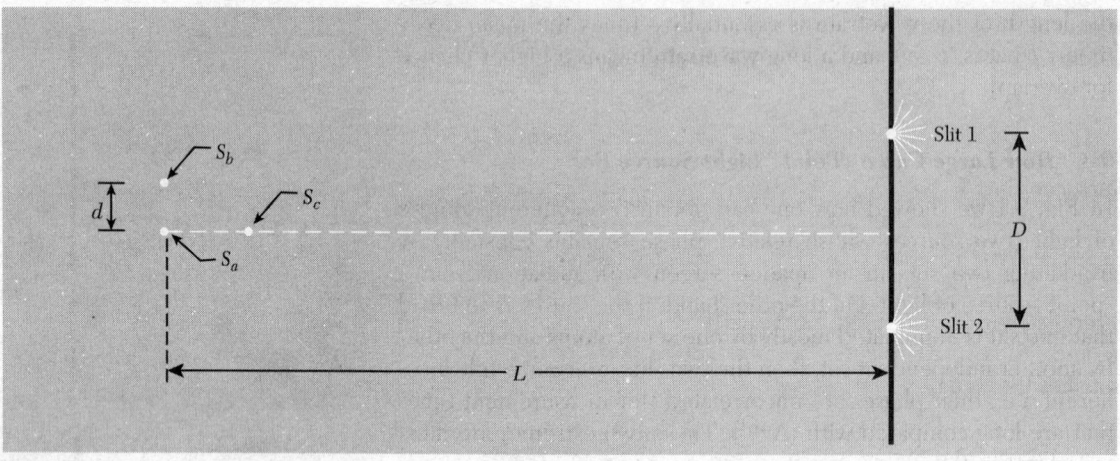

Fig. 9.9 Coherence. Slits 1 and 2 are driven by the three independent sources a, b, and c. Must the three sources be coalesced to a single point for slits 1 and 2 to be coherent?

Now consider the situation where both sources a and b are turned on (with c turned off). Sources a and b are independent sources having the same dominant frequency and bandwidth and the same average intensity. During any time interval short compared with $(\Delta \nu)^{-1}$, the amplitude and phase constant of each source remain constant. Suppose that for a given instant [an instant means a time interval short compared with the coherence time $(\Delta \nu)^{-1}$ but long enough to have at least one complete fast oscillation, so that we can tell what the amplitudes and phases are] it so happens that the amplitude of b is very small compared with that of a. Then to a good approximation the two slits are irradiated only by a, and the slit currents therefore have zero relative phase. Now let us wait a time long compared with the coherence time of sources a and b and look again.

Suppose that this time it so happens that the amplitudes of the oscillations of a and b are practically equal. In this case, the screen with the slits is irradiated by the two-source interference pattern that we have seen in Figs. 9.5, 9.6, and 9.7. The locations of the maxima and minima depend on the relative phase of sources a and b. The question of interest is whether or not the two slits 1 and 2 are still driven with zero relative phase. We know that the amplitude of the interference pattern changes sign when we go from one interference maximum to the next. [According to Eq. (13), Sec. 9.2, the amplitude $A(r, \theta)$ is proportional to the cosine of $\frac{1}{2}(\varphi_1 - \varphi_2) + \pi d \sin \theta / \lambda$. It thus changes sign when $d \sin \theta$ increases by an amount λ, as between successive interference maxima.] We see that both slits are driven at zero relative phase most of the time only if they are separated by much less than the separation x_0 between successive interference maxima of the two-source interference pattern. (Even when the slits are closely spaced, it may happen that a zero of the two-source pattern irradiating them falls between the two slits, in which case they are driven 180 deg out of phase. However, this happens a smaller and smaller fraction of the time as the slits get closer together.) Thus we need

$$D \ll x_0, \tag{33}$$

where x_0 is the spatial separation between successive maxima and is given according to Eq. (22), Sec. 9.2, by

$$x_0 = L\frac{\lambda}{d}. \tag{34}$$

Coherence condition. The "extended source" consisting of points a, b, and c therefore acts like an effective point source provided that it satisfies the *coherence condition*,

$$D \ll \frac{L\lambda}{d}, \tag{35}$$

i.e.,

$$d \ll \frac{L\lambda}{D}, \tag{36}$$

i.e.,

$$L \gg \frac{dD}{\lambda}, \tag{37}$$

where one or the other of these forms may be most appropriate depending on what parameters are experimentally variable. [You can verify Eq. (37) in an easy home experiment (see Home Exp. 9.20). In that experiment L is the variable.] The easiest way to remember the coherence condition is in the form

$$\boxed{dD \ll L\lambda,} \tag{38}$$

which says that the product of the two transverse widths d and D must be small compared with the product of the two longitudinal lengths L and λ.

If the source consists of a huge number of points between a and b, so that the source has width d, Eq. (38) applies to the entire source if it applies to the extreme points a and b (i.e., point sources closer together than d are coherent if those d apart are). Similarly, when we (later) consider several or many slits in a screen instead of just two, the coherence condition Eq. (38) can be applied to the entire array of slits with D taken as the separation of the outermost slits.

9.5   Angular Width of a "Beam" of Traveling Waves

A "beam" of traveling waves is a pattern of waves traveling in a given direction and having a finite lateral width. A flashlight beam of visible light and a radar beam of microwaves can each be made by putting a small source of electromagnetic radiation at the focal point of a parabolic reflector. The small source drives the electrons in the metallic surface of the reflector with just the proper phase relations so that the reflected radiation from all points of the surface interferes constructively along the direction of the beam. Another way to get a light beam is to reflect light from a small or distant source (the sun, for example) off a small plane mirror. Alternatively, we can use a hole in an opaque screen instead of a mirror. If the source is sufficiently far away and sufficiently small, the radiation incident on the mirror (or hole) can be approximated as a plane wave—i.e., one with all the radiation traveling in exactly the same direction. Then the mirror reflects "part of the plane wave." Similarly in the case of the small source at the focus of a parabolic mirror, if the source is sufficiently small and the mirror is a perfect paraboloid, the beam is (to a certain approximation) like a "segment of a plane wave," consisting of radiation all traveling in the same direction. All these considerations hold equally well for sound waves or water waves.

Angular width of a beam is diffraction limited. Now comes the interesting and very important question: Can one, by very careful design, make a beam of waves that is just like a "cross-sectional segment" of a plane wave, in the sense that all the waves are traveling in exactly the same direction, so that one has a perfectly parallel beam that will continue forever with the same width? *No.* No matter how small the point source at the focus of a perfect parabola, the radiation in the beam will not be perfectly parallel. If the "dominant" direction is along z and the spatial width of the beam (at a given value of z, say at the reflector) is D, then there will be an angular distribution of propagation directions with a "full width at half maximum intensity" of about λ/D. (We will show this below.) Similarly if we have a perfectly plane wave from a distant point source falling on a hole of width D (or a mirror of width D), the angular width of the transmitted beam is about λ/D. The angular width can only be zero if D is infinite (or if λ is zero). The angular width of the beam is said to be *diffraction limited.* In Fig. 9.10 we show some examples of beams. Note that if the original width of the beam is D, and if every attempt is made to make the beam as perfectly parallel as possible, the width W after the beam has traveled a large distance L is approximately the original width D plus L times the angular full width λ/D. For large enough L, we can neglect the original width D. Thus we have

Angular full width: $\boxed{\Delta\theta \approx \dfrac{\lambda}{D}}$, (39)

Beam width: $W \approx L\dfrac{\lambda}{D}.$ (40)

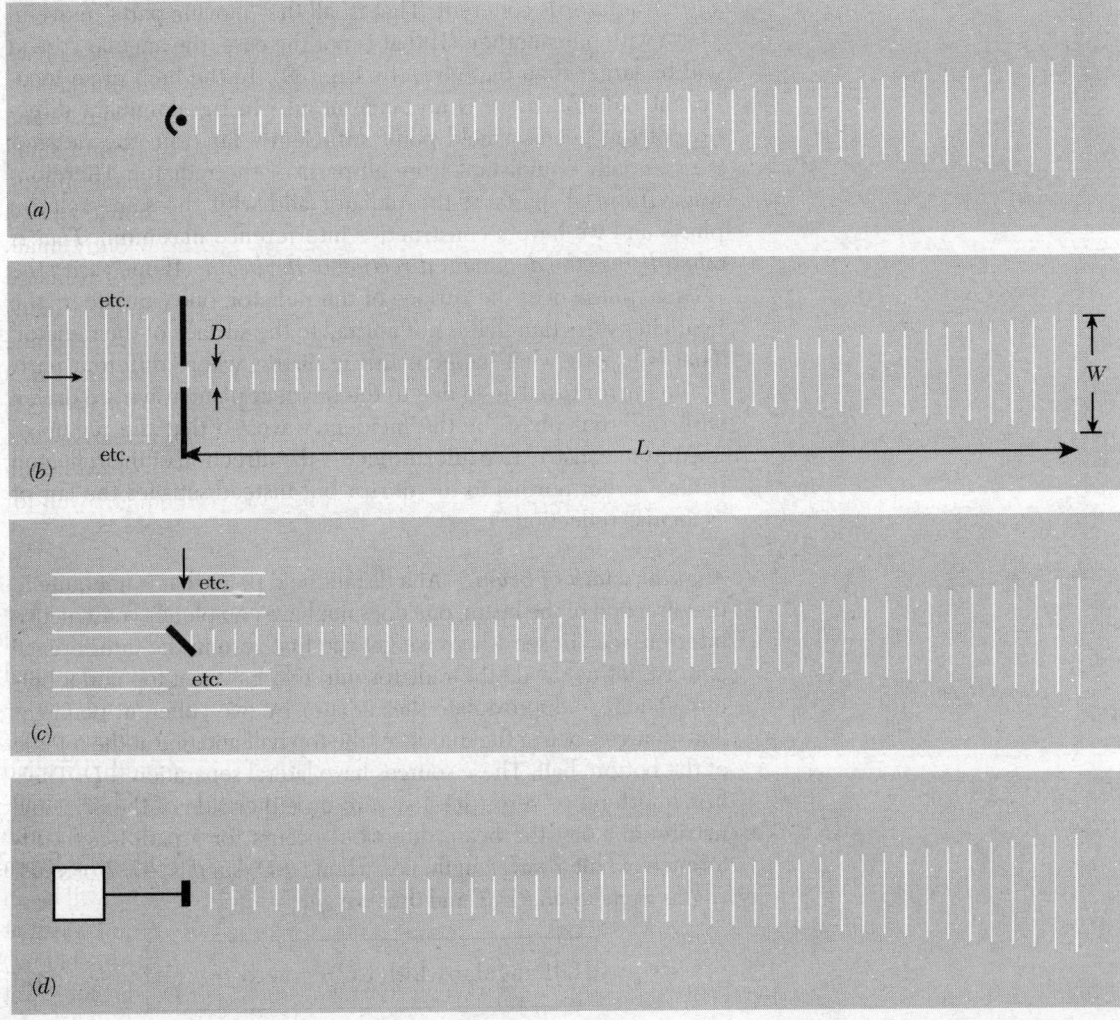

Fig. 9.10 Diffraction. Beam of width D has angular width ≈ λ/D and spreads by an amount W ≈ Lλ/D in traveling a distance L. (a) Beam made by point source and parabolic mirror. (b) Beam made by plane wave incident on hole in opaque screen. (c) Beam made by plane wave incident on plane mirror. (d) Beam emitted by plane radiator with all parts oscillating in phase.

Every one of the four sketches in Fig. 9.10 can be taken to represent either water waves, sound waves, or electromagnetic waves (visible light of wavelength 5×10^{-5} cm, for example, or microwaves of wavelength 10 cm, for another example).

A beam is an interference maximum.   Here is a crude derivation of Eq. (39). (In Sec. 9.6 we shall give an exact derivation.) The result is independent of the kind of waves and is independent of how the waves are produced. We might as well take the simplest source, which is probably a plane radiator as shown in Fig. 9.10*d*. For sound waves, this can be an oscillating piston *in free air*. For electromagnetic waves, it can be an oscillating sheet of charge of finite extent, for example a plane antenna array. In any case, *the* *entire radiator* is coherent. That is, all the "moving parts" move in phase with one another. (If that is not the case, the angular spread will be larger than that given by Eq. (39). In the limit of an incoherent radiator, there is no beam at all.) In the dominant direction of the beam, a field point sufficiently far from the radiator is essentially equidistant from all parts of the radiator. Therefore waves from all parts of the radiator add with the same relative phase and we have a constructive interference maximum. *That is what defines the dominant direction of the beam.* (If one varies the relative phase over the surface of the radiator, one can "steer" the beam in a direction that is not normal to the surface of the radiator. That is exactly what happens in Fig. 9.10c, where different parts of the mirror tilted at 45 deg to the incident plane wave are driven with different phase by the incident wave, so that the region of maximum constructive interference—the direction of the reflected beam—is not normal to the mirror but instead satisfies the law of "specular reflection.")

Angular width of beam. At a distant field point that is not quite in the direction of the beam, one does not have completely constructive interference. To see where we get the first zero in the interference pattern, let us divide the radiator into two halves, a top and a bottom. Then we approximate the radiator by two coherent point (or line) sources, one at the middle of the top half and one at the middle of the bottom half. These sources have lateral separation $\frac{1}{2}D$. Their first interference zero (the first zero on either side of the principal maximum along the beam direction) occurs for a path-length difference of half a wavelength, i.e., when $(\frac{1}{2}D)\sin\theta$ is $\frac{1}{2}\lambda$. For small angles we take $\sin\theta = \theta$, and thus we get

$$\text{Half angular width to first zero} = \frac{\lambda}{D}. \qquad (41)$$

This is shown in Fig. 9.11.

Where does the next maximum occur? If points 1 and 2 of Fig. 9.11 really were point (or line) sources, the next maximum would occur when the path length from source 2 to the field point exceeded that from source 1 by one wavelength. Indeed the top half and bottom half *are* in phase then, but they each contribute zero! That is because if you divide the top and bottom halves themselves into halves, so that the entire radiator is divided into four quarters, then the contribution of the first quarter is 180 deg out of phase with that of the second and cancels it; that of the third is 180 deg out of phase with that of the fourth and cancels it. Thus the first subsidiary maximum actually comes not when we have two halves with contributions differing in phase by 2π (since then we have four quarters with contributions differing in phase between

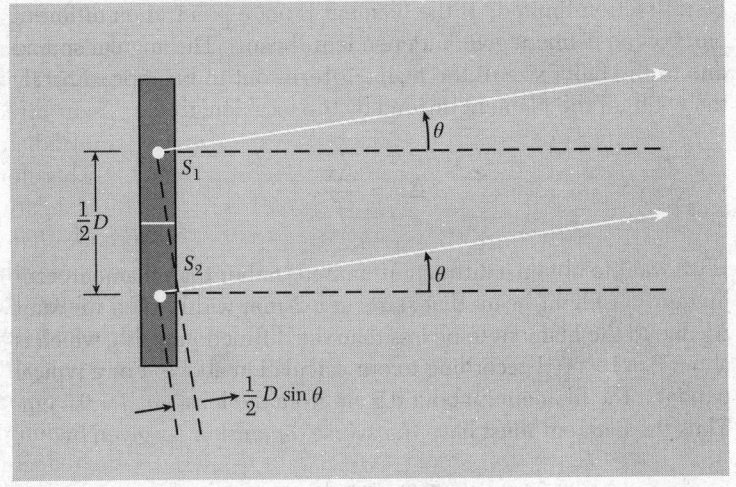

Fig. 9.11　Plane radiator. Source 1 represents the contributions from the top half, source 2 those from the bottom half.

successive quarters by π) but rather when we have *three* thirds of the radiator with adjacent thirds differing in phase by π from one another. *Two* of the thirds cancel one another, but the *third* third remains. Thus the amplitude for the first subsidiary maximum is smaller than that of the principal maximum by at least a factor of $\frac{1}{3}$ (it is actually smaller by more than that because of phase differences within the one-third contribution that is left). We see that the subsidiary maxima have small amplitude compared with the central maximum that gives the "beam" direction. When we study the exact pattern, we will find that the half angular width to the first zero is equal to the full angular width at *about* half maximum intensity, which is how we have defined the angular width of the beam in Eq. (39). Thus we have derived Eq. (39), roughly. (The exact result is given in Fig. 9.14, Sec. 9.6.)

Application: Laser beam versus flashlight beam

Suppose you have a diffraction-limited laser beam of diameter $D = 2$ mm, with wavelength 6000 Å. How much does the beam diameter increase in a distance of 15 m? The angular spread of the beam is

$$\Delta\theta \approx \frac{\lambda}{D} \approx \frac{6 \times 10^{-5}\,\text{cm}}{0.2} \approx 3 \times 10^{-4} \text{ rad.}$$

The angular spread times the distance $L = 15$ m $= 1500$ cm gives a spatial spread of $W \approx (1500)(3 \times 10^{-4}) \approx 0.5$ cm $= 5$ mm. (This can be nicely demonstrated in the classroom with a laser.) If you have a "penlight" type of flashlight with a beam of diameter 2 mm at the flashlight formed by a "point" filament at the focus of a lens, how small would the filament have to be for the flashlight beam to be diffraction-limited? If the filament is not a point, then different parts of the filament give "independent" beams. The angular spread due to the finite size of the filament turns out to be approximately the width of the filament divided by the focal length f:

$$\Delta\theta \approx \frac{\Delta x}{f}.$$

If we want to obtain a diffraction-limited (rather than filament-size-limited) flashlight beam that starts at a 2-mm width, then we want $\Delta\theta$ due to the filament to be less than the diffraction width, which is about 3×10^{-4} rad according to our calculation above. For a typical penlight, the filament is about 0.5 cm from the lens; i.e., $f \approx 0.5$ cm. Thus the filament must have transverse dimension Δx given by

$$\Delta x < f\Delta\theta \approx (0.5)(3 \times 10^{-4}) = 1.5 \times 10^{-4} \text{ cm}.$$

Such a small filament is hard to make.

9.6 Diffraction and Huygens' Principle

Difference between interference and diffraction. In Sec. 9.5 we discussed the angular width of a diffraction-limited beam. We gave a crude derivation of the diffraction pattern produced when an infinite plane wave strikes an aperture in an opaque screen (Fig. 10*b*) or a mirror (Fig. 10*c*) or is emitted by a plane radiator (Fig. 10*d*). In previous sections we discussed the interference pattern produced by two point or line sources. What is the difference between an interference pattern and a diffraction pattern? None, really. For historical reasons, the amplitude or intensity pattern produced by superposing contributions from a finite number of discrete coherent sources is usually called an *interference* pattern. The amplitude or intensity pattern produced by superposing the contributions from a "continuous" distribution of coherent sources is usually called a *diffraction* pattern. Thus one speaks of the interference pattern from two narrow slits, or the diffraction pattern from one wide slit, or the combined interference and diffraction pattern from two wide slits.

In Sec. 9.5 we assumed that the diffraction-limited beam produced when a plane wave is incident on an aperture in a screen (Fig. *10b*) is equivalent to that produced by a plane radiator having the size of the aperture, with all parts of the radiator oscillating in phase and with the same amplitude (Fig. *10d*). In the present section we shall seek to justify that assumed equivalence. In so doing, we shall find that the equivalence is not exact; it is a useful approximation that greatly simplifies the calculation of diffraction patterns. It only works if the aperture width is large compared with the wavelength. In that case, it works very well for calculating the radiation emitted at not too large angles from the beam direction and thus for calculating the intensity and amplitude sufficiently far downstream from the aperture or equivalent radiator. It is not of any use if you wish to know the fields inside the aperture itself. The calculation technique that makes use of this assumed equivalence is called *Huygens' construction.* We shall use it to calculate the diffraction pattern produced when a plane wave (produced, for example, by a distant point source) strikes a hole in an opaque screen.

How an opaque screen works. All electromagnetic radiation has its ultimate origin in oscillating charged particles. The total electric (and magnetic) field at any given point is a superposition of the waves produced by all the sources, i.e., all the oscillating charges. In the present problem, one of the sources is the distant point source that produces the plane wave incident on the screen. We shall call this the source S. Behind the opaque screen the total wave amplitude is zero (by hypothesis—that is what we mean by an opaque screen). This total wave is a superposition of the wave from S and the waves emitted by oscillating electrons in the material of the screen. That is, the screen doesn't just gobble up the incident wave from S. Its electrons are driven by the incident radiation (and also by the radiation emitted by the other electrons of the screen), and the superposition of *all* of the waves, i.e., from S and from all the electrons, gives zero behind the screen. If this seems strange to you, recall how it comes about that a static electric field is zero inside a good metallic conductor. The conductor doesn't eat up the external driving field. That field is still there inside the conductor, but charges move in the conductor (before static equilibrium is established) and come to rest on the surfaces until finally the *superposition* of the fields from the surface charges and the incident field gives zero total field inside the conductor. All electromagnetic fields come from charged particles, and such a "zero" field as that behind an opaque screen is the result of a superposition.

If you have a mental picture of the electric lines of force of a charged particle as being like a little stream of bullets sent out at the velocity of light from the point charge, you will get into trouble. Little bullets do not obey the superposition principle. They do not pass through one another without disturbance. Two bullets cannot superpose to give zero bullets. With this misleading mental picture, you will probably think of the effect of a metallic conductor on an electrostatic field as one of "stopping the bullets," like a sort of armor. That is also the way in which you might incorrectly think of a screen which is opaque to incident light, as a sort of armor that stops the light and eats it up, turning it into heat (if the screen is black) or bouncing the bullets back (if the screen is a shiny metal foil). This is a bad picture. If you have it, you will not be the first—but it is wrong. Get rid of it.

Shiny and black opaque screens.   There are two extremes in the varieties of opaque screen. At one extreme we can have a shiny opaque screen (such as an opaque piece of aluminum foil). The electrons in the metal are driven by the local electric field; consequently they emit electromagnetic waves. In the forward direction (the direction of the incident radiation), it turns out that the superposition of the incident wave and that from the electrons gives zero. In the backward direction, it gives a reflected wave. Far from any resonance, the motion of a given electron is entirely due to the elastic amplitude, and thus the velocity is 90 deg out of phase with the total electric field at its location; therefore no work is done on the electron during any complete cycle. (The electron "redirects" the radiation energy without permanently absorbing any energy.)

At the other extreme, we can have a black opaque screen (such as black cardboard or a microscope slide painted with a layer of "aquadag"—soot suspended in water). Again the electrons are driven by the incident radiation. They also suffer a resistive drag from the medium and are always at terminal velocity. Their radiation in the forward direction is 180 deg out of phase with the incident radiation and superposes with it to give zero (after sufficient thickness of screen). The velocity of a given electron is always in phase with the total electric force at its location, and consequently net work is done on the electron. The work done on the electron is transferred to the medium, which gets hotter. There is no net reflected wave—the contributions from different layers of the screen superpose to give zero in the backward direction.

Effect of a hole in an opaque screen.   Now let us proceed to cut a small hole (or slit) in our opaque screen. First let us mark the material to be removed. This will be called slit number 1, so we label the material to be removed plug 1. The material of the screen above and below plug 1 is labeled a (above) and b (below). The total field behind the screen, which is zero, is the superposition of the fields emitted by the source S and by the material from a, b, and plug 1. Thus, *before removing the material of slit* 1, we have

$$E = 0 = E_S + E_a + E_b + E_1. \tag{42}$$

This situation is shown in Fig. 9.12.

Now we remove the material that is plugging up slit 1. *Assume* that the motion of the electrons in the regions a and b is not changed by removal of the plug. (This is an approximation, because the

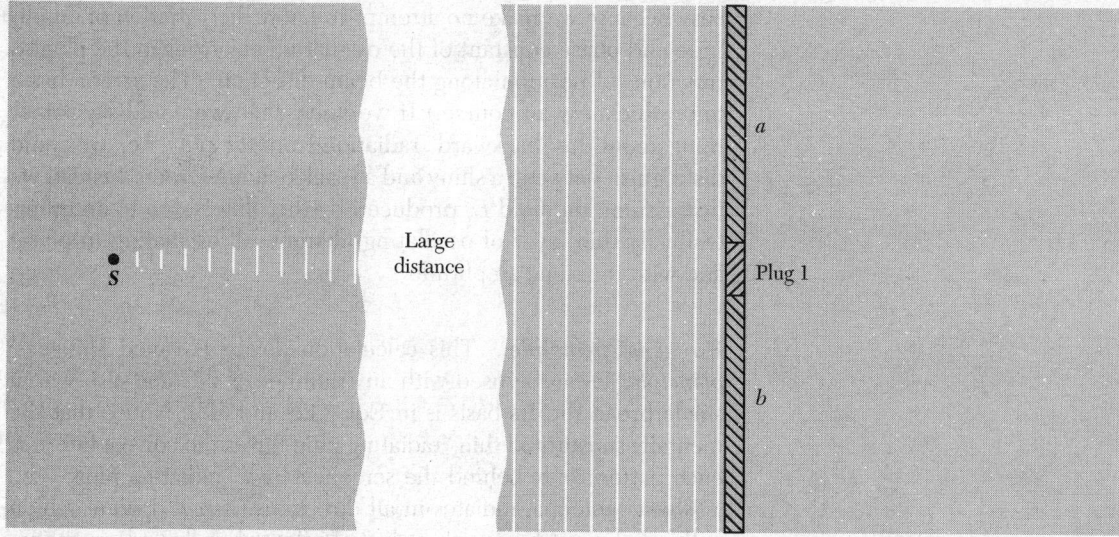

Fig. 9.12 *Plane waves from distant point source S are incident on opaque screen. Superposition of fields due to charges at S, a, b, and plug 1 gives zero behind the screen.*

electrons in regions *a* and *b* are driven by the total field at their locations, and that includes the fields radiated by the electrons that were in the plug. Those electrons in *a* and *b* within a few wavelengths of the edge of the slit will be most affected by removing the plug, because the radiation from a given electron falls off with increasing distance from the electron, so that the nearest neighbors are the most important.) With this assumption the total field behind the screen is no longer the superposition given by Eq. (42), which adds to zero. Instead it is that superposition minus the contribution from plug 1:

$$E = E_S + E_a + E_b$$
$$= (E_S + E_a + E_b + E_1) - E_1$$
$$\approx 0 - E_1$$
$$\approx -E_1. \tag{43}$$

We see that the remaining field, which is a superposition of contributions from the source *S* and the remaining material of the screen, *a* and *b*, is just the same (except for a minus sign) as that which *was* being emitted by the plug when it was in place. Thus we can find the field behind the screen by imagining that we substitute for the source and the screen with slit the simpler system consisting of just the plug by itself, with no source *S* and no remaining screen, and with the electrons in the plug all oscillating with the same phase and amplitude, as they actually were doing when the plug was in place. This gives us an easy way to calculate the interference patterns due to slits in an opaque screen. It is easy because we make no attempt to know the variation of amplitude and phase constant of the oscillating electrons in the plug as functions of position along the beam direction. (The screen has a finite thickness, of course.) If we knew that, we could say something about the "backward" radiation from the plug; i.e., we could distinguish between a shiny and a black opaque screen. Instead we just assume the field E_1 produced by the plug is due to an infinitesimally thin layer of oscillating charges, all oscillating in phase and with the same amplitude.

Huygens' principle. This calculation device is called *Huygens' principle*. It can be used with any number of slits and also with a single broad slit. Its basis is in Eqs. (42) and (43). Notice that the mentally substituted thin "radiating plug" gives the correct interference pattern only behind the screen. A *real* "radiating plug," i.e., a "sheet" antenna, radiates in all directions. A *real* opaque screen with a hole in it has much or little backward (reflected) radiation, depending on whether it is shiny or black. The Huygens' plug cannot be used to calculate the field to the left of the screen (taking the incident radiation to be coming from the left, as in the figure), because we neglected the phase and amplitude variation that occur between the front and back surfaces of the plug. That variation depends on whether the screen is shiny or black.

Another thing to notice is that in writing Eq. (43) we assumed E_a and E_b were the same with the plug in place as without it. This is only approximately true, as mentioned above. If one has, for example, a single wide slit and uses Huygens' construction to calculate the fields to the right of the screen and in the slit itself, one finds the following: If one is sufficiently far to the right of the screen and sufficiently near the forward direction, and if the screen is many wavelengths wide, then Huygens' construction gives a very good approximation to the right answer (as determined by experiment). If one is in the vicinity of the slit itself, then Huygens' construction gives a very poor approximation to the right answer. If you are in the slit, the most important moving charges in the remaining screen material are those nearest the edge of the slit, because they are closest. But these are just the ones most affected by the removal of the plug. The field pattern can be very complicated in the slit and especially near the edges of the slit, where the nearest oscillating charges dominate. You may ask, "Why not just solve the problem exactly?" This is very difficult. You must use Maxwell's equations in all the vacuum regions and in the material, specifying the properties of the material precisely, and fitting everything together at the boundaries. There are no general methods of finding solutions, and very few such problems have been solved exactly.

***Calculation of single-slit diffraction pattern using Huygens'
construction.*** We wish to calculate the diffraction pattern produced when a plane wave (emitted by a distant point source, say)
is incident on a slit. Using Huygens' construction, we mentally
replace the incident plane wave (or distant point source) and material of the screen by a slab of radiating material—the Huygens'
plug. Since we have a continuous distribution of oscillating charges
across the slab, we should perform an integration (a superposition)
over the contributions from infinitesimal elements of the slab.
Instead of an integration over a continuous distribution, we can
(and shall) consider a discrete sum over contributions from N identical equally spaced "antennas." In the limit that N goes to infinity,
we shall have a continuous distribution of radiating sources. (The
advantage of using N discrete sources rather than a continuous distribution is that we thereby obtain at the same time the solution for
the radiation pattern produced by N antennas or N narrow slits, for
arbitrary N from $N = 2$ to infinity.)

Let the total width of the single wide slit be D. Then D is the
width of the region that contains our linear array of N "Huygens'
antennas." Let the separation between adjacent antennas be d.
Then we have $D = (N - 1)d$. Suppose the incident plane wave is in
the $+z$ direction and the N slits are along x, as shown in Fig. 9.13.

Fig. 9.13 *N antennas, or N narrow
slits, with charges all oscillating in
phase.*

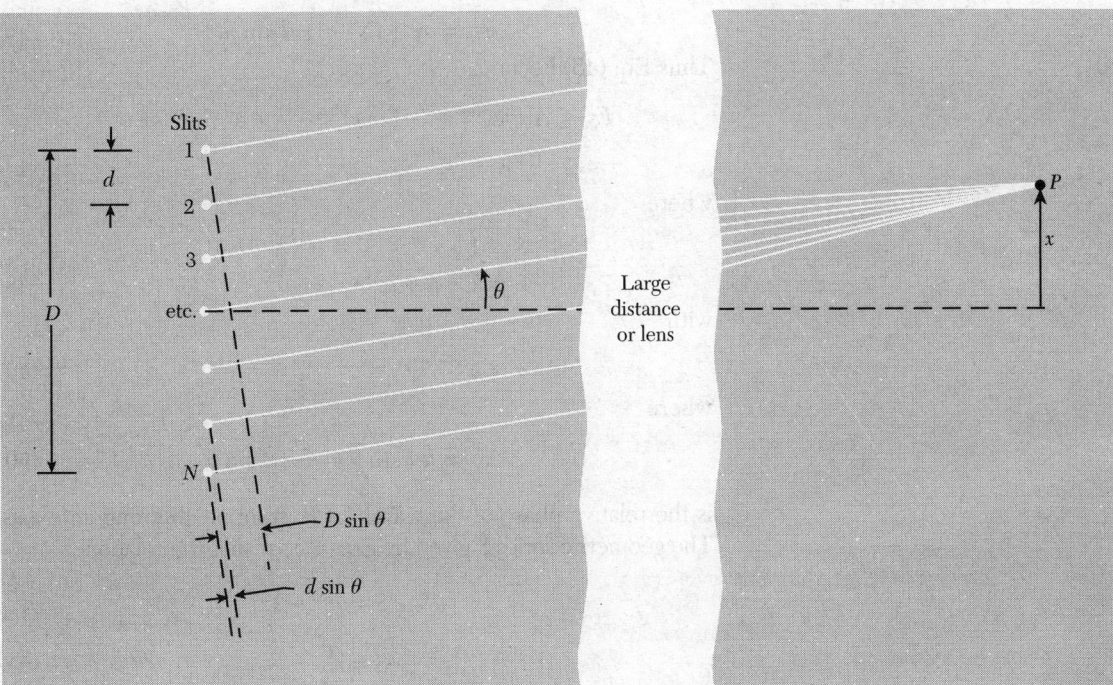

At a distant field point P, each antenna gives a contribution that has the same amplitude $A(r)$ (because P is distant enough so that in the dependence of amplitude on distance we can assume the distance is approximately the same for all the antennas). All the antennas oscillate in phase (by hypothesis). The electric field E at the point P is therefore given by the superposition

$$E = A(r) \cos (kr_1 - \omega t) + A(r) \cos (kr_2 - \omega t)$$
$$+ \cdots + A(r) \cos (kr_N - \omega t). \tag{44}$$

We wish to reexpress this .superposition of N outgoing traveling waves by a single outgoing traveling wave propagating from the average position of the array and having an amplitude that is modulated as a function of emission angle. (That is what we did when we considered the interference pattern of two point sources in Sec. 9.2. For $N = 2$ our present derivation should reproduce those results.) We can simplify the algebra by using complex numbers. The field E is the real part of the complex quantity E_c, where

$$E_c = A(r)e^{-i\omega t}(e^{ikr_1} + e^{ikr_2} + \cdots + e^{ikr_N}). \tag{45}$$

But according to Fig. 9.13,

$$r_2 = r_1 + d \sin \theta,$$
$$r_3 = r_1 + 2d \sin \theta,$$
$$\cdots\cdots\cdots\cdots\cdots\cdots$$
$$r_N = r_1 + (N - 1) \, d \sin \theta. \tag{46}$$

Thus Eq. (45) becomes

$$E_c = A(r)e^{-i\omega t}e^{ikr_1}(1 + e^{ik(r_2 - r_1)} + e^{ik(r_3 - r_1)} + \cdots)$$
$$= A(r)e^{-i\omega t}e^{ikr_1}S, \tag{47}$$

where

$$S \equiv 1 + e^{ik(r_2 - r_1)} + e^{ik(r_3 - r_1)} + \cdots$$
$$= 1 + a + a^2 + \cdots + a^{N-1}, \tag{48}$$

with

$$a \equiv e^{ik(r_2 - r_1)} = e^{ik(d \sin \theta)} = e^{i\Delta\varphi}, \tag{49}$$

where

$$\Delta\varphi = kd \sin \theta = \frac{2\pi}{\lambda} d \sin \theta \tag{50}$$

is the relative phase of the waves (at P) from neighboring antennas. The geometric series S given by Eq. (48) satisfies the relation

$$aS - S = a^N - 1,$$
$$S = \frac{a^N - 1}{a - 1} \tag{51}$$

$$= \frac{e^{iN\Delta\varphi} - 1}{e^{i\Delta\varphi} - 1}$$

$$= \frac{e^{i(1/2)N\Delta\varphi}}{e^{i(1/2)\Delta\varphi}} \frac{[e^{i(1/2)N\Delta\varphi} - e^{-i(1/2)N\Delta\varphi}]}{[e^{i(1/2)\Delta\varphi} - e^{-i(1/2)\Delta\varphi}]}$$

$$= e^{i(1/2)(N-1)\Delta\varphi} \frac{\sin\frac{1}{2}N\Delta\varphi}{\sin\frac{1}{2}\Delta\varphi}.$$

Then Eq. (47) becomes

$$E_c = A(r)e^{-i\omega t}e^{ik[r_1 + (1/2)(N-1)d\sin\theta]} \frac{\sin\frac{1}{2}N\Delta\varphi}{\sin\frac{1}{2}\Delta\varphi}$$

$$= A(r)e^{-i\omega t}e^{ikr} \frac{\sin\frac{1}{2}N\Delta\varphi}{\sin\frac{1}{2}\Delta\varphi}, \tag{52}$$

where the quantity

$$r \equiv r_1 + \tfrac{1}{2}(N-1)d\sin\theta$$

$$= r_1 + \tfrac{1}{2}D\sin\theta \tag{53}$$

gives the distance from P to the center of the array. Taking the real part
of Eq. (52), we obtain for the field at P

$$E(r, \theta, t) = \left[\frac{A(r)\sin\frac{1}{2}N\Delta\varphi}{\sin\frac{1}{2}\Delta\varphi} \right] \cos(kr - \omega t).$$

$$\equiv A(r, \theta)\cos(kr - \omega t). \tag{54}$$

Let us check that Eq. (54) gives the same result for $N = 2$ as our
earlier results, Eqs. (12) and (13), Sec. 9.2, using the identity $\sin 2x = 2\sin x \cos x$ with $x = \frac{1}{2}\Delta\varphi$:

$$E(r, \theta, t) = A(r) \frac{2\sin\frac{1}{2}\Delta\varphi \cos\frac{1}{2}\Delta\varphi}{\sin\frac{1}{2}\Delta\varphi} \cos(kr - \omega t)$$

$$= \left[2A\cos\tfrac{1}{2}\Delta\varphi \right] \cos(kr - \omega t),$$

which agrees with the earlier results.

Single-slit diffraction pattern. We let N go to infinity. We hold
D constant. The spacing d goes to zero. The relative phase shift $\Delta\varphi$
between the waves contributed by adjacent antennas goes to zero.
The total phase shift Φ between the contributions of the first and
Nth antennas at P is exactly $(N-1)\Delta\varphi$. This is approximately $N\Delta\varphi$
for N huge:

$$\Phi = (N - 1)\Delta\varphi = kD \sin\theta. \tag{55}$$

$$\Phi \approx N\Delta\varphi, \quad N \gg 1. \tag{56}$$

Thus the modulated amplitude in Eq. (54) becomes

$$A(r, \theta) = A(r) \frac{\sin\frac{1}{2} N\Delta\varphi}{\sin\frac{1}{2}\Delta\varphi} \approx A(r) \frac{\sin\frac{1}{2}\Phi}{\sin\left[\frac{1}{2}(\Phi/N)\right]}. \tag{57}$$

In the limit that N is sufficiently large, we can neglect all except the first term of the Taylor's series for $\sin\left|\frac{1}{2}(\Phi/N)\right|$ in Eq. (57):

$$\sin\frac{1}{2}\frac{\Phi}{N} \approx \frac{1}{2}\frac{\Phi}{N}, \tag{58}$$

$$A(r, \theta) = NA(r) \frac{\sin\frac{1}{2}\Phi}{\frac{1}{2}\Phi}. \tag{59}$$

We can make one further simplification. As N goes to infinity, we must let $A(r)$ go to zero in such a way that $NA(r)$ is constant, because we want the same contribution for a given infinitesimal element dx of the continuous array no matter how many antennas it contains. (Remember, we are using the antennas in a Huygens' construction.) We can eliminate specific reference to N and $A(r)$ in Eq. (59) by noticing that as θ goes to zero, Φ goes to zero, and the ratio $\sin\frac{1}{2}\Phi/\frac{1}{2}\Phi$ goes to unity:

$$\frac{\sin x}{x} = \frac{x - \frac{1}{6}x^3 + \cdots}{x} = 1 - \frac{1}{6}x^2 + \cdots$$
$$= 1 \text{ for } x = 0.$$

Thus $A(r, 0)$ equals $NA(r)$ times unity, according to Eq. (59). Finally we have

$$E(r, \theta, t) = A(r, 0)\left[\frac{\sin\frac{1}{2}\Phi}{\frac{1}{2}\Phi}\right]\cos(kr - \omega t), \tag{60}$$

with

$$\Phi = 2\pi\frac{D \sin\theta}{\lambda}. \tag{61}$$

The time-averaged energy flux has angular dependence (for fixed r)

$$I(r, \theta) = I_{max}\frac{\sin^2\frac{1}{2}\Phi}{(\frac{1}{2}\Phi)^2}, \tag{62}$$

as is easily seen from Eq. (60). The amplitude and intensity patterns of Eqs. (60) and (62) are plotted in Fig. 9.14.

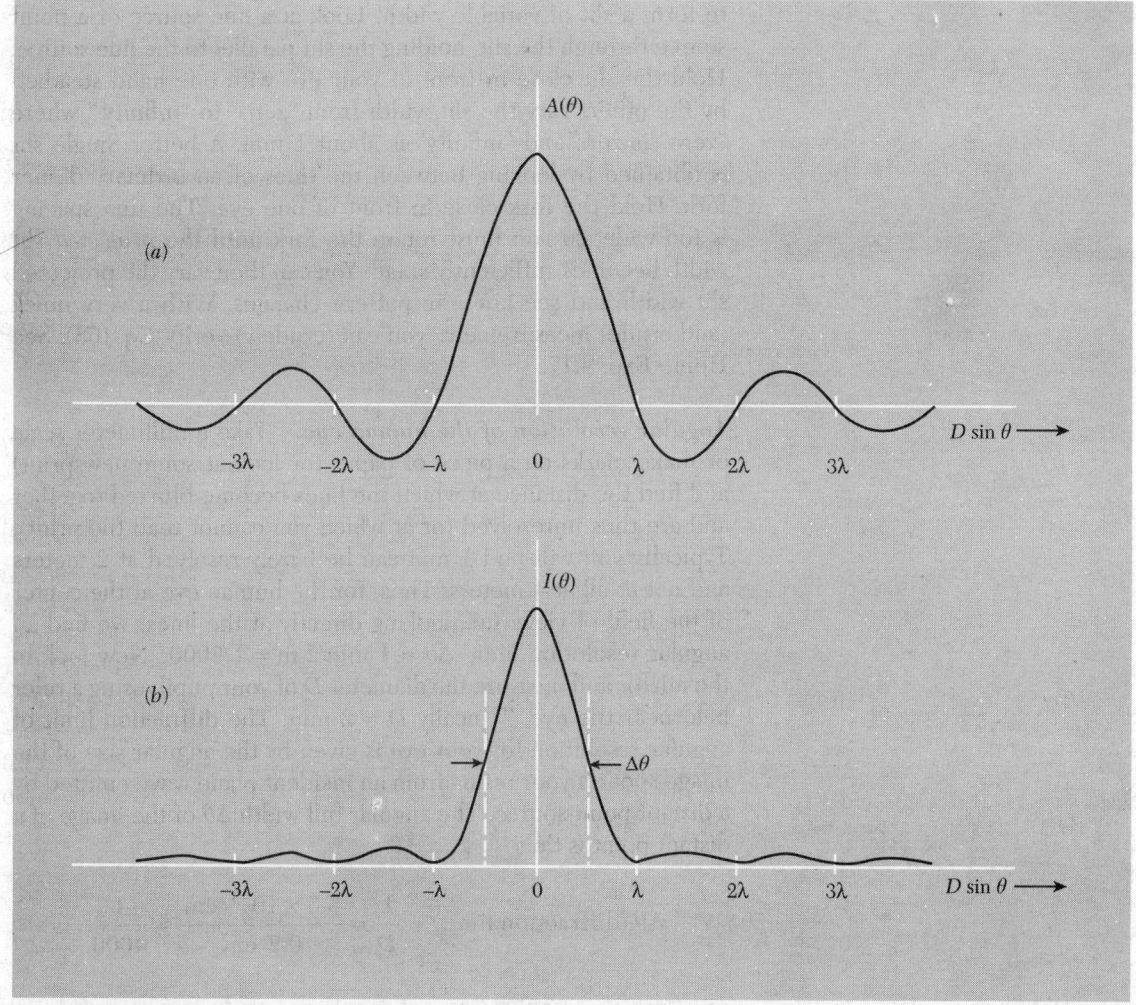

(a)

$A(\theta)$

$D \sin \theta \longrightarrow$

-3λ -2λ $-\lambda$ 0 λ 2λ 3λ

(b)

$I(\theta)$

$\Delta\theta$

$D \sin \theta \longrightarrow$

-3λ -2λ $-\lambda$ 0 λ 2λ 3λ

Angular width of a diffraction-limited beam. We have now justified the result given in Sec. 9.5 that a "beam" of width D has an angular full width $\Delta\theta$ approximately equal to λ/D. The precise shapes of the amplitude and intensity versus θ are given by Fig. 9.14. The main feature of the intensity plot is that the intensity is large only in an angular band roughly between $\theta = -\frac{1}{2}\lambda/D$ and $\theta = +\frac{1}{2}\lambda/D$:

$$\Delta\theta = \frac{\lambda}{D}. \qquad (63)$$

The simplest way to see a single-slit diffraction pattern is as follows: Tear off two little pieces of paper, each with a straight edge. Hold one piece in each hand with the straight edges parallel

Fig. 9.14 *Single-slit diffraction pattern. (a) Amplitude. (b) Intensity. The angular band $\Delta\theta$ extending from $-\frac{1}{2}\lambda/D$ to $+\frac{1}{2}\lambda/D$ corresponds approximately (for small angles) to the "full width at half intensity." The intensity is down by a factor $(2/\pi)^2 = 0.41$, rather than 0.5.*

to form a slit of variable width. Look at a line source or a point source through the slit, holding the slit parallel to the line source. Hold the slit close in front of your eye with one hand steadied by the other. Vary the slit width from "zero" to "infinity," where "zero" is zero and "infinity" is about 1 mm. A better Single slit is obtained by looking between the tines of an ordinary dinner fork. Hold the fork close in front of one eye. The tine spacing is too wide, so you must rotate the fork until the *projected* slit width becomes sufficiently small. You can then vary the projected slit width and see how the pattern changes. With a very quick (and crude) measurement, you can (crudely) verify Eq. (63). See Home Exp. 9.17.

Angular resolution of the human eye. Take a millimeter scale or make marks on a piece of paper (or look at some newsprint) and find the distance at which the lines become blurred together and are thus unresolved (or at which you cannot read the print). Typically you will find 1 mm can be barely resolved at 2 meters and not at all at 4 meters. Thus, for the human eye at the center of the field of view (i.e., looking directly at the lines) we find an angular resolution limit $\Delta\theta \approx 1$ mm/2 m = 1/2000. Now look in the mirror and measure the diameter D of your pupil, using a ruler held near the eye. Typically $D \approx 2$ mm. The diffraction limit of angular resolution for your eye is given by the angular size of the image spot on your retina from an incident plane wave emitted by a distant point source. The angular full width $\Delta\theta$ of the image of a distant point is thus

$$\Delta\theta \text{ (diffraction limit)} \approx \frac{\lambda}{D} \approx \frac{5.5 \times 10^{-5} \text{ cm}}{0.2 \text{ cm}} \approx \frac{1}{4000}.$$

Thus the brain (or mine anyway) likes to have the points separated by an angular separation of about twice the diffraction width before it sees them as resolved.

In order to verify that the (rough) agreement between eye resolution and diffraction width is not accidental, repeat the above experiment looking through a pinhole in a piece of paper (or opaque tape or foil or something). The pinhole should have diameter about 1 mm (assuming your pupil is about 2 mm). Does your angular resolution get worse? By a factor of 2?

Rayleigh's criterion. If two points have angular separation of one diffraction width λ/D, then according to Fig. 9.14b the maximum of intensity from one point will fall on the first minimum of the intensity pattern of the other point. In that case the two points are said to be *just resolved according to Rayleigh's criterion.*

The actual lateral width on your retina of the image of a distant point is given approximately by the focal length of the eye's lens times the angular width of the image. The focal length f is the inner diameter of the eye, about 3 cm (when a distant object is viewed). Thus the lateral width of the image spot for a distant point is roughly $f(\lambda/D) = 3 \times 5 \times 10^{-5}/0.2 = 8$ microns. The fact that you do about as well as the diffraction limit implies that the photoreceptors at the center of the retina (these are the ones called cones) are separated by no more than about 8 microns.

An astronaut orbiting at an elevation of about 150 miles once said that he could see the individual houses in the villages as they passed beneath him. Do you believe him?

Nomenclature: Fraunhofer diffraction and Fresnel diffraction. In our consideration of the diffraction pattern produced by a single slit or aperture we assumed we had an incident plane wave (from a distant point source S). We also assumed that we detected the radiation emitted from the slit at a given *angle*. This means we considered a superposition of waves traveling parallel to one another to the detection point P, and that either P is very far from the slit or that we use a lens (for example the lens of your eye) to focus the waves at P (for example on your retina). Diffraction observed under these two conditions—incident plane wave and diffracted wave emitted in a given direction—is called *Fraunhofer diffraction*. If no lenses are used, the point source S and detector P must each be in the "far zone" of the slit. To determine whether S (for example) is in the far zone, imagine that we pass a plane through the slit so that the plane is oriented perpendicular to the line of sight from S to the center of the slit. Consider the solid cone of all the straight lines from S that pass through all parts of the area of the slit. If these lines all intersect the plane described above at "practically the same" distance from S, then S is in the far zone of the slit. "Practically the same" distance means to within much less than one half-wavelength. In that case the radiation from S will be practically indistinguishable from a plane wave. An analogous criterion holds for the detection point P.

It is not difficult to show that for a slit of width D a point at distance L is in the far zone, provided

$$L\lambda \gg (\tfrac{1}{2} D \cos \theta)^2,$$

Where $\frac{1}{2} D \cos \theta$ is the projected half-width of the slit, as projected perpendicular to the line of sight from the slit to the point. If one or the other of these two conditions is not satisfied, i.e., if either the point source S or the detection point P is not in the far zone of the slit, then we have what is called *Fresnel diffraction* (which we shall not discuss in detail).

Fourier analysis of the transverse space dependence of a coherent source. The result Eq. (63) can be put in a different interesting form. Let us think of a single frequency component of the traveling wave. We may take this component to be exactly monochromatic. Then the bandwidth $\Delta\omega$ is zero. What about the propagation vector? The square of the propagation vector, k^2, is equal to ω^2/c^2 (for light in vacuum). Therefore k^2 must have a perfectly definite value if ω^2 does. But that does not mean that each component of **k** must have a definite value! Now, k^2 is the sum of the squares of its components:

$$k^2 = k_x^{\,2} + k_y^{\,2} + k_z^{\,2}, \tag{64}$$

where k_x gives the number of radians of phase per unit distance along $\hat{\mathbf{x}}$, k_y gives radians of phase per unit distance along $\hat{\mathbf{y}}$, and k_z gives radians of phase per unit distance along $\hat{\mathbf{z}}$. If the beam were a true plane wave traveling along $+z$ instead of a diffraction-limited beam, then k_x and k_y would be zero. For a Fourier component of the diffraction-limited beam that is traveling with a propagation vector in the xz plane and making a small angle θ with the z axis, we have $k_y = 0$, $k_z = k \sin\theta$, and $k_z = k \cos\theta$. For small angles θ, we can approximate $\sin\theta$ by θ and $\cos\theta$ by 1. Then we have for the x component

$$k_x \approx k\theta. \tag{65}$$

But we have already seen that the beam has an angular spread about the dominant direction z given by

$$\Delta\theta \approx \frac{\lambda}{D}. \tag{66}$$

Thus the spread in k_x is given by [combining Eqs. (65) and (66)]

$$\Delta k_x \approx k\Delta\theta \approx k\frac{\lambda}{D} = \frac{2\pi}{D},$$

or, writing Δx for the full width D of the beam in the x direction, we have

$$\boxed{\Delta k_x \Delta x \geq 2\pi.} \tag{67}$$

(The inequality reminds us that diffraction limit is only attained if the sources are coherent and all in phase.)

In fact we can be much more explicit. According to our Huygens' construction, we have a radiating slab consisting of sources distributed uniformly along x from (say) $x = -\frac{1}{2}D$ to $x = +\frac{1}{2}D$. All sources have the same strength and phase constant. A plot of source strength $f(x)$ versus x from $x = -\infty$ to $+\infty$ gives zero except in the region of width D centered at the origin. Thus it is a "square wave" in x. We should be able to Fourier-analyze it in terms of a superposition of sinusoidal space-dependent functions $\sin k_x x$ and $\cos k_x x$, just as we have Fourier-analyzed a square pulse in time in terms of $\sin \omega t$ and $\cos \omega t$. Now, in Eq. (6.95), Sec. 6.4, we found that the Fourier transform of a square pulse in time $f(t)$ with height $1/\Delta t$ and width Δt is given by

$$B(\omega) = \frac{1}{\pi} \frac{\sin \frac{1}{2} \omega \Delta t}{\frac{1}{2} \omega \Delta t}. \tag{68}$$

By analogy, a square pulse $f(x)$ in x of width D and height $1/D$ should have the Fourier transform

$$B(k_x) = \frac{1}{\pi} \frac{\sin \frac{1}{2} k_x D}{\frac{1}{2} k_x D}. \tag{69}$$

But

$$k_x D = kD \sin \theta = \Phi. \tag{70}$$

Therefore

$$B(k_x) = \frac{1}{\pi} \frac{\sin \frac{1}{2} \Phi}{\frac{1}{2} \Phi}. \tag{71}$$

By comparison of Eqs. (71) and (60), we see that the amplitude of the field detected at angle θ (which is given by k_x) is (aside from proportionality constants) the Fourier transform of the source strength at the slit (the square wave). At the slit, the oscillation amplitude is $f(x) \cos \omega t$, where $f(x)$ is the source strength (here a constant over the aperture of the slit). At distance r and direction θ, the traveling wave is obtained by replacing $\cos \omega t$ by $\cos (\omega t - kr)$ and replacing $f(x)$ by its Fourier transform $B(k_x)$. The other transverse dimension of the beam, y, satisfies a relation like Eq. (67), but with x replaced by y.

Important results of Fourier analysis. Recalling our previous results for Fourier analysis in terms of the longitudinal component of the wave vector, k_z, and for Fourier analysis in terms of the frequency, we can summarize all the results of Fourier analysis:

$$\begin{aligned} \Delta k_x \ \Delta x &\geq 2\pi \\ \Delta k_y \ \Delta y &\geq 2\pi \\ \Delta k_z \ \Delta z &\geq 2\pi \\ \Delta \omega \ \Delta t &\geq 2\pi \end{aligned} \tag{72}$$

Fourier analysis furnishes a powerful technique for calculating diffraction patterns. However, we shall not pursue that topic here (see Prob. 9.59).

Diffraction pattern for two wide slits. Make a double slit. (One
good method is to tape a piece of household aluminum foil on a
microscope slide so that it is stretched flat and tight against the
slide. Cut a slit in the foil with a razor blade guided by a straight
edge of another slide. The second slit should be as close to the first
as you can make it without ruining the first slit—a spacing of less
than $\frac{1}{2}$ mm is easily achieved.) Look at your line source both with
and without the red gelatin filter with the slit held close in front of
one eye. The narrowly spaced "interference fringes" correspond to
the double-slit interference pattern. Thus they have angular separation λ/d radians (using the small-angle approximation that sin θ
equals θ). On the same slide make a single slit of the same width as
those in the double slit (i.e., use the same razor blade and same type
of stroke, or simply make one of the two slits of the double slit longer than the other). Compare the single- and double-slit patterns.
Notice that *the double-slit pattern is modulated by the single-slit
pattern.* (See Fig. 9.15.) In fact it is usually rather difficult to see
any of the double-slit pattern except within the central maximum of
the single-slit modulation. (If you use the red filter and have a good
double slit, you may succeed.)

Here is the explanation for the appearance of the pattern. Each slit
gives an electric field at the detector (your retina) that has a certain
amplitude and a certain phase constant. The phase constant of the
contribution from one entire slit is the same as that from the differential contribution (the "antenna") at the center of the slit. That follows
from the fact that the wave has the factor cos $(kr - \omega t)$, where r is
the distance from the center of the slit to the detector. [See Eqs. (60)
and (53), Sec. 9.6.] The amplitude is proportional to sin $\frac{1}{2}\Phi/\frac{1}{2}\Phi$,
where Φ is the difference in phase of the contributions from opposite
edges of the slit. When there are two such slits separated by distance
d, each slit gives a contribution which, as far as phase is concerned,
is the same as that which one would get from a narrow slit located at
the center of the actual slit. As far as amplitude is concerned, there
is the factor sin $\frac{1}{2}\Phi/\frac{1}{2}\Phi$. Thus the pattern is just the two-slit pattern
found previously, except that the constant amplitude $A(r)$ contributed
by each slit is now replaced by a constant times sin $\frac{1}{2}\Phi/\frac{1}{2}\Phi$. In other
words, the double-slit pattern that one would get from two infinitely
narrow slits is *modulated* by being multiplied by sin $\frac{1}{2}\Phi/\frac{1}{2}\Phi$. Combining our previous results for the two-slit pattern [Eq. (13), Sec. 9.2]
with the modulation factor, we find that if the two slits are excited
with the same phase the radiation pattern is given by

$$E(\theta, t) = A(\theta) \cos (kr - \omega t), \tag{73}$$

$$A(\theta) = A(0) \cdot \left[\frac{\sin \frac{1}{2}\Phi}{\frac{1}{2}\Phi} \right] \cos \frac{1}{2}\Delta\varphi, \tag{74}$$

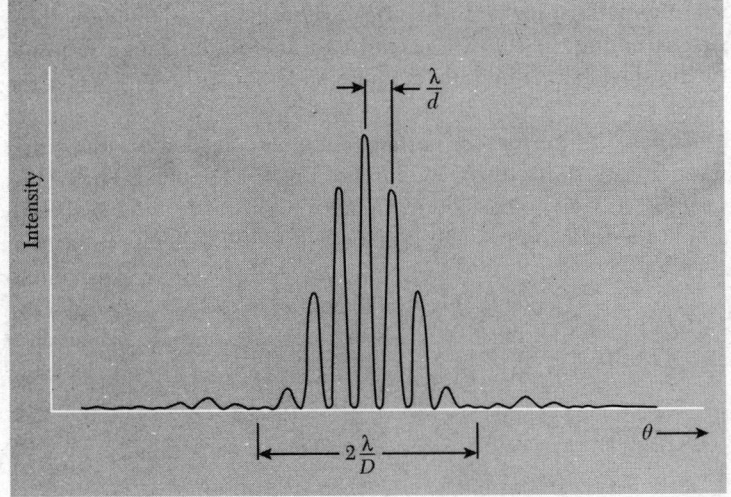

Fig. 9.15 *Double slit pattern. The slit separation d is four times the width D of each slit in this example. The expressions for the angular separation, λ/d, and for the full width between modulation zeroes, 2λ/D, use the small-angle approximation, sin θ = θ.*

$$\Phi = kD \sin \theta = 2\pi \frac{D \sin \theta}{\lambda}, \tag{75}$$

$$\Delta\varphi = kd \sin \theta = 2\pi \frac{d \sin \theta}{\lambda}, \tag{76}$$

where D is the slit width of each slit, d is the slit separation (center to center), and r is the distance from the observation point P to a point midway between the centers of the two slits. If D goes to zero, then the central maximum "covers the entire field of view," and we get the result of Sec. 9.2 for two narrow slits.

The intensity pattern $I(\theta)$ is proportional to the time average of the square of the electric field; i.e., according to Eqs. (73) and (74),

$$I(\theta) = I(0) \left[\frac{\sin \frac{1}{2}\Phi}{\frac{1}{2}\Phi} \right]^2 \left(\cos^2 \frac{1}{2}\Delta\varphi \right). \tag{77}$$

The factor $\cos^2 \frac{1}{2}\Delta\varphi$ gives the rapid angular variation characteristic of the two-slit pattern, with maxima separated in angle by λ/d. The factor $(\sin \frac{1}{2}\Phi/\frac{1}{2}\Phi)^2$ gives the single-slit modulation, with angular full width at approximately half intensity of λ/D, or angular full width between zeros on either side of the central maximum of $2\lambda/D$. By counting the number of "two-slit" fringes in the central maximum of the "single-slit modulation," you can estimate the ratio d/D for your double slit. The intensity pattern corresponding to Eq. (77) is plotted in Fig. 9. 15.

Diffraction pattern for many identical parallel wide slits. From our discussion of the case of two wide slits, it should be apparent that the pattern for many identical wide slits is easily obtained by first assuming the slits are narrow and then multiplying the result by the single-slit amplitude modulation factor $\sin \frac{1}{2}\Phi / \frac{1}{2}\Phi$.

Multiple-slit interference pattern.   Let us consider how the interference pattern for the N antennas of Fig. 9.13 depends on N. (We may just as well be considering N narrow slits as N antennas.) The amplitude for N narrow slits is given by Eq. (54), which we recopy:

$$E(r, \theta, t) = A(r, \theta) \cos (kr - \omega t), \tag{78}$$

$$A(r, \theta) = A(r) \frac{\sin \frac{1}{2} N \Delta\varphi}{\sin \frac{1}{2} \Delta\varphi}, \tag{79}$$

$$\Delta\varphi = 2\pi \frac{d \sin \theta}{\lambda}. \tag{80}$$

Principal maxima, central maximum, white light source.   The angles for which the denominator (and numerator) of Eq. (79) goes to zero are given by $\frac{1}{2}\Delta\varphi = 0, \pm \pi, \pm 2\pi$, etc. These are the angles for which the path length increment $d \sin \theta$ is zero, $\pm \lambda$, etc., corresponding to completely constructive interference between all N antennas. These are called *principal maxima*:

$$d \sin \theta = 0, \pm\lambda, \pm2\lambda, ..., m\lambda, \qquad m = 0, \pm1, \pm2, ... \tag{81}$$

The maximum at $\theta = 0$ is called the *central maximum* or *zeroth-order maximum*. Those with $m = \pm1$ are called *first-order* maxima, etc. The central maximum differs from all other principal maxima in one important respect, which is that all slits give contributions that are in phase *independent of wavelength*. Thus *for a white source the central maximum is white*. For all other principal maxima except the central one, the angle of the maximum depends on the wavelength, i.e., on the color.

   At a principal maximum, the amplitude of the superposition is just N times the amplitude contributed by each slit. That is physically obvious. It also follows from Eq. (79): For the central maximum, we have $\Delta\varphi = 0$. Then we use (taking $\frac{1}{2}\Delta\varphi = x$)

$$\frac{\sin Nx}{\sin x} = \frac{Nx - \frac{1}{6}(Nx)^3 + \cdots}{x - \frac{1}{6}x^3 + \cdots} = N \frac{[1 - \frac{1}{6}(Nx)^2 + \cdots]}{[1 - \frac{1}{6}x^2 + \cdots]}$$

$$= N \quad \text{for } x \to 0. \tag{82}$$

For the first-order maximum with $m = +1$, we can similarly show that the limit as x goes to π of $\sin Nx/\sin x$ is $\pm N$. To do that, we expand in terms of the small angle ϵ by which x differs from π:

$$x = \pi - \epsilon$$

$$\frac{\sin Nx}{\sin x} = \frac{\sin (N\pi - N\epsilon)}{\sin (\pi - \epsilon)} = (-1)^{N+1} \frac{\sin N\epsilon}{\sin \epsilon}.$$

(83)

As ϵ goes to zero, we get the limit of the ratio to be $(-1)^{N+1} N = \pm N$.

Angular width of a principal maximum. As N increases, the angular widths of the principal maxima decrease. The angular half-width from a principal maximum to the first zero on one side or the other of the maximum is given by inspection of Eq. (79). At a principal maximum, both the numerator and denominator are zero. When the argument $\frac{1}{2} N \Delta\varphi$ of the sinusoidal function in the numerator of Eq. (79) increases by π, the numerator is again zero. (The denominator is not then zero.) Thus *the phase increment $\Delta\varphi$ increases by $2\pi/N$ in going from a principal maximum to the first adjacent zero of amplitude.* That means *the path length increment $d \sin \theta$ increases by λ/N in going from a principal maximum to the first adjacent zero of amplitude.* Now, the path length increment between two successive principal maxima is λ. Thus we see that *the amplitude falls from maximum to zero in an interval in $\sin \theta$ that is N times narrower than the interval λ/d in $\sin \theta$ between successive principal maxima.*

For large N, or for N even (whether it is large or small), it is easy to see why the first zero (next to a principal maximum) occurs when the path length increment $d \sin \theta$ is λ/N. Suppose there are 6 antennas. The first zero occurs when the first three can be paired off with the last three in "canceling" pairs, with 1 canceled by 4, 2 by 5, and 3 by 6. The cancellation occurs when antennas 1 and 4 have $\frac{1}{2} \lambda$ path difference (and similarly for the other pairs). Then 1 and 2 have path difference $\lambda/6$, which is λ/N. When N is odd, this argument cannot be used because the antennas cannot be so paired off. In this case the easiest way to obtain this result "visually" rather than algebraically is to make a vector diagram in the complex plane of the contributing amplitudes. Then it is easy to see that the N complex amplitudes join to make a closed polygon, so as to give zero total amplitude, when $\Delta\varphi$ is $2\pi/N$. (Prob. 9.52.) In Fig. 9.16 we show how the interference pattern depends on N when the slit separation d is held fixed.

You can demonstrate the narrowing of the principal maxima when N increases from 2 to 3 as follows: With a razor blade make a triple slit in aluminum foil taped to a slide. Make two of the slits longer than the third, so that you can go from a double slit to a triple slit by a slight shift of the slide in front of one eye. After a half dozen tries, you may succeed (I did) in getting three decent slits having about the same separations, with d less than $\frac{1}{2}$ mm.

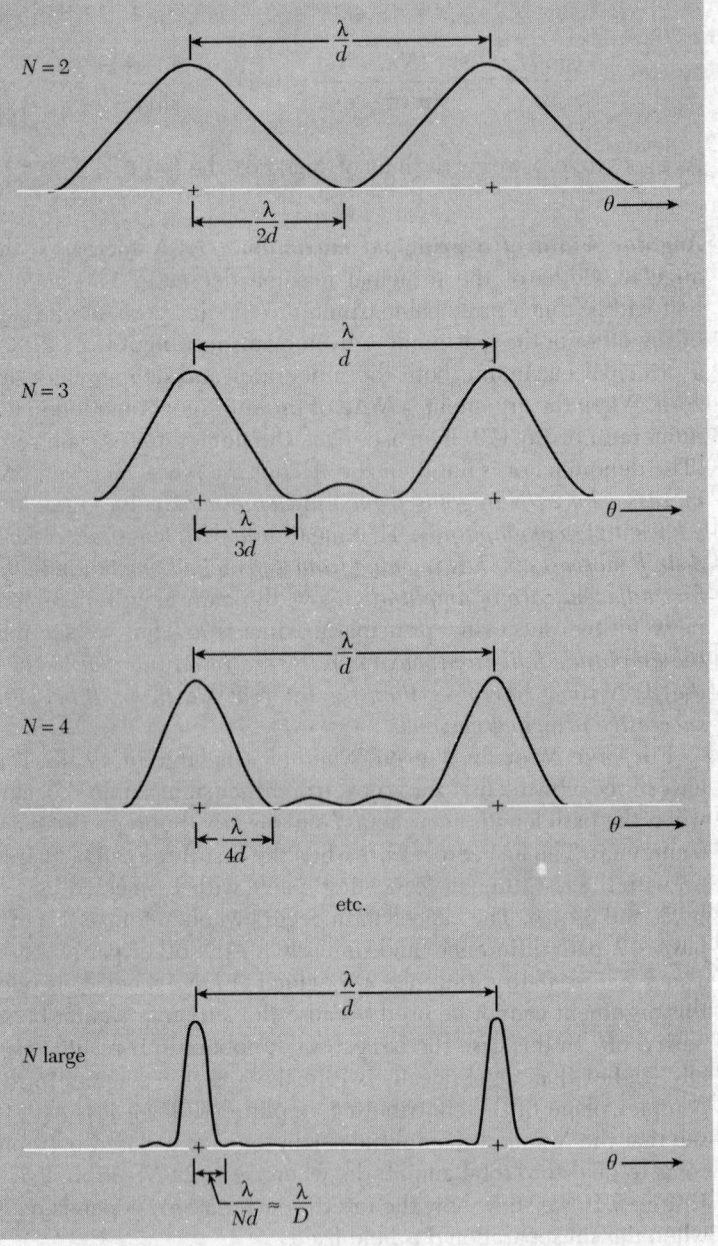

Fig. 9.16 *Multiple-slit interference pattern. Two principal maxima are shown. The angles are assumed to be small, so that sin θ = θ. For large N, each principal maximum has the shape of the single-slit diffraction pattern shown in Fig. 9.14b.*

(After each try, hold the array up to a light to inspect it. An ordinary cheap magnifying glass having magnification of 2 × or 3 × is helpful.) When you look at your line source with a double slit, the bright fringes look slightly broader than the "black" regions that separate the fringes. When you shift to the triple slit, the bright regions look narrower than the intervening black regions. Of course, if you do not have fairly evenly spaced and uniform slits, you will get patterns we have not discussed.

Transmission-type diffraction grating. Instead of N antennas or N slits in an opaque screen, we may have N parallel scratches on a piece of smooth glass or plastic of width D. If there were no scratches, the light would give a diffraction pattern corresponding to a single wide slit of width D. The scratches act like "antennas." They give an "N-scratch" interference pattern which is what we just obtained for N slits, with one exception. At the central maximum (at $0°$), we get contributions not only from the scratches but also from all the transparent material between the scratches. Therefore we expect the central maximum to be considerably brighter than the other principal maxima.

A line source of monochromatic light seen through a diffraction grating has at each principal maximum an intensity profile (plot of intensity versus angle) like that of the single-slit pattern of Fig. 9.14b.

The diffraction grating in your optics kit has this same design—it has 13,400 scratches per inch, which gives $d = 1.90 \times 10^{-4}$ cm., i.e., 1.9 microns. For green light with wavelength about 5500 Å, that is, 0.55 microns, how many principal maxima do you expect to find? According to Eq. (81), the principal maxima occur at values of $\sin \theta$ of 0, λ/d, etc. Of course $\sin \theta$ cannot be greater than 1. For our grating we have $d \approx 3.5\lambda$ when $\lambda = 0.55\ \mu$. Therefore if $\sin \theta = m\lambda/d$, we can have $m = 0, \pm 1, \pm 2,$ and ± 3, but not ± 4. Now look at a light bulb with your grating. The "straight ahead" bulb is the central maximum. All colors overlap at $\theta = 0$. The streaks of colors at the side are light-bulb images of various colors at angles corresponding to $d \sin \theta = \lambda$ for first order, 2λ for second order, etc. Do you see all three orders? (If you see four, somebody is wrong.) If you want to see the colors of an incandescent bulb as they really are, you should not use a large light bulb, because its size causes overlap of the various "colored light bulbs." Either put a narrow vertical slit over your light bulb (and hold the grating so it spreads the colors out horizontally) or, better, get a "display bulb" at any hardware or grocery store. (They have a clear glass envelope and a straight filament about three inches long, and they cost about 40 cents.)

You can measure d for your grating very easily, given that (for example) green light has wavelength 5500 Å. Just look at a light bulb and measure the angle in radians (or its sine or tangent) from the central maximum to "green," using your hand and arm or a ruler held at arm's length with the grating held close in front of one eye. Then use Eq. (81). Do you get $d \approx 3.5\lambda$? For further exploration of the properties of your grating, see the home experiments.

Diffraction by an opaque obstacle.   In Fig. 9.12 we showed a point source S and an opaque screen consisting of parts a, b, and 1. The (zero) field behind the screen was regarded as the superposition $E_S + E_a + E_b + E_1 = 0$. When plug 1 was removed, the field $E_S + E_a + E_b$ was taken to be the same as it was before removing the plug, i.e., to be equal to $-E_1$ (before the plug was removed). This gave the method of Huygens' construction for finding the diffraction pattern from the screen with plug 1 removed, i.e., from a screen with an aperture having the shape of the plug. Now we wish to consider what happens if we *leave the plug in place and remove the rest of the screen.* This will give the diffraction pattern of an opaque obstacle.

Before removing anything, we have $E_S + E_a + E_b + E_1 = 0$. Now remove a and b and assume that the motions of the electrons in plug 1 (the opaque obstacle) are unchanged. (That is an approximation, since these electrons are driven by radiation from the electrons in a and b as well as by radiation from S.) The field behind the plug is then $E_S + E_1$. In the region close behind the plug ("close" will be defined presently), the field should be essentially what it was when the entire screen was present, because the regions a and b were even then rather far away (compared to plug 1) and gave contributions small compared with $E_S + E_1$. Therefore the region close behind the plug should have essentially zero electric field. This is the "shadow" of the plug. It is caused by the fact that at a point close behind the screen the field (which is zero) is essentially due only to S and to the *nearby* charges, which in this case are those in plug 1. Thus close behind the screen the field E_1 superposes with E_S to give zero. Hence plug 1 is emitting a "part of a plane wave" in the same direction as the incident plane wave from the distant source S, with amplitude equal to that of the plane wave from S, and with a phase constant 180 deg out of phase with the plane wave so as to cancel it to zero in the superposition $E_S + E_1$. *This* is *the way the shadow* is *produced.* The opaque obstacle does not "eat up" the incident light; it radiates a beam of "negative-amplitude light" (i.e., negative with respect to the incident light) in the forward direction that combines with the incident wave to give zero close behind the obstacle.

How far downstream does a shadow extend?   Now, the plug does not emit a true plane wave of "negative-amplitude" light, because it has a finite width (or diameter) D. Instead it emits a "beam," with the direction of the beam the same as that of the plane wave E_S, but with a diffraction-limited angular spread in its direction with angular full width $\Delta\theta \approx \lambda/D$. By the time this beam has traveled a distance L from the screen (the plug), it has spread out in its lateral dimension by an amount W given roughly by $W \approx L\Delta\theta \approx L(\lambda/D)$. As the beam spreads out, its amplitude naturally decreases. (Each point charge in the plug gives a contribution that falls off as the inverse distance. Also, suppose the plug were radiating by itself; as its radiated energy spread over a larger area, its amplitude at any one point would have to decrease.) Only when its electric field amplitude is equal in magnitude (and opposite in sign) to that of the plane wave E_s can it cancel E_s to zero. Thus the shadow eventually disappears after a sufficient distance "downstream." Crudely speaking, we can say that the "negative-amplitude" light emitted by the driven plug is significantly weakened when the diffraction spreading of the beam has doubled its width. That gives us a crude "boundary distance" L_0, at which $D \approx W_0$. But since $W_0 \approx L_0(\lambda/D)$, we have

$$\boxed{L_0\lambda \approx D^2.} \tag{84}$$

Thus, for $L \ll L_0$, we expect to have a nice black shadow behind the obstacle, except near the edges (where the assumption that E_1 is unchanged when a and b are removed, breaks down badly). For $L \gg L_0$, we expect that it will be difficult to detect the effect of the obstacle at all, since its electric field contribution is then small compared with that of the plane wave E_s. In order to detect it easily, you can make use of the directional information; you can use a lens. The plane wave E_s will focus to a small spot in the focal plane, the spot size being given by $f\lambda/D_{lens}$, where D_{lens} is the lens diameter and f the focal length. The negative-amplitude light from the obstacle gives an image of width $f\lambda/D$. If D_{lens} is much larger than the obstacle size D, then the bright spot due to the plane wave obscures only a small region at the center of the image.

You may study the diffraction patterns of obstacles using a flashlight for a point source (with lens removed and reflector covered up) and pins and hairs for obstacles. One of the amazing results is the "bright spot" you see at the center of the shadow when you are at distances $L \gg L_0$. See Home Exp. 9.34.

Equation (84) can be tested with waves other than light waves. You can test it by putting an obstacle in the path of a beam of traveling water waves in a ripple tank or bathtub. The "shadow" is well defined for $L \ll L_0$ and gone for $L \gg L_0$. See Home Exp. 9.29.

9.7 *Geometrical Optics*

The name "geometrical optics" refers to the study of how light beams behave in optical instruments (which consist of various reflecting and refracting surfaces) in the approximation that one considers only the dominant directions of the beams and does not worry about the spreading of the beams due to diffraction. (The name "physical optics" is sometimes used to describe studies which take into account the wave nature of light and which therefore include interference and diffraction.) The basic "laws" of geometrical optics are *the law of specular reflection and Snell's law of refraction*. Of course both of these laws are actually due to the wave nature of light; each results from a particular constructive interference.

Specular reflection.   Whenever a plane wave is incident on a smooth flat material surface, *specular reflection* occurs; that is, (*a*) there is a reflected ray that lies in the plane of incidence (the plane containing the incident ray and the normal to the surface), and (*b*) the angle of reflection equals the angle of incidence (both angles are measured from the normal).

Specular reflection is due to constructive interference. The electrons in the material are driven by the incident wave. They reradiate. The reflected ray for specular reflection is the direction of a constructive interference maximum.

We can most easily understand this by considering our familiar linear antenna array. We let the antenna currents be driven by the electric field of an incident plane wave at nonnormal incidence, as shown in Fig. 9.17.

Now let us examine the distant radiation field due to the antenna currents alone. First consider the central interference maximum. It is easy to see that this occurs in the direction of propagation of the incident beam; antenna 1 is excited ahead of antenna 2 (in phase) and therefore radiates ahead of no. 2 by the same amount. At a distant point *P* the radiation from antennas 1 through *N* will be exactly in phase if the antenna-radiation propagation direction is the same as the incident propagation direction; a certain crest of the wave from antenna 1 has to travel farther than a crest from antenna *N*, but it started earlier by just the right amount.

It is obvious from the symmetry of the antenna array that the antennas excited as in Fig. 9.17 will form not only the central interference maximum "to the right" (in the figure) but also a corresponding maximum "to the left." This "image" maximum is the specularly reflected radiation. We see from Fig. 9.18 that the angle of reflection equals the angle of incidence.

Specular reflection from any smooth plane surface is due to constructive interference occurring in a manner completely analogous to that for the closely spaced antennas.

Nonspecular reflection from a regular array.   The central maximum and the specular reflection maximum are not the only interference maxima produced by the array of antennas shown in Figs. 9.17 and 9.18. In addition to these "zeroth"-order maxima for transmission and reflection there are also maxima for those directions for which the path difference from adjacent antennas to the detector is greater (or less), by an integer number of wavelengths, than the

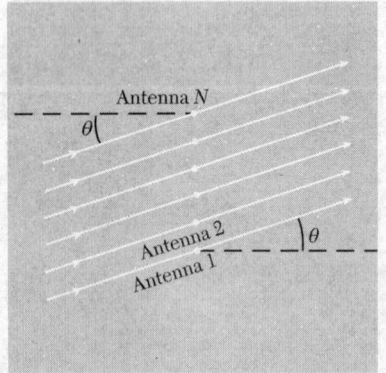

Fig. 9.17 *Antenna array driven by plane wave at nonnonnal incidence. The dotted line is the normal to the plane of the antennas. The arrows give the direction of propagation. The angle of incidence is θ.*

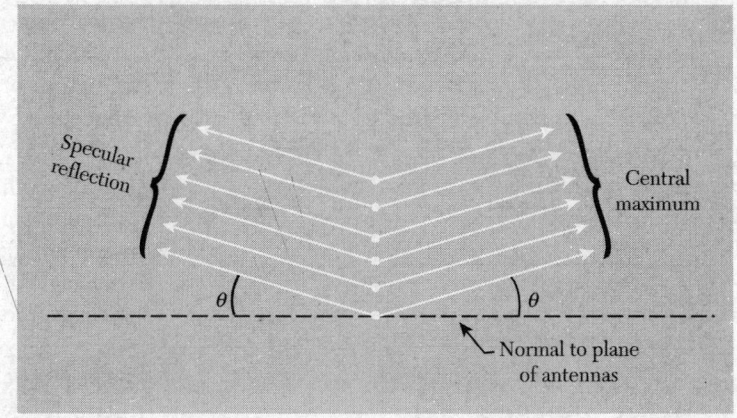

Fig. 9.18 Directions of interference maxima of antennas driven with phase relations of Fig. 9.17.

difference for a zeroth-order maximum. The interference pattern for transmitted waves (traveling to the right in Fig. 9.18) is simply that of an *N*-slit transmission diffraction grating, with incident light at nonnormal incidence. The interference pattern for the reflected waves is similar to that of the transmitted waves, of course, except that the reflected zeroth order (specular reflection) is not likely to be as bright as the transmitted zeroth order (central maximum). You may verify the existence of the interference pattern for reflected light from a regular array by using the transmission diffraction grating from your optics kit as a *reflection grating*, i.e., by holding it close to one eye and looking at the reflection of a point source. The zeroth-order (specular) reflection is easily identified, since it is "white." The nonzeroth-order reflection maxima are similar to the transmission maxima, for the same nonnormal angle of incidence.

If the spacing between neighboring antennas is less than one wavelength, then the only directions corresponding to completely constructive interference are those of the zeroth-order maxima, i.e., those corresponding to the central maximum and to specular reflection. When we study geometrical optics and optical instruments, we are usually considering visible light incident on glass or metal surfaces. The "driven antennas" are atoms in the surface, and are spaced about 10^{-8} cm apart. For visible light of wavelength about 5×10^{-5} cm, therefore, we can obtain only the zeroth-order maxima. (For x rays of wavelength less than 10^{-8} cm reflecting from the surface of a single crystal, one obtains the higher-order maxima.) Because we shall be considering optical instruments that use visible light, we assume from now on that we have only specular reflection.

Image of a point source in a mirror—virtual source and real source. The surfaces of constant phase for radiation from a point source are spheres. A sufficiently small region on one of these spheres can be approximated by a plane, and we can call the

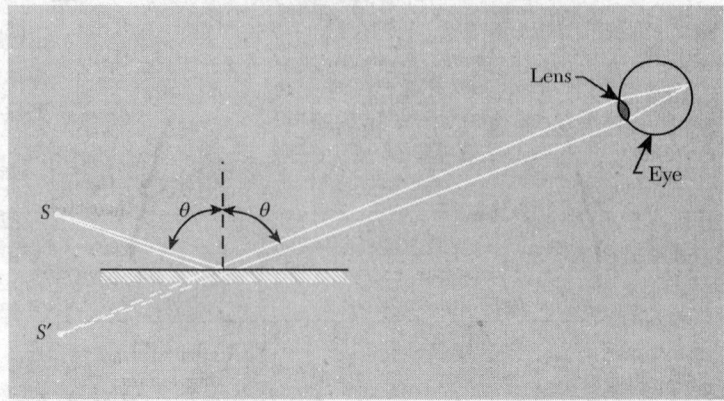

Fig. 9.19   Virtual point image S' of real point source S in plane mirror.

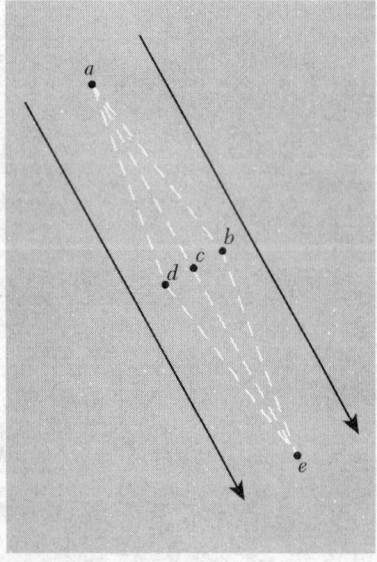

Fig. 9.20   Light beam propagating in glass. The arrows are along the propagation direction and give the width of the beam. The points a, b, c, d, and e are glass atoms.

(approximate) plane wave of radiation that passes through this small region a *ray*. In Fig. 9.19 we see a point source S viewed by means of a mirror. The radiation that enters the lens aperture (pupil) of the eye can be thought of as a " bundle of rays." Two of the rays are drawn in Fig. 9.19. Each ray is specularly reflected from the mirror. The light entering the eye appears to come from a point source S' located behind the mirror. The source S' is called a *virtual source* because there is not a real source of radiation at S', (The source S is called a *real source.*)

Refraction—Snell's law—Fermat's principle.   We have already given two derivations of Snell's law. One derivation used a simple geometric construction (Sec. 4.3). The other derivation used the fact that the number of wave crests per unit length along the boundary is the same on either side of the boundary (Sec. 7.2).Both of these derivations made use of plane waves. Since in geometrical optics one is always using rays, i.e., narrow beams of light, rather than plane waves, we shall give here a third derivation that uses a diffraction-limited beam rather than a plane wave. The spreading of the beam due to diffraction will not concern us, however, and we shall not show it.

First consider a beam propagating in a homogeneous piece of glass of index n, as shown very schematically in Fig. 9.20. Consider atom a, in the middle of the beam. It is driven by the beam. It radiates in all directions. Its radiation helps to drive atoms b, c, and d. Their radiation superposes to help drive atom e (which is also at the center of the beam). Now, the beam is the result of constructive interference. This means that for b and d lying sufficiently close on either side of c, all three of the atoms b, c, and d contribute with nearly the same phase at e, since all have been driven by a. In other words, the times for waves to travel at the phase velocity c/n from a to b to e, a to c to e, and a to d to e must all be nearly equal if a, c, and e are all along the path of the ray, and if b and d are sufficiently close to c. If that were not the case, the radiation from different driven atoms would not be superposing so as to maintain a beam by constructive interference.

Now, it is obvious from Fig. 9.20 that if a, c, and e are along the ray, then the neighboring paths $\overline{abe}$ and $\overline{ade}$ are slightly longer than the path $\overline{ace}$. What we mean by saying that they are nearly equal to $\overline{ace}$ is that if (for example) b has a small transverse displacement x from c, then the path $\overline{abe}$ exceeds $\overline{ace}$ by a quantity that is proportional to the square of the small quantity x, rather than proportional to the first power of x. Thus, in a Taylor's series expansion of the path length versus the parameter x, the first derivative vanishes (that term in the series is what gives a contribution linear in x).

Actually it is not path length but propagation time that matters. Thus we have the principle that a light beam propagates along a path such that the derivative of the propagation time with respect to x is zero, where x is a parameter that is zero for the path of the beam (like $\overline{ace}$) and not zero for a neighboring path (like $\overline{abe}$ or $\overline{ade}$). This condition says that *the propagation time along the beam is an extremum.* This is called *Fermat's Principle of Least Time*, or simply *Fermat's principle.*

We shall now use Fermat's principle to derive Snell's law. In Fig. 9.21 we show atom a in medium 1 and atom e in medium 2 (They are analogous to atoms a and e in Fig. 9.20). The point of intersection of the beam with the interface, labeled P, is variable. The path $\overline{aPe}$ has a segment $\overline{aP}$ that takes propagation time $t_1 = l_1 n_1/c$ and a segment $\overline{Pe}$ that takes propagation time $t_2 = l_2 n_2 / c$. The distances ct_1 and ct_2 are called the *optical path lengths* $n_1 l_1$ and $n_2 l_2$. The total optical path (o.p.) is a minimum if the total elapsed time is minimum. Thus we want to find the point P for which

$$\text{o.p.} \equiv n_1 l_1 + n_2 l_2 = \text{minimum}. \qquad (85)$$

From Fig. 9.21 we have

$$\text{o.p.} = n_1 (y_1^2 + x_1^2)^{1/2} + n_2 (y_2^2 + x_2^2)^{1/2}. \qquad (86)$$

Now let P move an infinitesimal distance away from its (as yet unknown) position which minimizes o.p. Let d(o.p.) be the change in o.p. due to this displacement. To find d(o.p.), we differentiate Eq. (86). The only variables are x_1 and x_2, since P remains on the interface. The sum of x_1 and x_2 is of course constant (since atoms a and e are fixed), so that the increment dx_2 is the negative of the increment dx_1 when P is displaced. Thus we have

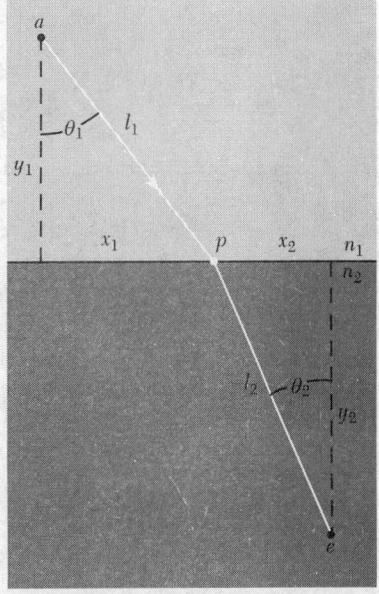

Fig. 9.21 *Refraction. The optical path length $n_1 l_1 + n_2 l_2$ varies, depending on the location of the point P. The actual path of the light ray that travels from a to e is found by varying the position of P so as to minimize the optical path length, according to Fermat's principle. In that case aPe is along the interference maximum, and is analogous to ace in Fig. 9.20.*

$$d(\text{o.p.}) = n_1\,dl_1 + n_2\,dl_2$$

$$= n_1\,d(y_1{}^2 + x_1{}^2)^{1/2} + n_2\,d(y_2{}^2 + x_2{}^2)^{1/2}$$

$$= \frac{n_1 x_1\,dx_1}{(y_1{}^2 + x_1{}^2)^{1/2}} + \frac{n_2 x_2\,dx_2}{(y_2{}^2 + x_2{}^2)^{1/2}}$$

$$= \frac{n_1 x_1}{l_1}\,dx_1 + \frac{n_2 x_2}{l_2}\,(-dx_1). \tag{87}$$

In writing Eq. (87), we have neglected the higher-order terms involving $dx_1{}^2$, $dx_1{}^3$, etc. Now we assume P is such that $\overline{aPe}$ is along the beam; then the first-order variation of o.p. with x_1 is zero, according to Fermat's principle. Then Eq. (87) gives

$$d(\text{o.p.}) = 0 = \left[\frac{n_1 x_1}{l_1} - \frac{n_2 x_2}{l_2} \right] dx_1,$$

i.e.,

$$n_1\,\frac{x_1}{l_1} = n_2\,\frac{x_2}{l_2},$$

i.e.,

$$n_1 \sin \theta_1 = n_2 \sin \theta_2, \tag{88}$$

which is Snell's law.

Now we shall consider some basic optical components.

Ellipsoidal mirror.   In Fig. 9.22 we see a hollow ellipsoid of revolution with a specularly reflecting inner surface and with a *point* source of light located at F, one of the two principal foci. From the definition of an ellipse, the distances from F to the other focus F' are the same for all paths (*except* for the direct path not involving a reflection). Therefore the focus F' is a region of complete constructive interference for radiation emitted by electrons in the surface that are driven by radiation from F. We say that the source at F is *imaged* at the point F'.

The image at F' is *not* a point; the phase of the resultant field at a point near F' is within about $\pm\pi$ of the phase at F' provided the point lies within a sphere with radius about $\lambda/4$ centered at F'. Therefore that is roughly the size of the image at F'.

Concave parabolic mirror.   Imagine that the focal point F and the focal length f of the ellipsoid in Fig. 9.22 are held fixed, but that the focal point F' is moved to the right; the ellipse is "stretched." If F' is moved infinitely far to the right, the ellipsoid degenerates into a paraboloid. Rays emitted from F then form a parallel beam (because they still focus at F', infinitely far away). This is shown in Fig. 9.23.

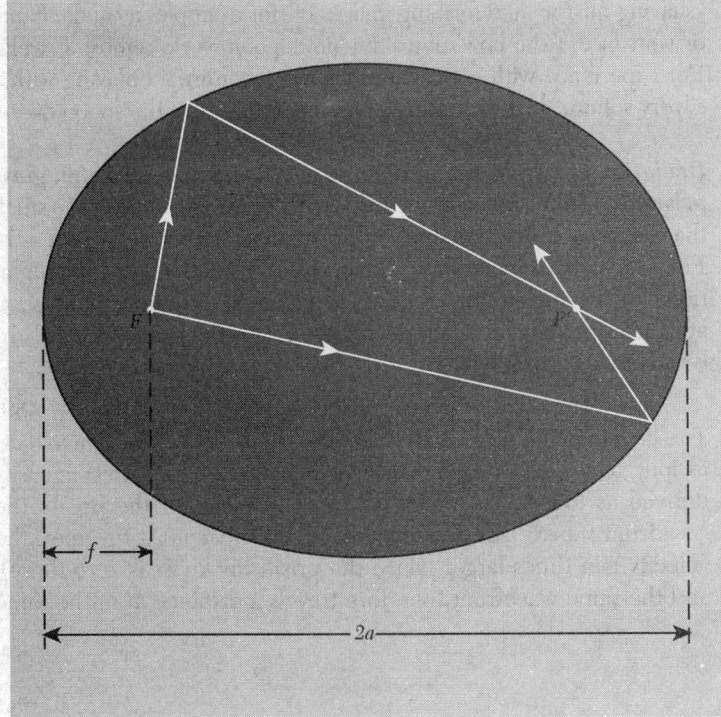

Fig. 9.22 Ellipsoidal mirror.

If the parabolic mirror aperture has a diameter D, then a point source at F does not form a perfectly parallel beam. The angular width of the interference maximum is $\Delta\theta \approx \lambda/D$. If D is "infinite," we get a perfect plane wave from the point source.

Conversely, an incident plane wave (perfectly well defined in angle) focuses to an image at F that is not a point unless D is infinite. The image has a width $\Delta x \approx f\Delta\theta \approx f\lambda/D$.

Concave spherical mirror.   A sphere is said to be "nestled" at the apex of a paraboloid if it is tangent to the paraboloid there and has the same radius as the radius of curvature of the paraboloid there. It is not difficult to show that the radius of such a nestled sphere is $2f$. See Fig. 9.24.

Spherical aberration.   For a small aperture diameter $D \ll 2f$, a spherical mirror is essentially "in contact" with an imagined nestled parabolic mirror. Then a point source at F forms an almost parallel beam. For large apertures, the deviation of the spherical surface from that of a paraboloid produces "spherical aberration." (See Fig. 9.24.)

For a discussion of image formation by concave mirrors, see PSSC, *Physics*, 2nd ed., Chap. 12 (D. C. Heath and Company, Boston, 1965). You can obtain experience with concave mirrors by getting a cheap

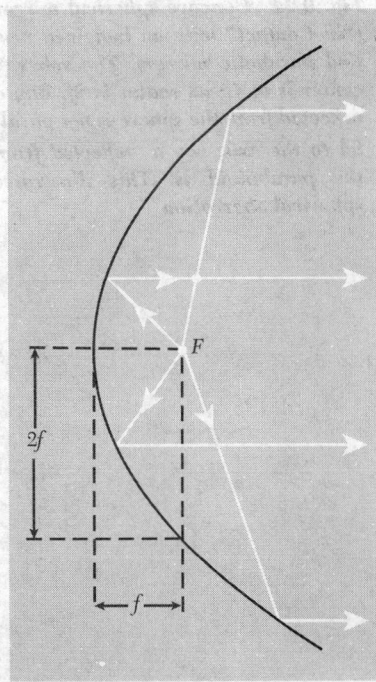

Fig. 9.23   Concave parobolic mirror.

"shaving mirror" and forming images of (for example) a candle flame or your face. (The bowl of a shiny new spoon works almost as well.) For experience with convex mirrors, we recommend playing with a silvery spherical Christmas tree ornament. (Or, turn the spoon over.)

Deviation of a light ray at near-normal incidence on a thin glass prism.   A "thin" prism is one for which the wedge angle a is so small that we can use the small-angle approximations $\sin a \approx a$, $\cos a \approx 1$. For near-normal incidence, we can also use small-angle approximations for the angle of incidence. Then a monochromatic plane wave at near-normal incidence is deviated "toward the base of the prism" by an angle δ given by

$$\delta = (n - 1)a. \tag{89}$$

The deviation δ is a constant, *independent of the angle of incidence*, as long as we stay near normal incidence. Equation (89) is easily derived as follows (See Fig. 9.25): At the base of the prism, the wavefront transverses the distance l at velocity c/n. At the apex, the velocity is n times larger (since the prism thickness is zero there), and the same wavefront therefore travels a distance nl in the same

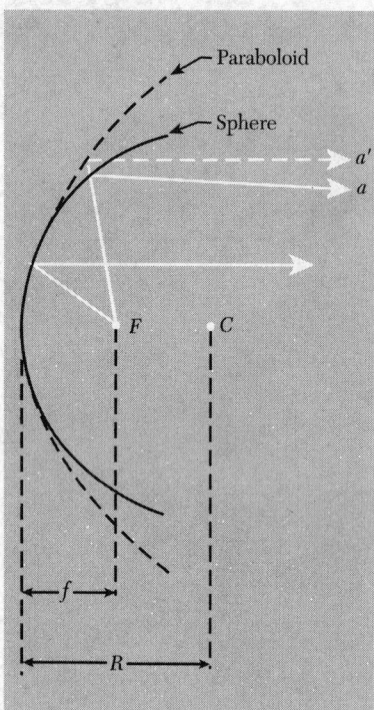

Fig. 9.24   Concave spherical mirror ("in Contact" with an imagined nestled parabolic mirror). The sphere's center is at C; its radius is 2f. Ray a reflected from the sphere is not parallel to the axis; ray a′ reflected from the paraboloid is. This illustrates spherical aberration.

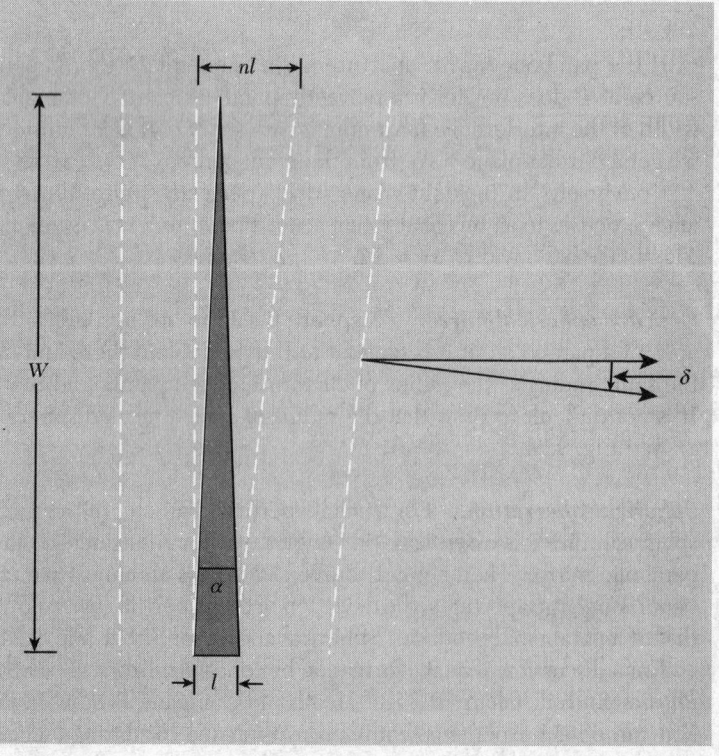

Fig. 9.25   Deviation by a thin prism.

time. Thus the wavefront is ahead by a distance $(n-1)l$ at the top. This distance divided by the width W of the prism is (for small angles) the angle of deviation $\delta = (n-1)(l/W) = (n-1)a$, which is Eq. (89).

Color dispersion of prism. As an example of a thin prism, suppose a is 30 deg (for which the small-angle approximation is still not too bad, for our purpose) and n is 1.50. Then the deviation is 15 deg, according to Eq. (89). That is actually the average deviation, because for typical glass with an average index of refraction of 1.5, blue light of wavelength 0.45 μ actually has index about 0.01 larger than red light of wavelength 0.65 μ. Therefore the blue light is deviated more than is the red light by about $0.01a$. For a of 30 deg, blue is deviated about 0.3 deg more than red. In radians, since 30 deg is about half a radian (1 rad = 57.3°), blue is deviated by about $\frac{1}{200}$ radian more than red. On a screen one meter beyond the 30-deg prism, blue is therefore separated from red by about $\frac{1}{2}$ cm. A prism spectrometer makes use of this dispersive effect of a glass prism to analyze spectra. In optical instruments involving glass lenses, dispersion leads to chromatic aberration—i.e., rays of different colors do not focus at the same places. One can avoid chromatic aberration in a telescope by using a parabolic mirror to gather the light to a focus rather than a refracting lens. (The law of specular reflection holds for all colors.) One can also eliminate chromatic aberration by using two kinds of glass with different dispersion. See Prob. 9.53.

Focusing of paraxial light rays by a thin lens. Suppose we have a glass lens in air with two convex spherical surfaces normal to a common axis of symmetry $\hat{\mathbf{z}}$. A light ray is incident from the left, traveling parallel to the axis of symmetry of the lens at distance $y = h$ from the axis. If the lens is "thin," we (by definition) neglect the variation of y as the ray passes through the lens; we also neglect the thickness compared with the focal length. To consider only "paraxial" rays means that we keep h small compared to the radii of curvature of the two surfaces, so that we can use small-angle approximations for all angles of interest.

Let us find the focal point F where a ray which is incident parallel to the symmetry axis crosses the symmetry axis after deviation by the lens, as shown in Fig. 9.26. We see that if the incident ray is focused at F it must have been deviated by the small angle

$$\delta = \frac{h}{f}. \tag{90}$$

Necessary condition for a focus. Thus we see that the necessary condition for the existence of a *common* focal point for all parallel paraxial incident rays is *that the deviation be linearly proportional*

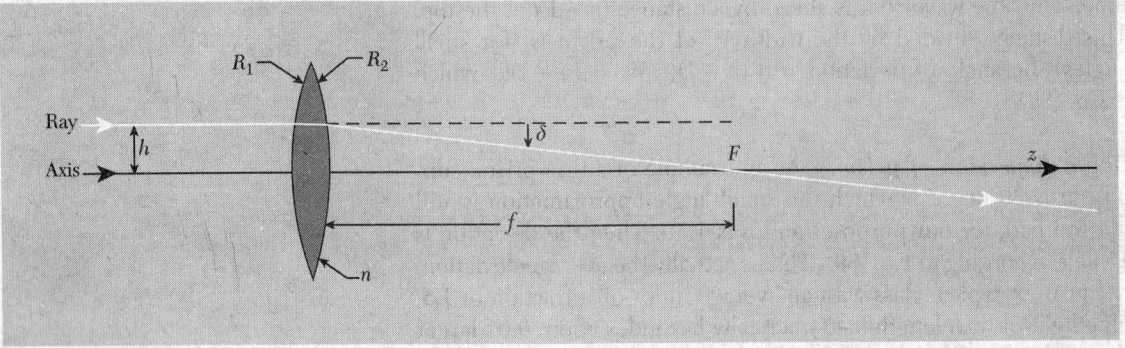

Fig. 9.26 Thin lens. Incident ray parallel to the axis.

to the displacement h of the ray from the axis. Thus if Eq. (90) is satisfied for all h (but always assuming small-angle deviations), then all parallel rays will be focused at the same distance f behind the lens. *This condition holds for any similar focusing problem*, for example for the focusing of a beam of charged particles by a magnetic lens.

It remains to be seen whether a thin lens with spherical surfaces satisfies Eq. (90) with f independent of h. This is seen as follows: As far as the ray in Fig. 9.26 is concerned, it could just as well have been deviated by an equivalent thin prism. The first surface is at angle h/R_1 to the vertical (where the ray strikes it). The second surface is at an angle h/R_2 to the vertical in the opposite sense. The equivalent prism angle a is therefore $hR_1^{-1} + hR_2^{-1}$. The deviation δ by the equivalent thin prism is $(n-1)a$, so that we have

$$\delta = (n-1)h(R_1^{-1} + R_2^{-1}). \qquad (91)$$

Lens-maker's formula. We see that Eq. (91) satisfies the condition for a focus, namely that δ be proportional to h; the focal length f is given by [see Eq. (90)]

$$\frac{1}{f} = (n-1)\left(\frac{1}{R_1} + \frac{1}{R_2}\right). \qquad (92)$$

Equation (92) is called the *lens-maker's formula*.

Focal plane. Now consider a bundle of parallel rays which are not parallel to the symmetry axis but rather make angle θ with the axis. The deviation of a thin prism is independent of the angle of incidence (for small angles). Therefore *a ray that strikes the lens at distance h from its center is deviated by $\delta = h/f$, independent of the angle of incidence*. That means that any parallel bundle focuses at a point in a plane, called the *focal plane*, a distance f behind the lens, and the lateral displacement in the plane of the point is $f\theta$ from the axis, as shown in Fig. 9.27.

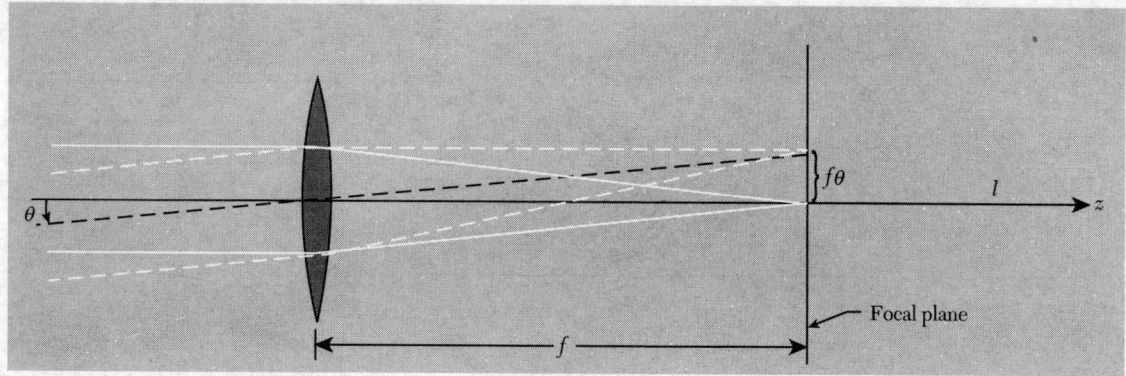

Fig. 9.27 *Focal plane.*

Real point image of a point object. We have found the point image of a parallel beam, i.e., a beam from an object point (source) infinitely far to the left. Let us now consider an object point o at distance p to the left of our converging lens and find its image I at distance q to the right. We let o be on the symmetry axis; then I will be on the axis. Now consider Fig. 9.28. It is obvious from the figure that if we start with a vector pointing from o in the $+\hat{\mathbf{z}}$ direction and then perform the rotations $+\theta_1, -\delta$, and $+\theta_2$, we are back to the $+\hat{\mathbf{z}}$ axis:

$$\theta_1 - \delta + \theta_2 = 0. \tag{93}$$

Thin-lens formula. But

$$\theta_1 = \frac{h}{p}, \qquad \theta_2 = \frac{h}{q}, \qquad \text{and } \delta = \frac{h}{f}.$$

(The deviation is always h/f, independent of angle of incidence.) Therefore Eq. (93) gives

$$\frac{h}{f} = \frac{h}{p} + \frac{h}{q},$$

i.e.,

$$\boxed{\frac{1}{p} + \frac{1}{q} = \frac{1}{f}.} \tag{94}$$

Equation (94) is called the *thin-lens formula.*

Lateral magnification. The angles of deviation of rays by a thin lens are unchanged if the lens is given a slight rotation about an axis through its center perpendicular to the plane of Fig. 9.28. Thus the ray from the object point through the center of the lens is still unde-viated, and the ray striking the lens at distance h from the center is deviated by h/f. Therefore the object and image points in Fig. 9.28

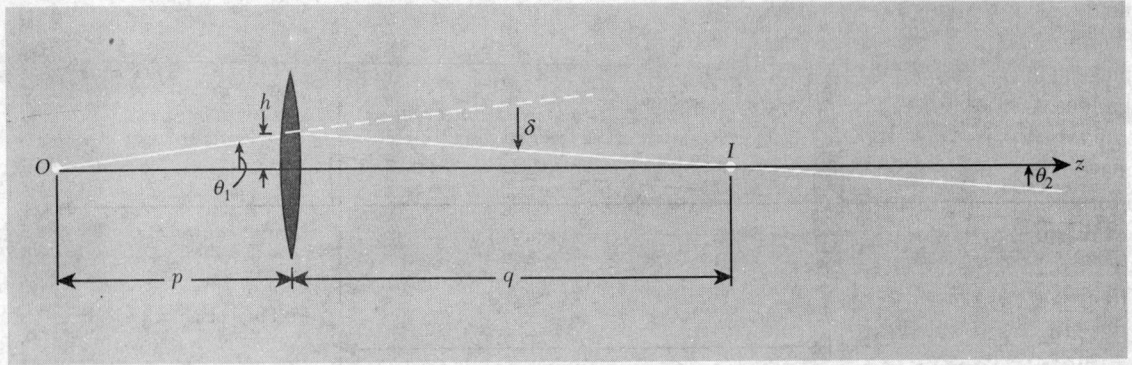

Fig. 9.28   *Real point image of a point object.*

are unchanged if the lens is given a slight rotation about its center. (On the other hand, if the lens is given a slight translation perpendicular to its axis, then the image point will be translated. The new location is obtained by the observation that the ray through the center of the lens is undeviated.) Instead of making a slight rotation of the lens about its center, suppose we hold the lens fixed and give the object point a slight upward translation perpendicular to the axis of the lens. The entire ray diagram can then be rotated about the center of the lens (because the deviations are independent of the angle of incidence, for near-normal incidence). Thus we see that if the object point is translated *up* an amount y, then the image point will be translated *down* an amount that is greater than y by the ratio of the "lever arms" q and p. One expresses this by saying that the lateral magnification is $-q/p$:

$$\text{Lateral magnification} = -\frac{q}{p}. \qquad (95)$$

The minus sign tells us that if the object point goes up the image point goes down. If the object is not a single point but an extended object, like a little arrow with head and tail, we see that the image is *inverted*.

Converging lens.   The lens shown in Fig. 9.28 is a *converging lens.* The image of an object that is at a distance greater than the focal length f from a converging thin lens is a *real inverted image.* The adjective "real" means that there really is light at the image. By contrast, an image in an ordinary plane mirror is "virtual"—there is no light behind the mirror surface.

Virtual image.   If the object point in Fig. 9.28 is at distance f to the left of the thin converging lens shown, then the deviation h/f of rays at distance h from the lens center is just such as to form a parallel beam to the right of the lens. If the object point is closer than f, then the deviation h/f is insufficient to direct the ray back toward the axis. Therefore the ray never crosses the axis again. Thus there is no real image. This ray seems to come from a "virtual" point to the left of the lens. One says that there is a *virtual image*. See Fig. 9.29. It is easy to show (we shall let you do it) that the location of the virtual image is still given by the thinlens formula, Eq. (94), provided we interpret a negative value of q to mean a distance measured to the left of the lens.

Diverging lens.   If a lens is thinner at the center than at the edges, it is a *diverging lens* (assuming it is a glass lens in air). If we think of the lens as consisting of thin prisms (as we did for the converging lens), then the apex of each prism is closer to the axis than the base is. Rays are deflected away from the lens axis (rather than toward it as in a converging lens). A parallel beam incident from the left gives a diverging beam that diverges from a *virtual focus* to the left of the lens, as shown in Fig. 9.30. It is easy to show (we shall let you do it) that all the formulas obtained for thin converging lenses can be used for thin diverging lenses if we give suitable interpretation to the meaning of negative quantities.

Fig. 9.29 Virtual point image of point object. The object distance p is less than the focal length f.

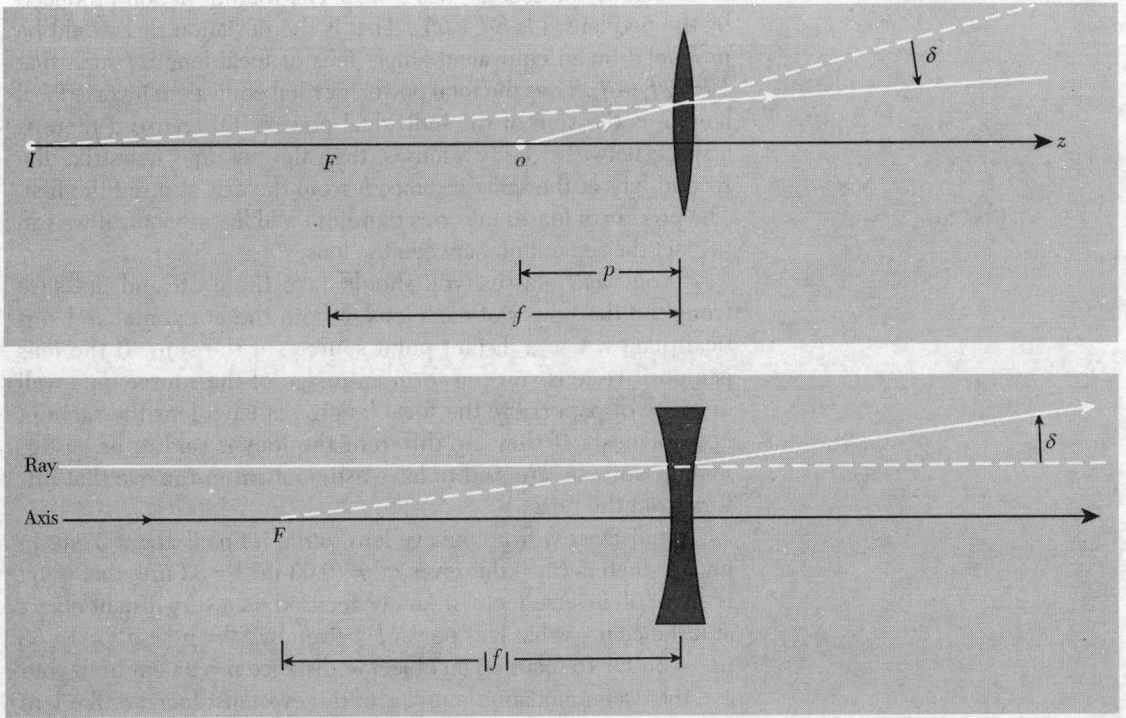

Fig. 9.30 Diverging lens.

Thus if we say that a diverging lens has a negative focal length, $f = -|f|$, we can use the thin-lens formula to relate object and image distances. For example, Fig. 9.30 corresponds to $p = +\infty$, $q = -|f|$, and $f = -|f|$ in the formula

$$p^{-1} + q^{-1} = f^{-1}.$$

Lens power in diopters. The inverse focal length in units of inverse meters is called the *lens power in diopters*. Thus a converging lens of focal length 50 cm has power +2 diopters (+2 D). A diverging lens of focal length –50 cm has power –2 D. The inverse focal length (the power) has the nice feature that it is *linear* in the following sense: If one thin lens is followed immediately by another, *the total power of the two thin lenses in contact is the sum of their individual powers.* That is easily seen as follows. The first lens deviates a ray *toward the axis* by an angle h/f_1, where f_1 is positive for a converging lens and negative for a diverging lens. If the second lens is located at the exit end of the first lens, then the ray does not have a chance to change its transverse distance h from the common axis of the two lenses. Therefore it is incident on the second lens at the same distance h as it was on the first lens. Therefore the deviation produced by the second lens is h/f_2. The total deviation produced by the two lenses is $h/f_1 + h/f_2$. That is the deviation that would be produced by an equivalent single lens of focal length f such that $1/f = 1/f_1 + 1/f_2$. Thus the total power, or total equivalent inverse focal length, is the sum of the individual powers. Of course if there is a space between the two lenses, then the ray does not strike the second lens at the same distance h from the axis as it did the first. The powers of lenses in series therefore add linearly only if we can neglect the separation between the lenses.

If you wear glasses, you should take them off and measure (roughly) the power of each lens in both the horizontal and vertical planes. Use a distant point source (or the sun). If the lens is a positive lens you can form an image of the source on a wall or piece of paper. Are the focal lengths of each lens the same in both planes? (If they are different the lens is said to be "astigmatic," and you are said to have astigmatism in the eye that differs from the norm.)

The distance q from the eye lens to the retina is about 3 cm. In inverse meters (m^{-1}) this gives $q^{-1} = (0.03 \text{ m})^{-1} = 33 \text{ m}^{-1}$, that is, q^{-1} is about 33 inverse meters. An eye focused on a very distant object at distance $p = \infty$ has lens power f^{-1} given by $f^{-1} = p^{-1} + q^{-1} = 0 + 33$ $\text{m}^{-1} = 33$ D. To focus on an object at distance $p = 25$ cm from your eye the accommodation muscles of the eye must increase the lens power by an amount $p^{-1} = (0.25 \text{ m})^{-1} = 4 \text{ m}^{-1} = 4$ D, giving a total of about 37 D. If you have sufficiently good accommodation muscles, you can instead increase your lens power by about 10 D and can then focus on an object at distance $p = (10 \text{ D})^{-1} = 0.1 \text{ m} = 10 \text{ cm}$. Then the object looks larger and you can see its details better. If you could bring it to within 1 cm of your eye and still focus an image on the retina, it would look 25 times larger than at 25 cm; correspondingly, you could resolve details 25 times smaller. No one has that much accommodation.

Simple magnifier. You can hold a small object at about 25 cm from your unaided eye and examine it without fatigue, if you have normal vision. If the height of the object is h (in cm), it subtends an angle $h/25$ (radians) at your eye, and this determines the size of the image on the retina. If you can bring the object closer, it will give a larger image on the retina. To maintain a clear (i.e., focused) image, the accommodation muscles must increase the lens power. That is difficult and tiring. Now use a lens of focal length f (cm). Hold the lens just in front of the eye. Bring the object closer. When the object is at the focal plane of the lens, each point on the object will give a parallel bundle of rays out of the lens and into your eye. This is easy for you to focus—your eye lens is relaxed. We shall let you show that the angular size of the object increases by a factor $25/f$ (assuming small angles so that you can use small-angle approximations). See Fig. 9.31. You can make a cheap magnifier (for about 50 cents) by getting a lens of focal length 2 or 3 cm and taping it on a microscope slide. (You can also buy one for about $1.00. See, for example, the catalog of Edmund Scientific Co., Barrington, N.J. 08007).

Pinhole magnifier. Take a piece of aluminum foil and make in it a pinhole of about $\frac{1}{2}$-mm diameter or less. Hold it close in front of your eye. Look at a light source. The "floaters" you see are the diffraction patterns of chains of cells in your eye. (They are not on the

Fig. 9.31 Simple magnifier. The power of the eye lens is supplemented by that of the magnifier. The object can be moved closer to the eye and consequently gives a larger image.

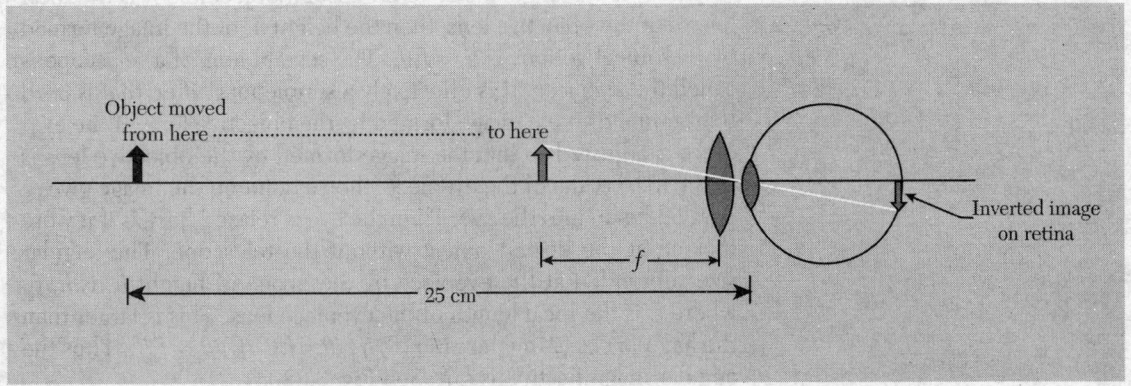

Object moved from here .... to here

f

25 cm

Inverted image on retina

surface, as you can see by trying to wipe them off by blinking.) Now look at a well-illuminated printed page through the pinhole. (If you wear glasses, take them off. You don't need them and they don't do any good.) Bring the page up closer and closer to your eye. Notice that the word you are looking at stays "in focus" and is magnified as it is brought closer! (It finally gets fuzzy because your pinhole is not small enough.) The magnification is easily calculated by a sketch like that of Fig. 9.31 with the lens replaced by a pinhole.

Do you really see things upside down?   Here is a way you can convince yourself that the image on your retina is inverted. Look through your pinhole at a broad light source. Hold a pencil point in front of the pinhole and look at its shadow on your retina. Everything behaves as expected. Now *reverse* the *order* and put the pencil point between the pinhole and your eye. Move the pencil and notice the direction of motion of the shadow! Now make a sketch and explain what's happening.

Exercising the pupils.   When you look at a broad source (like the sky) through your pinhole, you see a bright circle. That circle is the projection of your pupil on your retina. You can study the dilation and contraction of your pupil by covering and uncovering your *other* eye, the eye that is *not* looking through the pinhole! When you uncover the other eye so that light enters it, its pupil contracts. *So does the pupil of the eye looking through the pinhole!* You can easily see these "sympathetic" pupil contractions. Notice that it takes a time of the order of $\frac{1}{2}$ sec for the pupil to contract or dilate when the light intensity is suddenly changed.

Telescope.   A telescope consists of two lenses. The first is the "objective" lens, which forms a real image of a distant object. To a good approximation, the image is in the focal plane of the objective lens. If θ_0 is the angular size of the distant object and f_1 is the focal length of the objective lens, then the height h_1 of the image formed by the objective lens is $h_1 = f_1\theta_0$. The second lens of a telescope is called the *eyepiece*. It is effectively a simple magnifier. that is used to examine the real image formed by the objective lens. If the eyepiece is adjusted so that the image formed by the objective lens is in the focal plane of the eyepiece, then a point on the image gives a parallel beam into the eye. Then the eye is relaxed, just as if it were looking at the distant object without the telescope. The angular size subtended at the eyepiece by the image of height h_1 is h_1/f_2, where f_2 is the focal length of the eyepiece lens. This is larger than the angular size θ_0 by the ratio $(h_1/f_2)/\theta_0 = f_1\theta_0/f_2)/\theta_0 = f_1/f_2$. Thus the angular magnification is f_1/f_2. See Fig. 9.32.

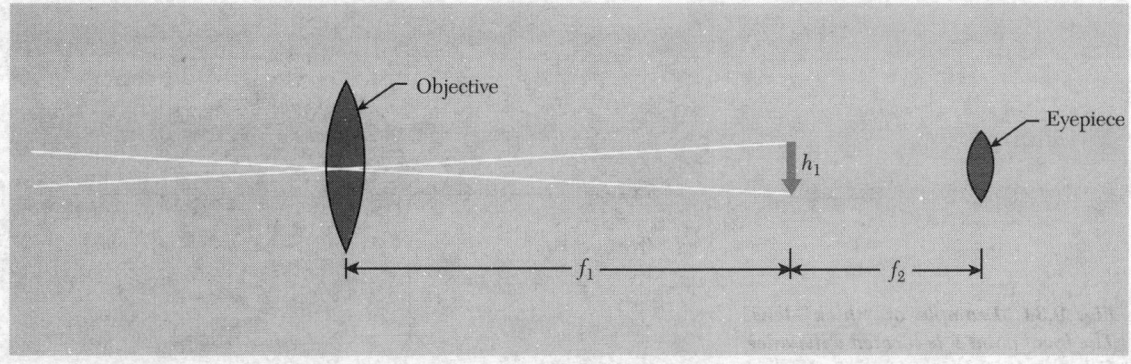

Fig. 9.32   Telescope.

Microscope.   A microscope is like a telescope in having an objective lens to form a real image of the object and an eyepiece to examine this image. The bug to be examined is nearly (but not exactly) at the focal plane of the objective. The image is formed a long distance L from the objective—say $L \approx 20$ cm. This distance is essentially the length of the barrel of the microscope. A bug of width x located at distance approximately f_1 from the objective gives a real image of width $h_1 = (L/f_1)x$ at the image point. This image is a distance f_2 from the eyepiece and subtends an angle h_1/f_2 there. If the bug were examined with the naked eye at distance 25 cm, it would subtend angle $x/25$ cm. Thus the magnification is $(h_1/f_2)/(x/25) \doteq 25L/f_1 f_2$. See Fig. 9.33.

Thick spherical or cylindrical lens.   A small glass baby-food jar makes a good cylindrical lens. (We recommend chocolate pudding. Eat the pudding, scrape off the label, and fill the clean jar with water or any other clear liquid.) In Fig. 9.34 we show the formation of an image of a parallel beam of light by such a lens.

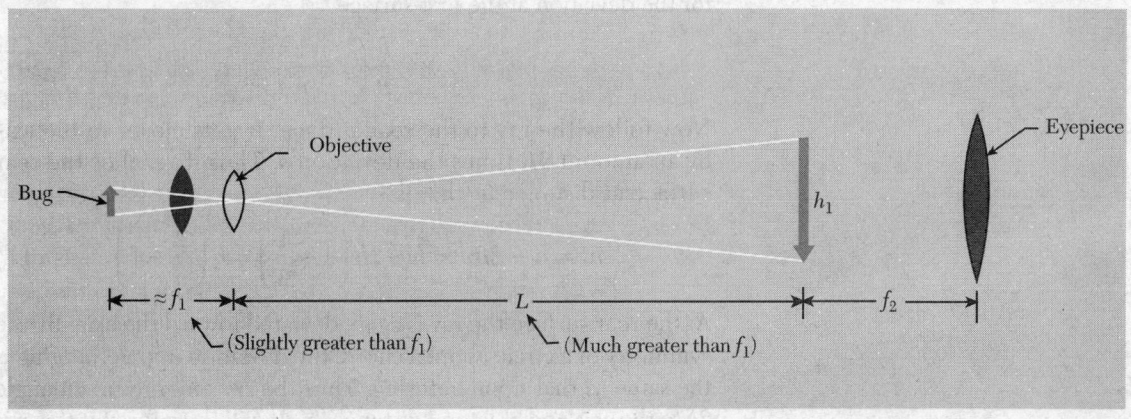

Fig. 9.33   Microscope.

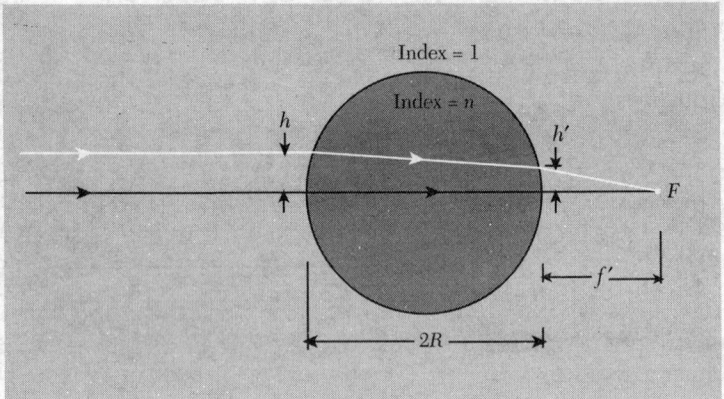

Fig. 9.34 *Example of "thick" lens.*
The focal point F is located a distance
f′ beyond the last surface. Indices of
re-fraction: air = 1, lens = n.

Deviation at a single spherical surface.   Let us trace parallel rays
through this lens. The ray through the center of the sphere or circle
is not deviated. The ray at transverse distance h from the center
line makes an angle of incidence θ_i given by $\theta_i = h/R$, for $h/R \ll 1$.
The deviation δ of this ray at the first surface is equal to the angle
of incidence θ_i minus the angle of refraction θ_r. For small angles,
Snell's law, $n_1 \sin \theta_1 = n_2 \sin \theta_2$, becomes $n_1 \theta_1 = n_2 \theta_2$. Then the *devia-*
tion toward the normal at one surface is given by

$$\delta = \theta_1 - \theta_2$$
$$= \theta_1 \left(1 - \frac{\theta_2}{\theta_1} \right)$$
$$= \theta_1 \left(1 - \frac{n_1}{n_2} \right). \tag{96}$$

Equation (96) is general (for small angles) and is useful for tracing
rays through complicated systems. In the present example, we find
for the deviation at the first surface

$$\delta = \frac{h}{R} \left(1 - \frac{1}{n} \right). \tag{97}$$

Now follow the ray to the rear surface. It gets closer to the axis
by an amount $2R$ times the deviation δ. Thus it reaches the rear
surface at distance h' that is

$$h' = h - 2R\delta = h - 2h \left(1 - \frac{1}{n} \right) = h \left(\frac{2}{n} - 1 \right). \tag{98}$$

At the rear surface the ray is again deviated toward the axis. By the
symmetry of a circle about a chord, the deviation upon emerging is
the same as that upon entering. Thus the ray emerges at an angle
2δ to the axis and at lateral distance h'. It will therefore hit the axis
at distance f' beyond the last surface, where

$$2\delta = \frac{h'}{f'}. \tag{99}$$

Equations (97), (98), and (99) give

$$f' = \frac{h'}{2\delta} = \frac{h\left(\dfrac{2}{n} - 1\right)}{\dfrac{2h}{R}\left(1 - \dfrac{1}{n}\right)} = \frac{R}{2}\frac{(2 - n)}{(n - 1)}. \tag{100}$$

You can use Eq. (100) and a jar to measure the index of refraction of water or of (for example) mineral oil. [Equation (100) holds for either a cylinder or a sphere.) See Home Exp. 9.42.

Leeuwenhoek's microscope. The world's first microscope was merely a tiny glass sphere. You can make one. (You can obtain tiny clear glass spheres by the pound from a chemical supply house. Make sure they are clear instead of translucent.) Here is how it works. Put the sphere right in front of your eye. Put the bug (to be looked at) at the focal point F of Fig. 9.34. A point on the bug gives a parallel beam of light entering the eye. Since it is a parallel beam, you can relax your accommodation muscles, and the beam will focus to a point on the retina. Another point on the bug will focus at another point on the retina. Let us calculate the magnification of this lens. Suppose the bug has lateral extension x_{bug}. Rays from the extremes of the bug through the center of the sphere are not deviated. That means that the angular size of the bug is x_{bug} divided by the distance from F to the center of the sphere:

$$\theta_{bug} = \frac{x_{bug}}{R + f'}. \tag{101}$$

This is the angle between the parallel beams that correspond to the images of the extremes of the bug on your retina and is therefore the angular size you "see" using the microscope. When you look at the bug without the microscope, you must hold the bug about 25 cm away to focus on it comfortably. The angular size of the bug is then $x_{bug}/25$ cm. The angular magnification M is therefore

$$M = \frac{25}{R + f'} = \frac{25}{R\left[1 + \dfrac{1}{2}\left(\dfrac{2 - n}{n - 1}\right)\right]} = \frac{50\text{ cm}}{R}\left(1 - \frac{1}{n}\right). \tag{102}$$

Thus, for example, if $R = 1$ mm and $n = \frac{3}{2}$ (glass), we get $M = 167$.

Scotchlite retrodirective reflector. If $n = 2$, then according to Eq. (98) a paraxial ray entering at transverse distance h strikes the rear surface of the sphere of Fig. 9.32 at distance $h' = 0$. Thus a parallel

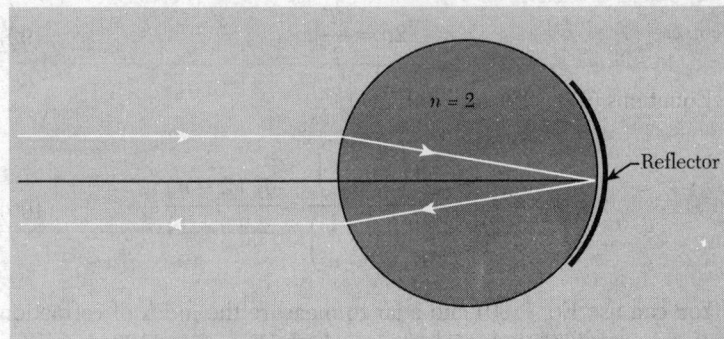

Fig. 9.35 Retrodirection of light by an ideal Scotchlite reflector having index n = 2.

beam is focused exactly at the rear surface. The beam is partly reflected and partly transmitted there. The reflected part is eventually directed back at 180 deg to the original direction, as is seen by inspection of Fig. 9.35. The transmitted light at the rear surface can be largely reflected back into the glass by covering the rear surface with a silvery reflector.

A reflecting material called Scotchlite, which uses this principle, can be obtained at any hardware store. It is used for bright road signs, among other things. Examine it with a magnifying glass. You will see that it consists of many tiny glass spheres embedded on a sticky silvery surface and then painted with clear-red shellac (for red Scotchlite) or something else for other effects. It turns out that the largest index that one can easily get with glass is about $n = 1.9$ This is close enough to 2 so that it works fairly well.

The "next generation" of the world's largest liquid hydrogen bubble chambers, now (1968) being designed, will (at least some of them will) use Scotchlite on the bottom of the chamber to retrodirect light rays toward their source. You can easily measure the retrodirective properties of Scotchlite. See Home Exp. 9.35.

Problems and Home Experiments

9.1   Near field and far field.   How far away should you be from a double slit of slit spacing 0.1 mm irradiated with visible light in order to use the far-field approximation *without* making use of a lens? How far should you be from two microwave antennas having spacing 10 cm and emitting 3-cm microwaves to use the far-field approximation?

9.2   A double slit of slit separation 0.5 mm is illuminated by a parallel beam from a helium-neon laser that emits monochromatic light of wavelength 6328 Å. Five meters beyond the slits is a screen. What is the separation of the interference fringes on the screen?

9.3 What is the "mean length" of the classical wavetrain (wave packet) corresponding to the light emitted by an atom with mean decay time 10^{-8} sec? In an ordinary gas-discharge source the atoms do not decay freely but rather have an effective coherence time $\approx 10^{-9}$ sec due to Doppler broadening and collision broadening. What is the length of the corresponding classical wavetrain?

9.4 If a "line" source of visible light is not really a line but has width 1 mm, how far must it be from a double slit which it illuminates in order for the two slits to be reasonably coherent? Assume the slit separation is $\frac{1}{2}$ mm.

9.5 How far away is an automobile when you can barely resolve the two headlights with your eye?

9.6 Venus has a diameter of about 8000 miles. When it is visible as a "morning star" (or "evening star"), it is about as far away as the sun, i.e., about 93 million miles. It looks "larger than a point" to the unaided eye. Are you seeing the true size of Venus?

9.7 Resolution of the eye. Tale two light bulbs of the same power (say 150 watt), one with a clear glass envelope and reasonably small filament (2.5 cm by 0.3cm), the other with a frosted bulb 7.5 cm in diameter. Find out by experiment how far away you must walk before the two lights have the same apparent size. (It will be a block or two.) At this same large distance. compare the apparent sizes of two frosted bulbs having the same actual size but differing in power by a factor of two or three. How do you explain the result? Why does Venus look larger than a point? (See Prob. 9.6.)

Home experiment

9.8 Diffraction grating moiré pattern. You need a white line source and two identical gratings. [The best line source (which you will need for many of these experiments) is a "display lamp." For example, one of about 40 watts with a 3-in. long straight filament in a clear glass envelope is available for about 40 cents in grocery and hardware stores. Your optics kit has only one grating. More gratings are available for about 25 cents each (or about 10 cents each in boxes of 100) from, e.g., Edmund Scientific Company, Barrington, N.J. 08007.] With the line source oriented vertically, look through one grating (hold it up close to one eye) and orient the grating so that the colors are spread out horizontally. Now superpose the second grating on the first. Carefully rotate it so as to superpose exactly the first-order images from the two gratings. With care, you will succeed (in a minute) in obtaining "black stripes" across the colored first-order image. Here is part of the explanation. The line spacing on the grating is d. Suppose the space

Home experiment

between the planes of the two gratings is s. Think of them as two picket fences superposed with a slight space between them or as two identical screens parallel to one another. At some angles, the grating scratches will lie one behind the other. At other angles, the scratches of one grating will lie (in projection) halfway between the scratches of the other. At these angles, the effective number of scratches per unit length (i.e., d^{-1}) is doubled. Now comes the physics: Why do you get the black stripes? Do they correspond to angles where the effective number of lines is "single" or "double"? Given the number of lines per cm, d^{-1}, for each grating, how can you determine the spacing s? Given s, how can you determine d?

Home experiment   **9.9   Silk-stocking diffraction pattern.**   You need a sheer silk (or nylon) stocking and a point source of white light. Although a reasonably distant street light will perhaps do for the point source, the best point source for this and other experiments is made from a 6-volt flashlight, for example a "camper" flashlight, with a bulb that has a filament about $\frac{1}{2}$ mm long. To get a good point source, remove the glass lens and cover the parabolic reflector with a piece of dark cloth or paper (with a hole cut for the bulb). Or simply look at the bulb from the side, out of the beam from the reflector. (*Note*: A "sealed beam" flashlight will not work!)

Look through the stocking at the point source. From the pattern you see, you could determine the average thread spacing and the number of sets of threads at different angles. Fold over many layers and look at the source again. The pattern of concentric circles you see is similar to an x-ray "powder diffraction pattern."

Home experiment   **9.10   Long-playing diffraction grating.**   Look at a white point source reflected at near-glancing incidence on a 33-rpm record. The record grooves make a good reflection grating. Measure crudely the wavelength of red and of green light using the record. Describe your method. How can you easily determine the location of the zeroth-order "specular" maximum?

9.11   Which side has the scratches?   One side of the plastic of your diffraction grating is smooth; the other side has the scratches. You can find out which side has the scratches by looking through it at a white source after rubbing one side of the grating with an oily finger; then clean it and try the other side. What is the explanation?

9.12   Consider nestled spherical and parabolic mirrors as shown in Fig. 9.24. Take the $+\hat{\mathbf{z}}$ direction to the right (along the axis of symmetry) and x transverse to z; take $x = z = 0$ at the apex of the mirrors.

(*a*) Show that the parabolic surface is given by

$$z = \frac{x^2}{4f}.$$

(*b*) Show that the spherical surface is given (for $x \ll f$) by

$$z = \frac{x^2}{4f} + \frac{x^4}{64f^3} + \cdots.$$

(c) Compare a spherical mirror with aperture of diameter D and with focal length f to a parabolic mirror with the same D and f. For the spherical mirror, consider the angular deviation $\delta\theta$ of the "worst" rays (near the rim of the aperture) due to spherical aberration. ($\delta\theta$ is the deviation from the $\hat{z}$ direction for rays from a point source). Show that $\delta\theta$ is less than the diffraction angular width $\Delta\theta \approx \lambda/D$ provided that

$$D < 4f \left(\frac{\lambda}{4f} \right)^{1/4}.$$

Thus (for example) for visible light and for focal length $f \approx 50$ in., a spherical mirror is about as good as a parabolic mirror provided the mirror diameter D is less than ~ 3.5 in.

9.13 A plane slab of glass of thickness t and index n is inserted between an observer's eye and a point source. Show that the point source appears to be displaced to a point closer to the observer by approximately $\left[(n-1)/n \right] t$. Use small-angle approximations.

9.14 A "corner reflector" consists of three plane mirrors joined so as to form an inside corner of a rectangular box. Show that a light beam that strikes a corner reflector is directed back at 180 deg to its original direction, independent of the angle of incidence, as long as it hits all three surfaces.

9.15 Show that a plane wave *normally* incident on one face of a wedge-shaped prism of angle A is deviated by an amount θ_{dev}, where

$$n \sin A = \sin (A + \theta_{dev}).$$

9.16 A diffraction-limited laser beam of diameter 1 cm is pointed at the moon. What is the diameter of the area illuminated on the moon? (The moon is 240,000 mi away.) Take the light wavelength to be 6328 Å. Neglect scattering in the earth's atmosphere.

9.17 Single slit diffraction pattern. Tape a piece of aluminum foil by its edges to a microscope slide. (The most convenient tape is Scotch translucent "magic mending tape.") Cut a single slit with a razor blade or sharp knife. Hold the slit close in front of one eye and look at a white line source. Estimate the angular full width of the central maximum by (for example) making marks on a piece of paper that is behind the line source to give a scale. Estimate the ratio of the wavelength of red light to that of green light, where these colors are given by your gelatin filters. Using the red filter, estimate the width of the razor cut, i.e., the width of your slit,

Home experiment

using the measured angular width of the diffraction pattern and assuming $\lambda \sim 6500$ Å. If you have a magnifying glass, you can lay your slit on a millimeter scale and estimate the slit width directly. How do the two results for the width compare?

Home experiment **9.18 Double-slit diffraction and interference pattern.** Make two parallel slits separated by $\frac{1}{2}$ mm or less, using the technique of Home Exp. 9.17. Make one slit about $\frac{1}{2}$ cm longer than the other, so that you can go quickly from the double-slit pattern to the single-slit pattern by displacing the slits slightly. You can thus see what part of the double-slit pattern is the "single-slit modulation," due to the nonzero width of the single slit. To see the effect of variable slit spacing d easily, cut one slit at a slight angle to the other, so that they cross in a "vee" shape. You should make many slits (it takes 10 sec to make one pair of slits—the tenth time you do it); some will be better than others. (Hold the slit up to a light and examine it to see why it is a bad one, if it is.)

Home experiment **9.19 Three-slit pattern.** Try this only after you have made a number of good double slits by the technique of Home Exps. 9.17 and 9.18. Cut a third slit parallel to the first two slits. Make the third slit not as long as the first two, so that you can quickly go from two to three slits by a slight translation. The important thing to try to see is the narrowing of the intensity maxima when the third slit is added. [You can get a very beautiful set of single, double, triple, and quadruple slits, along with slits of variable width and an array of up to 80 slits, all mounted on a single convenient slide called the Cornell Slitfilm Demonstrator, which is obtainable from The National Press, 850 Hansen Way, Palo Alto, Calif. List price is $1.50.]

Home experiment **9.20 Coherence—size of a "point" source or line source.** Use a single slit of known (estimated) width. Place the red gelatin filter over the source. Stand far enough from the source so that you obtain a sharp single-slit pattern. Now move closer to the source. Find the distance L at which the single-slit pattern "washes out." [It washes out at a distance at which different parts of the filament of the flashlight bulb (if that is your point source) become independent light sources and thus are incoherent for the resolution time of your eye, as discussed in Sec. 9.4.] Use your estimates of the sizes of the source and the slit and your measurement of the distance L at which the pattern washes out to estimate the wavelength of the light, using the relation derived in Sec. 9.4, $d(\text{source})\,D(\text{slit}) \approx L\lambda$.

Home experiment **9.21 Coherence—Lloyd's mirror, the "guaranteed coherent double slit."** If you hold an ordinary double slit in front of one eye and look at a broad source like the sky or a frosted light bulb, you will see no interference pattern. Why is that? We shall now design a double slit that will give a two-slit interference pattern even when you look at a frosted light bulb. First make a single slit by the technique of Home Exp. 9.17.

Now take a second microscope slide and place it edgewise to the first slide (the one with the slit) and parallel to the slit, so that the mirror image of the slit in the second slide is parallel to the first slit. Stick the second slide to the first with a big glob of putty or modeling clay (nondrying putty, e.g., Nu-Glaze glazing compound works well), so that you can easily wiggle the second slide to adjust it, but it will stay in place when you don't push on it. Adjust the mirror (the second slide) so as to get as narrow a separation between the slit and its "image slit" as you can manage—say $\frac{1}{2}$ mm. Do this by holding the assembly a foot or so from your head, so that you can easily focus your eyes on the "double" slit while you hold it in front of a bright background and adjust the mirror. When you have a good double slit, bring the assembly up close to one eye and focus the eye at a large distance (i.e., on the light source). Look for three or four "black streaks" parallel to the "coherent double slit." These are interference zeros due to destructive interference between the light coming from the real slit and that coming from the image slit. The image slit is of course always completely coherent with the real slit. (Why?) Because of the phase reversal in reflection, the slit currents and the "image slit currents" are 180 deg out of phase. Therefore the fringe at the plane of the mirror is "black"—an interference zero. Here is a question, to be answered both by experiment and by "theory": Are the " bright" lines between the "black" lines exactly as bright as the bright background that one sees with just the single slit? Brighter? Dimmer?

9.22 Paper-clip Lloyd's mirror. (See Home Exp. 9.21.) A paper clip illuminated by a light bulb gives a shiny narrow line source. Hold the clip against (and parallel to) the edge of a microscope slide used as a mirror. When you get a decent looking "coherent double slit" of separation less than $\frac{1}{2}$ mm, bring it up close to one eye and look for the dark interference bands discussed in Home Exp. 9.21. It takes a little more practice than the method of Home Exp. 9.21. The light must be at nearly grazing incidence on the mirror. Also the illumination should be arranged so that the light source doesn't blind you.

Home experiment

9.23 Two-dimensional diffraction patterns. (*a*) Look at a distant streetlight through a piece of ordinary window screen. Turn the screen sideways so that the projected wire separation is as small as you please. *Problem*: How far away must a streetlight of 20-cm diameter (frosted bulb) be in order to give coherent illumination over two neighboring wires of the screen?

(*b*) Look at a streetlight or your flashlight point source through various kinds of cloth—a silk handkerchief, nylon panties, an umbrella, etc.

(*c*) Look at a point source through two diffraction gratings of the type that you have in your kit. Rotate one grating so that its lines are perpendicular to those of the first. Notice that one gets some (rather faint) bright

Home experiment

spots at 45 deg to the two sets of lines. These spots are something new, not obtained by superposing *intensities* from the two gratings. Of course they *must* be due to superposition of amplitudes from the two sets of lines. Make a sketch and explain the origin of these "extra spots." The diffraction pattern produced by two crossed gratings is similar to the pattern produced by diffraction from a single crystal. You may have seen the movie made for Education Development Center (EDC, formerly ESI) by L. Germer, showing diffraction of a monoenergetic electron beam reflected from the surface of a single crystal. (The technique is easier for reflected than for transmitted waves, if one wants to look at a single crystal. Similarly, you can get reflection gratings from Edmund Scientific Co. They are like your transmission grating, except that the surface is lightly silvered to enhance reflection.)

Home experiment   **9.24 Diffraction grating–gelatin filter passbands.** Use your diffraction grating as follows to measure the wavelengths of the red and green passed by your filters. Put a line (or point) source right next to a wall or door. Make a mark on the wall about a foot to the side of the source. Look at the source through the grating, holding the filter over your grating (or put the filter over the source—but don't melt it!) Move closer and farther from the source until the color of interest appears to be superposed with your mark on the wall. Measure the appropriate distances and calculate λ. Thus calibrate the wavelengths transmitted by your red, green, and purple filters. Memorize the results. (Then you can use your filters and the grating to find the wavelengths of other colors when you wish to, without repeating the geometric measurement of this experiment.)

Home experiment   **9.25 Spectral lines.** Pour some table salt on a wet knife or spoon (one that you don't mind ruining). Set the knife in the flame of a gas stove. Look at the yellow flame through your diffraction grating (this is easiest at night in a darkened room). Notice that the first-order (and higher-order) images of the yellow sodium flame are as sharp and clear as the zeroth-order "direct" image. That is because the yellow light is a "spectral line" having narrow bandwidth. (Actually the yellow light from sodium is a "doublet" of two lines with wavelengths 5890 and 5896 Å.) Now look at a candle. In zeroth order, it does not look terribly different from the sodium flame; they are both yellow. But in the first-order diffraction image, the candle is very much spread out in color, whereas the sodium remains sharp. The "yellow" of the candle, which is due to hot particles of carbon, has a wavelength spectrum extending over (and beyond) the entire visible range.

Here are other convenient sources of sharp spectral lines; look at them through your grating:

Mercury vapor: Fluorescent lamps, mercury-vapor street lights, sunlamps. (A sunlamp is convenient in that it screws directly into an ordinary 110-volt AC socket. It is probably the cheapest source of mercury-vapor spectral lines; the cost is about $10.)

Neon: Many advertising signs. Neon has a profusion of lines; you see "many signs." A cheap broad monochromatic source is a G.E. bulb NE-34 which screws directly into a 110-volt AC socket (the cost is about $1.60). Others are a "circuit continuity tester," which plugs into any wall receptacle and which costs about $1 (at a hardware store), and a neon "night light."

Strontium: Strontium chloride salt (available at a chemical supply house for about 25 cents/oz); dissolve a little in a few drops of water and put it in the gas flame on your ruined spoon. The wavelength of the red line is a famous length standard.

Copper: Copper sulfate; availability and technique as for strontium chloride. It gives a beautiful green color.

Hydrocarbon: Look at your gas flame in the first-order spectrum. There are a sharp, clear blue image and a sharp, clear green image. The "blue" color of the flame is therefore due to one or more almost monochromatic spectral lines.

9.26 Monochromatic toilet paper. Burn a piece of toilet paper and look at it through your diffraction grating (held, as always, close in front of one eye). Notice the beautifully clear "first-order flame." This shows that the soft yellow light is almost monochromatic, with very little "white light" color spectrum due to hot carbon. The yellow that you see is the by now familiar (we hope) sodium doublet of wavelengths 5890 and 5896 Å.

Now that you recognize "sodium yellow," light an ordinary paper match and look at it with your grating. Most of the light is "hot carbon yellow," which is not really yellow but a complete "white" color spectrum. But look closely! In the yellow part of the hot carbon spectrum, down low next to the cardboard, where the flame is "blue" looking—below the blindingly bright hot carbon spectrum—do you see a crisp, clear little monochromatic match flame? If you don't, try again! Now burn other things and look. You may well conclude that everything is made of salt or is at least contaminated by it.

9.27 Fabry-Perot sodium fringes. The world's cheapest broad, almost monochromatic light source is obtained by burning a wad of toilet paper. You can use this source to, see Fabry-Perot fringes. Burn the paper. (The room should be dark—perhaps also you should have some water handy!) Look through the flame at the image of the flame at near-normal incidence in a piece of glass—a microscope slide or picture-frame glass. You will see finger-print-like fringes. If the glass is optically flat, the fringes will be circles centered on your eyeballs; in any case, you can see them easily. If you have a gas stove or bunsen burner, you can get a brighter monochromatic sodium source by sprinkling salt on a wet knife and immersing it in the flame. Then you can see the Fabry-Perot fringes even in the daytime. For a nice, steady, broad monochromatic source with which to look at the fringes, use the G.E. neon bulb NE-34.

Home experiment

Home experiment

9.28 Mailing-tube spectrometer—Fraunhofer lines. Use a mailing tube 45 cm to 60 cm. Mount your diffraction grating on one end. Mount
a single slit on the other end. The slit is best made with two single-edged
razor blades. Glue or tape one blade permanently in place; stick the other
on with nonhardening putty (glazing compound, obtainable in any hardware store), so that you can easily adjust it (narrower for better resolution,
wider for more light). Look at the spectra mentioned in Home Exp. 9.25.

Problem: Should you be able to resolve the sodium doublet (wavelengths
5890 and 5896 Å) with this spectrometer?

Ans. No—the line separation given by this grating is just about equal to
the image width due to diffraction in the pupil of your eye.

Can you resolve it by using a longer mailing tube?

Ans. No. There are two ways to improve the resolution. One is to get
a grating with smaller line spacing d. The other is to increase the number
of lines that are used, i.e., to increase the width D of grating used. With
the design above, D is the width of your pupil, about 2 mm. If you add a
telescope with an objective lens of diameter 2 cm, and if all the rays that
enter the objective lens get through the pupil of your eye, then with the
diffraction grating at the objective lens, your angular resolution, λ/D, is
improved tenfold.

With this simple spectrometer you can see the Fraunhofer lines in the
spectrum of the sun. Go outside on a sunny day. Lay a pile of half a dozen
sheets of white paper on the ground (more than one so that it is as "white
as possible"). Look at the sunlit paper with your spectrometer. Use a coat or
blanket to cover your head to keep out stray light; otherwise you will have
difficulty seeing the first-order spectrum. Also, use the edge of the tube
to "hide" the blindingly bright zeroth-order light. Adjust the slit to about
$\frac{1}{2}$ mm. Look for three or four or five dark lines crossing the continuous
spectrum of the sun. If you don't see anything, keep trying—adjust the slit
width for comfortable intensity. Another technique is to cover the slit with
several layers of waxed paper, use a very narrow slit, and look at the sky
near the sun, varying the intensity by how close you come to pointing the
spectrometer at the sun.

The dark Fraunhofer lines you see are absorption lines. Atoms in the
relatively cool outer gas mantle of the sun are driven by the continuous
spectrum emitted by the hot sun. Those frequencies that correspond to
natural resonances of the atoms excite the atoms. This takes energy out of
the continuous spectrum at the resonant frequency. The outer gas is actually opaque at those frequencies, so that the spectrum has corresponding
"black lines" at colors where the sunlight has been completely absorbed.
The easiest lines to see are some closely spaced lines in the yellow-green
due to iron, calcium, and magnesium; the H line in the blue-green due
to hydrogen; and several closely spaced lines in the blue due to hydrocar-
bons—similar to the emission lines you see with a gas flame. The sodium D

line is also present, but hard to see (for me at least). To see where to look for it, look at the sodium emission line by throwing salt on a gas flame. That is the color that is "missing" in the Fraunhofer spectrum. (See the color plate following p. 528.)

9.29 Diffraction of water waves. Illuminate a bathtub from above with an incandescent lamp that has a small filament in a clear envelope, so as to get sharp shadows. Generate traveling waves that are "straight waves"—the two-dimensional analog of plane waves— by jiggling a floating stick or board placed across the end of the tub. Float a coffee cup as an opaque obstacle. Estimate the distance downstream at which the "shadow" of the cup is "healed." Suppose that you didn't know the diameter of the cup. Determine this diameter (approximately) experimentally by multiplying the "healing length," L_0, by the wavelength of the water waves, λ, and taking the square root. (We assume you know where that formula comes from. See Sec. 9.6.) This is one way of finding the diameter of nuclei—by measuring their diffraction "cross section." (*Note*: It is rather difficult to measure the wavelength of the water waves with the crude technique we suggest. It is easier to shake the stick at a reproducible tempo (as fast as you can) and then measure the frequency. The wavelength can then be obtained from the dispersion relation for water waves, as tabulated in Sec. 4.2.) How does your cross-section measurement of the cup diameter compare with a direct measurement of it?

9.30 How wide is a "plane wave" from a distant point source? We have often said that the traveling wave from a distant point source is "like" a plane wave over a "limited region" transverse to the line of sight from the point source to the field point. How limited is the region? Suppose the source is at distance L and we wish to consider a circular plane region of radius R transverse to the line of sight from the source. How large can R be so that the phase at the center of the circle and that at the edge of the circle differ by less than $\Delta\varphi$ radians?

Ans. The phase at the center of the circle is ahead of that at the edge (the center is closer to the source) by an amount $\Delta\varphi = \pi R^2/L\lambda$. Thus the phase is "the same" over the entire plane of the circle to the extent that the area of the circle is small compared with $L\lambda$.

9.31 The world's largest parabolic radio antenna at present, at the National Radio Astronomy Observatory, Green Bank, West Virginia, is a paraboloid dish 91 m in diameter. What is its angular resolution in radians and in minutes of arc (the units used by astronomers) for the famous 21-cm radiation of hydrogen?

Ans. A point source will look like a volley ball at a distance of 91.0 m.

9.32 Telescope "exit pupil." Suppose you have a simple telescope consisting of an objective lens and an eyepiece. The angular magnification

Home experiment

is f_1/f_2, where f_1 and f_2 are the focal lengths of the objective and eyepiece lenses, respectively. Show that not all of the rays from a distant object which strike a very large diameter objective lens get into your eye and that in fact the "useful diameter" of the objective lens is about f_1/f_2 times the diameter of your eye pupil. Thus, in an eight-power telescope, if the exit beam is a parallel beam of width 4 mm (twice as wide as the pupil of your eye, so that your eye need not be perfectly lined up and also so that off-axis points in the field of view deliver all their light), the objective lens should be 32 mm in diameter. A larger diameter is a waste of objective lens.

Home experiment   **9.33 Eye-pupil size and mental activity.** If someone shows you a picture of a good-looking individual of the opposite sex, your eye-pupil diameter may increase by as much as 30%, according to Eckhard H. Hess, *Scientific American* p. 46 (April, 1965). This large a change is very easy to detect in your own pupil by using a pinhole in a piece of aluminum foil that covers one eye, with a bright source illuminating the pinhole, as discussed in Sec. 9.7. Perhaps by just thinking, you can vary your pupil size, depending on what you think about. Have someone read to you. (Concentrate on listening, not on the pupil size.)

Home experiment   **9.34 Diffraction by an opaque obstacle.** This experiment works well with a white point source consisting of a 6-volt "camper" flashlight with lens removed and reflector covered by dark cloth. (The filament size is about $\frac{1}{2}$ mm.) The source should be at least three meters from the obstacle, so that you get a decent "coherent plane wave" over an obstacle the size of a pin. The detecting "screen" is a microscope slide on which is stuck a layer of Scotch translucent magic mending tape. Make the shadow of the object fall on the screen with the screen held a foot in front of your face (or whatever distance you find comfortable for looking at the screen). Your eye should be almost in line with the light source and the image on the screen so as to take advantage of the large intensity scattered at small angles (about the forward direction) from the translucent screen. Aside from looking at the beautiful fringes, one purpose of the experiment is to explore (crudely) the concept of the "length of the shadow," L_0, given by $L_0\lambda \approx D^2$, where D is the width of the obstacle. Among other things, look at a pin (if the pin width is $\frac{1}{2}$ mm, then $L_0 \approx 50$ cm for visible light) and a human hair (yours). [Mine has width about $\frac{1}{20}$ mm. This gives $L_0 \approx \frac{1}{2}$ cm.]

First consider the pin. Place the screen 5 or 6 meters downstream from the pin. The diffraction image should then be large enough so that you do not need a magnifying glass. It may help to jiggle the screen slightly, so as to wash out the effect of the irregularities of the magic mending tape. Notice the famous bright spot at the center of the "shadow" of the pinhead and the bright line at the center of the pin shaft. Is the bright spot or line brighter or dimmer than the bright screen itself (at a point well outside the image)? Next examine the image of the pin with the screen at a distance of only 5 cm downstream from the pin. (You will need a magnifying glass, unless you have very good eyes.) Notice that the shadow is a nice solid black, with no bright spot in the center. That is because you are much closer than L_0. At the edges it shows fringes, as is expected from our discussion in Sec. 9.6.

Next consider the human hair. Put the screen immediately behind the hair (i.e., about 1 mm downstream). Look at the shadow with a magnifying glass. It should be nice and black, since L is small compared with L_0. Now go to a distance of a few centimeters. You should see nice fringes. Go to 5 or 6 meters downstream. This is several hundred times L_0. According to our discussion, the shadow should be practically "healed" and the image of the hair very difficult to see against the background of light from the source. Your eyes are very sensitive detectors of contrast, and you will see something. Look at other things, knife-edges, holes in aluminum foil, etc.

9.35 Scotchlite.   You can get a piece of red Scotchlite adhesive tape in a hardware store. It is used for decorating, safety reflectors, bubble chambers, etc. Look at it with a magnifying glass. Tape a piece on a wall and shine a flashlight beam on it, with the flashlight held right in front of your nose so you see the light that is retrodirected through 180 deg. Now gradually move the flashlight out to the side while still shining the beam on the tape. That way you can estimate the angular width of the retro-directed beam. Why do you expect some angular width; i.e., why doesn't it give perfect retrodirection?

Home experiment

9.36 Coherence and polarization.   Light is emitted from an unpolarized point source. First it passes through a linear polarizer with easy transmission axis at 45 deg to the x and y axes. Then it is incident on a double slit. Each slit is covered by a linear polarizer, one slit having the polarization axis along $\hat{\mathbf{x}}$, the other having it along $\hat{\mathbf{y}}$.

(*a*) Suppose you look at the interference pattern with the unaided eye. Do you expect to have the usual two-slit interference pattern? What *do* you expect?

(*b*) Next suppose you look at the interference pattern while holding a Polaroid linear polarizer in front of one eye. What do you expect to see? What happens as you rotate the polaroid in front of your eye?

(*c*) Now suppose you look at the pattern through a circular polarizer run backward as an analyzer. What pattern do you expect to see?

There are many nice variations you can make on this problem: (i) Put a right-handed circular polarizer over one slit and a left-handed circular polarizer over the other an repeat the above observations. (ii) Add a quarter- or half-wave plate just behind the slits, etc.

9.37 Double slit interferometer.   Suppose you cover one of two slits with a microscope slide and the other with nothing. If the slide has thickness 1 mm, show that monochromatic light of wavelength 5000 Å gets a retardation in one slit of about 1000 wavelengths relative to the other. If the double-slit pattern is not to wash out, the light must be fairly monochromatic.

How narrow a band of wavelengths (in angstroms) is required so that the relative phase shift of the two slits varies by less than 180 deg from one edge of the wavelength band to the other? How could you use this fact to measure the bandwidth of a spectral line? (What would you measure and plot versus what, and how would you obtain the bandwidth from the plot?)

Home experiment

9.38 Pinhole magnifier. Derive a formula for the magnification of a pinhole magnifier. Check the formula as follows: Make two marks 2 cm apart on one piece of paper; make two marks 2 mm apart on another. Put the pinhole over one eye and nothing over the other. Both pieces of paper should be illuminated from behind (at least, that is easiest to look at). With both eyes open, look with one eye through the pinhole at the 2 mm marks and with your other eye look at the 2 cm marks. Bring the 2 mm marks closer until you superpose the two sets of marks with your two eyes. Measure appropriate distances.

Home experiment

9.39 Floaters in the eye. Use a pinhole in a piece of aluminum foil illuminated by a broad source to study the floaters. When you are looking at one, try to wipe it off by blinking. Can it be wiped? "Roll" your eyes once or twice and then watch the floaters swirl! Now try to find if they are nearer to the pupil or to the retina: vary the distance of the pinhole from your eye. The size of the circle of light changes. (Make a sketch to aid in explaining why.) Any object at the same location as the pupil would change apparent size by the same ratio as does the projection of the pupil. (Why?) Anything at the retina (or near it) would not change apparent size at all. (Why?) What do the floaters do? Are they closer to the retina or to the pupil? Now try to estimate their length and diameter. To estimate the diameter, compare them with a human hair (yours) held in front of your pupil between the pinhole and the pupil. For this purpose, you need a very small pinhole, smaller than you can easily make with a pin. Crinkle up a piece of aluminum foil, then smooth it. Look for an accidental small pinhole. (You can tell if it is small—less light intensity gets through.) Now look at a hair. You should see its shadow and see nice diffraction fringes at its edge. Compare its size with that of a floater. Are they finer than a hair? (*Note*: A human hair has a diameter about $\frac{1}{20}$ mm, i.e., 50 μ (micron). A typical red blood cell has a diameter of 5 or 6 μ.)

Home experiment

9.40 Marbles. Get some clear glass marbles at a toy store. One of these can be used as a Leeuwenhoek magnifying glass. Put a point light source a meter or so away and focus it to a "point" with a marble. How far beyond the marble is the focal point? What is the index of refraction of the glass? (To put it differently, does the location of the focal point agree with the result derived in Sec. 9.7 if you take $n = 1.5$?) Look at something small. Measure the magnification using the technique of Home Exp. 9.38.

9.41 Planoconvex lens. A planoconvex lens is flat on one side and has a spherical (or cylindrical) surface on the other. Derive a formula for the location of the focal point for light incident on the plane side of the lens.

9.42 Measuring index of refraction of liquids. Use an empty glass jar, for example a baby-food jar. (You can also use a clear glass envelope from a broken light bulb.) With the jar full of clear liquid, upright, and illuminated from the side, one has a thick cylindrical lens as discussed in Sec. 9.7. With the jar lying on its side and half filled, it is a planoconvex lens, the flat side being the surface or the liquid. Illuminate it from above with a point or line source. Measure the location of the focal point. Use the appropriate formula to find the index of refraction. Try water, alcohol, mineral oil.

Home experiment

9.43 Satellite cameras. According to the newspapers, we now have a satellite carrying a camera that can resolve objects a foot in diameter. What must be the diameter of the lens if the satellite is at a height of 150 miles?

9.44 Inside-out lens. A baby-food jar full of air and immersed in water is a diverging lens. Use a fish tank with glass sides, or use an ordinary pan with a mirror to change a vertically downward flashlight beam into a horizontal beam. Put a little milk in the water so you can see the beam. A good pencil-sized beam is obtained from a flashlight covered by an opaque piece of cardboard with an off-center hole. (The flashlight bulb is usually irregular at the tip. Also, you don't want the direct light from the bulb, which falls off as the inverse square of distance, but the parallel beam from the parabolic reflector.) You can study lenses of air and mineral oil and glass using a suspension of milk in water to see the beam.

Home experiment

9.45 Color mixing. Your eye and brain do not Fourier-analyze light (the way your ear Fourier-analyzes sound). It takes some practice, but you can recognize the difference between a color due to monochromatic light and a color due to a mixture of wavelengths. Psychologically, "white" is a "color." However; your diffraction grating tells you it is the whole visible spectrum of wavelengths.

Home experiment

   (*a*) Look at things through your purple filter, which passes red and blue but absorbs green.

   (*b*) Look at two separated white light sources—line sources or incandescent bulbs—through your diffraction grating. Vary your distance from the two sources until the left-hand first-order spectrum of the right-hand bulb can be superposed with the right-hand first-order spectrum of the left-hand bulb. Then you cart superpose any two wavelengths and see what "psychological" color results. In order to be superposing two "pure" wavelengths, you should use two line sources (i.e., display lamps). One gets beautiful colors. Try it! (Joseph Doyle suggested this experiment.)

9.46   A point object is 2 meters from a positive lens of power 1 diopter. Where is the image? (The object is on the lens axis.)

9.47   A thin lens used as a magnifying glass has magnification 5. A second thin lens has magnification 7. When both lenses are used (one lens immediately following the other), what is the magnification of the magnifier? Is it 35? 12? 2?

9.48   Longitudinal magnification.   Show that if a point object on the axis of a thin positive lens is moved a distance dp along the axis, then the image moves in the same direction by an amount dq, where dq has magnitude q^2/p^2 times dp.

9.49   Depth of focus.   A point object at distance p is focused to a point image on a camera film located at distance q behind a thin lens of diameter D. Another point object at distance $p + \Delta p$ will not be in focus on the film. It will pass through its focus before (or after) reaching the mm and will give a "circle of confusion" on the film.

(a) Show that the diameter d of the circle of confusion on the film of an out-of-focus point is given by $d \approx D(q/p^2)\Delta p$. Thus for a given "tolerable" circle of confusion, i.e., a given value of d, and for given q and p, the "depth of focus" Δp is inversely proportional to the lens diameter D. Small D gives large depth of focus. The focal length divided by the diameter is called the "f-number." Thus large f-number means small diameter of the aperture "stop" on the lens and gives large depth of focus. For D equal to zero, we have a "pinhole camera"; then this formula would say that the depth of focus is infinite. You verify that to some extent when you use your pinhole magnifier and find that everything stays "in focus" from $p \approx 1$ cm to infinity, without your having to try to focus with your accommodation muscles.

(b) If D gets too small, we cannot neglect diffraction. Show that diffraction gives a circle of confusion $d \approx q\lambda/D$. Now suppose that you are not limited by the photographic film's "grain density" or anything else about the film. Also, suppose you are not limited by intensity (which might require large D). Define d^2_{av} as the sum of the squares of the two contributions to d, one from depth of focus and one from diffraction. Minimize d^2_{av} as a function of the lens diameter D, holding everything else constant. Show that, for given λ, p, and Δp, the least fuzzy image is obtained with D for which $D^2 = \lambda p^2/\Delta p$.

(c) Forget about diffraction. Suppose you are photographing two people at once; one is 4.5 m away, the other is 7.5 m away. The lens focal length is 10 cm. You want the circle of confusion on *both* people to be less than 1 mm in diameter "in the object space," i.e., 1 mm "on the person." Find the required f-number. (Use rough approximations; for example, take $p \approx 6.0$ m as an average.)

(d) Does diffraction make the image noticeably worse in the geometry of part (c)?

Ans. f-number ≈ 50. Diffraction contributes about as much as depth of focus to the fuzziness.

9.50 Radiation pattern of tuning fork—quadrupole radiation.

Hold a single vibrating tuning fork near your ear. Twirl it about its long axis (the axis of the handle) and listen to the maxima and minima of intensity. Hold the fork at one end of a tube that has been tuned for resonance with the fork. Slowly rotate the fork about its long axis. In 360 deg of rotation, you should find four angles of zero intensity and four angles of maximum intensity. Hold the fork so that you are at a zero of intensity. Without disturbing the relative position of the fork and tube, insert a piece of cardboard that closes half the end of the tube, giving one prong of the fork an open path down the tube but shielding the other. What happens? Why? Now strike two forks against one another, hold one at each end of the tube, and listen for the beats. When you have established the rhythm, twist one fork by 90 deg about its long axis, so that you go from one angle of maximum intensity to the next. The beats do not continue "on the beat." Just after the twisting, there are two beats when there should have been one, and then the beats continue on the offbeats of the original rhythm. Unless you have a good ear for keeping a steady beat, you may have trouble proving to yourself that the new beats are on the offbeats of the old, but you should have no difficulty in hearing the "extra" beat that comes when you rotate the fork. (It helps if you count "1 and 2 and 3 and ...," with the numbers on beat maxima and the "ands" on zeroes; then don't let your counting be influenced by what you hear when you twist the fork.)

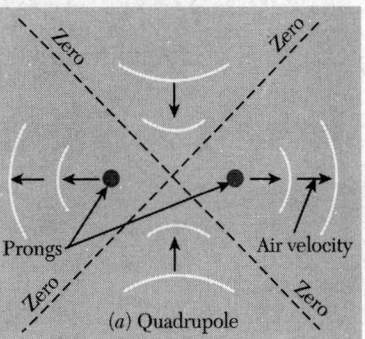

(a) Quadrupole

What is the explanation? Think of how the forks act on the air in their vicinity. When the prongs are separating, they are pushing air away from the outside of the prongs, giving an outward velocity to the pushed air. Meanwhile the region between the prongs develops a slight deficit of air because the prongs leave more space between them as they expand. Air rushes in from the sides to fill the deficit. The induced air velocity is thus outward for the air lying in the plane of the two prongs and inward for the air lying in the plane that passes between the two prongs. On the next half-cycle, the prongs are coming together; the air is sucked inward in the plane of the prongs and squeezed out into the plane between the prongs. Somewhere in between these directions, there must be a direction where the induced velocity is zero, i.e., where the velocity pattern has a *node*. This is the explanation for the four maxima and four minima when you rotate the fork. The radiation pattern is shown in part (a) of the figure at an instant when the prongs are moving outward. A radiation pattern such as this is called a *quadrupole radiation pattern*. If you had a single prong instead of two, the radiation pattern of maxima and minima and the relative phases would be those of a *dipole* radiation pattern. If you take one oscillating dipole (a sound wave dipole here, but the idea applies to radio waves or any other kind of waves) and superpose its radiation on that of an identical dipole which is slightly displaced in position and oscillating 180 deg out

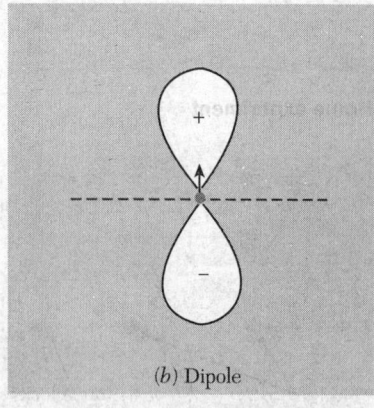

(b) Dipole

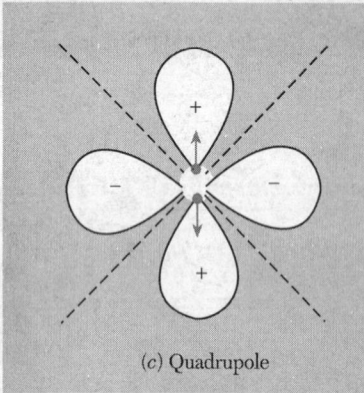

(c) Quadrupole

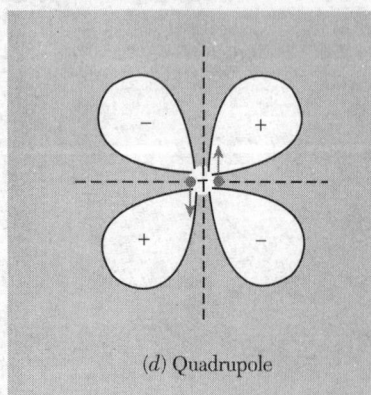

(d) Quadrupole

Problem 9.50 Radiation pattens of tuning fork. (a) Quadrupole. (b) Dipole. (c) Quadrupole. (d) Quadrupole.

of phase relative to the first, you get quadrupole radiation. Depending on the direction of the displacement relative to the pattern of waves from a single dipole, you can get many different quadrupole patterns. But they all have these common features: There are four "lobes" of strong intensity, where one gets constructive interference between the contributions from the two dipoles. The phases in neighboring lobes differ by 180 deg. There are nodes between lobes. (A dipole pattern has just two lobes and two nodes.) The figure shows one dipole pattern and two quadrupole patterns: Pattern (b) is a polar plot of the wave function for dipole radiation at a given instant. Pattern (c) is obtained by superposing two dipoles which are shifted slightly relative to one another along the direction of the dipole lobes and which oscillate 180 deg out of phase. This is the pattern of the tuning fork. Pattern (d) is obtained by superposing two dipoles which are shifted slightly relative to one another along the direction of the dipole nodes and which oscillate 180 deg out of phase.

9.51 Suppose you produce a radio beam with a plane radio transmitter antenna of area A_T. This beam is received by an antenna of area A_R located at a large distance D from the transmitter. Show that the transmitted power P_T and received power P_R are related approximately by

$$\frac{P_R}{P_T} \approx \frac{A_R A_T}{\lambda^2 D^2}.$$

Suppose the transmitting and receiving antennas each consist of a tapered microwave "horn" having an entrance aperture in the form of a square of edge length 3 meters. Suppose the microwave frequency is 1000 Mc and the distance between transmitter and receiver is 18.5 km. What is the ratio of received to transmitted power?

9.52 Interference pattern of N identical slits. The amplitude is given by Eq. (54), Sec. 9.6. Make a graphical representation of the sum of the corresponding complex amplitudes for an "arbitrary" value of $\Delta\varphi$ (the relative phase between contributions of neighboring slits). Make the graphical representation for the first zero adjoining a principal maximum; thus derive graphically the fact that $\Delta\varphi = 2\pi/N$ for this zero. Show from the graphical representation that the phase constanf for the superposition is the average of the phase constants of the first and last contributions.

9.53 Chromatic aberration correction. By using two different kinds of glass in a "compound" lens, you can get rid of some of the chromatic aberration. Rather than a lens, consider a thin prism. Design a thin compound prism from simple wedges with angles a_1 and a_2 that will give a certain desired deviation θ_0 at wavelength $\lambda_0 = 5500$ Å and will have zero rate of change of deviation with wavelength, given that the two types of glass have indices $n_1(\lambda)$ and $n_2(\lambda)$, where the functions $n_1(\lambda)$ and $n_2(\lambda)$ are known

Ans. $a_1 \dfrac{dn_1}{d\lambda} = a_2 \dfrac{dn_2}{d\lambda}.$

Now express the deviation θ of the prism for wavelength λ in terms of a Taylor's series in the quantity $\lambda - \lambda_0$. Carry the series only through $(\lambda - \lambda_0)^2$. How could you further reduce the chromatic aberration (given whatever you need)?

9.54   Seeing underwater.   Look at things underwater using a skin diver's face mask. Derive the fact that things appear to be about three-fourths as far away as they really are. An especially vivid demonstration is to look at someone in a swimming pool whose head is above water and whose body is below the surface. Immerse the lower half of your face mask in the water so that water level is at your eye level. Then you can look at the person's head by looking through the air above the water and can by a flick of your eyes look at the person's body through the water.

Home experiment

9.55   Underwater glasses.   When you put your face under water and try looking without a face mask, everything looks blurred, because the change of index of refraction in going from water to eye is not very great. As a simplification, assume there is no change in index. Also assume your eye lens has very little effect, as if all the focusing were done at the first air-to-eye interface. (This is a crude approximation. Actually, you can see underwater to same extent.) Assume the focal length of that first surface is 3 cm, and that a parallel beam of light in air is brought to a focus at the retina. When you look underwater, you lose that focusing action. Design glasses that can be worn underwater so as to enable you to see clearly. Use glass with index of refraction 1.5. Show that if the focal length when used underwater is 3 cm, then the focal length when used in air is about 1 cm. If one of these glass lenses is used as an ordinary magnifying glass what is its magnification? Suppose you use an ordinary glass marble for the lens. You want to form an image (of a parallel beam in water) 3 cm behind the rear surface of the marble. What should be the diameter of the marble?

Ans. about 1.7 cm. Now get a clear glass marble and try it. (Hold the marble up close to one eye.)

Home experiment

9.56   Interference in scattered light.   Here is a very simple way to get beautiful colored interference fringes. Rub a little ordinary talcum powder over an ordinary mirror. (You can also use flour, or dust, or you can simply breathe on the mirror so as to give condensed water vapor.) Stand back several meters from the mirror. Shine a small "penlight" flashlight in the mirror and look at the reflection of the light bulb. (Or use any flashlight, covering most of the reflector with your hand so as to have a source smaller than 1 cm or so; or use a candle at night.) Notice the fringes! Try different positions of the light source, moving the source closer to the mirror than your eyes and farther away. The fringes are produced by interference between the following two kinds of rays: The first type is a ray that passes through a grain of talcum powder, scatters, is specularly reflected from the mirror toward your eye, and is not scattered on the way back through the powder. The second kind is a ray that passes through the

Home experiment

powder without being scattered, reflects specularly from the mirror, and is scattered toward your eye from the *same* grain of powder. The powder grains are transparent. (They look white for the same reason that ground glass looks white.) The scattering is very nearly in the forward direction for either of the rays. Thus each of the two interfering rays goes through about the same thickness of transparent material in a given grain. Derive the observed result that the central fringe, meaning the one that apparently passes through the image of the point source, is always an interference maximum. For white light it is white. The fringes become colored only beyond a distance of several fringes on either side of this central fringe. The geometrical appearance of the fringes is not easily calculated. See A. J. de Witte, "Interference in Scattered Light," *Am. Jour. Phys.* **35**, 301 (April, 1967).

9.57   Stellar interferometers.   (*a*) A double slit followed by a lens and a photographic film can give angular resolution $\delta\theta \approx \lambda/d$, where λ is the wavelength of light and d is the slit separation. Thus one can detect the structure of astronomical objects emitting visible light provided they subtend an angle λ/d or more. Justify the preceding remark.

(*b*) When d gets to he about 30 cm or so, then the occurrence of turbulent "bubbles" of air in the earth's atmosphere having different index of refraction from that of surrounding air are sufficient to give relative phase shifts of order π for two air paths from the astronomical object to the two slits. (The air paths are then about 30 cm apart all the way through the atmosphere.) Show that the resulting limit on angular resolution is about 2 microradians for visible light at the earth's surface.

(*c*) Now let us replace the two optical slits by two radio antennas detecting waves of wavelength 30 cm. In place of the lens, which was used to bring to one place the light waves from the two slits in order to form an interfering superposition, let us either use coaxial cables or else rebroadcast (or redirect) the signal from each antenna through the air to a central receiving station. This station takes the place of the photographic emulsion. Show that in order to get the same resolution as for visible light with slit separation 30 cm, the radio antennas must have separation of about 180 km.

(*d*) Now, the size of the turbulent air bubbles is a few feet. Once the air paths are separated by many feet, the random phase shifts accumulated for the two paths in passing through the atmosphere are essentially independent of the path separation. For a first guess, therefore, you might suppose that the effect of the atmosphere on the two radio antennas separated by about 180 km would be about the same as that of the two slits for visible light separated by a few feet, and thus one might guess that atmospheric variation of index of refraction would wash out the angular resolution of the two radio antennas. It is indeed true that the index of refraction of air for the radio waves is not too different from that of the light. However, the corresponding relative *phase shift* for the 30-cm waves is many thousands of times smaller than for the light waves. Why is that the case?

(*e*) The radio interferometer does not suffer from atmospheric variation in index of refraction, and we can increase the separation between the antennas to much *more* than 180 Km, thus getting better resolution than with the interferometer that uses visible light. (Of course, the astronomical object must be emitting 30-cm waves as well as visible waves, if we are to detect it by both methods.) We can therefore contemplate a radio interferometer with, for example, one antenna in New York and one in California, giving a base line (separation of the antennas) of about 3000 km and a corresponding angular resolution 10^{-7} rad. Unfortunately there is a new problem, which is the variable phase shift of the radio waves sent by cable (due to temperature variations) or by broadcast from the two antennas to the central station at which the signals are to be superposed to give interference. The amount of air over our heads is equivalent to a uniform layer about 8 km thick with the same density as at sea level. The amount of air between New York or California and some central station in the Midwest is several hundred times more than that, and it won't work. What should we do? Here is an ingenious solution by N. Broten, et al., "Long Base-line Interferometry Using Atomic Clocks and Tape Recorders," *Science* **156**, 1592 (June 23, 1967): At each station there is an atomic clock, for example a hydrogen maser oscillator running at 1000 MHz (which is 10^9 cps). Such a clock may have a stability of 1 part in 10^{14}, which means that it drifts randomly in phase by only one cycle out of 10^{14} cycles. Show that such a clock is stable (drifts by less than one cycle) for a time of order 1 day.

(*f*) Suppose we wish to measure a star's radio waves at a central frequency $v_0 = 1000$ MHz (corresponding to 30-cm radio waves) with a bandwidth $\Delta v = 1$ MHz. The local oscillator at each station is then run at $v_0 = 1000$ MHz. At each station the local oscillator signal is superposed with the signal from the antenna. If the local oscillator provides a current $\cos \omega_0 t$, and if the antenna provides a current $A \cos (\omega t + \varphi)$, then the superposition of the two currents, *squared* and averaged over one fast cycle (at 1000 MHz), gives power $P = I^2 R$ with time dependence

$$P = 1 + A^2 + 2A \cos [(\omega_0 - \omega)t - \varphi],$$

times some constant. Justify that formula.

(*g*) If $v_0 t$ is known to better than 1 cycle for $t = 1$ day, then P is known to better than 1 cycle at the slower "beat" frequency $v_0 - v$ for $t = 1$ day. The oscillator frequency v_0 is set at the center of the desired band. Then the average of $v_0 - v$ through the band of signals detected is zero. The bandwidth Δv is about 1 MHz. This signal, P, with frequencies extending from zero up to about 1 MHz, is recorded on a tape recorder at each station. (Video tape recorders used in TV have sufficient bandwidth for this.) After each station has recorded for some time (of less than a day), the two tape recorders may be *carried*, by airplane for example, from the two antennas to the office of the physicist. The two tape recorders are then synchronized and played back together so as to superpose their signals. In order for the phase information not to be lost, it is necessary that neither tape recorder lose one cycle, or get out of step with the other by one cycle (one cycle at the oscillation frequency of the signal P on each recorder). This signal has frequency components from zero to about 1 MHz, since that is the original bandwidth. Therefore the two tapes must be synchronized to an accuracy better than 1 microsecond. Also, when the tapes were being made at their respective antennas, they had to have time markers put on them, so that the physicist would know at what times the incoming signals were simultaneous. The time markers on the magnetic tape must be accurate to better than 1 microsecond. The synchronization and time-marking of ordinary video tape recorders used in television to better than 1μ sec accuracy is *easy standard practice*! Thus one may contemplate a stellar interferometer consisting of a radio antenna in New York, mixed with a local hydrogen maser signal and recorded on magnetic tape, and a similar station in California. The interferometer sweeps across the sky as the earth turns. The phase "constant" φ in each antenna current is just the quantity kr, where r is essentially the distance from the antenna to the star. Thus antenna 1 in New York has current $A_1 \cos (\omega t + kr_1)$ due to a given star, and antenna 2 in California has current $A_2 \cos (wt + kr_2)$ due to the same star, for the same frequency component ω of the signal. Assume for simplicity that the antennas are identical in construction and that the received amplitudes are equal: $A_1 = A_2 = A$. Show that, when the taperecorder output currents P_1 and P_2 are superposed, the resultant current $P_1 + P_2$ is proportional to $1 + A^2 + 2A \cos \frac{1}{2} (\omega_0 - \omega)t \cos \frac{1}{2} k(r_1 - r_2)$. Now take this current and square and average over one cycle of $v_0 - v$. Show that the resulting time-averaged power is proportional to $(1 + A^2)^2 + A^2[1 + \cos k(r_1 - r_2)]$. The term $\cos k(r_1 - r_2)$ goes through one cycle every time California gets 15 cm closer to a point star while New York gets 15 cm farther from it (assuming λ is 30 cm). If there is a second star (or some internal structure), it will be resolved provided its value of $k(r_1 - r_2)$ differs from that of the first star by, for example, π, i.e., provided its angular separation from the first star is of order λ/d.

(*h*) There are other technical problems that we have not mentioned. For example, since there are many radio stars, we want each telescope to point only at the same small region of sky, so that there is only one star to worry about. How can we arrange that? (Assume each telescope has a diameter of 50 meters.)

(*i*) Here is another problem. We want to be able to recognize the "central" fringe in the interference pattern, so that we know the exact direction to the star. (*Hint:* When you look at a point source with a double slit and (for example) green light, it is not easy to pick out the central interference fringe. That is also true if you use red light. However, if you use white light, which includes both red and green, it is easy to pick out the central fringe.) Why is that? Show how you could use, at each antenna, two detected frequency bands (this is like using red and green light), mixing each with the same local oscillator signal, correlating the corresponding magnetic tapes for each frequency band separately, and finally mixing the resultant tape output signals from the two frequency bands.

The kind of radio-star interferometer just discussed may someday be used to give very precise measurements of the fluctuations in the rotational period of the earth. See T. Gold, *Science* **157,** 302 (July 21, 1967), and G. J. F. Mac-Donald, *Science* **157,** 304 (July 21, 1967).

9.58 Converting amplitude modulation to phase modulation in FM radio broadcasting. (This problem is closely coupled to Prob. 9.59.)

(*a*) An amplitude-modulated voltage has the form

$$V(t) = V_0[1 + a(t)] \cos \omega_0 t,$$

where ω_0 is the carrier frequency and $a(t)$ is the fractional amplitude modulation. A phase-modulated voltage has the form

$$V(t) = V_0 \cos [\omega_0 t + \varphi(t)],$$

where $\varphi(t)$ is the modulated (i.e., time-dependent) phase "constant." Show that the instantaneous angular frequency in this phase-modulated wave is given by $\omega(t) = \omega_0 + d\varphi(t)/dt$. Thus we could call this a frequency-modulated (FM) voltage rather than a phase-modulated voltage.

(*b*) The fractional amplitude modulation $a(t)$, or the modulated phase $\varphi(t)$, contains the information that is to be transmitted, for example, music. Let us Fourier-analyze the music and consider a single Fourier frequency component of frequency ω_m, where subscript m stands for "modulation." Then we replace $a(t)$ by $a_m \cos \omega_m t$. (We should also consider a term involving $\sin \omega_m t$, but we won't .) The amplitude-modulated voltage becomes

$$V(t) = V_0[1 + a_m \cos \omega_m t] \cos \omega_0 t = V_0 \cos \omega_0 t + V_0 a_m \cos \omega_m t \cos \omega_0 t.$$

This amplitude-modulated voltage is equivalent to a superposition of purely harmonic oscillations at the carrier frequency ω_0, the upper sideband frequency $\omega_0 + \omega_m$, and the lower sideband frequency $\omega_0 - \omega_m$. Justify the preceding remark by writing $V(t)$ explicitly as that superposition.

In AM radio these frequencies are broadast and are picked up by your radio antenna. Lightning and electric shavers also broadcast in the same frequency range and contribute to the amplitude modulation by suddenly increasing or decreasing the amplitude at a given frequency. This gives "static" and messes up the music. If we convert the amplitude-modulated voltage to a phase-modulated voltage, this static can be largely eliminated, because the lightning makes the mistake of changing the amplitude, and the FM receiver knows that that can't be part of the music, since the music has been broadcast at constant amplitude. The FM receiver can therefore arrange to filter out the sudden changes in amplitude, i.e., the static.

Here is how the AM voltage is converted to an FM (phase-modulated) voltage: The AM voltage is applied at the input of a bandpass filter which passes a narrow band that includes the carrier ω_0 but excludes the two sidebands $\omega_0 \pm \omega_m$. Once the carrier has been thus isolated from the two sidebands, it is given a quarter-cycle time delay (or advance) at the carrier frequency, i.e., it is shifted in phase by 90 deg. Then it is superposed once more with the two sidebands, which have been preserved without change. (We can also amplify or reduce the amplitude of the carrier voltage if we wish to, but we shall omit that.) Thus we can, for example, replace $\cos \omega_0 t$ by $\sin \omega_0 t$ in the carrier voltage $V_0 \cos \omega_0 t$. After phase-shifting and recombining, we obtain

$$V'(t) = V_0 \sin \omega_0 t + V_0 a_m \cos \omega_m t \cos \omega_0 t.$$

Now let us give the name $\varphi(t)$ to $a_m \cos \omega_m t$, which is also our original $a(t)$ for a given modulation frequency. For simplicity, let us assume that the magnitude of $a(t)$, i.e., of $\varphi(t)$, is small compared with unity. Then $\cos \varphi(t) \approx 1$, $\sin \varphi(t) \approx \varphi(t)$.

(c) Show that the above voltage can be written in the form

$$V'(t) = V_0 \sin [\omega_0 t + \varphi(t)], \varphi(t) = a_m \cos \omega_m t,$$

Thus we have found the trick for converting AM to FM (or vice versa): phase-shift the carrier wave by ±90 deg relative to the two sidebands. That beautiful invention of E. H. Armstrong in 1936 made commercial FM radio possible.

(d) Show that another way to convert AM to FM (or vice versa) is to shift either sideband by 180 deg, leaving the carrier and the other sideband unchanged.

9.59 Converting phase modulated light to amplitude-modulated light in a phasecontrast microscope.

(a) First consider an ordinary microscope. We are not primarily interested in the magnification now, so we let the magnification be unity, as follows. We put a glass microscope slide at $z = 0$, with the slide lying in the xy plane. We put a simple lens at $z = 2f$, where f is the focal length of the lens. We put a screen or photographic plate at $z = 4f$. Then the microscope slide is imaged at the screen, and the magnification is unity. Justify the preceding statement.

(b) Now we put an amoeba in a drop of water on the microscope slide and form its image on the screen. Unfortunately we cannot see the amoeba because it is transparent, with an index of refraction not very different from that of the water in which it is immersed. So we stain the amoeba with a dye. Now we can see it, but the dye killed it, and we wished to study its life processes.

The role of the dye was to modulate the amplitude of the light emerging at the $+z$ end of the amoeba, i.e., modulating it relative to the amplitude of the light that comes through the microscope slide where there is no amoeba. When there is no dye, the amplitude at the $+z$ end of the amoeba, at a given transverse position x, is the same as it would be if there were no amoeba. However the phase is different, because the light has passed through different amounts of amoeba depending on x, and the index of refraction of the amoeba is different from that of water. Thus suppose that

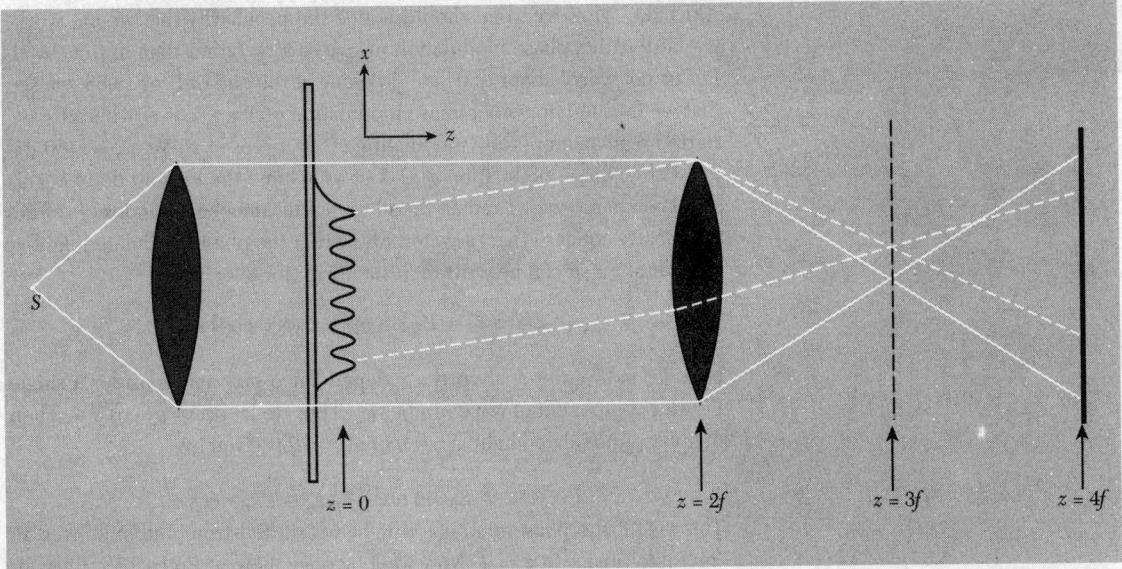

Problem 9.59 *Phase-contrast microscope. In this example we have chosen the magnification to be unity. The object plane is at $z = 0$. The image plane is at $z = 4f$. The focal plane of the objective lens is at $z = 3f$.*

the microscope slide is illuminated with a plane wave of monochromatic light traveling in the $+z$ direction, the light having been emitted by a point source S located at the focal point of a lens which then forms a parallel beam. (See the figure.)

Suppose that at $z = 0$, just downstream from the amoeba, the electric field when there is no amoeba is given for all x by

$$E(x, z, t) = E_0 \sin \omega t.$$

When the amoeba is present, there is a phase shift $\varphi(x)$ that depends on x. In that case the electric field in the light at $z = 0$ is given by

$$E(x, z, t) = E_0 \sin [\omega t + \varphi(x)].$$

Thus the unstained amoeba produces phase-modulated light. The lens at $z = 2f$ forms an image of the amoeba on the screen at $z = 4f$. The electric field at the screen is given by the same expression as at $z = 0$ (neglecting small losses, and not worrying about the inverting of the image, i.e., the replacing of x by $-x$). The time-averaged square of the electric field is then $\frac{1}{2}E_0^2$, independent of x, and thus we cannot see an image. Justify both parts of the preceding statement.

(c) The stain modulates the amplitude E_0 but kills the amoeba. Instead we wish to convert the phase-modulated light into amplitude-modulated light. How can we do it? Let us be guided by analogy with the conversion of an AM voltage to an FM voltage as discussed in Prob. 9.58. We want to work that problem backward, so to speak, and convert FM light to

AM light. (However, one should notice the peculiarity that we are working here with a phase modulation in space, $\varphi(x)$, rather than in time, $\varphi(t)$. Let us not worry about that yet!) Looking at the end of Prob. 9.58, we see that we finished up with a phase modulation $\varphi(t) = a_m \cos \omega_m t$ because we started with an amplitude modulation $a(t) = a_m \cos \omega_m t$. We are now starting with a phase modulation $\varphi(x)$. Let us assume the magnitude of $\varphi(x)$ is small compared with 1 radian for all x, i.e., the amoeba has index of refraction nearly equal to that of water. Show that the phase-modulated light at $z = 0$ or at $z = 4f$ can be written in the form (for $\varphi \ll 1$)

$$E(x, z, t) = E_0 \sin \omega t + E_0 \varphi(x) \cos \omega t.$$

Now let us Fourier-analyze the x dependence $\varphi(x)$ and consider a single Fourier component of wavenumber k_m. Thus we let $\varphi(x) = a_m \cos k_m x$. Then the phase-modulated light at $z = 0$ or at $z = 4f$ is given by

$$E(x, z, t) = E_0 \sin \omega t + E_0 a_m \cos k_m x \cos \omega t.$$

This is still the phase-modulated light we started from and still gives an "invisible" image at $z = 4f$. Now we look once more at Prob. 9.58. By analogy, let us call the contribution $E_0 \sin \omega t$ by the name "carrier light wave." Then we see that if we can shift the phase of the carrier light wave by 90 deg relative to the modulated light (the light with amplitude a_m), we will have produced AM light. Without worrying yet about how in the world it can be done, let us simply replace $\sin \omega t$ by $\cos \omega t$ in the carrier light in the above expression. Then we have for the light at the screen at $z = 4f$

$$\begin{aligned} E'(x, z, t) &= E_0 \cos \omega t + E_0 a_m \cos k_m x \cos \omega t \\ &= E_0[1 + a_m \cos k_m x] \cos \omega t \\ &= E_0[1 + a(x)] \cos \omega t. \end{aligned}$$

This amplitude-modulated light gives intensity proportional to the time-averaged square of the electric field, i.e., to $\frac{1}{2}E_0^2[1 + a(x)]^2$, which depends on x and thus shows us how the amoeba's thickness and internal index of refraction depends on x. Thus we can see the amoeba.

(*d*) There is only one "minor" problem left: How can we isolate the carrier light wave from the rest of the light, shift its phase by 90 deg relative to the rest of the light wave, and then recombine it (superpose it) with the rest of the wave at the screen all this to be accomplished between $z = 0$ and $z = 4f$? In the case of the conversion of AM voltage to FM voltage, the trick was to use a frequency bandpass filter to separate the carrier frequency $\omega = \omega_0$ from the sidebands $\omega = \omega_0 \pm \omega_m$. Therefore by analogy we must search for a *wavenumber bandpass filter* to separate the carrier wavenumber $k_z = k_0 = 0$ from the sidebands $k_x = k_0 \pm k_m$. This last statement is made more understandable if we write the phase-modulated electric field at $z = 0$ in the form

$$E(x, z, t) = E_0 \sin[\omega t + k_0 x] + \tfrac{1}{2}E_0 a_m \cos[\omega t - k_0 + k_m)x]$$
$$+ \tfrac{1}{2}E_0 a_m \cos[\omega t - k_0 - k_m)x],$$

where $k_0 = 0$ and where we have written the standing wave $\cos k_m x \cos \omega t$ as a superposition of two traveling waves with $k_x = +k_m$ and with $k_x = -k_m$. Justify this last formula. Thus we see that the phase-modulated oscillation at $z = 0$ produces three traveling waves. The carrier wave has $k_x = 0$; the modulations give one wave with $k_z = +k_m$ and one with $k_z = -k_m$. All three traveling waves have essentially the same value for k_z, this value being essentially ω/c, because we are assuming that k_z is small compared with k_z, i.e., that the waves all travel in essentially the z direction, and therefore that the magnitude $\omega/c = \sqrt{k_z^2 + k_x^2}$ of the propagation vector is essentially equal to k_z for all three waves. (We are omitting k_y from this discussion.)

(*e*) In the figure we show a glass microscope slide plus the k_m Fourier component of the x dependence of the thickness of the amoeba. The carrier wave is produced by the point source S. Its outermost rays are drawn with solid lines. The upper sideband with $k_x = +k_m$ is drawn with dotted lines. (The lower sideband with $k_x = -k_m$ is not shown.) The lens focuses these three traveling waves, each of which is almost a plane wave, to separate almost-point images in the focal plane of the lens, located at $z = 3f$. The rays continue on to the screen at $z = 4f$, where they again overlap. Notice that *at the focal plane of the lens* (at $z = 3f$) *the three waves are completely separated in space. That* is where we can work on the carrier without disturbing the sidebands! Thus we have a spatial filter to isolate a given k_x, in analogy with the time filter (Fourier-analyzing circuit) used to isolate a given ω. Now that we have the carrier wave isolated spatially from the sidebands at $z = 3f$, it should be easy to shift its phase by 90 deg without disturbing the sidebands. Invent a way to shift the phase of the carrier wave by 90 deg relative to the sidebands. The beautiful invention of the phase-contrast microscope was made by F. Zernicke in 1934.

We can describe the procedure we have just been through in a more abstract (and hence more general) way as follows: At $z = 0$ we were given the functional x dependence of the amplitude and phase constant of an oscillation $A(x) \cos [\omega t + \varphi(x)]$. [In the present example there was not amplitude modullation at $z = 0$, that is, $A(x)$ was constant. In other examples involving diffraction patterns, we have had $\varphi(x)$ constant instead.] We Fourier-analyzed the x dependence to find standing waves at $z = 0$. These acted as sources of traveling waves with known values of k_x and k_z. We then used a lens to convert a dependence on k_x (at $z = 0$) to a dependence on x (at the focal plane of the lens, a distance f beyond the lens), with different k_x focused at different x. The x dependence *at this focal plane* is thus equivalent to the k_x dependence, with a one-to-one correspondence between k_x and x. Thus the x dependence at this focal plane is equal to a constant times the *Fourier transform* of the x dependence at the object plane at $z = 0$. There is no other value of z for which that is true. When the waves finally reach the screen (the image plane), they have again the same x dependence as at the object plane (neglecting the replacement of x by $-x$, and neglecting the fact that the magnification may be different from unity). Thus, in going from the object plane to the focal plane behind the lens and then to the screen, the x dependence has gone from the x dependence at the object plane to the k_x dependence at the object plane and then to the x dependence at the object plane. In going from the k_x dependence at the object plane to the x dependence at the object plane, we are going through the *inverse Fourier transform*. Thus we may say that in the phase-contrast microscope we start with a given x dependence, Fourier-transform it, work on it (shift the phase of part of the Fourier transform, perhaps amplify or attenuate its amplitude also), and then perform the inverse Fourier transform. (If we leave the Fourier transform intact, i.e., put no phase shifter at the focal plane, then the final result is the same as the original x dependence.) Many remarkable effects can be achieved in this way, in what is sometimes called "Fourier-transform spectroscopy" or "focal-plane spectroscopy."

(*f*) Describe the conversion of AM voltage to FM voltage in the same general terms we have just used to describe the phase-contrast microscope.

(*g*) In our discussion we did not take into account the total width (in the x direction) of the amoeba, and of the carrier wave. Assume the carrier wave beam width is W and that the amoeba has total width ω. What is the effect of these widths on the x variation of intensity in the focal plane at $z = 3f$, i.e., how are our former results modified, if at all?

(*h*) Suppose that instead of shifting the phase of the carrier wave by 90 deg at the focal plane we completely remove the carrier by an opaque obstruction in the focal plane. What will be the x dependence of the intensity of the image in that case?

9.60 Two thin lenses in series. Given two thin lenses of power f_1^{-1} and f_2^{-1} arranged in series along a common axis, with separation s between the two lenses. Take both lenses to be positive lenses. (The results will hold in general, with suitable interpretations of signs.) Consider a ray parallel to the axis, at distance h from the axis, and incident on the first lens. Say the ray is incident from the left, and the lenses are in order 1, 2 from the left. The first lens deflects the ray toward the axis. Assume the ray hits the second lens before crossing the axis. Find the focal point F where the ray crosses the axis after leaving the second lens. Show that the location of F is independent of h (for small-angle approximations). Now define the location P (which stands for "principal plane") as follows: Extrapolate the incident ray forward (to the right) and the emergent ray (the one that goes through F) backward until they cross. They cross at the principal plane P. Let x be the distance of F to the right of the second lens. Let y be the distance of P to the left of the second lens. Then $x + y$ is the distance of the focal plane F to the right of the principal plane P. This distance is called the focal length f of the combination of the two lenses, considered as though it were a single thin lens located at the principal plane P. Find $x, y,$ and f in terms of $f_1, f_2,$ and s. Once you have found f and P for rays going from left to right, do the same for rays traveling from right to left. Are the focal lengths the same? Are the principal planes at the same place?

Ans. For rays incident from the left,

$$f^{-1} = f_1^{-1} + f_2^{-1} - sf_1^{-1}f_2^{-1};$$
$$x = (1 - sf_1^{-1})f;$$
$$y = sf_1^{-1}f.$$

9.61 Two lenses having $f = +20$ cm and $f_2 = +30$ cm are placed 10 cm apart. If an object 5 cm high is located 30 cm in front of the first lens, find (a) the location, (b) the orientation, and (c) the size of the final image. By the ray-tracing technique locate the position of the image on a diagram.

9.62 A distant green object is to be pictured with a pinhole camera in which the distance from pinhole to film is D. What should be the approximate diameter of the pinhole if the picture is to be of maximum sharpness?

Supplementary Topics

Supplementary Topics

1 "Microscopic" Examples of Weakly Coupled Identical Oscillators

First, read Sec. 1.5 on weakly coupled identical pendulums including the last paragraph, Esoteric Examples, which introduces this topic.

Here are some examples of weakly coupled oscillators taken from atomic physics and elementary-particle physics. In each case there are "two identical degrees of freedom" which are "weakly coupled," so that there are two "normal modes" with frequencies ω_1 and ω_2. However, we are not dealing here with macroscopic mechanical systems. Newton's laws do not suffice. The understanding of these "microscopic" systems requires quantum mechanics. Nevertheless, in the behavior of the microscopic systems we shall describe there is a great mathematical similarity to the behavior of two weakly coupled pendulums. The physical interpretation is very different, however. For the coupled pendulums, the square of the amplitude if a pendulum is proportional to the energy (kinetic plus potential) of that pendulum. The energy "flows" back and forth from one pendulum to the other at the beat frequency. For a system described by quantum mechanics, the square of the amplitude for a particular degree of freedom (actually the absolute square—the amplitudes are always complex quantities in quantum mechanics) gives the *probability* that that degree of freedom is "excited" (i.e., has *all* the energy). This probability "flows" back and forth from one degree of freedom to the other at the beat frequency, $\nu_1 - \nu_2$. The energy itself is "quantized"; it cannot subdivide to "flow." In the case of the pendulums, the total energy of both pendulums is constant. The corresponding fact in the microscopic systems is that the total probability that either one or the other degree of freedom be excited is constant. (This total probability is unity, as long as the system does not somehow lose the excitation energy.) The following two examples are famous; you will encounter them again when you study quantum mechanics.

Ammonia molecule. The ammonia molecule, NH_3, is composed of one nitrogen and three hydrogen atoms. (See Vol. II, page 316.) The three H atoms form an equilateral triangle. Call the plane of this triangle the H_3 plane. The N atom has two possible positions where it may vibrate, corresponding to the two pendulums a and b. One (position a) is on one side of the H_3 plane; the other (position b) is on the other side. The N atom cannot easily get from a to b, or vice versa, because there is a potential-energy "hill" or barrier between a and b. In classical mechanics (i.e., according to Newtonian mechanics, without quantum mechanics), a and b are positions of stable equilibrium, and an N atom vibrating at a can *never* get over to b. (In the pendulum analogy, this corresponds to removing the coupling spring. Then if pendulum a is vibrating and b is at rest, that condition will be preserved forever, neglecting friction.) However, quantum mechanics introduces a "coupling" between a and b, in that it allows "potential-barrier penetration." Suppose the molecule starts at time $t = 0$ in the quantum-mechanical state for which N is definitely at a. Then the initial probabilities are given by $|\psi_a|^2 = 1, |\psi_b|^2 = 0$ (i.e., unit probability for N to be vibrating at position a; zero probability for it to be at position b). However, one finds that this condition is not maintained. In fact, one finds (by solving the Schrödinger equation) that for this starting condition the probability of finding N at a, namely $|\psi_a|^2$, and that to find it at b, namely $|\psi_b|^2$, are given by

$$|\psi_a|^2 = \tfrac{1}{2}[1 + \cos(\omega_1 - \omega_2)t], \qquad (1a)$$

$$|\psi_b|^2 = \tfrac{1}{2}[1 - \cos(\omega_1 - \omega_2)t], \qquad (1b)$$

where ω_1 and ω_2 are the frequencies of the normal modes. Equations (1) are remarkably similar to Equations (1.99), Sec. 1.5. The total probability that N be at either one place or the other is of course unity, as we find by adding Eqs. (1a) and (1b).

Just as with coupled pendulums, an ammonia molecule can be "started up" so that it is in one normal mode (or the other). It turns out that if the molecule is in the mode with slightly higher frequency (call this mode 2; $\omega_2 > \omega_1$, then it is unstable. It tends to emit electromagnetic radiation and change from mode 2, called "the excited state," into mode 1, called "the ground state." This radiation can be detected. Its frequency is the beat frequency, $\nu_2 - \nu_1$, which has the value

$$\nu_{\text{beat}} = \nu_2 - \nu_1 \approx 2 \times 10^{10} \,\text{cps}.$$

This corresponds to a wavelength ($\lambda = c/\nu_{\text{beat}}$) of about 1.5 cm, which is in the typical "radar" or "microwave" range. If one sends a microwave beam of frequency 2×10^{10} cps through ammonia gas, some of the microwave photons induce transitions from the ground state (mode 1) to the excited state (mode 2). Thus energy is exchanged between the microwave beam and the gas by exciting the molecules. Similarly, an excited molecule may "decay" to its ground state, thus adding one photon to the microwave beam. This energy exchange between the microwave beam and the ammonia gas provides the basis for the "ammonia clock." Absorption of energy from the microwave beam "winds up" the clock. The "probability flow" from state a

to b and back at the beat frequency provides the "escapement mechanism" of the clock. The ammonia clock and its descendants provide the world's most accurate measurements of time.

Neutral K mesons. Another fascinating system with a behavior analogous to that of weakly coupled pendulums is that of the neutral K mesons, which are called "strange particles." They *are* very strange and are not yet understood completely. This system has two degrees of freedom, called the K^0 meson and the $\overline{K}^0$ meson, analogous to the two pendulums. They are "coupled" because either of them can interact with two pi mesons (among other things) via a "weak interaction." The pi mesons (or pions, for short) are analogous to the spring. There are therefore two normal modes, which are called the K_1^0 meson and the K_2^0 meson. Unlike the modes we have previously discussed, one of these modes (the K_1^0 mode) is strongly damped. The other is weakly damped. (Systems with damping are discussed in Chap. 3.) If the system starts out at $t = 0$ with unit probability of being in the K_1^0 mode, that probability decreases exponentially with time and is given by e^{-t/τ_1}. A similar (but much smaller) damping occurs for the K_2 mode. The "loss of probability" corresponding to the damping is the result of radioactive decay of the modes into other particles. For example, K_1^0 decays mostly into two pions, and τ_1 is the mean decay time for K_1^0.

If the system started at $t = 0$ with unit probability of being in the K^0 state (call that state a), and if there were no damping, then the probability at a later time of the system's being in that same state (K^0) would be given by Eq. (1a). The corresponding probability of finding the system in state b $\overline{K}^0$ would be given by Eq. (1b). Because of the damping, these formulas must be modified to

$$\left|\psi(K^0)\right|^2 = \tfrac{1}{4}\left[e^{-t/\tau_1} + e^{-t/\tau_2} + 2e^{-(1/2)(t/\tau_1 + t/\tau_2)} \cos(\omega_1 - \omega_2)t\right], \quad (2a)$$

$$\left|\psi(\overline{K}^0)\right|^2 = \tfrac{1}{4}\left[e^{-t/\tau_1} + e^{-t/\tau_2} - 2e^{-(1/2)(t/\tau_1 + t/\tau_2)} \cos(\omega_1 - \omega_2)t\right]. \quad (2b)$$

Note that when $\tau_1 = \tau_2 = \infty$ (no damping), Eqs. (2) are identical with Eqs. (1).

It is an interesting exercise to devise a damping mechanism that will damp only mode 1 and another that will damp only mode 2 for the weakly coupled pendulums. Then the equations for the energy in the pendulums will be like Eqs. (2) instead of Eqs. (1).

2 Dispersion Relation for de Broglie Waves

A de Broglie wave to describe a single particle of definite energy has the form

$$\psi(z, t) = A f(z) e^{-i\omega t}. \quad (1)$$

The probability of finding the particle in an interval dz at position z is $\left|\psi(z,t)\right|^2 dz$, which is independent of t. If the potential energy of the particle is constant, we have a "homogeneous medium," and $f(z)$ is then a sinusoidal function of kz:

$$\psi(z,t) = \left[A \sin kz + B \cos kz \right] e^{-i\omega t}. \tag{2}$$

The dispersion relation for a particle in a region of constant potential energy V is obtained by substituting $E = \hbar\omega$ and $p = \hbar k$ (the Bohr frequency condition and the de Broglie wavenumber relation) into the classical expression for the energy. For example, for nonrelativistic electrons of mass m, the classical relation between the energy E, momentum p, and potential V is given by

$$E = \frac{p^2}{2m} + V, \tag{3}$$

which gives the dispersion relation for the de Broglie waves as

$$\hbar\omega = \frac{\hbar^2 k^2}{2m} + V. \tag{4}$$

Electrons in a box.   As an example, take the electron to be confined to a one-dimensional "box" extending from $z = 0$ to $z = L$. Inside the box, we take $V = V_1$ (a constant). For z less than 0 or greater than L, we take $V(z)$ to be $+\infty$. Thus the electron is confined to the box. If this "bound electron" behaved like a classical particle, it could have any kinetic energy, given by

$$\frac{p^2}{2m} = E - V_1. \tag{5}$$

But a real electron is not a classical particle. The possible bound states of a real electron in the above "infinite potential well" are just the normal modes of the de Broglie waves of the electron; i.e., they are standing waves with frequency and wavelength related by Eq. (4).

Standing wave shapes like those of a violin string.   What is the sequence of wavenumbers k for the standing waves? The probability of finding the electron outside of the interval $0 \leq z \leq L$ is zero. Thus $\left|\psi(z,t)\right|^2$ is zero just outside the well. But $\psi(z,t)$ is a continuous function of z. Thus ψ must be zero at $z = 0$ and L. (These are the same boundary conditions as those one has for a homogeneous violin string fixed at $z = 0$ and L. The standing de Broglie waves therefore have exactly the same sequence of configurations as those of an ideal violin string.) Thus the boundary condition at $z = 0$ requires that $B = 0$ in Eq. (2), giving

$$\psi(z,t) = e^{-i\omega t} A \sin kz. \tag{6}$$

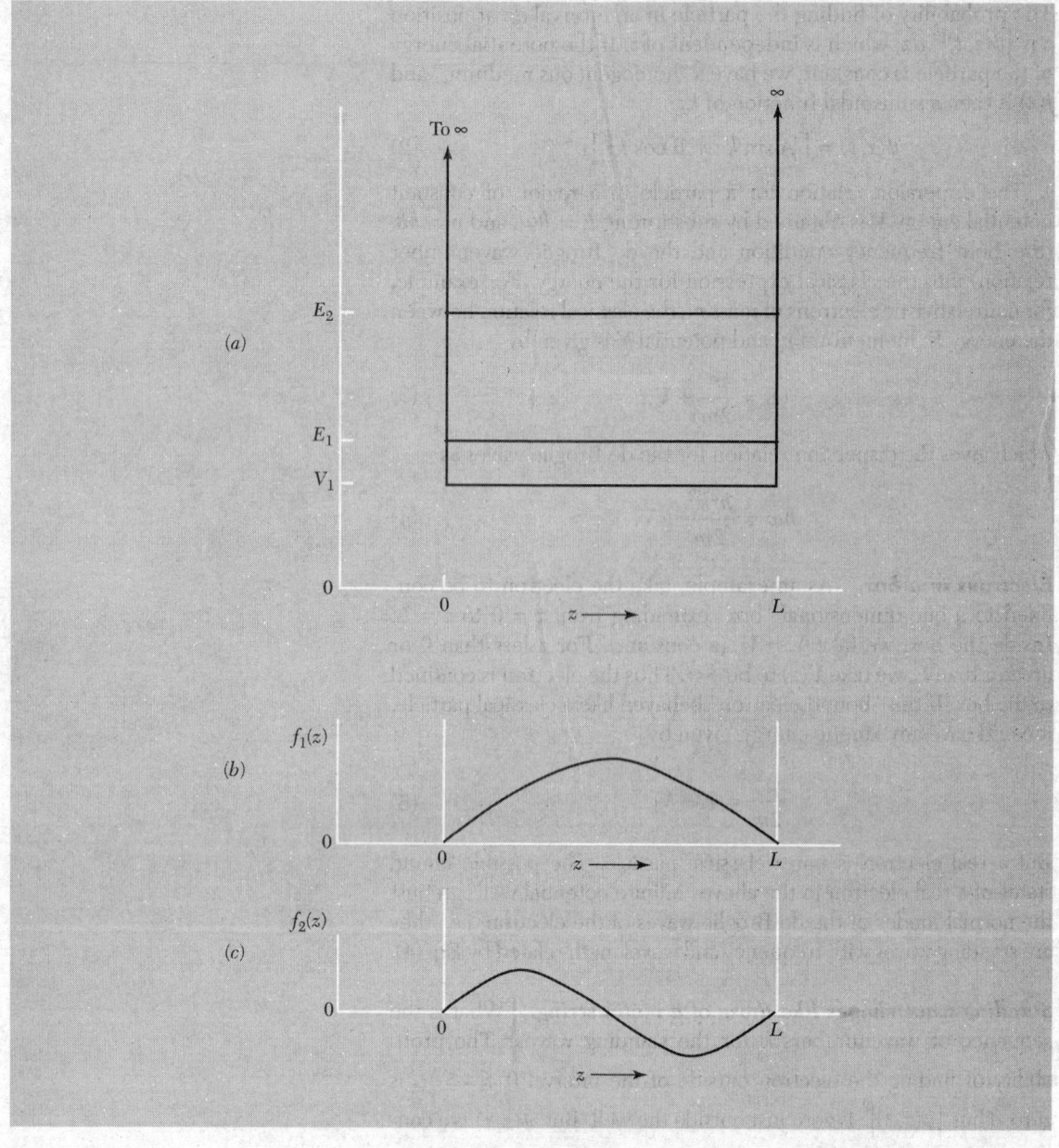

Fig. ST.1 *Electron bound in an infinite potential well. (a) Plot of V(z), with added horizontal lines E_1 and E_2 showing the energies of the first and second modes (the ground state and first excited state). The kinetic energy E_n-V_1 is proportional to n^2, so we have shown E_2-V_1 to be four times E_1-V_1. (b) Spatial dependence of ground-state wave function. (c) Spatial dependence of wave function of first excited state.*

The boundary condition at $z = L$ requires that $\sin kL = 0$. Thus the possible standing waves are given by $L =$ one half-wavelength, two half-wavelengths, etc.:

$$k_1 L = \pi, \quad k_2 L = 2\pi, \quad ..., \quad k_n L = n\pi, \quad ... \tag{7}$$

If the electron is in a single mode, the probability per unit length (so that we can drop the interval dz) of finding the particle at position z at time t is

$$\left|\psi(z, t)\right|^2 = \left|e^{-i\omega t} A \sin kz\right|^2 = \left|A\right|^2 \sin^2 kz. \tag{8}$$

This probability is independent of time, and the electron is said to be in a "stationary state." The probability that the electron is somewhere between $z = 0$ and L is unity. This gives the "normalization condition" on $\left|A\right|^2$:

$$1 = \int_0^L \left|\psi\right|^2 dz = \left|A\right|^2 \int_0^L \sin^2 kz \, dz = \tfrac{1}{2}\left|A\right|^2 L, \tag{9}$$

which determines $\left|A\right|$. Thus if

$$A = \left|A\right|e^{i\alpha} = \sqrt{\frac{2}{L}}e^{i\alpha},$$

Then

$$\psi(z, t) = \sqrt{\frac{2}{L}}e^{-i(\omega t - \alpha)} \sin kz, \tag{10}$$

where α is an undetermined phase constant.

The frequencies of the standing waves are given by the dispersion relation, Eq. (4):

$$\omega_n = \omega_0 + \frac{\hbar k_n^2}{2m}, \quad \omega_0 \equiv \frac{V_1}{\hbar}. \tag{11}$$

Thus the electron energy E is given by

$$E_n = V_1 + \frac{\hbar^2 k_n^2}{2m} = V_1 + \frac{\hbar^2 (n\pi/L)^2}{2m}, \quad \text{for } n = 1, 2, 3, ... \tag{12}$$

Standing wave frequencies unlike those of a violin string. Thus, although the shapes of the standing waves are like those of a violin string, the frequencies are not "harmonics" of the lowest mode frequency, as they are for a violin string. This is because the dispersion relation for de Broglie waves is very different from the dispersion relation for violin-string waves.

In Fig. ST.1 we plot the lowest mode (called the "ground state") and the second mode (called the "first excited state").

Inhomogeneous medium. If the potential function $V(z)$ is not a constant, independent of z, then the shapes of the de Broglie standing waves that correspond to the modes (states with definite wave frequencies, i.e., definite particle energies) are not sinusoidal in space. Thus there is no "dispersion relation" giving ω as a function of k, because the space dependence is not that of Eq. (2), and there is no single wavenumber k corresponding to frequency ω. Instead one must solve the Schrödinger differential wave equation to obtain $f(z)$. This is somewhat analogous to the case of the continuous string discussed in Sec. 2.3. There we found that the modes are sinusoidal in their space dependence only if the medium is homogeneous. For an inhomogeneous string, the space dependence of the standing waves obtained by solving the differential equation [Eq. (2.59), Sec. 2.3; we take the tension $T_0(z) = T_0 = $ constant, the mass density $\rho_0(z)$ not constant].

$$\frac{d^2 f(z)}{dz^2} = -\frac{\omega^2 \rho_0(z)}{T_0} f(z). \tag{13}$$

Similarly, for an inhomogeneous potential $V(z)$, the space dependence of the de Broglie standing waves is obtained by solving the Schrödinger equation, which is in this case

$$\frac{d^2 f(z)}{dz^2} = \frac{2m}{\hbar^2} \left[V(z) - \hbar\omega \right] f(z). \tag{14}$$

3 Penetration of a "Particle" into a "Classically Forbidden" Region of Space

The kinetic plus potential energy of a classical particle can be written (nonrelativistically) as

$$E = \frac{p^2}{2m} + V, \tag{1}$$

where $p^2/2m$ is the kinetic energy and V is the potential energy. Suppose that V is V_1 between $z = 0$ and $z = L$ and is $V_2(V_2 > V_1)$ from $z = L$ to $+\infty$ and from $z = 0$ to $-\infty$. Suppose that the classical particle is "bound" in the "potential well," described in Supplementary Topic 2. That is the case if the energy E of the particle lies somewhere between V_1 and V_2; then, if the classical particle somehow finds itself in the region between $z = 0$ and L, it can never get out. It bounces back and forth between the walls with momentum $p_z = \pm\sqrt{2m(E - V_1)}$, reversing the sign of p_z whenever it hits a wall. It cannot enter the regions where the potential is V_2, because in those regions the kinetic energy would be negative:

$$\frac{p^2}{2m} = E - V = E - V_2 = -(V_2 - E), \quad \text{for } E < V_2. \tag{2}$$

Of course, it makes no sense for a classical particle to have negative kinetic energy.

Real particles are not classical particles. They have wave properties as well as "particle" properties. The de Broglie relations $p = \hbar k$ and $E = \hbar\omega$ give the dispersion relation

$$\omega = \omega_0(z) + \frac{\hbar k^2}{2m}, \quad \text{for } \omega > \omega_0, \tag{3}$$

with

$$\omega_0(z) \equiv \frac{V(z)}{\hbar}.$$

Analogy with coupled pendulums. One can compare this with the dispersion relation for coupled pendulums (in the continuous approximation—see Sec. 3.5)

$$\omega^2 = \omega_0^2(z) + \frac{K^2 a^2}{M} k^2, \quad \text{for } \omega^2 > \omega_0^2, \tag{4}$$

with

$$\omega_0^2(z) \equiv \frac{g}{l}. \tag{5}$$

For the coupled pendulums, when ω is less than ω_0, the waves are not sinusoidal; they are exponential waves. The medium is said to be reactive. The dispersion relation becomes

$$\omega^2 = \omega_0^2 - \frac{K^2 a^2}{M} \kappa^2, \quad \omega^2 < \omega_0^2, \tag{6}$$

where κ is the attenuation constant and $\delta = 1/\kappa$ is the attenuation length. Similarly for de Broglie waves, when ω is less than ω_0, the dispersion relation becomes

$$\omega = \omega_0(z) - \frac{\hbar \kappa^2}{2m}, \quad \omega < \omega_0. \tag{7}$$

Thus the kinetic energy, $E - V$, is given in our example by

$$E - V_1 = \frac{\hbar^2 k_1^2}{2m}, \quad 0 \le z \le L \tag{8}$$

and

$$E - V_2 = -\frac{\hbar^2 \kappa_2^2}{2m}, \quad \text{other places.} \tag{9}$$

Hence, for positive kinetic energy of the particle, the corresponding de Broglie waves are sinusoidal (for a homogeneous medium) with wavenumber k_1, while for negative kinetic energy of the particle, the de Broglie waves are exponential waves with attenuation constant κ_2. The wave functions of the possible states of an electron bound in this "finite potential well" are very similar in shape to the "bound modes" of coupled pendulums described in Sec. 3.5. Thus the ground state wave function $f(z)$ is sinusoidal in the "positive kinetic energy" region (the dispersive region), with wavenumber such that kL is slightly less than π. At $z = 0$ and at $z = L$, this sinusoidal wave function joins smoothly to an exponential function which attenuates to zero at infinite distance from the dispersive region. (The two lowest stationary states are shown in Fig. ST.2.)

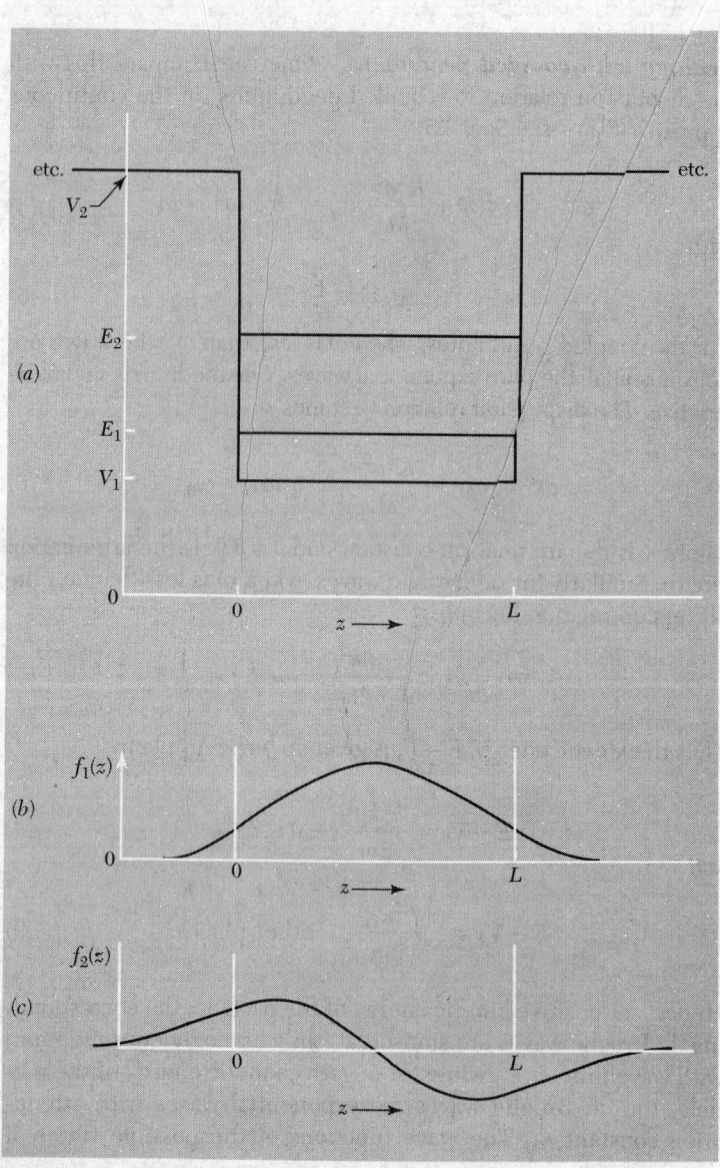

Fig. ST.2 Electron bound in a finite potential well. (a) Plot of V(z), with added horizontal lines E_1 and E_2 showing energies of ground state and first excited state. (b) Space dependence of ground state. (c) Space dependence of first excited state.

From the graph we see that the probability of finding the particle in the "classically forbidden" regions is not zero. For z less than zero, the probability is proportional to $\left|\exp\left[-\kappa_2(-z)\right]\right|^2$; for z greater than L it is proportional to $\left|\exp\left[-\kappa_2(z-L)\right]\right|^2$.

Notice that if V_2 goes to $+\infty$, then according to Eq. (9), κ_2 becomes infinite and the attenuation distance δ_2 goes to zero. This is the situation discussed in Supplementary Topic 2. For that case, we were able to write down immediately the wavenumbers of the allowed modes and then to obtain the corresponding energies from the dispersion relation. In the present example of a finite potential well, it takes more work to find the allowed values of k (inside the well) and κ (outside the well).

4   *Phase and Group Velocities for de Broglie Waves*

For a nonrelativistic electron of energy E in a constant potential V, the dispersion relation is (see Supplementary Topic 2)

$$\omega = \frac{\hbar k^2}{2m} + \frac{V}{\hbar}. \tag{1}$$

The phase velocity is

$$v_\varphi(k) = \frac{\omega}{k} = \frac{\hbar k}{2m} + \frac{V}{\hbar k}. \tag{2}$$

The velocity of the classical particle is p/m, that is, $\hbar k/m$. Thus Eq. (2) reads

$$v_\varphi(k) = \frac{1}{2}v\,(\text{particle}) + \frac{V}{p\,(\text{particle})}, \tag{3}$$

which is a peculiar relationship. Luckily, $v_\varphi(k)$ is not directly observable. The velocity of a quantum-mechanical particle is instead that of a "wave packet" made up of several neighboring values of k, not just one value. The velocity of propagation of a wave packet is given by the group velocity v_g. Thus, using Eq. (1), we find

$$v_g = \left(\frac{d\omega}{dk}\right)_0 = \left(\frac{\hbar k}{m}\right)_0, \tag{4}$$

where the subscript zero means one is to evaluate the derivative at the value of k at the center of the band Δk that forms the packet. Thus we see that $v_g = v\,(\text{particle})$ if we take the particle momentum to be $(\hbar k)_0$, corresponding to the center of the packet.

For a relativistic free particle, the relation between energy, momentum, and rest mass m is given by

$$E^2 = (mc^2)^2 + (cp)^2, \tag{5}$$

which gives the dispersion relation (using $E = \hbar\omega$ and $p = \hbar k$, which are relativistically correct)

$$\hbar^2\omega^2 = (mc^2)^2 + (\hbar ck)^2. \tag{6}$$

The phase velocity $v_\varphi = \omega/k$ the value $v_\varphi = \omega/k = E/p, = E/p$, which is c^2/v (particle) and is thus greater than c. The group velocity is

$$v_g = \frac{d\omega}{dk} = \frac{c^2k}{\omega} = \frac{c^2p}{E} = v \text{ (particle).} \tag{7}$$

The relation between the phase velocity, group velocity, and velocity of light is the same as that for radio waves in the ionosphere, namely $v_\varphi v_g = c^2$. That is because the dispersion relations are similar.

5 Wave Equations for de Broglie Waves

A harmonic de Broglie wave (i.e., a stationary state) in a region of constant potential has the form

$$\psi(z, t) = e^{-i\omega t}(Ae^{ikz} + Be^{-ikz}). \tag{1}$$

Thus

$$\frac{\partial \psi(z, t)}{\partial t} = -i\omega\psi(z, t); \tag{2}$$

$$\frac{\partial^2 \psi(z, t)}{\partial t^2} = -\omega^2\psi(z, t); \tag{3}$$

$$\frac{\partial \psi(z, t)}{\partial z} = e^{-i\omega t}(ikAe^{ikz} - ikBe^{-ikz}); \tag{4}$$

$$\frac{\partial^2 \psi(z, t)}{\partial z^2} = -k^2\psi(z, t). \tag{5}$$

For nonrelativistic particles, the dispersion relation is given by (see Supplementary Topic 2)

$$\hbar\omega = \frac{\hbar^2k^2}{2m} + V. \tag{6}$$

Multiplying Eq. (2) by $i\hbar$ and using Eqs. (5) and (6), we obtain

$$\frac{i\hbar\partial \psi(z, t)}{\partial t} = -\frac{\hbar^2}{2m}\frac{\partial^2 \psi(z, t)}{\partial z^2} + V\psi(z, t). \tag{7}$$

Equation (7) was derived using harmonic waves for a uniform potential, which gives a space dependence that is sinusoidal in kz.

There is no reason why one should demand that Eq. (7) hold when one has $V = V(z)$, i.e., when V is a function of position, but one can hope. That is what Schrödinger did. He guessed that perhaps Eq. (7) would hold even for a $V(z)$ which is not constant. Equation (7), with $V = V(z)$, is called the *Schrödinger equation* (more precisely, the one-dimensional, time-dependent Schrödinger equation). It works, for atomic physics.

When we cannot neglect relativistic effects, we cannot use Eq. (6), nor Eq. (7). For free relativistic particles, the relativistic dispersion relation is

$$\hbar^2 \omega^2 = \hbar^2 c^2 k^2 + (mc^2)^2. \tag{8}$$

Multiplying Eq. (8) by $-\hbar^{-2}\psi(z, t)$ and using Eqs. (3) and (5), we obtain

$$\frac{\partial^2 \psi(z, t)}{\partial t^2} = c^2 \frac{\partial^2 \psi(z, t)}{\partial z^2} - \frac{(mc^2)}{\hbar^2} \psi(z, t). \tag{9}$$

Equation (9) is called the *Klein-Gordon equation*. Notice that if we set $m = 0$ we get the classical wave equation for non dispersive waves of velocity c. This corresponds to the fact that a photon has zero rest mass.

6   Electromagnetic Radiation from a One-dimensional "Atom"

First, review Supplementary Topic 2. Consider the stationary states for an electron bound in a one-dimensional potential well with infinite "sides" located at $z = -L/2$ and $+L/2$. Suppose now that this bound electron happens to be in a superposition of the ground state and the first excited state:

$$\psi(z, t) = \psi_1(z, t) + \psi_2(z, t), \tag{1}$$

$$\psi_1(z, t) = A_1 e^{-i\omega_1 t} \cos k_1 z, \qquad k_1 L = \pi; \tag{2}$$

$$\psi_2(z, t) = A_2 e^{-i\omega_2 t} \sin k_2 z, \qquad k_2 L = 2\pi. \tag{3}$$

The probability (per unit length) of finding the electron at position z at time t is given by

$$\left| \psi(z, t) \right|^2 = \left| A_1 e^{-i\omega_1 t} \cos k_1 z + A_2 e^{-i\omega_2 t} \sin k_2 z \right|^2$$

$$= A_1^2 \cos^2 k_1 z + A_2^2 \sin^2 k_2 z$$
$$+ 2 A_1 A_2 \cos k_1 z \sin k_2 z \cos (\omega_2 - \omega_1) t \tag{4}$$

We see that the probability has a term which oscillates harmonically at the beat frequency between the two de Broglie frequencies ω_1 and ω_2. In fact, it is not difficult to perform the following integrations and obtain $\bar{z}$, the space-averaged value of z:

$$\int_{-L/2}^{L/2} |\psi|^2 dz = (A_1^2 + A_2^2)\frac{L}{2},$$

$$\int_{-L/2}^{L/2} z|\psi|^2 dz = \frac{16L^2}{9\pi^2} A_1 A_2 \cos{(\omega_2 - \omega_1)t},$$

$$\overline{z} = \frac{\int z|\psi|^2 \, dz}{\int |\psi|^2 \, dz} = \frac{32L}{9\pi^2}\frac{A_1 A_2}{A_1^2 + A_2^2} \cos{(\omega_2 - \omega_1)t},$$

i.e.,

$$\overline{z} = (0.36L)\frac{A_1 A_2}{A_1^2 + A_2^2} \cos{(\omega_2 - \omega_1)t}. \tag{5}$$

Why the radiation frequency is the beat frequency.   If the electron has a charge $q = -e$, it will radiate electromagnetic radiation of the frequency at which it oscillates. We see from Eq. (5) that the average position of the charge oscillates at the beat frequency $\omega_2 - \omega_1$. Therefore the radiation frequency is the beat frequency between the two stationary states involved in the "transition":

$$\omega_{\text{rad}} = \omega_2 - \omega_1. \tag{6}$$

7   Time Coherence and Optical Beats

One can obtain interference between waves of different frequency. That holds for optical phenomena as well as other phenomena. Suppose we have two light waves 1 and 2 producing electric fields $\mathbf{E}_1$ and $\mathbf{E}_2$, both polarized (let us say) along $\hat{\mathbf{x}}$. (Thus we can drop vector signs.) The total field at fixed z is the superposition of E_1 and E_2. Using a complex field $E_c(t)$, we have

$$E_c(t) = E_1 e^{i\omega_1 t} e^{i\varphi_1} + E_2 e^{i\omega_2 t} e^{i\varphi_2}. \tag{1}$$

The energy flux, which can be measured by a photomultiplier (whose output current is proportional to the incident energy flux), is proportional to the average of $E^2(t)$ over one period T of the "fast" oscillations at the average frequency:

$$\left\langle E^2(T) \right\rangle = \tfrac{1}{2}\left| E_c(t) \right|^2$$

$$= \tfrac{1}{2}\left\{ E_1^2 + E_2^2 + 2E_1 E_2 \cos{\left[(\omega_1 - \omega_2)t + (\varphi_1 - \varphi_2)\right]} \right\}. \tag{2}$$

Thus one may hope to measure a photomultiplier current that varies at the relatively slow beat frequency $\omega_1 - \omega_2$. What are the requirements on the bandwidths? Remember that our simple point of view is to think of the amplitudes and phase constants as slowly varying in an unpredictable manner, with (for example) φ_1 drifting randomly by an amount of order 2π in a coherence time interval which is the inverse of the bandwidth of oscillation 1:

$$t_{1(\text{coh})} \approx (\Delta \nu_1)^{-1}. \tag{3}$$

$$t_{2(\text{coh})} \approx (\Delta \nu_2)^{-1}. \tag{4}$$

It is clear that if we are to observe beats, the individual components must keep their phases roughly constant for the duration of a beat period. Thus, to observe beats, we require that both coherence times be long compared with the beat period, i.e., that both bandwidths be small compared with the beat frequency:

$$\begin{aligned}\Delta \nu_1 &< \left|\nu_1 - \nu_2\right| \\ \Delta \nu_2 &< \left|\nu_1 - \nu_2\right|\end{aligned} \qquad \text{(for observable beats).} \tag{5}$$

In addition, you need to be able to detect variations in the photomultiplier current at the beat frequency. It also helps to be ingenious. The experiment has been done, and it is a beauty. †

8 Why Is the Sky Bright?

We learned why the sky is blue when we considered in Sec. 7.5 the color dependence of the scattering of light from individual air molecules. Here is an argument that seems to show that the sky should be invisible: Consider a given monochromatic component of sunlight. The electric field drives a given air molecule. Each oscillating electron of the air molecule radiates waves in all directions, some of which travel to the eye of a given observer. But for a given molecule (call it 1) there is another (No. 2) that is one half-wavelength farther from the observer. If both molecules are driven with same amplitude and phase constant, their waves should superpose to give zero at the position of the observer. For scattering near 90 deg, we can obviously satisfy these phase and amplitude conditions, provided the number of air molecules per unit volume is large enough so that there is nearly always a molecule "No.2" for every molecule "No. 1." (For scattering near zero deg, molecules that are a half-wavelength farther from the observer are excited half a period earlier; therefore they do not give destructive interference.) For air at STP, the number density is about 3×10^{19} cm^{-3}. Therefore a cube of edge length 5×10^{-5} cm (the wavelength of blue light) contains about 4×10^6 molecules, i.e., about 100 molecules along each edge of a cube with edge length equal to one wavelength. That would seem to be more than sufficient to give almost completely destructive interference, even when we take into account the fact that the air density falls off exponentially with height above the surface of the

---

† A. T. Forrester, R. A. Gudmundsen, and P. O. Johnson, "Photoelectric Mixing of Incoherent Light," *Phys. Rev.* **99**, 169 (1955).

earth. Therefore we arrive at the prediction that the part of the sky which corresponds to scattering through about 90 deg should be "black" rather than bright blue!

That prediction obviously completely contradicts experience. In fact, the intensity observed is very nearly that which would be predicted on the basis of the scattering from individual air molecules, taking into account the number density of molecules and adding the *intensities* contributed by independent molecules. For some reason, the predicted destructive interference does not occur. Why?

Here is another related fact: If instead of air we use glass or clean water, the predicted destructive interference for 90-deg scattering *does* indeed occur. That is why a flashlight beam proceeds through clean water with negligible loss of intensity (except for the spreading of the beam due to diffraction). Now, the amount of air above the surface of the earth is equivalent in weight and (approximately) in number of molecules to about 33 ft of water. Nevertheless the amount of 90-deg scattering experienced by a flashlight beam in traversing 33 ft of clean water is very small compared with that experienced by sunlight in traversing the atmosphere. In the case of water, one adds the *amplitudes* for 90-deg scattering, and one gets the expected destructive interference. In the case of air, it is the *intensities* which seem to add. Why?

The answer lies in the *uniformity* of the spacing of the water molecules as compared with that of the air molecules. (It has nothing to do with the difference between an air molecule and a water molecule: water vapor behaves like gaseous air; liquid air behaves like liquid water.) The water molecules are "in contact" and have very uniform spacing. There is always "guaranteed" to be a molecule "No. 2" to cancel the contribution of a given molecule "No. 1" (in the superposition of their radiation fields at the position of the observer). In the case of air, it is only *on the average* that there is a molecule "No. 2" for every molecule "No. 1"; sometimes there is, and sometimes there isn't. *The fluctuations* (in uniformity of number density of air molecules) *destroy the coherence*. The "expected" destructive interference of *amplitudes* for 90-deg scattering does not occur. Instead (as is always the case for incoherent sources) the total intensity is the sum of the *intensities* from the contributing sources.

Here is a simplified derivation: Consider a tiny region in space, called region 1. Now pick. another region (called No. 2) of the same size, situated at the same distance from the sun, and situated one half-wavelength farther from the observer than region 1. (We are considering a monochromatic component of the sunlight.) Suppose each of these two regions is tiny compared with a wavelength. Then all the molecules in region 1 are driven in phase. Each contributes a field $\mathbf{E}_1$ at the observer's position. If there are n_1 molecules in region 1 at a given time, then the field at the observer due to these molecules is $n_1\mathbf{E}_1$. Similarly, the field due to region 2 is $n_2\mathbf{E}_2$. The total field due to these two regions is the superposition $\mathbf{E} = n_1\mathbf{E}_1 + n_2\mathbf{E}_2$. Because of the fact that the two regions are driven in phase and are half a wavelength apart along the direction from the observer, we have that $\mathbf{E}_2$ is $-\mathbf{E}_1$. Thus at a given instant we have

$$\mathbf{E} = n_1\mathbf{E}_1 + n_2\mathbf{E}_2 = (n_1 - n_2)\mathbf{E}_1. \tag{1}$$

The field $\mathbf{E}_1$ is the field radiated by one driven air molecule; For this field, we can write (dropping the vector notation, because we are not interested in the polarization)

$$E_1 = A_1 \cos(\omega t + \varphi). \tag{2}$$

Therefore the field contributed by the two regions 1 and 2 is given by

$$E = A \cos(\omega t + \varphi), \tag{3}$$

where

$$A = (n_1 - n_2)A_1. \tag{4}$$

What is the average or "expected" value of the amplitude A? Sometimes n_1 is larger than n_2; sometimes it is smaller. On the average n_1 and n_2 are equal; therefore on the average A is zero. *If n_1 and n_2 remained fixed at their average values, E would always be zero, and we would get no scattering at 90 deg. But this is not the case*, as we shall see.

Now let us look at the intensity of scattered radiation. The intensity is proportional to the square of the radiated electric field. Let us average over one period of oscillation. (The period is about 10^{-15} sec; n_1 and n_2 do not change during that short time interval.) Then the scattered intensity is proportional to the square of the amplitude A. Aside from uninteresting constants, we have that

$$\text{Intensity} = A^2 = (n_1 - n_2)^2 A_1{}^2. \tag{5}$$

Now we consider the effect of the fluctuations in $n_1 - n_2$. If we average over a long enough time interval, so that out regions 1 and 2 have had time to "sample" the constantly varying number density, we see that the time-averaged intensity from the two regions is just the average of $(n_1 - n_2)^2$ times the intensity we would get from a single molecule that remained in region 1. Using the letter I for the time-averaged intensity from the two regions, we have

$$I = \overline{(n_1 - n_2)^2}I_1, \tag{6}$$

where I_1 is the intensity from a single molecule (that remains in region 1), and where the bar denotes the time averaging. Now, the average value of n_1, $\overline{n}_1$, equals the average value of n_2, $\overline{n}_2$. There-fore we can write

$$(n_1 - n_2)^2 = \left[(n_1 - \overline{n}_1) - (n_2 - \overline{n}_2)\right]^2$$

$$= (n_1 - \overline{n}_1) + (n_2 - \overline{n}_2)^2 - 2(n_1 - \overline{n}_1)(n_2 - \overline{n}_2). \tag{7}$$

Taking the average, we find

$$\overline{(n_1 - n_2)^2} = \overline{(n_1 - \overline{n}_1)^2} + \overline{(n_2 - \overline{n}_2)^2}$$

$$- 2\overline{(n_1 - \overline{n}_1)(n_2 - \overline{n}_2)}. \tag{8}$$

So far, everything we have written applies both to air and water. Now we come to the crucial difference. For air, the two regions 1 and 2 are completely "independent," in the sense that the fluctua-tions in n_1 are (at a given instant) independent of those in n_2. That is because there is no direct influence of molecules in region 1 upon those in region 2. (For water, that is not the case. All the molecules are touching. If you want to push a molecule into one side of region 1, you have to make room for it by pushing one out the other side. You will even be pushing molecules through region 2 when you do.) Therefore *for air* the average of the product of $(n_1 - \overline{n}_1)$ and $(n_2 - \overline{n}_2)$ is just the product of the two independent average values:

$$\overline{(n_1 - \overline{n}_1)(n_2 - \overline{n}_2)} = \overline{(n_1 - \overline{n}_1)} \cdot \overline{(n_2 - \overline{n}_2)} = (\overline{n}_1 - \overline{n}_1)(\overline{n}_2 - \overline{n}_2) = 0. \tag{9}$$

(The crucial step was in recognizing that for air the fluctuation in n_1 is independent of that in n_2.) Next we evaluate the mean square fluctuations of n_1 and n_2 about their average values. In the case of the air, there is "plenty of space" in region 1 (or region 2); there is no crowding of molecules. The fact that there may happen to be an excess of molecules in region 1 at a given instant has no influence on whether an additional molecule can get in. In that case, it turns out (as you will learn in Vol. V) that the number of molecules in region 1 (or region 2) obeys a probability distribution function (called "the Poisson distribution") for which the mean square deviation of n_1 from its average value is equal to the average value itself:

$$\overline{(n_1 - \overline{n}_1)^2} = \overline{n}_1, \qquad \overline{(n_2 - \overline{n}_2)^2} = \overline{n}_2. \tag{10}$$

This relationship holds for the air molecules. However, it does *not* hold for the water molecules, because the presence of a small excess strongly inhibits the entrance of any additional molecules. Instead, we have for water

$$\overline{(n_1 - \overline{n}_1)^2} \ll \overline{n}_1, \qquad \overline{(n_2 - \overline{n}_2)^2} \ll \overline{n}_2. \tag{11}$$

The time-averaged intensity from the two regions is thus given for air by

$$
\begin{aligned}
I &= \overline{(n_1 - n_2)^2 I_1} \\
&= \overline{(n_1 - \overline{n}_1)^2} I_1 + \overline{(n_2 - \overline{n}_2)^2} I_1 + 0 \\
&= \overline{n}_1 I_1 + \overline{n}_2 I_1 \\
&= \overline{n}_1 I_1 + \overline{n}_2 I_2.
\end{aligned}
\tag{12}
$$

This intensity is just the sum of the *intensities* contributed by the air molecules in region 1 plus those contributed by the molecules in region 2. For water, we have instead

$$
I = \overline{(n_1 - n_2)^2} I_1 \ll \overline{n}_1 I_1 + \overline{n}_2 I_2.
\tag{13}
$$

If n_1 and n_2 were always exactly equal, we would have "perfectly rigid and uniform water" which would give zero intensity.

By a very simple and ingenious experiment, R. W. Wood has demonstrated that the intensity of light scattered from air at 90 deg is proportional to the number of contributing molecules, as predicted by Eq. (12). You can easily repeat his experiment. See for example the description in M. Minnaert, *Light and Colour*, paragraphs 172 and 174 (Dover Publications, Inc., New York, 1954).

9   *Electromagnetic Waves in Material Media*

Our discussion will be more general than that in the main text. We shall not avoid discussing the absorptive part of the dielectric constant nor shall we avoid the use of complex numbers.

Maxwell's equations.   We start by writing Maxwell's equations in their most general form (in esu):

$$
\nabla \cdot \mathbf{B} = 0
\tag{1}
$$

$$
\vec{\nabla} \cdot \mathbf{E} = \frac{\rho_{\text{tot}}}{\epsilon_0} = \frac{\rho_{\text{free}}}{\epsilon_0} - \frac{1}{\epsilon_0} \vec{\nabla} \cdot \mathbf{P}
\tag{2}
$$

$$
\begin{aligned}
\vec{\nabla} \times \vec{\mathbf{B}} &= \mu_0 \mathbf{J}_{\text{total}} + \frac{1}{c^2} \frac{\partial \mathbf{E}}{\partial t} \\
&= \mu_0 \mathbf{J}_{\text{free}} + \left(\mu_0 \vec{\nabla} \times \overline{\mathbf{M}} + \mu_0 \frac{\partial \vec{\mathbf{P}}}{\partial t} \right) + \frac{1}{c^2} \frac{\partial \mathbf{E}}{\partial t}
\end{aligned}
\tag{3}
$$

$$
\vec{\nabla} \times \vec{\mathbf{E}} = -\frac{\partial \vec{\mathbf{B}}}{\partial t}.
\tag{4}
$$

[For Eq. (1), see Vol. II, p. 360, Eq. (10.1); for Eq. (2), see Vol. II, p. 332, Eq. (9.57); for Eq. (3), see Vol. II, p. 343, Eq. (9.79), which holds when **M** is zero, and p. 390, Eq. (10.50), which holds when $\partial \mathbf{P}/\partial t$ and $\partial \mathbf{E}/\partial t$ are zero; for Eq. (4), see Vol. II, p. 243, Eq. (10.30).]

An alternative way to write Eqs. (1) through (4) is as follows:

$$\nabla \cdot \mathbf{B} = 0 \tag{5}$$

$$\vec{\nabla} \cdot \left\{ \epsilon_0 \vec{\mathbf{E}} + \vec{\mathbf{P}} \right\} = \rho_{\text{free}} \tag{6}$$

$$\vec{\nabla} \times \left\{ \frac{1}{\mu_0} \vec{\mathbf{B}} - \vec{\mathbf{M}} \right\} = \frac{\partial}{\partial t} \left\{ \epsilon_0 \vec{\mathbf{E}} + \vec{\mathbf{P}} \right\} + \mathbf{J}_{\text{free}} \tag{7}$$

$$\vec{\nabla} \times \vec{\mathbf{E}} = -\frac{\partial \vec{\mathbf{B}}}{\partial t} \tag{8}$$

The combination $\epsilon_0 \vec{\mathbf{E}} + \vec{\mathbf{P}}$ is called $\mathbf{D}$. The combination $\dfrac{\vec{\mathbf{B}}}{\mu_0} - \vec{\mathbf{M}}$ is called $\mathbf{H}$:

$$\epsilon_0 \vec{\mathbf{E}} + \vec{\mathbf{P}} \equiv \vec{\mathbf{D}}, \qquad \frac{1}{\mu_0} \vec{\mathbf{B}} - \vec{\mathbf{M}} \equiv \vec{\mathbf{H}}. \tag{9}$$

We shall avoid using the symbols $\mathbf{D}$ and $\mathbf{H}$, however.

Linear isotropic medium. The force on a point charge q located at a given point x, y, z, at a given time t is given by

$$\mathbf{F} = q\vec{\mathbf{E}} + q(\vec{\mathbf{v}} \times \vec{\mathbf{B}}), \tag{10}$$

where $\mathbf{E}$ and $\mathbf{B}$ are the instantaneous local fields. In discussing "continuous" media we use the average force per unit charge, averaged over a small volume element, to define the space-averaged values of $\mathbf{E}$ and $\mathbf{B}$. We consider these fields to act on an "average" charge, whose charge and velocity are averages over the volume element, and correspond to the charge and current densities in the volume element.

The forces on the charges and currents in the medium are due to the fields $\mathbf{E}$ and $\mathbf{B}$ in the medium. These forces modify the charge and current distributions and contribute to $\mathbf{P}$ and $\mathbf{M}$. The medium is said to be *isotropic* if the polarization $\mathbf{P}$ is along $\pm \mathbf{E}$ and the magnetization $\mathbf{M}$ is along $\pm \mathbf{B}$. This also implies that $\mathbf{P}$ is zero when $\mathbf{E}$ is zero and $\mathbf{M}$ is zero when $\mathbf{B}$ is zero. It also implies that (for example) P_x depends only on E_x and not on E_y or E_z. (In some crystals, if you push on the electrons of the atoms with a force proportional to $\mathbf{E}$, their displacement, which is the source of $\mathbf{P}$, is not along $\mathbf{E}$ because the constraint forces of the crystal make it easier for the electrons to move in some directions than in others.) Thus for an isotropic medium we have, for example,

$$P_x = \chi E_x + a E_x^2 + \beta E_x^3 + \cdots. \tag{11}$$

For sufficiently weak fields the quadratic and higher terms in Eq. (11) are negligible. This is the case for ordinary intensities of electromagnetic fields in ordinary matter. (For sufficiently strong fields such as can be produced with a pulsed ruby laser the nonlinear contributions to **P** can be detected and studied.) A medium is said to be *linear* if we can neglect the terms aE_x^2, βE_x^3, etc., in Eq. (11). We see that "linearity" is a property not just of the medium but also 'of the intensity of the fields present.

Definitions of χ, χ_m, ϵ, and μ for static fields. For a linear isotropic medium the electric susceptibility χ and magnetic susceptibility χ_m are defined for time-independent fields as follows:

$$P_x(x, y, z) = \chi(x, y, z)E_x(x, y, z) \tag{12}$$

$$M_x(x, y, z) = \frac{\chi_m}{\mu} B_x(x, y, z). \tag{13}$$

The permittivity of the material ϵ and magnetic permeability μ are defined by

$$\epsilon_0 E_x + \vec{P}_x = \epsilon E_x \tag{14}$$

$$\frac{1}{\mu_0} B_x - M_x = \frac{1}{\mu} Bx. \tag{15}$$

Combining these definitions we have

$$\epsilon = \epsilon_0 \left(1 + \chi\right) \tag{16}$$

$$\mu = \mu_0 \left(1 + \chi_m\right). \tag{17}$$

[For Eq. (14), see Vol. II, p. 323, Eq. (9.38). To get Eq. (15), see Vol. II, p. 393, Eq. (10.55) for the equation $\mathbf{M} = \chi_m \mathbf{H}$, and p. 390, Eq (10.52) for the definition $\mathbf{H} \equiv \frac{1}{\mu_0} \vec{\mathbf{B}} - \mathbf{M}$. The further definition $\mathbf{H} = \mathbf{B}/\mu$ gives Eq. (15).]

Susceptibilities for time-dependent fields. We wish to extend these linear relations so that they hold for time-dependent fields in a linear isotropic medium. We might hope that once we had measured (for example) χ for static electric fields, we could simply generalize Eq. (12) so as to write $P_x(x, y, z, t) = \chi E_x(x, y, z, t)$, where χ is the value obtained from the static measurements. That hope is forlorn, as we shall see. In general it is necessary to Fourier-analyze the fields into various frequency components. The electric and magnetic susceptibilities depend on frequency. Thus there is no "overall" χ that can be factored from a sum of contributions to **P** from different frequencies.

Once we have found that the susceptibilities depend on frequency, we might expect that we could generalize Eq. (12) by writing

$$P_x(x, y, z, \omega, t) = \epsilon_0 \chi(x, y, z, \omega)E_x(x, y, z, \omega t), \tag{18}$$

with a similar expression for M_x. However, we shall find that even Eq. (18) is an oversimplification, because it implies that P_x is proportional to E_x at every instant, i.e., that P_x is *in phase* (except for a possible minus sign) with E_x. More generally, we must include the possibility of a component of P_x that is *in quadrature* with E_x (i.e., is ± 90 deg out of phase with E_x). We shall find that the part of P_x which is in phase with E_x does not lead to absorption of electromagnetic energy by the medium. We shall therefore call the in-phase part of P_x the "elastic" or "dispersive" part. The part of P_x that is in quadrature with E_x gives energy absorption and will be called the "absorptive" part of P_x. We can write $P_x(x, y, z, \omega t)$ as the sum of an elastic and an absorptive part. For a linear isotropic medium the elastic part is proportional to $E_x(x, y, z, \omega t)$ with a proportionality constant $\chi_{el}(x, y, z, \omega)$. The absorptive part can be taken to be proportional $E_x(x, y, z, \omega t - \frac{1}{2}\pi)$, with a proportionality constant $\chi_{ab}(x, y, z, \omega)$:

$$P_x(x, y, z, \omega t) = \chi_{el}(x, y, z, \omega)E_x(x, y, z, \omega t)$$
$$+ \chi_{ab}(x, y, z, \omega)E_x(x, y, z, \omega t - \tfrac{1}{2}\pi). \quad (19)$$

Let us consider a given location and drop x, y, z from the notation. Suppose that at that location we have

$$E_x(\omega t) = E_0 \cos(\omega t - \varphi). \quad (20)$$

Then Eq. (19) gives

$$P_x(\omega, t) = \epsilon_0 \chi_{el} E_x(\omega t) + \epsilon_0 \chi_{ab} E_x(\omega t - \pi), \quad (21)$$

i.e.,

$$P_x(\omega t) = \epsilon_0 \chi_{el} E_0 \cos(\omega t - \varphi) + \chi_{ab} E_0 \sin(\omega t - \varphi). \quad (22)$$

Simple model of linear isotropic medium. Assume that in a small neighborhood of a given fixed location the medium consists of N neutral "atoms" per unit volume. Each atom consists of a particle (an "electron") having mass M, charge q (with q algebraic, i.e., unspecified as to sign) bound by a spring of spring constant $M\omega_0^2$ to a much heavier "nucleus" having a charge equal in magnitude and opposite in sign to q. (We also include the case where ω_0 is zero. In that case we have a neutral "plasma.") We neglect the comparatively tiny motion of the nucleus and hence neglect its contribution to **P**. The atom has no magnetic moment, and no magnetic moment is induced by magnetic fields. Thus the magnetization is zero. We neglect the fluctuations and irregularities in the motion of individual particles and assume that every particle acts like a fictitious "average" particle. It is assumed that the particle M is acted on by the spring, by the electric field $E_x(\omega t)$ at its location, and by a "damping force" which takes into account the loss of energy from the particle to its neighbors by collisions (or by radiation). We neglect the force

$q(\mathbf{v}/c) \times \mathbf{B}$ on the particle M compared with the force $q\mathbf{E}$. This is because we assume there are no static magnetic fields present, and because we assume v/c is always tiny. (This is the case even for the intense electric fields produced by a pulsed ruby laser.) Therefore for the x component of the motion of q we have

$$M\ddot{x} = -M\omega_0^2 x - M\Gamma\dot{x} + qE_x, \tag{23}$$

with

$$E_x(\omega t) = E_0 \cos(\omega t - \varphi). \tag{24}$$

The damping force $-M\Gamma\dot{x}$ represents transfer of energy from the oscillating charge to the medium. This energy is no longer in either the electromagnetic field components of frequency ω nor is it in the oscillation energy of M at frequency ω, but is instead in the form of translational and rotational energy of the atoms, and also of "random" vibrations at other frequencies. It is called *heat*.

In writing Eq. (24) we are assuming that the amplitude E_0 and phase constant φ depend only on the equilibrium position of the charge q and not on the instantaneous displacement $x(t)$ of q from its equilibrium position. We are therefore assuming that the amplitude of vibration of q is very small compared with the wavelength of the electromagnetic waves that give the space and time dependence of E_x. Otherwise we would have to include an x dependence in E_0 and φ.

We shall assume that the "local field" E_x which appears in Eq. (23) for the motion of our "average" charge q is the same as the space-averaged field E_x which appears in Eq. (21). This is nearly the case for a gas and for certain crystals. (In many crystals the electric field felt by a given charge is dominated by a close neighbor. In general the average local field is not the same as the space-averaged field.)

According to Sec. 3.2, the steady-state solution of Eq. (23) has the form

$$x(t) = A_{el} \cos(\omega t - \varphi) + A_{ab} \sin(\omega t - \varphi),$$

where $A_{el} \cos(\omega t - \varphi)$ is the elastic component of the displacement x, i.e., the part in phase with the driving force, and $A_{ab} \sin(\omega t - \varphi)$ is the absorptive part of the displacement, i.e., the part in quadrature with the driving force. The elastic and absorptive amplitudes are given by

$$A_{el} = \frac{qE_0}{M} \frac{(\omega_0^2 - \omega^2)}{(\omega_0^2 - \omega^2)^2 + \Gamma^2\omega^2} \tag{25}$$

$$A_{ab} = \frac{qE_0}{M} \frac{\Gamma\omega}{(\omega_0^2 - \omega^2)^2 + \Gamma^2\omega^2}. \tag{26}$$

The polarization P_x is the number density N times the dipole moment qx corresponding to the displacement x of q from its equilibrium position. Thus we have

$$P_x(t) = Nqx(t) \qquad (27)$$

i.e.,

$$P_x(t) = NqA_{el} \cos(\omega t - \varphi) + NqA_{ab} \sin(\omega t - \varphi), \qquad (28)$$

i.e.,

$$P_x(\omega t) = \frac{NqA_{el}}{E_0} E_x(\omega t) + \frac{NqA_{ab}}{E_0} E_x\left(\omega t - \tfrac{1}{2}\pi\right). \qquad (29)$$

By comparison of Eq. (29) with Eq. (21) we find

$$\chi_{el} = \frac{NqA_{el}}{\epsilon_0 E_0} = \frac{Nq^2}{\epsilon_0 M} \frac{(\omega_0^2 - \omega^2)}{(\omega_0^2 - \omega^2)^2 + \Gamma^2\omega^2} \qquad (30)$$

$$\chi_{ab} = \frac{NqA_{ab}}{\epsilon_0 E_0} = \frac{Nq^2}{\epsilon_0 M} \frac{\Gamma\omega}{(\omega_0^2 - \omega^2)^2 + \Gamma^2\omega^2}. \qquad (31)$$

Use of complex quantities in Maxwell's equations. Maxwell's equations do not contain the square root of -1. Neither do any observable quantities such as $\mathbf{E}$ or $\mathbf{B}$ or $\mathbf{P}$ or $\mathbf{M}$. However, we can greatly simplify the algebra used to describe electromagnetic waves in media *where there is absorption* by making use of complex numbers.

When absorption can be neglected, Eq. (21) reduces to the simpler form $P_x(\omega t) = \epsilon_0 \chi(\omega) E_x(\omega t)$, where $\chi(\omega)$ is χ_{el}. This is the form of Eq. (18), which in turn is similar to the linear relation that holds for static fields, Eq. (12). In that case the definitions of dielectric constant and magnetic permeability given by Eqs. (12) through (17) can be used for time-dependent fields.

When absorption cannot be neglected, Eq. (18) must be replaced by the more complicated expression given by Eq. (21). That is because we must include the "in quadrature" as well as the "in phase" components of $\mathbf{P}$ (and similarly of $\mathbf{M}$) when we cannot neglect absorption. In that case we have to keep track separately of $\mathbf{E}(\omega t)$, $\mathbf{E}\left(\omega t - \tfrac{1}{2}\pi\right)$, $\mathbf{B}(\omega t)$, $\mathbf{B}\left(\omega t - \tfrac{1}{2}\pi\right)$, and the corresponding polarizations and magnetizations that are in phase or in quadrature with $\mathbf{E}(\omega t)$ and $\mathbf{B}(\omega t)$.

A very neat way to accomplish this "bookkeeping" is to make use of complex quantities which we call $\mathbf{E}$, $\mathbf{B}$, $\mathbf{P}$, and $\mathbf{M}$, with the understanding that *the actual physical fields are the real parts of these "complex fields."*

The time dependence of each of the complex fields is taken to be of the form $\exp(-i\omega t)$, where the minus sign is the convention used in optics. [In electrical engineering the usual convention is $\exp(+i\omega t)$. In quantum mechanics the convention is always $\exp$

($-i\omega t$).] Thus we introduce the complex quantity E_x (at a given location) given by

$$E_x(\omega t) = E_0 e^{i\varphi} e^{-i\omega t} = E_0 \cos(\omega t - \varphi) - i E_0 \sin(\omega t - \varphi). \quad (32)$$

The physical field corresponding to the complex field E_x is the real part of E_x, and is thus equal to $E_0 \cos(\omega t - \varphi)$ according to Eq. (32).

The simplification that comes from using complex time dependence $\exp(-i\omega t)$ is the result of the fact that a 90-deg phase shift is merely equivalent to multiplying by i:

$$e^{-i[\omega t - (1/2)\pi]} = e^{i(1/2)\pi} e^{-i\omega t} = i e^{-i\omega t}.$$

Thus

$$E_x(\omega t - \tfrac{1}{2}\pi) = i E_x(\omega t). \quad (33)$$

Complex susceptibility. Whether or not we are using complex fields, the physical polarization is related to the physical electric field by the linear relation (for a linear isotropic medium)

$$P_x(\omega t) = \epsilon_0 \chi_{el} E_x(\omega t) + \epsilon_0 \chi_{ab} E_x(\omega t - \pi/2), \quad (34)$$

where all quantities are real and hence physical. We now make use of complex $E_x(\omega t)$ given by Eq. (32) and reinterpret (34) with P_x and E_x complex (χ_{el} and χ_{ab} are still real):

$$P_x(\omega t) = \epsilon_0 \chi_{el} E_x(\omega t) + \epsilon_0 \chi_{ab} E_x\left(\omega t - \frac{\pi}{2}\right)$$

$$= \epsilon_0 \chi_{el} E_x(\omega t) + i \chi_{ab} \epsilon_0 E_x(\omega t)$$

i.e.,

$$P_x(\omega t) = \chi(\omega) \epsilon_0 E_x(\omega t), \quad (35)$$

where

$$\chi(\omega) = \chi_{el} + i \chi_{ab}. \quad (36)$$

The physical polarization along x is the real part of the complex quantity P_x given by Eq. (35). It involves both the real part χ_{el} and the imaginary part χ_{ab} of the complex susceptibility $\chi_{el} + i\chi_{ab}$. (Of course, χ_{el} and χ_{ab} are both real quantities.) For example [taking $\varphi = 0$ in Eq. (32)] we have

$$E_x = E_0 e^{-i\omega t} = E_0 \cos \omega t - i E_0 \sin \omega t \quad (37)$$

$$P_x = \chi \epsilon_0 E_x = (\chi_{el} + i\chi_{ab}) \epsilon_0 (E_0 \cos \omega t - i E_0 \sin \omega t)$$

$$= \epsilon_0 \chi_{el} E_0 \cos \omega t + \epsilon_0 \chi_{ab} E_0 \sin \omega t + i \cdot (\text{imaginary part}). \quad (38)$$

The real part of P_x as given by Eq. (38), the real part of E_x, as given by Eq. (37), and the real quantities χ_{el} and χ_{ab} satisfy Eq. (34), which holds for the physical (hence real) fields.

Complex dielectric constant. Because we have introduced complex fields E_x and P_x we have obtained the very simple expression

$P_x = \epsilon_0 \chi E_x$ given by Eq. (35), in place of the more complicated expression, Eq. (34). The price has been that we now have complex susceptibility $\chi(\omega)$, given by Eq. (36). Because Eq. (35) is similar in form to Eq. (12) (which holds for static fields), we can extend the definitions given in Eqs. (12) through (17) so that they hold for time-dependent fields. This means that when we cannot neglect absorption we must use complex dielectric constant ϵ_r and complex magnetic permeability if we want Eqs. (12) through (17) to hold. Then according to Eqs. (16) and (36) we have

$$\epsilon = \epsilon_0 \left(1 + \chi_e \right) = \epsilon_0 \left(1 + \chi_{el} + i\chi_{ab} \right) \tag{39}$$

Thus
$$\epsilon_r = \left(1 + \chi_e \right) = \left(1 + \chi_{el} + i\chi_{ab} \right).$$

$$\epsilon_r = \operatorname{Re} \epsilon_r + i \operatorname{Im} \epsilon_r,$$

where

$$\operatorname{Re} \epsilon_r = 1 + \chi_{el} \tag{40}$$

$$\operatorname{Im} \epsilon_r = \chi_{ab}. \tag{41}$$

For $\omega = 0$ all quantities reduce to their static values.

Complex dielectric constant for the simple model of linear isotropic medium. For our simple model we have $\mathbf{M} = 0$. Hence $\chi_m = 0$ and $\mu = 1$ according to Eqs. (13), (15), and (17). The electric susceptibility has real (i.e., elastic) and imaginary (i.e., absorptive) parts given by Eqs. (30) and (31). Thus Eq. (39) gives

$$\begin{aligned} \epsilon_r &= 1 + \frac{Nq^2}{\epsilon_0 M} \cdot \frac{\left(\omega_0{}^2 - \omega^2 \right)}{(\omega_0{}^2 - \omega^2)^2 + \Gamma^2 \omega^2} \\ &+ i \frac{Nq^2}{\epsilon_0 M} \cdot \frac{\Gamma \omega}{(\omega_0{}^2 - \omega^2)^2 + \Gamma^2 \omega^2}. \end{aligned} \tag{42}$$

We may remark that once we have decided to use complex numbers the solution of the equation of motion of q, Eq. (23), becomes quite simple:

$$\ddot{x} + \Gamma \dot{x} + \omega_0{}^2 x = \frac{q}{M} E_x = \frac{q}{M} E_0 e^{-i\omega t}, \tag{43}$$

with E_0 complex. Now let $x = x_0 \exp\left(-i\omega t\right)$. Then $\dot{x} = -i\omega x$ and $\ddot{x} = -\omega^2 x$. Substitution in Eq. (43) gives

$$(-\omega^2 - i\omega\Gamma + \omega_0{}^2)x = \frac{q}{M} E_x,$$

i.e.,

$$x(\omega t) = \frac{q}{M} \cdot \frac{1}{(\omega_0{}^2 - \omega^2) - i\omega\Gamma} E_x(\omega t). \tag{44}$$

Then the complex susceptibility is given by

$$\chi(\omega) = \frac{P_x}{\epsilon_0 E_x} = \frac{Nqx}{\epsilon_0 E_x} = \frac{Nq^2}{\epsilon_0 M(\omega_0{}^2 - \omega^2) - i\omega\Gamma}. \tag{45}$$

The complex dielectric constant ϵ_r is given by

$$\epsilon_r = 1 + \chi = 1 + \frac{Nq^2}{\epsilon_0 M} \cdot \frac{1}{(\omega_0^2 - \omega^2) - i\omega\Gamma}. \qquad (46)$$

You can easily check that Eqs. (46) and (42) are equivalent, by multiplying numerator and denominator of $\epsilon_r - 1$ [in Eq. (46)] by $(\omega_0^2 - \omega^2) + i\omega\Gamma$, so as to be able to write ϵ_r as the sum Re $\epsilon_r + i$ Im ϵ. Sometimes it is more convenient to leave ϵ_r in the form of Eq. (46).

Maxwell's equations for linear isotropic medium.   We start with the general Maxwell's equations given by Eqs. (5) through (8). We then assume the linear relations between P_x, and E_x, and between M_x, and B_x, given by Eqs. (12) through (17). These relations hold with real quantities only if we take $\omega = 0$. They hold for arbitrary frequency ω only if we take all quantities to be complex, as we have seen. We thus obtain Maxwell's equations that relate the complex fields **B** and **E** (whose real parts are the physical fields):

$$\nabla \cdot \mathbf{B} = 0 \qquad (47)$$

$$\vec{\nabla} \cdot \vec{D} = \rho_{\text{free}} \qquad (48)$$

$$\vec{\nabla} \times \vec{H} = \vec{J}_{\text{free}} + \frac{\partial \vec{D}}{\partial t} \qquad (49)$$

$$\vec{\nabla} \times \vec{E} = -\frac{\partial \vec{B}}{\partial t}. \qquad (50)$$

For the general case of frequency-dependent ϵ and μ these equations all refer to a given frequency ω. Since the physical ρ_{free} and $\mathbf{J}_{\text{free}}$ may each have parts proportional to both cos ωt and sin ωt, they will generally be the real parts of complex quantities that appear in the above equations. Of course in the special case of a medium where ϵ and μ do not depend on frequency, all quantities are real.

Maxwell's equations for neutral homogeneous linear isotropic medium.   In Eqs. (48) and (49) the relative electrical permittivity and magnetic permeability are complex functions of frequency ω and also functions of x, y, z, since we have not assumed that all places in the medium have the same properties. For example, in our simple model we can take the number density to be a function of position, $N = N(x, y, z)$. We now consider the especially simple and important case where the medium is *homogeneous*, which means that μ and ϵ do not depend on $x, y,$ and z. With these assumptions ϵ and μ are constants in Eqs. (48) and (49). We also assume that the medium is *neutral*, by which we mean that ρ_{free} and $\mathbf{J}_{\text{free}}$ are both zero. (Our simple model consists of a neutral gas or amorphous solid or plasma.) Then Maxwell's Eqs. (47) through (50) become

$$\nabla \cdot \mathbf{B} = 0 \tag{51}$$

$$\nabla \cdot \mathbf{E} = 0 \tag{52}$$

$$\vec{\nabla} \cdot \mathbf{B} = \mu\epsilon \frac{\partial \mathbf{E}}{\partial t} \tag{53}$$

$$\vec{\nabla} \cdot \mathbf{E} = -\frac{\partial \vec{\mathbf{B}}}{\partial t}. \tag{54}$$

Notice that if we set $\mu = 1$ and $\epsilon = 1$ we obtain Maxwell's equations for vacuum. For the cases we are interested in, μ and ϵ are in general complex, so that $\mathbf{E}$ and $\mathbf{B}$ are complex. For example, for our simple model, $\mu = 1$ and ϵ is complex. Then both $\mathbf{E}$ and $\mathbf{B}$ are complex, and the physical fields are their real parts.

Wave equation. Equations (51) through (54) are first-order linear differential equations. Equations (53) and (54) are "coupled" equations, relating $\mathbf{B}$ and $\mathbf{E}$. We can obtain second-order uncoupled equations as follows. Take the curl of Eq. (53) and then use Eq. (54):

$$\vec{\nabla} \times (\vec{\nabla} \times \vec{\mathbf{B}}) = \mu\epsilon \frac{\partial}{\partial t} (\vec{\nabla} \times \vec{\mathbf{E}}) = -\mu\epsilon \frac{\partial^2 \vec{\mathbf{B}}}{\partial t^2}. \tag{55}$$

Similarly take the curl of Eq. (54) and then use Eq. (53):

$$\vec{\nabla} \times (\vec{\nabla} \times \vec{\mathbf{E}}) = -\frac{\partial}{\partial t} (\vec{\nabla} \times \vec{\mathbf{B}}) = -\mu\epsilon \frac{\partial^2 \vec{\mathbf{E}}}{\partial t^2}. \tag{56}$$

Now apply the vector identity [Appendix Eq. (39)]

$$\nabla \times (\nabla \times \mathbf{C}) \equiv \nabla(\nabla \cdot \mathbf{C}) - \nabla^2 \mathbf{C} \tag{57}$$

to the left side of Eq. (55) and likewise to Eq. (56) and use the fact that $\nabla \cdot \mathbf{E}$ and $\nabla \cdot \mathbf{B}$ are both zero. That gives

$$\nabla^2 \mathbf{B} - \mu\epsilon \frac{\partial^2 \vec{\mathbf{B}}}{\partial t^2} = 0, \ \nabla^2 \vec{\mathbf{E}} - \mu\epsilon \frac{\partial^2 \vec{\mathbf{E}}}{\partial t^2} = 0. \tag{58}$$

Equations (58) actually consist of six separate equations each of the form

$$\nabla^2 \psi(x, y, z, t) - \mu\epsilon \frac{\partial^2 \psi(x, y, z, t)}{\partial t^2} = 0, \tag{59}$$

where $\psi(x, y, z, t)$ stands for any one of the six quantities E_x, E_y, E_z, B_x, B_y and B_z.

For the special case that ϵ and μ are real and positive and independent of frequency, Eq. (59) is the classical wave equation for nondispersive waves. That is the case for vacuum, where we have $\mu = \epsilon = 1$. We are interested in the general case of a homogeneous neutral isotropic linear medium where ϵ and μ are complex and depend on frequency. In that case we take **E** and **B** as complex quantities with time dependence $\exp(-i\omega t)$ Thus for all six quantities represented by $\psi(x, y, z, t)$ we have

$$\psi(x, y, z, t) = \varphi(x, y, z)e^{-i\omega t} \tag{60}$$

$$\frac{\partial^2 \psi}{\partial t^2} = -\omega^2 \psi. \tag{61}$$

Substituting Eq. (61) into Eq. (59) and canceling $(-i\omega t)$ gives the differential equation satisfied by the space dependence $\varphi(x, y, z)$:

$$\nabla^2 \varphi(x, y, z) + k^2 \varphi(x, y, z) = 0, \tag{62}$$

where we define the complex constant k^2 by

$$k^2 = \mu\epsilon\omega^2. \tag{63}$$

Complex index of refraction.   We further define the complex constant n^2, called the *square of the complex index of refraction*, by

$$n^2 = \frac{\mu\epsilon}{\mu_0 \epsilon_0}. \tag{64}$$

Thus

$$k^2 = n^2 \frac{\omega^2}{c^2} = \mu\epsilon\omega^2 \cdot \tag{65}$$

Notice that since ϵ and μ are complex, so is k^2 and so is n^2. We can take the square root of k^2 or of n^2. The square root of a complex number is a complex number. Thus we have a complex k and a complex index of refraction.

Plane-wave solutions.   The general solution of Eq. (62) can be written as a superposition of terms of the form

$$\varphi(x, y, z) = e^{ik \cdot r} = \exp i(k_x x + k_y y + k_z z), \tag{66}$$

where

$$k_x{}^2 + k_y{}^2 + k_z{}^2 = k^2 = n^2 \frac{\omega^2}{c^2} = \mu\epsilon\omega^2. \tag{67}$$

Then the general solution of Eq. (59) can be written as a superposition of plane traveling waves of the form

$$\psi(x, y, z, t) = e^{-i(\omega t - k \cdot r)}, \tag{68}$$

with k^2 complex.

Plane waves propagating along z. As a special case we consider the case where only k_z differs from zero. Then the general solution has a plane wave propagating along $+z$ and one propagating along $-z$:

$$\psi(z, t) = [A^+ e^{+ikz} + A^- e^{-ikz}]e^{i\omega t}, \qquad (69)$$

where $+k$ and $-k$ are the two square roots of k^2 and A^+ and A^- are complex constants. Since we want $\exp[i(kz - \omega t)]$ to represent a wave propagating in the $+z$ direction, we let k be that square root of k^2 that has a positive real part, provided k has a real part. If k is a pure imaginary, we call k that square root of k^2 that equals $+i|k|$.

Relation between E and B for plane wave. Equation (69) must hold for any one of the six quantities $E_x, E_y, E_z, B_x, B_y, B_z$, since all these quantities satisfy the wave equation Eq. (59). In obtaining that second-order wave equation we have discarded some of the information contained in the Maxwell's first-order equations. We shall now go back to Maxwell's equations and incorporate all the information. From $\nabla \cdot \mathbf{B} = 0$ and $\nabla \cdot \mathbf{E} = 0$ we conclude that B_z and E_z are constant (for $\mathbf{k}$ along the z axis). Since we are not considering the special case of zero frequency, that constant is zero. We thus have only E_x, E_y, B_x, and B_y to consider. For simplicity we consider only linear polarization with E_x different from zero and E_y equal to zero. Then according to Eq. (69) we have

$$E_x(z, t) = (E^+ e^{ikz} + E^- e^{-ikz})e^{-i\omega t}, \qquad (70)$$

where E^+ and E^- are complex constants. From Maxwell's Eqs. (53) and (54) we then find that B_x is zero and that B_y and E_x are related by

$$\frac{\partial B_y}{\partial z} = +\mu\epsilon \frac{\partial E_x}{\partial t} \qquad (71)$$

$$\frac{\partial B_y}{\partial t} = -\frac{\partial E_x}{\partial z}.$$

Using the fact that B_y has a form given by Eq. (69) and using Eqs. (71), we find

$$B_y(z, t) = \frac{n}{c}\Big[E^+ e^{ikz} - E^- e^{-ikz} \Big]e^{-i\omega t}. \qquad (72)$$

Thus if we are given E_x [see Eq. (70)], then B_y is completely determined [by Eq. (72)]. Similar results are obtained if we consider E_y to be nonzero. The general results are that for components propagating in the $\pm\hat{\mathbf{z}}$ direction, $\mathbf{B}$ and $\mathbf{E}$ are related by

$$\mathbf{B}^+ = +\hat{\mathbf{z}} \times \left(\frac{n}{c} \mathbf{E}^+ \right), \qquad \mathbf{B}^- = -\hat{\mathbf{z}} \times \left(\frac{n}{c} \mathbf{E}^- \right), \qquad (73)$$

where the superscripts refer to propagation along $+\hat{\mathbf{z}}$ or $-\hat{\mathbf{z}}$. In all these relations n and k are generally complex.

Numerical example of complex index of refraction. Suppose we have a medium with $\mu = 1.0$ and $\epsilon = 1 + i\sqrt{3}$, at a given frequency ω. Then

$$n^2 = 1 + i\sqrt{3} = 2\exp\left(i\frac{1}{3}\pi\right). \tag{74}$$

$$n = \sqrt{2}\exp i\frac{\pi}{6} = \sqrt{2}\left(\frac{1}{2}\sqrt{3} + \frac{1}{2}i\right) = 1.225 + 0.707i$$

$$k = n\frac{\omega}{c} = 1.225\frac{\omega}{c} + 0.707i\frac{\omega}{c}.$$

Suppose the wave is linearly polarized along x and is propagating in the $+z$ direction. Then $E^- = 0$. We take $E^+ = E_0$, where E_0 is real. Then

$$E_x = E_0 e^{i(kz-\omega t)} = E_0 e^{-0.707(\omega/c)z} e^{i\omega[1.225z/c)-t]}$$

$$B_y = \frac{n}{c}E_x = \frac{\sqrt{2}}{c}E_x \exp\left(i\frac{\pi}{6}\right).$$

In this example the wave is propagating in the $+z$ direction. Its wavelength (distance over which phase increases by 2π at fixed t) is $(1.225)^{-1}$ times its vacuum wavelength. The wave amplitude decreases exponentially with distance. The magnetic field is $\sqrt{2}$ times larger in magnitude than the electric field, and lags it by 60 deg in phase angle.

Reflection and transmission of plane waves. Suppose media 1 and 2 are different homogeneous media having the plane $z = 0$ as an interface. Medium 1 occupies all space with negative z, medium 2 occupies all space with positive z. A plane wave is generated by a source at $z = -\infty$. This gives an incident wave traveling in the $+z$ direction in medium 1. The discontinuity generates a reflected and a transmitted wave. For simplicity we consider only normal incidence. Let the incident wave be linearly polarized along x and have complex amplitude unity for E_x. Let R_{12} and T_{12} be the complex amplitudes of the reflected and transmitted E_x. Thus we have

$$E_x(1) = 1 \cdot e^{i(k_1 z - \omega t)} + R_{12} e^{-i(k_1 z + \omega t)} \tag{75}$$

$$E_x(2) = T_{12} e^{i(k_2 z - \omega t)}, \tag{76}$$

where E_x (1) is the total (i.e., incident plus reflected) field E_x in medium 1, E_x (2) is the total (i.e., transmitted) field E_x in medium 2, and R_{12} and T_{12} are unknown complex constants to be determined. We can use Eq. (72) to find B_y in both media, once is E_x is known:

$$B_y(1) = \frac{n_1}{c} e^{i(k_1 z - \omega t)} - \frac{n_1}{c} R_{12} e^{-i(k_1 z - \omega t)}, \tag{77}$$

$$B_y(2) = \frac{n_2}{c} T_{12} e^{i(k_2 z - \omega t)}. \tag{78}$$

Boundary conditions at $z = 0$. Since we have a discontinuity at $z = 0$, we should not use Maxwell's equations for a homogeneous medium when we consider the region immediately bordering the plane $z = 0$. Instead we use Maxwell's Eqs. (47) through (50) for a linear isotropic medium. We assume both media are neutral and that there are no surface charges or currents on the plane of discontinuity. The two Maxwell equations of interest are those involving the curl:

$$\vec{\nabla} \times (\mathbf{B}/\mu) = \frac{\partial(\epsilon\vec{\mathbf{E}})}{\partial t} = -i\omega\epsilon\vec{\mathbf{E}} \tag{79}$$

$$\vec{\nabla} \times \vec{\mathbf{E}} = \frac{-\partial\vec{\mathbf{B}}}{\partial t} = +i\omega\vec{\mathbf{B}} \tag{80}$$

where $\mathbf{E} = \hat{\mathbf{x}}E_x$ and $\mathbf{B} = \hat{\mathbf{y}}B_y$ for our problem. According to Stokes' theorem, any vector $\mathbf{C}$ satisfies

$$\int (\nabla \times \mathbf{C}) \cdot d\mathbf{A} = \oint \mathbf{C} \cdot d\mathbf{l}, \tag{81}$$

where $d\mathbf{A}$ is an element of surface area and $d\mathbf{l}$ is a line element along the contour that encloses the area. We apply Eq. (81) to $\mathbf{C} \equiv \hat{\mathbf{y}}(B_y/\mu)$ using a contour that goes out along the $+y$ direction on one side of the plane $z = 0$ and returns along the $-y$ direction on the other side of the plane, with tiny separation ∇z between these two legs of the contour. As. ∇z goes to zero, the area enclosed by the .contour goes to zero. Hence the surface integral on the left side of Eq. (81) goes to zero, provided $\nabla \times \mathbf{C}$ is not infinite. (It isn't.) Therefore the contour integral on the right side of Eq. (81) is zero. Therefore the component of $\mathbf{C}$ tangential to the boundary is the same on each side of the boundary. Thus we find that the tangential component of $\mathbf{B}/\mu$ is the same on either side of the boundary; it is "continuous" at $z = 0$. Similarly the tangential component of $\mathbf{E}$ is continuous at $z = 0$.

Continuity of E_x at $z = 0$ gives [using Eqs. (75) and (76)]

$$1 + R_{12} = T_{12}. \tag{82}$$

Continuity of $H_y = B_y/\mu$ at $z = 0$ gives [using Eqs. (77) and (78)]

$$\frac{n_1}{\mu_1}(1 - R_{12}) = \frac{n_2}{\mu_2} T_{12}. \tag{83}$$

Defining the characteristic impedance (aside from a proportionality constant) as

$$Z = \frac{\mu}{n} = \frac{\mu}{\sqrt{\epsilon\mu}} = \sqrt{\frac{\mu}{\epsilon}} \tag{84}$$

and solving Eqs. (82) and (83) we obtain

$$R_{12} = \frac{Z_2 - Z_1}{Z_2 + Z_1}, \quad T_{12} = 1 + R_{12}. \tag{85}$$

For the special case where the magnetic permeability μ is unity we have $Z = n^{-1}$. Then Eq. (85) becomes

$$R_{12} = \frac{n_1 - n_2}{n_1 + n_2}, T_{12} = 1 + R_{12}. \tag{86}$$

For the special case where medium 1 is vacuum with $n_1 = 1$ and medium 2 is a medium with complex index $n = n_R + in_I$, Eq. (86) gives

$$R_{12} = \frac{1 - n}{1 + n} = \frac{(1 - n_R) - in_I}{(1 + n_R) + in_I} \equiv |R| \exp i\varphi. \tag{87}$$

The amplitude of the reflected wave is $|R|$ times that of the incident wave. The time dependence $\exp(-i\omega t)$ of the incident wave becomes $\exp(-i\omega t + i\varphi)$ for the reflected wave, so that φ is the phase *lag* introduced by the reflection. The fractional intensity is $|R_{12}|^2$, where according to Eq. (87)

$$|R_{12}|^2 = \frac{(1 - n_R)^2 + n_I^2}{(1 + n_R)^2 + n_I^2}. \tag{88}$$

Example: Simple model of dispersion relation of conductor

We assume we can apply our simple model. We take the spring constant $M\omega_0^2$ to be zero. That means that the "average charges" obey the equation of motion

$$\ddot{x} + \Gamma\dot{x} = \frac{q}{M} E_x. \tag{89}$$

We first consider a steady electric field turned on suddenly at $t = 0$. The velocity $\dot{x}$ builds up exponentially with time until it reaches "terminal velocity" given by setting $\ddot{x} = 0$ in Eq. (89). For a "DC" (constant) field E_x turned on at $t = 0$ the solution of Eq. (89) is

$$\dot{x} = \frac{qE_x}{\Gamma M}(1 - e^{-\Gamma t}), \quad \dot{x} = \frac{qE_x}{\Gamma M} \quad \text{when } t \gg \Gamma^{-1}. \tag{90}$$

Thus Γ, which has dimensions of frequency, gives the "rate of reaching terminal velocity." Putting it differently, Γ^{-1} is the mean relaxation time of "transient" currents when the electric field is suddenly changed to a new constant value.

"Purely resistive" frequency range. For "small" ω (meaning for ω small compared to Γ), the charges will always be at essentially the terminal velocity appropriate to the instantaneous field E_x. In that case the phase relation between $\dot{x}$ and E_x is practically the same as for zero frequency (i.e., for DC). The medium is then said to be *purely resistive*. Then Eq. (90) gives

$$\dot{x}(t) = \frac{qE_x(t)}{\Gamma M}, \qquad \omega \ll \Gamma. \tag{91}$$

The current density $\mathbf{J}_x$ is then proportional to E_x. (This is called *Ohm's law*.) The "purely resistive" conductivity σ_{DC} is related to Γ as follows:

$$\mathbf{J}_x = Nq\dot{x} = Nq\left(\frac{qE_x}{\Gamma M}\right) \equiv \sigma_{\mathrm{DC}}E_x, \tag{92}$$

i.e.,

$$\sigma_{\mathrm{DC}} = \frac{Nq^2}{\Gamma M}, \qquad \omega \ll \Gamma. \tag{93}$$

For the general case of arbitrary frequency the velocity $\dot{x}$ will have not only a component in phase with E_x, as for the DC case, but will also have a component in quadrature with E_x. Then we use complex quantities, all having time dependence $\exp(-i\omega t)$. The steady-state solution of Eq. (89) is easily obtained. [Set $\omega_0 = 0$ in Eq. (44). The complex conductivity $\sigma(\omega)$ is then given by

$$\mathbf{J}_x = Nq\dot{x} = Nq(-i\omega x) = -i\omega P_x = -i\omega\epsilon_0\chi E_x \equiv \sigma(\omega)E_x. \tag{94}$$

Thus

$$\sigma(\omega) = -i\omega\epsilon_0\chi = -i\omega\epsilon_0(\chi_{\mathrm{el}} + i\chi_{\mathrm{ab}}) = \epsilon_0\omega\chi_{\mathrm{ab}} - i\omega\epsilon_0\chi_{\mathrm{el}}. \tag{95}$$

We see that if $\sigma(\omega)$ is real, $\dot{x}$ is in phase with E_x, and σ is proportional to the absorptive electric susceptibility.

Rather than separate $\chi(\omega)$ or $\sigma(\omega)$ into their real and imaginary parts, it is less cumbersome to write them with complex denominator as in Eq. (45). Thus we have [setting $\omega_0 = 0$ in Eq. (45)]

$$\chi(\omega) = \frac{Nq^2}{\epsilon_0 M} \cdot \frac{1}{-\omega^2 - i\omega\Gamma} \tag{96}$$

$$\sigma(\omega) = -i\omega\epsilon_0\chi(\omega) = \frac{Nq^2}{M} \cdot \frac{i\omega}{\omega^2 + i\omega\Gamma}. \tag{97}$$

In the limit $\omega \ll \Gamma$ we can neglect ω^2 compared with $\omega\Gamma$, so that in the *purely resistive* or DC limit we have

$$\chi(\omega) = i\frac{Nq^2}{\epsilon_0 M}\frac{1}{\omega\Gamma}, \qquad \omega \ll \Gamma \tag{98}$$

and

$$\sigma(\omega) = \frac{Nq^2}{M\Gamma} = \sigma(0) = \sigma_{\mathrm{DC}}, \qquad \omega \ll \Gamma. \tag{99}$$

We see that for the purely resistive frequency range $0 \lesseqgtr \omega \ll \Gamma$, the conductivity $\sigma(\omega)$ is real and is equal to its DC (zero frequency) value $\sigma(0)$. The velocity $\dot{x}$ is then in phase with E_x.

The complex electric susceptibility $\chi(\omega)$ is pure imaginary for $\omega \ll \Gamma$ according to Eq. (98). The complex square of the index of refraction, n^2, is then given for $\omega \ll \Gamma$ by

$$n^2 = 1 + \chi = 1 + i \frac{Nq^2}{\epsilon_0 M} \frac{1}{\omega \Gamma} = 1 + i \frac{\omega_p^{\,2}}{\omega \Gamma}, \qquad (100)$$

where

$$\omega_p^{\,2} \equiv \frac{Nq^2}{\epsilon_0 M}. \qquad (101)$$

There are two limiting cases of a "purely resistive medium" with qualitatively different physical characteristics.

Case 1: "Dilute resistive medium"

That means ω, Γ, and ω_p satisfy the relations

$$\omega_p \ll \Gamma, \qquad \frac{\omega_p^{\,2}}{\Gamma} \ll \omega \ll \Gamma. \qquad (102)$$

Then according to Eq. (100) we have

$$n = \left[1 + i \frac{\omega_p^{\,2}}{\omega \Gamma} \right]^{1/2} \approx 1 + \frac{1}{2} i \frac{\omega_p^{\,2}}{\omega \Gamma}, \qquad (103)$$

with neglect of higher-order terms. Then

$$k = n \frac{\omega}{c} = \frac{\omega}{c} + i \frac{1}{2} \frac{\omega_p^{\,2}}{c \Gamma} = \frac{\omega}{c} + i \frac{\sigma_{dc}}{2 \epsilon_0 c}, \qquad (104)$$

where we used Eqs. (101) and (93) in the last equality. The real part of k is equal to ω/c, just as in vacuum. The imaginary part is much smaller than the real part. The imaginary part of k gives an exponential attenuation of a traveling plane wave. The mean attenuation length is large compared with one wavelength. The *intensity* of the plane wave is proportional to the absolute square of the complex amplitude. Therefore it is exponentially attenuated with distance by a factor $\exp(-2k_I z)$, where k_I is the imaginary part of k. The distance $d \equiv (2k_I)^{-1}$ in which the *intensity* is attenuated by a factor e^{-1} is given by Eq. (104):

$$\frac{1}{d} \equiv 2k_I = \frac{\sigma_{dc}}{\epsilon_0 c}, \qquad i.e., \frac{\rho_{dc}}{d} = \frac{1}{\epsilon_0 c}. \qquad (105)$$

The "resistance per square" of a square slab of dilute resistive medium having slab thickness d and edge lengths L is the DC resistivity

divided by d. That equals $\dfrac{1}{\epsilon_0 c} = \dfrac{1}{8.85 \times 10^{-12} \times 3 \times 10^8} \approx 377\Omega$
ohms per square, according to Eq. (105). You may recall that 377 ohms per square is the characteristic impedance for "perfect termination" of an electromagnetic plane wave. (See Chap. 5.) Of course the wave is not absorbed in just one exponential decay length d of the intensity. However practically none is reflected, and it is all eventually absorbed.

More precisely, since the real part of n is essentially unity and the imaginary
part is small compared with unity, the fractional reflected intensity for a plane wave normally incident from vacuum is given by

$$|R|^2 = \frac{(n_R - 1)^2 + n_I^2}{(n_R + 1)^2 + n_I^2} \approx \frac{0 + n_I^2}{2^2 + n_I^2} \approx \frac{n_I^2}{4} \ll 1. \qquad (106)$$

Using Eqs. (103) and (105) this becomes

$$|R|^2 \approx \frac{1}{16}\left(\frac{\omega_p^2}{\omega\Gamma}\right)^2 = \left(\frac{\lambdabar}{4d}\right)^2 \ll 1, \qquad (107)$$

where $\lambdabar \equiv c/\omega$ is the "reduced" wavelength in vacuum.

Case 2: "Dense resistive medium"

That means we have the relations

$$\omega \ll \Gamma, \qquad \omega \ll \omega_p, \qquad \omega\Gamma \ll \omega_p^2. \qquad (108)$$

Then n^2 is essentially pure imaginary, according to Eq. (100). When we take the square root of n^2, we use the fact that the square root of i is $[\exp{(i\tfrac{1}{2}\pi)}]^{1/2} = \exp{(i\tfrac{1}{4}\pi)}$, which is $2^{-(1/2)}(1 + i)$. That gives

$$n = \left[i\frac{\omega_p^2}{\omega\Gamma}\right]^{1/2} = \left(\frac{\omega_p^2}{2\omega\Gamma}\right)^{1/2}(1 + i) = |n|\frac{(1 + i)}{\sqrt{2}} \qquad (109)$$

Then

$$k = n\frac{\omega}{c} = \sqrt{\frac{\omega}{c}}\left(\frac{\omega_p^2}{2c\Gamma}\right)^{1/2}(1 + i)$$

$$= \sqrt{\frac{\omega}{c}}\left(\frac{\alpha_{dc}}{2\epsilon_0 c}\right)^{1/2}(1 + i), \qquad (110)$$

Then the real and imaginary parts of k are equal. Each is large compared to the vacuum value of k (that is, ω/c). The mean penetration distance for the amplitude, k_I^{-1}, is small compared with the vacuum wavelength. It then turns out that a plane wave incident from vacuum to a dense resistive medium is reflected with practically no absorption. That is because the penetration distance is so small that relatively few charges feel any electric field. Those which *do* feel it are at terminal velocity, in phase with E_x, and hence are absorbing energy. However they are so few that the wave "escapes" with little loss of intensity.

More precisely, the fractional reflected intensity is given by

$$|R|^2 = \frac{(n_R - 1)^2 + n_I^2}{(n_R + 1)^2 + n_I^2} \approx \frac{|n|^2 - 2n_R}{|n|^2 + 2n_R} = \frac{|n|^2 - \sqrt{2}\,|n|}{|n|^2 - \sqrt{2}\,|n|}$$

$$\approx 1 - \frac{2\sqrt{2}}{|n|} = 1 - 2\sqrt{2}\left(\frac{\omega\Gamma}{\omega_p{}^2}\right)^{1/2} \qquad (111)$$

Thus $|R|^2 \approx 1$ since $\omega\Gamma \ll \omega_p{}^2$.

The *e*-fold attenuation length for the intensity, $d \equiv (2k_I)^{-1}$, is given by

$$d = \lambdabar\sqrt{\frac{\omega\Gamma}{2\omega_p{}^2}} \ll \lambdabar.$$

Although this d is small compared with the wavelength, it is greater by a factor $(\lambdabar/2d)$ than the thickness that gives 377 ohms per square for DC. Therefore the impedance is small compared with that which gives perfect termination. This is why the sign of E_x reverses upon reflection.

We see that there is a great qualitative difference between a dilute and a dense resistive medium. A dilute resistive medium is essentially "black." It is almost completely absorbing. By contrast, a dense resistive medium acts like a "lumped" impedance that is very small. It gives almost complete reflection.

Finally, we must remember that our descriptive phrases are merely names for the conditions expressed by the inequalities (102) and (108). These names omit the important fact that a given conductor has different properties, depending on the frequency. For example, according to Eq. (108), any conductor acts like a dense resistive medium if ω is sufficiently small. On the other hand, a conductor cannot be a "dilute" resistive medium at any frequency unless it has $\Gamma \gg \omega_p$. If that condition is satisfied, it is a dilute resistive medium only in the frequency range given by Eq. (102).

Purely elastic frequency range.   The equation of motion for a single average charge is given by Eq. (89). For complex time dependence $\exp(-i\omega t)$ this equation can be written

$$-i\omega\dot{x} + \Gamma\dot{x} = \frac{q}{M}E_x. \qquad (112)$$

The purely resistive frequency range just considered was that for which we could neglect ω compared with Γ. The purely elastic range is that for which ω is very large compared with Γ. Thus for a *purely elastic* frequency range we have

$$\dot{x} = \frac{iq}{\omega M} E_x, \qquad \omega \gg \Gamma. \tag{113}$$

Then the velocity is in quadrature with the force and no net work is done on the charge in one cycle. There is no absorption. The complex conductivity is a pure imaginary given by [using Eq. (113)]

$$J_x = Nq\dot{x} = i \frac{Nq^2}{\omega M} E_x \equiv \sigma(\omega) E_x,$$

i.e.,

$$\sigma(\omega) = i \frac{Nq^2}{\omega M}, \qquad \omega \gg \Gamma. \tag{114}$$

[See also Eq. (97), with neglect of $\omega\Gamma$ compared with ω^2.]

The square of the complex index of refraction, n^2, is given by

$$n^2 = 1 + \chi = 1 - \frac{Nq^2}{\epsilon_0 M \omega^2} = 1 - \frac{\omega_p^2}{\omega^2}, \qquad \omega \gg \Gamma. \tag{115}$$

There are two qualitatively different purely elastic frequency ranges.

Case 1: Dispersive frequency range

That means we have

$$\Gamma \ll \omega_p \lesssim \omega. \tag{116}$$

Then according to Eq. (115) we have

$$0 \leq n^2 < 1, \tag{117}$$

i.e.,

$$0 \leq n \leq 1. \tag{118}$$

Thus for a conductor in its dispersive frequency range the index n is real and lies between 0 and 1. The medium is transparent. There is no absorption. The phase velocity is greater than c. The fractional reflected intensity is $(n-1)^2/(n+1)^2$.

Case 2: Reactive frequency range

That means

$$\Gamma \ll \omega \lesssim \omega_p. \tag{119}$$

Then Eq. (115) gives

$$-\frac{\omega_p^2}{\Gamma^2} \ll n^2 \leq 0. \tag{120}$$

Thus n^2 is negative and n is a pure imaginary:

and
$$n = i|n| = i\left[\frac{\omega_p^2}{\omega^2} - 1\right]^{1/2},$$

$$k = n\frac{\omega}{c} = i\frac{\omega}{c}|n| = i|k|.$$

A plane wave in a reactive medium has the form

$$E_x = \left[A^+ e^{-|k|z} + A^- e^{+|k|z}\right] e^{-i\omega t}.$$

If the medium extends to $z = +\infty$ then A^- is zero. Thus a plane wave incident from vacuum to such a medium must be totally reflected without absorption. More precisely, the fractional reflected intensity is given by

$$|R|^2 = \frac{(n_R - 1)^2 + n_I^2}{(n_R + 1)^2 + n_I^2} \approx \frac{1 + n_I^2}{1 + n_I^2} = 1.$$

In the main text we avoided discussion of complex index of refraction and complex wave number k by avoiding discussion of absorptive media. For the reactive frequency range we used the symbol κ in place of what we are now calling the magnitude of complex k, $|k|$, for the reactive range. For the dispersive range we used k, which corresponds to our present complex k when it is real.

Summary of properties of conductors. We can now summarize the properties of any conductor (to the extent that our simple model works):

(i) For sufficiently low frequency any conductor is a dense resistive medium. It then gives practically complete reflection with very little absorption.

(ii) For sufficiently high frequency any conductor is a dispersive medium. It is then transparent.

Conductors can be roughly divided into three classes, according to the relative magnitude of the rate of reaching terminal velocity, Γ, and the plasma oscillation frequency, ω_p.

(i) A conductor with $\Gamma \gg \omega_p$ has a frequency range for which it is a dilute resistive medium. In that range it can absorb a wave without reflection. Such a conductor has no purely reactive frequency range. Thus such a conductor cannot give total reflection at any frequency.

(ii) A conductor with $\Gamma \ll \omega_p$ has a frequency range for which it is a purely reactive medium. In that range it can give total reflection without absorption. It has no frequency range for which it is a dilute resistive medium. It can therefore never absorb a plane wave without reflection.

(iii) A conductor with $\Gamma \approx \omega_p$ has no frequency range for which it is a dilute resistive medium, nor does it have a range for which it is a purely reactive medium. Of course, it still has the general property that for sufficiently low ω it is a dense resistive medium and for sufficiently high ω it is transparent.

Application: Solid silver

Assume solid silver can be approximated by our simple model. The movable charges are the "conduction electrons," which are supplied by the "valence electrons" of silver atoms. The valence is unity. The atomic weight is 107.9 gm/mole. The mass density is 10.5 gm/cm³. Avogadro's number is 6×10^{23} per mole. Then N is $(6 \times 10^{23})(10.5)/(107.9) = 5.8 \times 10^{22}$. Assuming M and q are the mass and charge of a free electron we find

$$\omega_p = \sqrt{\frac{Ne^2}{\epsilon_0 M}} = 1.36 \times 10^{16} \, \text{rad/sec.}$$

The resistivity ρ_{dc} is 1.59×10^{-6} ohm cm. Then the rate of reaching terminal velocity, Γ, is given by

$$\Gamma = \frac{Ne^2}{\epsilon_0 M \sigma_{dc}} = \omega_p^2 \rho_{dc} = 2.7 \times 10^{13} \, \text{sec}^{-1}.$$

We see that, for solid silver, $\Gamma \ll \omega_p$. For $\omega \ll 2.7 \times 10^{13}$ rad/sec silver is a dense resistive medium, according to the model. (This is the case for microwaves, for example.) For $\omega \gg 2.7 \times 10^{13}$ rad/sec it is purely elastic. For the purely elastic range having $\omega < 1.36 \times 10^{16}$ rad/sec it is purely reactive. (That range includes visible light.) For the purely elastic range having $\omega > 1.36 \times 10^{16}$ rad/sec it is transparent. (That is the far ultraviolet and x-ray region.) Of course, real silver does not follow this model exactly. (For one thing, we have neglected contributions from the "bound" electrons.)

Application: Graphite

We assume the valence is 4, density is 2.0, atomic weight is 12. Then the simple model gives

$$\omega_p = 0.36 \times 10^{17} \, \text{rad/sec.}$$

The resistivity ρ_{DC} is 1.57×10^{-15} statohm cm. This gives

$$\Gamma = 1.6 \times 10^{17} \text{ sec}^{-1}.$$

For $\omega \ll 1.6 \times 10^{17}$ rad/sec graphite is purely resistive, according to the model. For $\omega \ll 8 \times 10^{15}$ rad/sec it is a dense resistive medium. For $8 \times 10^{15} \ll 1.6 \times 10^{17}$ it is a dilute resistive medium. Since that range only covers a factor of 20 in frequency, both inequalities cannot be well satisfied, so graphite is not very dilute at any frequency, and hence is not "completely black" at any frequency. Graphite has no reactive frequency range. For $\omega \gg 1.6 \times 10^{17}$ it is transparent, according to the model.

Let us predict the reflectivity $|R|^2$ for visible light on idealized graphite. For green light of vacuum wavelength 5500 Å we have $\omega = 2(3.14)(3 \times 10^{10})/(5.5 \times 10^{-5}) = 3.42 \times 10^{15}$ rad/sec. This is not in the "dense resistive medium" frequency range, given by $\omega \ll 8 \times 10^{15}$. Thus we do not expect nearly 100% reflectivity. Neither do we expect very small reflectivity. We have

$$n^2 = \epsilon_r = \epsilon_{rR} + i\epsilon_{rI},$$

$$\epsilon_R = 1 + \frac{\omega_p^2(\omega_0^2 - \omega^2)}{(\omega_0^2 - \omega^2) + \Gamma^2\omega^2} = 1 - \frac{\omega_p^2}{\omega^2 + \Gamma^2}$$

$$= 1 - \frac{(36)^2}{(3.42)^2 + (160)^2} = 0.951$$

$$\epsilon_I = \frac{\omega_p^2 \Gamma \omega}{(\omega_0^2 - \omega^2)^2 + \Gamma^2\omega^2} = \frac{\omega_p^2(\Gamma/\omega)}{\omega^2 + \Gamma^2}$$

$$= \frac{160}{3.42} \frac{(36)^2}{(3.42)^2 + (160)^2} = 2.36$$

$$n^2 = 0.951 + 2.36i = 2.55 \ \exp i\varphi,$$

where

$$\varphi = \tan^{-1} \frac{2.36}{0.951} \approx 68 \text{ deg}.$$

Then

$$n = \sqrt{2.55} \ \exp\left(i \tfrac{1}{2} \varphi\right) = 1.60(\cos 34° + i \sin 34°) = 1.33 + i0.90.$$

Then

$$|R|^2 = \frac{(n_R - 1)^2 + n_I^2}{(n_R + 1)^2 + n_I^2} = \frac{(.33)^2 + (.90)^2}{(2.33)^2 + (.90)^2} = 0.15.$$

Thus according to the model a sheet of polished graphite reflects about 15% of the intensity of normally incident visible green light.

Appendix

A.1 Taylor's Series

We assume that $f(x)$ can be written as an infinite series of the form

$$f(x) = c_0 + c_1(x - x_0) + c_2(x - x_0)^2 + c_3(x - x_0)^3 + \dots, \qquad (1)$$

where the c's are constants. Then $f(x)$ is said to be expressed as "an expansion at point x_0. To find c_0, we set $x = x_0$; then all terms on the right-hand side vanish except the first. Thus $c_0 = f(x_0)$. To find c_1, we differentiate Eq. (1) once with respect to x and then set $x = x_0$. All terms vanish except the c_1 term; thus we find $c_1 = (df/dx)_0$, where the subscript zero means that df/dx is evaluated at $x = x_0$. Similarly,

$$(d^m f/dx^m)_0 = m!\,c_m, \qquad (2)$$

and Eq. (1) becomes

$$f(x) = f(x_0) + (x - x_0)\left(\frac{df}{dx}\right)_0 + \frac{(x - x_0)^2}{2!}\left(\frac{d^2 f}{dx^2}\right)_0$$
$$+ \frac{(x - x_0)^3}{3!}\left(\frac{d^3 f}{dx^3}\right)_0 + \cdots \qquad (3)$$

A.2 Commonly Used Series

sin x and cos x. We use $d(\sin x)/dx = \cos x$, $d(\cos x)/dx = -\sin x$, $\cos(0) = 1$, $\sin(0) = 0$, and $x_0 = 0$ in Eq. (3) to obtain

$$\sin x = x - \frac{x^3}{3!} + \frac{x^5}{5!} - \cdots, \qquad (4)$$

$$\cos x = 1 - \frac{x^2}{2!} + \frac{x^4}{4!} - \cdots. \qquad (5)$$

Exponential e^{ax}. We use $d(e^{ax})/dx = ae^{ax}$, $e^0 = 1$, and $x_0 = 0$ in Eq. (3) to obtain

$$e^{ax} = 1 + ax + \frac{a^2 x^2}{2!} + \frac{a^3 x^3}{3!} + \frac{a^4 x^4}{4!} + \cdots. \qquad (6)$$

sinh x and cosh x. These functions may be defined by $d(\sinh x)/dx = \cosh x$, $d(\cosh x)/dx = \sinh x$, $\sinh(0) = 0$, $\cosh(0) = 1$, and Eq. (3) with $x_0 = 0$ to give

$$\sinh x = x + \frac{x^3}{3!} + \frac{x^5}{5!} + \cdots, \qquad (7)$$

$$\cosh x = 1 + \frac{x^2}{2!} + \frac{x^4}{4!} + \cdots. \qquad (8)$$

Relationships involving the exponential. If we set $a = +1$ in Eq. (6) and compare with Eqs. (7) and (8), and if we then do the same for $a = -1$, we find

$$e^x = \cosh x + \sinh x, \tag{9}$$

$$e^{-x} = \cosh x - \sinh x, \tag{10}$$

which may be solved to obtain

$$\cosh x = \frac{e^x + e^{-x}}{2}, \tag{11}$$

$$\sinh x = \frac{e^x - e^{-x}}{2}. \tag{12}$$

If we set $a = +i \equiv + \sqrt{-1}$ in Eq. (6), we obtain

$$e^{ix} = 1 + ix - \frac{x^2}{2!} - \frac{ix^3}{3!} + \frac{x^4}{4!} + \frac{ix^5}{5!} - \frac{x^6}{6!} + \cdots. \tag{13}$$

Similarly, if we set $a = -i$ in Eq. (6), we obtain

$$e^{-ix} = 1 - ix - \frac{x^2}{2!} + \frac{ix^3}{3!} + \frac{x^4}{4!} - \frac{ix^5}{5!} - \frac{x^6}{6!} + \cdots. \tag{14}$$

By adding or subtracting Eqs. (13) and (14) and comparing the results with Eqs. (4) and (5), we obtain

$$\frac{e^{ix} + e^{-ix}}{2} = \cos x, \tag{15}$$

$$\frac{e^{ix} - e^{-ix}}{2i} = \sin x, \tag{16}$$

which may be solved to give

$$e^{ix} = \cos x + i \sin x, \tag{17}$$

$$e^{-ix} = \cos x - i \sin x. \tag{18}$$

tan x. We use $\tan x \equiv \sin x/\cos x$, $d(\sin x)/dx = \cos x$, and $d(\cos x)/dx = -\sin x$ to obtain $d(\tan x)/dx = (\cos x)^{-2}$, $d^2(\tan x)/dx^2 = 2 \sin x (\cos x)^{-3}$, $d^3(\tan x)/dx^3 = 2(\cos x)^{-2} - 6 \sin^2 x (\cos x)^{-4}$, etc. Then we use $x_0 = 0$ in Eq. (3) to obtain

$$\tan x = x + \frac{x^3}{3} + \frac{2x^5}{15} + \cdots. \tag{19}$$

Appendix

Binomial series $(1 + x)^n$. We use $d(1+x)^n/dx = n(1+x)^{n-1}, d^2(1+x)^n/dx^2 = n(n-1)(1+x)^{n-2}, d^3(1+x)^n/dx^3 = n(n-1)(n-2)(1+x)^{n-3}$, etc., and Eq. (3) with $x_0 = 0$ to obtain

$$(1 + x)^n = 1 + nx + \frac{n(n-1)x^2}{2!} + \frac{n(n-1)(n-2)x^3}{3!} + \cdots. \quad (20)$$

Equation (20) holds for any n, positive or negative, and for any x, positive or negative, as long as x satisfies the relation $x^2 < 1$.

A.3 *Superposition of Harmonic Functions*

The following superpositions of N harmonic functions are encountered in wave phenomena:

$$\begin{aligned} u(t) = \cos \omega_1 t + \cos (\omega_1 + a)t + \cos (\omega_1 + 2a)t \\ + \cdots + \cos [\omega_1 + (N-1)a]t; \end{aligned} \quad (21)$$

$$\begin{aligned} u(z) = \cos kz + \cos (kz + \beta) + \cos (kz + 2\beta) \\ + \cdots + \cos [kz + (N-1)\beta]. \end{aligned} \quad (22)$$

These are both of the form

$$\begin{aligned} u = \cos \theta_1 + \cos (\theta_1 + \gamma) + \cos (\theta_1 + 2\gamma) \\ + \cdots + \cos [\theta_1 + (N-1)\gamma]. \end{aligned} \quad (23)$$

We wish to find a convenient expression for Eq. (23). We note that u can be written as the real part of v, where

$$v = e^{i\theta_1} + e^{i(\theta_1 + \gamma)} + e^{i(\theta_1 + 2\gamma)} + \cdots + e^{i[\theta_1 + (N-1)\gamma]} = e^{i\theta_1}S, \quad (24)$$

where S is a geometric series of N terms, given by

$$S = 1 + a + a^2 + a^3 + \cdots + a^{N-1}, \quad \text{with } a = e^{i\gamma}. \quad (25)$$

Multiply S by a. Then subtract S from aS, term by term, to get

$$aS - S = a^N - 1, \quad (26)$$

i.e.,

$$\begin{aligned} S &= \frac{a^N - 1}{a - 1} = \frac{e^{iN\gamma} - 1}{e^{i\gamma} - 1} = \frac{e^{(1/2)iN\gamma}}{e^{(1/2)i\gamma}} \frac{(e^{(1/2)iN\gamma} - e^{-(1/2)iN\gamma})}{(e^{(1/2)i\gamma} - e^{-(1/2)i\gamma})} \\ &= e^{(1/2)i(N-1)\gamma} \frac{\sin \frac{1}{2} N\gamma}{\sin \frac{1}{2} \gamma}, \end{aligned} \quad (27)$$

where we used Appendix Eq. (16) in the last step. Inserting Eq. (27) into Eq. (24) gives

$$v = e^{i[\theta_1 + (1/2)(N-1)\gamma]} \frac{\sin \frac{1}{2} N\gamma}{\sin \frac{1}{2}\gamma}. \quad (28)$$

Taking the real part, we obtain the desired result,

$$u = \cos \left[\theta_1 + \frac{1}{2} (N-1)\gamma \right] \frac{\sin \frac{1}{2} N\gamma}{\sin \frac{1}{2} \gamma}. \quad (29)$$

The result, Eq. (29), can be put in another useful form. In Eq. (23) θ_1 is the argument of the first term, and the argument of the last term, which we shall call θ_2, is given by

$$\theta_2 \equiv \theta_1 + (N-1)\gamma. \tag{30}$$

The average of the first and last arguments θ_1 and θ_2 is then

$$\theta_{av} \equiv \tfrac{1}{2}(\theta_1 + \theta_2) = \tfrac{1}{2}\theta_1 + \tfrac{1}{2}\theta_1 + \tfrac{1}{2}(N-1)\gamma. \tag{31}$$

Thus the first factor in Eq. (29) is just $\cos\theta_{av}$. Using this and the fact that γ is $(\theta_2 - \theta_1)/(N-1)$ [according to Eq. (30)], we write Eq. (29) in the form

$$u = \cos\theta_{av}\,\frac{\sin\left[\tfrac{1}{2}N(\theta_2 - \theta_1)/(N-1)\right]}{\sin\left[\tfrac{1}{2}(\theta_2 - \theta_1)/(N-1)\right]}. \tag{32}$$

Equation (29) emphasizes the increment γ between the arguments of successive terms in the sum, Eq. (23). Equation (32) is equivalent to Eq. (29) but emphasizes the first and last contributions, θ_1 and θ_2, and their average. Note that $\cos\theta_{av}$ is a harmonic oscillation of the same form as each of the terms in the superposition, Eq. (23); but instead of amplitude unity, it has amplitude $A(\theta_1, \theta_2, N)$ given by the expression

$$A(\theta_1, \theta_2, N) = \frac{\sin\left[\tfrac{1}{2}N(\theta_2 - \theta_1)/(N-1)\right]}{\sin\left[\tfrac{1}{2}(\theta_2 - \theta_1)/(N-1)\right]}. \tag{33}$$

The most compact expression of our result is then

$$u = A(\theta_1, \theta_2, N)\cos\theta_{av}. \tag{34}$$

The case $N = 2$ corresponds for oscillations in time [Eq. (21)] to the phenomenon of "beats" and for oscillations in space [Eq. (22)] to the two-slit interference pattern. For oscillations in time, larger N gives "modulations," which produce a "pulse" behavior in $u(t)$ in the limit as $N \to \infty$. For oscillations in space, larger N gives the multiple-slit interference pattern, and the limit $N \to \infty$ gives the single-slit diffraction pattern due to a single slit that is many wavelengths wide.

A.4   Vector Identities

We shall use A, B, and C to stand for scalar functions of x, y, and z, that is, $A(x, y, z)$, $B(x, y, z)$, and $C(x, y, z)$. Similarly $\mathbf{A}$, $\mathbf{B}$, and $\mathbf{C}$ stand for vector functions of x, y, and z. Thus $\mathbf{A}$ means $\hat{\mathbf{x}}\,A_x(x, y, z) + \hat{\mathbf{y}}\,A_y(x, y, z) + \hat{\mathbf{z}}\,A_z(x, y, z)$, where $\hat{\mathbf{x}}$, $\hat{\mathbf{y}}$, and $\hat{\mathbf{z}}$ are unit vectors. We want to learn how to work with the object ∇ (pronounced "del") that is both a vector and a "take the derivative" operator. The trick is to write equations involving del so as to satisfy both its vector aspects and its "take the derivative" aspects. For example, in

$$\nabla(AB) = (\nabla A)B + A(\nabla B) = B\nabla A + A\nabla B, . \qquad (35)$$

the first equality comes from the rule for differentiation of a product: first B is held constant, then A. The second equality gets rid of the parentheses, because, by convention, del differentiates only what is to the right of it . We can temporarily symbolize this by writing del as ∇_a when it is to operate only on $\mathbf{A}$ (or A) and ∇_b when it is to operate on B (or $\mathbf{B}$). In that way we take care of the product differentiation rule by adding subscripts. Then we move the operators and vectors around so that quantities not to be differentiated are "safely" on the left side of del, making sure that we satisfy the vector rules as we do so. Finally we erase the subscripts a and b. Thus

$$\nabla(AB) = \nabla_a(AB) + \nabla_b(AB) = B\nabla_a A + A\nabla_b B = B\nabla A + A\nabla B. \quad (36)$$

Similarly

$$\nabla \times (\mathbf{AB}) = \nabla_a \times (\mathbf{AB}) + \nabla_b \times (\mathbf{AB}) = B\nabla_a \times \mathbf{A} - \mathbf{A} \times \nabla_b B$$

$$= B\nabla \times \mathbf{A} - \mathbf{A} \times \nabla B. \qquad (37)$$

After some practice, you don't need to write the intermediate steps.

Now we wish to find an identity for $\nabla \times (\nabla \times \mathbf{C})$. We assume you know the identity

$$\mathbf{A} \times (\mathbf{B} \times \mathbf{C}) = \mathbf{B}\,(\mathbf{A} \cdot \mathbf{C}) - \mathbf{C}(\mathbf{A} \cdot \mathbf{B}) \quad \text{("back minus cab" rule)} \quad (38a)$$

$$= \mathbf{B}\,(\mathbf{A} \cdot \mathbf{C}) - (\mathbf{A} \cdot \mathbf{B})\mathbf{C}. \qquad (38b)$$

We can use this rule, substituting del for $\mathbf{A}$ and del for $\mathbf{B}$. We must be careful to keep both dels on the left side of $\mathbf{C}$, because both dels are supposed to differentiate $\mathbf{C}$. Thus we cannot use Eq. (38a); we must use Eq. (38b). Then we get

$$\nabla \times (\nabla \times \mathbf{C}) = \nabla(\nabla \cdot \mathbf{C}) - (\nabla \cdot \nabla)\mathbf{C}. \qquad (39)$$

In terms of x, y, and z components, Eq. (39) means

$$[\nabla \times (\nabla \times \mathbf{C})]_x = \frac{\partial(\nabla \cdot \mathbf{C})}{\partial x} - \nabla^2 C_x, \qquad (40)$$

with similar expressions for y and z, where

$$\nabla^2 \equiv \frac{\partial^2}{\partial x^2} + \frac{\partial^2}{\partial y^2} + \frac{\partial^2}{\partial z^2}. \qquad (41)$$

Supplementary Reading

General References

(The parenthetical cross reference at the end of an entry refers to a chapter or problem in Vol. III of the Berkeley Physics Series.)

American Institute of Physics, Selected Reprints, *Polarized Light* (American Institute of Physics, New York, 1963). Reprints of 18 articles chosen for their interest and importance.

American Institute of Physics, Selected Reprints, *Quantum and Statistical Aspects of Light* (American Institute of Physics, New York, 1963). Includes the Brown and Twiss experiment mentioned in the text.

Arthur H. Benade, *Horns, Strings and Harmony* (Anchor Books, Science Study Series S 11, Doubleday & Company, Inc., Garden City, N.Y., 1960). A delightful book by a flute –playing physicist.

George H. Darwin, *The Tides, and Kindred Phenomena in the Solar System* (W. H. Freeman & Company, San Francisco, 1962). This popular classic, written in 1898, has fascinating descriptions of seiches in Lake Geneva, tidal "bores," the past and future history of the earth and moon as deciphered from the tides, etc.

Donald G. Fink and David M. Lutyens, *The Physics of Television* (Anchor Books, Science Study Series S 8, Doubleday & Company, Inc., Garden City, N.Y., 1960).

Winston Kock, *Sound Waves and Light Waves* (Anchor Books, Science Study Series S 40, Doubleday & Company, In c., Garden City, N.Y., 1965).

E. H. Land, "Some Aspects of the Development of Sheet Polarizers," *J. Opt. Soc. Am.* **41,** 957 (1951).

M. Minnaert, *Light and Colour in the Open Air* (Dover Publications, Inc., New York, 1954). A classic full of "outdoor home experiments." (A general reference, but see also Chap. 8 and Supplementary Topic 8.)

Physical Science Study Committee, *Physics*, 2nd ed. (D. C. Heath & Company, Boston, Mass., 1965).

John R. Pierce, *Electrons and Waves* (Anchor Books, Science Study Series S 38, Doubleday & Company, Inc., Garden City, N.Y., 1964). A fine introduction to electronics and communication by a physicist who has made notable contributions to both of these arts.

William A. Shurcliff and Stanley S. Ballard, *Polarized Light* (Momentum Book 7, D. Van Nostrand Company, Inc., Princeton, N.J., 1964). Fascinating and dramatic examples of the production and use of polarized light in many branches of physics.

Ivan Simon, *Infrared Radiation* (Momentum Book 12, D. Van Nostrand Company, Inc., Princeton, N.J., 1966).

Alex G. Smith and Thomas D. Carr, *Radio Exploration of the Planetary System* (Momentum Book 2, D. Val) Nostrand Company, Inc., Princeton, N.J., 1964).

Elizabeth A. Wood, *Crystals and Light, An Introduction to Optical Crystallography* (Momentum Book 5, D. Van Nostrand Company, Inc., Princeton, N.J., 1964). Paperback; this is a beautiful little book describing the study of crystals, use of polarization microscopes, etc. The piece of Polaroid pasted inside the book for home experiments actually started me on my binge of inventing home experiments for Vol. III of the Berkeley Physics Series.

Specific References

Reinhard Beer, "Remote Sensing of Planetary Atmospheres by Fourier Spectroscopy," *The Physics Teacher*, p. 151 (April 1968). (See Prob. 6.33.)

G. L. Berge and G. A. Seielstad, "The Magnetic Field of the Galaxy," *Scientific American*, p. 46 (June 1965). (See Chap. 8.)

G. R. Bird and M. Parrish, Jr., *J. Opt. Soc. Am.* **50,** 886 (1960). Evaporation of gold onto a plastic diffraction grating. (See Chap. 8.)

N. Broten, et al., "Long Base-line Interferometry Using Atomic Clocks and Tape Recorders," *Science* **156,** 1592 (June 23, 1967). (See Prob. 9.57.)

R. Hanbury Brown and R. O. Twiss, "The Question of Correlation between Photons in Coherent Light Rays," *Nature* **178,** 1447 (1956). (See Chap. 9.)

B. A. Burgel, "Dispersion, Reflection, and Eigenfrequencies on the Wave Machine," *Am. J. Phys.* **35,** 913 (1967). (See Prob. 4.14.)

W. Calvert, R. Knecht, and T. Van Zandt, "Ionosphere Explorer I Satellite: First Observations from the Fixed-Frequency Topside Sounder," *Science* **146,** 391 (Oct. 16, 1964). (See Chap. 4.)

D. D. Coon, *Am. J. Phys.* **34,** 240 (1966). (See Chap. 7.)

A. de Maria, D. Stetser, and W. Glenn, Jr., "Ultrashort Light Pulses, "*Science* **156,** 1557 (June 23, 1967). (See Prob. 6.23.)

A. J. de Witte, "Interference in Scattered Light," *Am. J. Phys.* **35,** 301 (April 1967). (See Prob. 9.56.)

Rene Dubos, *Pasteur and Modern Science* (Anchor Books, Doubleday & Company, Inc., Garden City, N.Y., 1960). (See Chap. 8.)

R. Feynman, *The Feynman Lectures on Physics*, vol. I, chap. 33 (Addison Wesley, Reading, Mass., 1963). (See Chap. 8.)

A. T. Forrester, R. A. Gudmundsen, and P. O. Johnson, "Photoelectric Mixing of Incoherent Light," *Phys. Rev.* **99,** 1691 (1955). (See Chap. 1 and Supplementary Topic 7.)

J. M. Fowler, J. T. Brooks, and E. D. Lambe, "One-dimensional Wave Demonstration," *Am. J. Phys.* **35,** 1065 (1967). (See Chap. 4.)

Martin Gardner, *The Ambidextrous Universe* (Basic Books, Inc., Publishers, New York, 1964.) An account of handedness in living organisms and in the weak decay interactions of elementary particles. (See Chap. 8.)

J. A. Giordmaine, "The Interaction of Light with Light," *Scientific American*, p. 38 (April 1964). (See Prob. 1.13.)

T. Gold, "Radio Method for the Precise Measurement of the Rotation Period of the Earth," *Science* **157,** 302 (July 21, 1967). (See Prob. 9.57.)

Eckhard H. Hess, "Attitude and Pupil Size," *Scientific American*, p: 46 (April 1965). (See Prob. 9.3.3.)

J. Lovelock , D. Hitchcock, P. Fellgett, J. and P. Connes, L. Kaplan, and J. Ring, "Detecting Planetary Life from Earth," *Science Journal*, p. 56 (April 1967). (See Prob. 6.33.)

G. J. F. MacDonald , "Implications for Geophysics of the Precise Measurement of the Earth's Rotation," *Science* **157,** 304 (July 21, 1967). (See Prob. 9.57.)

J. S. Mayo, "Pulse-Code Modulation," *Scientific American,* p. 102 (March 1968). (Sec. 6.2.)

R. Pfleegor and L. Mandel, "Interference of Independent Photon Beams," *Phys. Rev.* **159,** 1084 (1967). (See Cha p. 9)

F. Press and D. Harkrider, "Air-Sea Waves from the Explosion of Krakatoa, "*Science* **154,** 1325 (Dec. 9, 1966). (See Prob. 6.12.)

S. J. Smith and E. M. Purcell. "Visible Light from Localized Surface Charges Moving across a Grating," *Phys. Rev.* **92,** 1069 (1953). (See Prob. 7.28.)

J. R. Tessman and J. T. Finnell, Jr., *Am .J. Phys.* **35,** .523 (1967). (See Chap. 7)

Karl von Frisch, *Bees. Their Vision, Chemical Sense. and Language* (Cornell University Press, Ithaca, N.Y., 1950). (See Chap. 8.)

Index

Reference to a specific problem is indicated by p and the number in parentheses following the page citation.

Useful Constants

Speed of light in vacuum°	$C = 2.997925 \times 10^{10}$ cm/sec $= 3 \times 10^8$ m/s
Fundamental charge	$e = 1.6 \times 10^{-19} C$
Planck's constant	$h = 6.6 \times 10^{-34}$ J-sec.
"Reduced" Planck's constant	$\hbar = h/2\pi = 1.0 \times 10^{-34}$ J-sec.
Electron rest mass	$m_e = 9.1 \times 10^{-31}$ kg.
Proton rest mass	$m_p = 1.7 \times 10^{-27}$ kg.
Gravitational constant	$G = 6.7 \times 10^{-11}$ kg^{-1} m^{3}s^{-1}
Acceleration of gravity at sea level	$g \approx 9.8$ m/s^2
Bohr radius	$a_0 = 0.5 \times 10^{-10}$ m.
Avogadro's number	$N_0 = 6.0 \times 10^{23}$ mole^{-1}
Boltzmann's constant	$k = 1.4 \times 10^{-23}$ J/degree Kelvin
Standard temperature	$To = 273$ degree Kelvin
Standard pressure	$p_0 = 1$ atm $= 1.01 \times 10^5$ N/m^2
Molar volume at S.T.P.	$V_0 = 22.4 \times 10^{-5}$ m^3/mole
Thermal energy kT at S.T.P.	$kT_0 = 3.8 \times 10^{-21}$ J
Density of air at S.T.P.	$\rho_0 = 1.3 \times 10^2$ kg/m^3
Speed of sound in air at S.T.P.	$v_0 = 3.32 \times 10^2$ m/s
Sound impedance of air at S.T.P.	$Z_0 = 428$ (N/m^2)/(m/sec)
Standard sound intensity	$I_0 = 10^{-2}$ W/m^2
Factor of ten in intensity	$= 1$ bel $= 10$ db
One fermi (F)	$= 10^{-15}$ m
One angstrom unit Å)	$= 10^{-10m}$
One micron (μ)	$= 10^{-4}$ m
One hertz (Hz)	$= 1$ cycle per second (cps)
Wavelength of one-electron-volt photon	$= 1.24 \times 10^{-4}$ cm ≈ 12345 Å
One electron volt (ev)	$= 1.6 \times 10^{-19}$ J.
One watt (W)	$= 1$ joule/sec
One coulomb (coul)	
One volt (V)	
One ohm (Ω)	
Thirty ohms	
Impedance per square of vacuum for electromagnetic waves	
One farad (F)	
One henry (H)	

°In converting from practical units to electrostatic units we have approximated the velocity of light ar 3.00×10^{10} cm/sec. Wherever a 3 appears, a more accurate conversion factor can be obtained by using the more accurate value of c. Similarly wherever 9 appears, it is more accurately $(2.998)^2$.

Recommended Unit Prefixes

Multiples and Submultiples	Prefixes	Symbols
10^{12}	tera	T
10^{9}	giga	C
10^{6}	mega	M
10^{3}	kilo	k
10^{2}	hecto	h
10	deka	da
10^{-1}	deci	d
10^{-2}	centi	c
10^{-3}	milli	m
10^{-6}	micro	μ
10^{-9}	nano	n
10^{-12}	pico	p

Useful Identities

$$\cos x + \cos y = [2 \cos \tfrac{1}{2}(x - y)] \cos \tfrac{1}{2}(x + y)$$
$$\cos x - \cos y = [-2 \sin \tfrac{1}{2}(x - y)] \sin \tfrac{1}{2}(x + y)$$
$$\sin x + \sin y = [2 \cos \tfrac{1}{2}(x - y)] \sin \tfrac{1}{2}(x + y)$$
$$\sin x - \sin y = [2 \sin \tfrac{1}{2}(x - y)] \cos \tfrac{1}{2}(x + y)$$
$$\cos (x \pm y) = \cos x \cos y \mp \sin x \sin y$$
$$\sin (x \pm y) = \sin x \cos y \pm \sin y \sin x$$
$$\cos 2x = \cos^2 x - \sin 2 x$$
$$\sin 2x = 2 \sin x \cos x$$
$$\cos^2 x = \tfrac{1}{2}(1 + \cos 2x)$$
$$\sin^2 x = \tfrac{1}{2}(1 - \cos 2x)$$
$$\sin x = x - \tfrac{1}{6}x^3 + \dots$$
$$\cos x = 1 - \tfrac{1}{2}x^2 + \dots$$
$$(1 + x)^n = 1 + nx + \tfrac{1}{2}n(n - 1)x^2 + \dots; x2 < 1.$$
$$\cos \theta_1 + \cos (\theta_1 + \gamma) + \cos (\theta_1 + 2\gamma) + \dots + \cos [\theta_1 + (N - 1)\,\gamma] = \cos \left[\theta_1 + \tfrac{1}{2}(N - 1)\,\gamma\right] \frac{\sin \tfrac{1}{2} N \gamma}{\sin \tfrac{1}{2}\gamma}$$

The Electromagnetic Spectrum

Common Name of Electromagnetic Radiation	Practical Units°		Order of Magnitude		
	λ	$hv, v, v/c$	$\lambda(cm)$	$v(Hz)$	$hv(ev)$
Bremasstrahlung x rays (maximum energy) from:					
Stanford electron linac	0.067 F	18 Gev	10^{-14}	10^{24}	10^{10}
Typical electron synchrotron	4 F	300 Mev	10^{-13}	10^{23}	10^{9}
Gamma rays:					
Neutral pi-meson decay $\pi^0 \rightarrow 2\gamma$	19 F	67 Mev	10^{-12}	10^{22}	10^{8}
Decay of excited nucleus	100 F	10 Mev	10^{-11}	10^{21}	10^{7}
X ray (excited atoms or electron bremsstrahlung)	0.1 Å	100 kev	10^{-9}	10^{19}	10^{5}
Ultraviolet light (excited tomas)	100 Å	100 ev	10^{-6}	10^{16}	10^{2}
Visible light: Dark blue visibility limit	3900 Å	2.5 ev	10^{-5}	10^{15}	10^{1}
Blue of mercury vapor stree light	4358 Å	22,940 cm^{-1}			
Green of mercury vapor street light	5461 Å	18,310 cm^{-1}			
Yellow of mercury vapor street light	5770 Å	17,330 cm^{-1}			
Helium-neon laser red light	6328 Å	15,800 cm^{-1}			
Visible light: Dark red visibility limit Infrared	7600 Å	1.6 ev	10^{-4}	10^{14}	10^{0}
Dominant heat radiation ($hv \approx 3kT$) from:					
Surface of sun (T ≈ 6000°K)	1 μ	1 ev	10^{-4}	10^{14}	10^{0}
Room temperature ($T \approx 300°K$)	20 μ	15,000 GHz	10^{-3}	10^{13}	10^{-1}
Universal primeval fireball (3°K)	2 mm	150 GHz	10^{-1}	10^{11}	10^{-2}
Microwaves and radio waves:					
Ammonia clock	1.5 cm	20 GHz	10^{0}	10^{10}	10^{-3}
Radar (S band)	10 cm	3 GHz	10^{1}	10^{9}	10^{-4}
Interstellar hydrogen line	21 cm	1.5 GHz	10^{1}	10^{9}	10^{-4}
Ultra-high-frequency (UHF) † TV carrier	37 cm	800 Mc	10^{1}	10^{9}	10^{-4}
	75 cm	400 Mc	10^{2}	10^{8}	10^{-5}
Ordinary TV carrier (VHF)†	1.5–5.5 m	210 to 55 Mc	10^{2}	10^{8}	10^{-5}
Commercial FM radio carrier (VHF)†	2.8–3.4 m	108 to 88 Mc	10^{2}	10^{8}	10^{-5}
10 m	30 Mc	10^{3}	10^{7}	10^{-6}	
Amateur bands (HF)†	100 m	3 Mc	10^{4}	10^{6}	10^{-7}
Commercial AM radio carrier (MF)†	200 m	1500 kc	10^{4}	10^{6}	10^{-7}
600 m	500 kc	10^{5}	10^{5}	10^{-8}	
Audio frequencies (VLF) †	10 km	30 kc	10^{6}	10^{4}	10^{-9}
10^4 km†	30 cps	10^{9}	10^{1}	10^{-12}	

° The "practical" unit are those most commonly used by experiments. When different field overlap or when technology changes rapidly, different names and different units may be used for the same frequency region. *For examples:* When x ray are detected by photon countries, it is natural to use energy units (ev, kev, Mev); when they are detected by crystal diffraction, it is natural to use length units (Å). *Another example:* Lasers are now developed mostly by electrical engineers, who tend to use frequency units (Mc or MHz, GHz, etc.) where spectroscopists might tend to use wavelength units (Å, μ, etc.).

† U = ultra, H = high, F = frequency, V = very, M = medium, L = low.

‡ Wavelength of audio-frequency radio waves, not of sound waves in air.